Edited by
Paul Seidel

Applied Superconductivity

Related Titles

Bezryadin, A.

Superconductivity in Nanowires

Fabrication and Quantum Transport

2012

ISBN: 978-3-527-40832-0

Also available in digital formats

Buckel, W. and Kleiner, R.

Supraleitung

Grundlagen und Anwendungen7., aktualisierte und erweiterte Auflage

2012

ISBN: 978-3-527-41139-9

Ireson, G.

Discovering Superconductivity – An Investigative Approach

2012

ISBN: 978-1-119-99141-0

Also available in digital formats

Waser, R. (ed.)

Nanoelectronics and Information Technology

Advanced Electronic Materials and Novel Devices; Third, Completely Revised and Enlarged Edition

2012

ISBN: 978-3-527-40927-3

Kalsi, S.S.

Applications of High Temperature Superconductors to Electric Power Equipment

2011

ISBN: 978-0-470-16768-7

Also available in digital formats

Bhattacharya, R., Paranthaman, M. (eds.)

High Temperature Superconductors

2010

ISBN: 978-3-527-40827-6

Also available in digital formats

Padamsee, H.

RF Superconductivity

Science, Technology, and Applications

2009

ISBN: 978-3-527-40572-5

Also available in digital formats

Hansen, R.C.

Electrically Small, Superdirective, and Superconducting Antennas

2006

ISBN: 978-0-471-78255-1

Also available in digital formats

Edited by Paul Seidel

Applied Superconductivity

Handbook on Devices and Applications

Volume 1

Verlag GmbH & Co. KGaA

The Editor

Prof. Dr. Paul Seidel
Friedrich–Schiller-Universität Jena
Institut für Festkörperphysik
AG Tieftemperaturphysik
Helmholtzweg 5
D-07743 Jena
Germany

Cover
Illustration and assembly: Grafik-Design Schulz, Fußgönheim
Cavity: Courtesy DESY
Deserializer: IOP Publishing. Reproduced with permission. All rights reserved. M. H. Volkmann *et al*, 2013 *Supercond. Sci. Technol.* **26**, 015002
Superconductor cable: Courtesy Nexans Deutschland GmbH

All books published by **Wiley-VCH** are carefully produced. Nevertheless, authors, editors, and publisher do not warrant the information contained in these books, including this book, to be free of errors. Readers are advised to keep in mind that statements, data, illustrations, procedural details or other items may inadvertently be inaccurate.

Library of Congress Card No.: applied for

British Library Cataloguing-in-Publication Data
A catalogue record for this book is available from the British Library.

Bibliographic information published by the Deutsche Nationalbibliothek
The Deutsche Nationalbibliothek lists this publication in the Deutsche Nationalbibliografie; detailed bibliographic data are available on the Internet at <http://dnb.d-nb.de>.

© 2015 Wiley-VCH Verlag GmbH & Co. KGaA, Boschstr. 12, 69469 Weinheim, Germany

All rights reserved (including those of translation into other languages). No part of this book may be reproduced in any form – by photoprinting, microfilm, or any other means – nor transmitted or translated into a machine language without written permission from the publishers. Registered names, trademarks, etc. used in this book, even when not specifically marked as such, are not to be considered unprotected by law.

Print ISBN: 978-3-527-41209-9
ePDF ISBN: 978-3-527-67066-6
ePub ISBN: 978-3-527-67065-9
Mobi ISBN: 978-3-527-67064-2
oBook ISBN: 978-3-527-67063-5

Cover Design Grafik-Design Schulz, Fußgönheim, Germany
Typesetting Laserwords Private Limited, Chennai, India
Printing and Binding Markono Print Media Pte Ltd, Singapore

Printed on acid-free paper

Contents to Volume 1

Contents to Volume 2

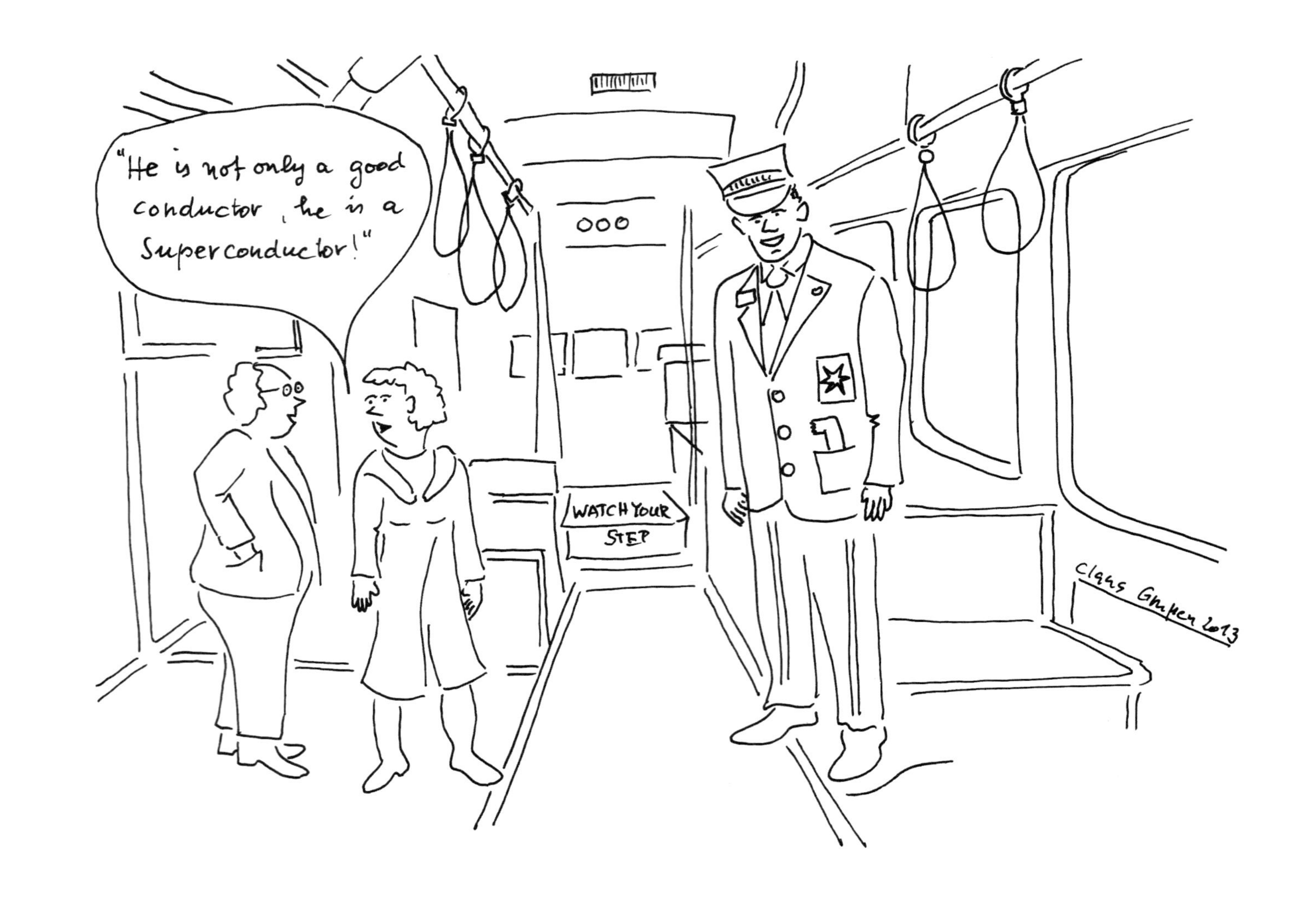
"He is not only a good conductor, he is a Superconductor!"
WATCH YOUR STEP
Claus Grupen 2013

Preface

During the celebrations of the 100 years of superconductivity in 2011, many times the question came up about real applications and the commercial impact of superconducting materials. Actually, since already the Applied Superconductivity Conference 1998 themed "Superconductivity coming to market" had promised a positive answer to this long-standing question, we felt that the present situation should be evaluated and summarized. There exist a lot of very good textbooks on basics of superconductivity as well as some monographs concerning special applications for specialists in detail like the *The SQUID Handbook* edited by John Clarke and Alex Braginski. The collections of articles like "Engineering Superconductivity" edited by Peter Lee in 2001, reflecting the status up to 1999, and "Applied Superconductivity" edited by Bernhard Seeber, already published in 1998, were first steps in the direction to discuss and introduce the applications for a wider audience. Nevertheless, today, they no longer represent the topical situation and latest developments. Thus, the immense progress in applications of superconductivity will be demonstrated within this new handbook on a level which covers the range from popular aspects for students and beginners till details for specialists. Because of the finite size for a two-volume book, the basic knowledge on superconductivity had to be reduced to a minimum required for understanding the main part on applications. The historical development is not reflected in detail but sometimes with respect to the actual status in order to demonstrate the speed of progress. For historic details, we refer the reader to the book *100 Years of Superconductivity* edited by Horst Rogalla and Peter H. Kes, which also covers many historic aspects of applications.

This handbook wants to demonstrate that applied superconductivity has a rising impact in science and industry. The breathtaking development within the last 20 years involved a large number of different fields, for example, in medicine, geophysics, high-energy physics, and power engineering. Thus, not all examples and details can be given here, but the references will guide the reader to additional sources. The dynamics of the development of superconductivity, especially in materials and technologies toward applications is astonishing. Applications even in niches like radio frequency (RF) filters for mobile communication are a strong forcing mechanism in this development. As one example for the rapid development, the rise in critical current densities in high-T_c superconductors from 10^3 A cm^{-2} in the beginning about 20 years ago till today's second generation of YBCO-coated conductors with some 10^6 A cm^{-2} should be mentioned. This progress is the basis for many magnet or power applications. But this example also illustrates the problem of production and availability of big amounts of superconducting materials adapted to the requirements needed for all the possible projects. A production of the coated conductors fixed by pre-contracts between producers and costumers as has been done for low-T_c superconducting cables for accelerators like CERN or fusion reactors like International Thermonuclear Experimental Reactor (ITER) may be a good practical way to cope with this problem. There will be a rising importance of superconducting materials, technologies, and devices due to their superior properties in comparison with well-established commercial standards. The aspects of saving of energy and reduction of pollution open new possibilities for applied superconductivity as expressed in the topic of the ASC 2014 "Race to energy efficiency."

In future, there will be many systems where the user or costumer takes the advantages of superconductivity sometimes even without knowing that there is some superconducting component like in medical magnetic resonance imaging (MRI) (MRT in German). The applications of superconductors will be additionally forced by the progress in cooling technologies with cryogen-free or cryocooler solutions. I hope that this handbook will help to enhance the understanding of the immense potential of superconductivity for applications in many fields.

The handbook is organized in the following parts:

1) Fundamentals of superconductivity and main related effects will be given only in a way to understand the main part of the book.
2) Superconducting materials will be introduced, but besides some overview in the fundamentals, there will be detailed contributions only on materials relevant for applications now and in near future.
3) Technology, preparation, and characterization concerning bulk materials, single crystals, thin films, Josephson junctions, wires and tapes, as well as cooling technology will be discussed with respect to the parameters and conditions needed for applications.
4) The main part consists of eight extended chapters on different application fields, including engineering aspects as well as main important parameters and interesting details up to examples for real applications.
5) In the summary and outlook, we try to forecast the development of the present main applications within the next 20 years.

Finally, I like to thank all contributors, the referees, and the staff of Wiley-VCH, especially Vera Palmer, Ulrike Werner, and Nina Stadthaus, as well the staff of Laserwords, especially Madhubala Venkatesan, for their excellent contributions and stimulating cooperation.

Jena
September 15, 2014

Paul Seidel

List of Contributors

Marie-Cécile Alvarez-Hérault
Domaine Universitaire
G2ELAB (Grenoble Institute of Technology, UJF, CNRS)
Ense3, 11, rue des Mathématiques - BP 46
38402 Saint Martin d'Hères Cedex
France

Solveig Anders
Leibniz Institute of Photonic Technology
Department Quantum Detection
Albert-Einstein-Street 9
D-07745 Jena
Germany

Tolga Aytug
Oak Ridge National Laboratory
Chemical Sciences Division
PO Box 2008 MS6100
Oak Ridge
TN 37831-6100
USA

Robert Bach
University of Applied Science South Westphalia
Department of Electrical Engineering
Lübecker Ring 2
D-59494 Soest
Germany

Mikhail Belogolovskii
National Academy of Sciences of Ukraine
Donetsk Institute for Physics and Engineering
Department of the Theory of Dynamic Properties of Complex Systems
Street R. Luxemburg 72
83114 Donetsk
Ukraine

Sergey A. Belomestnykh
Collider-Accelerator Department
Bldg. 911B, Brookhaven National Laboratory
P.O. Box 5000
Upton
NY 11973-5000
USA

and

Stony Brook University
Department of Physics and Astronomy
Stony Brook
NY 11794
USA

Jörn Beyer
Physikalisch-Technische Bundesanstalt (PTB)
Cryophysics and Spectrometry
Abbestr. 2-12
D-10587 Berlin
Germany

Joachim Bock
Nexans SuperConductors GmbH
Chemiepark Knapsack
D-50351 Hürth
Germany

Luca Bottura
CERN TE-MSC, M24500
CH-1211 Geneva, 23
Switzerland

Audrius Brazdeikis
University of Houston
Department of Physics and Texas Center for Superconductivity
Houston
TX 77004
USA

Wolf-Rüdiger Canders
Technische Universität Braunschweig
Institut für Elektrische Maschinen
Antriebe und Bahnen
Postfach 3329
D-38023 Braunschweig
Germany

Claudia Cantoni
Oak Ridge National Laboratory
Chemical Sciences Division
PO Box 2008 MS6100
Oak Ridge
TN 37831-6100
USA

James R. Claycomb
Houston Baptist University
Department of Mathematics and Physics
7502 Fondren Road
Houston
TX 77074
USA

Roberto Cristiano
CNR Istituto SPIN - Superconductors
Innovative Materials and Devices
UOS - Napoli
80125 Napoli
Italy

Jonathan A. Demko
LeTourneau University
School of Engineering and Engineering Technology
2100 South Mobberly Avenue
Longview
TX 75607
USA

Jean-Luc Duchateau
CEA/IRFM
Institute for Magnetic Fusion Research
13108 St Paul lez Durance Cedex
France

Andreas Erb
Bayerische Akademie der Wissenschaften
Walther-Meissner-Institut für Tieftemperaturforschung
Walther-Meissner-Str. 8
D-85748 Garching
Germany

Robert L. Fagaly
Quasar Federal Systems
5754 Pacific Center Blvd.
Suite 203
San Diego
CA 92121
USA

Pascal Febvre
University of Savoie
IMEP-LAHC
Campus Scientifique
73376 Le Bourget du Lac Cedex
France

Herbert C. Freyhardt
University of Houston
Texas Center for Superconductivity
202 UH Science Center
Houston
TX 77204-5002
USA

Ludwig Fritzsch
Leibniz Institute of Photonic Technology
Department Quantum Detection
Albert-Einstein-Street 9
D-07745 Jena
Germany

Günter Fuchs
Leibniz-Institut für Festkörper-und Werkstoffforschung (IFW) Dresden
Department Superconducting Materials
Postfach 270116
D-01171 Dresden
Germany

Camille Gandioli
Domaine Universitaire
G2ELAB (Grenoble Institute of Technology, UJF, CNRS)
ENSE3
38402 Saint Martin d'Heres
France

Flavio Gatti
INFN and Università di Genova
Dipartimento di Fisica
Via Dodecaneso 33
16146 Genova
Italy

Rene Geithner
Helmholtz Institute Jena
Fröbelstieg 3
D-07743 Jena
Germany

Michael A. Green
Lawrence Berkeley National Laboratory
Engineering Division
M/S 46-0161, 1 Cyclotron Road
Berkeley
CA 94720
USA

and

FRIB Michigan State University
640 South Shaw
East Lansing
MI 48824
USA

Francesco Grilli
Karlsruhe Institute of Technology
Institute for Technical Physics
Hermann-Von Helmholtz-Platz 1
D-76344
Eggenstein-Leopoldshafen
Germany

Claus Grupen
Siegen University
Faculty for Science and Engineering
Emmy-Noether-Campus
Walter-Flex-Straße 3
D-57068 Siegen
Germany

Nouredine Hadjsaid
Domaine Universitaire
G2ELAB (Grenoble Institute of Technology, UJF, CNRS)
ENSE3
38402 Saint Martin d'Heres
France

Seungyong Hahn
Massachusetts Institute of Technology
Francis Bitter Magnet Laboratory, Plasma Science and Fusion and Center
170 Albany Street
Cambridge
MA 02139
USA

Eric Hellstrom
Florida State University
Department of Mechanical Engineering
National High Magnetic Field Laboratory
Applied Superconductivity Center
2031 E. Paul Dirac Dr.
Tallahassee
FL 32310
USA

Dagmar Henrich
Karlsruhe Institute of Technology
Department of Electrical Engineering and Information Technology
Institute of Micro- und Nanoelectronic Systems
Hertzstraße 16
D-76187 Karlsruhe
Germany

and

Oxford Instruments Omicron NanoScience
Limburger Straße 75
D-65232, Taunusstein-Neuhof
Germany

Roland Hott
Karlsruhe Institute of Technology
Institute of Solid State Physics
Hermann-von-Helmholtz-Platz 1
D-76021 Karlsruhe
Germany

John R. Hull
Boeing
Advanced Physics Applications
P.O. Box 3707, MC 2T-50
Seattle
WA 98124-2207
USA

Yukikazu Iwasa
Massachusetts Institute of Technology
Francis Bitter Magnet Laboratory
Plasma Science and Fusion and Center
170 Albany Street
Cambridge
MA 02139
USA

Quanxi Jia
Los Alamos National Laboratory
Center for Integrated Nanotechnologies
MPA-CINT, MS K771
Los Alamos
NM 87545
USA

Jianyi Jiang
National High Magnetic Field Laboratory
Applied Superconductivity Center
2031 E. Paul Dirac Dr.
Tallahassee
FL 32310
USA

Andreas Kade
Gemeinnützige Gesellschaft mbH
ILK Dresden
Institut für Luft- und Kältetechnik
Bertolt-Brecht-Allee 20
D-01309 Dresden
Germany

Gunter Kaiser
Gemeinnützige Gesellschaft mbH
ILK Dresden
Institut für Luft- und Kältetechnik
Bertolt-Brecht-Allee 20
D-01309 Dresden
Germany

Swarn Singh Kalsi
Consulting Engineer
Kalsi Green Power Systems, LLC
46 Renfield Drive
Princeton
NJ 08540
USA

Neeraj Khare
Indian Institute of Technology Delhi
Physics Department
Hauz Khas
New Delhi 110016
India

John Kirtley
Stanford University
Applied Physics
476 Lomita Mall
McCullough Bldg 139
Stanford
CA 94305
USA

Reinhold Kleiner
Universität Tübingen
Physikalisches Institut and
Center for Collective Quantum
Phenomena in LISA$^+$
Auf der Morgenstelle 14
D-72076 Tübingen
Germany

Johannes Kohlmann
Physikalisch-Technische
Bundesanstalt (PTB)
Quantum Electronics
Bundesallee 100
D-38116 Braunschweig
Germany

Gernot Krabbes
Leibniz-Institut für
Festkörper-und
Werkstoffforschung (IFW)
Dresden
Department Superconducting
Materials
Postfach 270116
D-01171 Dresden
Germany

Hans-Joachim Krause
Forschungszentrum Jülich
Institute of Bioelectronics, Peter
Grünberg Institute (PGI-8)
Wilhelm-Johnen-Str.
D-52425 Jülich
Germany

Helmut Krauth
Bruker EAS
Ehrichstraße 10
D-63450 Hanau
Germany

Jürgen Kunert
Leibniz Institute of Photonic
Technology
Department Quantum Detection
Albert-Einstein-Street 9
D-07745 Jena
Germany

Jürgen Lisenfeld
Karlsruhe Institute of
Technology (KIT)
Physikalisches Institut
Wolfgang-Gaede-Straße 1
D-76131 Karlsruhe
Germany

Cesar Luongo
Jefferson Laboratory
600 Kelvin Drive, Suite 3
Newport News
VA 23606
USA

Doris Maier
Institut de RadioAstronomie
IRAM
300, Rue de la Piscine
38406 St. Martin d'Heres
France

Robert McDermott
University of Wisconsin
Department of Physics
1150 University Avenue
Madison
WI 53706
USA

Hans-Georg Meyer
Leibniz Institute of Photonic
Technology
Department Quantum Detection
Albert-Einstein-Street 9
D-07745 Jena
Germany

Donald L. Miller
Northrop Grumman
Corporation
Electronic Systems
PO Box 1521, Mail Stop 3B10
Baltimore
MD 21203
USA

Antonio Morandi
University of Bologna
Department of Electrical
Electronic and Information
Engineering
Viale Risorgimento 2
40136 Bologna
Italy

Michael Mück
ez SQUID Mess- und
Analysegeräte
Herborner Strasse 9
D-35764 Sinn
Germany

Oleg Mukhanov
HYPRES Inc.
175 Clearbrook Road
Elmsford
NY 10523
USA

Davide Nardelli
Columbus Superconductors,
S.p.A.
Via delle Terre Rosse 30
16133 Genova
Italy

Hannes Nowak
JenaSQUID GmbH & Co. KG
Münchenroda 29
D-07751 Jena
Germany

Gregor Oelsner
Leibniz Institute of Photonic
Technology
Department Quantum Detection
Albert-Einstein-Street 9
D-07745 Jena
Germany

Thomas Ortlepp
CiS Research Institute for
Microsensor Systems and
Photovoltaics GmbH
Konrad-Zuse-Street 14
D-99099 Erfurt
Germany

and

Ilmenau University of
Technology
Microelectronics and
nanoelectronic systems
PO Box 10 05 65
D-98684 Ilmenau
Germany

Hasan S. Padamsee
Cornell University
Laboratory for Elementary
Particle Physics
153 Sciences Drive
Ithaca
NY 14853-5001
USA

and

Fermilab
P.O. Box 500, MS 316
Batavia
IL 60510-5011
USA

Ilaria Pallecchi
CNR-SPIN and University of
Genova
Dipartimento di Fisica
Via Dodecaneso 33
16146 Genova
Italy

Mariappan P. Paranthaman
Oak Ridge National Laboratory
Chemical Sciences Division
PO Box 2008 MS6100
Oak Ridge
TN 37831-6100
USA

Giovanni P. Pepe
CNR Istituto SPIN -
Superconductors
Innovative Materials and Devices
UOS - Napoli
80125 Napoli
Italy

and

University of Naples Federico II
Department of Physics
Via Cinthia
Naples
80126 Monte Sant'Angelo
Italy

Werner Prusseit
THEVA Dünnschichttechnik
GmbH
Rote-Kreuz Str. 8
D-85737 Ismaning
Germany

John X. Przybysz
Northrop Grumman
Corporation
Electronic Systems
1550 West Nursery Road
Mail Stop C425
Linthicum
MD 21090
USA

Marina Putti
CNR-SPIN and University of
Genova
Dipartimento di Fisica
Via Dodecaneso 33
16146 Genova
Italy

Lucio Rossi
CERN—European Organization
for Nuclear Research
Technology Department
385 Route de Meyrin
1217 Meyrin
Switzerland

Hannes Rotzinger
Karlsruher Institut für
Technologie
Physikalisches Institut
Wolfgang-Gaede-Straße 1
D-76131 Karlsruhe
Germany

Steven T. Ruggiero
University of Notre Dame
Department of Physics
225 Nieuwland Science Hall
Notre Dame
IN 46556
USA

Jean-Claude Sabonnadiere
Domaine Universitaire
G2ELAB (Grenoble Institute of
Technology, UJF, CNRS), ENSE3
38402 Saint Martin d'Heres
France

Klaus Schlenga
Bruker EAS
Ehrichstraße 10
D-63450 Hanau
Germany

Matthias Schmelz
Leibniz Institute of Photonic
Technology
Department Quantum Detection
Albert-Einstein-Street 9
D-07745 Jena
Germany

Gunar Schroeder
Gemeinnützige Gesellschaft
mbH
ILK Dresden
Institut für Luft- und
Kältetechnik
Bertolt-Brecht-Allee 20
D-01309 Dresden
Germany

Thomas Schurig
Physikalisch-Technische
Bundesanstalt (PTB)
Cryophysics and Spectrometry
Abbestr. 2-12
D-10587 Berlin
Germany

Paul Seidel
Friedrich Schiller University Jena
Institute of Solid State Physics
Helmholtzweg 5
D-07743 Jena
Germany

Tengming Shen
Fermi National Accelerator
Laboratory
Magnet Systems Department
Wilson Street & Kirk Road
Batavia
IL 60510
USA

Michael Siegel
Karlsruhe Institute of
Technology
Department of Electrical
Engineering and Information
Technology
Institute of Micro- und
Nanoelectronic Systems
Hertzstraße 16
D-76187 Karlsruhe
Germany

Frederic Sirois
Polytechnique Montreal
C.P. 6079, succ. centre-ville
Montreal, QC, H3C 3A7
Canada

Liliana Stan
Los Alamos National Laboratory
Center for Integrated
Nanotechnologies
MPA-CINT, MS K771
Los Alamos
NM 87545
USA

and

Argonne National Laboratory
Center for Nanoscale Materials
9700 South Cass Avenue,
Building 440
Argonne
IL 60439–4806
USA

Ronny Stolz
Leibniz Institute of Photonic
Technology
Department Quantum Detection
Albert-Einstein-Street 9
D-07745 Jena
Germany

Keiichi Tanabe
International Superconductivity
Technology Center
Superconductivity Research
Laboratory
2-11-19 Minowa-cho
Kohoku-ku
Yokohama
Kanagawa 223-0051
Japan

Saburo Tanaka
Toyohashi University of
Technology
Tempaku-cho
441-8580 Toyohashi
Aichi
Japan

Pascal Tixador
Domaine Universitaire
G2ELAB (Grenoble Institute of
Technology, UJF, CNRS), ENSE3
38402 Saint Martin d'Heres
France

Hannes Toepfer
Technische Universität Ilmenau
Theoretische Elektrotechnik
PF 10 05 65
D-98684 Ilmenau
Germany

Masayoshi Tonouchi
Osaka University
Institute of Laser Engineering
2-6 Yamada-Oka
Suita-city
Osaka 565-0871
Japan

Matteo Tropeano
Columbus Superconductors,
S.p.A.
Via delle Terre Rosse 30
16133 Genova
Italy

Wolfgang Vodel
Friedrich Schiller University Jena
Institute of Solid State Physics
Helmholtzweg 5
D-07743 Jena
Germany

and

Helmholtz Institute Jena
Fröbelstieg 3
D-07743 Jena
Germany

Huabing Wang
National Institute for Materials Science (NIMS)
Superconducting Properties Unit
1-2-1 Sengen
Tsukuba 3050047
Japan

Martin Weides
Karlsruher Institut für Technologie
Physikalisches Institut
Wolfgang-Gaede-Straße 1
D-76131 Karlsruhe
Germany

Frank N. Werfel
Adelwitz Technologiezentrum GmbH (ATZ)
Naundorfer Street 29
D-04860 Torgau
Germany

Martin N. Wilson
33 Lower Radley
OX14 3AY Abingdon
United Kingdom

Stuart C. Wimbush
Victoria University of Wellington
Robinson Research Institute
PO Box 600
69 Gracefield Road
Lower Hutt 5010
Wellington 6140
New Zealand

Thomas Wolf
Karlsruhe Institute of Technology
Institute of Solid State Physics
Hermann-von-Helmholtz-Platz 1
D-76021 Karlsruhe
Germany

Roger Wördenweber
Forschungszentrum Jülich GmbH
Peter Grünberg Institute (PGI-8) and JARA-Fundamentals of Future Information Technology
Leo-Brandt-Straße
D-52425 Jülich
Germany

Jarek Wosik
University of Houston
Department of Electrical and Computer Engineering, and Texas Center for Superconductivity
Houston
TX 77004
USA

Alexander B. Zorin
Physikalisch-Technische
Bundesanstalt
Quantenelektronik
Bundesallee 100
D-38116 Braunschweig
Germany

and

Moscow State University
Skobeltsyn Institute of Nuclear
Physics
119899 Moscow
Russia

Gertrud Zwicknagl
Technische Universität
Braunschweig
Institut für Mathematische
Physik
Mendelssohnstraße 3
D-38106 Braunschweig
Germany

1
Fundamentals

1.1
Superconductivity

1.1.1
Basic Properties and Parameters of Superconductors[1)]
Reinhold Kleiner

1.1.1.1 Superconducting Transition and Loss of DC Resistance
In the year 1908, Kamerlingh-Onnes [3], director of the Low-Temperature Laboratory at the University of Leiden, had achieved the liquefaction of helium as the last of the noble gases. At atmospheric pressure, the boiling point of helium is 4.2 K. It can be reduced further by pumping. The liquefaction of helium extended the available temperature range near the absolute zero point and Kamerlingh-Onnes was able to perform experiments at these low temperatures.

At first, he started an investigation of the electric resistance of metals. At that time, the ideas about the mechanism of the electric conduction were only poorly developed. It was known that it must be electrons being responsible for charge transport. Also one had measured the temperature dependence of the electric resistance of many metals, and it had been found that near room temperature the resistance decreases linearly with decreasing temperature. However, at low temperatures, this decrease was found to become weaker and weaker. In principle, there were three possibilities to be discussed:

1) The resistance could approach zero value with decreasing temperature (James Dewar, 1904).
2) It could approach a finite limiting value (Heinrich Friedrich Ludwig Matthiesen, 1864).
3) It could pass through a minimum and approach infinity at very low temperatures (William Lord Kelvin, 1902).

In particular, the third possibility was favored by the idea that at sufficiently low temperatures the electrons are likely to be bound to their respective atoms. Hence, their free mobility was expected to vanish. The first possibility, according to which

1) Text and figures of this chapter are a short excerpt from monographs [1, 2].

Applied Superconductivity: Handbook on Devices and Applications, First Edition.
Edited by Paul Seidel.
© 2015 Wiley-VCH Verlag GmbH & Co. KGaA. Published 2015 by Wiley-VCH Verlag GmbH & Co. KGaA.

the resistance would approach zero value at very low temperatures, was suggested by the strong decrease with decreasing temperature. Initially, Kamerlingh-Onnes studied platinum and gold samples, since at that time he could obtain these metals already with high purity. He found that during the approach of zero temperature the electric resistance of his samples reached a finite limiting value, the so-called residual resistance, a behavior corresponding to the second possibility discussed above. The value of this residual resistance depended upon the purity of the samples. The purer the samples, the smaller the residual resistance. After these results, Kamerlingh-Onnes expected that in the temperature range of liquid helium, ideally, pure platinum or gold should have a vanishingly small resistance. In a lecture at the Third International Congress of Refrigeration 1913 in Chicago, he reported on these experiments and arguments. There he said: "*Allowing a correction for the additive resistance I came to the conclusion that probably the resistance of absolutely pure platinum would have vanished at the boiling point of helium*" [4]. These ideas were supported further by the quantum physics rapidly developing at that time. Albert Einstein had proposed a model of crystals, according to which the vibrational energy of the crystal atoms should decrease exponentially at very low temperatures. Since the resistance of highly pure samples, according to the view of Kamerlingh-Onnes (which turned out to be perfectly correct, as we know today), is only due to this motion of the atoms, his hypothesis mentioned above appeared obvious.

In order to test these ideas, Kamerlingh-Onnes decided to study mercury, the only metal for which he hoped at that time that it can be extremely purified by means of a multiple distillation process. He estimated that at the boiling point of helium he could barely just detect the resistance of the mercury with his equipment, and that at still lower temperatures it should rapidly approach zero value. The initial experiments carried out by Kamerlingh-Onnes together with his coworkers, Gerrit Flim, Gilles Holst, and Gerrit Dorsman, appeared to confirm these concepts. At temperatures below 4.2 K, the resistance of mercury, indeed, became immeasurably small. During his further experiments, he soon recognized that the observed effect could not be identical to the expected decrease of resistance. The resistance change took place within a temperature interval of only a few hundredths of a degree and, hence, it resembled more a resistance jump than a continuous decrease.

Figure 1.1.1.1 shows the curve published by Kamerlingh-Onnes [5]. He commented himself: "*At this point* (slightly below 4.2 K) *within some hundredths of a degree came a sudden fall not foreseen by the vibrator theory of resistance, that had framed, bringing the resistance at once less than a millionth of its original value at the melting point. … Mercury had passed into a new state, which on account of its extraordinary electrical properties may be called the* superconductive state" [4].

In this way also the name for this new phenomenon had been found. The discovery came unexpectedly during experiments, which were meant to test some well-founded ideas. Soon it became clear that the purity of the samples was unimportant for the vanishing of the resistance. The carefully performed experiment had uncovered a new state of matter.

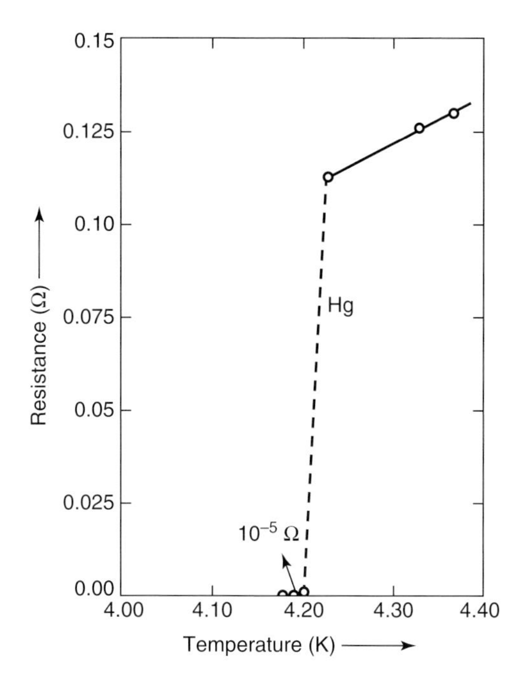

Figure 1.1.1.1 Superconductivity of mercury. (From [1], after Ref. [5].)

Today we know that superconductivity represents a widespread phenomenon. In the periodic system of the elements superconductivity occurs in many elements. Here, at atmospheric pressure, niobium is the element with the highest "transition temperature" or "critical temperature" T_c of about 9 K. Eventually, thousands of superconducting compounds have been found, and this development is by no means closed.

The vanishing of the DC electric resistance below T_c is not the only unusual property of superconductors. An externally applied magnetic field can be expelled from the interior of superconductors except for a thin outer layer ("ideal diamagnetism" or "Meissner–Ochsenfeld effect"). This happens for type-I superconductors for field below the so-called critical field B_c, and for type-II superconductors below the lower critical field B_{c1}. For higher fields, type-II superconductors can concentrate the magnetic field in the form of "flux tubes." Here the magnetic flux[2] is quantized in units of the "magnetic flux quantum" $\Phi_0 = 2.07 \cdot 10^{-15}$ Wb. The ideal diamagnetism of superconductors was discovered by Meissner and Ochsenfeld in 1933. It was a big surprise, since based on the induction law one would

2) The magnetic flux Φ through a loop of area F, carrying a perpendicular and spatially homogeneous flux density B is given by $\Phi = B \cdot F$. In the following, we denote B simply by "magnetic field." In the general case of an arbitrarily oriented and spatially inhomogeneous magnetic field $\boldsymbol{B}$, one must integrate over the area of the loop, $\Phi = \int_F \boldsymbol{B}\, d\boldsymbol{f}$. The unit of magnetic flux is weber (Wb), the unit of the magnetic field is tesla (T). We have 1 Wb = 1 T m^2. If a loop is placed at a large distance around the axis of an isolated flux tube, we have $\Phi = \Phi_0$.

have only expected that an ideal conductor conserves its interior magnetic field and does not expel it.

The breakthrough of the theoretical understanding of superconductivity was achieved in 1957 by the theory of Bardeen, Cooper, and Schrieffer ("BCS theory") [6]. They recognized that at the transition to the superconducting state, the electrons pairwise condense into a new state, in which they form a coherent matter wave with a well-defined phase, following the rules of quantum mechanics. Here the interaction of the electrons is mediated by the "phonons," the quantized vibrations of the crystal lattice. The pairs are called *Cooper pairs*. In most cases, the spins of the two electrons are aligned antiparallelly, that is, they form spin-singlets. Also, at least in most cases, the angular momentum of the pair is zero (s-wave). The theory also shows that at nonzero temperatures, a part of the electrons remain unpaired. There is, however, an energy gap Δ which separates these unpaired "quasiparticles" from the Cooper pairs. It requires the energy 2Δ to break a pair.

For more than 75 years, superconductivity represented specifically a low-temperature phenomenon. This changed in 1986, when Bednorz and Müller [7] discovered superconductors based on copper oxide.

This result was highly surprising for the scientific community, also because already in the middle 1960s, Matthias and coworkers had started a systematic study of the metallic oxides. They searched among the substances based on the transition metal oxides, such as W, Ti, Mo, and Bi [8]. They found extremely interesting superconductors, for example, in the Ba–Pb–Bi–O system, however, no particularly high transition temperatures.

During the turn of the year 1986–1987, the "gold rush" set in, when it became known that the group of Shigeho Tanaka in Japan could exactly reproduce the results of Bednorz and Müller. Only a few weeks later, transition temperatures above 80 K were observed in the Y–Ba–Cu–O system [9]. During this phase, new results more often were reported in press conferences than in scientific journals. The media anxiously followed this development. With superconductivity at temperatures above the boiling point of liquid nitrogen ($T = 77$ K), one could envision many important technical applications of this phenomenon.

Today we know a large series of cuprate "high-temperature superconductors." Here the mostly studied compounds are $YBa_2Cu_3O_7$ (also "YBCO" or "Y123") and $Bi_2Sr_2CaCu_2O_8$ (also "BSCCO" or "Bi2212"), which display maximum transition temperatures around 90 K. Some compounds have transition temperatures even above 100 K. The record value is carried by $HgBa_2Ca_2Cu_3O_8$, having at atmospheric pressure a T_c value of 135 K and at a pressure of 30 GPa, a value as high as $T_c = 164$ K. Figure 1.1.1.2 shows the evolution of the transition temperatures since the discovery by Kamerlingh-Onnes. The jump-like increase due to the discovery of the copper-oxides is particularly impressive.

In Figure 1.1.1.2, we have also included the metallic compound MgB_2, as well as the iron pnictides.

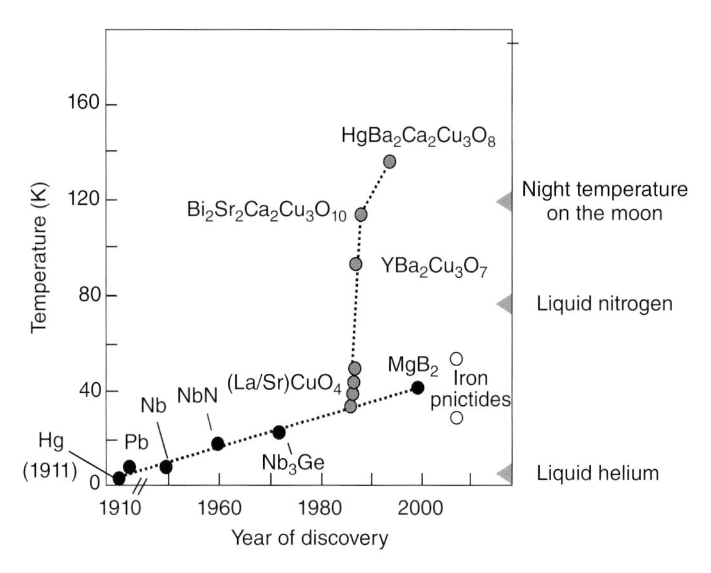

Figure 1.1.1.2 Evolution of the superconducting transition temperature since the discovery of superconductivity. (From [2], after Ref. [10].)

For MgB_2, surprisingly, superconductivity with a transition temperature of 39 K was detected only in 2000, even though this material has been commercially available for a long time [11]. Also, this discovery had a great impact in physics, and many essential properties of this material have been clarified in the subsequent years. It turned out that MgB_2 behaves similarly as the "classical" metallic superconductors, however with two energy gaps. The discovery of the iron pnictides in 2008 [12] had a similar impact. These are compounds like $LaFeAsO_{0.89}F_{0.11}$ or $Ba_{0.6}KFe_2As_2$, with transition temperatures of up to 55 K. The iron pnictides contain layers made of FeAs as the basic building block, in analogy to the copper oxide layers in the cuprates.

Many properties of the high-temperature superconductors (in addition also to other superconducting compounds) are highly unusual. For example, the Cooper pairs in the cuprates have an angular momentum of $2\hbar$ (d-wave) and the coherent matter wave has $d_{x^2-y^2}$ symmetry. For the d-wave symmetry, the energy gap Δ disappears for some directions in momentum space. More than 25 years after their discovery, it is still unclear how the Cooper pairing is accomplished in these materials. However, it seems likely that magnetic interactions play an important role.

Another important issue is the maximum current which a superconducting wire or tape can carry without resistance, the so-called critical current. We will see that the property "zero resistance" is not always fulfilled. When alternating currents are applied, the resistance can become finite. Also for DC currents, the critical current is limited. It depends on the temperature and the magnetic field, and also on the type of superconductor used and the geometry of the wire. It is a big challenge to fabricate conductors in a way that hundreds or even thousands of amperes can be carried without or at least with very low resistance.

1.1.1.2 Ideal Diamagnetism, Flux Quantization, and Critical Fields

It has been known for a long time that the characteristic property of the superconducting state is that it shows no measurable resistance for direct current. If a magnetic field is applied to such an *ideal conductor*, permanent currents are generated by induction, which screen the magnetic field from the interior of the sample. For that reason, a permanent magnet can levitate when placed on top of an ideal conductor. This effect is demonstrated in Figure 1.1.1.3.

What happens if a magnetic field $\boldsymbol{B}_a$ is applied to a *normal conductor* and if subsequently by cooling below the transition temperature T_c ideal conductivity is reached? At first, in the normal state at the application of the magnetic field, eddy currents are flowing because of induction. However, as soon as the magnetic field has reached its final value and does not change anymore with time, these currents decay, and finally the magnetic fields within and outside the superconductor become equal. If now the ideal conductor is cooled below T_c, this magnetic state simply remains, since further induction currents are generated only during *changes* of the field. Exactly this is expected, if the magnetic field is turned off below T_c. In the interior of the ideal conductor, the magnetic field remains conserved. Hence, depending upon the way in which the final state, namely a temperature below T_c and an applied magnetic field $\boldsymbol{B}_a$, has been reached, within the interior of the ideal conductor we have completely different magnetic fields.

Accordingly, a material with the only property $R = 0$, for the same external variables T and $\boldsymbol{B}_a$, could be transferred into completely different states, depending upon the previous history. Therefore, for the same given thermodynamic variables, we would not have just one well-defined superconducting phase, but, instead, a continuous manifold of superconducting phases with arbitrary shielding currents, depending upon the previous history. However, the existence of a manifold of superconducting phases appeared so unlikely, that also before 1933 one referred to only a single superconducting phase even without an experimental verification.

As a matter of fact, a superconductor behaves quite different than an ideal electric conductor. Again we imagine that a sample is cooled below T_c in the presence of an applied magnetic field. If this magnetic field is very small, one finds that

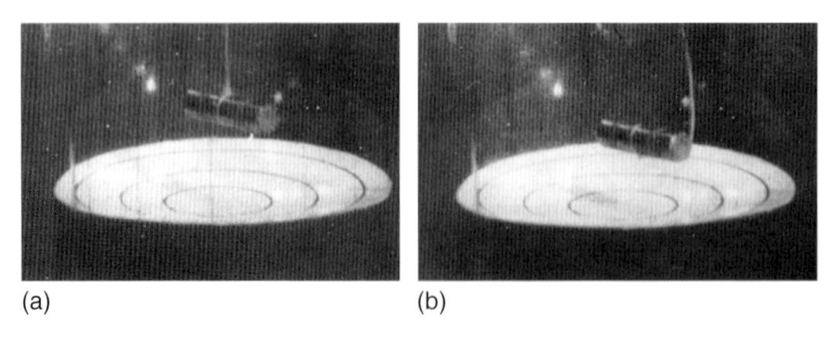

Figure 1.1.1.3 The "levitated magnet" for demonstrating the permanent currents, which are generated in superconducting lead by induction during the lowering of the magnet. (a) Starting position and (b) equilibrium position.

the field is completely expelled from the interior of the superconductor except for a very thin layer at the sample surface. In this way, one obtains an ideal diamagnetic state, independent of the temporal sequence in which the magnetic field was applied and the sample was cooled.

This ideal diamagnetism has been discovered in 1933 by Meissner and Ochsenfeld [13] for rods made of lead or tin. This expulsion effect, similar as the property $R=0$, can be nicely demonstrated using the "levitated magnet." In order to show the property $R=0$, in Figure 1.1.1.3 we have lowered the permanent magnet toward the superconducting lead bowl, generating in this way by induction the permanent currents. For demonstrating the Meissner–Ochsenfeld effect, we place the permanent magnet into the lead bowl at $T > T_c$ (Figure 1.1.1.4a) and then cool further down. The field expulsion appears at the superconducting transition, the magnet is repelled from the diamagnetic superconductor, and it is raised up to the equilibrium height (Figure 1.1.1.4b). In the limit of ideal magnetic field expulsion, the same levitation height is reached as in Figure 1.1.1.3.

Above, we had assumed that the magnetic field applied to the superconductor would be "small." Indeed, one finds that the ideal diamagnetism only exists within a finite range of magnetic fields and temperatures, which, furthermore, also depends upon the sample geometry.

Next we consider a long, rod-shaped sample where the magnetic field is applied parallel to the axis. For other shapes, the magnetic field often can be distorted. One finds that there exist two different types of superconductors:

- The first type, referred to as *type-I superconductors* or *superconductors of the first kind*, expels the magnetic field up to a maximum value B_c, the critical field. For larger fields, superconductivity breaks down, and the sample assumes the normal-conducting state. B_c depends on the temperature and reaches zero at T_c. Pure mercury or lead are examples of a type-I superconductor.
- The second type, referred to as *type-II superconductors* or *superconductors of the second kind*, shows ideal diamagnetism for magnetic fields smaller than the

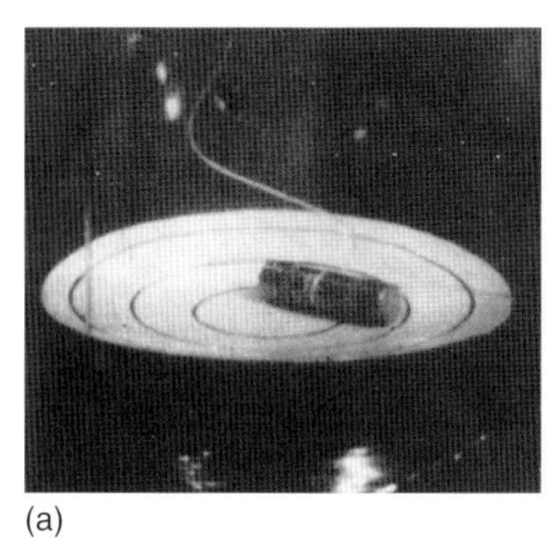
(a)

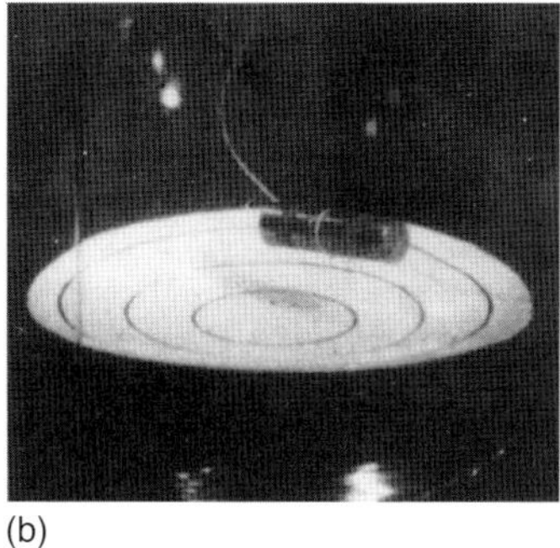
(b)

Figure 1.1.1.4 "Levitated magnet" for demonstrating the Meissner–Ochsenfeld effect in the presence of an applied magnetic field. (a) Starting position at $T > T_c$ and (b) equilibrium position at $T < T_c$.

"lower critical magnetic field" B_{c1}. Superconductivity completely vanishes for magnetic fields larger than the "upper critical magnetic field" B_{c2}, which often is much larger than B_{c1}. Both critical fields reach zero at T_c. This behavior is found in many alloys, but also in the high-temperature superconductors. In the latter, B_{c2} can reach even values larger than 100 T, depending on the direction the field is applied relative to the CuO layers.

What happens in type-II superconductors in the "Shubnikov phase" between B_{c1} and B_{c2}? In this regime, the magnetic field only partly penetrates into the sample. Now shielding currents flow within the superconductor and concentrate the magnetic field lines, such that a system of flux lines, also referred to as *Abrikosov vortices*, is generated. In an ideal homogeneous superconductor, in general, these vortices arrange themselves in form of a triangular lattice. In Figure 1.1.1.5, we show schematically this structure of the Shubnikov phase. The superconductor is penetrated by magnetic flux lines, each of which carries a magnetic flux quantum and is located at the corners of equilateral triangles. Each flux line consists of a system of circular currents, which in Figure 1.1.1.5 are indicated for two flux lines. These currents together with the external magnetic field generate the magnetic flux within the flux line and reduce the magnetic field between the flux lines. Hence, one also talks about flux vortices. With increasing external field $\boldsymbol{B}_a$, the distance between the flux lines becomes smaller.

The first experimental proof of a periodic structure of the magnetic field in the Shubnikov phase was given in 1964 by a group at the Nuclear Research Center in Saclay using neutron diffraction [14]. However, they could only observe a basic period of the structure. Real images of the Shubnikov phase were generated by Essmann and Träuble [15] using an ingenious decoration technique. In

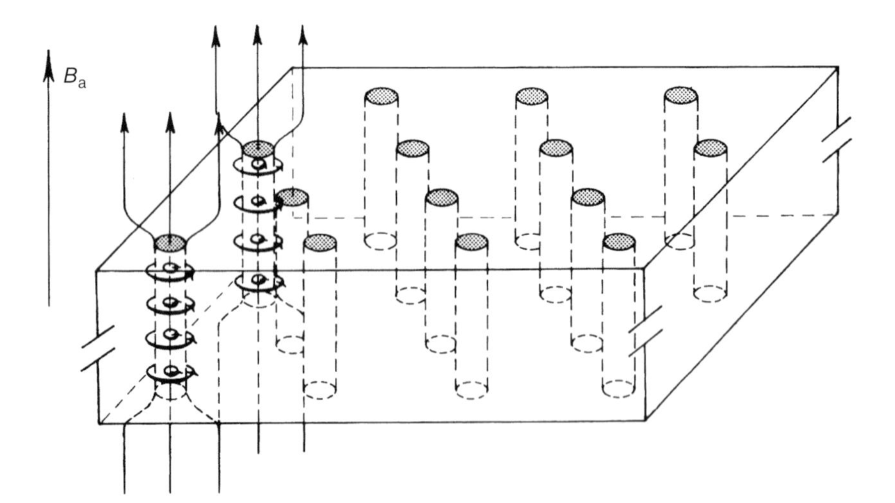

Figure 1.1.1.5 Schematics of the Shubnikov phase. The magnetic field and the supercurrents are shown only for two flux lines.

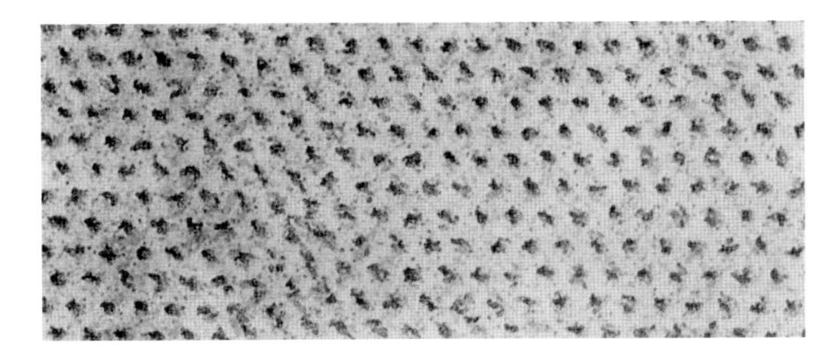

Figure 1.1.1.6 Image of the vortex lattice obtained with an electron microscope following the decoration with iron colloid. Frozen-in flux after the magnetic field has been reduced to zero. Material: Pb + 6.3 at.% In; temperature: 1.2 K; sample shape: cylinder 60 mm long, 4 mm diameter; magnetic field B_a parallel to the axis. Magnification: 8300× (Reproduced by courtesy of Dr. Essmann.)

Figure 1.1.1.6, we show a lead–indium alloy as an example. These images of the magnetic flux structure were obtained as follows: above the superconducting sample iron atoms are evaporated from a hot wire. During their diffusion through the helium gas in the cryostat, the iron atoms coagulate forming iron colloids. These colloids have a diameter of <50 nm, and they slowly approach the surface of the superconductor. At this surface, the flux lines of the Shubnikov phase exit from the superconductor. In Figure 1.1.1.6, this is shown for two flux lines. The ferromagnetic iron colloid is collected at the locations, where the flux lines exit from the surface, since here they find the largest magnetic-field gradients. In this way, the flux lines can be decorated. Subsequently, the structure can be observed in an electron microscope. The image shown in Figure 1.1.1.6 was obtained in this way. Such experiments convincingly confirmed the vortex structure predicted theoretically by Abrikosov.

The question remains if the decorated locations at the surface indeed correspond to the ends of the flux lines carrying only a single flux quantum. In order to answer this question, we just have to count the number of flux lines and also have to determine the total flux, say, by means of an induction experiment. Then we find the value of the magnetic flux of a flux line by dividing the total flux Φ_{tot} through the sample by the number of flux lines. Such evaluations exactly confirmed that in highly homogeneous type-II superconductors each flux line contains a single flux quantum $\Phi_0 = 2.07 \cdot 10^{-15}$ T m^2.

Today, apart from neutron diffraction and decoration, there are a number of different methods for imaging magnetic flux lines. We will not go into detail but mention that the methods often supplement each other and provide valuable information about superconductivity.

Flux quantization, in integer multiples of Φ_0, also occurs in a superconducting ring. This has been demonstrated very nicely in pioneering experiments by Doll and Näbauer [16] in Munich and by Deaver and Fairbank [17] in Stanford.

1.1.1.3 The Origin of Flux Quantization, London Penetration Depth and Ginzburg–Landau Coherence Length

Next we will deal with the conclusions to be drawn from the quantization of the magnetic flux in units of the flux quantum Φ_0.

For atoms we are well used to the appearance of discrete states. For example, the stationary atomic states are distinguished due to a quantum condition for the angular momentum appearing in multiples of $\hbar = h/2\pi$. This quantization of the angular momentum is a result of the condition that the quantum mechanical wave function, indicating the probability for finding the electron, be single-valued. If we move around the atomic nucleus starting from a specific point, the wave function must reproduce itself exactly if we return to this starting point. Here, the phase of the wave function can change by an integer multiple of 2π, since this does not affect the wave function. We can have the same situation also on a macroscopic scale. Imagine that we have an arbitrary wave propagating without damping in a ring with radius R. The wave can become stationary, if an integer number n of wavelengths λ exactly fits into the ring. Then we have the condition $n\lambda = 2\pi R$ or $kR = n$, using the wave vector $k = 2\pi/\lambda$. If this condition is violated, after a few revolutions the wave disappears due to interference.

Applying these ideas to the Cooper pair matter wave propagating around the ring[3] one obtains [1, 2]:

$$n \cdot \frac{h}{q} = \mu_0 \lambda_L^2 \oint \boldsymbol{j}_s d\boldsymbol{r} + \Phi \qquad (1.1.1.1)$$

Equation (1.1.1.1) represents the so-called fluxoid quantization. The integral has to be taken along some closed contour inside the superconductor and Φ is magnetic flux penetrating this contour. The expression on the right-hand side denotes the "fluxoid." The quantity

$$\lambda_L = \sqrt{\frac{m}{\mu_0 q^2 n_s}} \qquad (1.1.1.2)$$

is the London penetration depth (q: charge; m: particle mass; n_s: particle density; μ_0: permeability) and $\boldsymbol{j}_s$ is the super-current density.

In many cases, the super-current density and, hence, the path integral on the right-hand side of Eq. (1.1.1.1) is negligibly small. This happens in particular if we deal with a thick-walled superconducting cylinder or with a ring made of a type-I superconductor. Because of the Meissner–Ochsenfeld effect, the magnetic field is expelled from the superconductor. The shielding super-currents only flow near the surface of the superconductor and decay exponentially toward the interior, as we will discuss further below. We can place the integration path, along which Eq. (1.1.1.1) must be evaluated, deep into the interior of the ring. In this case, the integral over the current density is exponentially small, and we obtain in good

3) The moving wave is connected with the motion of the center of mass of the pairs, which is identical for all pairs.

approximation

$$\Phi \approx n \cdot \frac{h}{q} \tag{1.1.1.3}$$

This is exactly the condition for the quantization of the magnetic flux, and the experimental observation $\Phi = n \cdot (h/2|e|) = n \cdot \Phi_0$ clearly shows that the superconducting charge carriers have the charge $|q| = 2e$. The sign of the charge carriers cannot be found from the observation of the flux quantization, since the direction of the *particle* current is not determined in this experiment. In many superconductors, the Cooper pairs are formed by electrons, that is, $q = -2e$. On the other hand, in many high-temperature superconductors, we have hole conduction, similar to that found in p-doped semiconductors. Here we have $q = +2e$.

From Eq. (1.1.1.1), one can also show [1, 2] that for a solid material which is superconducting everywhere in its interior only $n = 0$ is possible. Then one arrives at

$$\mu_0 \lambda_L^2 \oint \boldsymbol{j}_s d\boldsymbol{r} = -\Phi = -\int_F \boldsymbol{B} d\boldsymbol{f} \tag{1.1.1.4}$$

Using Stokes's theorem, this condition can also be written as

$$\boldsymbol{B} = -\mu_0 \lambda_L^2 \,\mathrm{curl}\, \boldsymbol{j}_s \tag{1.1.1.5}$$

Equation (1.1.1.5) is the second London equation. It is one of two fundamental equations with which the two brothers F. London and H. London in 1935 [18] constructed a successful theoretical model of superconductivity. From Eq. (1.1.1.5), after some math one finds

$$\Delta \boldsymbol{B} = \frac{1}{\lambda_L^2} \boldsymbol{B} \tag{1.1.1.6}$$

Δ is the Laplace operator, $\Delta f = (\partial^2 f/\partial x^2) + (\partial^2 f/\partial y^2) + (\partial^2 f/\partial z^2)$, which in Eq. (1.1.1.6) must be applied to the three components of $\boldsymbol{B}$.

Equation (1.1.1.6) produces the Meissner–Ochsenfeld effect, as we can see from a simple example. For this purpose, we consider the surface of a very large superconductor, located at the coordinate $x = 0$ and extended infinitely along the (x,y)-plane. The superconductor occupies the half-space $x > 0$ (see Figure 1.1.1.7). An external magnetic field $\boldsymbol{B}_a = (0, 0, B_a)$ is applied to the superconductor. Owing to the symmetry of our problem, we can assume that within the superconductor only the z-component of the magnetic field is different from zero and is only a function of the x-coordinate. Equation (1.1.1.6) then yields for $B_z(x)$ within the superconductor, that is, for $x > 0$:

$$\frac{d^2 B_z(x)}{dx^2} = \frac{1}{\lambda_L^2} B_z(x) \tag{1.1.1.7}$$

This equation has the solution

$$B_z(x) = B_z(0) \cdot \exp\left(-\frac{x}{\lambda_L}\right) \tag{1.1.1.8}$$

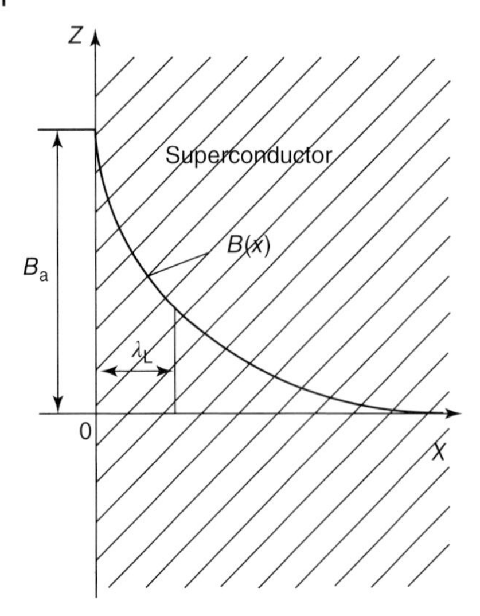

Figure 1.1.1.7 Decrease of the magnetic field within the superconductor near the planar surface.

which is shown in Figure 1.1.1.7. Within the length λ_L the magnetic field is reduced by the factor $1/e$, and the field vanishes deep inside the superconductor.

We note that Eq. (1.1.1.7) also yields a solution increasing with x: $B_z(x) = B_z(0) \cdot \exp(+x/\lambda_L)$. However, this solution leads to an arbitrarily large magnetic field in the superconductor and, hence, is not meaningful.

From Eq. (1.1.1.2), we can obtain a rough estimate of the London penetration depth with the simplifying assumption that one electron per atom with the free-electron mass m_e contributes to the super-current. For tin, for example, such an estimate yields $\lambda_L = 26$ nm. This value deviates only little from the measured value, which at low temperatures falls in the range 25–36 nm. More values for λ_L are listed in Table 1.1.1.1 together with a number of other parameters that will be introduced in this chapter. All numbers should be taken just as rough guidelines since they depend strongly on the sample purity. For some materials, λ_L, as well as the other quantities, depend strongly on the crystal orientation. These materials are often layered structures. The London penetration depth can be very large when the magnetic field is applied parallel to the layers.

Only a few nanometer away from its surface the superconducting half-space is practically free of the magnetic field and displays the ideal diamagnetic state. The same can be found for samples with a more realistic geometry, for example, for a superconducting rod, as long as the radii of curvature of the surfaces are much larger than λ_L and the superconductor is also much thicker than λ_L. Then on a length scale of λ_L, the superconductor closely resembles a superconducting half-space. Of course, for an exact solution Eq. (1.1.1.6) must be solved.

The London penetration depth depends upon temperature. From Eq. (1.1.1.2) we see that λ_L is proportional to $1/n_s^{1/2}$. We can assume that the number of electrons combined to Cooper pairs decreases with increasing temperature and

Table 1.1.1.1 Critical temperature and zero temperature values of the energy gap, the coherence length, and the upper critical field. Numbers vary strongly in the literature and thus should be taken as a rough guide only. For Pb and Nb, the critical field rather than B_{c2} is quoted. (*ab*) and (*c*), respectively refer to in-plane and out-of-plane properties; "*max.*" indicates the maximum energy gap.

Material	T_c (K)	Δ (meV)	ξ_{GL} (nm)	λ_L (nm)	B_c, B_{c2} (T)
Pb	7.2	1.38	51–83	32–39	0.08 (B_c)
Nb	9.2	1.45	40	32–44	0.2 (B_c)
NbN	13–16	2.4–3.2	4	250	16
Nb_3Sn	18	3.3	4	80	24
Nb_3Ge	23.2	3.9–4.2	3–4	80	38
NbTi	9.6	1.1–1.4	4	60	16
$YBa_2Cu_3O_7$	92	15–25 (*max., ab*)	1.6 (*ab*) 0.3 (*c*)	150 (*ab*) 800 (*c*)	240 (*ab*) 110 (*c*)
$Bi_2Sr_2CaCu_2O_8$	94	15–25 (*max., ab*)	2 (*ab*) 0.1 (*c*)	200–300 (*ab*) >15 000 (*c*)	>60 (*ab*) >250 (*c*)
$Bi_2Sr_2Ca_2Cu_3O_{10}$	110	25–35 (*max., ab*)	2.9 (*ab*) 0.1 (*c*)	150 (*ab*) >1000 (*c*)	40 (*ab*) >250 (*c*)
MgB_2	40	1.8–7.5	10 (*ab*) 2 (*c*)	110 (ab) 280 (*c*)	15–20 (*ab*) 3 (*c*)
$Ba_{0.6}K_{0.4}Fe_2As_3$	38	4–12	1.5 (*ab*) $c > 5$ (*c*)	190 (*ab*) 0.9 (*c*)	70–235 (*ab*) 100–140 (*c*)
$NdO_{0.82}F_{0.18}FeAs$	50	37	3.7 (*ab*) 0.9 (*c*)	190 (*ab*)*c* >6000 (*c*)	62–70 (*ab*) 300 (*c*)

vanishes at T_c. Above the transition temperature, no stable Cooper pairs should exist anymore.[4] Hence, we expect that λ_L increases with increasing temperature and diverges at T_c. Correspondingly, the magnetic field penetrates further and further into the superconductor until it homogeneously fills the sample above the transition temperature.

How can one measure the London penetration depth? In principle, one must determine the influence of the thin shielding layer upon the diamagnetic behavior. This has been done using several different methods. For example, one can measure the magnetization of plates which are thinner and thinner [19]. As long as the thickness of the plate is much larger than the penetration depth, one will observe nearly ideal diamagnetic behavior. However, this behavior becomes weaker, if the plate thickness approaches the range of λ_L. Another method uses spin-polarized muons, which, by varying their kinetic energy, are implanted in different depths from the surface. The spin of the muon precesses in the local magnetic field and, by measuring the electron that are emitted upon its decay, it is possible to determine the precession frequency and thus the local magnetic field [20]. For determining the temperature dependence of λ_L, only relative

4) Here we neglect thermal fluctuations by which Cooper pairs can be generated momentarily also above T_c.

measurements are needed. One can determine the resonance frequency of a cavity fabricated from a superconducting material. The resonance frequency sensitively depends on the geometry. If the penetration depth varies with the temperature, this is equivalent to a variation of the geometry of the cavity and, hence, of the resonance frequency, yielding the change of λ_L [21].

A strong interest in the exact measurement of the penetration depth, say, as a function of temperature, magnetic field, or the frequency of the microwaves for excitation, arises because of its dependence upon the density of the superconducting charge carriers. It yields important information on the superconducting state and can serve as a sensor for studying superconductors.

What causes the difference between type-I and type-II superconductivity and the generation of vortices? From the assumption of a continuous superconductor, we have obtained the second London equation and the ideal diamagnetism. In type-I superconductors, this state is established, as long as the applied magnetic field does not exceed a critical value. At higher fields, superconductivity breaks down. For a discussion of the critical magnetic field, we must treat the energy of a superconductor more accurately. This is done in the framework of the Ginzburg–Landau theory. Here, one can see that it is the competition between two energies, the energy gain from the condensation of the Cooper pairs and the energy loss due to the magnetic field expulsion, which causes the transition between the superconducting and the normal-conducting state.

At small magnetic fields, the Meissner phase is also established in type-II superconductors. However, at the lower critical field, vortices appear within the material. Turning again to Eq. (1.1.1.1), we see that the separation of the magnetic flux into units[5] of $\pm 1\Phi_0$ corresponds to states with the quantum number $n = \pm 1$. Here, the superconductor cannot display continuous superconductivity anymore, for which case $n = 0$ was the only possibility. Instead, we must assume that the vortex axis is not superconducting, similar to the ring geometry.

A more accurate treatment of the vortex structure based on the Ginzburg–Landau theory shows that the magnetic field decreases nearly exponentially with the distance from the vortex axis on the length scale λ_L. Hence, we can say that the flux line has a magnetic radius of λ_L.

Second, on a length scale ξ_{GL}, the Ginzburg–Landau coherence length, the density n_s of the Cooper pairs vanishes as one approaches the vortex axis. Depending on the superconducting material, this length ranges between 0.1 nm and a few hundred nanometers; see also Table 1.1.1.1. Similar to the London penetration depth, it is temperature dependent, in particular close to T_c. We also mention here that there is also a coherence length associated with the distance over which the two electrons forming the Copper pairs are correlated. This is the BCS coherence length $\xi_0 = \hbar v_F / k_B T_c$, where v_F is the Fermi velocity.

Why does each vortex carry exactly one flux quantum Φ_0? Again we must look at the energy of a superconductor. Essentially, we find that in a type-II superconductor it is energetically favorable if it generates an interface superconductor/normal

5) The sign must be chosen according to the direction of the magnetic field.

conductor above the lower critical magnetic field. Therefore, as many of these interfaces as possible are generated. This is achieved by choosing the smallest quantum state with $n = \pm 1$, since in this case the maximum number of vortices and the largest interface area near the vortex axis is established.

Now we can estimate the lower critical field B_{c1}. Each flux line carries a flux quantum Φ_0, and at least one needs a magnetic field $B_{c1} \approx \Phi_0$/(cross-sectional area of the flux line) $\approx \Phi_0/(\pi\lambda_L^2)$ for generating this amount of flux. With a value of $\lambda_L = 100\,\text{nm}$ one finds $B_{c1} \approx 65\,\text{mT}$. From the Ginzburg–Landau theory, one obtains an expression which differs from our simple estimate by a factor of $(\ln\kappa + 0.08)/4$, with $\kappa = \lambda_L/\xi_{GL}$. This factor is on the order of unity for not too small values of κ.

For increasing magnetic field, the flux lines are packed closer and closer to each other, until near B_{c2} their distance is about equal to the Ginzburg–Landau coherence length ξ_{GL}. For a simple estimate of B_{c2}, we assume a cylindrical normal-conducting vortex core. Then superconductivity is expected to vanish, if the distance between the flux quanta becomes equal to the core diameter, that is, at $B_{c2} \approx \Phi_0/(\pi\xi_{GL}^2)$. An exact theory yields a value smaller by a factor of 2. In fact, often one uses the corresponding relation for determining ξ_{GL}. We further note that, depending on the value of ξ_{GL}, B_{c2} can become very large. With the value $\xi_{GL} = 2\,\text{nm}$, one obtains a field larger than 80 T. Such high values of the upper critical magnetic field are reached or even exceeded in the high-temperature superconductors.

Table 1.1.1.1 lists B_{c2} for several superconductors. In the table, we have also listed the critical field of Nb and Pb. Pure single crystals of these materials are type-I superconductors. It should be noted, however, that in most practical cases, due to a reduced mean free path, the electrons can travel without scattering, the coherence length is smaller, and λ_L is larger, making these materials type-II.

At the end of this section, we wish to ask how the permanent current and zero resistance, the key phenomena of superconductivity, can be explained in terms of the macroscopic wave function. From the second London equation (1.1.1.5), with the use Maxwell's equations one obtains

$$\boldsymbol{E} = \mu_0 \lambda_L^2 \dot{\boldsymbol{j}}_s \tag{1.1.1.9}$$

This is the first London equation. For a temporally constant super-current, the right-hand side of Eq. (1.1.1.9) is zero. Hence, we obtain current flow without an electric field and zero resistance.

Note that the relation $\boldsymbol{E} \propto \dot{\boldsymbol{j}}_s$ is similar to that of an inductor, $U_L \propto \dot{I}_L$. We can thus understand one of the reasons why an alternating current will produce a finite resistance. At nonzero temperatures, a part of the electrons in the superconductor is unpaired (quasiparticles). In the presence of an alternating electric field, both quasiparticles and Cooper pairs are accelerated and a nonzero resistance appears which grows with increasing frequency.

Equation (1.1.1.9) also indicates that in the presence of a DC electric field, the super-current density continues to increase with time. For a superconductor this seems reasonable, since the superconducting charge carriers are accelerated more and more due to the electric field. On the other hand, the super-current density cannot increase up to infinity. Therefore, additional energy arguments are needed for finding the maximum super-current density which can be reached. This can be treated in the framework of the Ginzburg–Landau theory and yields the so-called pair-breaking critical current density.

We could have derived the first London equation also from classical arguments, if we note that for current flow without resistance the superconducting charge carriers cannot experience (inelastic) collision processes. Then, in the presence of an electric field, we have the force equation $m\dot{\boldsymbol{v}} = q\boldsymbol{E}$. We use $\boldsymbol{j}_s = qn_s\boldsymbol{v}$ and find $\boldsymbol{E} = (m/q^2n_s)\dot{\boldsymbol{j}}_s$. The latter equation can be turned into Eq. (1.1.1.9) using the definition (1.1.1.2) of the London penetration depth.

Finally, we briefly mention here that the well-defined phase of the superconducting matter wave is responsible for interference effects as they appear in Josephson junctions and in superconducting quantum interference devices (SQUIDs). It turns out that, in a Josephson junction, a sandwich consisting of two superconducting electrodes separated by a very thin barrier, there is a super-current across the barrier which varies sinusoidally with the difference δ of the phases of the matter wave of the two electrodes. If there is a voltage drop U across the barrier this phase differences increases in time, with the time derivative of δ given by $\dot{\delta} = 2\pi U/\Phi_0$. In SQUIDs two Josephson junctions are integrated in a superconducting loop. Here, the maximum super-current that can be sent across the two junctions varies sinusoidally with the magnetic flux threading the loop. The modulation period is given by the flux quantum. Details will be given in the corresponding chapters.

1.1.1.4 Critical Currents

We have already mentioned that a superconductor can carry only a limited electric current without resistance. The existence of a critical current is highly important for technical applications of superconductivity. In type-II superconductors, we have materials which can remain still superconducting also for technically interesting magnetic fields. However, for applications it is also important that these superconductors still can transport sufficiently high electric currents without resistance also in high magnetic fields. As we will see, here we are confronted with a problem, which has been solved only with the so-called hard superconductors.

Before we turn to the special features in type-I and type-II superconductors, we want to briefly look at the magnitude of the critical super-current density in the ideal case of a thin and homogeneous superconducting wire. This pair-breaking critical current density j_{cp}, which can be reached under most favorable conditions, can be treated within the Ginzburg–Landau theory. We consider a homogeneous superconducting wire having a diameter which is smaller than the London penetration depth λ_L and the Ginzburg–Landau coherence length ξ_{GL}. We find

$$j_{cp} = \frac{2}{3}\sqrt{\frac{2}{3}}B_{cth} \cdot \frac{1}{\mu_0\lambda_L} \tag{1.1.1.10}$$

B_{cth} is the so-called thermodynamical critical field which for a type-I superconductor under certain conditions equals the critical field B_c. For a type-II superconductor, it can be related to the upper critical field via $B_{cth} = B_{c2}/\sqrt{2}\kappa$, with the Ginzburg–Landau parameter $\kappa = \lambda_L/\xi_{GL}$.

If for B_{cth} we take a value of 1 T and for λ_L a value of 100 nm, we obtain for j_{cp} a value of about $4.3 \cdot 10^8$ A cm^{-2}.

With respect to type-I superconductors, we consider a wire with circular cross-section carrying a current I. The wire is assumed much thicker than the London penetration depth. At sufficiently small currents, the superconducting wire resides in the Meissner phase. In this phase, the interior of the superconductor must remain free of magnetic flux. However, this also means that the interior cannot carry an electric current, since otherwise the magnetic field of the current would exist. From this, we conclude that also the current passing through a superconductor is restricted to the thin surface layer, into which the magnetic field can penetrate in the Meissner phase. The external currents applied to a superconductor are referred to as *transport currents*, in contrast to the shielding currents appearing in the superconductor as circulating currents. The total current is given by the integral of the current density over the cross-sectional area.

Already in 1916, Silsbee [22] proposed the hypothesis, that in the case of "thick" superconductors, that is, for superconductors with a fully developed shielding layer, the critical current is reached exactly, when the magnetic field of the current at the surface attains the value B_{cth}. This hypothesis has been confirmed perfectly. In other words, it means that the magnetic field and the current density at a surface with a well-developed shielding layer are strongly correlated. The critical value of the current density is associated with a certain critical field, namely B_{cth}, where it is completely irrelevant, if the current density is due to shielding currents or to a transport current.

Because of the validity of the Silsbee hypothesis, it is very simple to calculate the critical currents of wires with circular cross-section from the critical fields. The magnetic field at the surface of such a wire carrying the current I is given by

$$B_0 = \mu_0 \frac{I}{2\pi R} \tag{1.1.1.11}$$

where B_0 is the field at the surface, I is the transport current, R is the wire radius, and $\mu_0 = 4\pi \cdot 10^{-7}$ V s (A m)$^{-1}$.

The only requirement is cylinder symmetry of the current distribution. The radial dependence of the current density is arbitrary. According to Eq. (1.1.1.11), the critical field of about 30 mT at 0 K – the value for the critical field of tin – corresponds to a critical current $I_{c0} = 75$ A. This critical current increases only proportionally to the wire radius, since the total current only flows within the thin shielding layer.

We can also find an average critical *current density* at the surface. In this case, we replace the exponentially decaying current density by a distribution, in which the full current density at the surface remains constant to a depth λ_L, the penetration depth, and then abruptly drops to zero.[6] Based on this argument, for the tin wire at 0 K, we obtain a critical current density

$$j_{c0} = \frac{I_{c0}}{2\pi R \lambda_L(0)} = 7.9 \cdot 10^7\ \mathrm{A\,cm^{-2}} \tag{1.1.1.12}$$

where $R = 0.5$ mm, $\lambda_L(0) = 3 \cdot 10^{-6}$ cm, $I_{c0} = 75$ A.

This critical current density is similar to the critical pair-breaking current density of a thin wire of Sn. It would allow very high transport currents, if the shielding effect, leading to the restriction of the current to a thin surface layer, can be avoided. Such substances have been developed in form of the hard superconductors.

Using Silsbee's hypothesis, we can also calculate the critical currents of a superconductor in an external magnetic field. One only has to add the vectors of the external field and of the field of the transport current at the surface. The critical current density is reached, when this resulting field attains the critical value.

Next we turn to the type-II superconductors which differ in an important fundamental point from the type-I superconductors. For small magnetic fields and, hence, also for small transport currents, the type-II superconductors reside in the Meissner phase. In this phase, they behave like type-I superconductors, that is, they expel the magnetic field and the current into a thin surface layer. A difference to the type-I superconductors first appears when the magnetic field at the surface exceeds the value B_{c1}. Then the type-II superconductor must enter the Shubnikov phase, that is, flux lines must penetrate into the superconductor.

One finds that in the Shubnikov phase, an "ideal," that is, perfectly homogeneous, type-II superconductor has a finite electric resistance already at very small transport currents. On the other hand, in type-II superconductors containing a large amount of defects, we can observe very large super-currents. These are the "hard superconductors."

With respect to an ideal type-II superconductor, we consider a rectangular plate, carrying a current parallel to the plane of the plate and kept in the Shubnikov phase due to a magnetic field $B_a > B_{c1}$ oriented perpendicular to the plate (Figure 1.1.1.8).

As the first important result of such an experiment, one finds that under these conditions the transport current I is distributed over the total cross-section of the plate, that is, it is not completely restricted anymore to a thin surface layer. After the penetration of the magnetic flux into the superconducting sample, the transport current can flow also within the interior of the superconductor. The transport current, say, along the x-direction, also passes through the vortices, that is, through regions, where a magnetic field is present. This causes a Lorentz force between the vortices and the current. In the case of a current along a wire of length L in a perpendicular magnetic field B_a, the absolute magnitude of this force is $F = I \cdot L \cdot B$. It is oriented perpendicular to B and to the current (here given by the wire axis). Since the transport current is spatially fixed by the boundaries of the

6) Since the penetration depth is only a few 10^{-6} cm, for macroscopic wires we always have $R \gg \lambda_L$. Therefore, for our considerations, the surface of the wire can be treated as a plane.

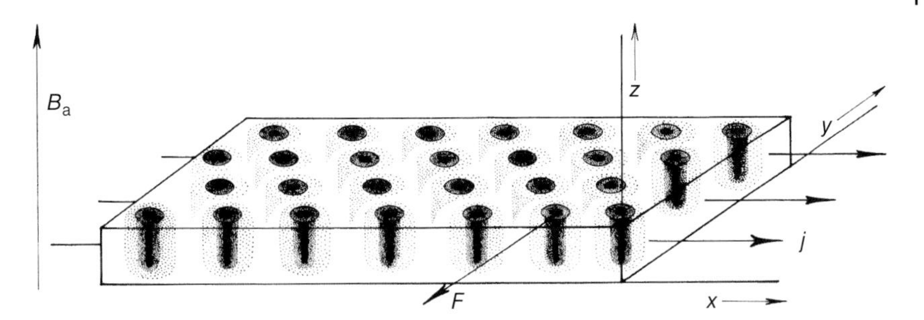

Figure 1.1.1.8 Shubnikov phase in the presence of a transport-current density *j*. The flux lines experience a force *F* driving them along the *y*-direction. The magnetic-field distribution around the flux lines is indicated by the hatching.

plate, under the influence of the Lorentz force the vortices must move perpendicular to the current direction and to the magnetic field, that is, perpendicular to their own axis [23]. For ideal type-II superconductors, in which the free motion of the vortices is possible, this vortex motion should appear already at arbitrarily small forces and, hence, at arbitrarily small transport currents. However, the vortex motion across the superconductor causes dissipation, that is, electric energy is changed into heat. This energy can only be taken from the transport current by means of an electric voltage appearing along the sample. Hence, the sample shows electric resistance.

Therefore, ideal type-II superconductors are useless for technical applications, say, for building magnets, in spite of their high critical field B_{c2}. Finite critical currents in the Shubnikov phase can only be obtained if the vortices in some way are bound to their locations.

Such pinning of the vortices can indeed be achieved by incorporating suitable "pinning centers" into the material. In the simplest way, we can understand the effect of pinning centers by means of an energy consideration. The formation of a vortex requires a certain amount of energy. This energy is contained, say, in the circulating currents flowing around the vortex core. We see that, for the given conditions, a vortex is associated with a certain amount of energy per unit length, that is, the longer the flux line the larger is also the energy needed for its generation. We denote this energy by ε^*. It can be estimated from the lower critical field B_{c1}, above which magnetic flux starts to penetrate into a type-II superconductor. The resulting gain in expulsion energy suffices for generating the vortices in the interior. For simplicity, we consider again a long cylinder in a magnetic field parallel to its axis, that is, a geometry with zero demagnetization coefficient. At B_{c1}, the penetration of the magnetic flux results in n flux lines per unit area. Each flux line carries just one flux quantum Φ_0. This requires the energy

$$\Delta E_F = n \cdot \varepsilon^* \cdot L \cdot F \tag{1.1.1.13}$$

where n is the number of flux lines per unit area, ε^* is the energy per unit length of vortex, L is the sample length, and F is the sample cross-section.

The gain in magnetic expulsion energy is

$$\Delta E_{\mathrm{M}} = B_{\mathrm{c1}} \cdot \Delta M \cdot V \tag{1.1.1.14}$$

where ΔM is the change of the magnetization of the sample and $V = L{\cdot}F$, is the sample volume.

ΔM can be expressed in terms of the penetrated flux quanta. We have

$$\Delta M = \frac{n \cdot \Phi_0}{\mu_0} \tag{1.1.1.15}$$

This yields for the gain in expulsion energy

$$\Delta E_M = \frac{1}{\mu_0} B_{\mathrm{c1}} \cdot n \cdot \Phi_0 L \cdot F \tag{1.1.1.16}$$

If both energy changes are being set equal ($\Delta E_{\mathrm{F}} = \Delta E_{\mathrm{M}}$), from the definition of B_{c1} we obtain

$$n \cdot \varepsilon^* \cdot L \cdot F = \frac{1}{\mu_0} \cdot B_{\mathrm{c1}} \cdot n \cdot \Phi_0 \cdot L \cdot F \tag{1.1.1.17}$$

and hence

$$\varepsilon^* = \frac{1}{\mu_0} \cdot B_{\mathrm{c1}} \cdot \Phi_0 \tag{1.1.1.18}$$

From our knowledge of the vortex energy ε^*, we can easily understand the pinning effect of normal precipitates. If a vortex can pass through a normal-conducting inclusion, its length within the superconducting phase and thereby its energy are

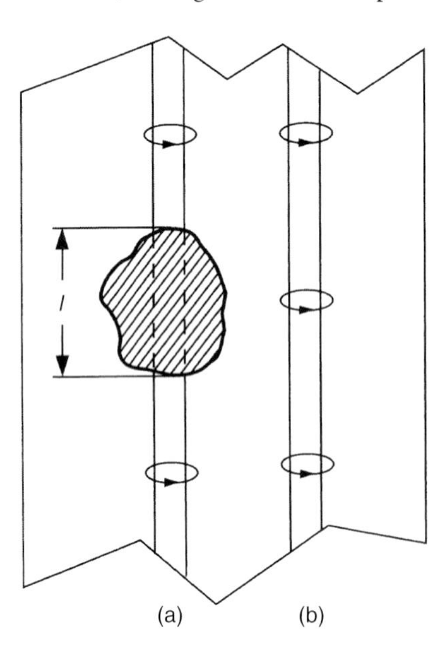

Figure 1.1.1.9 Pinning effect of normal-conducting precipitates. In location (a), the effective length of the vortex is shorter compared to location (b), since in the normal-conducting region there are no circulating currents.

reduced. In Figure 1.1.1.9, this is schematically indicated. The hatched region indicates the normal inclusion. A vortex in location (a) has an energy smaller by the amount $\varepsilon^{*}\cdot l$ compared to one in location (b). This means that we must supply the energy $\varepsilon^{*}\cdot l$ to the vortex, in order to move it from (a) to (b). Hence, a force is needed to effect this change in location.

If there are many pinning centers, the vortices will attempt to occupy the energetically most favorable locations. As shown in Figure 1.1.1.10, they will also bend in order to reach the minimum value of the total energy. The length increment caused by the bending must be overcompensated by the effective shortening within the normal-conducting regions. In a vortex lattice, as it is generated in the Shubnikov phase, in the total energy balance we must take into account also the repulsive forces acting between the flux lines.

In principle, also other pinning centers, say, lattice defects, can be understood in the same way. Every inhomogeneity of the material, which is less favorable for superconductivity, acts as a pinning center, with the completely normal state representing the limiting case. For example, superconducting precipitates, however, with a lower transition temperature in general act as pinning centers. We will not

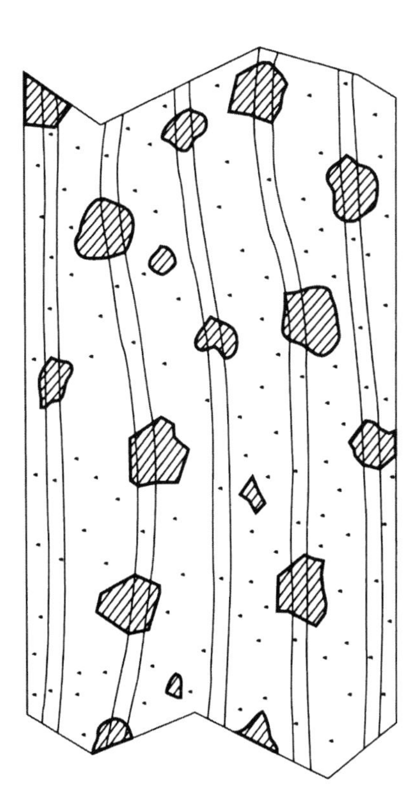

Figure 1.1.1.10 Vortex configuration in a hard superconductor. The hatched regions represent pinning centers. The dots indicate atomic defects.

go into details but mention details can be very complicated. It is still a high art to obtain superconductors which sustain a large transport current.

The effect of the pinning centers can also be described in terms of an energy landscape. Now the pinning center represents a potential well of depth E_p. The vortex is located at its most favorable position, similar to a ball at the lowest point of a bowl. If the ball is to be displaced from this location, one needs a force in order to supply the increase of the potential energy. For removing the ball from its most favorable location, we must supply the energy needed to lift the ball out of the bowl. Usually, in a material there exist many pinning centers, which are irregularly distributed and which have different energy depths E_p. If the superconductor is cooled below T_c in a magnetic field, the vortices will quickly occupy the potential wells, instead of generating a regular triangular lattice. At best, we have a distorted lattice, or in the extreme case even a glassy state [24].

The deviation of an individual vortex from its ideal location within the triangular vortex lattice depends not only on the depth of the potential wells but also on the configuration of all other vortices, because of the repulsive interaction between them. An energetically highly unfavorable arrangement of the vortices will be changed quickly because of the thermal fluctuations. These fluctuations can provide the energy difference ΔE, needed for leaving the potential well, with a probability $w = \exp(-\Delta E/k_B T)$. In this case, the thermodynamic fluctuations can reduce the depth of the potential well, or they can supply the missing energy to the vortex. At low temperatures and for large values of ΔE, this probability can become very small, such that the state with the lowest energy cannot be occupied anymore. Furthermore, because of the interaction between the vortices, ΔE can approach infinity. In this case, we deal with the state of the vortex glass, which experiences no changes anymore within finite times.

Next, we want to discuss the effect of the pinning centers during the current transport in superconducting wires or thin films. We have seen that an ideal type-II superconductor in the Shubnikov phase cannot carry a current perpendicular to the direction of the magnetic field without dissipation. However, in a *real* superconductor, the vortices are never completely freely mobile. There is always a perhaps very small force necessary in order to tear the vortices off the pinning centers which are practically always present. The strength of the pinning forces acting on the individual vortices will have a certain distribution about an average value F_H. Also, the whole vortex lattice will affect the pinning forces due to collective effects. However, for simplicity, we will only speak of a single pinning force F_H.

As long as the Lorentz force F_L is smaller than the pinning force F_H, the vortices cannot move. Therefore, also in every real type-II superconductor in the Shubnikov phase, we will be able to observe current flow without dissipation. If the transport current exceeds its critical value at which $F_L = F_H$, the vortex motion sets in, and electric resistance appears.[7] We see that the critical current

7) If the pinning forces acting on the individual vortices are different, initially the most weakly pinned vortices will start moving, resulting in only a relatively small resistance. With increasing current, their number and, hence, the sample resistance will approach a certain limiting value.

is a measure of the force F_H, with which the vortices are pinned at energetically favored locations.

By means of a systematic study of the hard superconductors, one has been able to develop empirically quite useful materials.

To return to the effect of levitation – applications of this effect are discussed in Sections 4.1 and 4.2 – let us now consider a hard superconductor which is cooled in the field of a permanent magnet. We assume that this field is well above B_{c1}.

Quite in contrast to a standard permanent magnet, but also in contrast to an ideal type-I or type-II superconductor, the hard superconductor will try to keep the field in its interior at the value, at which it was cooled down. After they are pinned, the flux lines do not move anymore as long as the maximum pinning force of the pinning centers is not exceeded. If a hard superconductor is cooled down within a certain distance above a permanent magnet, an *attractive* force is active, if the superconductor is moved away from the magnet. In the same way, a *repulsive* force is active, if the superconductor is moved closer to the permanent magnet. The same applies in the case of any arbitrary directions of the movement. As soon as the external field changes, the hard superconductor generates shielding currents in such a way, that the field (or the vortex lattice) remains unchanged in its interior. Therefore, a hard superconductor, including a loading weight, can not only float above a magnet but also hang freely below a magnet, or placed at an arbitrary angle. This effect is demonstrated in Figure 1.1.1.11 [25]. In this case, properly prepared little blocks of $YBa_2Cu_3O_7$ were mounted within a toy train, and the blocks were cooled down within a certain working distance from the magnets, forming the "train tracks." The train can move along the track practically without friction, since the magnetic field keeps its value along this direction.

When, say in a magnet, the magnetic field is swept between two large values $\pm B_{max}$, vortices are forced to enter and leave the superconductor once the

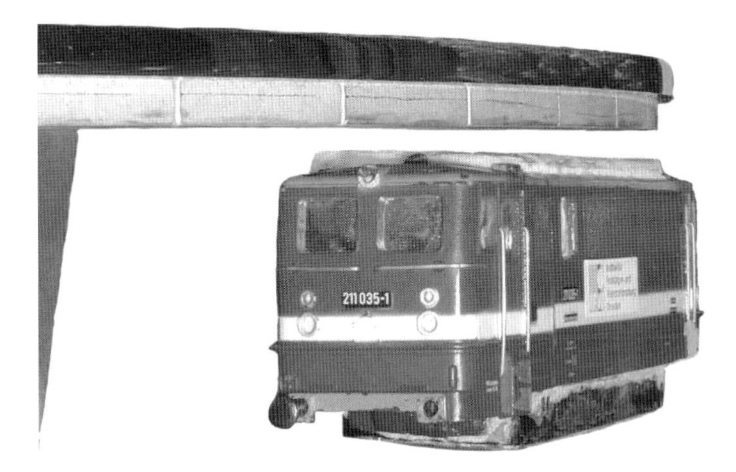

Figure 1.1.1.11 Hanging toy train [25]. (Institut für Festkörper- und Werkstoffforschung, Dresden. From [1].)

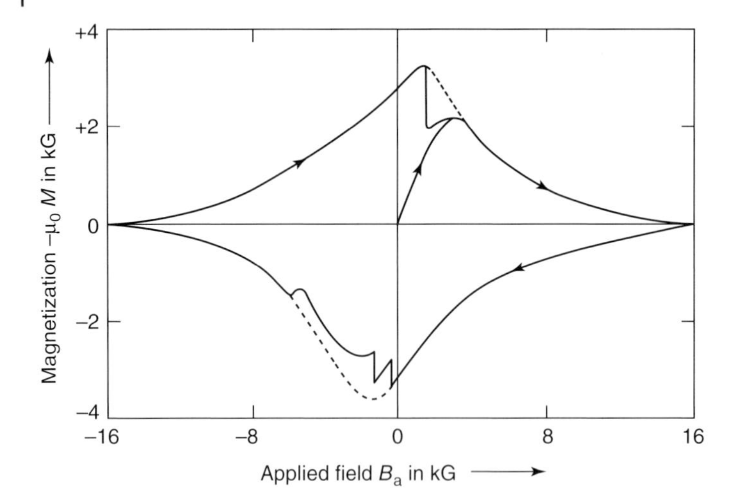

Figure 1.1.1.12 Complete magnetization cycle of a Pb–Bi alloy [26]. The jumps on the solid curve are due to jumps of magnetic flux lines. The dashed curve is expected if there were no such jumps (1 kG = 0.1 T). (From [1].)

pinning force is surpassed. This leads to a hysteresis in the magnetization of the superconductor, see Figure 1.1.1.12. Accompanied with this are, for example, hysteresis losses for alternating magnetic fields.

We want to conclude our discussion of the critical currents in superconductors with a few general remarks. We have seen that the mechanism of pair-breaking results in an intrinsic maximum super-current density. However, in the cases which are technically relevant, the critical current of a superconductor is determined by *extrinsic* properties. On the one hand, the latter properties in the form of pinning centers in the Shubnikov phase only allow a finite super-current, and on the other hand, for example, in the form of grain boundaries in high-temperature superconductors, represent weak regions in the material strongly reducing the maximum super-current. The question, if a new material, say, the iron pnictides, finds interesting technical applications depends on the concrete problems and often can be answered only after a long development period.

In summary, in this chapter, we have seen that the main ingredient of the superconducting state is that electron pairs (Cooper pairs) form a macroscopic matter wave. For conventional superconductors, as described by the BCS theory, the electrons interact via phonons. There are also unconventional superconductors like the cuprates where the pairing mechanism is not yet clear. The well-defined phase of the matter wave leads us to the ideal diamagnetism at not too large fields and to the vortex state in type-II superconductors. Interference effects of coupled matter waves are the basis of the physics of Josephson junctions and of SQUIDs. We have further introduced important length scales like the London penetration depth λ_L (the scale over which magnetic fields decay inside the superconductor), the Ginzburg–Landau coherence length ξ_{GL} (the scale over which the amplitude of the matter wave and thus the Cooper pair density varies), and the BCS coherence length ξ_0 (the scale over which the two partners of a Cooper pair are correlated). We also mentioned that unpaired electrons are separated by an energy gap Δ from a Cooper pair and we have seen that there is a maximum field, as well as a maximum current a superconductor can carry.

References

1. Buckel, W. and Kleiner, R. (2004) *Superconductivity, Fundamentals and Applications*, 2nd edn, Wiley-VCH Verlag GmbH.
2. Buckel, W. and Kleiner, R. (2012) *Supraleitung, Grundlagen und Anwendungen*, 7th German edn, Wiley-VCH Verlag GmbH.
3. Kamerlingh-Onnes, H. (1908) *Proc. R. Acad. Amsterdam*, **11**, 168.
4. Kamerlingh-Onnes, H. (1913) *Commun. Leiden*, Suppl. Nr. **34**.
5. Kamerlingh-Onnes, H. (1911) *Commun. Leiden*, **120b**.
6. Bardeen, J., Cooper, L.N., and Schrieffer, J.R. (1957) *Phys. Rev.*, **108**, 1175.
7. Bednorz, J.G. and Müller, K.A. (1986) *Z. Phys. B*, **64**, 189.
8. Raub, C.J. (1988) *J. Less-Common Met.*, **137**, 287.
9. (a) Wu, M.K., Ashburn, J.R., Torng, C.J., Hor, P.H., Meng, R.L., Gao, L., Huang, Z.J., and Chu, C.W. (1987) *Phys. Rev. Lett.*, **58**, 908; (b) Zhao, Z.X. (1987) *Int. J. Mod. Phys. B*, **1**, 179.
10. Kirtley, J.R. and Tsuei, C.C. (1996) *Spektrum der Wissenschaften*, German edition of Scientific American, p. 58, Oktober 1996.
11. Nagamatsu, J., Nakagawa, N., Muranaka, T., Zenitani, Y., and Akimitsu, J. (2001) *Nature*, **410**, 63.
12. Takahashi, H., Igawa, K., Arii, K., Kamihara, Y., Hirano, M., and Hosono, H. (2008) *Nature*, **453**, 376.
13. Meissner, W. and Ochsenfeld, R. (1933) *Naturwissenschaften*, **21**, 787.
14. (a) Cribier, D., Jacrot, B., Madhav Rao, L., and Farnoux, B. (1964) *Phys. Lett.*, **9**, 106; (b) see also:Gorter, C.J. (ed.) (1967) *Progress Low Temperature Physics*, vol. **5**, North Holland Publishing Comp, Amsterdam, p. 161 ff.
15. Essmann, U. and Träuble, H. (1967) *Phys. Lett.*, **24 A**, 526 and *J. Sci. Instrum.* (1966) **43**, 344.
16. Doll, R. and Näbauer, M. (1961) *Phys. Rev. Lett.*, **7**, 51.
17. Deaver, B.S. Jr., and Fairbank, W.M. (1961) *Phys. Rev. Lett.*, **7**, 43.
18. (a) London, F. and London, H. (1935) *Z. Phys.*, **96**, 359; (b) London, F. (1937) *Une conception nouvelle de la supraconductivite*, Hermann and Cie, Paris.
19. Lock, J.M. (1951) *Proc. R. Soc. London, Ser. A*, **208**, 391.
20. Jackson, T.J., Riseman, T.M., Forgan, E.M., Glückler, H., Prokscha, T., Morenzoni, E., Pleines, M., Niedermayer, C., Schatz, G., Luetkens, H., and Litterst, J. (2000) *Phys. Rev. Lett.*, **84**, 4958.
21. Pippard, A.B. (1950) *Proc. R. Soc. London, Ser. A*, **203**, 210.
22. Silsbee, F.B. (1916) *J. Wash. Acad. Sci.*, **6**, 597.
23. Gorter, C.J. (1962) *Phys. Lett.*, **1**, 69.
24. Blatter, G., Feigel'man, M.V., Geshkenbein, V.B., Larkin, A.I., and Vinokur, V.M. (1994) *Rev. Mod. Phys.*, **66**, 1125.
25. Schultz, L., Krabbes, G., Fuchs, G., Pfeiffer, W., and Müller, K.-H. (2002) *Z. Metallkd.*, **93**, 1057.
26. (a) Campbell, A.M., Evetts, J.E., and Dew Hughes, D. (1964) *Philos. Mag.*, **10**, 333; (b) Evetts, J.E., Campbell, A.M., and Dew Hughes, D. (1964) *Philos. Mag.*, **10**, 339.

1.1.2
Review on Superconducting Materials

Roland Hott, Reinhold Kleiner, Thomas Wolf, and Gertrud Zwicknagl

1.1.2.1 Introduction

The discovery of superconductivity was the result of straightforward research to see how low one can go concerning the electrical resistance of metals: studies on alloys and temperature-dependent measurements had evidenced that it could be decreased by reducing the density of impure atoms as well as by lowering temperature. Mercury offered the best low-impurity perspectives – Kamerlingh Onnes had built up in Leiden a unique cryogenic facility: the jump to apparently zero resistivity that he observed here in 1911 below 4 K came nevertheless as a big surprise [1]. He soon extended the list of superconducting (SC) materials by tin (3.7 K) and lead (7.2 K), and his Leiden successors found thallium (2.4 K) and indium (3.4 K) [2]. Meißner successfully continued the search through the periodic table until 1930 with tantalum (4.2 K), thorium (1.4 K), titanium (0.4 K), vanadium (5.3 K), and niobium, the element with the highest critical temperature, $T_c = 9.2$ K [3] (Figure 1.1.2.1). The extension to binary alloys and compounds in 1928 by de Haas and Voogd was fruitful with SbSn, Sb_2Sn, Cu_3Sn, and Bi_5Tl_3 [4]. Bi_5Tl_3 and, shortly afterwards, a Pb–Bi eutectic alloy established first examples of critical magnetic field values B_{c2} in the tesla range, which revived hope for high-field persistent current SC electromagnets as already envisioned by Kamerlingh Onnes.

After 1930, SC materials research fell more or less asleep until Matthias and Hulm started in the early 1950s a huge systematic search which delivered a number of new compounds with $T_c > 10$ K as well as technically attractive $B_{c2} > 10$ T: NbTi ($T_c = 9.2$ K) and the A15 materials were the most prominent examples. Matthias

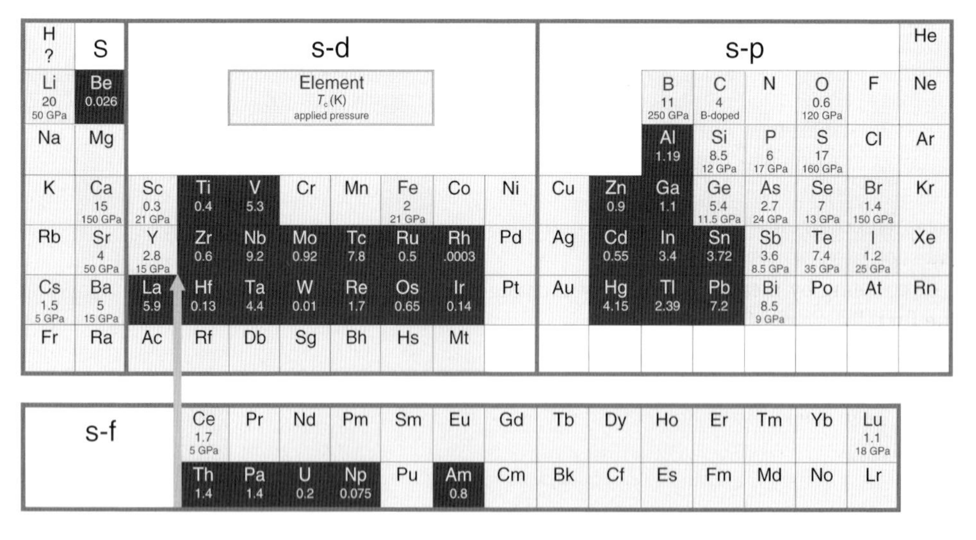

Figure 1.1.2.1 Periodic table with the distribution and T_c [K] of the chemical elements for which superconductivity has been observed with or without application of pressure [1, 5, 6].

condensed his huge practical knowledge from his heroic preparation of some 3000 different alloys into "rules" on how to prepare "good" superconductors: high crystal symmetry, high density of electronic states at the Fermi level, no oxygen, no magnetism, no insulators! [7].

In spite of his inofficial sixth rule "Stay away from theorists!" in 1957 the *Bardeen–Cooper–Schrieffer (BCS) theory* [8] brought the desperately awaited breakthrough of theoretical solid state physics to a microscopic explanation of superconductivity. The key idea of BCS theory is that in metals even a tiny attractive interaction between the conduction electrons results in the formation of bound electron pair states ("Cooper pairs") which are no longer obliged to obey the Fermi–Dirac statistics which enforced the electrons to occupy high kinetic energy single particle states due to the Pauli principle. The energy gain of the SC state with respect to the normal state does not result from the small binding energy of the pairs, but it is the condensation energy of the pairs merging into a macroscopic quantum state which can be measured as an energy gap for electron excitations into single particle states. Although the BCS theory was derived from the physical idea of attractive electron–phonon coupling, the model-based weak pair coupling theory as its mathematical kernel is well applicable to other pairing mechanisms. BCS theory had an impact not only on solid-state physics but also on elementary particle physics where it was further developed to the Higgs mechanism of mass generation [9].

In 1979, in violation of another Matthias rule, superconductivity was discovered in the magnetic material $CeCu_2Si_2$ as the first representative of *heavy-fermion (HF) superconductors* [10] where magnetism is suspected as mechanism responsible for the Cooper pairing: in these intermetallic compounds, the electronic degrees of freedom which are responsible for superconductivity are directly linked with magnetic moments of partially filled f-shells of lanthanide or actinide atoms. The superconductivity below a typical $T_c \sim 1$ K seems to arise here from the delicate balance between the localized magnetic moments which try to imprint their magnetic signature on the shielding conduction electrons, and the conduction electrons which try to neutralize these magnetic moments by spin flipping, for example, via Kondo effect [11].

The search for *organic superconductors* had been boosted in the 1960s by the idea that conductive polymer chains with polarizable molecular groups may provide for electrons running along the polymer chains a highly effective Cooper pair coupling by means of an energy exchange via localized excitons [12]. Since the first discovery of an organic superconductor in 1980 [13] $T_c > 10$ K has been achieved [14]. However, the origin of superconductivity has turned out to be far from the suggested excitonic mechanism. Electric conduction stems here from π-electrons in stacked aromatic rings forming one-dimensional or two-dimensional (2D) delocalized electron systems. This restriction of the effective dimensionality and strong Coulomb repulsion effects push the systems toward metal-insulator, magnetic, and SC transitions [15].

The Mermin–Wagner theorem [16] that long-range order cannot exist in two dimensions at finite temperature due to strong fluctuations seemed to restrict superconductivity to the physical dimension $d = 3$. The *cuprate high-temperature superconductors (HTS)* [17] proved in 1986 that the limiting case $d = 2 + \varepsilon$ ($\varepsilon \to 0$), that is, basically 2D CuO layer-oriented superconductivity with slight SC coupling to neighboring CuO layers, can even be enormously beneficial for SC long-range order [18]. The problem for a theoretical description of cuprate HTS within BCS theory and its extensions [19] is not the high T_c of up to 138 K under normal pressure [20], far above the pre-HTS record of 23 K [21]. There is no theoretical argument why a textbook phonon BCS superconductor should not achieve such a high T_c: in the McMillan–Rowell formula [22], the commonly used theoretical T_c approximation, T_c, depends in a very sensitive way on the involved material parameters. The HTS T_c range is readily accessible with a still reasonable parameter choice [23]. The real problem is that, in contrast to the "deep" Fermi sea of quasi-free electrons in classical metals where the Cooper-pair condensed electrons amount only to a small part of the valence electron system ($k_B T_c \ll E_{Fermi}$), in these layered cuprate compounds there is only a "shallow" reservoir of charge carriers ($k_B T_c \sim E_{Fermi}$) which first have to be introduced in the insulating antiferromagnetic (AF) stoichiometric parent compound by appropriate doping. The thus generated normal-conducting state corresponds to a "bad metal" in which Coulomb correlations strongly link the charge and spin degrees of freedom. The BCS recipe to express the SC wavefunction in terms of the normal-metal single particle states does not work here anymore since the macroscopic many-particle wavefunction is thoroughly changing in the superconductive transition: additional electronic degrees of freedom come into play which had not been accessible before in the normal-conducting state. This transition process still lacks a satisfactory theoretical description [24, 25]. Nevertheless, the SC instability in cuprate HTS, as well as in the structurally and chemically related layered cobaltate and ruthenate compounds, is believed to stem predominantly from a magnetic and not from a phononic interaction as in the case of the classical metallic superconductors where magnetism plays only the role of an alternative, intrinsically antagonistic long-range order instability.

Fullerides (C_{60}, C_{70}, ...) discovered in 1985 are a third modification of elementary carbon. The superconductivity in C_{60} induced by doping and intercalation of alkali-metal atoms, with T_c values up to 33 K at normal pressure [26], followed soon as another surprise. In spite of the high T_c, superconductivity can be explained by BCS theory based on intramolecular phonons [27]. *Borides* were investigated with respect to high-T_c superconductivity already in the 1950s: the rationale was the BCS T_c-formula [22] where a high characteristic phonon frequency, as provided by the light boron atoms, was predicted to be particularly helpful. In the 1990s, the borocarbide superconductors RE Ni_2B_2C with T_c up to 16.5 K [28] fulfilled this promise at least halfway. However, phonons are here apparently only one of the contributing superconductivity mechanisms: additional magnetism due to localized RE^{3+} 4f-electrons is here weakly interacting with the SC 3d-electrons of the Ni_2B_2 layers. The huge surprise came in 2001 with the discovery of superconductivity up to $T_c = 40$ K in MgB_2, a compound which was well known since the 1950s and which was in 2001 already commercially available in quantities up to metric tons [29]. In spite of the high T_c, a phononic mechanism is here highly plausible. As a new feature, multiband superconductivity, that is, the coherent coupling of the Cooper-pair instabilities of several Fermi surfaces [30], is essential for the theoretical description of the SC properties [31]. This multiband mechanism plays an even more dominant role in *iron-based superconductors* [32] where the scattering between the Fermi surfaces of up to five Fe-derived electronic bands is apparently the origin of a complicated magnetic superconductive coupling mechanism [33]. $T_c = 26$ K observed for LaFeAsO in 2008 [34] started a "gold rush" where T_c was immediately pushed up to 55 K [35].

The Matthias rules paradigm has thus changed completely: layered materials with strong electronic correlations – that is where you can expect new high-T_c superconductors!

1.1.2.2 Cuprate High-Temperature Superconductors

Cuprate HTS have played an outstanding role in the scientific and technological development of superconductors due to the enormous efforts made to cope with the challenges due to the plethora of preparational degrees of freedom and the inherent tendency toward inhomogeneities and defects in combination with the very short SC coherence lengths of the order of the dimensions of the crystallographic unit cell.

The structural element of HTS compounds related to the location of mobile charge carriers are stacks of a certain number $n = 1, 2, 3, \ldots$ of CuO_2 layers "glued" on top of each other by means of intermediate Ca layers (see Figure 1.1.2.2) [36]. Counterpart of these *active blocks* of $(CuO_2/Ca/)_{n-1}CuO_2$ stacks are *charge reservoir blocks* $EO/(AO_x)_m/EO$ with $m = 1, 2$ monolayers of a quite arbitrary oxide AO_x "wrapped" on each side by a monolayer of alkaline earth oxide EO with E = Ba, Sr (see Figure 1.1.2.2b). The HTS structure results from alternating stacking of these two block units. The general chemical formula

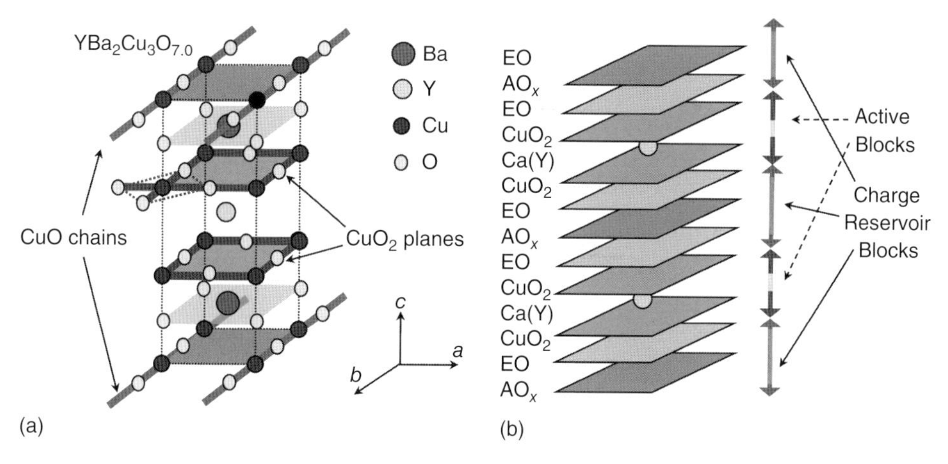

Figure 1.1.2.2 (a) Crystal structure of $YBa_2Cu_3O_7$ ("YBCO") [39]. (b) General structure of a cuprate HTS A-$m2(n-1)n$ ($A_mE_2Ca_{n-1}Cu_nO_{2n+m+2+y}$) for $m = 1$.

$A_mE_2Ca_{n-1}Cu_nO_{2n+m+2+y}$ (see Figure 1.1.2.2b) is conveniently abbreviated as **A-*m*2(*n* − 1)*n*** [37] (e.g., $Bi_2Sr_2Ca_2Cu_3O_{10}$: Bi-2223) neglecting the indication of the alkaline earth element (see Figure 1.1.2.1). The family of all $n = 1, 2, 3, \ldots$ representatives with common AO_x are referred to as *A-HTS*, for example, Bi-HTS. The most prominent compound $\mathbf{YBa_2Cu_3O_7}$ (see Figure 1.1.2.2a), the first HTS discovered with a critical temperature T_c above the boiling point of liquid nitrogen [38], traditionally abbreviated as "**YBCO**" or "**Y-123**" ($\mathbf{Y_1Ba_2Cu_3O_{7-\delta}}$), fits into this classification scheme as a modification of Cu-1212 where Ca is completely substituted by Y.

The following scenario (see Figure 1.1.2.3) applies to *hole-doping* [40] as well as to *electron-doping* of all HTS [41]: the undoped compounds are AF insulators up to a critical temperature T_N well above 300 K, with alternating spin orientations of the hole states that are localized around the Cu atoms in the CuO_2 layers. Adding charge carriers by doping relaxes the restrictions of spin alignment: T_N decreases and the insulator turns into a "bad metal." At low temperature, the electric transport shows a dramatic change within a small doping range from an insulating to a SC behavior [42]. For $La_{2-x}Sr_xCuO_4$, this happens at a critical hole concentration $x = 0.05$ in the CuO_2 planes (see Figure 1.1.2.2). On stronger doping, superconductivity can be observed up to an increasingly higher critical temperature T_c until the maximum T_c is achieved for "optimal doping" ($x \approx 0.16$ for $La_{2-x}Sr_xCuO_4$). On further doping, T_c decreases again until finally ($x \geq 0.27$ for $La_{2-x}Sr_xCuO_4$) only normal conducting behavior is observed.

The rationale that the phenomenon of superconductivity in HTS can be conceptually reduced to the physics of the CuO_2 layers [44] has evolved to a more and more 2D view in terms of CuO_2 *planes*. The superconductive coupling

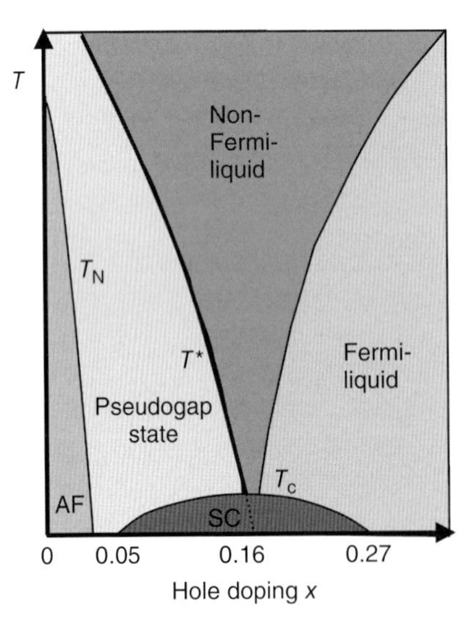

Figure 1.1.2.3 HTS temperature-doping phase diagram with the interplay of antiferromagnetism (AF) and superconductivity (SC) [40, 43].

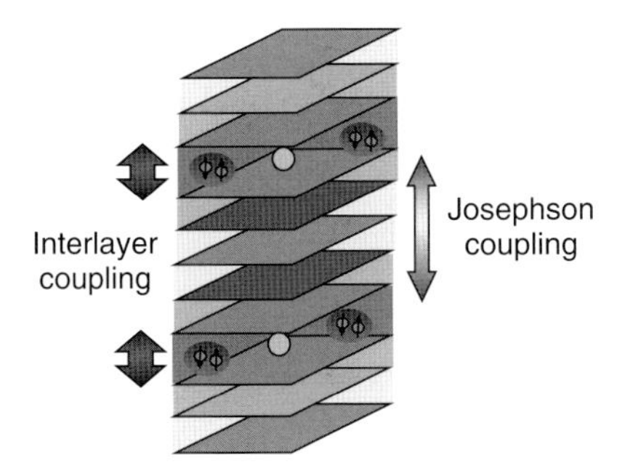

Figure 1.1.2.4 Hierarchy of the superconductive coupling in cuprate HTS.

between these planes within a given $(CuO_2/Ca/)_{n-1}CuO_2$ stack ("interplane coupling") is much weaker than the intraplane coupling, but still much stronger than the superconductive coupling between the $(CuO_2/Ca/)_{n-1}CuO_2$ stacks which can be described as Josephson coupling (see Figure 1.1.2.4).

HTS are extreme type-II superconductors [45] with $\lambda > 100$ nm and $\xi \sim 1$ nm. The quasi-2D nature of superconductivity in HTS leads to a pronounced anisotropy of the SC properties with much higher super-currents along the CuO_2 planes than in the perpendicular direction [46] and a corresponding anisotropy of the magnetic penetration depth λ, for example, $\lambda_{ab} = 750$ nm and $\lambda_c = 150$ nm in optimally doped YBCO [47] (the indices refer to the respective orientation of the magnetic field). Material imperfections of the dimension of the coherence length which are required as pinning centers preventing the flux flow of magnetic vortices are easily encountered in HTS due to their small coherence lengths, for example, for optimally doped YBCO $\xi_{ab} = 1.6$ nm, $\xi_c = 0.3$ nm for $T \rightarrow 0$ K [48] which are already comparable to the lattice parameters (YBCO: $a = 0.382$ nm, $b = 0.389$ nm, $c = 1.167$ nm [39]). The high T_c in combination with the small value of coherence volume $(\xi_{ab})^2\xi_c \sim 1$ nm^3 allows large thermally induced magnetic fluctuations in the SC phase at temperature close to T_c, an effect which could be completely neglected in classical superconductors [4]. Moreover, since technical superconductor materials consist of a network of connected grains, already small imperfections at the grain boundaries with spatial extensions of the order of the coherence length lead to a substantial weakening of the SC connection of the grains and thus to "weak-link behavior" of the transport properties which has to be avoided in technical conductor materials [49]. On the other hand, this has been widely exploited for the fabrication of HTS Josephson junctions [50].

The low ξ_c, that is, the weak superconductive coupling between the $(CuO_2/Ca/)_{n-1}CuO_2$ stacks, may lead for c-axis transport to an intrinsic Josephson effect within the unit cell even for perfectly single-crystalline materials [51]. If the thickness of the charge reservoir blocks $EO/(AO_x)_m/EO$ in-between these stacks is larger than ξ_c, vortices are here no longer well defined due to the low Cooper pair density (see Figure 1.1.2.5). This leads to a quasi-disintegration of the vortices into stacks of *pancake vortices* which are much more flexible entities

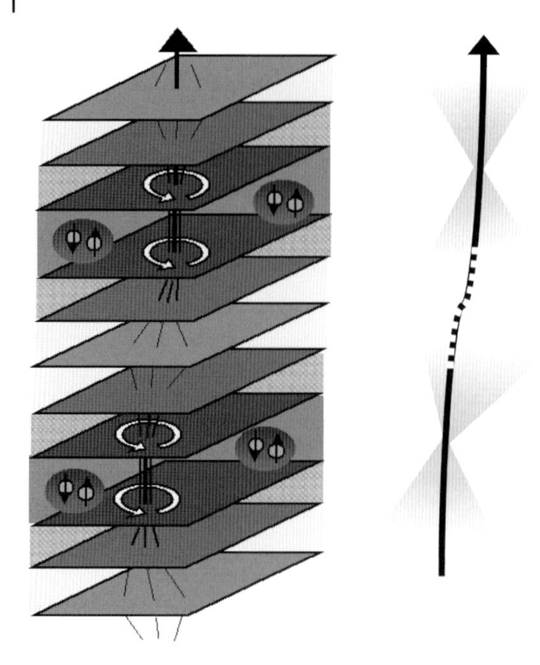

Figure 1.1.2.5 Quasi-disintegration of magnetic vortex lines into "pancake" vortices [52].

than the continuous quasi-rigid vortex lines in conventional superconductors. The effects described in the preceding two paragraphs combine to reduce the *irreversibility field* $B_{irr}[T]$, the tolerable limit for magnetic fields with respect to SC transport, in cuprate HTS substantially below the thermodynamical critical field $B_{c2}[T]$, a distinction which was more or less only of academic interest in the case of classical superconductor.

Besides these intrinsic obstacles for the transport of super-current in single-crystalline HTS materials, there are additional hurdles since HTS materials are not a homogeneous continuum but rather a network of linked grains (see Figure 1.1.2.6). The mechanism of crystal growth is such that material that cannot be fitted into the lattice structure of the growing grains is pushed forward into the growth front with the consequence that in the end all remnants of secondary phases and impurities are concentrated at the boundaries in-between the grains. Such barriers impede the current transport even if they consist of only a few atomic layers and have to be avoided by careful control of the growth process, in particular of the composition of the offered material. Another obstacle for super-currents in HTS (which is not only detrimental for transport currents but also enables the fabrication of HTS Josephson junctions) is misalignment of the grains: exponential degradation of the super-current transport is observed as a function of the misalignment angle due to the d-symmetry of the SC order parameter [53] and even more due the build-up of charge inhomogeneities [54].

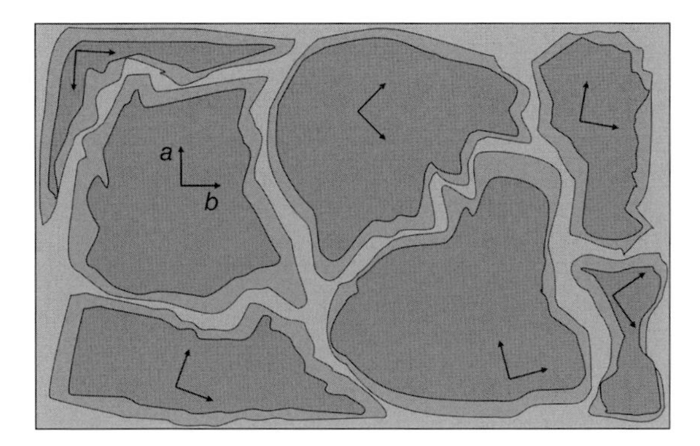

Figure 1.1.2.6 Schematic HTS microstructure with differently oriented single crystal grains separated by secondary phase regions. Oxygen depletion and thus T_c reduction may occur at grain boundaries.

1.1.2.3 **Other Oxide Superconductors**

The discovery of superconductivity in the *bismuthate* $BaPb_{1-x}Bi_xO_3$ in 1975 with a rather high $T_c \sim 13$ K in those days for $x \sim 0.25$ [55] raised great interest in the mechanism of superconductivity in this at that time quite exotic oxide compound with a low density of states at the Fermi level. The cuprate HTS soon chased away that exotic touch in spite of the rise of T_c to >30 K in $Ba_{1-x}K_xBiO_3$ (BKBO; $x \sim 0.35$) in the middle of the HTS bonanza days [56]: tunneling showed clean gap structures consistent with weak-to-moderate coupling BCS theory [57]. In the parent compound $BaBiO_3$, a three-dimensional (3D) charge-density wave (CDW) arrangement of $Bi^{(4-\delta)+}O_6$ and $Bi^{(4+\delta)+}O_6$ octahedra ($|\delta| \ll 1$) creates a gap at the Fermi level and leads to an insulating electric behavior. K or Pb doping suppresses this CDW by means of the random occupation of the A position with Ba and K or Pb ions in a simple pseudo-cubic ABO_3 solid solution structure [58]. Furthermore, this doping introduces hole carriers and finally results in a metal-insulator transition at a critical doping level $x_c \sim 0.35$. The maximum T_c occurs for slightly higher doping. On further doping, T_c rapidly decreases and finally disappears at the K solubility limit $x \sim 0.65$. The SC pairing mechanism is apparently related to the structural and concomitant electronic 3D CDW instability.

The extensive search for other SC transition metal oxides following the discovery of the cuprate HTS came in 1994 across strontium *ruthenate* (Sr_2RuO_4), a layered perovskite with an almost identical crystal structure as the cuprate HTS $La_{2-x}Sr_xCuO_4$ ("LSCO"), albeit only with a $T_c \sim 1.5$ K [59]. In both materials, the conduction electrons stem from partially filled d-bands (of the Ru or Cu ions, respectively) that are strongly hybridized with oxygen p-orbitals. In contrast to the nearly filled Cu 3d-shell in cuprate HTS with only one hole state, in Sr_2RuO_4, in the formal oxidation state of the ruthenium ion Ru^{4+} four electrons are left in the 4d-shell. The closely related ferromagnetic material $SrRuO_3$ shows the inherent tendency of Ru^{4+} toward ferromagnetism. Hence, in analogy with the cuprate HTS, where on doping the AF ground state of the parent compounds seems to "dissolve" in spin-singlet Cooper pairs in a d-wave orbital channel, it was suggested that the superconductivity in Sr_2RuO_4 is brought about by spin-triplet pairing where the Ru ions "release" parallel-spin, that is, triplet Cooper pairs in p-wave or even higher odd order angular orbital channels.

$RuSr_2GdCu_2O_8$ (Ru-1212) is a *ruthenate-cuprate hybrid* containing both CuO_2 and RuO_2 layers. It fits into the elucidated cuprate HTS layer structure scheme (see Figure 1.1.2.2b) substituting the Ca of the canonical 1212-HTS structure (or the Y in YBCO, or the RE in RE-123, respectively) by Gd to render CuO_2/Gd/CuO_2 stacks, separated by a SrO "wrapping layer" from the RuO_2 layers as "charge reservoir layers." Like rare earth borocarbides (see Chapter 7), Ru-1212 and some other closely related rutheno-cuprate compounds display ferromagnetism and superconductivity coexisting on a microscopic scale [60], with $T_{Curie} \sim 135$ K and T_c up to 72 K for Ru-1212. The CuO_2/Gd/CuO_2 stacks are believed to be responsible for the superconductivity, whereas the (ferro)magnetic ordering arises from the RuO_2 layers. A clear intrinsic Josephson effect shows that the material acts as a natural superconductor–insulator–ferromagnet–insulator–superconductor superlattice [61].

Cobaltates made in 2003 their entry into the SC zoo. $T_c = 4.5$ K has been achieved in hydrated sodium cobaltate $Na_{0.3}CoO_2 \cdot 1.4\ H_2O$ [62]. Na provides here the doping. The intercalation of water increases the separation between the CoO_2 layers and seems to be essential for the onset of superconductivity: $Na_{0.3}CoO_2 \cdot 0.6\ H_2O$, with the same formal Co oxidation state but substantially less separation between the CoO_2 layers is not SC [63]. A major difference compared to cuprate HTS is the triangular lattice geometry of the CoO_2 layers which introduces magnetic frustration into the Co spin lattice, in contrast to the square lattice geometry of Cu ions in cuprate HTS which favors unfrustrated AF spin orientation.

β-Pyrochlore oxide superconductors AOs_2O_6 with A = K, Rb, Cs, and respective $T_c = 9.6, 6.3, 3.3$ K [64] have a triangle-based crystal structure, which is in principle even more subject to magnetic frustration [65]. The A ion sits here in a cage formed by the surrounding OsO_6 tetrahedra (see Figure 1.1.2.7) [66]. Anomalous phonons are observed as an anharmonic oscillation ("rattling" motion) of the A ion cage [67]. Intriguingly, the rattling motion participates in the SC properties

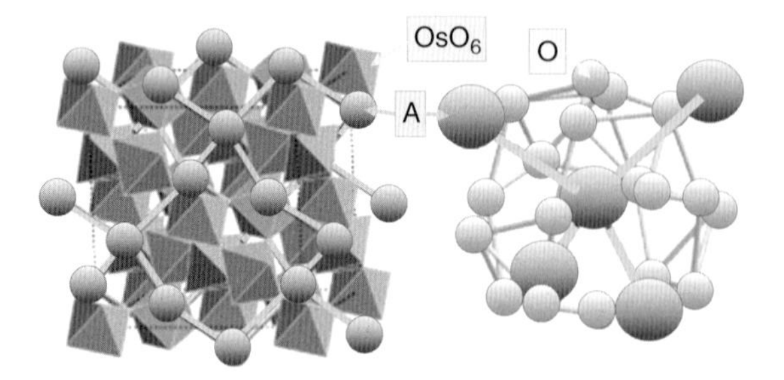

Figure 1.1.2.7 Crystal structure of the β-pyrochlore oxides AOs_2O_6. The A atom is located in an oversized atomic cage made of OsO_6 octahedra and can move with a large excursion along the four [72] directions pointing to the neighboring A atoms in adjacent cages [68].

[68, 69]. Comparing the three compounds, T_c rises with increasing magnitude of the rattling motion and electron–phonon coupling [70], while the density of state decreases [71], opposite to what is expected from BCS theory.

1.1.2.4 Iron-Based Superconductors

The 2D layer structure of iron-based superconductor families ("Fe–Sc") [32, 73] bears close resemblance to the cuprate HTS structure: the transition element atoms (Fe/Cu) are arranged in a quadratic lattice (Figure 1.1.2.8b) and apparently provide the SC mechanism. Instead of the Cu–Cu bonding via O atoms sitting halfway in-between next-nearest Cu atoms in the cuprate HTS, in Fe–Sc the Fe–Fe bonding happens via tetrahedrally arranged P, As, Se, or Te atoms above and underneath the Fe plane and affects the second-nearest Fe neighboring atoms as well (see Figure 1.1.2.8a). The Fe atoms form thus with the (P/As/Se/Te) atoms a network of regular pyramids with alternating upward/downward orientation. For both SC families, optimum T_c is observed for the most symmetric arrangement of these layer geometries, that is, for flat CuO_2 layers [74] and for regular $Fe(P/As/Se/Te)_4$ tetrahedra [75].

A huge difference is the replacibility of the transition metal atoms: in cuprate HTS, 10% substitution of Cu atoms by Zn, the rightward neighboring atom in the periodic table, suppresses superconductivity completely. In $BaFe_{2-x}Co_xAs_2$, the introduction of Co into the Fe layers even introduces superconductivity by the concomitant electron doping, for example, up to $T_c = 24$ K for Co concentration $x = 0.06$ [76]: Fe–Sc apparently tolerate considerable disorder in the Fe planes. Another huge difference: the undoped, "parent" compounds of cuprate HTS are AF insulators whereas Fe–Sc derive from magnetic metal compounds [7]. Electronic correlations in Fe–Sc are certainly weaker than in cuprate HTS, but electron–orbital selective correlation mechanisms in the Fe atoms introduce here novel basic physics [77].

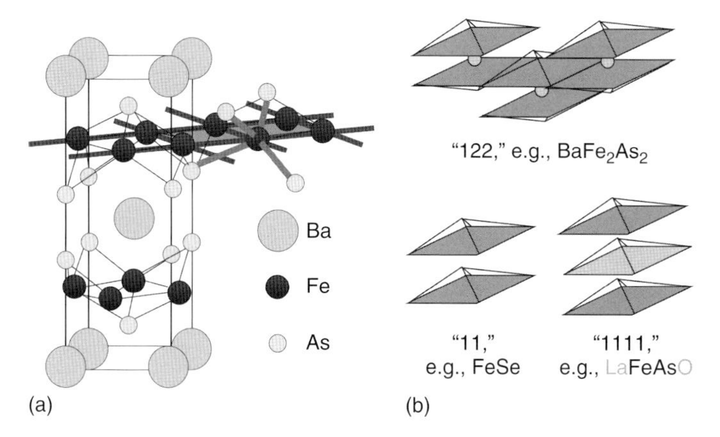

Figure 1.1.2.8 (a) Crystal structure of $BaFe_2As_2$. (b) Schematic structure of the 122, 11, and 1111 Fe–Sc indicating the up/down-orientation of the Fe(P/As/Se/Te) pyramids.

Just like HTS, Fe–Sc are extreme type-II superconductors with $\lambda > 100$ nm [78] and $\xi \sim 1$ nm [79]. For cuprate HTS, the discrepancy between the distance $d \sim 1$ nm in between the SC $(CuO_2/Ca/)_{n-1}CuO_2$ stacks [47] and the coherence length $\xi_c = 0.3$ nm perpendicular to the planes leads to the discussed weak-link behavior. For Fe–Sc, the situation is a bit more benign with the reported $d = 0.86$ nm/$\xi_c = 0.6$ nm for Nd-1111, $d = 0.65$ nm/$\xi_c = 1.5$ nm for Ba-122, and $d = 0.6$ nm/$\xi_c = 0.6$ nm for FeSe [79]. The almost isotropic magnetic field behavior, for example, for Ba-122 with B_{c2} [$T = 0$ K] estimates of ~50 T and 40 T for magnetic fields parallel and perpendicular to the planes, respectively [79], is not related to the supposed isotropic $s_{+/-}$ SC order parameter but stems from the fact that the Fe–Sc are apparently Pauli-limited: the Zeeman splitting of the electronic single-particle states makes it energetically favorable that the Cooper-pairs split into the constituent up- and down-spin states at fields below the "orbital limit" $B_{c2} = \Phi_0/(2\pi\xi^2)$ given by the magnetic flux quantum Φ_0 and the product "ξ^2" of the SC coherence lengths perpendicular to the field.

The restrictions with respect to the crystalline alignment of neighboring grains appear to be much less severe for Fe–Sc than for Cu-HTS: in both cases, for small misalignment angles α, the critical current density J_c is observed to remain more or less constant up to a critical angle α_c, followed by an exponential decrease $J_c \sim e^{-\alpha/\alpha_0}$ for larger α. However, for Fe–Sc the recently reported values $\alpha_c \approx 10°$ and $\alpha_0 \approx 15°$ [80] indicate a much less stringent texture requirement than for cuprate HTS ($\alpha_c \approx 3°–5°$, $\alpha_0 \approx 3°/5°$). Moreover, it is not clear if this granularity is already an intrinsic limit: the progress achieved in 2011 [80] compared to 2009 [81] gives rise to hope for further substantial improvement.

1.1.2.5 Heavy Fermion Superconductors

HF systems are stoichiometric lanthanide or actinide compounds whose qualitative low-temperature behavior in the normal state closely parallels the one well known from simple metals. The key features are the specific heat which varies approximately linearly $C \sim \gamma T$, the magnetic susceptibility which approaches a temperature independent constant $\chi(0)$, and the electrical resistivity which increases quadratically with temperature $\rho(T) = \rho_0 + AT^2$. However, the coefficient $\gamma \sim 1$ J mol^{-1} K^{-2} as well as $\chi(0)$ are enhanced by a factor of 100–1000 as compared to the values encountered in ordinary metals while the Sommerfeld–Wilson ratio $[\pi(k_B)^2\ \chi(0)]/[3(\mu_B)^2\gamma]$ is of order unity. The large enhancement of the specific heat is also reflected in the quadratic temperature coefficient A of the resistivity $A \sim \gamma^2$. These features indicate that the normal state can be described in terms of a Fermi liquid [82]. The excitations determining the low-temperature behavior correspond to heavy quasiparticles whose effective mass m^* is strongly enhanced over the free electron mass m. The characteristic temperature T^* which can be considered as a fictitious Fermi temperature or, alternatively, as an effective band width for the quasiparticles is of the order 10–100 K. Residual interactions among the heavy quasiparticles lead to instabilities of the normal Fermi liquid state. A hallmark of these systems is the competition or coexistence of various different cooperative phenomena

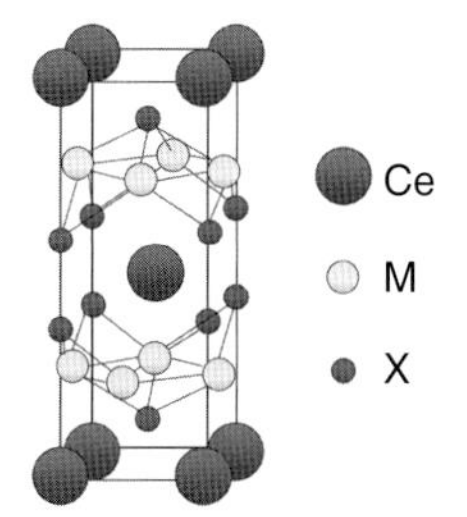

Figure 1.1.2.9 Conventional unit cell of CeM_2X_2 (M = Cu, Ni, Ru, Rh, Pd, Au, ...; X = Si, Ge) and URu_2Si_2.

which results in highly complex phase diagrams. Of particular interest are the SC phases which typically form at a critical temperature $T_c \leq 2$ K [83]. $PuCoGa_5$ with $T_c \sim 18.5$ K is up to now the only "high-T_c" HF representative [84].

The discovery of superconductivity in $CeCu_2Si_2$ ($T_c = 1.5$ K; see Figure 1.1.2.9) [10] forced condensed-matter physicists to revise the generally accepted picture of the electrons occupying the inner shells of the atoms. Traditionally, the corresponding states were viewed as localized atomic-like orbitals which are populated according to Hund's rules in order to minimize the mutual Coulomb repulsion. This leads to the formation of local magnetic moments which tend to align and which are weakly coupled to the delocalized conduction electrons. The latter were viewed as "free" fermions which occupy coherent Bloch states formed by the valence orbitals of the atoms. Usually, Cooper pairs which characterize a SC phase are broken by magnetic centers. The damaging effect of 4f- and 5f-ions was well established by systematic studies of dilute alloys. In stark contrast, in $CeCu_2Si_2$ [10] the magnetic degrees of freedom of the partially filled f-shells must generate superconductivity since the non-f reference compound $LaCu_2Si_2$ remains normal.

During the past decade, it became clear that there are different routes to heavy fermion behavior [85] where the magnetic degrees of freedom of the partially filled f-shells form a strongly correlated paramagnetic Fermi liquid with an effective Fermi energy of the order of 1 – 10 meV [86]. In Ce- and Yb-based compounds, the heavy quasiparticles with predominantly 4f-character arise through the Kondo effect in the periodic lattice [87]. For the actinide compounds, increasing experimental evidence points toward a dual character of the 5f-electrons with some of them being delocalized forming coherent bands while others stay localized reducing the Coulomb repulsion by forming multiplets [85, 88]. In Pr skutterudites, on the other hand, the quasiparticles are derived from the conduction states whose effective masses are strongly renormalized by low-energy excitations of the Pr 4f-shells [89]. It is generally agreed that the pairing interaction in HF superconductors is of electronic origin.

In the past decade, superconductivity at ambient pressure was found in the Ce-based HF superconductors CeM_mIn_{3+2m} (M = Ir or Co; $m = 0, 1$) [90]. The most prominent member of this family is $CeCoIn_5$, which has a relatively high $T_c = 2.3$ K (see Figure 1.1.2.10) [91]. Of fundamental interest is the discovery of HF superconductivity in $CePt_3Si$ ($T_c = 0.75$ K; see Figure 1.1.2.11) [92] which crystallizes in a lattice without inversion symmetry [93]. Highly promising systems with tailored violation of local inversion symmetry are artificial superlattices consisting of the

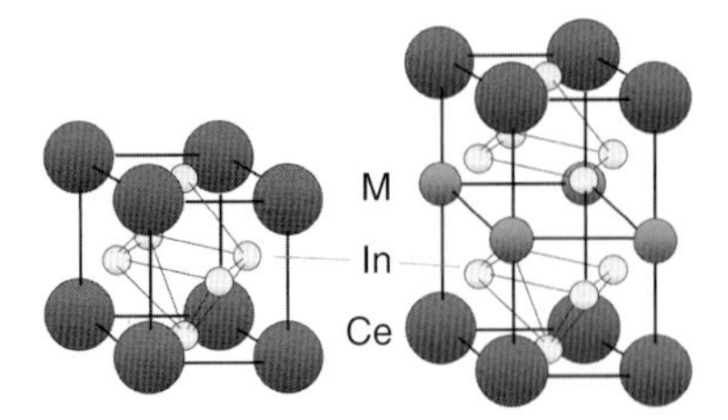

Figure 1.1.2.10 Unit cell of $CeIn_3$ and $CeMIn_5$ (M = Co, Ir).

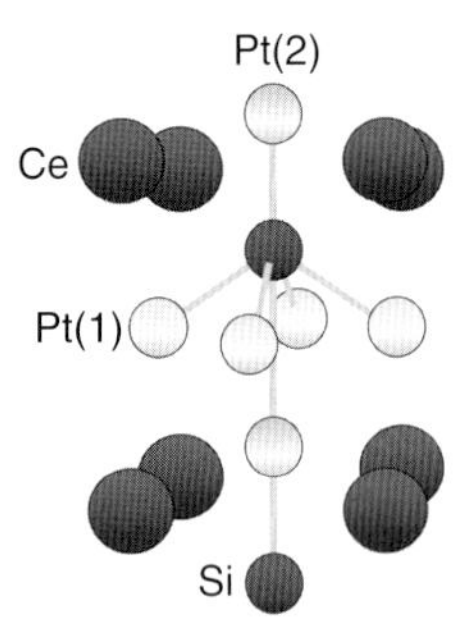

Figure 1.1.2.11 Crystal structure of $CePt_3Si$.

HF superconductor $CeCoIn_5$ and its conventional metallic counterpart $YbCoIn_5$ [94]. It remains a major challenge to give reasons for the apparent asymmetry between Ce- and Yb-based HF systems, that is, to explain why there is a great variety of Ce-based HF superconductors but only one weak Yb-based HF superconductor β-$YbAl_4B$ [95]. HF superconductivity is found more frequently in intermetallic actinide-compounds than in lanthanide-compounds. This may be related to the different nature of heavy quasiparticles in actinide-compounds where the 5f-electrons have a considerable, though orbitally dependent, degree of delocalization. The genuine Kondo mechanism is not appropriate for heavy quasiparticle formation as in lanthanide-compounds. This may lead to more pronounced delocalized spin fluctuations in U-compounds which mediate unconventional Cooper pair formation. AF order, mostly with small moments of the order 10^{-2} μ_B is frequently found to envelop and coexist with the SC phase.

UPt_3 (see Figure 1.1.2.12) [96] exhibits triplet pairing. It sticks out as the most interesting case of unconventional superconductivity with a multicomponent order parameter whose degeneracy is lifted by a symmetry-breaking field due to a small moment AF order. In contrast, in UPd_2Al_3 (see Figure 1.1.2.13) [97] superconductivity coexists with large moment antiferromagnetism. Probably spin singlet pairing is realized. There is experimental evidence for a new kind of magnetic pairing mechanism mediated by propagating magnetic exciton modes. The sister compound UNi_2Al_3 [98] is an example of coexistence of large moment antiferromagnetism with a SC triplet order parameter. In URu_2Si_2 [99], the SC order parameter symmetry is still undetermined. The interest in this compound is focused more on the enveloping phase with a "hidden" order parameter presumably of quadrupolar type or an "unconventional" spin density wave (SDW)

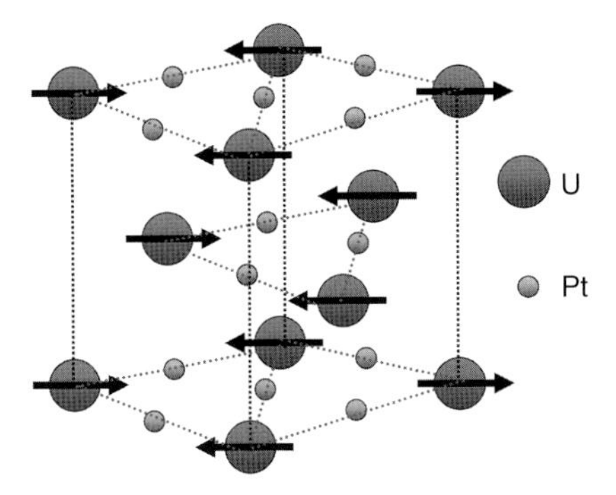

Figure 1.1.2.12 Crystal structure of UPt_3 and AF magnetic structure ($T < T_N = 5.8\,K$).

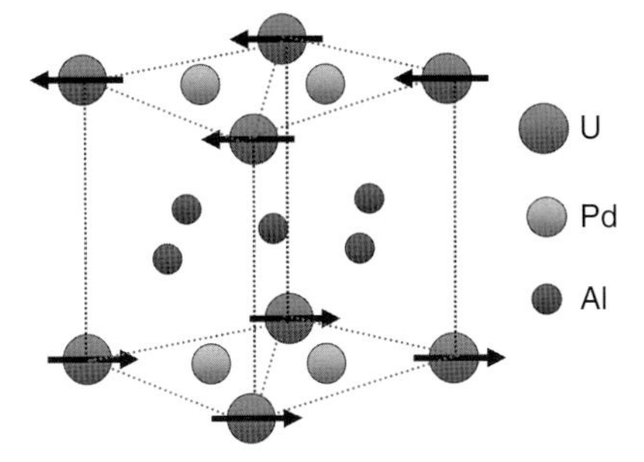

Figure 1.1.2.13 Conventional unit cell of UPd_2Al_3 and simple AF magnetic structure.

[100]. The oldest cubic U-HF superconductor UBe_{13} [101] and its thorium alloy $U_{1-x}Th_xBe_{13}$ is also the most mysterious one. While for the pure system there is a single SC phase of yet unknown symmetry, in the small Th concentration range two distinct phases exist which either may correspond to two different SC order parameters or may be related to a coexistence of superconductivity with a SDW phase. In UGe_2 [102], ferromagnetism and superconductivity coexist. Due to the ferromagnetic polarization the triplet gap function contains only equal spin pairing.

The possibility of coexisting ferromagnetism and superconductivity was first considered by Ginzburg [103] who noted that this is only possible when the internal ferromagnetic field is smaller than the thermodynamic critical field of the superconductor. Such a condition is hardly ever fulfilled except immediately below the Curie temperature T_C where coexistence has been found in a few superconductors with local moment ferromagnetism and $T_C < T_c$ such as $ErRh_4B_4$ and $HoMo_6S_8$. If the temperature drops further below T_C, the internal ferromagnetism molecular field rapidly becomes larger than H_{c2} and superconductivity is destroyed. The reentrance of the normal state below T_C has indeed been observed in the above compounds.

The transuranium-based superconductors $PuCoGa_5$ ($T_c = 18.5\,K$) [84], $PuRhGa_5$ ($T_c = 8.7\,K$) [104], and $NpPd_5Al_2$ ($T_c = 4.9\,K$) [105] are all unconventional superconductors at ambient pressure with the highest transition

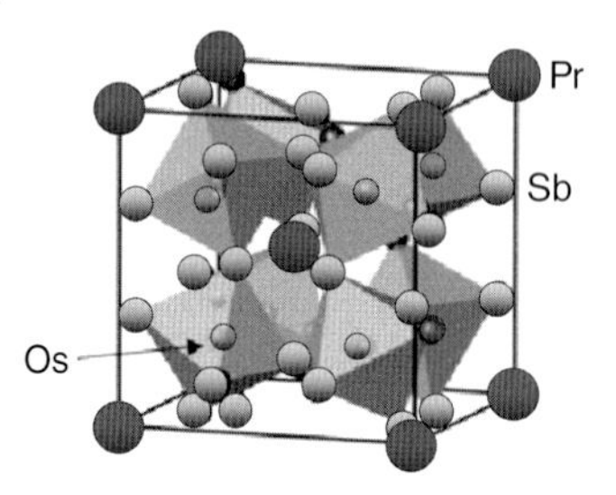

Figure 1.1.2.14 Cubic crystal structure of the filled skutterudite RT_4X_{12}. The T atoms are located in the center of the X octahedra.

temperatures T_c among all the HF superconductors. The HF superconductor $PrOs_4Sb_{12}$ [106] is potentially of similar interest as UPt_3 because it represents the second example of multiphase superconductivity [107] with a critical temperature $T_c = 1.85$ K. The skutterudites RT_4X_{12} (R = alkaline earth, rare earth or actinide; T = Fe, Ru, or Os and X = P, As, or Sb) show a cage structure where large voids formed by tilted T_4X_{12} octahedrons can be filled with R atoms (see Figure 1.1.2.14). They are, however, rather loosely bound and are therefore subject to large anharmonic oscillations ("rattling") in the cage.

1.1.2.6 **Organic and Other Carbon-Based Superconductors**

Carbides, for example, NbC ($T_c = 12$ K) [108, 109] and MoC ($T_c = 14.3$ K) [109, 110], were among the first discovered compound superconductors and contended in these early days of superconductivity with nitrides and borides for the highest T_c.

Theoretical speculations of superconductivity in organic compounds [12] were met for a long time with total disbelief from the experimental side, for example, from B. Matthias. Things changed when immediately after Matthias's death in 1980 superconductivity was discovered below 0.9 K in the compound $(TMTSF)PF_6$ under a hydrostatic pressure of 12 kbar, with the organic molecule TMTSF (tetra-methyl-tetra-selenium-fulvalene; see Figure 1.1.2.15) [13]. Meanwhile, a number of TMTSF-based superconductors with $T_c \sim 1$ K have been found, for example, $(TMTSF)_2ClO_4$ which becomes SC at 1 K already under normal pressure conditions [111]. The organic molecules are stacked here on top of each other (see Figure 1.1.2.15). The general chemical formula is $(TMTSF)_2X$ where X denotes an electron acceptor such as PF_6, ClO_4, AsF_6, or TaF_6. In the normal state, the TMTSF compounds have a relatively large electric conductivity along the stacks, but only a small conductivity perpendicular to the stacks, thus forming nearly one-dimensional (normal) conductors. The TMTSF compounds are type-II superconductors with highly anisotropic properties. For example, in $(TMTSF)_2ClO_4$ along the stacks the Ginzburg–Landau coherence length is about 80 nm, whereas along the two perpendicular directions of the crystal axes it is about 35 and 2 nm, respectively. The latter value is of the same order of magnitude as the lattice constant along the *c*-axis. Hence, the compound represents a nearly 2D superconductor [72].

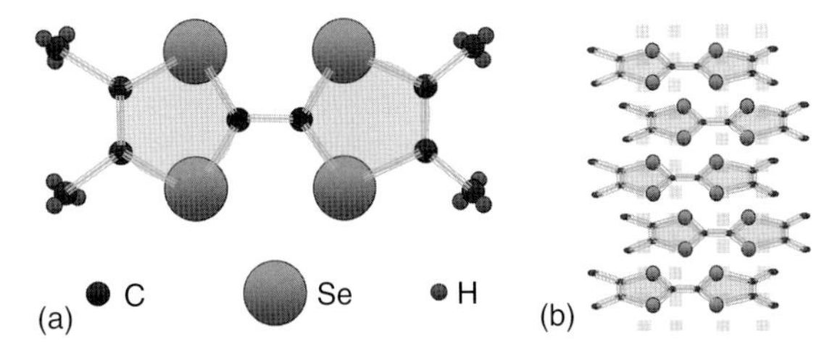

Figure 1.1.2.15 (a) Structure of the organic molecule tetra-methyl-tetra-selenium-fulvalene (TMTSF). (b) Stack arrangement of the molecules forming one-dimensional conduction channels.

Another important class of organic superconductors, often exhibiting T_c well above 1 K, is based on the bis-ethylene-dithia-tetra-thiafulvalene molecule, abbreviated as "BEDT-TTF" or "ET." $(BEDT\text{-}TTF)_2Cu[N(CN)_2]Br$ becomes SC at 11.2 K [14], $(BEDT\text{-}TTF)_2Cu(NCS)_2$ at 10.4 K. The ET-compounds are also highly anisotropic. However, in contrast to the TMTSF-compounds, in the normal state they form 2D layered structures with a large electric conductivity in two dimensions. Like the TMTSF-based materials, the ET-compounds are type-II superconductors, with very short out-of-plane coherence lengths. These compounds thus also represent SC layered structures making them in many respects similar to HTS. The pairing mechanism of the organic superconductors is at present still unclear as well. At least some compounds appear to be d-wave superconductors, in the compound $(TMTSF)_2PF_6$ one may even deal with a spin-triplet superconductor [112].

The quasi-2D organic superconductors are prime candidates for exhibiting the long-sought Fulde–Ferrell–Larkin–Ovchinnikov ("FFLO/LOFF") phases [113]. When the magnetic field is applied parallel to the conducting planes, the orbital critical field is strongly enhanced and superconductivity is Pauli limited. First thermodynamic evidence for the formation of a FFLO/LOFF state was found in κ-$(BEDT\text{-}TTF)_2Cu(NCS)_2$ [114]. The angle-dependence of the formation of the FFLO/LOFF state was demonstrated in [115].

In 1994 superconductivity was found in *boron carbides* [116] (see the Chapter 7), in 2004 in *diamond* with T_c up to 4 K when doped with boron [6] and up to 11.4 K in thin films [117]. For yttrium and rare earth carbide compounds T_c as high as 18 K [118] was reported. Superconductivity in a graphite intercalation compound was first observed in 1965 [119] on KC_8 which exhibits very low critical temperature $T_c = 0.14$ K [120]. Later, several ternary graphite intercalation compounds revealed higher T_c of 1.4 K for $KHgC_8$ [121] and 2.7 K for $KTl_{1.5}C_4$ [122]. Recently, the discovery of high critical temperatures in graphite intercalation compounds YbC_6 ($T_c = 6.5$ K) [123], CaC_6 ($T_c = 11.5$ K) [124], and $Li_3Ca_2C_6$ ($T_c = 11.15$ K) [125] has renewed the interest in this family of materials [123].

Fullerides are compounds of the form A_3C_{60} which may become SC on the admixture of alkali atoms or of alkaline earth atoms [126]: Rb_3C_{60} has a value of T_c of 29.5 K, the present record under pressure is held by Cs_3C_{60} with $T_c = 40$ K [27, 127]. The crystal structure of the fullerides is face-centered cubic, with

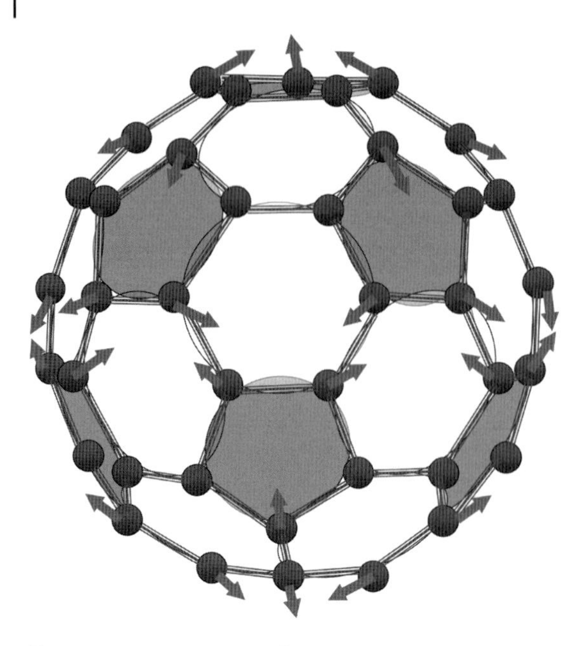

Figure 1.1.2.16 Structure of the C_{60} molecule. The arrows indicate one of the intramolecular phonon modes which are believed to be mainly responsible for the SC pairing [27].

the alkali atoms occupying interstitial sites between the large C_{60} molecules. Fullerides are BCS-like s-wave superconductors. Intramolecular C_{60} phonons (see Figure 1.1.2.16) seem to contribute the most important part of the pairing interactions [27]. However, for body-centered cubic A15-structured Cs_3C_{60} (which is not SC at ambient pressure), an apparently purely electronic transition to a SC state with T_c up to 38 K can be induced by pressure, where the T_c dependence on pressure cannot be described by BCS theory in terms of the induced changes of anion packing density [128].

Recent experiments on alkali-doped *picene* and *dibenzopentacen,* hydrocarbon molecules made up of an assembly of five and seven fused benzene rings, respectively, reported superconductivity up to T_c of 18 [129] and 33 K [130], respectively. A linear increase of T_c with the number of constituent benzene rings is suspected. However, the fabrication process cannot be controlled yet sufficiently to achieve single-phase preparation. This holds true even more for *carbon nanotubes* where long-standing speculations on superconductivity [131] have now been confirmed experimentally for the case of double-wall carbon nanotubes (DWNTs) with resistively measured $T_c = 6.8$ K [132].

1.1.2.7 **Borides and Borocarbides**

Rare earth borocarbide superconductors have provided the first example of a homogeneous coexistence of superconductivity and ferromagnetism for all temperatures below T_c: the two antagonistic long-range orders are carried by different species of electrons that interact only weakly through contact exchange

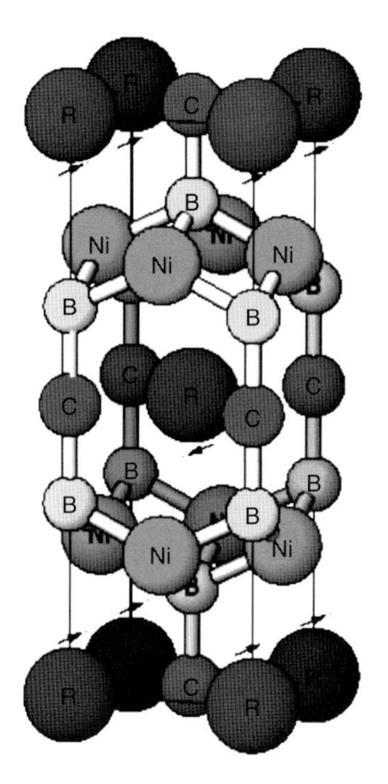

Figure 1.1.2.17 Structure of the layered transition metal borocarbides RNi_2B_2C.

interaction leading to a small effect of the local moment molecular field on the SC conduction electrons. The nonmagnetic rare earth borocarbides have extremely large gap anisotropy ratios $\Delta_{max}/\Delta_{min} \geq 100$ [133]. Surely, the standard electron–phonon mechanism has to be supplemented by something else, perhaps anisotropic Coulomb interactions to achieve this "quasi-unconventional" behavior in borocarbides.

The SC class of layered transition metal borocarbides RNi_2B_2C (nonmagnetic R = Y, Lu, Sc; magnetic R = lanthanide elements in a R^{3+} state; see Figure 1.1.2.17) was discovered in 1994 [116, 134–136]. The crystal structure consists of R C rock-salt-type planes separated by Ni_2B_2 layers built from NiB_4 tetrahedra and stacked along the *c*-axis. More general structures with more than one R C layer are possible [135]. The nonmagnetic borocarbides have relatively high T_c values of around 15 K. There is evidence that the SC mechanism is primarily of the electron–phonon type, although this cannot explain the large anisotropy of the SC gap. At first sight, the layered structure is similar to the HTS cuprates. However, unlike the copper oxide planes the NiB_2 planes show buckling. As a consequence, the electronic states at the Fermi level in the borocarbides do not have quasi-2D $d_{x^2-y^2}$ character and, therefore, have much weaker correlations excluding the possibility of AF spin-fluctuation-mediated superconductivity.

The discovery of superconductivity in MgB_2 (see Figure 1.1.2.18) in early 2001 with $T_c \sim 40$ K, came as a huge surprise since this simple material was

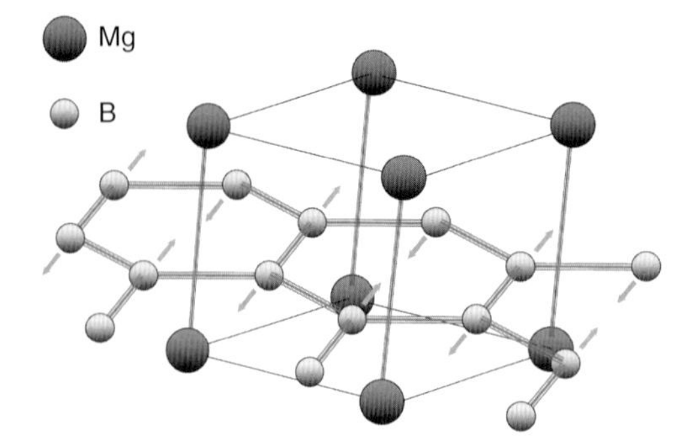

Figure 1.1.2.18 Hexagonal crystal structure of MgB_2. The arrows indicate the B-phonon mode which presumably introduces the strongest SC coupling.

known since the early 1950s and had simply been missed in the systematic research for superconductivity [137]. Since no atomic d- or f-shells are involved in the conduction electron system of this binary compound of light elements Coulomb correlation do not play a role. The simple crystal structure consisting of graphite-like B-layers with intercalated Mg favors conduction along these layers and a respective superconductive and normal state anisotropy, but it does not introduce a reduction of the effective dimensionality, as in the case of organic superconductors due to the stacking of isolated aromatic rings. The coupling of the conduction electrons to a particular boron phonon mode (see Figure 1.1.2.18) was identified right from the start as basic origin of superconductivity in MgB_2 [138]. The observation of two energy gaps (at 1.8 and 6.8 meV [139, 140]) and the considerable superconductive anisotropy as large as 6–9 [141] challenged a more thorough theoretical investigation which explained these findings in terms of two-band superconductivity [30] on the basis of the large anharmonicity of the involved phonon mode and a refined treatment of its coupling with the different sheets of the electronic conduction band [139, 142].

References

1. (a) Buckel, W. and Kleiner, R. (2013) *Supraleitung – Grundlagen und Anwendungen*, Wiley-VCH Verlag GmbH, Weinheim; (b) Buckel, W. and Kleiner, R. (2004) *Superconductivity – Fundamentals and Applications*, Wiley-VCH Verlag GmbH, Weinheim.
2. Goodstein, D. and Goodstein, J. (2000) *Phys. Perspect.*, **2**, 30.
3. Huebener, R. and Lübbig, H. (2011) *Die Physikalisch-Technische Reichsanstalt*, Vieweg+Teubner-Verlag, Wiesbaden.
4. (a) Fossheim, K. and Sudbo, A. (2004) *Superconductivity – Physics and Applications*, John Wiley & Sons, Ltd; (b) van Aubel, E., de Haas, W. J., and Voogd, J. (1928) *Comm. Leiden*, Nr 193c.
5. (a) Geballe, T.H. (2001) *Science*, **293**, 223; (b) Buzea, C. and Yamashita, T. (2001) *Supercond. Sci. Technol.*, **14**, R115.
6. Ekimov, E.A. *et al.* (2004) *Nature*, **428**, 542.
7. Mazin, I.I. (2010) *Nature*, **464**, 183.
8. Bardeen, J., Cooper, L.N., and Schrieffer, J.R. (1957) *Phys. Rev.*, **108**, 1175.

9. Cooper, L.N. and Feldman, D. (eds) (2011) *BCS: 50 Years*, World Scientific Press, Singapore.
10. Steglich, F. *et al.* (1979) *Phys. Rev. Lett.*, **43**, 1892.
11. Pfleiderer, C. (2009) *Rev. Mod. Phys.*, **81**, 1551.
12. Little, W.A. (1964) *Phys. Rev. A*, **134**, 1416.
13. Jérome, D., Mazaud, A., Ribault, M., and Bechgaard, K. (1980) *J. Phys. Lett. (Paris)*, **41**, L95.
14. Saito, G. *et al.* (1991) *Physica B*, **169**, 372.
15. Wosnitza, J. (2012) *Crystals*, **2**, 248.
16. Mermin, N.D. and Wagner, H. (1966) *Phys. Rev. Lett.*, **17**, 1133 and 1307.
17. Bednorz, J.G. and Müller, K.A. (1986) *Z. Phys. B*, **64**, 189.
18. Chu, C.W. (2011) in *BCS: 50 Years* (eds L.N. Cooper and D. Feldman), World Scientific Press, Singapore.
19. (a) Schrieffer, J.R., Scalapino, D.J., and Wilkins, J.W. (1963) *Phys. Rev. Lett.*, **10**, 336; (b) Eliashberg, G.M. (1960) *Sov. Phys. JETP*, **11**, 696.
20. Putilin, S.N. *et al.* (2001) *Physica C*, **338**, 52.
21. Gavaler, J.R. (1973) *Appl. Phys. Lett.*, **23**, 480.
22. McMillan, W.L. and Rowell, J.M. (1965) *Phys. Rev. Lett.*, **14**, 108.
23. Zeyer, R. and Zwicknagl, G. (1990) *Z. Phys. B*, **78**, 175.
24. Lee, P.A., Nagaosa, N., and Wen, X.-G. (2006) *Rev. Mod. Phys.*, **78**, 17.
25. Maier, T.A., Poilblanc, D., and Scalapino, D.J. (2008) *Phys. Rev. Lett.*, **100**, 237001.
26. Tanigaki, K. *et al.* (1991) *Nature*, **352**, 222.
27. Gunnarsson, O. (1997) *Rev. Mod. Phys.*, **69**, 575.
28. Canfield, P.C., Gammel, P.L., and Bishop, D.J. (1998) *Phys. Today*, **51**, 40.
29. Nagamatsu, J. *et al.* (2001) *Nature*, **410**, 63.
30. (a) Suhl, H., Matthias, B.T., and Walker, L.R. (1959) *Phys. Rev. Lett.*, **3**, 552; (b) Moskalenko, V. (1959) *Fiz. Met. Metalloved.*, **8**, 503; (c) Binnig, G., Baratoff, A., Hoenig, H.E., and Bednorz, J.G. (1980) *Phys. Rev. Lett.*, **45**, 1352.
31. Geerk, J. *et al.* (2005) *Phys. Rev. Lett.*, **94**, 227005.
32. Stewart, G.R. (2011) *Rev. Mod. Phys.*, **83**, 1589.
33. Hirschfeld, P., Korshunov, M.M., and Mazin, I.I. (2011) *Rep. Prog. Phys.*, **74**, 124508.
34. Kamihara, Y., Watanabe, T., Hirano, M., and Hosono, H. (2008) *J. Am. Chem. Soc.*, **130**, 3296.
35. Ren, Z.-A. *et al.* (2008) *Europhys. Lett.*, **83**, 17002.
36. R. Hott & T. Wolf, Cuprate high temperature superconductors. *Applied Superconductivity. Handbook on Devices and Applications* (ed. P. Seidel), Wiley-VCH Verlag GmbH, Weinheim (2015).
37. Yamauchi, H. and Karppinen, M. (2000) *Supercond. Sci. Technol.*, **13**, R33.
38. Wu, M.K. *et al.* (1987) *Phys. Rev. Lett.*, **58**, 908.
39. Harshman, D.R. and Mills, A.P. Jr., (1992) *Phys. Rev. B*, **45**, 10684.
40. Orenstein, J. and Millis, A.J. (2000) *Science*, **288**, 468.
41. Onose, Y., Taguchi, Y., Ishizaka, K., and Tokura, Y. (2001) *Phys. Rev. Lett.*, **87**, 217001.
42. Ando, Y. *et al.* (2001) *Phys. Rev. Lett.*, **87**, 017001.
43. (a) Tallon, J.L. and Loram, J.W. (2001) *Physica C*, **349**, 53; (b) Varma, C.M. (1997) *Phys. Rev. B*, **55**, 14554; (c) Alff, L. *et al.* (2003) *Nature*, **422**, 698.
44. Anderson, P.W. (1987) *Science*, **235**, 1196.
45. Brandt, E.H. (2001) *Phys. Rev. B*, **64**, 024505.
46. (a) Grasso, G. and Flükiger, R. (1997) *Supercond. Sci. Technol.*, **10**, 223; (b) Sato, J. *et al.* (2001) *Physica C*, **357-360**, 1111; (c) Okada, M. (2000) *Supercond. Sci. Technol.*, **13**, 29.
47. Saxena, A.K. (2010, 2012) *High-Temperature Superconductors*, Springer-Verlag, Berlin, Heidelberg.
48. Plakida, N.P. (1995) *High-Temperature Superconductivity*, Springer-Verlag, Berlin, Heidelberg.
49. Larbalestier, D., Gurevich, A., Feldmann, D.M., and Polyanskii, A. (2001) *Nature*, **414**, 368.
50. (a) Hilgenkamp, H. and Mannhart, J. (2002) *Rev. Mod. Phys.*, **74**, 485; (b)

Gross, R. *et al.* (1997) *IEEE Trans. Appl. Supercond.*, **7**, 2929.
51. Kleiner, R., Steinmeyer, F., Kunkel, G., and Müller, P. (1992) *Phys. Rev. Lett.*, **68**, 2394.
52. Brandt, E.H. (1995) *Rep. Prog. Phys.*, **58**, 1465.
53. (a) Tsuei, C.C. and Kirtley, J.R. (2000) *Rev. Mod. Phys.*, **72**, 969; (b) Tsuei, C.C. and Kirtley, J.R. (2002) *Physica C*, **367**, 1.
54. Graser, S. *et al.* (2010) *Nat. Phys.*, **6**, 609.
55. (a) Sleight, A.W., Gillson, J.L., and Bierstedt, P.E. (1975) *Solid State Commun.*, **17**, 27; (b) Merz, M. *et al.* (2005) *Europhys. Lett.*, **72**, 275.
56. Mattheiss, L.F., Gyorgy, E.M., and Johnson, D.W. Jr., (1988) *Phys. Rev. B*, **37**, 3745.
57. Hellman, E.S. and Hartford, E.H. Jr., (1995) *Phys. Rev. B*, **52**, 6822.
58. (a) Cava, R.J. *et al.* (1988) *Nature*, **332**, 814; (b) Pei, S. *et al.* (1990) *Phys. Rev. B*, **41**, 4126; (c) Klinkova, L.A. *et al.* (2003) *Phys. Rev. B*, **67**, 140501.
59. Maeno, Y., Rice, T.M., and Sigrist, M. (2001) *Phys. Today*, **54**, 42.
60. Bernhard, C. *et al.* (1999) *Phys. Rev. B*, **59**, 14099.
61. Nachtrab, T. *et al.* (2004) *Phys. Rev. Lett.*, **92**, 117001.
62. Takada, K. *et al.* (2003) *Nature*, **422**, 53.
63. Foo, M.L. *et al.* (2003) *Solid State Commun.*, **127**, 33.
64. (a) Yonezawa, S. *et al.* (2004) *J. Phys. Condens. Matter*, **16**, L9; (b) Yonezawa, S., Muraoka, Z., and Hiroi, Z. (2004) *J. Phys. Soc. Jpn.*, **73**, 1655; (c) Yonezawa, S., Muraoka, Y., Matsushita, Y., and Hiroi, Z. (2004) *J. Phys. Soc. Jpn.*, **73**, 819; (d) Hiroi, Z., Yonezawa, S., and Muraoka, Y. (2005) *J. Phys. Soc. Jpn.*, **74**, 3399.
65. Ong, N.P. and Cava, R.J. (2004) *Science*, **305**, 52.
66. Schuck, G. *et al.* (2006) *Phys. Rev. B*, **73**, 144506.
67. Kunes, J., Jeong, T., and Pickett, W.E. (2004) *Phys. Rev. B*, **70**, 174510.
68. Nagao, Y. *et al.* (2009) *J. Phys. Soc. Jpn.*, **78**, 064702.
69. Hattori, K. and Tsunetsugu, H. (2010) *Phys. Rev. B*, **81**, 134503.
70. Hiroi, Z., Yamaura, J., Yonezawa, S., and Harima, H. (2007) *Physica C*, **460-462**, 20.
71. Sainz, R. and Freeman, A.J. (2005) *Phys. Rev. B*, **72**, 024522.
72. Mansky, P.A., Danner, G., and Chaikin, P.M. (1995) *Phys. Rev. B*, **52**, 7554.
73. (a) Johnston, D.C. (2010) *Adv. Phys.*, **59**, 803; (b) Mandrus, D., Sefat, A.S., McGuire, M.A., and Sales, B.C. (2010) *Chem. Mater.*, **22**, 715.
74. Chmaissem, O. *et al.* (1999) *Nature*, **397**, 45.
75. Lee, C.-H. *et al.* (2008) *J. Phys. Soc. Jpn.*, **77**, 083704.
76. (a) Hardy, F. *et al.* (2010) *Europhys. Lett.*, **91**, 47008; (b) Sefat, A.S. *et al.* (2008) *Phys. Rev. Lett.*, **101**, 117004.
77. Lanata, N. *et al.* (2013) *Phys. Rev. B*, **87**, 045122.
78. Prozorov, R. and Kogan, V.G. (2011) *Rep. Prog. Phys.*, **74**, 124505.
79. Putti, M. *et al.* (2010) *Supercond. Sci. Technol.*, **23**, 034003.
80. Katase, T. *et al.* (2011) *Nat. Commun.*, **1419** (2), 409.
81. Lee, S. *et al.* (2009) *Appl. Phys. Lett.*, **95**, 212505.
82. Fulde, P., Keller, J., and Zwicknagl, G. (1988) *Solid State Physics, Advances in Research and Applications*, Vol. 41, Academic Press, p. 1.
83. Thalmeier, P. and Zwicknagl, G. (2004) *Handbook on the Physics and Chemistry of Rare Earths*, Vol. 34, Elsevier; cond-mat/0312540.
84. Sarrao, J.L. *et al.* (2002) *Nature*, **420**, 297.
85. Fulde, P., Thalmeier, P., and Zwicknagl, G. (2006) *Solid State Physics, Advances in Research and Applications*, Vol. 60, Academic Press, p. 1.
86. Zwicknagl, G. (1992) *Adv. Phys.*, **41**, 203.
87. Ernst, S. *et al.* (2011) *Nature*, **474**, 362.
88. (a) Zwicknagl, G., Yaresko, A.N., and Fulde, P. (2002) *Phys. Rev. B*, **63**, 081103; (b) Zwicknagl, G., Yaresko, A.N., and Fulde, P. (2003) *Phys. Rev. B*, **68**, 052508.
89. Zwicknagl, G., Thalmeier, P., and Fulde, P. (2009) *Phys. Rev. B*, **79**, 115132.

90. (a) Thompson, J.D. *et al.* (2001) *J. Magn. Magn. Mater.*, **226-230**, 5; (b) Petrovic, C. *et al.* (2001) *Europhys. Lett.*, **53**, 354.
91. Petrovic, C. *et al.* (2001) *J. Phys. Condens. Matter*, **13**, L337.
92. Bauer, E. *et al.* (2004) *Phys. Rev. Lett.*, **92**, 027003.
93. Kimura, N. and Bonalde, I. (2012) in *Non-Centrosymmetric Superconductors*, Lecture Notes in Physics, Vol. 84, Chapter 2 (eds E. Bauer and M. Sigrist), Springer, Heidelberg, pp. 35–79.
94. Nakatsuji, S. *et al.* (2008) *Nat. Phys.*, **4**, 603.
95. Shishido, H. *et al.* (2010) *Science*, **327**, 980.
96. Stewart, G.R., Fisk, Z., Willis, J.O., and Smith, J.L. (1984) *Phys. Rev. Lett.*, **52**, 679.
97. Geibel, C. *et al.* (1991) *Z. Phys. B*, **84**, 1.
98. Geibel, C. *et al.* (1991) *Z. Phys. B*, **83**, 305.
99. Palstra, T.T.M. *et al.* (1985) *Phys. Rev. Lett.*, **55**, 2727.
100. (a) Schmidt, A.R. *et al.* (2010) *Nature*, **465**, 570; (b) Aynajian, P. *et al.* (2010) *Proc. Natl. Acad. Sci. U.S.A.*, **107**, 10383; (c) Yuan, T., Figgins, J., and Morr, D.K. (2012) *Phys. Rev. B*, **86**, 035129; (d) Shibauchi, T. and Matsuda, Y. (2012) *Physica C*, **481**, 229; (e) Das, T. (2012) *Sci. Rep.*, **2**, 596; (f) Pourret, A. *et al.* (2013) *J. Phys. Soc. Jpn.*, **82**, 034706.
101. Ott, H.R., Rudigier, H., Fisk, Z., and Smith, J.L. (1983) *Phys. Rev. Lett.*, **80**, 1595.
102. Saxena, S.S. *et al.* (2000) *Nature*, **406**, 587.
103. Ginzburg, V. (1957) *JETP*, **4**, 153.
104. Wastin, F. *et al.* (2003) *J. Phys. Condens. Matter*, **15**, S2279.
105. Aoki, D. *et al.* (2007) *J. Phys. Soc. Jpn.*, **76**, 063701.
106. Bauer, E.D. *et al.* (2002) *Phys. Rev. B*, **65**, 100506.
107. Izawa, K. *et al.* (2003) *Phys. Rev. Lett.*, **90**, 117001.
108. Meissner, W. and Franz, H. (1930) *Z. Phys.*, **65**, 30.
109. Mourachkine, A. (2004) *Room-Temperature Superconductivity*, Cambridge International Science Publishing, Cambridge.
110. Matthias, B.T. and Hulm, J.K. (1952) *Phys. Rev.*, **87**, 799.
111. Bechgaard, K. *et al.* (1981) *Phys. Rev. Lett.*, **46**, 852.
112. Lee, I.J. *et al.* (2002) *Phys. Rev. Lett.*, **88**, 017004.
113. (a) Zwicknagl, G. and Wosnitza, J. (2011) in *BCS: 50 Years* (eds L.N. Cooper and D. Feldman), World Scientific Press, Singapore; (b) Casalbuoni, R. and Nardulli, G. (2004) *Rev. Mod. Phys.*, **76**, 263; (c) Beyer, R. and Wosnitza, J. (2013) *Low Temp. Phys.*, **39**, 225.
114. Lortz, R. *et al.* (2007) *Phys. Rev. Lett.*, **99**, 187002.
115. Beyer, R. *et al.* (2012) *Phys. Rev. Lett.*, **109**, 027003.
116. (a) Cava, R.J. *et al.* (1994) *Nature*, **367**, 252; (b) Cava, R.J. *et al.* (1994) *Nature*, **367**, 146.
117. Takano, Y. (2006) *Sci. Technol. Adv. Mater.*, **7**, S1.
118. (a) Amano, G. *et al.* (2004) *J. Phys. Soc. Jpn.*, **73**, 530; (b) Nakane, T. *et al.* (2004) *Appl. Phys. Lett.*, **84**, 2859; (c) Kim, J.S. *et al.* (2007) *Phys. Rev. B*, **76**, 014516.
119. Hannay, N.B. *et al.* (1965) *Phys. Rev. Lett.*, **14**, 225.
120. Koike, Y., Tanuma, S., Suematsu, H., and Higuchi, K. (1980) *J. Phys. Chem. Solids*, **41**, 1111.
121. Pendrys, L.A. *et al.* (1981) *Solid State Commun.*, **38**, 677.
122. Wachnik, R.A. *et al.* (1982) *Solid State Commun.*, **43**, 5.
123. Emery, N. *et al.* (2008) *Sci. Technol. Adv. Mater.*, **9**, 44102.
124. (a) Weller, T.E. *et al.* (2005) *Nat. Phys.*, **1**, 39; (b) Emery, N. *et al.* (2005) *Phys. Rev. Lett.*, **95**, 087003.
125. Emery, N. *et al.* (2006) *J. Solid State Chem.*, **179**, 1289.
126. (a) Hebard, A. *et al.* (1991) *Nature*, **350**, 600; (b) Fleming, R.M. *et al.* (1991) *Nature*, **352**, 787.
127. (a) Pennington, C.H. and Stenger, V.A. (1996) *Rev. Mod. Phys.*, **68**, 855; (b) Buntar, V. and Weber, H.W. (1996) *Supercond. Sci. Technol.*, **9**, 599.
128. Takabayashi, Y. *et al.* (2009) *Science*, **323**, 1585.

129. Mitsuhashi, R. *et al.* (2010) *Nature*, **464**, 76.
130. Xue, M. *et al.* (2012) *Sci. Rep.*, **2**, 389.
131. (a) Tang, Z.K. *et al.* (2001) *Science*, **292**, 2462; (b) Bockrath, M. (2006) *Nat. Phys.*, **2**, 155.
132. Shi, W. *et al.* (2012) *Sci. Rep.*, **2**, 625.
133. Izawa, K. *et al.* (2002) *Phys. Rev. Lett.*, **89**, 137006.
134. Nagarajan, R. *et al.* (1994) *Phys. Rev. Lett.*, **72**, 274.
135. Hilscher, G. and Michor, H. (1999) *Studies in High Temperature Superconductors*, Vol. 28, Nova Science Publishers, pp. 241–286.
136. Müller, K.-H. and Narozhnyi, V.N. (2001) *Rep. Prog. Phys.*, **64**, 943.
137. (a) Grant, P. (2001) *Nature*, **411**, 532; (b) Campbell, A.M. (2001) *Science*, **292**, 65.
138. (a) Kortus, J. *et al.* (2001) *Phys. Rev. Lett.*, **86**, 4656; (b) Bohnen, K.-P., Heid, R., and Renker, B. (2001) *Phys. Rev. Lett.*, **86**, 5771.
139. Choi, H.J., Roundy, D., Sun, H., Cohen, M.L., and Louie, S.G. (2002) *Nature*, **418**, 758.
140. Canfield, P.C. and Crabtree, G. (2003) *Phys. Today*, **56**, 34.
141. Bud'ko, S.L., Canfield, P.C., and Kogan, V.G. (2002) *Physica C*, **382**, 85.
142. (a) Kong, Y., Dolgov, O.V., Jepsen, O., and Andersen, O.K. (2001) *Phys. Rev. B*, **64**, 020501; (b) Mazin, I.I. *et al.* (2004) *Phys. Rev. B*, **69**, 056501; (c) Choi, H.J. *et al.* (2004) *Phys. Rev. B*, **69**, 056502.

1.2
Main Related Effects

1.2.1
Proximity Effect
Mikhail Belogolovskii

1.2.1.1 Introduction

The proximity effect (hereafter as PE) manifests itself as a mutual induction of physical properties from one material into an adjacent one across their interface. This definition means that the structure studied consists, at least, of three main elements: two materials with distinct characteristics and the interface whose transparency determines an effectiveness of the discussed phenomenon. In the most known N/S (normal/superconductor) example, due to the adjacent superconductor (S), electron pairs permeate into the neighboring normal (N) metal and, conversely, through the N/S interface, normal electrons leak into the superconductor partly destroying superconducting correlations in the S-side of the bilayer ([1]; Chapter 6).

Two prototypical examples of the N/S sandwich, namely, that with a low transparent interface (the so-called tunneling limit) as well as a direct NS contact, were studied, for the first time, in two seminal papers by McMillan published in 1968 [2, 3]. Taking into account incoherent single-particle scatterings from one layer to another treated within the tunnel-barrier approximation, he succeeded to explain such proximity-induced features as the enhancement of the Bardeen–Cooper–Schrieffer (BCS) potential in the N film, induced mini-gap Ω_N, qualitative changes of electronic densities of states in the adjacent metals, and so on [2]. The tunneling PE model was widely used to explain experimental data, for example, those for Nb/Pb sandwiches [4]. Note that the barrier in this model need not to be directly related to an insulating layer since reflection of quasiparticle states can be associated, for example, with different electronic structures of two metals in contact. In the second paper by McMillan [3], the analysis was based on a specific elastic phase-coherent transfer process across the NS interface by which normal current is converted into a super-current (in particular, at energies below the superconducting energy gap Δ_S, when direct single particle transmissions are forbidden). It is known as the *Andreev scattering*: an electron (hole) incident on the interface from the N side is elastically retroreflected into a hole (electron) from a spin-reversed band which is traveling in the opposite direction to the incoming charge. Starting from a step-like BCS pairing potential, McMillan [3] recalculated it and found that the new potential, which is nearly self-consistent, is roughly half Δ_S at the interface, very quickly (exponentially) approaches the bulk value of the energy gap Δ_S into the S-side, and drops rapidly into the N-side.

Generally, an advanced PE theory should be based on the microscopic Gor'kov equations and on a realistic treatment of the N/S interface. But even the oversimplified McMillan's models, which took into account two different aspects of the very complicated problem, are able to provide deep insight into the underlying

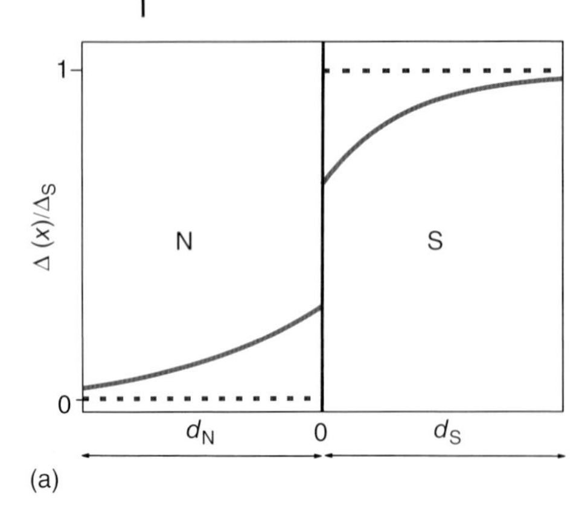

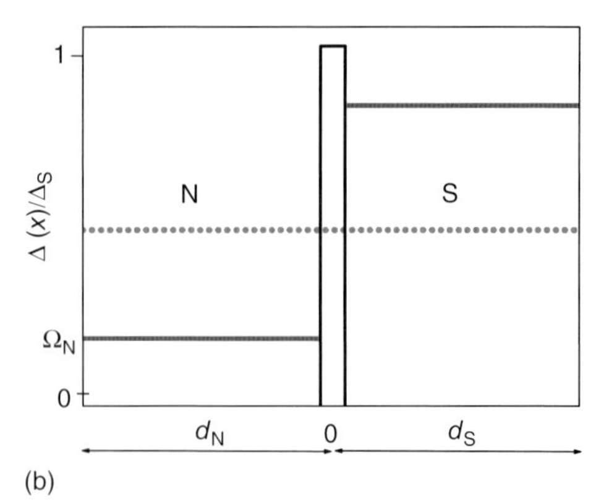

Figure 1.2.1.1 Sketch of the spatial dependence of the energy gap value (solid lines) in a proximity N/S sandwich with comparatively thick (a) and very thin (b) N (with a weak attractive electron–electron interaction) and S layers; Δ_S is the BCS potential in an isolated S metal, d_N and d_S are thicknesses of N and S layers, respectively. In (a), the dashed line demonstrates a step-like approximation for the BCS pairing potential. In (b), N and S films are separated by a potential barrier; if its height is comparatively low, we get a so-called Cooper limit shown with a dotted line.

physics and to explain, at least, qualitatively the main part of experimental data. According to them, principal spatial behavior of the order parameter $\Delta(x)$ in a planar N/S sandwich for extremely thick and extremely thin proximity-coupled S and N layers has the form shown in Figure 1.2.1.1a,b, respectively. The extent of superconducting correlations in the N part of the bilayer is determined by its structure and geometry: in a thin N layer, the proximity-induced mini-gap Ω_N is uniform, whereas in a thick N film the correlations extend over some distance determined by the energy of the electrons E relative to the Fermi level E_F. One of the signatures of the PE is the modification of local electronic densities of states $N(\varepsilon)$ in N and S parts, $\varepsilon = E - E_F$. The case of very thin N and S layers with one single gap Ω_N, two different peaks corresponding to Ω_N, and a new (corrected) value of the gap in a superconductor is shown in Figure 1.2.1.2.

Although the main conclusions of the McMillan's papers remain valid to date, see reviews by Gilabert [5], Pannetier and Courtois [6], and Klapwijk [7], a substantial body of novel results and new developments has contributed to the present level of PE understanding. One of the main advances of the last two decades has been a comprehension idea about the key role of the Andreev reflection in the PE. Other novel aspects of the problem relate new nonsuperconducting and superconducting materials which were unknown or unused in previous experiments with proximized bilayers, an effect of the interface between superconducting and nonsuperconducting films, which in some cases is not limited to the penetrability of superconducting correlations but can be a source of unexpected interfacial phenomena, and so on.

In the following, we attempt to present a simple introduction into the PE with the aim to explain fundamentals of the phenomenon and at the same time to

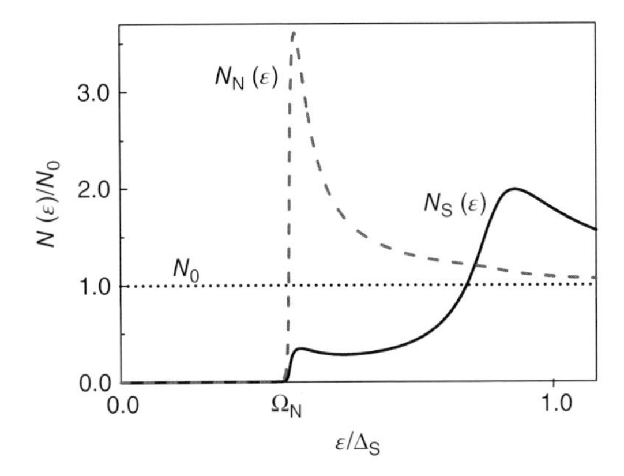

Figure 1.2.1.2 Sketch of normalized electronic densities of states versus energy on the normal $N_N(\varepsilon)$ and superconductor $N_S(\varepsilon)$ sides of an S/N sandwich; N_0 = const is the normal-state density of states of the sample, Δ_S is the BCS potential in an isolated S metal, and Ω_N is the minigap induced in the proximized N layer.

deliver current understanding of the general physical picture to a reader. In contrast to previous PE reviews, where the main attention was paid to basic effects in N/S sandwiches, we discuss all types of superconducting hybrid structures, composed from constituents of fundamentally different electronic structures, and present some novel devices and experiments in order to show PE perspectives for practical applications. The ideas underlying the two McMillan's papers will form a basis for our discussion.

The overview is structured as follows. In order to put the superconducting proximity phenomenon into a broader context, we commence with a simplest version of the PE by considering a relatively trivial case of a metal–insulator (MI) contact. In spite of diverse nature and very different values of superconducting and dielectric gaps, we demonstrate a few similar features exhibited by MI and NS sandwiches. A summary of the works on N/S bilayers, the most often studied PE samples, is presented in the next section. Then we deal with a specific PE arising between a ferromagnetic (F) metal and a superconductor in contact, explain the origin of the singlet-triplet conversion and the long-range PE in F/S hybrid bilayers. We finish the chapter by reviewing last ideas concerning the discussed phenomenon, in particular those relating contacts of superconductors with PE-affected topological insulators, a new class of quantum materials which, due to time-reversal symmetry, relativistic effects, and the inverted band structure, are insulating in bulk and completely metallic with a Dirac-like spectrum at their surface.

1.2.1.2 **Metal–Insulator Contact**

In order to be able to gain an insight into the main ideas of superconducting PEs, we start with a more simple system made of a metal and an insulator in contact. Owing to the wave nature of an electron, there is a finite probability to find the charge in a classically forbidden region adjacent to a metal film. It is well known that the probability exponentially decays into the insulating film and, when the film thickness is nanoscale, an electron can be found at the outer side of the potential barrier. The quantum-mechanical tunneling phenomenon in solid-state systems realized by separating two conductors with a very thin insulator is known from 1950s and now constitutes a basis upon which such devices as Josephson junctions, scanning tunneling microscope, and others are operating.

Less known is that fits of the rectangular barrier model to experimental data often lead to unphysically small values of extracted parameters, especially, barrier heights [8]. The origin of the discrepancy can be, in particular, related to the presence of a high density of extra metal-induced states in the gap energy range [9]. What is important for superconductor–insulator interfaces is that, according to Choi *et al.* [10], these states are strongly localized at the MI interface by inevitably random fluctuations in the electronic potential and so can produce paramagnetic spins. It can explain the origin of magnetic flux noise in superconducting quantum interference devices (SQUIDs) with a power spectrum $\sim 1/f$ (f is the frequency) which limits the decoherence time of superconducting flux-sensitive qubits [10]. Spin-flip scattering of conduction electrons by local magnetic moments, possibly located at metal–oxide interfaces, was revealed in granular Al films in the vicinity of the metal-to-insulator transition [11]. The most surprising finding was the coexistence of enhanced superconducting properties in Al granules with surface magnetic moments which raises a question about the mechanism of superconductivity in such films.

Notice that electrons trapped into the localized states in the near-surface region in an insulator can tunnel into the adjacent metal when an electric field is applied to the interface (in a superconductor it occurs only when the energy gained by an electron is larger than the energy gap). Such a model was proposed by Halbritter [12] to explain loss mechanisms in superconducting niobium cavities. Usually, this effect is negligible but in the case of Nb it can be important due to the presence of conducting Nb oxides. When the radio frequency (RF) field in the cavities is raised, a longitudinal electric field penetrates the insulator and stimulates electrons to tunnel to the superconductor and to return back when the field is lowered. The losses due to interface tunnel exchange are an example of the interface-induced effect which should be more pronounced in the case of small grain sizes.

Let us now look at the NI proximity problem from a perspective of elastic scattering processes at the interface (see Figure 1.2.1.3a). An electron (hole) incident on the interface from the N side is retroreflected into an electron (hole) of the same energy E and the same absolute value of momentum but traveling in the opposite direction to the incoming charge. Let us consider an ideal three-dimensional planar structure with two normal metallic electrodes, two insulating layers I which are so thick that electrons can tunnel across them very rarely, and a normal-metal nanometer-thin N′ interlayer of the thickness d_N. In the NIN′IN structure, an electron is spatially confined in the direction x normal to the layers but remains free to move in the parallel direction. It results in the creation of a two-dimensional electron gas at quasi-bound quantized electron states.

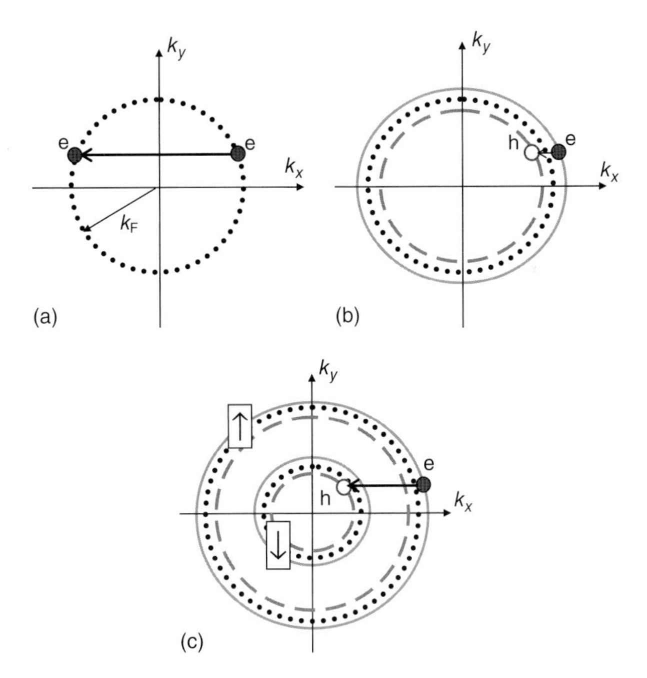

Figure 1.2.1.3 Scattering processes at NI (a), NS (b), and FS (c) interfaces. The filled and empty circles denote the quasielectron (e) and quasihole (h) states in the nonsuperconducting layer. The Fermi surface with the Fermi wave vector k_F, the constant-energy electron surface, and the constant-energy hole surface are shown with dotted, solid, and dashed lines, respectively; arrows in the (c) panel denote the spin direction of the corresponding energy band.

The electron transmission coefficient (transparency) D through the NIN′IN structure is sharply peaked about certain energies corresponding to virtual resonant levels in the quantum well of the thickness d_N. The energies can be easily found from the demand of coherent superposition of scattered electron waves. According to the Bohr–Sommerfeld quantization rule, the electron wave-function phase shift acquired during the electron "round-trip" inside the N′ interlayer (shown in Figure 1.2.1.4a) $2k_x d_N = 2\pi j$, where k_x is the x-component of the electron wave vector, j is integer. When two insulating I layers are atomically thin and identical, an electron with one of the resonant energies can cross the NIN′IN system without being reflected (the resonant tunneling effect). Although the latter phenomenon is well known from elementary quantum mechanics, we want to stress that the peaked structure appears just due to the presence of insulating layers adjacent to the N′ interlayer and that energy locations of strong maxima and minima in its spectrum can be obtained by very simple quantum-phase arguments which will be applied further to NS and FS cases. The second remark concerns a strong effect of barrier inhomogeneities on electron tunneling near resonance energies (see [13] and references therein).

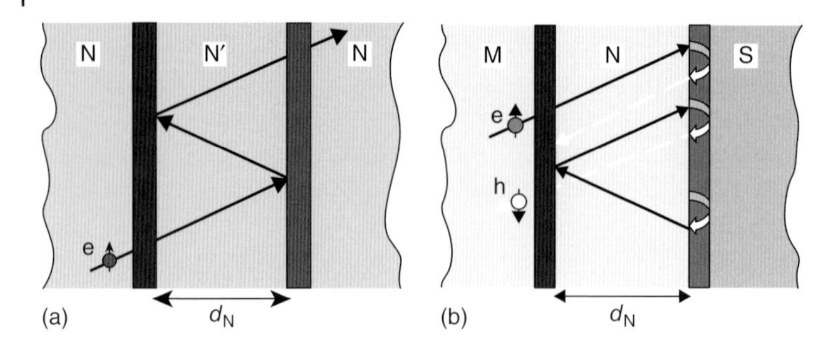

Figure 1.2.1.4 Backscattering processes within a normal interlayer are decomposed into a series of alternating reflections from tunneling barriers in a double-barrier NIN′IN structure (a) and those by the superconducting pair potential and a barrier potential in an MIN′IS junction (b).

Resuming this subsection, we would like to attract the attention to three different aspects of the NI PE which are common for the cases discussed below as well: (i) leakage of itinerant electrons from a metal into an insulator, (ii) existence of energy regions with a strongly suppressed density of states in the excitation spectrum of a spatially confined N′ interlayer due to backscattering of electron waves from the N′I interface, and (iii) specific properties of the MI interface which appears only when the surfaces of the two separated materials are brought into a contact. Generally, the three features are only different manifestations of the same phenomenon and, in principle, should be described self-consistently within a unified formalism for the inhomogeneous solid-state structures. Keeping in mind the analogy (and the difference as well) between I and S layers, we can now transfer to the superconducting PE.

1.2.1.3 **Normal Metal–Superconductor Contact**

Even the simplest hybrid NS heterostructure with an ideal interface reveals several interesting phenomena. The energy scale in the NS bilayer is characterized by a superconducting gap energy of the order of 1 meV, which is much less than a N-metal electron band width (~1 eV and more). That is why the probability of the normal backscattering from a clean NS interface is extremely small and a quasi-classical approximation, see the review in [14], applied to the Green-function formalism serves usually as a starting point for theoretical works dealing with a quasi-ballistic NS problem when elastic quasielectron-into-quasihole transformations of Bogoliubov quasiparticles and inverse (with a missing charge of $2e$ absorbed into the superconducting ground state as a Cooper pair) occur at the NS interface – see Figure 1.2.1.3b. At voltage bias $V = 0$, the differential conductance $G(V) = dI(V)/dV$ of an ideal NS bilayer is determined only by pair transferring processes and is as large as twice normal-state conductance, whereas in the very low transparency limit $G(0)$ of the N/S contact is vanishing at $T = 0$. The current across the junction with N and S layers divided by an insulating barrier does not increase until the electron volt reaches the superconducting gap Δ_S [15]. At $V = \Delta_S/e$, the differential conductance $G(V)$ of the trilayer exhibits a peak.

Proximity-induced changes in the electronic density of states $N_N(\varepsilon)$ of a normal metal are directly reflected in the $G(V)$ characteristic of a planar tunnel junction formed by a normal counter-electrode M, an isolator I, and the NS contact. Thus, the tunneling experiment (real or imagined) can serve as a probe of the local electronic density of states at the tunneling surface weighted by the angular distribution of tunneling electrons [3]. Considering MINS junction within the Bohr–Sommerfeld quantization condition, we can easily understand the origin of the coherent peak in the spectrum of a superconductor and its modification for the N layer of the thickness $d_{N'}$. The characteristic energy $\overline{E}$ of a bound state formed in the N interlayer corresponds to the coherent superposition of reflected quasiparticle waves. To find it, we suppose a planar geometry of the MINS structure (Figure 1.2.1.4b) where the trajectories of quasiparticles in the electron branch and those in the hole branch are denoted by solid and dashed lines, respectively. A charge velocity component perpendicular to the layers changes its sign in the normal reflection from the MI interface, whereas all velocity components change their signs in the Andreev backscattering from the NS interface. The reflection coefficients can be calculated from the boundary conditions of the wave functions using the fact that in the discussed geometry transverse components of the wave vectors are unchanged. In this case, each Andreev reflection within the energy gap contributes an additional phase shift $\chi^{\mathrm{eh(he)}}(\boldsymbol{k}) = -\arccos(\varepsilon/|\Delta(\boldsymbol{k})|) \mp i\Phi(\boldsymbol{k})$, where $|\Delta(\boldsymbol{k})|$ and $\Phi(\boldsymbol{k})$ are absolute value and phase of the complex superconducting order parameter (see, e.g., [16]). Adding the phases accumulated along an electron "round-trip" in the N interlayer with two subsequent Andreev reflections (Figure 1.2.1.4b), for an s-wave superconducting electrode, we get the following expression for the phase shift: $k_x^{\mathrm{e}} d_n + \chi^{\mathrm{eh}} - k_x^{\mathrm{h}} d_n + \pi - k_x^{\mathrm{h}} d_n + \chi^{\mathrm{he}} + k_x^{\mathrm{e}} d_n + \pi = (4\varepsilon/\hbar v_{\mathrm{F}})(d_{\mathrm{N}}/\cos\theta) - 2\arccos(\varepsilon/\Delta_{\mathrm{S}})$, where θ is the incident angle, the x-axis is normal to interfaces. The bound-state energies $\overline{\varepsilon} = \overline{E} - E_{\mathrm{F}}$ can be found requiring the total phase shift to be an integer multiple of 2π. Hence, the lowest bound level follows from the relation $\overline{\varepsilon} = \hbar v_{\mathrm{F}} \arccos(\overline{\varepsilon}/\Delta_{\mathrm{S}})\cos\theta/2d_{\mathrm{N}}$. For a clean system which is translationally invariant in the y and z directions, contributions from long path lengths ($\theta \to \pi/2$) result in no gap in the excitation spectrum [3]. But this conclusion is valid for infinite samples. If the volume of the N part of a mesoscopic N/S bilayer is finite, we get a sharp gap in the excitation spectrum like that in the PE tunneling model [2] or in the dirty (diffusive) limit when electrons experience a huge number of elastic scatterings on impurities during the way from one surface to another. In the latter case, the size limiting the phase coherence length for electron-like and hole-like quasiparticles traversing a diffusive trajectory is the elastic mean free path $l_{\mathrm{e}} = \sqrt{D\tau_{\mathrm{e}}}$, here D is the diffusion constant in the normal metal, and τ_{e} is the elastic scattering time. If so, then in the previous formula for $\overline{\varepsilon}$, we should replace the average time $\sim d_{\mathrm{N}}/v_{\mathrm{F}}$ for the motion of a quasiparticle across a clean N film with $\tau_{\mathrm{e}} \sim d_{\mathrm{N}}^2/D$. When $\overline{\varepsilon} \ll \Delta_{\mathrm{S}}$, the minigap in a dirty N layer is thus expected to be approximately $\overline{\varepsilon} \approx \pi\hbar D/4d_{\mathrm{N}}^2$ (compare with the numerically exact expression $\overline{\varepsilon} \approx 0.78\hbar D/d_{\mathrm{N}}^2$ obtained in [17]).

In d-wave superconductors like YBCO, dissimilar values of $\Phi(\boldsymbol{k})$ seen by a quasiparticle moving along different scattering trajectories (Figure 1.2.1.4b) brings to the Bohr–Sommerfeld quantization rule an additional phase shift $\delta\Phi(\boldsymbol{k})$. The most dramatic effect of $\delta\Phi(\boldsymbol{k}) = \pi$ occurs for the tunneling direction [110], converting destructive interference at $\bar{\varepsilon} = 0$ in the s-wave case into constructing interference (see, e.g., [16]). Owing to the phase conjugation between electrons and holes at the Fermi level, it survives even in the presence of a normal interlayer when tunneling across the barrier in MINS devices is specular and a strong zero-bias peak in the $G(V)$ characteristic of a MINS junction is one of the main signatures of the d-wave symmetry of the order parameter in an S electrode.

Notice that the stepwise approximation for the pair potential used above and known as a *rigid-boundary condition* for superconducting bilayers is not self-consistent. According to Likharev [18], in real N/S junctions the deviation of the self-consistent solution from the step-like function strongly decreases when the interface resistivity is much bigger than that of metal electrodes. From the Andreev-scattering view, the penetration of superconducting correlations into a normal material is limited by dephasing between electron and hole wave functions $\delta\varphi = (k_x^{\mathrm{e}} - k_x^{\mathrm{h}})l$ which grows with increasing distance l from the N/S boundary into the N-side. The relevant length scale governing superconducting correlations in a clean N-metal is given by the temperature-dependent coherence length $\xi_{\mathrm{T}} = \hbar v_{\mathrm{F}}/(2\pi k_{\mathrm{B}} T)$.

Since practically used films are usually full of different imperfections, the charge scatterings inside them render the Green function to be isotropic in space. The quasi-classical Eilenberger equations can be further simplified to the form which is more appropriate for numerical calculations and is known as *Usadel equations*, see the review in [14]. Related boundary conditions for an interface between two superconductors were derived by Kupriyanov and Lukichev [19]. The validity of Usadel equations depends on the superconductor ($l_{\mathrm{e}} \ll \xi_0$) as well as on the normal metal ($l_{\mathrm{e}} \ll d_{\mathrm{N}}, \xi_{\mathrm{T}}$), where $\xi_{\mathrm{T}} = \sqrt{\hbar D/(2\pi k_{\mathrm{B}} T)}$ is the temperature-dependent decay length in the dirty N layer. Note that the validity of the clean-limit solution is restricted to $l_{\mathrm{e}}, \xi_{\mathrm{T}} \gg d_{\mathrm{N}}$ and to $l_{\mathrm{e}} \gg d_{\mathrm{N}} \exp(2d_{\mathrm{N}}/\xi_{\mathrm{T}})$ if $\xi_{\mathrm{T}} \ll d_{\mathrm{N}}$ [20].

Specific properties of proximity-coupled NS bilayers can be useful for practical purposes since they provide an additional internal degree of freedom which can offer new abilities to design and tune such characteristics as the critical temperature, the energy gap, and the shape of corresponding temperature dependencies. The most straightforward application consists in the control over T_{c}, which is determined by PEs and can, therefore, be tuned by varying individual layer thicknesses. As an example, we consider NS sandwiches where the thicknesses d_{S} and d_{N} are somewhat larger than corresponding coherence lengths [21]. Then the bilayer critical temperature as a function of d_{N} follows an exponential law $T_{\mathrm{c}} = \widetilde{T}_{\mathrm{c}} - \mathrm{const} \cdot \exp(-\alpha d_{\mathrm{N}})$ where the decay constant α increases with the decrease of the electron mean free path in the superconductor (see the related experimental data in Figure 1.2.1.5). The possibility to tune T_{c} with a PE was realized, in particular, in bolometers [22] and superconducting screening ground planes [23]. In the first case, the absorber and the thermistor were Ti–Al–Ti trilayers with thicknesses ~50 nm chosen to achieve T_{c} of 380 mK. In order to produce superconducting screening planes with tunable T_{c}'s between 4 and 7 K,

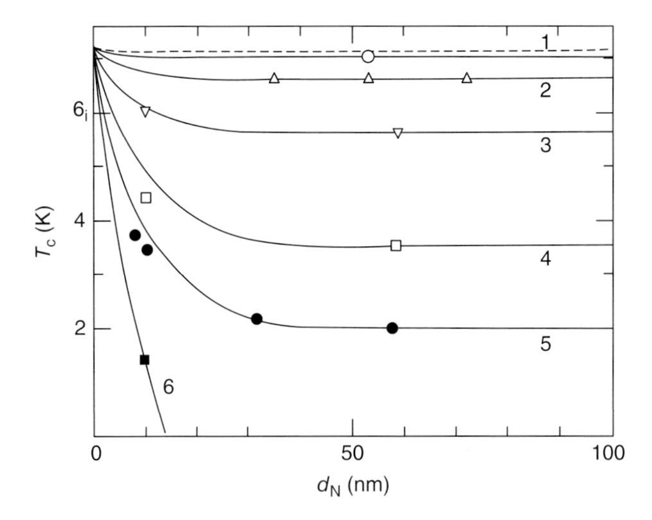

Figure 1.2.1.5 Transition temperature of Pb/Cu bilayers versus the Cu-film thickness d_N; the thickness of the Pb films was 100 nm (1), 50 nm (2), 30 nm (3), 15 nm (4), 10 nm (5), and 7 nm (6). (Adapted from [1], Figure 6.6. Reproduced with permission of John Wiley & Sons.)

Kain *et al.* [23] used Pb/Ag bilayers. In agreement with the theory, in samples with good electrical contact between two layers, the measured T_c values followed well the Cooper model of the PE when superconducting characteristics are averaged over the bilayer (see Figure 1.2.1.1b).

The main goal of the work by Lacquaniti *et al.* [24] was to design an intrinsically shunted Josephson device resistant to temperature fluctuations above 4.2 K and based on the niobium technology. A strong temperature dependence of critical current I_c in conventional Nb-based superconductor–insulator–superconductor (SIS) Josephson junctions appears when they are employed above 5 K. The problem comes from a steep-like behavior of the $I_c(T)$ curve above $T \approx 0.5T_{cs}$, where T_{cs} is the S-film critical temperature. In order to obtain more gentile curve, the authors replaced one of the superconducting electrodes with an N/S bilayer. It has been shown experimentally that with increasing temperature the superconducting order parameter in a normal layer of a thickness d_N comparable with that of the S film d_S first rapidly decreases and then at $T \approx T^*$ becomes flat up to a nearest vicinity of the bilayer T_c. Changing the ratio d_N/d_S, the authors modified the T^* value and were able to engineer thermal stability of the Josephson superconductor–normal conductor–isolator–superconductor (SNIS) devices.

1.2.1.4 Ferromagnetic Metal–Superconductor Contact

Coexistence of such antagonistic phenomena as superconductivity and ferromagnetism is a long-standing problem in solid-state physics. Originally, it was believed that they are mutually exclusive, but more recently it was found that they can coexist under certain circumstances giving rise to novel combined effects. One of the possibilities to observe the interplay between itinerant electron ferromagnetism (F) and superconductivity is to put the two metallic films into a contact. Besides a fundamental physics interest, such bilayers hold important potential for applications in spintronics devices [25] as well as for recovery of a mechanism of the high-T_c phenomenon which typically occurs in the vicinity of a magnetic instability.

The main changes caused by replacing the N film in the MINS heterostructure with an F layer are related to the difference in their electronic structures. Since the charge reflected at the F/S interface in an MIFS sample is created in the electron density of states with a spin opposite to that of the incident quasiparticle, the scattering strength (and, as a result, the F/S bilayer resistance) strongly depends on the spin imbalance in the ferromagnet. Its theoretical analysis is usually based on the Stoner model of metallic ferromagnetism where charge carriers with opposite spins occupy rigidly shifted bands with the difference in energy equal to $2\varepsilon_{ex}$ (see Figure 1.2.1.3c). In this case, Fermi wave vectors for electron spin-up $k_F^\uparrow$ and spin-down $k_F^\downarrow$ bands are different: $\delta k^{(F)} = k_F^\uparrow - k_F^\downarrow = 2\varepsilon_{ex}/\hbar v_F =$ const (their energy dependence may be not taken into account for $\varepsilon \leq \Delta \ll E_F$). For example, in Ni the averaged over Fermi surface s-subband values are $k_F^\uparrow = 5.1\ \text{nm}^{-1}$ and $k_F^\downarrow = 4.7\ \text{nm}^{-1}$ [26]. Thus, the semiclassical quantization condition for the Andreev bound state in an F interlayer looks as follows: $k_{Fx}^\uparrow d_F + \chi^{eh} - k_{Fx}^\downarrow d_F + \pi - k_{Fx}^\downarrow d_F + \chi^{he} + k_{Fx}^\uparrow d_F + \pi = 2\delta k_x^{(F)} d_F - 2\arccos(\overline{\varepsilon}/\Delta_S) = 2\pi j$, where d_F is the thickness of the F interlayer. As the distance from the F/S interface increases, the phase shift $\delta k_x^{(F)} d_F$ grows continuously and for $d_F \approx 2\pi\hbar v_F/4\varepsilon_{ex}$ is equal to π. This phase shift produces oscillations of the superconducting order parameter

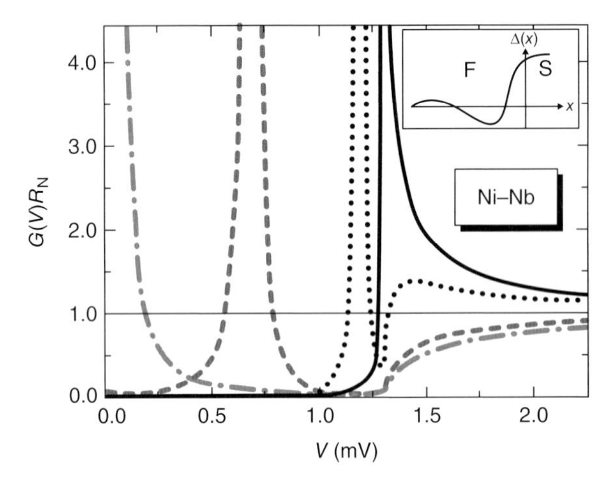

Figure 1.2.1.6 Calculated differential conductance-versus-voltage dependencies for a tunneling junction with a Ni/Nb bilayer; $d_F = 0$ nm (solid line), 1.0 nm (dotted line), 2.5 nm (dashed line), and 4.0 nm (dashed-dotted line), R_N is the resistance of the junction in a normal state; the Nb energy gap was set to 1.3 meV; temperature $T = 0$. The inset demonstrates schematic behavior of the superconducting order parameter in the F/S bilayer. The continuity of $\Delta(x)$ at the interface implies the absence of the potential barrier. In the general case, the jump of the order parameter occurs at $x = 0$ like in Figure 1.2.1.1. (Adapted from [26]. Reproduced with permission of World Scientific Publishing Company.)

on a length scale $\sim 1/\delta k^{(F)} = \hbar v_F/2\varepsilon_{ex}$ (see the inset in Figure 1.2.1.6). The states corresponding to a positive sign of the order-parameter real part are called *0-states* and those corresponding to the negative sign are known as *π-states*. In general, the overall character of the differential conductance for finite thicknesses of the F interlayer remains similar to that for MINS junctions, but now the measurable shift of the peak in $G(V)$ can be obtained for relatively small values of d_F. The effect of increasing d_{Ni} from 0 to 5 nm in transport characteristics of Al–I–Ni/Nb samples measured by SanGiorgio *et al.* [27] well agree with related numerical simulations by Belogolovskii *et al.* [26] including an anomalous double-peak structure for extremely thin Ni films (Figure 1.2.1.6).

In addition to the spatial oscillations of the pair wavefunction, spin-singlet Cooper pairs are fragile under the exchange potential which affects differently electrons with opposite spins. As in N metals, the penetration of superconducting correlations into a ferromagnetic metal is limited by dephasing between electron and hole wave functions but in the case of an F layer, it is much stronger. That is why for a direct FS bilayer, the penetration depth is atomically short (the only exception is the clean limit $l_e \gg \xi_0$ at $k_B T \ll \varepsilon_{ex}$ when it is limited only by elastic impurity scattering and typically exceeds the oscillation period). In general, the decay value depends on the presence of elastic scatterings as well as on the strength of the exchange field (see the review by Golubov *et al.* [14]). In the dirty limit at $T = 0$, the decay length exactly coincides with the oscillation period and for sufficiently strong diffusive ferromagnetic materials like Fe, Co, Ni even at finite temperatures $\xi_F = \sqrt{\hbar D/(2\varepsilon_{ex})}$. But in the general case, especially for weak ferromagnets, when the temperature and the exchange field are equally important, the period of the Cooper potential oscillations and their decay do not coincide. Moreover, when the temperature is going down, the decay length increases whereas the oscillations period decreases. Because of it, the temperature variation can be used as a good tool for engineering the phase shift of the superconducting order parameter along the F layer from 0 to π [28].

The new burst of interest to the F/S problem is associated with the so-called long-range equal spin-triplet pairing state, a new type of proximity-induced superconducting state in ferromagnets which can be realized in the presence of a magnetically inhomogeneous F/S interface [29]. Spin-dependent phase shifts which are acquired by electrons penetrating through the interface can induce spin-triplet s-wave Cooper pairs in ferromagnets. The latter ones should have the odd-frequency symmetry to satisfy the requirement from the Fermi–Dirac statistics of electrons and be not suppressed by the exchange potential. If so, they are able to deepen into the ferromagnet at distances of the order of the decay length ξ_T for a normal metal. The novel aspect of the PE consists in the fact that by introducing a low-transparent magnetic interface, we can enhance the penetration depth into the F layer from nanometers for singlet pairing to microns (at least, at very low temperatures) for triplet one. The relative fraction of odd-frequency pairs to even-frequency pairs depends sensitively on such junction parameters as its geometry, the interface transparency, and so on. Notice also that the inhomogeneous magnetization needed for the transformation does not necessarily be intrinsic. For example, it can be introduced by two misaligned magnetic layers like in the experiment by Khare *et al.* [30].

Let us now discuss a few practically useful ideas from a long list of striking proposals relating the PE in systems with S and F layers. As was explained above, in FS contacts the PE manifests itself in the damped oscillatory behavior of superconducting correlations induced in a ferromagnet. As a result, for certain thicknesses of the F layer and temperatures, the order parameter in a superconductor–ferromagnet–superconductor (SFS) junction may become positive at one S electrode and negative at the other S electrode. In this situation, one gets a π Josephson junction with the spontaneous π-shift of the phase difference in its ground state. Its properties are indeed unusual. For example, when connecting the S electrodes with a superconducting wire, one may expect the spontaneous super-current circulating in the loop, passing clockwise or counterclockwise. It has been experimentally demonstrated that the π-junction improves the performance and simplifies the design of classical and quantum circuits. An idea to introduce π-junctions as passive phase shifters in rapid single flux quantum (RSFQ) circuits was proposed by Ustinov and Kaplunenko [31]. It permits (i) to use only conventional junctions to carry flux quanta, and thus to provide logical functionality, (ii) to substitute by π-junctions the relatively large geometrical inductance of storing a single-flux quantum RSFQ cells and, hence, to reduce greatly the size of RSFQ circuits, (iii) to operate in the passive mode which has some advantages over the active regime. In a φ-junction, a generalization of the π-junction with a doubly degenerate ground state, the Josephson phase takes the values $+\varphi$ or $-\varphi$ $(0 < \varphi < \pi)$ [32]. This device is, in fact, a phase battery providing an arbitrary phase shift and being closed into a ring is able to self-generate a fractional flux $\varphi\Phi_0/(2\pi)$, where Φ_0 is the magnetic flux quantum.

Moreover, superconducting devices with a magnetic interlayer can be useful for the solution of other problems of superconducting electronics and so are able to realize its high performance potential. Let us provide some examples. The first problem relates the lack of high-capacity superconducting random access memory [33]. Magnetic Josephson junctions, that is, superconducting structures with incorporated magnetic layer(s), were suggested to perform both data storage and readout functions. By applying magnetic field pulses (e.g., by current pulses through a superconducting write line), the F-layer can be magnetized in two opposite directions. To discriminate these directions, a read current bias is applied through the junction inducing a reference magnetic field. Depending on F-layer magnetization, this field either adds to or subtracts from F-layer magnetic field effectively forming two possible magnetic states with high or low magnetizations corresponding to low ("1") and high ("0") critical currents, respectively [33]. Another problem relates a superconducting three-terminal device which could switch and amplify electric signals like a semiconducting transistor. One of the most promising designs is a double-barrier $S_1IS_2IS_3$ structure based on the tunneling injection of nonequilibrium quasiparticles through a thin insulating layer into the middle S_2 layer and detection of the resultant energy gap suppression by the second junction. While possessing essential transistor-like characteristics, the "symmetric" $S_1IS_2IS_3$ device has a drawback detrimental for its implementation, namely, lack of isolation between the input and output terminals. Nevirkovets and Belogolovskii [34] proposed to block parasitic back-action of the acceptor junction by a few-nanometer thick inhomogeneous ferromagnetic film inserted between the barrier and the middle S_2 layer. At last, the long-range PEs in ferromagnets pave the way for the controlled creation of completely spin-polarized super-currents which would necessarily have to be triplet. Such superconducting spintronics devices are, in particular, ideal candidates for quantum computing.

1.2.1.5 New Perspectives and New Challenges

A new step in the development of the PE that started several years ago is related to the search of Majorana fermions (see the review by Beenakker [35]). These hypothetical elementary particles which are often described as "half fermions" since they are the only fermionic particles expected to be their own antiparticles, have not been identified in the nature yet but can exist as quasiparticle excitations in solid-state systems. A great variety of strategies has been put forward in recent years to engineer Majorana fermions in different condensed-matter platforms. One of the most promising proposals is to use proximity-induced superconductivity in the surface of topological insulators, a new phase of matter where conduction of electrons occurs only on the surfaces due to strong spin–orbit coupling, which inverts the order of conduction and valence bands. Three-dimensional topological insulators have surface electron states with massless Dirac cones in which the spin of an electron is locked perpendicular to its momentum in a chiral spin structure where electrons with opposite momenta have opposite spins. When such a material is brought in contact with a conventional spin-singlet, s-wave S layer, superconductivity is induced on its surface, with a nondegenerate state at the Fermi level ($\varepsilon = 0$), in the middle of the superconducting gap. This specific midgap state following from the electron–hole symmetry is just the Majorana fermion. A key probe to detect it experimentally is the tunneling differential conductance which should show a peak at zero-voltage bias. Several groups have already reported zero-bias anomalies in corresponding PE devices (see, in particular, [36]). Another interesting prediction is a 4π-periodic Josephson effect [35]. At the moment it is not clear whether or not the observed anomalous features reflect a Majorana bound state signature. But, if the presence of Majorana fermions is proved, it will provide a fundamentally new way to store and manipulate quantum information, with possible applications in a quantum computer [35].

The progress in the PE research and arising challenges are not limited by this exotic field. Whereas the main fundamental features of the N/S PE (at least, for traditional superconductors) are clear, less is known about such structures very far from equilibrium as well as about nonlocal correlations in a proximized N layer (see [37, 38]). The problem of the competition between electron–electron interactions and the superconducting PE remains largely unexplored. Recent experiments [39] have shown that graphene provides a useful experimental platform to investigate it. Up to now, it is not clear why sometimes the T_c of a superconductor is increased upon attachment to a nonsuperconducting material. For example, in the work by Katzir *et al.* [40], it was shown that the T_c of thin Nb films increases by up to 10% when they are chemically linked to gold nanoparticles using ~3 nm long disilane linker molecules.

Even more questions are arising for contacts between superconducting and magnetic materials. While much of the work on F/S PE is focused on the penetration of the superconducting order parameter into the ferromagnet, very little was done to understand the penetration of the magnetization into the adjacent S metal which is often called *the inverse proximity effect*. The first theoretical interpretation of the problem arising in F/S bilayers was done by Krivoruchko and Koshina [41]. Direct experimental observation of the inverse PE in Ni/Pb and Co–Pd/Al bilayers was presented in the work by Xia *et al.* [42] where it was shown that the magnetization in a ferromagnetic film induces a magnetization in a superconducting film that is much smaller and opposite in sign.

Although singlet superconductivity and ferromagnetism are mutually exclusive in homogeneous bulk materials, magnetization noncollinearity is expected to enhance T_c. Zhu *et al.* [43] observed a nonmonotonic enhancement of superconductivity with the increase of magnetic noncollinearity in a related F/S sandwich. An interest in PE contacts formed by superconductors and antiferromagnets (AFs) has increased when it was found that at low temperatures iron-based superconductors can be intrinsically phase separated into antiferromagnetic and superconducting regions. Corresponding theoretical predictions [44] have not been yet checked experimentally.

1.2.1.6 **Summary**

The current understanding of the relationship between the PE in the old sense, as a leakage of superconductivity into a normal metal, and the Andreev backscattering at the NS interface is that they are only two sides of the same coin. Whereas this statement helped to understand the basic observations for N/S bilayers, it is not so in the case of F/S structures. From the experimental side, one of the most significant recent developments has been a striking phenomenon of generation of odd-frequency spin-triplet s-wave pairs in F/S devices by spin-mixing due to inhomogeneous magnetization or spin-dependent potential. New physics is yet to be captured in the theoretical treatments of F/S systems.

Discovery of topological superconductivity has paved the way for the novel states of quantum matter which are not only of a fundamental significance but have potential practical implications as well. The resemblance of the new exotic phase with strong spin–orbit coupling to the already studied spin-triplet superconductivity without it [45] permits to reveal the essential physics of the PE in the novel materials using the previous knowledge about spin-triplet superconductors. In our opinion, experimental and theoretical efforts aimed to create topological superconductors, gapped phases of fermionic quantum matter whose zero-energy states can be associated with Majorana quasiparticles, will certainly be a major research theme in the PE field for the nearest future.

References

1. Buckel, W. and Kleiner, R. (2004) *Superconductivity. Fundamentals and Applications*, 2nd edn, John Wiley & Sons, Ltd, Chichester.
2. McMillan, W.L. (1968) Tunneling model of the superconducting proximity effect. *Phys. Rev.*, **175** (2), 537–542.
3. McMillan, W.L. (1968) Theory of superconducting–normal-metal interface. *Phys. Rev.*, **175** (2), 559–568.
4. Seidel, P. and Richter, J. (1980) Theoretical investigation of the current-voltage characteristics of superconducting niobium-lead tunnel junctions. *Phys. Status Solidi B*, **98** (1), 189–197.
5. Gilabert, A. (1977) Effet de proximité entre un metal normal et un supraconducteur. *Ann. Phys.*, **2** (4), 203–252.
6. Pannetier, B. and Courtois, H. (2000) Andreev reflection and proximity effect. *J. Low Temp. Phys.*, **118** (5-6), 599–615.
7. Klapwijk, T.M. (2004) Proximity effect from the Andreev perspective. *J. Supercond.*, **17** (5), 593–611.
8. Miller, C.W. and Belyea, D.D. (2009) The impact of barrier height distributions in tunnel junctions. *J. Appl. Phys.*, **105** (9), 094505-1–094505-5.
9. Louie, S.G. and Cohen, M.L. (1976) Electronic structure of a metal-semiconductor interface. *Phys. Rev. B*, **13** (6), 2461–2469.
10. Choi, S.K., Lee, D.-H., Louie, S.G., and Clarke, J. (2009) Localization of metal-induced gap states at the metal–insulator interface: origin of flux noise in SQUIDs and superconducting qubits. *Phys. Rev. Lett.*, **103** (19), 197001-1–197001-4.
11. Bachar, N., Lerer, S., Hacohen-Gourgy, S., Almog, B., and Deutscher, G. (2013) Kondo-like behavior near the metal-to-insulator transition of nanoscale granular aluminum. *Phys. Rev. B*, **87** (21), 214512-1–214512-4.
12. Halbritter, J. (2005) Transport in superconducting niobium films for radio frequency applications. *J. Appl. Phys.*, **97** (8), 083904-1–083904-12.
13. Knauer, H., Richter, J., and Seidel, P. (1977) A direct calculation of the resonance tunneling in metal–insulator–metal tunnel junctions. *Phys. Status Solidi A*, **44** (1), 303–312.
14. Golubov, A.A., Kupriyanov, M.Y., and Il'ichev, E. (2004) The current-phase relation in Josephson junctions. *Rev. Mod. Phys.*, **76** (2), 411–469.
15. Blonder, G.E., Tinkham, M., and Klapwijk, T.M. (1982) Transition from metallic to tunneling regimes in superconducting microconstrictions: excess current, charge imbalance, and supercurrent conversion. *Phys. Rev. B*, **25** (7), 4515–4532.
16. Belogolovskii, M., Grajcar, M., Kúš, P., Plecenik, A., Beňačka, Š., and Seidel, P. (1999) Phase-coherent charge transport in superconducting heterocontacts. *Phys. Rev. B*, **59** (14), 9617–9626.
17. Pilgram, S., Belzig, W., and Bruder, C. (2000) Excitation spectrum of mesoscopic proximity structures. *Phys. Rev. B*, **62** (18), 12462–12467.
18. Likharev, K.K. (1979) Superconducting weak links. *Rev. Mod. Phys.*, **51** (1), 101–159.
19. Kupriyanov, M.Y. and Lukichev, V.F. (1988) Influence of boundary transparency on the critical current of "dirty" SS'S structures. *Zh. Eksp. Teor. Fiz.*, **94** (6), 139–149 (*Sov. Phys. JETP*, **67**(6), 1163–1168).
20. Belzig, W., Bruder, C., and Fauchère, A.L. (1998) Diamagnetic response of a normal-metal–superconductor proximity system of arbitrary impurity concentration. *Phys. Rev. B*, **58** (21), 14531–14540.
21. De Gennes, P.G. (1964) Boundary effects in superconductors. *Rev. Mod. Phys.*, **36** (1), 225–237.
22. Gildemeister, J.M., Lee, A.T., and Richards, P.L. (1999) A fully lithographed

voltage-biased superconducting spider-web bolometer. *Appl. Phys. Lett.*, **74** (6), 868–870.

23. Kain, B., Khan, S.R., and Barber, R.P. Jr., (2002) Tuning the transition temperature of superconducting Ag/Pb films via the proximity effect. *Physica C*, **382** (4), 411–414.
24. Lacquaniti, V., De Leo, N., Fretto, M., Sosso, A., Boilo, I., and Belogolovskii, M. (2012) Advanced four-layered Josephson junctions for digital applications. *Physics Procedia*, **36**, 100–104.
25. Žutić, I., Fabian, J., and Das Sarma, S. (2004) Spintronics: fundamentals and applications. *Rev. Mod. Phys.*, **76** (2), 323–410.
26. Belogolovskii, M., De Leo, N., Fretto, M., and Lacquaniti, V. (2011) Quantum size effects in superconducting junctions with a ferromagnetic interlayer. *Int. J. Quantum Inf.*, **9** (Suppl. 01), 301–308.
27. SanGiorgio, P., Reymond, S., Beasley, M.R., Kwon, J.H., and Char, K. (2008) Anomalous double peak structure in superconductor/ferromagnet tunneling density of states. *Phys. Rev. Lett.*, **100** (23), 237002-1–237002-4.
28. Ryazanov, V.V., Oboznov, V.A., Rusanov, A.Y., Veretennikov, A.V., Golubov, A.A., and Aarts, J. (2001) Coupling of two superconductors through a ferromagnet: evidence for a π junction. *Phys. Rev. Lett.*, **86** (11), 2427–2430.
29. Bergeret, F.S., Volkov, A.F., and Efetov, K.B. (2005) Odd triplet superconductivity and related phenomena in superconductor–ferromagnet structures. *Rev. Mod. Phys.*, 77 (4), 1321–1373.
30. Khare, T.S., Khasawneh, M.A., Pratt, W.P. Jr.,, and Birge, N.O. (2010) Observation of spin-triplet superconductivity in Co-based Josephson junctions. *Phys. Rev. Lett.*, **104** (13), 137002-1–137002-4.
31. Ustinov, A.V. and Kaplunenko, V.K. (2003) Rapid single-flux quantum logic using π- shifters. *J. Appl. Phys.*, **94** (8), 5405–5407.
32. Sickinger, H., Lipman, A., Weides, M., Mints, R.G., Kohlstedt, H., Koelle, D., Kleiner, R., and Goldobin, E. (2012) Experimental evidence of a φ Josephson junction. *Phys. Rev. Lett.*, **109** (10), 107002-1–107002-5.
33. Larkin, T.I., Bol'ginov, V.V., Stolyarov, V.S., Ryazanov, V.V., Vernik, I.V., Tolpygo, S.K., and Mukhanov, O.A. (2012) Ferromagnetic Josephson switching device with high characteristic voltage. *Appl. Phys. Lett.*, **100** (22), 222601-1–222601-4.
34. Nevirkovets, I.P. and Belogolovskii, M.A. (2011) Hybrid superconductor–ferromagnet transistor-like device. *Supercond. Sci. Technol.*, **24** (2), 024009-1–024009-8.
35. Beenakker, C.W.J. (2013) Search for Majorana fermions in superconductors. *Annu. Rev. Condens. Matter Phys.*, **4**, 113–136.
36. Mourik, V., Zuo, K., Frolov, S.M., Plissard, S.R., Bakkers, E.P.A.M., and Kouwenhoven, L.P. (2012) Signatures of Majorana fermions in hybrid superconductor–semiconductor nanowire devices. *Science*, **336** (6084), 1003–1007.
37. Kauppila, V.J., Nguyen, H.Q., and Heikkilä, T.T. (2013) Nonequilibrium and proximity effects in superconductor–normal metal junctions. *Phys. Rev. B*, **88** (7), 075428-1–075428-7.
38. Noh, T., Davis, S., and Chandrasekhar, V. (2013) Nonlocal correlations in a proximity-coupled normal metal. *Phys. Rev. B*, **88** (2), 024502-1–024502-7.
39. Deon, F., Sopic, S., and Morpurgo, A.F. (2014) Tuning the influence of microscopic decoherence on the superconducting proximity effect in a graphene Andreev interferometer. *Phys. Rev. Lett.*, **112** (12), 126803-1–126803-5.
40. Katzir, E., Yochelis, S., Zeides, F., Katz, N., Kalcheim, Y., Millo, O., Leitus, G., Myasodeyov, Y., Shapiro, B.Y., Naaman, R., and Paltiel, Y. (2012) Increased superconducting transition temperature of a niobium thin film proximity coupled to gold nanoparticles using linking organic molecules. *Phys. Rev. Lett.*, **108** (10), 107004-1–107004-5.
41. Krivoruchko, V.N. and Koshina, E.A. (2002) Inhomogeneous magnetism induced in a superconductor at a superconductor–ferromagnet interface. *Phys. Rev. B*, **66** (1), 014521-1–014521-6.
42. Xia, J., Shelukhin, V., Karpovski, M., Kapitulnik, A., and Palevski, A.

(2009) Inverse proximity effect in superconductor–ferromagnet bilayer structures. *Phys. Rev. Lett.*, **102** (8), 087004-1–087004-4.

43. Zhu, L.Y., Liu, Y., Bergeret, F.S., Pearson, J.E., te Velthuis, S.G.E., Bader, S.D., and Jiang, J.S. (2013) Unanticipated proximity behavior in ferromagnet-superconductor heterostructures with controlled magnetic noncollinearity. *Phys. Rev. Lett.*, **110** (17), 177001-1–177001-5.

44. Krivoruchko, V.N. and Belogolovskii, M.A. (1995) Proximity-induced superconductivity in an antiferromagnetic exchange field. *Phys. Rev. B*, **52** (13), 9709–9713.

45. Nagai, Y., Nakamura, H., and Machida, M. (2013) Quasiclassical treatment and odd-parity/triplet correspondence in topological superconductors. *J. Phys. Soc. Jpn.*, **83** (5), 053705-1–053705-4.

1.2.2
Tunneling and Superconductivity

Steven T. Ruggiero

1.2.2.1 Introduction

Superconducting and related tunneling is a rich and broad field involving investigations of the basic nature of superconductivity and other phenomena, and includes a wide variety of unique electronic devices. Tunneling is a powerful tool for uncovering the properties of the metals and insulators comprising tunnel systems, and can reveal the excitation spectra of various types of species incorporated into tunnel systems. Since electrons are injected from one conductor into another through an insulating barrier, we are performing spectroscopy [1]. And because of the physical nature of tunnel junctions, it is straightforward to accurately measure the energy of the tunneling electrons.

1.2.2.2 Normal/Insulator/Normal Tunnel Junctions

Consider a simple tunnel junction comprising two normal metals separated by a thin (~1–2 nm) insulating barrier, a normal-metal/insulator/normal-metal (NIN) junction as shown in Figure 1.2.2.1. Depicted is the semiconductor model (Figure 1.2.2.1a) for a system comprising two normal metals with states filled to the Fermi level, offset by the energy eV, where V is the externally applied bias voltage. In this case, the current–voltage ($I-V$) characteristic, Figure 1.2.2.1b, is linear for small voltages compared to the tunnel barrier height of the insulator, $\phi \sim 1$ eV, and rises rapidly as the applied voltage approaches ϕ as:

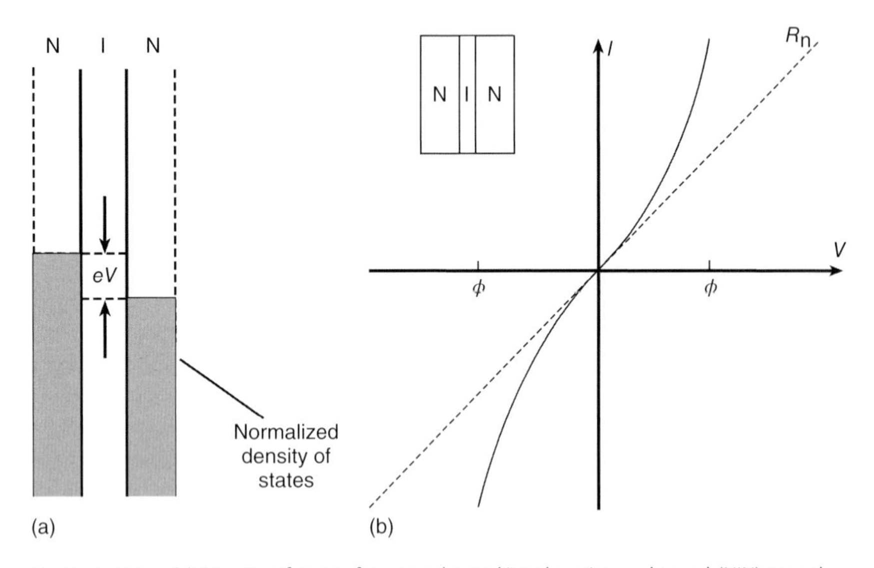

Figure 1.2.2.1 (a) Density of states for normal-metal/insulator/normal-metal (NIN) tunnel junction. (b) Current–voltage characteristics for a NIN tunnel junction.

$$I = \left(\frac{1}{R_n}\right)(V + aV^3) \quad (1.2.2.1)$$

where a is a barrier-dependent scaling parameter. Otherwise, the tunneling characteristics are featureless.

1.2.2.3 Normal/Insulator/Superconducting Tunnel Junctions

If we make one of the materials in a tunnel system superconducting, we have a normal-metal/insulator/superconductor (NIS) junction, with $I-V$ characteristics markedly changed at low voltage compared to the featureless NIN systems. At applied bias voltages typically on the order of $\sim 1-10$ mV, the energy gap in the superconducting density of states will be strongly manifest in the tunnel characteristics. We can see this effect by again viewing tunneling in the so-called semiconductor model (Figure 1.2.2.2a). Here, the tunneling process is depicted as the convolution of the tunneling densities of state a superconductor, with an energy gap Δ, and a normal metal, where the zero-temperature superconducting energy gap is given by

$$\begin{aligned}\Delta(0) &= \pi e^{-\gamma} k_B T_c \\ &= 1.764\ k_B T_c\end{aligned} \quad (1.2.2.2)$$

and $\gamma = 0.5772\ldots$ is the Euler–Mascheroni constant.

In this case, the tunnel current is given by

$$I_{NS} = \frac{1}{eR_n}\int_{-\infty}^{\infty} \frac{|E|}{[E^2 - \Delta^2]^{\frac{1}{2}}}[f(E) - f(E + eV)]dE \quad (1.2.2.3)$$

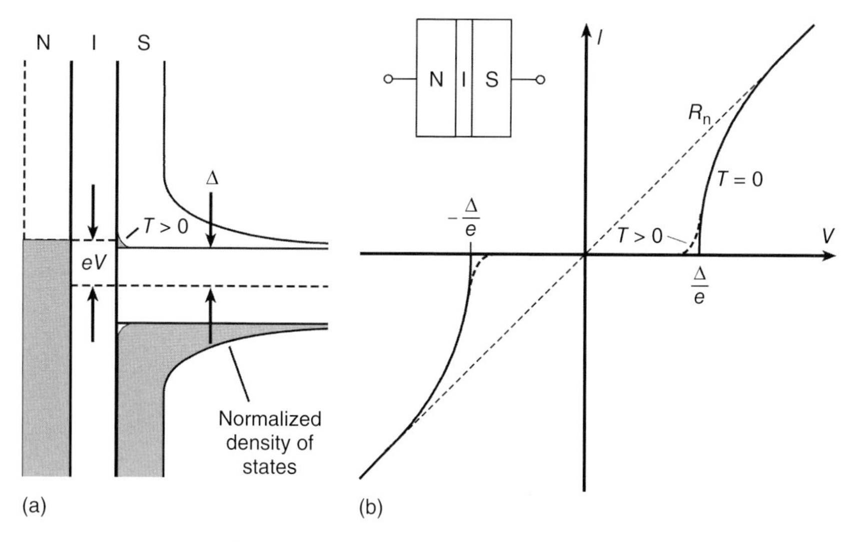

Figure 1.2.2.2 (a) Density of states for a normal-metal/insulator/superconductor (NIS) tunnel junction. For $T > 0$, some electrons are thermally excited across (twice) the superconducting energy gap Δ. (b) Current–voltage characteristics for NIS tunnel junction.

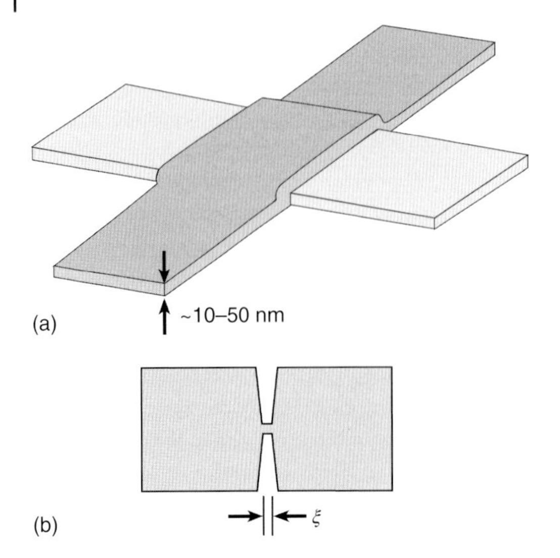

Figure 1.2.2.3 Shown on (a) is a tunnel junction, in its simplest form comprising a "sandwich" of a thin-film base electrode, a thin insulating barrier, and a top electrode and (b) is a weak-link Josephson system. Here, macroscopic superconducting elements are coupled by a small bridge or point contact, on the order of or smaller than the superconducting coherence length ξ.

where $f(E)$ is the Fermi function. The resultant $I-V$ characteristics are shown in Figure 1.2.2.2b. Giaever [2] observed the superconducting energy gap by creating NIS junctions of the form $Al/Al_2O_3/Pb$. This was done by vapor depositing a thin film of Al, allowing it to oxidize to create a $\sim 1-2$ nm thick insulating tunnel barrier, and finally depositing a cross stripe of Pb (and in a separate experiment In). These so-called "sandwich" junctions (see Figure 1.2.2.3) were measured at temperatures above the critical temperature of the Al stripe. Countless experiments on a variety of superconducting materials followed from this seminal work. Barriers can also be formed "artificially" by the direct deposition of insulating or semiconducting materials [3].

1.2.2.4 Superconductor/Insulator/Superconducting Tunnel Junctions

In the case where both materials are superconducting, the tunneling behavior becomes yet more interesting (Figure 1.2.2.4). For superconductor/insulator/superconductor (SIS) systems, the tunnel current is given by the expression:

$$I_{SS} = \frac{1}{eR_n} \int_{-\infty}^{\infty} \frac{|E|}{[E^2 - \Delta_1^2]^{\frac{1}{2}}} \frac{|E + eV|}{[(E + eV)^2 - \Delta_2^2]^{\frac{1}{2}}} [f(E) - f(E + eV)] dE \quad (1.2.2.4)$$

where the integral is assumed to exclude values of E, where $|E| < \Delta_1$ and $|E + eV| < \Delta_2$. If we follow the characteristics along the voltage axis, for quasiparticle tunneling, we see a current rise to V/R_n when we reach a bias potential $V = (\Delta_1 + \Delta_2)/e$, where R_n is the normal tunnel resistance in the absence of

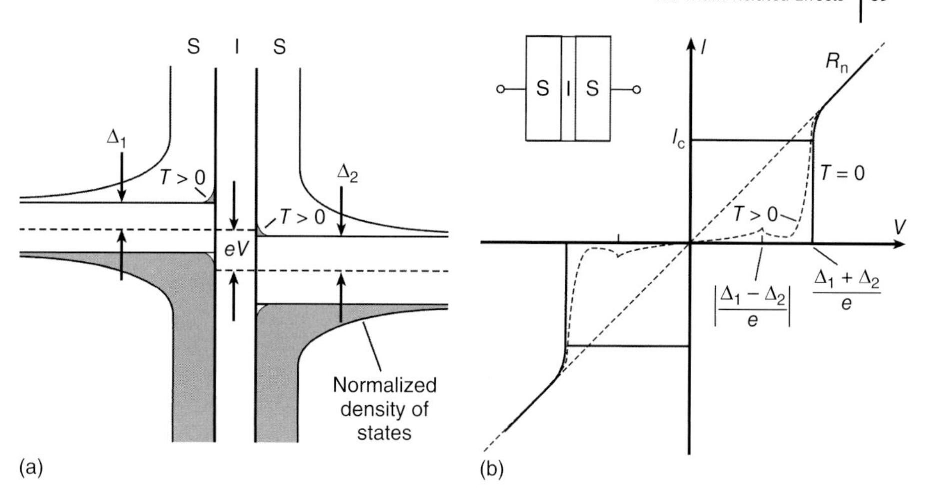

Figure 1.2.2.4 (a) Density of states for a superconductor/insulator/superconductor (SIS) tunnel junction. (b) Current–voltage characteristics for an SIS tunnel junction.

superconductivity. At finite temperature, an increase in current will also be observed at a bias voltage of $V = |\Delta_1 - \Delta_2|/e$.

There is also a rich set of phenomenology associated with the second, zero-voltage Josephson branch of the I–V characteristics. Here, the current rises – at zero voltage identically – until a maximum sustainable or critical current, I_c, is reached. This is a manifestation of the DC Josephson effect [4], where Cooper pairs tunnel through the barrier. This phenomenon was observed by Anderson and Rowell [5] using Sb/Sb-oxide/Pb tunnel junctions prepared using the technique of Giaever as noted earlier.

Josephson effects can also be observed in so-called weak-link systems (Figure 1.2.2.3), where a nanoscale-size superconducting element connects macroscopic-scale superconductors, as with a so-called point contact. The critical current is given in general by Ambegaokar and Baratoff [6]

$$I_c R_n = \frac{\pi[\Delta_1(T) + \Delta_2(T)]}{4e} \tanh \frac{\Delta_1(T) + \Delta_2(T)}{4k_B T} \tag{1.2.2.5}$$

At $T = 0$, then, both tunnel junctions and – in the dirty limit [7] – weak links will have a critical value given by:

$$I_c R_n = \frac{\pi[\Delta_1(0) + \Delta_2(0)]}{4e} \tag{1.2.2.6}$$

For temperatures in the vicinity of T_c, we have:

$$I_c R_n = \left(\frac{2.34\pi k_B}{e}\right)(T_c - T)$$
$$\approx (T_c - T)\ 635\,\mu\text{V}\,\text{K}^{-1} \tag{1.2.2.7}$$

We note that this temperature dependence is strictly applicable in the case of tunnel junctions (the topic of this chapter). Other Josephson devices – such

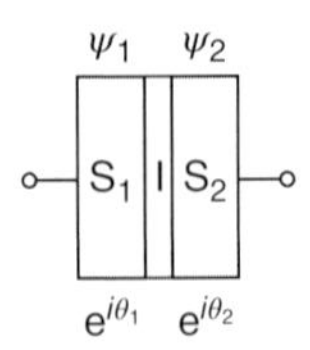

Figure 1.2.2.5 Two superconductors separated by an insulating tunnel barrier.

as weak links and superconductor/constriction/superconductor (ScS), superconductor/normal-metal/superconductor (SNS), superconductor/insulator/normal-metal/superconductor (SINS), superconductor/insulator/normal-metal/insulator/superconductor (SINIS), superconductor/ferromagnet/superconductor (SFS), and so on systems – will have a temperature dependence of $I_c R_n$ that can differ considerably from this classic Ambegaokar–Baratoff result.

To better understand the origin of Josephson behavior in tunnel junctions, let us consider two superconductors separated by an insulating barrier, as sketched in Figure 1.2.2.5. Each superconducting condensate can be described by a single wavefunction as $\psi_i = \sqrt{n_i} e^{i\theta_i}$, where n is the density of superconducting electrons and θ is a phase factor. If a potential energy $U_2 - U_1 = 2eV$ exists between the superconductors, then the applicable coupled Schroedinger equations can be written as:

$$
\begin{aligned}
i\hbar\frac{\partial\psi_1}{dt} &= U_1\psi_1 + K\psi_2 \\
i\hbar\frac{\partial\psi_2}{dt} &= U_2\psi_2 + K\psi_1
\end{aligned} \tag{1.2.2.8}
$$

where K is a coupling constant. Defining $\Delta\theta = \theta_1 - \theta_2$, this leads to a current flow of:

$$I = I_c \sin \Delta\theta \tag{1.2.2.9}$$

where $I_c = 2K(n_1 n_2)^{1/2}/\hbar$. We also have

$$\frac{\partial(\Delta\theta)}{\partial t} = \frac{2eV}{\hbar} \tag{1.2.2.10}$$

Equations (1.2.2.9) and (1.2.2.10) represent the DC and AC Josephson effects. Thus, for a fixed applied voltage difference, V, there is a steadily increasing phase difference

$$\Delta\theta(t) = \Delta\theta(0) + \frac{2eV}{\hbar}t \tag{1.2.2.11}$$

Therefore, the Josephson current will oscillate at a frequency

$$f = \frac{2eV}{h} \tag{1.2.2.12}$$

This result is remarkable in that it is true exactly, and that the appropriate charge is indeed $2e$, the charge of a Cooper pair. The Josephson frequency $2e/h = 483.6$ GHz mV^{-1}. When a Josephson junction is irradiated with radio-frequency waves of frequency f, a series of steps will appear in the DC I–V

characteristics at voltages

$$V_n = \frac{nhf}{2e} \tag{1.2.2.13}$$

These Shapiro steps [8] are the basis of Josephson-junction-based voltage standards because the frequency of microwaves can be very accurately determined, see Chapter 7. $N \times M$ arrays of microwave-irradiated Josephson junctions can produce so-called "giant" Shapiro steps of voltage

$$V_n = N\frac{nhf}{2e} \tag{1.2.2.14}$$

for arrays N junctions long in the direction of the current flow. This situation can also be reversed to produce voltage-tunable microwave radiation sources [9]. While typically producing very low rf power, such arrays are suitable for applications such as local oscillators for low-noise superconductor-based mixers.

We note finally that for the high-temperature superconductors, a separate class of internal, so-called intrinsic Josephson effects has been discussed in the literature, related to interlayer coupling [10].

1.2.2.5 Superconducting Quantum Interference Devices (SQUIDs)

We can also create a very interesting device by placing two Josephson junctions in a ring as shown in Figure 1.2.2.6a. Here, we have two critical currents of I_{c1} and I_{c2}, so the total current is

$$I = I_{c1}\sin(\Delta\theta_1) + I_{c2}\sin(\Delta\theta_2) \tag{1.2.2.15}$$

If we apply a magnetic field, then the flux in the ring will then be given as

$$\begin{aligned}\Phi &= \int \boldsymbol{B}\cdot d\boldsymbol{S}\\ &= \oint \boldsymbol{A}\cdot d\ell\\ &= \frac{\Phi_0}{2\pi}\oint (\boldsymbol{\nabla}\theta)\cdot d\ell\end{aligned} \tag{1.2.2.16}$$

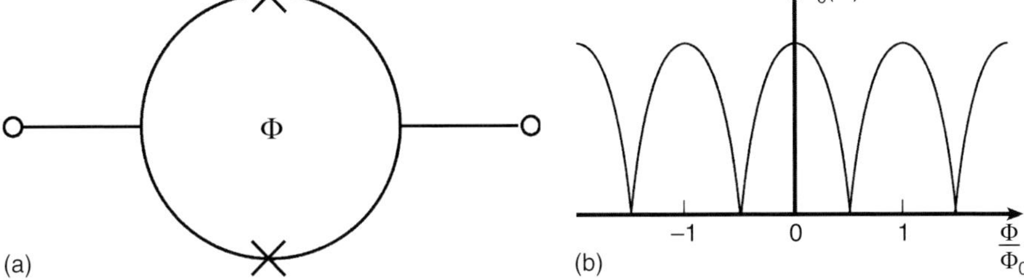

Figure 1.2.2.6 (a) A superconducting quantum interference device (SQUID) comprising two Josephson junctions in a loop. (b) Critical current as a function of applied flux for a SQUID.

where $\Phi_0 = h/2e = 2.068 \times 10^{-15}$ T m^2 is the magnetic flux quantum. Assuming we have chosen a specific loop contour, we can then say

$$\Delta\theta = 2\pi\frac{\Phi}{\Phi_0}(\text{mod } 2\pi) \tag{1.2.2.17}$$

For a system with identical junctions, we can write

$$\begin{aligned} I &= I_c \sin(\Delta\theta_1) + I_c \sin(\Delta\theta_2) \\ &= I_c \sin\left[\Delta\theta + \frac{\pi\Phi}{\Phi_0}\right] + I_c \sin\left[\Delta\theta - \frac{\pi\Phi}{\Phi_0}\right] \\ &= 2I_c \sin(\Delta\theta) \cos\frac{\pi\Phi}{\Phi_0} \end{aligned} \tag{1.2.2.18}$$

So the critical current will be modulated by the applied flux as

$$I_c(\Phi) = I_0 \left|\cos\frac{\pi\Phi}{\Phi_0}\right| \tag{1.2.2.19}$$

as depicted in Figure 1.2.2.6b. Because the flux quantum is so small, this means that superconducting quantum interference devices (SQUIDs) represent the most sensitive systems for the detection of magnetic flux, permitting measurements of absolute magnetic fields as small as $\sim 10^{-15}$ T.

Note that for a single junction, the application of a magnetic field leads to a Fraunhofer dependence of the critical current for the case of homogeneous current density. This behavior is analogous to single-slit optical diffraction. For further discussion and examples, see Chapter 9. Other references to Josephson effects can be found in texts on the subject [11].

1.2.2.6 Phonon Structure

Basic information can be obtained about the superconducting electrodes of tunnel junctions by studying the details of their $I-V$ characteristics. Of course the tunneling characteristics immediately supply the superconducting energy gap (or gaps) of the materials involved. In addition, we can learn about the phonon structure of phonon-mediated superconductors. The superconducting density of states, embodied earlier in our tunneling equations, can be expressed as:

$$N_S(E) = N(0)\text{Re}\left[\frac{E^2}{E^2 - \Delta^2(E)}\right]^{\frac{1}{2}} \tag{1.2.2.20}$$

Here, we allow for the fact that the energy gap is not a structureless constant but has both real and imaginary parts, the latter of which correspond to damping by the creation of phonons, especially in the vicinity of energies $E = h\nu_{\text{phonon}}$ [12]. This effect can be readily observed as small deviations in the tunnel conductance as noted by Giaever *et al.* [13]. More detailed analysis can provide $\alpha^2 F(\omega)$, the electron–phonon coupling strength times the phonon density of states [14].

Obtaining the phonon density of states from tunneling falls into a larger category of inelastic electron tunneling spectroscopies (IETS), another important example of which is the examination of the vibrational spectra of molecular absorbates. This is done by introducing molecular species into tunnel structures to create metal/insulator/absorbate/metal (MIAM) systems. This process can be as simple as first depositing an aluminum film, allowing it to oxidize to form a tunnel barrier, exposing the barrier to molecules in vapor form, and completing the junction with a compatible counter-electrode metal. As discussed by Hipps and Mazur [15], MIAMs can be employed to explore the vibrational and electron spectroscopic information of the metal electrodes (magnons and phonons, the latter as noted above), insulator, and absorbate. MIAMs have been applied to the study of surface chemistry and catalysis, adhesion and corrosion, molecular vibrational spectroscopy, and orbital-mediated tunneling.

1.2.2.7 **Geometrical Resonances**

Phenomena related to electron interference effects can also be explored. These effects have been observed as oscillations in the tunnel characteristics in systems of the type SIN′S and/or SIS′, where N′ and S′ vary in thickness. The observation of these effects may require especially clean films. For sub-gap energies, multiple Andreev reflections [16] and de Gennes–Saint-James bound states [17] can give rise to oscillations in the tunnel conductance which are nonperiodic in energy. For energies above the superconducting gap, geometric resonances can involve Tomasch oscillations [18] in superconducting electrodes and McMillan–Rowell oscillations [19] in normal electrodes. The bias-voltage spacings for Tomasch and McMillan–Rowell oscillations generally scale as $\Delta V = hv_{FS}/2ed_S$ ($\Delta_S \ll e\Delta V$) and $\Delta V = hv_{FN}/4ed_N$. Thus, the Fermi velocity and thicknesses of the superconducting and normal layers govern the period of the conductance oscillations.

1.2.2.8 **Scanning Tunneling Microscopy**

Scanning tunneling microscopy (STM) is a technique of vast importance and applicability. Both superconducting and normal-metal tips with atomic-level sharpness can be positioned within tunneling distance above surfaces. As the tip is rastered over a surface, a two-dimensional picture of its tunneling density of states can be created. As noted by De Lozanne [20], this information can be coupled with other powerful surface-scanning techniques such as atomic force microscopy (AFM), magnetic force microscopy (MFM), Hall effect (SHPM, scanning Hall probe microscopy), SQUID, microwave, near-field optical, or magneto-optic microscopies. These techniques have been notably useful in exploring the properties of the high-temperature superconductors.

1.2.2.9 **Charging Effects**

Charging effects can be observed in tunneling if either one or more of the tunnel electrodes has an ultra-small capacitance [21, 22] or else nano-size elements [23] are otherwise incorporated into the tunnel structure. To observe charging effects, it must also be generally true that $E_c = e^2/2C > k_B T$ and that the charging energy exceed the thermal energy, where C is the capacitance of the nano-element(s). It

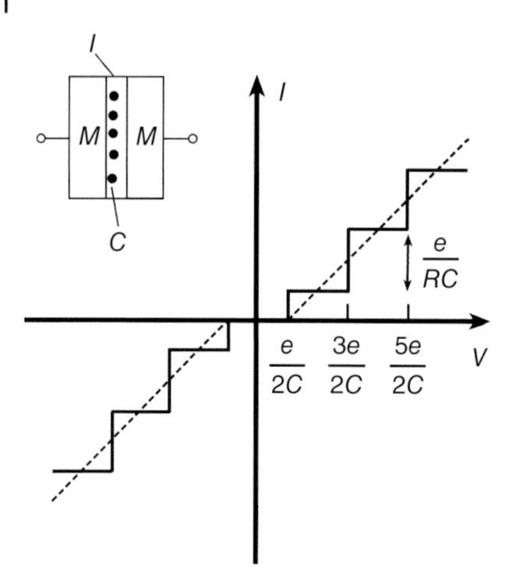

Figure 1.2.2.7 Shown is the coulomb staircase for tunneling into nano-droplets of metal incorporated into the insulating barrier structure of a metal/insulator/metal (M/I/M) tunnel system.

must also be true that $R > h/4e^2$ and that the effective resistance of the measured system exceed the quantum resistance.

In the case of nano-sized elements, tunnel junctions can be formed which incorporate metal particles $\sim 1-10$ nm in size into the tunnel barrier, which have charging energies in the vicinity of $\sim$10 meV, as shown in Figure 1.2.2.7 [23]. Such junctions can exhibit both the "Coulomb blockade," a gap-like overall offset in the $I-V$ characteristics of voltage $e/2C$, and a series of steps of width e/C and height e/RC, which is the "Coulomb staircase."

References

1. Wolf, E.L. (1985) *Principles of Electron Tunneling Spectroscopy*, Oxford University Press, Inc., New York.
2. Giaever, I. (1960) Energy gap in superconductors measured by electron tunneling. *Phys. Rev. Lett.*, **5**, 147–148.
3. Ruggiero, S.T. (1988) Artificial tunnel barriers, in *Superconducting Devices* (eds S.T. Ruggiero and D.A. Rudman), Academic Press, New York.
4. Josephson, B.D. (1962) Possible new effects in superconductive tunneling. *Phys. Lett.*, **1**, 251–253; *Adv. Phys.*, (1965), **14**, 4191.
5. Anderson, P.W. and Rowell, J.M. (1963) Probable observation of the Josephson superconducting tunneling effect. *Phys. Rev. Lett.*, **10**, 230–232.
6. Ambegaokar, V. and Baratoff, A. (1963) Tunneling between superconductors. *Phys. Rev. Lett.*, **10**, 486–489; erratum (1963) **11**, 104.
7. Tinkham, M. (1996) *Introduction to Superconductivity*, McGraw-Hill, New York.
8. Shapiro, S. (1963) Josephson currents in superconducting tunneling: the effect of microwaves and other observations. *Phys. Rev. Lett.*, **11**, 80–82.
9. Benz, S.P. and Burroughs, C.J. (1991) Coherent emission from two-dimensional Josephson junction arrays. *Appl. Phys. Lett.*, **58**, 2162–2164.

10. Kleiner, R., Steinmeyer, F., Kunkel, G., and Müller, P. (1992) Intrinsic Josephson effects in $Bi_2Sr_2CaCuO_8$ single crystals. *Phys. Rev. Lett.*, **68**, 2394–2397.
11. (a) Barone, A. and Paterno, G. (1982) *Physics and Applications of the Josephson Effect*, John Wiley & Sons, Inc., New York; (b) Likharev, K.K. (1986) *Dynamics of Josephson Junctions and Circuits*, Gordon and Breach, Philadelphia, PA; (c) van Duzer, T. and Turner, C.W. (1999) *Principles of Superconductive Devices and Circuits*, 2nd edn, Prentice Hall, Upper Saddle River, NJ; (d) Orlando, T.P. and Delin, K.A. (1991) *Foundations of Applied Superconductivity*, Addison-Wesley, Reading, MA.
12. (a) Schrieffer, J.R., Scalapino, D.J., and Wilkins, J.W. (1963) Effective tunneling density of states in superconductors. *Phys. Rev. Lett.*, **10**, 336–339; (b) Scalapino, D.J., Schrieffer, J.R., and Wilkins, J.W. (1966) Strong-coupling superconductivity. I. *Phys. Rev.*, **148**, 263–279.
13. Giaever, I., Hart, H.R. Jr., and Megerle, K. (1962) Tunneling into superconductors at temperatures below 1°K. *Phys. Rev.*, **126**, 941–948.
14. McMillan, W.L. and Rowell, J.M. (1969) Tunneling and strong-coupling superconductivity, in *Superconductivity* (ed. R.D. Parks), Marcel Dekker, New York.
15. Hipps, K.W. and Mazur, U. (2002) Inelastic electron tunneling spectroscopy, in *Handbook of Vibrational Spectroscopy* (eds J.M. Chalmers and P.R. Griffiths), John Wiley & Sons, Ltd, Chichester.
16. Andreev, A.F. (1964) The thermal conductivity of the intermediate state in superconductors. *Zh. Eksp. Theor. Fiz.*, **46**, 1823–1836; *Sov. JETP*, (1964), **19**, 1228-1231.
17. de Gennes, P.D. and Saint-James, D. (1963) Elementary excitations in the vicinity of a normal metal-superconducting metal contact. *Phys. Lett.*, **4**, 151–152.
18. Tomasch, W.J. (1965) Geometrical resonance in the tunneling characteristics of superconducting Pb. *Phys. Rev. Lett.*, **15**, 672–675.
19. Rowell, J.M. and McMillan, W.L. (1966) Electron interference in a normal metal induced by superconducting contacts. *Phys. Rev. Lett.*, **16**, 453–456.
20. De Lozanne, A. (1999) Scanning probe microscopy of high-temperature superconductors. *Supercond. Sci. Technol.*, **12**, R43–R56.
21. Fulton, T.A. and Dolan, G.J. (1987) Observation of single-electron charging effects in small tunnel junctions. *Phys. Rev. Lett.*, **59**, 109–112.
22. Averin, D.V. and Likharev, K.K. (1991) Single electronics: correlated transfer of single electrons and cooper pairs in systems of small tunnel junctions, in *Mesoscopic Phenomena in Solids* (eds B.L. Altshulter, P.A. Lee, and R.A. Web), Elsevier, Amsterdam.
23. Barner, J.B. and Ruggiero, S.T. (1987) Observation of the incremental charging of Ag particles by single electrons. *Phys. Rev. Lett.*, **59**, 807–810.

1.2.3
Flux Pinning

Stuart C. Wimbush

1.2.3.1 Introduction

The amount of DC electrical current able to be transported without loss (or, in practice, below some small but measurable voltage drop, typically quoted as 1 V over 10 km) by a superconducting wire under a given set of operating conditions (temperature, magnetic field) is the parameter that ultimately determines its technological applicability, and this critical current in turn is determined by the immobilization, or pinning, of magnetic flux lines within the superconductor. The origin of the magnetic flux may be either an applied magnetic field, as exists in motors or generators, or the field generated by the transport current itself (termed the *self-field*) as is the case, for example, in power transmission cables or transformers. Consequently, the existence of a pinning-limited critical current is unavoidable. Likewise, in superconducting electronic devices and in bulk superconductors used as permanent magnets, the prevention of flux motion through pinning is critical in order to reduce noise or to effectively trap an applied magnetic field. This section examines the mechanisms by which magnetic flux lines can be pinned, the different types of pinning centers that have been employed to engineer high critical current superconductors, the measurements able to provide experimental information on flux pinning, and the state-of-the-art flux pinning presently achievable in second generation coated conductors.

1.2.3.2 Flux Lines, Flux Motion, and Dissipation

Flux lines form within a type II superconductor because it is energetically favorable for the material to allow a magnetic field to locally penetrate the bulk to form a mixed state comprising quantized threads of magnetic flux encircled by regions of superconducting material through which spontaneous supercurrents flow so as to screen the field from the bulk of the material, thereby forming a vortex (Figure 1.2.3.1). This enables the superconductivity as a whole to persist to a much higher field than would be possible if the Meissner state of complete flux expulsion were to be maintained (cf. Section 1.1.1), and is the case precisely because the spatial extent of the magnetic penetration in the type II materials exceeds the extent of the disruption to the superconducting state $\kappa = \lambda/\xi > 1/\sqrt{2}$. Thus, by allowing a small volume of material to revert to the normal state, the resulting penetrating magnetic field can be screened over a large volume of superconducting material. The upper critical field $B_{c2} = \Phi_0/2\pi\xi^2$ at which superconductivity ceases can be expressed intuitively as the field at which the non-superconducting regions overlap, resulting in an entirely normal state material. However, from a technological point of view, both this limiting field and the depairing current that provides an absolute upper limit to the current-carrying capacity of the material are supplanted by stricter limits governed not by the thermodynamic phase transition

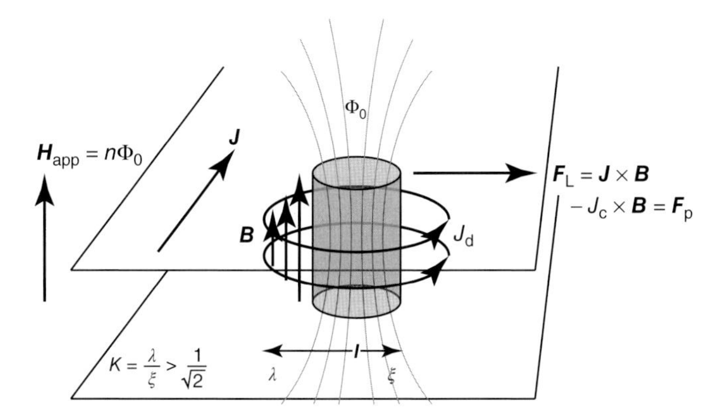

Figure 1.2.3.1 A quantized line of magnetic flux penetrates a type II superconducting slab subjected to an applied magnetic field exceeding its lower critical field. Spontaneous circulating supercurrents form a vortex that screens the field from the bulk of the superconductor. Under the influence of an imposed transport current, the unpinned flux line experiences a force analogous to the Lorentz force that causes it to move, dissipating energy. For this to be prevented, the flux line must be pinned by a countering force that inhibits dissipative motion up to some critical value of the transport current.

from the superconducting to the normal state but rather by energy dissipation within the superconductor caused by motion of the flux lines.

Flux motion occurs as a result of a force analogous to the Lorentz force $\boldsymbol{F}_L = \boldsymbol{J} \times \boldsymbol{B}$ being exerted on the flux lines under the influence of an imposed transport current $\boldsymbol{J}$. The moving flux lines are accelerated until the retarding force due to viscous flow, proportionally opposed to their velocity $\boldsymbol{v}$, matches the Lorentz force. The spatially varying magnetic field of the moving flux lines induces an electric field $\boldsymbol{E} = -\boldsymbol{v} \times \boldsymbol{B}$ in the direction of the transport current. This electric field acts on the normal electrons within the superconductor, dissipating energy through ohmic losses. Since the magnitude of the electric field is likewise proportional to the velocity of the moving flux lines, it will also be proportional to the transport current ($\boldsymbol{E} = \rho_f \boldsymbol{J}$), and the response is therefore indistinguishable from an ohmic resistance. Macroscopically, while remaining in the superconducting state and conducting current through the transport of Cooper pairs, the superconductor will nonetheless exhibit a resistance, termed the *flux flow resistance*. It is therefore the DC supercurrent that can be maintained in the absence of this resistance that dictates the range of practical operation of the superconductor, and this is termed the *critical current*, J_c.

Such a zero-resistance DC transport current can only be established if some means of preventing flux motion is available. Otherwise, the smallest transport current will generate a self-field of some degree, which will penetrate the superconductor in the form of flux lines as soon as it exceeds the lower critical field. Those flux lines will flow unimpeded through the superconductor under the influence of the Lorentz force, generating a flux flow resistance and dissipating energy. Such a situation arises in high-quality single crystals, which have extremely low

J_c values. To overcome this, a force termed the *pinning force* must counter the Lorentz force that is acting to excite the flux lines into motion. The opposition to vortex motion from pinning forces provides a mechanism for the enhancement of J_c, and also introduces our second limiting parameter in the form of the irreversibility field B_{irr} beyond which the number of flux lines present within the superconductor exceeds its capability to effectively pin them all, with the result that the magnetization becomes reversible (flux flowing freely into and out of the superconductor) and J_c drops to zero. Increasing flux pinning in technological materials is an essential endeavor in order to increase the achievable current and field ranges and thus the performance of superconducting machines and devices. The amount of pinning can be increased either by promoting naturally occurring sources of pinning or by introducing new sources, so-called artificial pinning centers.

1.2.3.3 Sources of Flux Pinning

Flux pinning may best be described as the effect of spatial inhomogeneities within the superconductor upon the flux line lattice that forms when mutually repulsive vortex–vortex interactions (resulting from the Lorentz force between the circulating supercurrent J_d of one vortex and the magnetic flux Φ_0 of another) are taken into account. Such inhomogeneities may arise naturally, for example, through local variations in material density, elasticity or electron–phonon coupling strength and the existence of crystalline defects, or they may be introduced artificially through doping, microstructural modification, or the incorporation of foreign bodies. A change in the material density is equivalent to a change in the (chemical) pressure, and will give rise to a local change in the superconducting transition temperature, T_c, as may strain fields and variations in the electron–phonon coupling. Genuine material doping may act to combine several of these effects. The pinning that results is termed *δT_c pinning* [1]. A similar effect occurs upon the inclusion of non-superconducting material, which provides a region of suppressed T_c that may be highly localized or rather extended, depending on the nature of the inclusion; however, such an inclusion also constitutes a material defect that will have a scattering effect on the charge carriers, acting to vary the electronic mean free path, l (δl pinning, formerly termed *$\delta\kappa$ pinning* [2]). In all cases, the result is the same: the creation of a lower energy (preferred) site for vortex occupation.

For maximum effect, the spatial variations must occur on the length scale of either ξ or λ, depending on the nature of the interaction; any larger and the inhomogeneity will not be seen by the superconductor as an inhomogeneity but rather as a distinct phase, any smaller and the effectiveness will be reduced due to the proximity effect. That said, small-scale modifications such as atomic substitutions can create larger-scale inhomogeneities through alterations of the electronic structure or the creation of strain fields, and so on. It is the extent of the inhomogeneity, not the extent of the modification, that matters.

A useful distinction can be drawn between pinning forces operating over the length scale of ξ, termed (vortex) *core pinning*, and those operating over a length scale of λ, termed *magnetic pinning*. To the former category belong the majority of artificial pinning centers successfully employed to date as well as all common growth defects, while the latter category comprises such natural sources as extended surfaces and pores where the flux entirely enters or exits the superconductor as well as pinning due to vortex–vortex interactions. While vortex–vortex interactions do act to pin vortices, it is important to note that this can only occur if at least some of the vortices are pinned by another source; otherwise, the entire lattice will simply slip through the material. Where some vortices are strongly pinned and others are not, shearing as well as melting of the flux line lattice becomes an effective depinning mechanism. Inherently, magnetic artificial pinning centers will combine both of these types, exerting a core pinning influence through their localized suppression of superconductivity as well as a magnetic pinning effect resulting from their magnetization (see Section 1.2.3.8).

By definition, all of these spatial inhomogeneities constitute defects in the crystal lattice of the superconductor, and so flux pinning is inextricably linked to the defect structure (often termed the *defect landscape*) of the material. As a consequence, defect engineering of technologically relevant materials forms an extensive field of endeavor. It has become common to catalog the available types of pinning defects, which is nothing more than a list of possible crystal defects, and a useful classification is to enumerate them in terms of their dimensionality, as illustrated in Figure 1.2.3.2.

Zero-dimensional (point) defects include foreign (impurity) atoms, atomic substitutions, and vacancies (particularly when referred to the non-stoichiometric oxygen content of the cuprates). These are effective in high temperature superconductors (HTS) due to the short coherence length of the materials, meaning that a variation in stoichiometry over even a single atomic site can have a sufficiently wide sphere of influence to locally suppress the superconducting order parameter.

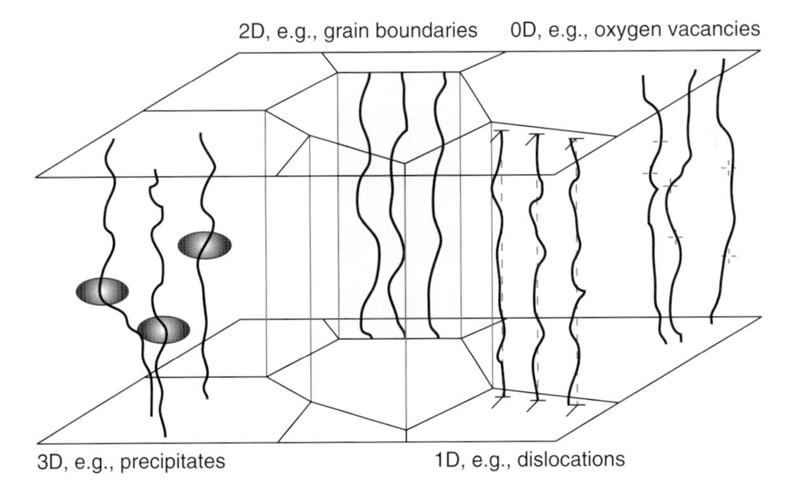

Figure 1.2.3.2 Schematic illustration of the pinning of flux lines by crystalline defects of varying dimensionality.

One-dimensional (line or columnar) defects include dislocations and artificial defects such as the damage tracks resulting from heavy ion, neutron, or proton irradiation. They are a particularly interesting class from the point of view of flux pinning due to their congruence with the form of the flux line itself. Correlation between the defects and the flux lines can lead to extremely strong pinning, with the flux line being pinned along its entire length. As an example, screw dislocations arise naturally during the growth of yttrium barium copper oxide (YBCO) thin films, forming part of the growth mode of the material and contributing strongly to its high J_c even before microstructural modification. A heavy focus is applied to engineering artificial pinning centers of this type in an attempt to smooth out the naturally occurring J_c anisotropy of the HTS materials. Since high-energy irradiation techniques are impractical for industrial production, the emphasis is on self-assembly processes of second-phase inclusions, although low-energy ion implantation methods have been employed with some success in introducing clusters of atoms to act as volume pinning centers [3].

Two-dimensional (planar) defects include grain boundaries, twin planes, stacking faults, and antiphase boundaries. Grain boundaries are important pinning centers in low temperature superconductors (LTS) and MgB_2, while twin planes and stacking faults are particularly prevalent in second generation coated conductors due to the orthorhombic structure of the YBCO crystal promoting twinning and the close proximity of the equilibrium phase formation conditions of Y123 and Y124 encouraging stacking fault formation to accommodate compositional variations. The low angle grain boundaries in HTS are better viewed as "dislocation fences" of one-dimensional defects than as continuous planar defects, due to the short coherence lengths. Antiphase boundaries tend to be of limited effectiveness due to their restricted spatial extent, tendency to heal out during growth by combination with stacking faults, and the fact that superconductivity is only somewhat suppressed by them. Planar defects can also provide correlated pinning, although with the risk of channeling of the flux lines along the planes [4].

Three-dimensional (volume) defects generally constitute secondary (impurity) phases, precipitates, or inclusions as well as voids (porosity). The similarity to point defects is apparent, although volume defects can also be effective in high coherence length materials such as LTS or at temperatures close to T_c. The majority of work on artificial pinning centers focuses on the intentional introduction of non-superconducting secondary phases intended to promote pinning. The requirements on these are that they should remain segregated from the superconducting matrix, and that they should be of a size appropriate to provide effective pinning while not consuming too great a volume fraction of the material (which then becomes unavailable for supercurrent transport). The most effective pinning arises from inclusions the size of ξ, and in HTS materials, this implies nano-engineering.

1.2.3.4 **Flux Pinning in Technological Superconductors**

Having enumerated the available sources of pinning in general terms, we now look, by way of example, at the specifics of flux pinning in the technological superconductors. Only a small subset of the many known superconductors has been developed for commercial application: Nb–Ti and Nb_3Sn of the LTS materials and Bi-2223 and YBCO of the HTS materials, with MgB_2 now also fighting to enter the fold. Each of these has decidedly different flux pinning characteristics.

The ductile alloy Nb–Ti is formed into a wire by repeated bundling, drawing, and annealing of rods of the material embedded within a stabilizing Cu matrix to form a filamentary conductor. In this material, pinning occurs when flux lines interact with dislocation tangles created during the drawing process. These regions of dense dislocations decrease the electronic mean free path, resulting in δl pinning. The microstructure of the filaments comprises elongated Nb–Ti grains with non-superconducting Ti precipitates lying along the filaments. The radial Lorentz force therefore drives flux lines across the grain boundaries and through the non-superconducting precipitates.

The A15-structure compound Nb_3Sn is initially prepared in a similar way, this time using rods of pure Nb placed within a bronze (Cu–Sn) matrix. This is drawn to again produce a filamentary conductor which is then heated to allow the Nb to post-react with Sn leached from the bronze. Production in this sequence is necessary due to the brittleness of the resulting intermetallic phase limiting subsequent processing. The post-reaction process results in columnar grains of superconductor oriented perpendicular to the filament. Consequently, the Lorentz force drives some of the flux lines along, rather than across, the grain boundaries, causing the flux line lattice to shear and thereby reducing the effectiveness of the pinning.

In the high temperature superconductors, the critical current is limited by two effects. The first, due to the small coherence length, results in any large angle grain boundary forming a weak link, severely limiting current flow [5]. However, careful materials processing has today virtually eliminated this type of grain boundary from technological materials, and since this effect is not pinning related, it shall not be considered further here. Once weak links have been eliminated, the critical current is again determined by flux pinning.

The "first generation" HTS material, Bi-2223, is fabricated by a powder-in-tube technique, whereby the precursor powder is packed into Ag tubes which are bundled and drawn before reaction to form the superconducting phase. A filamentary conductor again results, with a final rolling deformation into a tape serving to induce the texture required to eliminate weak links. In contrast, the "second generation" HTS material based on YBCO must be fabricated as a thin film "coated" conductor in order to achieve the required texturing, due to its less anisotropic crystal structure preventing mechanical texturing. In both of these layered, textured materials, a form of pinning termed *intrinsic* pinning arises due to the intrinsic inhomogeneity of the material in passing through the crystal planes commonly considered to be insulating that separate the superconducting CuO_2 planes. This form of pinning is only effective for flux lines (and therefore applied fields) lying in the plane of the material, leading to a sharp increase in the critical current for high fields (where only the most prevalent pinning centers are still effective) applied in-plane at low temperatures (where the coherence length is small, comparable to the planar spacing, and the flux lines are rigid).

In seeking to improve the performance of the second generation YBCO conductors, so-called "artificial" pinning centers have been introduced. In contrast to the microstructural modifications applied to the formation of the other classes of superconducting wire to maximize performance, these are non-superconducting secondary (impurity) phases intentionally introduced with the aim of providing an enhancement in J_c through pinning that outweighs their detriment through the reduction in the superconducting cross-section of the wire. The most successful, effective, and widely studied artificial pinning center for second generation coated conductors identified to date is the perovskite dielectric $BaZrO_3$. Its structural similarity to YBCO as well as its oxidic nature and the fact that it adds only a single, benign, element into the mix are all points in its favor. Zr was already known not to substitute into the YBCO lattice, while the material as a whole was well known to be compatible with YBCO since it was commonly used as a crucible for single crystal growth. It was also expected that its high melting point would correspond to slow growth kinetics, resulting in the desirably small size of inclusions critical for effective pinning in HTS. The initial report [6] of $BaZrO_3$ incorporation in an YBCO thin film in the form of nanoparticles of size ~10 nm provided an immediate J_c enhancement of up to a factor 5 in samples grown on both single crystal and technical substrates across the entire field range. Furthermore, in the case of *in situ* film growth techniques such as pulsed laser deposition and chemical vapor deposition, the strain created by the epitaxial

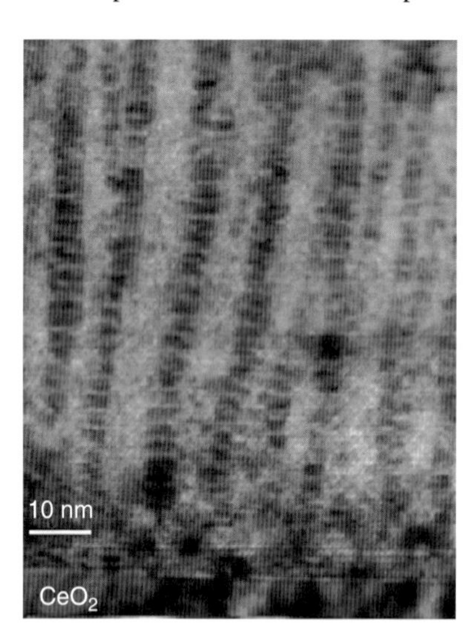

Figure 1.2.3.3 Epitaxial strain-induced self-assembly of individual $BaZrO_3$ nanoparticles in YBCO to form chains aligned with the film growth direction (the so-called "bamboo" microstructure), acting as one-dimensional correlated artificial pinning centers. (Adapted from Ref. [7]. Reproduced with permission of AIP Publishing LLC.)

growth of the lattice-mismatched nanoparticles within the YBCO matrix leads to a self-assembly of chains of nanoparticles into nanocolumns aligned with the growth direction (Figure 1.2.3.3), resulting in a much coveted enhancement in pinning correlated with that direction acting to balance out the anisotropic pinning due to the planar structure of the superconductor. In contrast, where *ex situ* film growth occurs, as is the case for chemical solution deposition, this epitaxial strain-induced self-assembly cannot occur, and instead a general reduction in J_c anisotropy associated with dispersed nanoparticle pinning is obtained. Thus, even for the same sample composition, ultimately, it is the microstructure of the particular sample that determines the pinning response.

Little further progress has been made in terms of the discovery of improved pinning species since this earliest attempt at artificial pinning center creation. However, an additional contender has emerged in the form of $BaHfO_3$ [8]. The need for further improvement arises in the move to thicker films, where it is observed that the nanocolumns of $BaZrO_3$ bend or "splay" as growth proceeds, reducing the effectiveness of their correlated pinning. $BaHfO_3$ nanocolumns, in contrast, maintain their orientation throughout the film thickness as well as being smaller in size due to a further increased melting point. Hf is similarly inert to Zr in YBCO. Some also suggest that the enhancements due to $BaHfO_3$ pinning are maintained to lower temperatures (see Section 1.2.3.7), although this is disputed with some questioning whether $BaHfO_3$ will prove truly superior to $BaZrO_3$, or whether they are in fact just very similar.

MgB_2 wires are also prepared by a powder-in-tube method with both pre-reaction (*ex situ*) and post-reaction (*in situ*) methods presently being employed. Multiband superconductivity in MgB_2 allows for a large enhancement in the upper critical field to be achieved through carbon doping in place of boron, commonly achieved through the addition of malic acid [9], enabling improved in-field performance through flux pinning. The primary pinning mechanism in MgB_2 is through the dense three-dimensional network of grain boundaries resulting from the solid state processing. The grain size, and thereby the effectiveness and density of this pinning, is tailored through modifications to the processing parameters. Pinning by nanoparticle artificial pinning centers is also commonly employed, with the most effective addition to date being SiC [10], which acts to combine the benefits of nanoparticle addition in the form of Mg_2Si with carbon doping during the reaction process.

1.2.3.5 Experimental Determination of Pinning Forces

It is a relatively straightforward matter to determine experimentally the magnitude of the pinning force $|\boldsymbol{F}_p(T, B)| = J_c(T, B)B$ in a given sample under a range of operating conditions. Fietz and Webb [11] were the first to show that the pinning force values so obtained scale rather simply with both temperature and field for a variety of Nb–Ti alloys of widely varying κ. Kramer [12] extended this analysis to other superconductors, most notably Nb_3Sn, and ultimately Dew-Hughes [13] generalized the function to the form commonly employed today for all superconductor materials:

$$F_p = J_c B \propto B_{irr}^{p+q} b^p (1-b)^q \quad \text{with } b = \frac{B}{B_{irr}} \tag{1.2.3.1}$$

where the temperature dependence is entirely contained within the temperature dependence of the irreversibility field B_{irr}. The increasing b^p part of the curve describes the increase in total pinning force as the density of pinned flux lines increases with the applied field until at the peak a "matching field" is reached where all available strong pinning sites are occupied. Beyond this, the further increasing applied field results in the $(1 - b)^q$ diminishment in the pinning force as the superfluid density decreases. The details of the distinct scaling laws thus derived (different values of p and q) have been linked to specific mechanisms of flux pinning thought to be operating, and explained through the similar scaling of materials properties related to those mechanisms. The pinning force curves of different technological materials are found to take on characteristically different forms (Figure 1.2.3.4), and it is therefore considered likely that the curves contain definitive information regarding the contribution of the particular material microstructure to flux pinning. For example, the $b(1 - b)$ form observed for Nb–Ti is associated with transverse depinning of flux lines from the interface of superconducting and non-superconducting regions, while the $b^{1/2}(1 - b)^2$ form common to Nb_3Sn is identified with longitudinal shearing of the flux line lattice. In HTS materials, more extreme forms of the behavior are observed, peaking at much lower reduced field values and diminishing more rapidly in field.

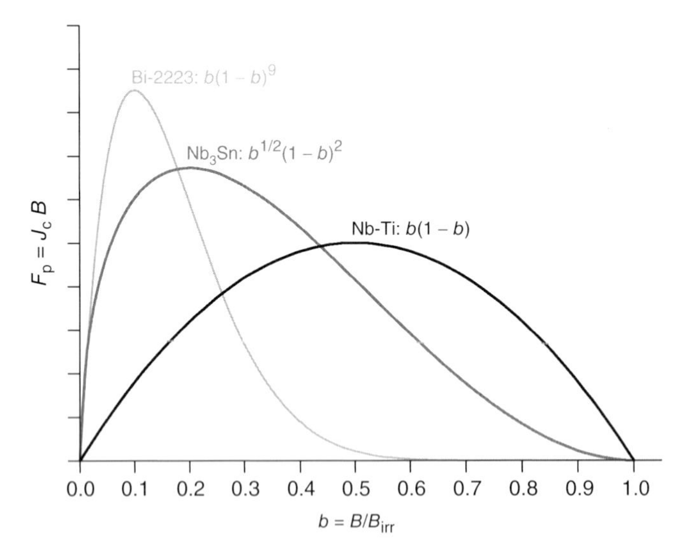

Figure 1.2.3.4 Schematic pinning force plots (as a function of the reduced applied field) representative of the pinning commonly observed in different technological superconductors. The characteristically different functional form of the curves for the different materials is expressed in terms of the generalized Kramer formula, Eq. (1.2.3.1).

1.2.3.6 Regimes of Flux Motion

We have already described the low-pinning limiting case of flux flow, where the Lorentz force fully overcomes the pinning force and the only resistance to flux motion is the viscous drag on the flux lines. This is the inevitable consequence of a sufficiently high current or a sufficiently high field, and is also the regime that most commonly arises in close to perfect single crystals. At the other extreme, we may consider the idealized case of extremely strong flux pinning or a close to negligible transport current. In this regime, the pinning forces dominate, although even here, thermal activation, particularly at the typical operating temperatures of the HTS materials, will lead to so-called "thermally activated flux flow" in which individual flux lines will hop statistically from one pinning site to the next. (It is important to note that the persistent current experiments of S. C. Collins at MIT, in which supercurrents were shown to remain undiminished over a period of years, were performed on *type I* superconductors, where flux penetration does not occur.)

In an intermediate regime, the combination of a significant transport current and thermal activation leads to a directed flux creep of particular significance in the HTS materials. Here, not only the higher operation temperatures but also the reduced core energy (related to ξ^3) stemming from the small coherence length mean that even at temperatures as low as 10 K, flux motion, and consequent dissipation due to creep can be significant. In LTS, in contrast, flux creep can be held to a manageably low level, allowing the operation, for example, of highly stable gigahertz class NMR magnets formed from type II LTS materials, a possibility that does not exist for HTS.

These three regimes of flux motion are illustrated schematically in Figure 1.2.3.5.

1.2.3.7 Limitations on Core Pinning Efficacy

We have seen how the depairing current is supplanted by the depinning current as the limiting factor governing the operation of technological superconductors. Nonetheless, the depairing current (at which the kinetic energy of the charge carriers constituting the current exceeds the binding energy of the Cooper pairs), as the more fundamental limit, is still held up as a goal for efforts aimed at increasing flux pinning. However, since the two are determined by entirely different mechanisms, there is no reason to suppose that their values should coincide. Indeed, it could have been the case that depairing was the limiting mechanism for performance, with depinning only occurring at higher transport currents than could ever be achieved. However, it has been recognized for some time [14, 15] that this is not the case, although sight of this fact appears to have been lost in regard to the HTS materials [16].

If the depairing current were to flow, it would produce a Lorentz force per unit length of flux line given by:

$$f_{\mathrm{d}} = J_{\mathrm{d}}\Phi_0 = \frac{4B_{\mathrm{c}}\Phi_0}{3\sqrt{6}\mu_0\lambda} \tag{1.2.3.2}$$

In the case of core (ξ) pinning, the maximum pinning force is obtained when each vortex is pinned along its entire length. This requires an idealized microstructure

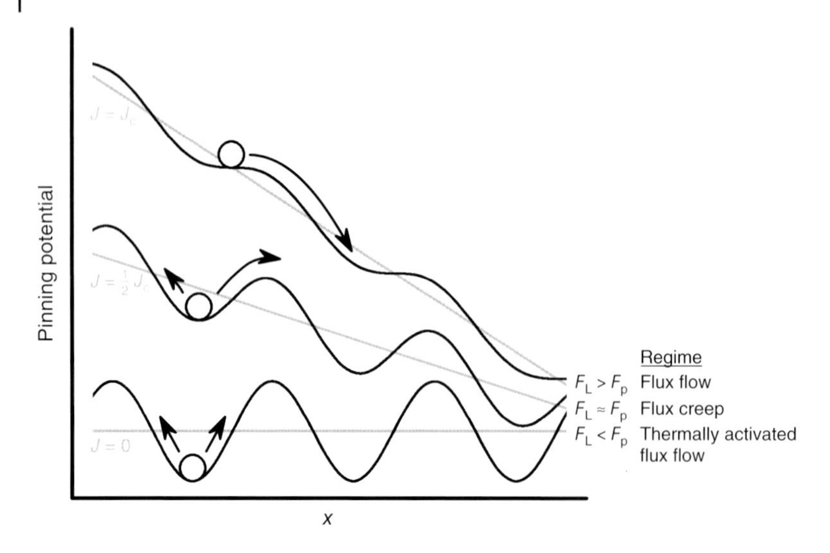

Figure 1.2.3.5 Schematic pinning potentials for different regimes of flux motion. A generic pinning site is represented as a parabolic potential well, superimposed upon a series of which is a linear potential gradient due to the Lorentz force arising from the transport current. When the Lorentz force overcomes the pinning force, the flux lines flow freely. Below this, thermally activated regimes of flux creep and flux flow occur.

comprising cylindrical non-superconducting regions of radius $\sim\xi$ oriented parallel to the applied field in an arrangement corresponding to that of the flux line lattice. If this were achieved, the saving in condensation energy per unit length of flux line would be

$$\varepsilon_{\text{core}} = -\frac{1}{2}\frac{B_{\text{c}}^2}{\mu_0}\pi\xi^2 \tag{1.2.3.3}$$

producing a potential well over a length scale $\sim\xi$ that would result in a pinning force per unit length of flux line of

$$f_{\text{p}}^{\text{core}} = -\nabla\varepsilon_{\text{core}} \approx \frac{1}{2}\frac{B_{\text{c}}^2}{\mu_0}\pi\xi \tag{1.2.3.4}$$

Using $B_{\text{c}} = \Phi_0/2\sqrt{2}\pi\lambda\xi$, the ratio of the optimal core pinning force to the depairing Lorentz force is thus

$$\frac{f_{\text{p}}^{\text{core}}}{f_{\text{d}}} = \frac{1}{2}\frac{B_{\text{c}}^2}{\mu_0}\pi\xi\frac{3\sqrt{6}\mu_0\lambda}{4B_{\text{c}}\Phi_0} = \frac{3\sqrt{3}}{16} \approx 0.32 \tag{1.2.3.5}$$

A more detailed calculation by Matsushita [17] taking into account the precise geometry of the flux line brings this down slightly to a value of 0.28, confirming that the maximum critical current achievable through core pinning is around 30% of the depairing current. In practice, such an idealized microstructure is impossible to achieve, would occupy a significant fraction of the sample volume with non-superconducting material (reducing the effective J_c), and in any case would be ideal for only a single value of magnetic field applied in a particular direction. Consequently, actual critical currents in the commercial LTS materials lie around one-tenth of the optimal value, or one-thirtieth of the depairing current.

HTS materials are surprisingly similar in their performance, and it is perhaps not surprising then that performance gains through pinning modification have stalled. For YBCO at 77 K, 0 T, $J_d \approx 30\,\mathrm{MA\,cm^{-2}}$ [16]. Therefore, core pinning alone cannot be expected to achieve a J_c higher than about $9\,\mathrm{MA\,cm^{-2}}$, which is close to what has been observed (and never exceeded) experimentally [18]. At low fields, J_c lies within an order of magnitude of the depairing current, better than has been achieved in LTS. In-field, there remains room for improvement ($J_c(B)$ decays more rapidly than $J_d(B)$), but the challenge is one of obtaining an extremely high density of correctly spaced near-perfect pinning centers, and any further gains to be had must lie in increasing the irreversibility field through improved pinning at high fields (>3 T). At low temperatures, the requirement for perfection becomes more stringent as the flux lines become more rigid, possibly explaining the seeming ineffectiveness of presently engineered core pins at the temperatures (20–30 K) of interest for high-performance in-field applications. Hard-won performance gains at 77 K vanish completely when the same sample is cooled to 20 K. Indeed, a pinning-engineered sample that performs better at 77 K than its counterpart may in fact be found to perform *worse* at 20 K [19]. Presently, there is no known species of artificial pinning center that can reliably be said to improve low temperature performance, and this is unquestionably the next great challenge for defect engineering of HTS materials. Performance tweaks are still being achieved, but generally performance lies around where we can expect it to reach, between one-tenth and one-hundredth of the depairing current. Any further substantial pinning gain must be based on an alternative pinning mechanism.

1.2.3.8 Magnetic Pinning of Flux Lines

Magnetic (λ) pinning offers a hitherto untapped opportunity to raise the depinning critical current beyond what can be achieved through core pinning alone, and to attain values closer to the depairing current. The simplest example of magnetic pinning arises when a high density of flux lines is introduced, and they form themselves into a lattice, the so-called vortex glass phase. Vortex–vortex interactions typically encourage this to be hexagonal in shape, and if the lattice is sufficiently rigid, it suffices to pin a single vortex by core pinning and others nearby will be held in place by those interactions. However, it is equally possible to conceive of artificial magnetic pinning in which the magnetic interaction of an appropriate artificial pinning center is utilized to provide a magnetic pinning contribution *in addition to* its core pinning effect. The slow take-up of magnetic pinning lies in the challenge of incorporating ferromagnetic material into the microstructure of the superconductor without detriment to the superconductivity through pair breaking interactions.

The claim is often made that magnetic pinning offers an advantage over core pinning through the greatly increased strength of the magnetic Zeeman energy term compared to the condensation energy:

$$\varepsilon_{\text{mag}} = -\frac{1}{2}\int_A \boldsymbol{M} \cdot \boldsymbol{B}\ dA = \frac{1}{2}M\Phi_0 \tag{1.2.3.6}$$

which is limited only by the magnetization of the pinning center, and might be expected to reach a value several orders of magnitude greater than the core pinning energy for a strong ferromagnet. However, it must be remembered that due to the magnetic interaction occurring over the length scale of λ instead of ξ, the pinning *force* may in fact not be much greater, or may even be less than the core pinning force. The magnetic pinning force depends on the ratio of the magnetization of the pinning site to the thermodynamic critical field, which is of the order of 1 T in the technological materials, very similar to that of available strong ferromagnets:

$$f_{\text{p}}^{\text{mag}} = -\nabla\varepsilon_{\text{mag}} \approx \frac{M\Phi_0}{2\lambda} = 2\sqrt{2}\frac{\mu_0 M}{B_{\text{c}}}f_{\text{p}}^{\text{core}} \tag{1.2.3.7}$$

where we have again made use of the relation $B_{\text{c}} = \Phi_0/2\sqrt{2}\pi\lambda\xi$.

Nonetheless, the possibility of magnetic pinning remains attractive not only because its effects are *additive* to those of core pinning, but also due to the potential existence of more exotic interaction mechanisms such as field compensation effects [20] or a reduction in the Lorentz force experienced by the pinned flux line [21] that may serve to lift the depinning limit to J_{c} altogether.

1.2.3.9 Flux Pinning Anisotropy

In general, the macroscopic pinning force resulting from a given sample microstructure depends not only on the applied field, but also on the field angle with respect to the sample, and attempts have been made to provide similar scaling rules for the angle dependence of the pinning as for the field dependence. The most popular of these was introduced by Blatter *et al.* [22] who proposed the field angle scaling law:

$$\widetilde{B} = \varepsilon_\theta B \quad \text{with} \quad \varepsilon_\theta^2 = \cos^2\theta + \varepsilon^2\sin^2\theta \tag{1.2.3.8}$$

for a field applied at an angle θ to the c-axis, where $\varepsilon^2 = m_{ab}/m_c$ is the electronic mass anisotropy of the superconductor. By scaling the applied field values in this way, dependent on their angle, and choosing an appropriate value of ε, the pinning force curves for different field angles can be made to coincide (Figure 1.2.3.6). In the usual interpretation, the pinning force variation described by the universal curve is ascribed to electronic mass anisotropy effects, while the deviations are explained by microstructure-related pinning, in this case a significant additional in-plane component resulting from intrinsic pinning due to the layered structure [23]. Under this interpretation, the ε value that provides the most consistent scaling of the data is directly associated with the electronic mass anisotropy of the superconductor [24]; however, where unfeasibly low ε values have been obtained from such an analysis, the concept of an *effective* electronic mass anisotropy has been introduced where the reduction in anisotropy is attributed to

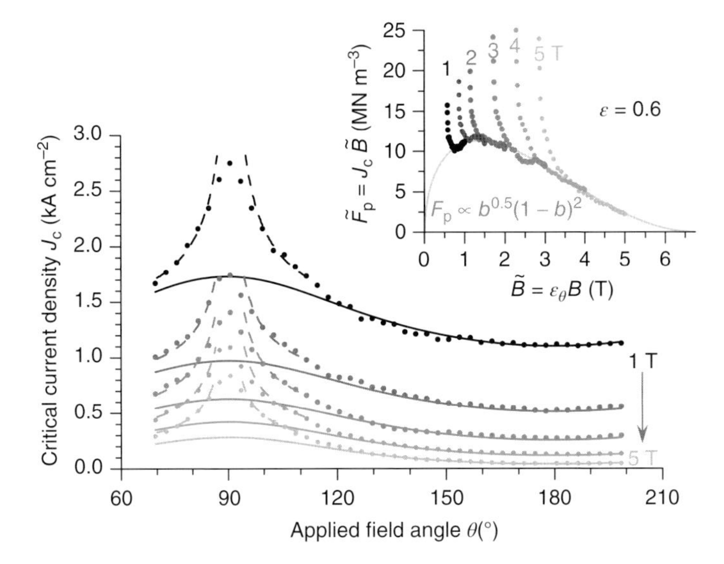

Figure 1.2.3.6 Blatter scaling of $J_c(\theta)$ for a clean $Ba(Fe_{1-x}Co_x)_2As_2$ film at 15 K under various applied fields, showing the experimental data and the angular J_c variation ascribed to electronic anisotropy (solid lines) and intrinsic pinning (broken lines). The inset shows the combined pinning force plot of the scaled datasets. Deviations from the scaled universal curve are due to correlated pinning. (Data from Ref. [23].)

a quasi-isotropic pinning force resulting from nanoscale inhomogeneous strain fields that act to inhibit Cooper pair formation [25].

1.2.3.10 **Maximum Entropy Treatment of Flux Pinning**

To date, the most comprehensive framework proposed for modeling the general experimental results of flux pinning studies is a method based around a maximum entropy derivation of the effects of statistical populations of pinning defects [26]. A consistent mathematical approach is able to provide formal derivations of the commonly used empirical relations:

$$J_c(t) \propto (1-t)^p \quad \text{with} \quad t = \frac{T}{T_c} \tag{1.2.3.9}$$

$$F_p(b) \propto b^\alpha (1-b)^{\beta-1} \quad \text{with} \quad b = \frac{B}{B_{irr}} \tag{1.2.3.10}$$

arising from the Ginzburg–Landau theory [27] and initially proposed by Kramer [12], respectively. In particular, it provides the generalization of the fixed exponents $p = 3/2$, $\alpha = 1/2$, and $\beta = 3$ occurring in those theories that is commonly applied empirically without justification. For the geometrical dependence of J_c on the applied field angle, it provides three fundamental components that are summed to represent different statistically significant combinations of pinning defect populations within the sample:

$$\text{Uniform } J_c(\psi) = \frac{J_0}{\pi} \tag{1.2.3.11a}$$

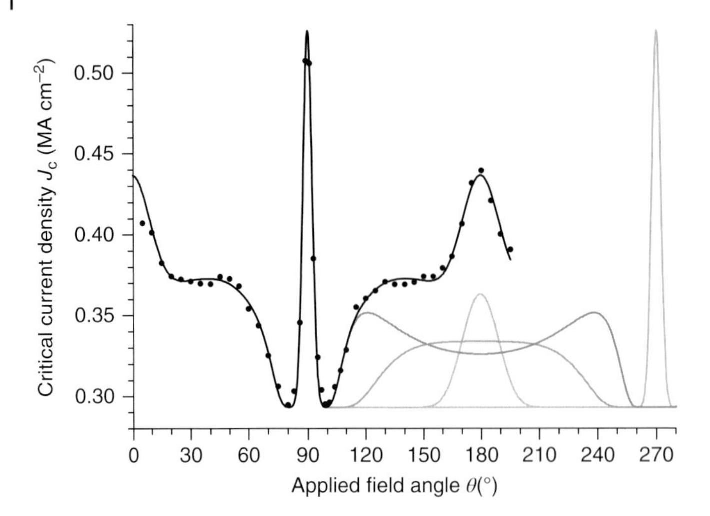

Figure 1.2.3.7 Maximum entropy modeling of $J_c(\theta)$ for a YBCO coated conductor at 65 K, 6 T, showing the experimental data and the resulting fit on the left, and the individual pinning components on the right. In addition to a uniform component, this dataset features a distinct *ab*-centered angular Gaussian describing the intrinsic pinning and three statistically distinguishable *c*-axis angular Gaussian contributions to the out-of-plane pinning.

$$\text{Angular Gaussian } J_c(\psi) = \frac{J_0}{\sqrt{2\pi}\sigma\sin^2\psi} \exp\left(-\frac{1}{2\sigma^2\tan^2\psi}\right) \quad (1.2.3.11b)$$

$$\text{Angular Lorentzian } J_c(\psi) = \frac{J_0}{\pi} \frac{\gamma}{\cos^2\psi + \gamma^2\sin^2\psi} \quad (1.2.3.11c)$$

An example of the application of this modeling to typical results obtained on pinning-optimized YBCO coated conductors is shown in Figure 1.2.3.7. It must be noted that the method of the previous section would fail to provide any meaningful interpretation of this dataset, where microstructure-based pinning completely dominates the response.

The same maximum entropy functions have been shown equally capable of describing the in-plane (variable Lorentz force) $J_c(\phi)$ variation as the out-of-plane $J_c(\theta)$ variation, and are also able to accurately model the effects of oblique defect structures such as those obtained on inclined substrates or through heavy ion irradiation. A further development of the same mathematical framework has been shown to have physical significance in describing vortex channeling along planar pinning defects.

References

1. Blatter, G., Feigel'man, M.V., Geshkenbein, V.B., Larkin, A.I., and Vinokur, V.M. (1994) Vortices in high-temperature superconductors. *Rev. Mod. Phys.*, **66**, 1125–1388.
2. Hampshire, R.G. and Taylor, M.T. (1972) Critical supercurrents and the pinning of

vortices in commercial Nb–60 at% Ti. *J. Phys. F: Met. Phys.*, **2**, 89–106.

3. Matsui, H., Ogiso, H., Yamasaki, H., Kumagai, T., Sohma, M., Yamaguchi, I., and Manabe, T. (2012) 4-fold enhancement in the critical current density of $YBa_2Cu_3O_7$ films by practical ion irradiation. *Appl. Phys. Lett.*, **101**, 232601.
4. Palau, A., Durrell, J.H., MacManus-Driscoll, J.L., Harrington, S., Puig, T., Sandiumenge, F., Obradors, X., and Blamire, M.G. (2006) Crossover between channeling and pinning at twin boundaries in $YBa_2Cu_3O_7$ thin films. *Phys. Rev. Lett.*, **97**, 257002.
5. Dimos, D., Chaudhari, P., and Mannhart, J. (1990) Superconducting transport properties of grain boundaries in $YBa_2Cu_3O_7$ bicrystals. *Phys. Rev. B*, **41**, 4038–4049.
6. MacManus-Driscoll, J.L., Foltyn, S.R., Jia, Q.X., Wang, H., Serquis, A., Civale, L., Maiorov, B., Hawley, M.E., Maley, M.P., and Peterson, D.E. (2004) Strongly enhanced current densities in superconducting coated conductors of $YBa_2Cu_3O_{7-x}$ + $BaZrO_3$. *Nat. Mater.*, **3**, 439.
7. Yamada, Y., Takahashi, K., Kobayashi, H., Konishi, M., Watanabe, T., Ibi, A., Muroga, T., Miyata, S., Kato, T., Hirayama, T., and Shiohara, Y. (2005) Epitaxial nanostructure and defects effective for pinning in $Y(RE)Ba_2Cu_3O_{7-x}$ coated conductors. *Appl. Phys. Lett.*, **87**, 132502.
8. Hänisch, J., Cai, C., Stehr, V., Hühne, R., Lyubina, J., Nenkov, K., Fuchs, G., Schultz, L., and Holzapfel, B. (2006) Formation and pinning properties of growth-controlled nanoscale precipitates in $YBa_2Cu_3O_{7-d}$/transition metal quasi-multilayers. *Supercond. Sci. Technol.*, **19**, 534.
9. Kim, J.H., Zhou, S., Hossain, M.S.A., Pan, A.V., and Dou, S.X. (2006) Carbohydrate doping to enhance electromagnetic properties of MgB_2 superconductors. *Appl. Phys. Lett.*, **89**, 142505.
10. Dou, S.X., Soltanian, S., Horvat, J., Wanga, X.L., Zhou, S.H., Ionescu, M., Liu, H.K., Munroe, P., and Tomsic, M. (2002) Enhancement of the critical current density and flux pinning of MgB_2 superconductor by nanoparticle SiC doping. *Appl. Phys. Lett.*, **81**, 3419–3421.
11. Fietz, W.A. and Webb, W.W. (1969) Hysteresis in superconducting alloys—temperature and field dependence of dislocation pinning in niobium alloys. *Phys. Rev.*, **178**, 657–667.
12. Kramer, E.J. (1973) Scaling laws for flux pinning in hard superconductors. *J. Appl. Phys.*, **44**, 1360–1370.
13. Dew-Hughes, D. (1974) Flux pinning mechanisms in type II superconductors. *Philos. Mag.*, **30**, 293.
14. Hampshire, D.P. (1998) A barrier to increasing the critical current density of bulk untextured polycrystalline superconductors in high magnetic fields. *Physica C*, **296**, 153–166.
15. Dew-Hughes, D. (2001) The critical current of superconductors: an historical review. *Low Temp. Phys.*, **27**, 713–722.
16. Sarrao, J. (ed) (2006) *Basic Research Needs for Superconductivity*, US Department of Energy Office of Science.
17. Matsushita, T. (2007) *Flux Pinning in Superconductors*, Springer.
18. Foltyn, S.R., Civale, L., MacManus-Driscoll, J.L., Jia, Q.X., Maiorov, B., Wang, H., and Maley, M. (2007) Materials science challenges for high-temperature superconducting wire. *Nat. Mater.*, **6**, 631–642.
19. Selvamanickam, V., Yao, Y., Chen, Y., Shi, T., Liu, Y., Khatri, N.D., Liu, J., Lei, C., Galstyan, E., and Majkic, G. (2012) The low-temperature, high-magnetic-field critical current characteristics of Zr-added $(Gd,Y)Ba_2Cu_3Ox$ superconducting tapes. *Supercond. Sci. Technol.*, **25**, 125013.
20. Moshchalkov, V.V., Golubovic, D.S., and Morelle, M. (2006) Nucleation of superconductivity and vortex matter in hybrid superconductor/ferromagnet nanostructures. *C. R. Phys.*, **7**, 86–98.
21. Blamire, M.G., Dinner, R.B., Wimbush, S.C., and MacManus-Driscoll, J.L. (2009) Critical current enhancement by Lorentz force reduction in superconductor–ferromagnet nanocomposites. *Supercond. Sci. Technol.*, **22**, 025017.

22. Blatter, G., Geshkenbein, V.B., and Larkin, A.I. (1992) From isotropic to anisotropic superconductors: a scaling approach. *Phys. Rev. Lett.*, **68**, 875–878.
23. Hänisch, J., Iida, K., Haindl, S., Kurth, F., Kauffmann, A., Kidszun, M., Thersleff, T., Freudenberger, J., Schultz, L., and Holzapfel, B. (2011) J_c scaling and anisotropies in Co-doped Ba-122 thin films. *IEEE Trans. Appl. Supercond.*, **21**, 2887–2890.
24. Civale, L., Maiorov, B., Serquis, A., Willis, J.O., Coulter, J.Y., Wang, H., Jia, Q.X., Arendt, P.N., MacManus-Driscoll, J.L., Maley, M.P., and Foltyn, S.R. (2004) Angular-dependent vortex pinning mechanisms in $YBa_2Cu_3O_7$ coated conductors and thin films. *Appl. Phys. Lett.*, **84**, 2121–2123.
25. Llordés, A., Palau, A., Gázquez, J., Coll, M., Vlad, R., Pomar, A., Arbiol, J., Guzmán, R., Ye, S., Rouco, V., Sandiumenge, F., Ricart, S., Puig, T., Varela, M., Chateigner, D., Vanacken, J., Gutiérrez, J., Moshchalkov, V., Deutscher, G., Magen, C., and Obradors, X. (2012) Nanoscale strain-induced pair suppression as a vortex-pinning mechanism in high-temperature superconductors. *Nat. Mater.*, **11**, 329–336.
26. Long, N.J. (2013) Maximum entropy distributions describing critical currents in superconductors. *Entropy*, **15**, 2585–2605.
27. Ginzburg, V.L. and Landau, L.D. (1950) On the theory of superconductivity. *Zh. Eksp. Teor. Fiz.*, **20**, 1064; In English in Landau, L.D. (1965) *Collected Papers*, Pergamon Press, Oxford, p. 546.

1.2.4
AC Losses and Numerical Modeling of Superconductors

Francesco Grilli and Frederic Sirois

1.2.4.1 Introduction

Type-II superconductors can carry DC current without dissipation, but they do exhibit energy dissipation when they carry AC current or when they are subjected to AC magnetic field. This is because the magnetic field penetrates in the form of discrete flux lines (or vortices) that get pinned to the superconductor material; when there is a change of magnetic field (as in an AC cycle), the flux distribution inside the superconductor material has to rearrange: the movement of magnetic flux induces an electric field, which in turn creates dissipation because this electric field induces currents in the normal conducting regions associated with the core of each vortex. Dissipation occurs whenever there is a variation of the magnetic flux, so the term *AC losses* is generally used for all the situations where the magnetic field changes over time, for example, during the current ramp of a magnet. This kind of energy dissipation is referred to as *hysteresis loss*.[1)]

Technical superconductors are composed of several materials, including metallic and sometimes magnetic parts: as a consequence, they are affected by additional loss contributions (such as eddy current, resistive, coupling, and magnetic losses), which can become important and in some cases largely exceed the hysteretic losses. In multifilamentary superconductors, coupling losses are caused by the current induced by external magnetic fields and flowing from one filament to the other via the normal metal in between; effective ways to reduce them include filament twisting and resistive barriers around the filaments. Eddy current losses can be reduced by increasing the stabilizer's resistivity.

1.2.4.2 General Features of AC Loss Characteristics

Throughout this chapter, we maintain the distinction between *transport* and *magnetization* losses to identify the dissipation caused by transport current and external magnetic field, respectively. This distinction is merely of practical nature because the mechanism responsible for the hysteresis losses inside superconductors (i.e., the movement of magnetic flux) is the same in both cases.

Hysteresis losses strongly depend on the amplitude of the current or the applied field. The transport losses typically increase with the third or fourth power of the applied current (depending on the superconductor's shape) for currents below I_c, then they increase even more rapidly due to flux-flow dissipation. At sufficiently high currents, some of the current starts flowing in the metallic parts of the conductor, giving rise to a resistive contribution. The magnetization losses too increase rapidly with the amplitude of the field (third or fourth power), then when the field fully penetrates the superconductor, they increase less rapidly, typically

1) Other loss mechanisms occur in superconductors, such as the response of normal electrons and the losses associated to the Meissner state. However, they are important only at very high frequencies and at extremely low fields, and they will not be addressed here.

with the first power of the field. More details on the dependence of the hysteresis losses on the current and field amplitude are given in Section 1.2.4.4 for different geometries.

Generally, the hysteresis losses (per cycle) have a feeble dependence on frequency, and the observed frequency dependence of measured losses is usually due to the eddy current or resistive losses occurring in the metal parts. A model for loss dissociation was proposed in Ref. [1], where the dissipated power is split into resistive, hysteretic, and eddy current contributions, each with a different dependence on the frequency:

$$P_{\text{tot}} = \underbrace{P_{\text{res}}}_{\text{RI}^2_{\text{rms}}} + \underbrace{P_{\text{hyst}}}_{fQ_{\text{hyst}}} + \underbrace{P_{\text{eddy}}}_{\propto f^2} \tag{1.2.4.1}$$

An example of the identification of the three components on measured data is shown in Figure 1.2.4.1a. In general, the dominance of one component on the others depends on several factors, related to the properties of the materials and on the operating conditions (frequency and amplitude of the field and current). Sometimes other loss components become important: for example, coupling losses or losses in magnetic parts. Figure 1.2.4.1b shows an example where the losses in the ferromagnetic substrate of a coated conductor are the major loss components in a significant current interval.

The variation of the critical current density J_c inside the superconductor affects the AC losses. Owing to the pinning mechanisms in type-II superconductors, J_c depends on the local magnetic flux density and on its orientation, sometimes in a very complicated fashion. This local reduction of J_c can influence the shape of the

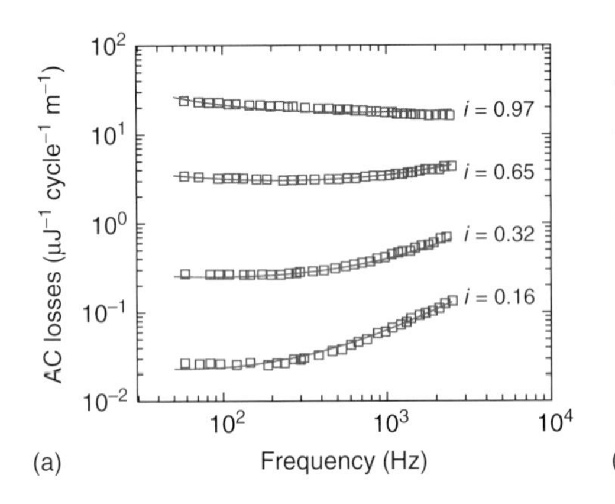

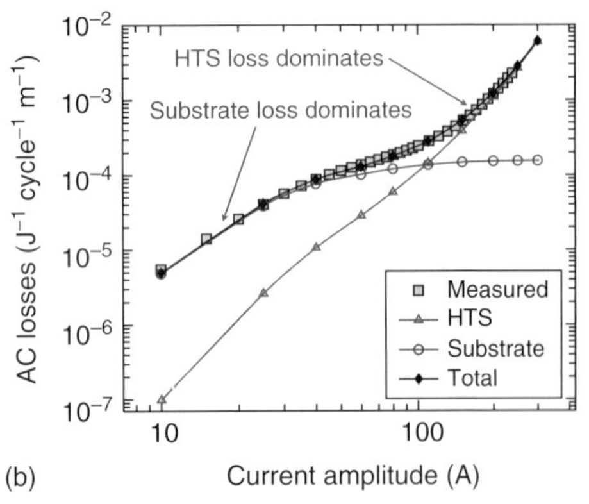

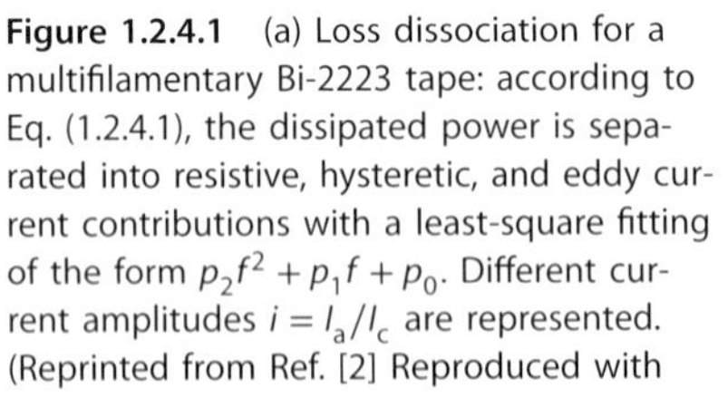

Figure 1.2.4.1 (a) Loss dissociation for a multifilamentary Bi-2223 tape: according to Eq. (1.2.4.1), the dissipated power is separated into resistive, hysteretic, and eddy current contributions with a least-square fitting of the form $p_2 f^2 + p_1 f + p_0$. Different current amplitudes $i = I_a/I_c$ are represented. (Reprinted from Ref. [2] Reproduced with permission of Ecole Polytechnique Fédérale de Lausanne.) (b) Transport losses of a yttriumbarium copper oxide (YBCO) coated conductor with ferromagnetic substrate: the losses in the substrate are the dominating components for currents <100 A. The different components are calculated by finite-element simulations.

AC loss characteristics of the whole tape, as do local nonuniformities of J_c caused by the manufacturing process.

1.2.4.3 **Measuring AC Losses**

This section describes the main methods used to measure AC losses in various circumstances (transport, magnetization, and their combination) and for different types of samples (short pieces of tape and large assemblies). Before proceeding, a brief comment on the terminology used here: in the following, *calorimetric* method indicates a method based on measuring the local increase of temperature caused by AC loss dissipation; this must be distinguished from the *boil-off* method (sometimes called *calorimetric* as well in the literature), which consists in measuring the amount of evaporated coolant.

1.2.4.3.1 **Transport Losses**

The goal is to measure the variation of magnetic flux caused by the AC transport current. The standard technique for measuring transport losses in straight tapes consists in soldering a pair of voltage taps on the tape, shaping them in the form of a loop, and measuring the voltage (see Figure 1.2.4.2a). It can be easily shown that, for the case of a sinusoidal transport current, only the voltage component in phase with the current generates dissipation [3]. The measurements are therefore developed to extract such component: this can be done, for example, by means of a lock-in amplifier. The voltage taps measure the sum of the voltage drop between the contact points and the voltage corresponding to the rate of variation of the magnetic flux in the loop. In practice, one is usually interested in measuring losses for currents smaller than the critical current I_c, and for such subcritical currents the voltage drop is zero. The question is therefore to determine what flux variation one measures.

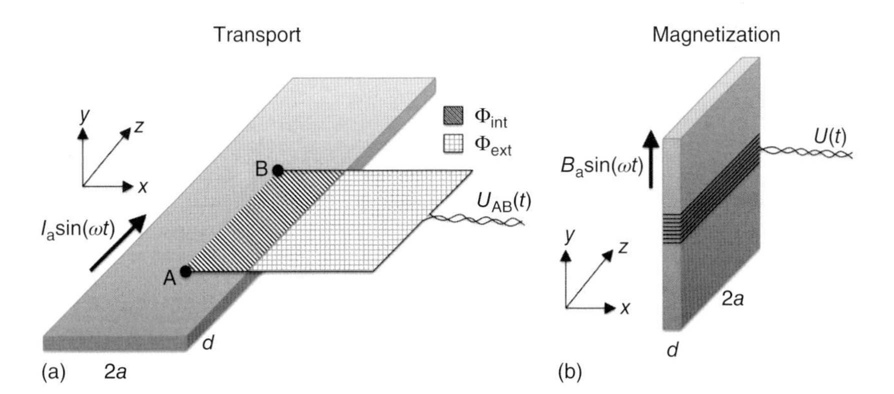

Figure 1.2.4.2 (a) Setup for measuring transport AC losses by electrical technique: for a tape with noncircular cross-section, the measuring loop needs to extend outside the tape. (b) Setup for measuring magnetization losses of a rectangular tape in parallel external field: in perpendicular field, the recorded loss value needs to be multiplied by a factor accounting for the demagnetizing effect.

In general, the voltage loop catches two types of fluxes, inside and outside the superconductor, which we can name Φ_{int} and Φ_{ext}, respectively (see Figure 1.2.4.2a). The ideal case is represented by a wire with circular cross-section because the flux lines are circular and concentric and do not cross the wire's surface: one could therefore think of placing the voltage loop on the surface of the wire, in this way catching all the relevant flux variations because the voltage associated with Φ_{ext} would be purely inductive (i.e., in quadrature with the transport current). Real tapes have elliptical or rectangular cross-section, and in those cases, things are more complicated because both Φ_{int} and Φ_{ext} have in-phase components. One must therefore extend the measuring loop outside the tape [4]. In principle, a loop extending far away from the sample would make sure that all the flux is caught; at the same time, however, a larger loop is also more prone to catch spurious signals and noise, which could affect the measured loss value. Clem calculated that a distance of three times the sample's half-width is a good compromise: the loop is sufficiently close to the sample and it provides loss values within 5–10% of the expected value [5]. More details on the influence of the position of the voltage taps and of the measured voltage signal on AC losses can be found in Refs. [6] and [7].

A completely different approach to measure transport AC losses is represented by the boil-off method, which consists in measuring the amount of coolant that evaporates as a consequence of the energy dissipation in the superconductor [8]. The main advantage is that it can be used with differently shaped and sized samples and it does not present the problem (typical of the electrical method) of recording the correct electric signal. This is particularly useful in the case of complex tape assemblies, such as cables and coils: tapes are closely packed together and soldering voltage taps and extracting the true loss signal can be quite problematic. The boil-off method is the standard measuring technique for low-temperature superconductors, especially for coils and cables. Its application to high-temperature superconductors cooled with liquid nitrogen is generally more difficult because of the very small volume of generated gas compared to the case of helium [9, 10].

1.2.4.3.2 Magnetization Losses

The standard technique for measuring the magnetization losses of a superconducting sample consists in placing the sample in a varying uniform magnetic field and measuring its magnetization by means of a pickup coil wound around the sample (see Figure 1.2.4.2b). The signal measured by the pickup coil is proportional to the energy dissipation in the sample. The ideal case is represented by an infinite slab in parallel field, where the field on the superconductor's surface is known and is equal to the applied field. A tape with rectangular cross-section with the field parallel to its longitudinal direction (as the one schematically shown in Figure 1.2.4.2b) is a good approximation of the slab configuration. In this case, the loss value is simply given by $Q = (\mu_0 N d 2a)^{-1} \int_0^{1/f} B_a(t) U(t) dt$, where N is the number of turns of the pickup coil, d and $2a$ are the thickness and width of the tape, respectively, and f is the frequency of the field. If the infinite slab approximation is not valid, for example, in the case of magnetic field applied perpendicular to the tape's face, a calibration constant c taking into account the distortion of the field lines (which depends on the tape's shape and on the pickup coil's position) must be added in the expression for Q.

The calibration constant can be determined by a calorimetric method, and in general, it depends on field amplitude and frequency. A detailed description of the calibration process, including a discussion on the influence of the pickup coil position, can be found in [11].

In fact, the calorimetric method itself constitutes an alternative method for measuring the magnetization losses [12, 13] and it can be used for measuring the transport losses as well. It provides a direct measurement of the AC losses because it measures the dissipation in terms of temperature increase of the sample; compared to the electromagnetic method, it is not prone to electromagnetic disturbances, since the measured signal simply records a temperature increase; however, it requires a more complicated hardware setup (thermometers, insulations); in addition, the acquisition of data points is much slower, especially at liquid nitrogen temperatures, when one has to detect small temperature increases in a thermally noisy environment, which makes measurement repetition and data averaging necessary. As a rule of thumb, at the liquid nitrogen temperature, the calorimetric method is around 10 times slower than the electromagnetic method.

The calorimetric method can however be used to calibrate the AC loss values measured with the electromagnetic method, in particular to calculate the calibration factor c by which the measured signal needs to be multiplied in order to take into account the differences from the ideal case of a slab in parallel field. One does not need to wind the pickup coil literally around the sample, but can wind it on a fixed frame built to accommodate different samples with the same size. However, differently shaped or sized samples still require different pickup coils.

An alternative measuring approach that overcomes all the problems mentioned above is the so-called calibration-free method [14]. The method is based on the observation that the losses in the sample constitute a fraction of the power supplied to the whole system by the AC source; this fraction is generally small, but can be detected by building a symmetric system consisting of two pickup coils, subjected to the same external uniform magnetic field, connected in series, but wound in opposite directions (see Figure 1.2.4.3a,b). In the absence of a sample, the system is perfectly balanced and the measured voltage $U(t)$ is zero. When the sample is inserted inside one of the measuring coils, the symmetry is broken, the total voltage induced in the two coils by the external field cancels out because of their opposite winding directions, and the measured signal gives directly the loss of the sample, irrespective of its shape and dimensions. This method has the big advantage that the same experimental setup can be used to measure samples of different shapes and sizes, without the need of building *ad hoc* pickup coils. An alternative setup to measure the magnetization AC losses is presented in [15].

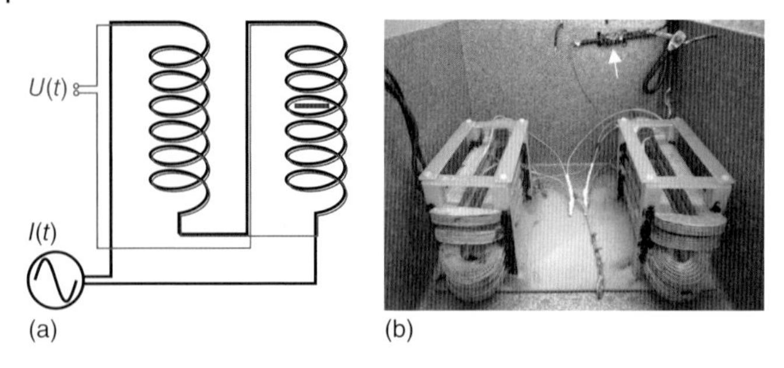

Figure 1.2.4.3 Conceptual design (a) and practical realization (b) of the calibration-free method. In the circuit, the rectangle represents the sample; in the picture, the arrow indicates the Rogowski coil used to extract the reference phase signal.

1.2.4.3.3 Combination of Transport and Magnetization AC Losses

In real applications, superconductors carry current while being also subjected to external magnetic fields, for example, produced by the adjacent turns in a coil. In this case, due to the complex interaction of the self-field and of the external field inside the superconductor, the AC losses are different from the simple sum of the losses occurring in the current-only and field-only cases.

Ashworth and Suenaga [16, 17] performed pioneering calorimetric measurements on Bi-2223 tapes. The calorimetric method, measuring the temperature increase associated with the AC losses, is not affected by the complex interaction of the electromagnetic fields. Its main limitation is the low speed of data acquisition and low accuracy for small currents and fields. Methods based on measuring the sample's magnetization by electromagnetic methods have also been developed [18, 19], even for the case when current and field are not in phase [20, 21].

1.2.4.4 Computing AC Losses

1.2.4.4.1 Analytical Computation

In the framework of the critical state model originally proposed by Bean [22], the loss density of a superconducting slab of width $2a$ in a parallel magnetic field of amplitude H_a can be easily calculated as [8][2]:

$$Q_M = \begin{cases} \frac{2}{3}\mu_0 H_a^2 \left(\frac{H_a}{H_p}\right) & H_a \leq H_p \\ 2\mu_0 H_a^2 \left[\frac{H_p}{H_a} - \frac{2}{3}\left(\frac{H_p}{H_a}\right)^2\right] & H_a \geq H_p \end{cases} \quad (\text{J cycle}^{-1}\,\text{m}^{-3}) \tag{1.2.4.2}$$

where $H_p = J_c a$ is the field for which the slab is fully penetrated. The loss density increases rapidly (proportionally to H_a^3) for $H_a < H_p$ and more slowly (proportionally to H_a) for $H_a > H_p$. Norris [23] derived the formulas for the transport AC losses of superconductors with elliptical and infinitely thin cross-sections carrying

2) Here and in the following expressions for AC losses, *amplitude* is intended as the peak value of a sinusoidal oscillation; this explains the slight difference between formulas (1.2.4.2) and the formulas given in Ref. [8], which are written in terms of the peak-to-peak amplitude.

current $i = I/I_c$:

$$Q_T = \frac{I_c^2 \mu_0}{\pi} \begin{cases} (1-i)\ln(1-i) + (2-i)\frac{i}{2} & \text{ellipse} \\ (1-i)\ln(1-i) + (1+i)\ln(1+i) - i^2 & \text{thin strip} \end{cases} \quad (\text{J cycle}^{-1}\,\text{m}^{-1}) \tag{1.2.4.3}$$

Formulas for magnetization losses of a thin strip of width $2a$ and *sheet* current density j_c in a field of amplitude H_a perpendicular to the strip's surface were independently derived by Brandt and Indenbom [24] and Zeldov *et al.* [25][3]:

$$Q_M = 4\mu_0 a^2 j_c H_a g\left(\frac{H_a}{H_c}\right) \quad (\text{J cycle}^{-1}\,\text{m}^{-1}) \tag{1.2.4.4}$$

with $g(x) = (2/x)\ln\cosh x - \tanh x$ and $H_c = j_c/\pi$. The losses are proportional to the square of the width, which means that a practical way of reducing them is by making narrower conductors: a strip cut into N filaments has losses N times lower, provided that the filaments are electromagnetically uncoupled, for example, by means of twisting or transposition. In addition, similarly to the case of an infinite slab, the curve of the magnetization losses of a thin strip presents a change of slope: from $Q_M \sim H_a^4$ at low fields to $Q_M \sim H_a$ at high fields. A similar change of slope, related to the full penetration of the field in the superconductor, is observed in other geometries too.

Owing to their simplicity and applicability to conductor geometries found in practice, formulas (1.2.4.2)–(1.2.4.4) are very often used to estimate the losses of superconducting tapes and wires. Other analytical expressions have been derived for certain tape arrangements, like tape arrays and stacks. An exhaustive review of the analytical models for superconductors can be found in [26].

While useful for a quick estimation of AC losses, analytical models suffer from a number of limitations that affect their accuracy and applicability. For example, most analytical models are based on the critical state approach and as a consequence they cannot take into account the intrinsic frequency dependence of hysteresis losses nor current densities exceeding J_c, see for example, the expression of Eq. (1.2.4.3), which diverges for $I = I_c$. Also taking into account nonuniform fields and currents with arbitrary temporal evolution is a very difficult task to perform analytically. These and other limitations can be overcome by numerical methods, which can account for virtually any arbitrary geometry and excitation.

1.2.4.4.2 **Numerical Computation**

Many approaches exist for computing AC losses numerically, most of them summarized in the open-access review article [27]. We can divide them in two broad categories: *differential methods*, based on partial differential equations (PDEs), such as the finite-element method (FEM), and *integral methods*, based on the use of Green's function to transform the PDEs into integral equations. Each approach

3) Owing to the different utilized notation and approach, the expression for the losses has a different form in the two original papers, but they are in fact equivalent. Here, we utilize the one from [24].

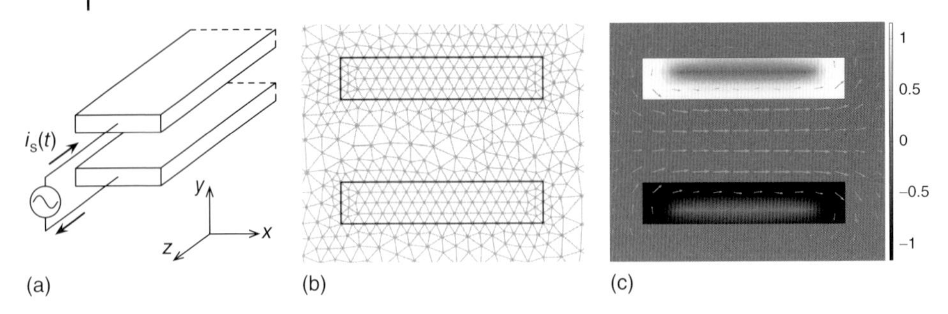

Figure 1.2.4.4 (a) Model of a two-conductor arrangement carrying antiparallel transport currents; (b) discretization in triangular mesh of a 2D cross-section of the geometric model; (c) plot of a typical solution (gray scale: J_z/J_c; arrows: H_x and H_y).

has advantages and drawbacks. In what follows, only the principles common to most numerical techniques are presented.

The first step for numerically computing AC losses consists in building a geometric model that represents the device under consideration, and then in discretizing it in a mesh of *elements* (i.e., domains of simple geometrical shape, typically triangles or quadrilaterals) that are compatible with the numerical method to be used (see, for example, Figure 1.2.4.4a,b). By using an integral method, the meshing of nonconducting regions can be avoided, although at the expense of a computational cost that grows approximately with the cube of the number of elements, as opposed to a roughly linear increase with differential methods [28].

Once geometry and mesh are created, one must choose a numerical method and a formulation, the latter being based on the variable one wants to solve for. There are many possible choices for electromagnetic variables: $H, E, A-V, T-\Omega$, J, and variants or combinations of these. Campbell reviewed many of the possible formulations in [29].

Regardless of the choice, the formulation must satisfy Maxwell's equations $\nabla \cdot B = 0$, $\nabla \times E = -\partial B/\partial t$, and $\nabla \times H = J$ (displacement current term $-\partial D/\partial t$ neglected), with the constitutive relationships $E = \overline{\rho}(J)J$ and $B = \overline{\mu}(H)H$, where $\overline{\rho}(J)$ and $\overline{\mu}(H)$ are, in the general case, nonlinear tensors, but very often they can be taken as scalars, especially in 2D problems where the H and J components are perpendicular.

Note that by neglecting the displacement current term, we obtain a diffusion like equation, which is in principle simpler to solve than the classical "wave equation," characterized by the presence of a second-order time derivative. This approximation is well justified in most practical cases, especially near power frequencies. It also explains why we can ignore the $\nabla \cdot D$ Maxwell equation. However, the resulting diffusion problem is highly nonlinear because of the $\overline{\rho}(J)$ and $\overline{\mu}(H)$ terms. The former represents the nonlinearity of the superconductor (for which one usually takes $\mu = \mu_0$ even near H_{c_1}, which is usually lower than the fields of practical interest), and the latter generally accounts for the nonlinear behavior of ferromagnetic parts, ρ being generally taken as constant in this case.

In the simplified 2D case with J_z perpendicular to H_x and H_y (see Figure 1.2.4.4a for axis definition), one can use an empirical power-law model to describe the E–J constitutive relationship near the critical current density, that is,

$$E = \rho(J)J = E_0 \left(\frac{|J|}{J_c} \right)^n \text{sign}(J) \Rightarrow \rho = \frac{E_0}{J_c} \left(\frac{|J|}{J_c} \right)^{n-1} \tag{1.2.4.5}$$

where n is the power law exponent and J_c is the critical current density of the material. Both J_c and n depend in general on the local field B, the temperature T, and possibly also on the position. Whether to consider these model refinements depends on the operating conditions of the considered problem. In addition, other E–J constitutive equations can be used.

Once the problem is fully defined, it has to be discretized according to the chosen numerical method. In all cases, this operation results in a generally large system of equations to be solved numerically. The numerical solution obtained is a piecewise approximation of the continuous problem, and this approximation converges toward the exact solution as the discretization is refined. Figure 1.2.4.4c shows an example of a solution obtained with the FEM.

The numerical solution of the problem is not straightforward though. Since the problem is systematically nonlinear, static or time harmonic solutions are not possible, and a time transient simulation must be performed, which is usually delicate and may result in divergence of the solver if the time-stepping algorithm is not robust enough. An adaptive time solver able to handle differential algebraic equation systems is typically preferred over simple basic methods, although it is possible to succeed with any methods if one is willing to use small time steps and thus wait long times. One can avoid these problems by using methods based on the critical state model instead of the smooth current–voltage characteristics shown in Eq. (1.2.4.5) [30, 31]; these methods are computationally faster and can be preferable when flux creep is not a concern.

The final step for numerically computing AC losses simply consists in performing post-processing operations on the obtained numerical solution. The fundamental quantity to retrieve is J (see Figure 1.2.4.4c for instance), from which one can compute the electric field E and the local power density $p(t) = \int_\Omega E \cdot J \; d\Omega$, where Ω is the cross-section (2D) or volume (3D) of the superconducting domain in which one wants to compute AC losses. For example, in a 2D case like that illustrated in Figure 1.2.4.4, one has $J = J_z\ (x, y, t)$, and using Eq. (1.2.4.5) to express E in terms of J, one obtains:

$$Q = \int_{t_0}^{t_0+T_p} dt \int_\Omega \rho(J) J^2 d\Omega \quad (\text{J cycle}^{-1}) \tag{1.2.4.6}$$

where T_p is the period of the AC signal ($T_p = 1/f$), and t_0 is an initial time for starting the integration, chosen in a region where the $\rho(J)J^2$ waveform has reached a steady state. Additional post-processing computations might be required if J_c or n is a function of B or any other parameter.

Expression (1.2.4.6) is very general and includes all losses in the domain Ω. However, other approaches are possible for computing the AC losses, namely using global quantities such as the current and voltage in each conductor, or using magnetic quantities (see Ref. [27] for details).

References

1. Stavrev, S., Dutoit, B., Nibbio, N., and Le Lay, L. (1998) Eddy current self-field loss in Bi-2223 tapes with a.c. transport current. *Physica C*, **307**, 105–116.
2. Stavrev, S. (2002) Modelling of high temperature superconductors for ac power applications. PhD thesis. Ecole Polytechnique Fédérale de Lausanne.
3. Rabbers, J.J. (2001) AC loss in superconducting tapes and coils. PhD thesis, University of Twente.
4. Campbell, A.M. (1995) AC losses in high Tc superconductors. *IEEE Trans. Appl. Supercond.*, **5** (2), 682–687.
5. Clem, J.R., Benkraouda, M., and McDonald, J. (1996) Penetration of magnetic flux and electrical current density into superconducting strips and disks. *Chin. J. Phys.*, **34** (2-11), 284–290.
6. Yang, Y., Hughes, T., Beduz, C., Spiller, D., Scurlock, R., and Norris, W. (1996) The influence of geometry on self-field AC losses of Ag sheathed PbBi2223 tapes. *Physica C*, **256**, 378–386.
7. Klinčok, B., Gömöry, F., and Pardo, E. (2005) The voltage signal on a superconducting wire in AC transport. *Supercond. Sci. Technol.*, **18**, 694–700.
8. Wilson, M.N. (1983) *Superconducting Magnets*, Clarendon Press, Oxford.
9. Okamoto, H., Sumiyoshi, F., Miyoshi, K., and Suzuki, Y. (2006) The nitrogen boil-off method for measuring AC losses in HTS coils. *IEEE Trans. Appl. Supercond.*, **16** (2), 105–108.
10. Murphy, J.P., Mullins, M.J., Barnes, P.N., Haugan, T.J., Levin, G.A., Majoros, M., Sumption, M.D., Collings, E.W., Polak, M., and Mozola, P. (2013) Experiment setup for calorimetric measurements of losses in HTS coils due to AC current and external magnetic fields. *IEEE Trans. Appl. Supercond.*, **23** (3), 4701505.
11. Schmidt, C. (2008) Ac-loss measurement of coated conductors: the influence of the pick-up coil position. *Physica C*, **468** (13), 978–984.
12. Schmidt, C. (2000) Calorimetric ac-loss measurement of high *Tc*-tapes at 77 K, a new measuring technique. *Cryogenics*, **40**, 137–143.
13. Ashworth, S. and Suenaga, M. (2001) Local calorimetry to measure ac losses in HTS conductors. *Cryogenics*, **41** (2), 77–89.
14. Šouc, J., Gömöry, F., and Vojenčiak, M. (2005) Calibration free method for measurement of the AC magnetization loss. *Supercond. Sci. Technol.*, **18**, 592–595.
15. Grilli, F., Ashworth, S.P., and Stavrev, S. (2006) AC loss characteristics of stacks of YBCO coated conductors. *Mater. Res. Soc. Symp. Proc.*, **946**, 0946-HH10-06.
16. Ashworth, S.P. and Suenaga, M. (1999) Measurement of ac losses in superconductors due to ac transport currents in applied ac magnetic fields. *Physica C*, **313**, 175–187.
17. Ashworth, S.P. and Suenaga, M. (1999) The calorimetric measurement of losses in HTS tapes due to ac magnetic fields and transport currents. *Physica C*, **315**, 79–84.
18. Rabbers, J., ten Haken, B., Gömöry, F., and ten Kate, H.H.J. (1998) Self-field loss of BSCCO/Ag tape in external AC magnetic field. *Physica C*, **300**, 1–5.
19. Jiang, Z. and Amemiya, N. (2004) An experimental method for total AC loss measurement of high Tc superconductors. *Supercond. Sci. Technol.*, **17** (3), 371–379.
20. Nguyen, D.N., Sastry, P., Zhang, G.M., Knoll, D.C., and Schwartz, J. (2005) AC loss measurement with a phase difference between current and applied magnetic field. *IEEE Trans. Appl. Supercond.*, **15** (2), 2831–2834.
21. Vojenčiak, M., Šouc, J., Ceballos, J., Gömöry, F., Pardo, E., and Grilli, F. (2006) Losses in Bi-2223/Ag tape at simultaneous action of AC transport and AC magnetic field shifted in phase. *J. Phys.: Conf. Ser.*, **43**, 63–66.

22. Bean, C.P. (1962) Magnetization of hard superconductors. *Phys. Rev. Lett.*, **8** (6), 250–252.
23. Norris, W. (1970) Calculation of hysteresis losses in hard superconductors carrying ac: isolated conductors and edges of thin sheets. *J. Phys. D Appl. Phys.*, **3**, 489–507.
24. Brandt, E.H. and Indenbom, M. (1993) Type-II-superconductor strip with current in a perpendicular magnetic field. *Phys. Rev. B*, **48** (17), 12893–12906.
25. Zeldov, E., Clem, J., McElfresh, M., and Darwin, M. (1994) Magnetization and transport currents in thin superconducting films. *Phys. Rev. B*, **49** (14), 9802–9822.
26. Mikitik, G., Mawatari, Y., Wan, A., and Sirois, F. (2013) Analytical methods and formulas for modeling high temperature superconductors. *IEEE Trans. Appl. Supercond.*, **23** (2), 8001920.
27. Grilli, F., Pardo, E., Stenvall, A., Nguyen, D.N., Yuan, W., and Gömöry, F. (2013) Computation of losses in HTS under the action of varying magnetic fields and currents. *IEEE Trans. Appl. Supercond.*, **20**, 1379–1382.
28. Sirois, F., Roy, F., and Dutoit, B. (2009) Assessment of the computational performances of the semi-analytical method (SAM) for computing 2-D current distributions in superconductors. *IEEE Trans. Appl. Supercond.*, **19** (3), 3600–3604.
29. Campbell, A.M. (2011) An introduction to numerical methods in superconductors. *J. Supercond. Novel Magn.*, **24**, 27–33.
30. Pardo, E., Gömöry, F., Šouc, J., and Ceballos, J.M. (2007) Current distribution and ac loss for a superconducting rectangular strip with in-phase alternating current and applied field. *Supercond. Sci. Technol.*, **20** (4), 351–364.
31. Campbell, A.M. (2007) A new method of determining the critical state in superconductors. *Supercond. Sci. Technol.*, **20**, 292–295.

2
Superconducting Materials

2.1
Low-Temperature Superconductors

2.1.1
Metals, Alloys, and Intermetallic Compounds
Helmut Krauth and Klaus Schlenga

2.1.1.1 Introduction
Superconductivity was first observed in relatively pure metallic elements. The hope for immediate applications in electrical engineering and electrical energy technology was soon disappointed, as elemental superconductors are very sensitive to magnetic fields and can therefore not carry significant currents. Most metals are Type I superconductors trying to shield a magnetic field completely from its interior (Meissner state), until superconductivity breaks down suddenly. Only three metals, among them Nb, exhibit Type II superconductivity tolerating higher magnetic fields by allowing flux to penetrate in the form of flux lines (Shubnikov phase). Consequently, the only application-relevant superconductor metal is Nb.

Alloying usually changes metals into Type II conductors and/or enhances the upper critical magnetic field further. In addition, intermetallic compounds of specific composition and crystal structure, for example, A15 structure, show much improved properties including higher T_c than alloys. Defect-free Type II superconductors still cannot carry significant current. The flux lines move under the Lorentz force exerted on them; this results in losses and heat generation. Therefore, the microstructure must be optimized by introduction of defects to pin the flux lines, leading to so-called *hard superconductors*.

Mainly electromagnetic stability considerations require that application-relevant conductors take the form of multifilamentary wires, with the superconducting filaments embedded in a normal conducting matrix.

The goal of conductor fabrication for application is therefore the production of wires with many kilometer length and high current carrying capacity in such an electromagnetic stable configuration. At least in part incidentally, pure Nb, the

Applied Superconductivity: Handbook on Devices and Applications, First Edition.
Edited by Paul Seidel.
© 2015 Wiley-VCH Verlag GmbH & Co. KGaA. Published 2015 by Wiley-VCH Verlag GmbH & Co. KGaA.

Nb-based alloy NbTi, and the Nb-based A15-type compound Nb_3Sn are virtually the only superconductor materials industrially produced in large quantities.

2.1.1.2 **Type I and Type II Superconductor Elements and High-Field Alloys**

We first deal with the superconductivity of elemental metals and of metallic alloys. Superconductivity in these materials is phonon-mediated and the Bardeen–Cooper–Schrieffer (BCS) theory applies. The relationships given below were derived by this microscopic theory and especially from the more macroscopic phenomenological Ginzburg–Landau–Abrikosov–Gorkov (GLAG) theory [1–3].

2.1.1.2.1 **Fundamental Superconductor Properties**

Among the primary superconductor properties of a material is the critical temperature, T_c, below which superconductivity exists. All superconducting metals and alloys have $T_c \leq 10\,\mathrm{K}$. To provide sufficient performance the operational temperature T_{op} must be significantly lower than T_c, for example, 4 K or even 2 K. The next important question is how the superconductor performs in a magnetic field B (Throughout this chapter we will express the magnetic field H in terms of the vacuum induction B. Similarly the magnetization will also be given in tesla.). By fundamental thermodynamic considerations comparing the Cooper pair condensation energy with the magnetic field energy one gets an estimation of the thermodynamic critical field B_{cth} (0 K) as

$$B_{cth}(0) \approx \sqrt{\mu_0 \cdot \frac{\gamma}{2}} \cdot T_c \tag{2.1.1.1}$$

Here, γ is the Sommerfeld coefficient of the electronic specific heat $c_E = \gamma \cdot T$, connecting to BCS via $\gamma \propto N(E_F)$.The temperature dependence of B_{cth} empirically is as follows:

$$B_{cth}(T) \approx B_{cth}(0) \cdot (1 - t^2) \quad \text{with} \quad t = \frac{T}{T_c} \tag{2.1.1.2}$$

Below the borderline defined by Eq. (2.1.1.2) the superconductor is in Meissner state and expels the magnetic field completely. Above this line it is in the normal conducting state. Most metals behave in this way, but there are some with different behaviors. The performance is governed by the Ginzburg–Landau (GL) parameter $\kappa(T)$, microscopically defined by $\kappa = \lambda/\xi$, with λ the penetration depth of the magnetic field and ξ the coherence length of the Cooper pairs. Below $\kappa < 1/\sqrt{2}$, the material behaves as described above ("Type I superconductors").

Above $\kappa > 1/\sqrt{2}$ ("Type II superconductors"), it is energetically more efficient that the magnetic field enters as flux lines, each of them carrying one flux quantum $\Phi = h/2e$ above a lower critical field B_{c1}. The superconductor is in the mixed state, also called the *Shubnikov phase*. Above an upper critical field B_{c2} the material is normal conducting. Intuitively, at B_{c1} the field is high enough to create an isolated flux line and at B_{c2} the whole cross-section is filled with the normal conducting cores of the flux line. According to [1], one gets the following correlations between

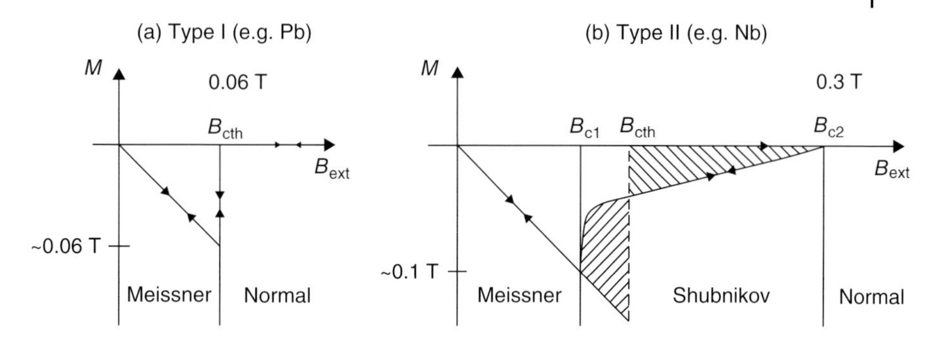

Figure 2.1.1.1 Magnetization curves of Type I and Type II superconductors at 4.2 K (demagnetization factor $n = 0$). (a) Type I expels an external magnetic field completely up to B_{cth}. (b) Type II allows the magnetic flux to penetrate above B_{c1} in the form of quantized flux lines until, at B_{c2}, the cross-section is completely normal. The shaded areas should be equal according to the thermodynamic definition of B_{cth}.

upper, lower, and thermodynamic critical fields and the microscopic properties with, for simplicity, assuming validity of GLAG for the whole temperature range:

$$B_{c2}(T) = \frac{\Phi_0}{2\pi\xi^2(T)} = \sqrt{2} \cdot \kappa(T) \cdot B_{cth}(T) \tag{2.1.1.3}$$

$$B_{c1}(T) \approx \frac{\Phi_0}{4\pi\lambda^2(T)} \cdot \ln\kappa(T) = \frac{\ln\kappa(T)}{\sqrt{2}\kappa(T)} \cdot B_{cth}(T) \quad \text{for} \quad \kappa \gg 1 \tag{2.1.1.4}$$

These equations allow estimating the critical fields with Eq. (2.1.1.4) being less accurate than Eq. (2.1.1.3) because the value of B_{c1} depends on the details of modeling a flux line.

As can be seen, B_{c2} increases linearly with κ, while B_{c1} decreases. The resulting, reversible magnetization curves of Type I and Type II superconductors are compared in Figure 2.1.1.1.

2.1.1.2.2 **Elemental Superconductors and Their Applications**

Among the many metallic elements exhibiting Type I superconductivity those with the highest performance for (T_c, $B_{cth}(0)$) are Pb (7.2 K, 0.08 T), Ta (4.4 K, 0.1 T), Hg (4.2 K, 0.04 T), and Sn (3.7 K, 0.03 T). These parameters are obviously too small for applications with magnetic fields. The three metals exhibiting Type II superconductivity are Nb (9.2 K), Tc (7.8 K), and V (5.3 K). Tc is a radioactively unstable element, such that Nb remains as the favorite metal for applications.

Indeed Nb is used in many applications such as sensor technologies, metrology, digital electronics, and radio frequency (RF) devices. An important large-scale application are RF cavities for particle accelerators.

For high RF performance very pure Nb is needed. This Nb is produced industrially in large quantities by multiple e-beam remelting Nb metal for purification by removing low melting point metals and interstitial gases. The purity of these qualities is characterized by RRR (residual resistivity ratio) values of 250–400.

This high purity Nb exhibits $B_{c1}(0) = 0.18\,\mathrm{T}$, $B_{cth}(0) = 0.2\,\mathrm{T}$, $B_{c2}(0) = 0.4\,\mathrm{T}$, and $\kappa(0) \approx 1.4$.

Nb cavities are usually operated at 2 K. This results in a maximum RF operational field of ~0.15 T, which in turn gives, for usual cavity geometries, an accelerating electrical field of ~40 MV m^{-1}, in the range of presently planned accelerator projects [4].

2.1.1.2.3 The Effect of Alloying

Alloying of metals leads to an increase of κ and therefore enhances B_{c2}. This can be explained on the basis of GLAG in the following way. A further important characteristic length is the electron mean free path l^* of the normal conducting electrons. If $l^* \gg \xi$, the superconductor is called *clean*. If on the other hand, $l^* \ll \xi$, one is in the dirty limit. For alloys, κ can be separated into a clean part κ_{cl} and a dirty part [2, 5]. Accordingly, κ of an alloy is

$$\kappa(T_c) \approx \kappa_{cl} + 2.4 \cdot 10^6 \cdot \rho_n \cdot \sqrt{\gamma} \quad \text{(SI units)} \tag{2.1.1.5}$$

A high normal resistivity ρ_n reflects the influence of decreased l^*. In the dirty limit, $\kappa \gg 1$ and κ_{cl} can be neglected.

According to BCS, κ increases for these materials about linearly when going to lower temperature. By assuming $\kappa(0) \approx 1.2 \cdot \kappa(T_c)$ it follows from Eqs (2.1.1.2) and (2.1.1.3):

$$B_{c2}(0) \approx 3.1 \cdot 10^3 \cdot \rho_n \cdot \gamma \cdot T_c \quad \text{in the dirty limit} \quad \text{(SI units)} \tag{2.1.1.6}$$

A wealth of superconducting alloys is known today. Two Nb-based alloys are showing the best superconductor properties: NbZr with T_c up to ~11 K and NbTi with a typical T_c of ~9.3 K, depending on the exact composition. Despite its higher T_c, NbZr was abandoned mainly because of its inferior characteristics during wire manufacturing. Therefore, we will concentrate in the following on NbTi.

Starting with Nb, T_c remains about constant with increasing Ti content, whereas ρ_n increases significantly. According to measurements, B_{c2} increases up to a Ti content of ~50 wt% and then decreases as T_c decreases (Larbalestier in Foner, [6]).

Today the standard alloy composition as starting material for wire production is NbTi47–48 wt%. The relevant properties are $T_c \approx 9.3\,\mathrm{K}$, $\rho_n \approx 65 \cdot 10^{-8}\,\Omega\mathrm{m}$, and $\gamma \approx 10^3\,\mathrm{J\,m^{-3}\,K^{-2}}$. With these values one gets from Eq. (2.1.1.5) $\kappa \sim 50$. This is in fair agreement with $\lambda \sim 160\,\mathrm{nm}$ and $\xi \sim 4\,\mathrm{nm}$, resulting in $\kappa \sim 40$. From Eqs (2.1.1.6) and (2.1.1.3), we get $B_{c2} \sim 18 - 20\,\mathrm{T}$. As the Pauli paramagnetism of the conduction electrons is fairly strong in NbTi, the energy of the normal conducting ground state and thus the energy difference to the superconducting state is depressed. Therefore, all critical fields, including B_{c2}, are reduced, leading to qualitative agreement with the observed value of about 15 T. Using Eq. (2.1.1.4) B_{c1} can be estimated to be ~0.02 T.

Further alloying of NbTi, for example, with Ta, Ti, or Hf can lead to an enhancement of B_{c2} by ~1 T especially at low temperature of 2 K [7] but this fact has not led so far to a commercial product because of the difficulties of consistent industrial fabrication of the alloys.

2.1.1.3 Superconducting Intermetallic Compounds

With the alloys described, the high-field properties could considerably be enhanced, but T_c remains at the maximum at ~10 K. Fortunately, there exist groups of material with much higher T_c. As we stay in this chapter within the BCS-type superconductors, the most important family of material is made up of the A15-type intermetallic compounds [8]. The highest T_c values among the many A15-type superconductors were found in Nb_3Ge (23 K), Nb_3Ga (20 K), Nb_3Al (19 K), Nb_3Sn (18 K), V_3Si (17 K), and V_3Ga (16 K). T_c is highest near stoichiometry and decreases on both sides.

As can be seen in the Nb–Sn phase diagram (Suenaga in Foner, [6]) of Figure 2.1.1.2, the A15 phase is stable in a composition range between 19 and 25.5 at.% Sn. Measurements on single crystals and bulk materials with different Sn content showed with increasing Sn content an approximately linear increase of T_c from 6 to 18 K with a saturation effect near stoichiometry (Flükiger in

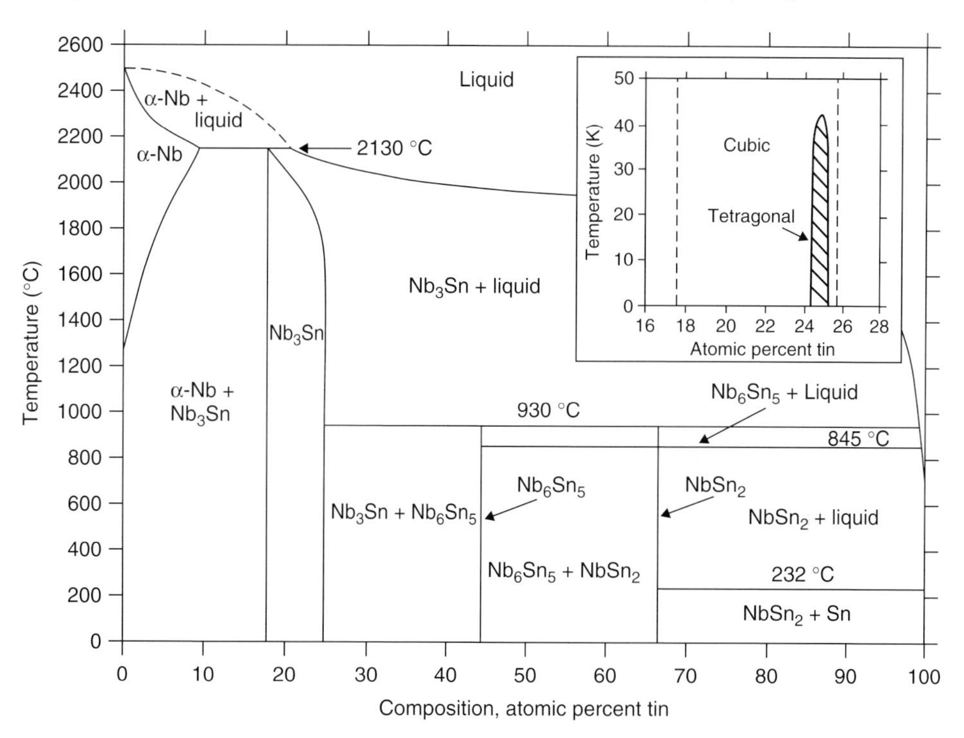

Figure 2.1.1.2 Pure binary phase diagram of Nb–Sn. Besides the A15 phase the two stable line compounds, $NbSn_2$ and Nb_6Sn_5, exist. All three phases form at temperatures below 930 °C. Above 930 °C, Nb_3Sn forms exclusively, but with a too large grain size. High-temperature part from Suenaga in Foner, [6]. Low-temperature insert according to Flükiger in Heller, [9].

Heller, [9], probably a signature of the martensitic transformation seen in the low temperature inset in Figure 2.1.1.2). In parallel, ρ_n decreases by more than an order of magnitude due to ordering effects and γ increases by a factor 2. Also in parallel, B_{c2} increases from about 5 to 30 T followed by a fast drop to 20–25 T near stoichiometry. An explanation of this performance can be as follows: stoichiometric Nb_3Sn is a clean superconductor with intrinsically high $\kappa_{cl}(0) \approx 35$ due to intrinsically small ξ ($\lambda = 124$ nm, $\xi = 3.6$ nm, and $l^* = 10$ nm). From Eq. (2.1.1.3), we estimate $B_{c2}(0) \sim 25$ T for the clean case. By doping, technically preferentially with Ta or Ti, the fast drop in B_{c2} can be prevented by increasing ρ_n again such that $B_{c2}(0) \sim 30$ T as long as T_c remains about constant. The optimum amounts of doping are 1–2 at.% Ti and 3–4 at.% Ta. It has been suggested that the different amounts are due to the fact that Ta occupies Nb sites, whereas Ti occupies Sn sites. An additional beneficial effect of doping is that the spontaneous tetragonal transformation shown in the low temperature phase diagram of Nb_3Sn (Figure 2.1.1.2) is suppressed.

The phase diagrams of the other A15 superconductors look somewhat different. The most similar one concerns V_3Ga and to some extent V_3Si, whereas for Nb_3Ge, Nb_3Ga, and Nb_3Al the stoichiometric composition is not stable at medium temperatures (Flükiger in Foner, [6]). We will come back to this in the manufacturing section.

2.1.1.4 Pinning in Hard Type II Superconductors

An ideal Type II superconductor in a magnetic field subjected to a transport current would immediately show losses because the flux lines would start moving under the Lorentz force exerted and the involved electrical fields would lead to losses. To avoid this, the flux lines have to be pinned. Each discontinuity of the lattice can in principle act as pinning center according to the energy reduction for the flux line residing in it. It is one of the major tasks of conductor manufacturing to modify the microstructure in such a way as to introduce efficient pinning centers during fabrication and herewith to optimize the critical current density.

There are three major areas to be investigated. At first, one has to look at the basic interactive forces between a material defect and a flux line. Secondly, a summation theory must be developed to get the volume pinning force F_p. The summation theories reach from direct summation valid for isolated flux lines to the limited pinning of a stiff flux line lattice in high fields, leading to shearing as the mechanism for de-pinning. Some basic work in the field is given in [10–12], a comprehensive treatment of the topic can be found in [13].

We will concentrate on scaling laws and their practical importance. In accordance with the behavior of many materials, the volume pinning force $F_p(B, T)$ can be written as

$$F_p \equiv J_c \cdot B \approx C \cdot s(\varepsilon) \cdot \frac{B_{c2}(T)^m}{\kappa(T)^n} \cdot g(h) \quad h = \frac{B}{B_{c2}(T)} \tag{2.1.1.7}$$

C is a normalization constant and the term $s(\varepsilon)$ describes the strain sensitivity of the superconductor (see Section 6.2). The magnetic field-dependent factor $g(h)$

can often be written as

$$g(h) = h^p \cdot (1-h)^q \quad \text{with a maximum at } h_{\text{max}} = \frac{p}{(p+q)} \tag{2.1.1.8}$$

In case of ideal scaling, all values of the normalized pinning force $f_p(B, T) = F_p/F_{p\,max}$ collapse onto a single curve. A good scaling for a material is indicative of a single and temperature-independent pinning mechanism. On the other hand, a changing shape or non-scaling with temperature indicates a change of mechanism. The exponents (p, q) are characteristic for different types of pinning centers:

- (1, 1) with $F_{P_{max}}$ @$h = 0.5$: very fine and dense normal conducting precipitations as is valid for optimized NbTi. The maximum pinning force at 5 T is relatively small with ~15 GN m^{-3}. But due to the partial "matching" or "synchronization" between flux line lattice and pinning centers, F_p decreases only linearly for high fields.
- (1/2, 1) with $F_{P_{max}}$ @$h = 0.33$: large precipitations or dislocation networks, as for NbTi at the beginning of the optimization process.
- (1, 2) with $F_{P_{max}}$ @$h = 0.2$ is valid for Nb_3Sn with grain boundaries as the only important pinning centers. For Nb_3Sn, F_p has its maximum also at ~5 T and reaches ~50 GN m^{-3} for bronze route (BR) conductors. The relatively large grain size of ~100 nm leads to partially unpinned flux lines at high field (separation of flux lines at 12 T is only about 12 nm). De-pinning by shearing of the flux line lattice is supposed to be responsible for the quadratic decrease of F_p at high field. This explains also the "saturation effect," that is, reducing the grain size by lower temperature treatment is only effective up to medium fields and not at very high field.

Due to flux pinning, hard superconductors exhibit an irreversible magnetization curve as shown in Figure 2.1.1.3. Its shape is dominated by pinning effects and the vertical full width of the magnetization loop can be expressed as

$$2\Delta M = k \cdot J_c \cdot d_m \tag{2.1.1.9}$$

with d_m the magnetically measured dimension (e.g., diameter) of the sample and k a geometrical factor, reflecting demagnetization effects. In a multifilamentary conductor, the value of d_m may be significantly higher than the nominal geometrical d_f, for example, in case of touching or proximity coupled filaments. Magnetization measurements may be used also to determine the irreversibility field B_{irr} as the field value at which the loop closes due to vanishing pinning which for low temperature superconductor (LTS) is only somewhat smaller than B_{c2}.

The reversible fraction of magnetization (including B_{c1}) is usually masked because of the much larger irreversible part, but starts to become detectable with NbTi/Cu multifilamentary wires in the very low field region. In this field regime also several small-scale effects become important. For the filaments one observes field penetration effects as λ is no more negligible compared to $d_f/2$ and also surface pinning adds to bulk pinning. More importantly, Cu becomes superconducting due to the proximity effect. At 4 K and 0.3 T (a typical accelerator injection field), this starts to be important with Cu at a filament separation

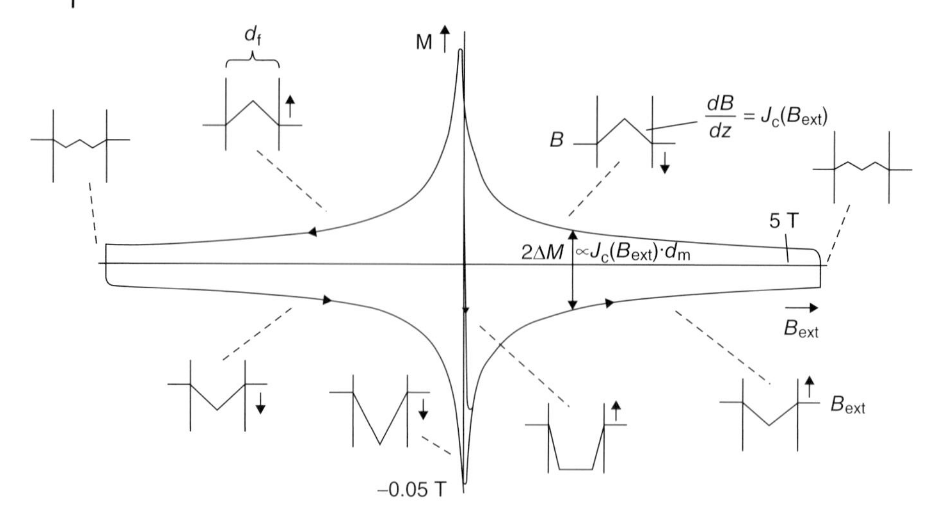

Figure 2.1.1.3 Magnetization of the NbTi volume of a wire with filament diameter $d_f \approx 6\,\mu m$ in a perpendicular field. The vertical width of the loop is proportional to $J_c \cdot d_m$. The field at which the loop closes represents an option to define a practical B_{c2}, or, more exact, the irreversibility field B_{irr}. The small insets show the penetration of B into a filament according to Bean's critical state model.

$s < 0.5\,\mu m$. To allow non-deformed (non-sausaged) filaments after complete area reduction s/d must be ~0.15, corresponding to a filament diameter >3–4 μm. For this reason, a (paramagnetic) CuMn matrix is chosen for fast-cycled fields of accelerators [14]. This reduces the critical coupling distance by a factor of ~3.

2.1.1.5 Design Principles of Technical Conductors

Virtually all practical superconductors based on LTS materials are multifilamentary wires with the superconducting filaments imbedded in a normal conducting matrix. How this comes about will be described in the following.

2.1.1.5.1 Electromagnetic Considerations

A superconductor suited for high-field application consists of a composite wire containing simultaneously normal metal and superconductor material. The behavior of these composites follows of course the Maxwell equations, but is dominated by the material properties of the superconductor. This leads to special design requirements [15].

Reasons for Filamentization If a cylindrical piece of a hard superconductor is exposed to a perpendicular magnetic field, a magnetic energy proportional to $(\Delta M)^2 \propto (J_c \cdot d_m)^2$ is stored due to the screening currents. Any energy input due to a small distortion will increase the temperature and therefore reduce J_c. This changes the flux distribution in the filaments, leading to further temperature increase. The condition that this does not lead to an avalanche, called *flux jump*, is size dependent and can, under adiabatic condition, be written as

$$\mu_0 \cdot (J_c \cdot (B_{op}) \cdot d_m)^2 < 12 \cdot C_v \cdot (T(B_{op}) - T_{op}) \tag{2.1.1.10}$$

The value of the numerical pre-factor can change with the details of the model. The small volumetric specific heat C_v at operational temperature implicates that even a small energy amount can lead to a complete flux jump and finally a breakdown of superconductivity and transition to normal conducting state (quench). This is especially true for NbTi. For Nb_3Sn recovery can take place because of the higher T_c (partial flux jump).

Highest possible J_c of course is the major goal of conductor development and fabrication, such that d_m must be limited to ensure stability. If we take NbTi as an example, with $J_c(4.2\,K, 1\,T) = 10^{10}\,A\,m^{-2}$, this leads to a maximum stable filament diameter $d_m = d_f \sim 50\,\mu m$. For 6 T, the corresponding value is ~200 μm. Conservative magnet designers choose therefore $d_f \approx 30 - 50\,\mu m$, whereas others have demonstrated that values of 100 μm and above can be safe, especially in case of high Cu ratio providing additional dynamic stabilization.

For BR Nb_3Sn, stability was never a problem because the individual filaments are inherently small as the diffusion reaction to form the A15 phase leads to thin layers. d_f is therefore typically 3–5 μm. During reaction treatment, the filament volume is growing, but, if the filament separation is properly chosen, d_m remains of the same order as d_f by avoiding touching and growing together of filaments. This is drastically changed for the new high J_c conductor variants, where the high J_c together with a large magnetic diameter d_m leads to flux jumps. From Eq. (2.1.1.10), a stability limit of ~20 μm at ~2 T follows.

In addition to stability-related limitations, the magnetic field profile to be generated by a transport current gets distorted by the higher order type of magnetic field generated due to the magnetization currents. In the case of NbTi, the technology to make wire with <20 μm filaments is available. For the dipole and quadrupole magnets of Large Hadron Collider (LHC), a filament diameter of 5–7 μm had been chosen.

For the next generation of accelerator magnets, continued Nb_3Sn conductor development is still in progress. The present manufacturing technology is capable of producing conductors with $J_{cnonCu} = 2.5 - 3 \cdot 10^9\,A\,m^{-2}$ and $d_m = 40 - 50\,\mu m$, whereas magnet designers aim for $> 3 \cdot 10^9\,A\,m^{-2}$ and <20 μm, respectively [16].

Even finer filaments are needed if the magnetic fields are to be fast-cycled and the hysteresis loop is passed frequently. The hysteresis losses generated are proportional to the area of the loop and to the frequency. For the NbTi magnets of the LHC upgrade injector and the Heavy Ion Facility FAIR, the design foresees NbTi filaments of ~2.5 μm diameter in a CuMn matrix.

At this point, a historical remark seems to be appropriate. During the 1970/1980s, the feasibility of superconductor equipment for electrical power technology was extensively investigated and technically suitable NbTi conductors were successfully developed [17]. At technical frequencies of 50/60 Hz, the filament diameter for the LTS known then must be in the sub-micrometer range.

As is well known, these development activities were discontinued essentially because of economical reasons.

Multifilamentary Wires A single filament is obviously not suited for application as it can carry only very limited current. Technical conductors consist therefore of many filaments embedded in a normal conducting matrix, preferentially high conductivity copper. Depending on the application requirements (filament diameter and critical current), a wire contains from ~20 to ~100 000 filaments. The function of the matrix is manifold: it allows current redistribution, for example, during current sweep or due to an interrupted filament. It carries away any heat developed and damps magnetic field changes, for example, caused by mechanical energy release or an incipient flux jump. It is beyond the scope of this article to discuss all details here. Only the major implications of the presence of the matrix shall be mentioned here.

First of all, the matrix couples the filaments electromagnetically. A composite with parallel filaments in a changing magnetic field would not work because it would still act more or less as a large filament. Therefore, the composite must be twisted around its own axis resulting in filaments spiraling around the wire axis with a pitch L_p. This partly decouples the filaments by periodically changing the sign of an external field-induced voltage. The twist length cannot be made arbitrarily small due to mechanical reasons, $L_p \geq 5 \cdot D_w$ with D_w the diameter of the filamentary area of the wire. Therefore, a certain coupling remains, leading to coupling currents which partly have to cross the matrix, where these currents decay with a time constant $\tau \propto L_p^2/\rho_\perp$, with $\rho_\perp$ the transverse resistance of the composite. Depending on application requirements, a highly resistive matrix or transverse barriers (CuNi, CuMn for NbTi, or, inherently, CuSn in Nb_3Sn) have to be implemented in the composite.

It is important to realize that twisting does (partly) decouple the filaments in a transverse field but of course not against the self-field of a transport current. Considerations on the stability of the self-field profile follow the same line as above in the subsection on filamentization. In the self-field case, the involved field energy is now $\propto (\lambda \cdot J_c \cdot D_w)^2$ where λ is the filling factor of the superconductor in the matrix. Self-field instability does limit the stable wire diameter for NbTi with high λ to below around 2 mm. It became important also for Nb_3Sn through the emergence of high J_c conductors limiting the stable diameter of these wires to below ~1 mm for accelerator magnet application. This can be different for other magnet configurations.

Quench Protection In addition to dynamic stabilization, the copper in a composite wire also fulfills the function of conductor protection in case that quenching cannot be prevented. As the superconductor material in the normal state exhibits inherently a large resistance and the current flow cannot be interrupted instantly, it would be prone to burn out and the device, for example, the magnet coil would be destroyed. The Cu can for short time carry the current and if, for example, an adequate quench detection system detects the normal conducting zone and triggers a fast discharge of the magnet, it can be protected from damage.

2.1.1.5.2 **Mechanical Properties**

As a consequence of the high operational currents I_{op} and the high magnetic fields B_0 in high-field magnets, significant forces and stresses are generated. For a circular loop with radius R, the tensional hoop stress is $\sigma = J_{op} \cdot B_0 \cdot R$. In some cases, this stress is the limiting factor for J_{op} rather than J_c. Long straight parts on the other hand, as in accelerator magnets, can lead to high compressive stresses that must be tolerated by the conductor.

NbTi is a mechanically friendly material with high yield strength (~600 MPa) and the superconducting properties are not sensitive to stress and strain. By combining in the very final manufacturing step an annealing heat treatment with a small percentage of cold reduction (see below), RRR and strength of the stabilizer Cu can be adjusted, such that the Cu does contribute to the wire strength while still fulfilling its stabilizing function.

In Nb_3Sn wires, the final fabrication step is the reaction heat treatment such that the Cu matrix is fully annealed and soft. The brittle A15 phase in the filaments and the other residual components have to carry the mechanical load. In large magnets, the conductor must therefore be externally or internally reinforced. In addition, the superconducting properties of Nb_3Sn are very strain sensitive. This aspect will be discussed in the section on Nb_3Sn.

2.1.1.5.3 **Co-Workability and Compatibility of Wire Components**

The fabrication of a multifilamentary composite wire requires hot and cold co-working of different materials, limiting the possible material combinations. As an example, high conductivity Al cannot replace Cu as it is too soft and the composite would break apart during area reduction.

Additionally many interface areas are present in a multifilamentary wire. Especially during heat treatment, diffusion takes place at the interfaces, destroying the desired material property of the individual component. Diffusion barriers must therefore be included. Examples are Nb barriers around each filament in fine filament NbTi conductors to prevent formation of TiCu intermetallic particles. These hard particles will not co-deform during subsequent cold-working steps deteriorating the filament quality substantially ("sausaging") below ~15 μm. Other examples are the Ta or Nb barriers in Nb_3Sn conductors to prevent Sn contamination of the stabilizing Cu during final reaction heat treatment.

An important aspect are also differences in thermal contraction properties as they lead to internal strains in the composite introduced due to cooldown to operational temperature. This is especially important for Nb_3Sn because the superconducting properties are highly strain dependent.

2.1.1.5.4 Cost Aspects

Usually, cost of a material is given in cost per weight or cost per volume. This remains the case for the pre-materials used in conductor fabrication. This part of the cost is mostly out of control for the conductor manufacturer.

Another metric is more relevant for an electrical conductor, and especially for superconductors. In this metrics, conductor cost is calculated as the cost per performance, that is, conductor cost per current it can carry over a given length. This is measured, for example, in euro per kiloampere meter.

In these units, Cu costs about 20–30€/kAm. In contrast to this, NbTi conductor cost is ~1€/kAm for an application at 5 T, depending on the complexity of the wire. Nb_3Sn is with ~10€/kAm at 10 T about one order of magnitude more expensive. Both of course must be operated at 4.2 K or close to that, whereas for Cu removing of the ohmic heat is required. Thus, not only investment cost but also operational cost must be taken into account. Simplified, it follows that essentially large magnets or high-field magnets up to ~23.5 T [18] are made from LTS superconductors.

2.1.1.6 Wire Manufacturing Routes and Properties

The fabrication of multifilamentary wires involves hot- and cold-working processes with an unusually high overall area reduction (Hillmann in Foner, [6], Krauth [19]). This leads to demanding requirements on the purity of the used materials and on the cleanliness during all fabrication steps.

Owing to the relatively good compatibility of NbTi and Cu, the manufacturing is relatively straightforward, if certain rules are obeyed. In contrast, all routes for Nb_3Sn conductor production involve even more complex material combinations, leading to more complicated processes and to stringent technical consequences and, finally, to the described higher cost.

2.1.1.6.1 NbTi Wires

NbTi/Cu multifilamentary wires represent the largest application of superconductors. In terms of material quantity, we talk about several hundreds of NbTi per year. All fabrication and quality assurance processes are well established (Hillmann in Foner, [6], Cooley, [7]).

Conductor Fabrication NbTi is virtually a perfect material to produce a multifilamentary superconductor wire. Despite its relatively poor T_c, it exhibits large critical current densities even up to 10 T at 4.2 K. It is perfectly well workable together with the stabilizer material Cu. The most important steps will be described with the help of the phase diagram of the Nb–Ti system shown in Figure 2.1.1.4.

The first step is the melting of the NbTi alloy under vacuum with a nominal composition of 47–48 wt% Ti from electrodes assembled from purified Nb and Ti with well-controlled impurity content. Additional remelting steps are applied to assure sufficient homogeneity. Special care must be applied because of the large separation of solidus and liquidus. Hard particles in the bulk and on the surface must be avoided. The final ingot must exhibit small grain size and hardness to ensure

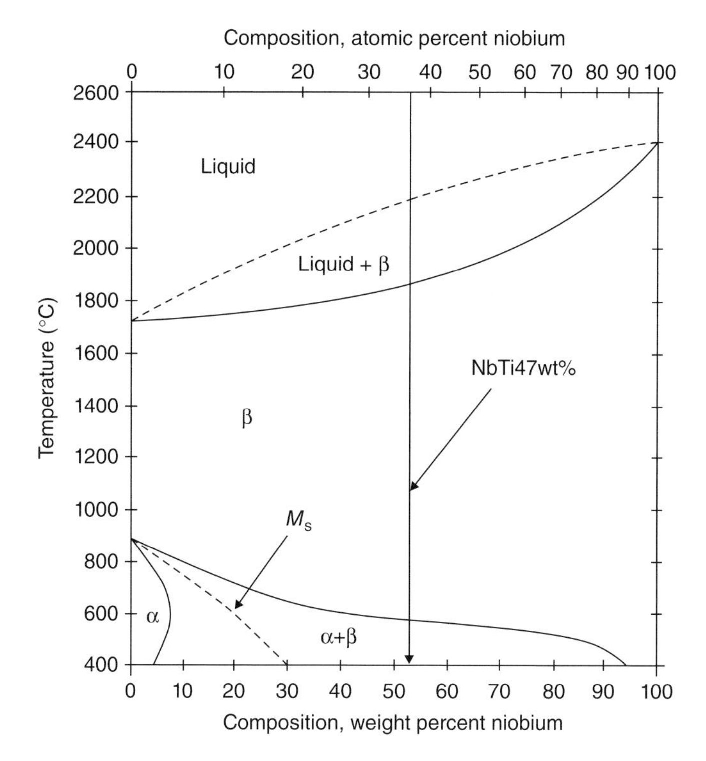

Figure 2.1.1.4 Phase diagram of NbTi [7] showing the large separation of liquidus and solidus, the high temperature β phase, and the two-phase α + β region.

good workability. The ingot is worked down to a size needed for billet assembly. At some intermediate step, a so-called *β-quench* is performed. This step includes heating, for example, to 800 °C and fast cooling to room temperature to freeze the superconducting β-phase.

Conductor production starts with the assembly of a composite billet (typically 250–350 mm diameter) from NbTi and Cu. The design of the final wire determines how this billet looks like. One conductor route starts with a monofilament billet consisting of a NbTi core, usually a Nb sheet wrapped around the NbTi bar acting as a diffusion barrier against formation of hard Cu–Ti intermetallic particles, and an outer Cu can. After warm extrusion the monofilament rod is cold drawn, worked into hexagonal shape, and cut into length. The hexagonal rods are stacked into a Cu can to form a (first) multifilamentary billet.

An alternative route for the multifilamentary billet is gun drilling of holes into a Cu bar and filling of the holes with NbTi rods. A third method is putting round NbTi rods into hexagonal Cu tubes with round hole and assembly in a Cu can.

The applied method is chosen according to technical feasibility and cost. When applying the first route, billets with up to several 1000 filaments can be manufactured. After again warm extrusion an additional restacking and extrusion process can follow, especially if more than 10 000 filaments are needed.

The last extrusion is performed at a temperature as low as allowed by press capacity and by the need of bond formation. Low temperature is desirable to avoid complete recrystallization and to partly retain the cold-working microstructure.

The most efficient pinning centers in NbTi are deformed α-Ti precipitations with small dimensions and high density. The precipitations are generated by heat treatments in the two-phase region typically between 380 and 420 °C. During wire drawing, the α-Ti precipitations are also being deformed. Thus, wire drawing includes a thermomechanical treatment adjusted to optimize J_c under operational conditions. A typical sequence of steps is as follows:

- Pre-strain of $\varepsilon_p \sim 5$ to get the preferred precipitation morphology,
- Three to six heat treatments with intermediate strains $\varepsilon_i \sim 1.0 - 1.5$, and
- final strain of $\varepsilon_f \sim 4 - 5$.

The logarithmic area reduction ε is given by $\varepsilon = 2 \ln d_1/d_2$. During final processing, the pinning force dependency on reduced field h is changing shape. Initially, the α-precipitations are relatively large with a size of 100–200 nm and the pinning force has its maximum near $h \approx 0.3$. With increasing strain, F_p increases for all fields and the maximum shifts to $h \approx 0.5$. Above the optimum ε_f, F_p decreases again as the effectiveness of the pinning centers reduces.

The maximum of F_p increases linearly with the volumetric density of α-precipitations. At about 20 vol%, one reaches $F_p \approx 15\,\text{GN}\,\text{m}^{-3}$ at ~5 T corresponding to $J_c = 3000\,\text{A}\,\text{mm}^{-2}$ at 5 T, a typical value of state-of-the-art commercial wires. Microstructural investigations have shown that in the optimized state the precipitations represent a high density of pinning centers and have taken the form of curled ribbons with a thickness of 1–2 nm and a separation of 5–10 nm. This compares with a coherence length of 4 nm and a flux line separation of 22 nm at 5 T.

The high volume percent of α-Ti precipitate reduces the Ti content of the β-NbTi matrix and therefore the basic superconducting properties T_c and B_{c2} of the superconductor somewhat with respect to the properties of the starting alloy. But usually this is not explicitly taken into account.

Conductor Examples and Properties The design of multifilamentary conductors depends strongly on the application. A few cross-sections of wires for different applications are shown in Figure 2.1.1.5.

The NbTi critical current density of typical conductors at 4.2 K is shown in Figure 2.1.1.6. The fast drop of J_c above 10 T limits the operating field to ~10 T. Cooling down to 2 K shifts the high-field part by about 3 T, thus increasing the accessible field appreciably.

Also shown in the figure is the non-Cu critical current density of BR Nb_3Sn conductors (see below).

J_c Scaling Law Scaling laws for J_c (often expressed in terms of F_p) for wires are intended to describe the critical surface formed by T_c, B_{c2}, J_c, and ε. Starting

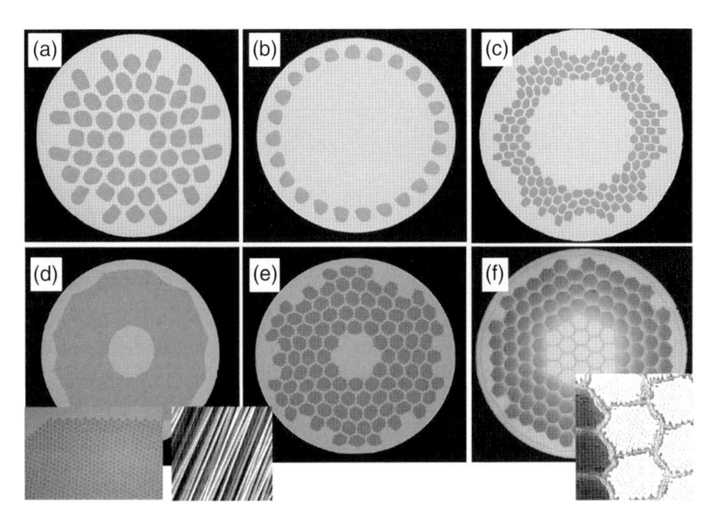

Figure 2.1.1.5 Cross-section of multifilamentary NbTi wires with different Cu/NbTi area values α. Standard wires: (a) for NMR ($\alpha = 1.3$, $D_w \approx 0.5-1$ mm), and (b) for MRI ($\alpha = 6.5$, $D_w \approx 1-2$ mm). Custom-made strands: (c) strand for W7X CiCC ($\alpha = 2.6$, $D_w = 0.57$ mm), (d) strand for LHC ($\alpha = 1.25$, $D_w = 0.74$ mm) with 6360fil à 6 μm (see insets), (e) double-stack strand for LHC ($\alpha = 1.65$, $D_w = 1.065$ mm) with 8670fil à 7 μm, (f) mixed Cu/CuNi matrix strand for pulsed-field coils ($D_w \sim 1$ mm, $d_f \sim 10$ μm, $\tau \sim 100$ μs). The close-up shows some of the CuNi barriers.

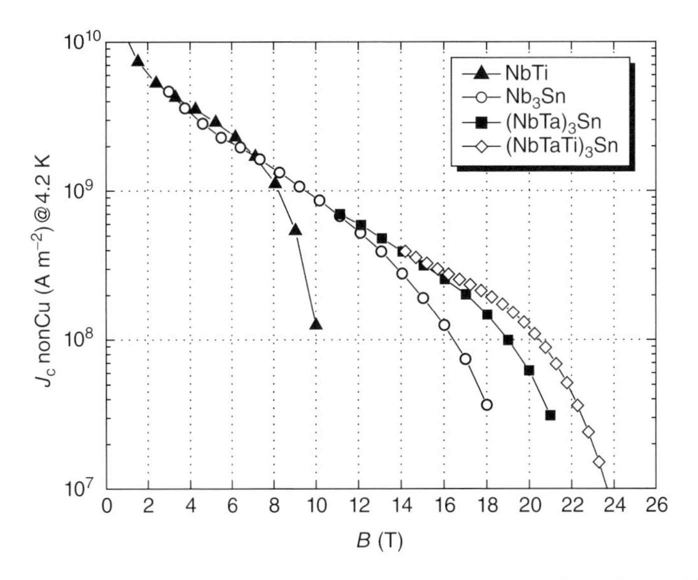

Figure 2.1.1.6 Non-Cu critical current density up to date NbTi and typical bronze route Nb_3Sn wires at 4.2 K.

with measurements at, for example, 4.2 K they shall allow estimating J_c at conditions not easily accessible for experiments. Of course, scaling depends on the manufacturing process and is therefore not universal.

The general form of a scaling law is given by Eq. (2.1.1.7). The term $s(\varepsilon)$ can be neglected for NbTi. A simplified formulation was proposed in [20]

$$F_p(B,T) \equiv J_c(B,T) \cdot B = C_{NbTi} \cdot (1 - t^{1.7})^n \cdot h^p \cdot (1-h)^q \tag{2.1.1.11}$$

In comparison with Eq. (2.1.1.7) the term with the explicit temperature dependence of κ is omitted and an empirically determined relation $B_{c2} = B_{c2}(0) \cdot (1 - t^{1.7})$ was chosen for $B_{c2}(T)$. Fitting parameters are n, p, q, and C_{NbTi}. Instead of B_{c2} also B_{irr} may be chosen.

Comparison of this law with experimental results for nominally optimized wires from different sources and different Ti content showed the following. The maximum of F_p always occurs at $h \sim 0.4–0.5$. But the fitting parameters deviate from the theoretical ones and show substantial scatter with n between 1.8 and 2.3, p between 0.6 and 0.9, and q between 0.8 and 1.1. This demonstrates that $J_c(B,T)$ can reasonably well be described by a scaling law of type (Eq. (2.1.1.11)), but that, not unexpectedly, alloy composition and process-dependent variations of the scaling parameters over a relatively wide range exist.

2.1.1.6.2 Nb_3Sn

A15 materials are very brittle and conductors cannot be manufactured in a way as described for NbTi. Initially, tape-type conductors were produced by coating of wide thin substrates, as is now the technique for some high temperature superconductors (HTSs). Magnets built with these tapes exhibited severe flux jumps and quenches, especially in a magnetic field with components perpendicular to the broad face of the tape, as expected.

Multifilamentary wire production requires availability of suitable precursor materials allowing A15 formation by a diffusion reaction. With Nb–Sn, the formation of exclusively the A15 phase is made possible by the presence of Cu. Without Cu other stable phases ($NbSn_2$, Nb_6Sn_5) according to Figure 2.1.1.2 would also form. Cu additions destabilize these undesired phases and allow the direct diffusion path as shown in Figure 2.1.1.7 at 700 °C or even below (Suenaga in [6]).

For V_3Ga, a similar process was developed using the V–Cu–Ga system. But, in addition to the somewhat inferior superconductor properties, Ga is too expensive for commercial conductors. In case of Nb_3Ge, Nb_3Ga, and Nb_3Al, the stoichiometric composition is not included in the A15 phase field at equilibrium. In case of Nb_3Al, a nonequilibrium process including a fast-quenching process has therefore been developed to form near-stoichiometric A15. But a corresponding industrial process is not yet implemented.

Production Routes Several production routes have initially been tried out, including many modifications [21]. We concentrate here on the industrially implemented ones [9]: BR, the internal tin process (IT), and a powder-in-tube method (PiT) [22]. In the first two routes, Nb is added to the billets as rods, and

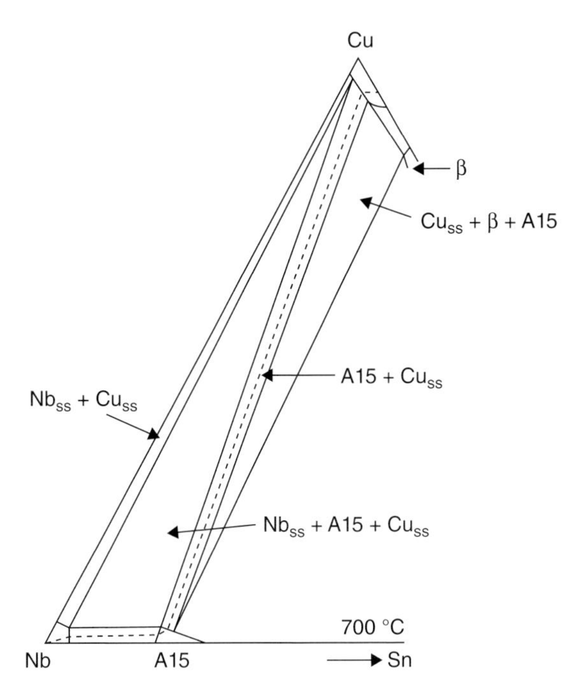

Figure 2.1.1.7 Ternary phase diagram of the Nb–Sn–Cu system at 700 °C (Suenaga in Foner, [6]) showing the tie line allowing the formation of the A15 phase by a diffusion reaction process from CuSn.

in PiT as tubes. Ternary alloys are produced typically by using an NbTa7.5 wt% alloy, or by adding Ti in the Sn source as Ti co-diffuses with Sn.

In BR, the Sn source is the CuSn matrix in which the filaments are imbedded. Unfortunately only about 15–16 wt% Sn dissolves in Cu. As we will see this limits the achievable critical current density, but has also several beneficial effects. In IT, the filaments are imbedded in a Cu-matrix with distributed sources of high Sn content. During reaction heat treatment first a bronze matrix is formed and then the Sn is transported from the source through the matrix to the filaments.

The slow process of A15 layer formation in these processes requires that the final filament diameter is about ~3–5 μm. As a consequence, a large number of filaments are needed, such that a double stacking technique is applied. In both routes, the first stacking billet is warm extruded. For IT, the second stacking contains sub-elements with the inserted low-melting Sn source, and can therefore only be cold-worked. This results in delicate working processes because of the much different hardness of the composite components in connection with limited metallic bonding. In case of BR also the final billet is a full size (>100 kg) warm extrusion billet. This allows production of long conductors as needed, for example, for NMR magnets.

In both cases, the Cu for stabilization and quench protection is included as an outer Cu shell. A diffusion barrier is added between the multifilamentary area and

Cu to protect the high conductivity Cu from contamination during reaction heat treatment. Ta may be preferential because it is perfectly inert and has no effect on the magnetization properties of the wire, while Nb will react to a large hollow filament increasing magnetization and therefore alternating current (AC) losses. However, this needs to be compared with cost considerations for materials and compatibility with manufacturing process.

In the tube type processes Nb (alloy) tubes are filled either with a ductile high Sn alloy or with a powder containing high Sn components. Here, we concentrate on the process based on a $NbSn_2$ powder source with Cu additions. In this process, A15 forms much faster than in above-described processes via bronze [22]. Therefore, a layer thickness of 10 μm can be achieved.

Common to all production routes the A15 phase is formed at final diameter of the wire in the temperature range of 650 to ~700 °C by a diffusion/reaction process. From this, it follows that in principle all compositions between 19 and 25.5 at.% are present. Accordingly, T_c and B_{c2} show a process-dependent distribution of significant width. Also the grain size (50 – 150 nm) and morphology (globular or columnar), both determining J_c, vary across the A15 layer. All superconducting properties are appropriately averaged values.

As a consequence of the complex configuration of Nb_3Sn wires we have to distinguish between the physically relevant (average) critical current density in the A15 layer J_{cA15} and the application-relevant current density J_{cnonCu}, averaged over the total wire area except the stabilizer Cu (and of course the overall or engineering wire critical current density J_{ceng}).

Wires with Standard Critical Current Density BR conductors represent the standard for applications such as NMR, because of the high performance with respect to filament quality, vanishing residual resistance along their length and in superconducting joints, properties relevant for persistent mode operation. In addition, they can be produced from extrusion billets in long unit lengths and with large cross-sections, because they show excellent warm and cold-working properties.

The large area fraction of bronze needed to provide enough Sn for Nb_3Sn formation limits the non-Cu critical current density, but is beneficial for keeping the filaments separated even after reaction, such that magnetically effective filament diameters remain at ~5 μm. Figure 2.1.1.8 shows cross-sections of NMR-type BR conductors and Figure 2.1.1.6 shows their typical J_{cnonCu}. Ternary compounds $(NbTa)_3Sn$ or $(NbTi)_3Sn$ (not shown in Figure 2.1.1.6) are now standard, whereas the binary compound is obsolete. The observed increase of J_c in the quaternary alloy is presumed to originate from speeding up the diffusion in Ta-doped compounds and therefore from the increase of the percentage of high Sn content A15.

For the nuclear fusion project ITER the coils are pulsed or are subjected to pulsed fields such that AC loss specifications have to be fulfilled. This is automatically the case for BR.

In case of IT wires, the design flexibility allows increasing the filamentary area such that the specifications can also be fulfilled.

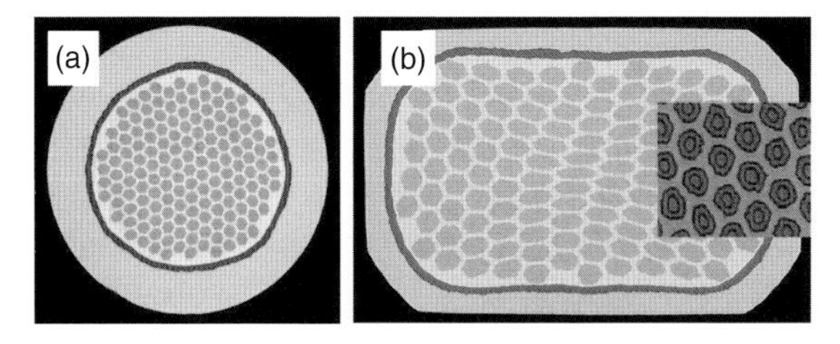

Figure 2.1.1.8 BR Nb_3Sn wires for NMR with Ta barrier: (a) round cross-section and (b) rectangular cross-section. The inset shows partially reacted filaments (~4 μm).

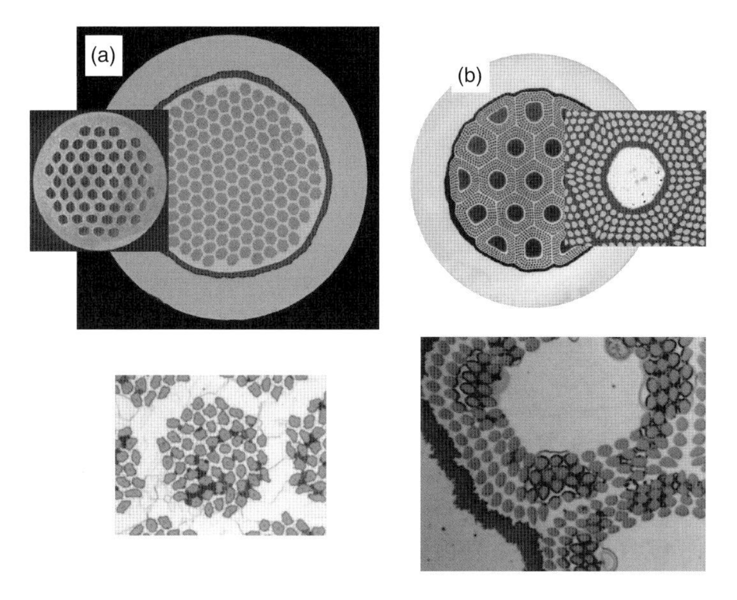

Figure 2.1.1.9 Low ac loss Nb_3Sn strands for ITER. (a) BR wire cross-section with inset showing first stack extrusion. Bottom: reacted filaments. (b) IT wire cross-section with close-up of a first stack extrusion rod. Bottom: reacted filaments. (Pictures courtesy Mark Glajchen, Oxford Superconducting Technology, OST.)

In order to get the needed Nb_3Sn strand quantities and to include the worldwide manufacturing capabilities both, BR and IT, strands are used in the ITER magnets. Figure 2.1.1.9 shows examples of ITER BR and IT strands.

Wires with High Critical Current Density Some applications, such as accelerator magnets for High Energy Physics require a much higher current density than can be delivered by the above conductor types. At present, the current densities of the level required by this application can be achieved by two routes.

The first route is an IT process with minimized Cu amount in the filamentary area, acting as diffusion channels for the Sn. But during reaction the individual filaments of a sub-element are growing together and act magnetically as a single

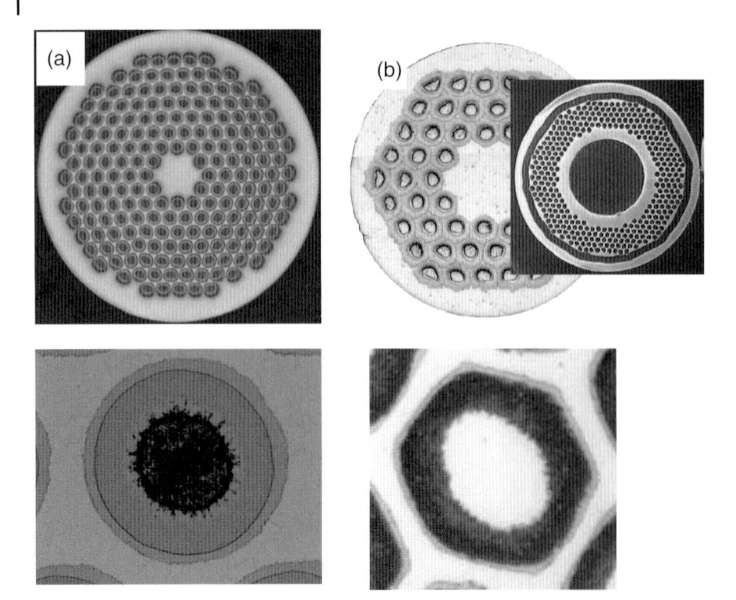

Figure 2.1.1.10 High J_c Nb_3Sn for high-field accelerator magnets. (a) PiT wire cross-section. Bottom: reacted tubular filament with residual Nb. (b) IT ("RRP", restacked rod process) wire cross-section with extruded sub-element. Bottom: reacted sub-element. (Pictures courtesy Mark Glajchen, Oxford Superconducting Technology, OST.)

hollow filament similar to PiT. Each sub-element is surrounded by a diffusion barrier (typically Nb) to protect the Cu in between from poisoning with Sn.

The other type of wires is produced by the PiT process described above. The Nb tubes filled with $NbSn_2$ correspond to an IT sub-element and the final reacted strands look to some extent very similar. Figure 2.1.1.10 shows wire cross-sections of typical high J_c Nb_3Sn wires.

In both cases, typical reaction heat treatment temperatures are between 660 and 680 °C for 50–100 h.

Achieved J_c values are $J_{\mathrm{cnonCu}} \approx 2.5 - 3 \cdot 10^9\,\mathrm{A\,m^{-2}}$ at 4.2 K and 12 T. In Figure 2.1.1.11, this is shown in comparison with BR conductors. The factor ~4 seen there has two origins, each accounting for about half of the effect: first, the intrinsic critical current density is improved by the larger amount of near-stoichiometric A15 and, second, the loss in "real estate" due the matrix is about halved.

In these wires, stability against both flux jump and self-field effects has drastically suffered. Conductors with $d_m \sim 50\,\mu\mathrm{m}$ show flux jump activity for $B \leq 1\,\mathrm{T}$. In medium field, self-field instabilities are observed if $D_w \geq 1\,\mathrm{mm}$.

The ultimate design and manufacturing limits for both routes still have to be explored. IT seems to offer somewhat higher J_c, whereas PiT presently allows smaller filament diameters and better control of high residual resistivity of the surrounding Cu by avoiding Sn penetration through the barrier. This is important to support dynamic stabilization.

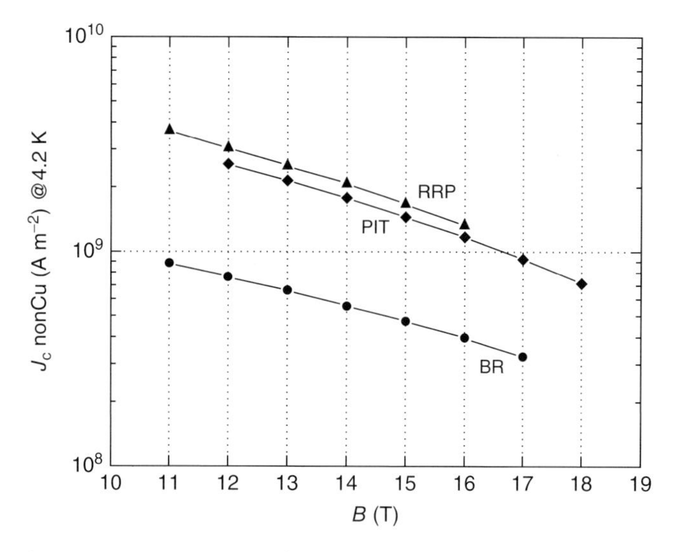

Figure 2.1.1.11 Non-Cu J_c of high J_c strands at 4.2 K, PiT (diamonds) and RRP (triangles), in comparison with bronze route (BR) (dots).

Scaling Laws Including Strain Sensitivity Other than in NbTi, for A15 superconductors all technically important parameters T_c, B_{c2}, and J_c are strain dependent. In addition, the A15 phase is very brittle and is sensitive to crack formation. Strain sensitivity must therefore be included in the scaling laws for J_c. The sensitivity is very high for Nb_3Sn compared with, for example, Nb_3Al. Therefore, there remains a strong incentive for trying to produce Nb_3Al wires despite the involved difficulties (Takeuchi in Heller, [9]).

The stress sensitivity was shown to originate from softening of the phonon spectrum and the related change of superconducting properties. The deviatoric (i.e., non-hydrostatic) strain contribution turned out to be the most relevant part.

The sensitivity function $s(\varepsilon)$ reads:

$$T_c(\varepsilon) = T_{cm}(0) \cdot s(\varepsilon)^{\frac{1}{3}} \tag{2.1.1.12}$$

This definition of $s(\varepsilon)$ was chosen in such a way that the practically more important behavior of B_{c2} can be written as

$$B_{c2}(T, \varepsilon) = B_{c2}(T, 0) \cdot s(\varepsilon) \tag{2.1.1.13}$$

For the case pf axial loading a practical formulation was developed by Ekin, for example, in Foner, [6]

$$s(\varepsilon_{ax}) = 1 - a \cdot |\varepsilon_{ax}|^u = 1 - a \cdot |\varepsilon_{app} - \varepsilon_m|^u \tag{2.1.1.14}$$

For Nb_3Sn, Ekin determined, for example, $u = 1.7$ and $a = 900$ in compression and $a = 1250$ in tension. The actual law depends on production route and detailed conductor design and must be determined experimentally.

In a composite wire without externally applied strain ε_{app}, the A15 phase is in compressive pre-strain, typically $\varepsilon_m \approx -0.15\%$ to -0.3%, because of the differential contraction of the wire components during cooling from reaction to operation temperature. After applying an opposite tensional strain, the resulting total strain is zero and $s(\varepsilon)$ has its maximum.

The additional mechanical brittleness of the A15 phase is responsible for the fact that a tensional strain above 0.3–0.5% leads to an irreversible degradation by crack formation, depending on conductor type. Strain sensitivity and especially fracture sensitivity seems to be increased in the high current density material.

Several unified scaling laws for J_c and F_p according to Eq. (2.1.1.7) were proposed and tested for validity. Here, we will present a relatively recent relationship [23]:

$$F_p(B,T,\varepsilon) = J_c(B,T,\varepsilon) \cdot B = C_{NbSn} \cdot s(\varepsilon) \cdot (1 - t^{1.52}) \cdot (1 - t^2) \cdot h^{0.5} \cdot (1 - h)^2 \tag{2.1.1.15}$$

In this relationship $h = B/\{B_{c2m}(0) \cdot s(\varepsilon) \cdot (1 - t^{1.52})\}$ and $t = T/\{T_{cm} \cdot s(\varepsilon)^{1/3}\}$.

The normalizing parameter C_{NbSn} and the bulk averages $B_{c2m}(0)$ and $T_{cm}(0)$, $p(\sim 0.5)$ and $q(\sim 2)$, and especially the strain dependencies have to be determined experimentally. Besides axial stress also transverse stresses reduce J_c and must be regarded, especially, for example, in accelerator magnets.

Also the fracture sensitivity must be carefully taken into account during all design considerations to avoid irreversible degradation due to mechanical overstressing during coil fabrication and operation. The existence and high performance of hundreds of persistent mode NMR magnets using BR conductors proves that both, strain and fracture sensitivity, can be managed successfully. It is still a big challenge to demonstrate this also for high-field accelerator magnets and especially for large fusion magnets using cable-in-conduit conductors (CiCCs).

2.1.1.7 **Built-Up and Cabled Conductors**

The size and herewith the current-carrying capacity of single wire is limited for several reasons. As described, self-field related instability is a physics-based limiting factor for the outer diameter of the filamentary area depending on the current density λJ_c. This results in a maximum diameter of a few millimeter depending on filling factor λ (or equivalently Cu : NbTi ratio α) for NbTi wire and <1 mm for the high current density Nb_3Sn strands to be used in accelerator magnets.

For NbTi, larger diameters than 2 mm start to lead to decreasing critical current density because of limited space for cold-work during the optimization process.

Finally, of course the material output is limited by the initial billet size. This varies, in terms of mass, from 250–300 kg for NbTi to 30–60 kg for the high current Nb_3Sn variants.

2.1.1.7.1 **Wire-in-Channel (WiC)**

The wire-in-channel (WiC) technique (Figure 2.1.1.12) is applied for NbTi conductors with very high Cu content larger than about 6, for example, for magnetic resonance imaging (MRI). In this design, a low Cu content round multifilamentary wire is soldered into a groove of a Cu profile. One advantage compared to an

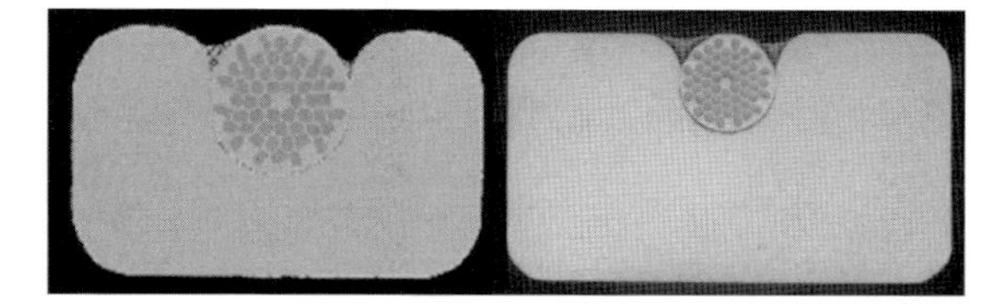

Figure 2.1.1.12 Two wire-in-channel conductors for MRI with different overall Cu content. (Typical dimensions 1.1 mm × 1.7 mm to 2.2 mm × 4.2 mm and $\alpha \approx 6-20$.)

equivalent monolith wire is the small diameter of the current-carrying part with the possibility to apply large total strain and to fully optimize the critical current density. Very long unit length of 30–50 km can be produced, depending on wire diameter. In addition, the Cu profile can be separately produced with a cheaper process than using the superconductor process.

2.1.1.7.2 Cabled Conductors

The conductors described so far can carry currents of up to several 100 A. For critical currents in the kiloampere to multi-kiloampere range, conductors cabled from many wires ("strands") are needed. In order to enforce equal current distribution, the cables must be fully transposed. In contrast to twisting, where the filaments remain on the same radius, with transposition each strand takes each radial location of the cross-section within a transposition length. Examples are flat or keystoned Rutherford cables for accelerator magnets, Rutherford cables co-extruded with high purity Al for particle detector magnets, and internally cooled cabled conductors (CiCC) for nuclear fusion. These conductors will be described in the respective application chapters.

2.1.1.8 Concluding Remarks

More than 100 years after their discovery the low T_c superconducting elements, alloys and intermetallic compounds still play a major role in the applied superconductivity. The metal Nb allows use in accelerator RF cavities with acceleration voltages of ~40 MV m^{-1}. The alloy NbTi is the workhorse of magnet technology up to fields of ~10 T, whereas the intermetallic compound Nb_3Sn extends the field up to beyond 20 T. Today many thousands of NbTi-based whole body MRI systems are installed worldwide, as well as high-resolution spectroscopic NMR systems using Nb_3Sn coils with fields up to 23.5 T [18]. Only with these materials large accelerators and storage rings, and nuclear fusion systems have become possible.

Acknowledgments

Finally, the authors would like to thank their colleagues V. Abaecherli, B. Sailer, M. Thoener, and M. Wanior during preparation of this manuscript.

References

1. Tinkham, M. (1996) *Introduction to Superconductivity*, Dover Publications, New York.
2. de Gennes, P.G. (1999) *Superconductivity of Metals and Alloys*, Westview Press.
3. Buckel, W. and Kleiner, R. (2004) *Superconductivity: Fundamentals and Applications*, Wiley-VCH Verlag GmbH, Weinheim.
4. Saito, K. (2003) Theoretical Critical Field in RF Application, *https//accelconf.web.cern.ch/accelconf/srf2003/papers/moo02.pdf* (accessed 20 May 2013).
5. Goodman, B.B. (1966) Type II superconductors. *Rep. Prog. Phys.*, **39** (2), 445.
6. Foner, S. and Schwartz, B.B. (eds) (1981) *Superconductor Material Science: Metallurgy, Fabrication and Applications*, Plenum Press, New York and London.
7. Cooley, L., Lee, P., and Larbalestier, D. (2001) Conductor processing of low Tc materials: the alloy Nb-Ti, in *Handbook of Superconducting Materials*, vol. **1** (eds D.A. Cardwell and D.S. Ginley), IOP Publishing, Bristol.
8. Dew Hughes, D. (1975) Superconducting A-15 compounds: a review. *Cryogenics*, **15** (8), 435.
9. Heller, R. (ed) (guest ed.) (2008) Low-Tc superconducting materials. *Cryogenics*, **48**, 7–8.
10. Ullmaier, H. (1975) *Irreversible Properties of Type II Superconductors*, Springer, Berlin.
11. Campbell, A.M. and Evetts, J.E. (1972) Flux vortices and transport currents in Type II superconductors. *Adv. Phys.*, **21** (90), 199.
12. Kramer, E.J. (1975) Microstructure-critical current relationships in hard superconductors. *J. Electron. Mater.*, **4** (5), 839.
13. Matsushita, T. (2010) *Flux Pinning in Superconductors*, Springer, Berlin.
14. Collings, E.W. *et al.* (1990) Magnetic studies of proximity-effect coupling in a very closely spaced fine-filament NbTi/CuMn composite superconductor. *Adv. Cryog. Eng. Mater*, **36**, 231.
15. Wilson, M.N. (1983) *Superconducting Magnets*, Clarendon Press, Oxford.
16. Bottura, L. and Godeke, A. (2012) Superconducting materials and conductors: fabrication and limiting parameters. *Rev. Accel Sci. Technol.*, **05**, 25.
17. Dubots, P. *et al.* (1985) NbTi wires with ultra-fine filaments for 50–60Hz use: influence of the filament diameter upon losses. *IEEE Trans. Magn.*, **21**, 177.
18. Roth, G. (2012) in *Superconducting Magnets for NMR in 100 Years of Superconductivity* (eds H. Rogalla and P.H. Kes), CRC Press, Boca Raton, FL.
19. Krauth, H. (1998) Conductors for DC applications, in *Handbook of Applied Superconductivity* (ed B. Seeber), CRC Press, Boca Raton, FL.
20. Bottura, L. (2000) A practical fit for the critical surface of NbTi. *IEEE Trans. Appl. Supercond.*, **10**, 1054.
21. Suenaga, M. and Clark, A.F. (eds) (1980) *Filamentary A15 Superconductors*, Plenum Press, New York and London.
22. Veringa, H. *et al.* (1983) Growth kinetics and characterization of superconducting properties of multifilament materials made by the ECN powder method. *IEEE Trans. Magn.*, **19**, 773.
23. Godeke, A. *et al.* (2006) A general scaling relation for the critical current density. *Supercond. Sci. Technol.*, **19**, R100.

2.1.2
Magnesium Diboride

Davide Nardelli, Ilaria Pallecchi, and Matteo Tropeano

2.1.2.1 Introduction

Magnesium diboride MgB_2, is a simple binary compound that is well known since the 1950s [1], however its high-T_c superconductivity had been hidden until its discovery in 2001 (see Figure 2.1.2.1) [2]. At that time, several groups around the world were searching for ternary or quaternary compounds that were rich in light elements (Li, B, C, Mg) and Jun Akimitsu's group at Aoyama Gakuin University in Japan was exploring the Ti–B–Mg ternary phase diagram [3]. In early January 2001, the group reported their results at a meeting in Sendai, Japan. The first paper was published in March 2001, setting off a rush of experimental and theoretical work to confirm and explore that remarkably high T_c value for a binary intermetallic compound.

In few months, four key papers were published, describing crucial properties and pointing to the unique potential of MgB_2 for the applications, starting of course from its very simple structure, made of inexpensive elements, and its relatively high T_c: (i) the remarkable absence of weak-links [5], which means no need of high-degree texturing to transport high superconducting current; (ii) the possibility of enhancing the critical current density and the irreversibility field in thin films [6]; (iii) the enhancement of the high magnetic field critical current

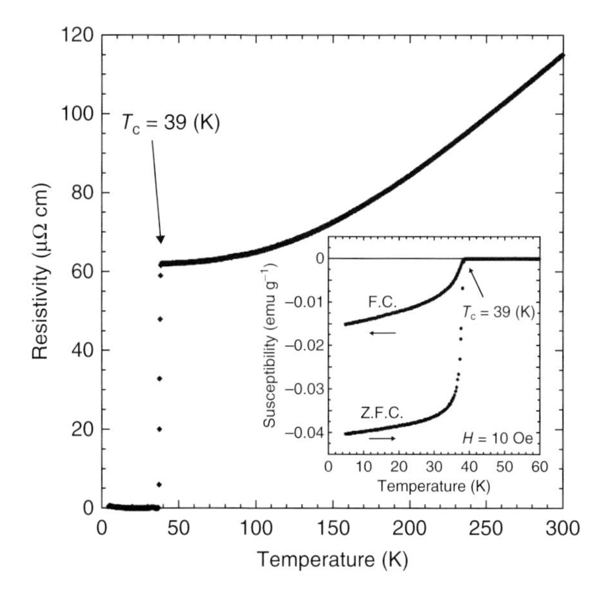

Figure 2.1.2.1 Temperature dependence of resistivity (main panel) and magnetic susceptibility (inset), as first reported by Jun Akimitsu's group [2]. (Figure reproduced with permission from Ref. [4] (Copyright 2011 by Oldenbourg Verlag).)

density by proton irradiation [7]; (iv) the possibility of making superconducting wires by the simple powder-in-tube (PIT) technique [8].

To date, intensive research has clarified all the fundamental physical mechanisms that determine MgB_2 superconducting and normal state behavior. Its peculiarity is mainly due to the presence of two bands and two energy gaps, which influence its properties. Moreover, most of MgB_2 properties depend sensitively on intraband and interband scattering rates, so that in principle they can be tailored at will by suitable chemical substitutions and irradiation.

Excellent reviews on MgB_2 have been published so far. The first review paper appeared as early as in 2001 [9], whereas other more recent papers have focused on general aspects [4, 10, 11], theoretical perspectives [12], critical current density [13], thin films [14], and wires and tapes [15–17]. In the following, a general overview on MgB_2 is presented, with main focus on the properties that are most relevant for current transport applications and hints also to electronics applications, which are remarkable as well, but addressed in other sections of this book. In Section 2.1.2.2, we describe intrinsic and extrinsic properties of MgB_2; among the former ones, we briefly review band structure, phonon spectrum, critical fields, and characteristic length, and among the latter ones, the effects of doping and impurities, and the critical current density in polycrystalline sample. In Section 2.1.2.3, we give some information about the phase diagram and the thermodynamic aspects of MgB_2 reaction synthesis and we describe the most used preparation methods for bulk, single crystal, and thin film MgB_2 samples. In Section 2.1.2.4, the main applications of MgB_2 are reviewed, namely wires and tapes, coated conductors, and electronics applications such as superconducting quantum interference devices (SQUIDs) and Josephson junctions. Finally, we draw an outlook on the perspectives of MgB_2 and on the edge of improvement of the performances of MgB_2-based systems.

2.1.2.2 Intrinsic and Extrinsic Properties of MgB_2

Magnesium diboride MgB_2 is a binary compound with very simple AlB_2-type *crystal structure*, characterized by lattice parameters $a = 3.08$ Å and $c = 3.51$ Å and described by the space group $P6/mmm$. In this structure, shown in Figure 2.1.2.2a, the B layers are alternately stacked with Mg layers. The B layers form a two-dimensional honeycomb lattice, where each B atom is surrounded by three other B atoms, forming an equilateral triangle with B–B distance equal to 1.78 Å. The Mg layers form a triangular lattice, with Mg–Mg distance just equal to the lattice parameter a. Each Mg atom is at the center of a hexagonal prism of B atoms, with B–Mg distance of 2.5 Å. This AlB_2-type structure has been observed to be unaffected by variations of temperature and applied external pressure.

The *band structure* of MgB_2 [19] is described by two σ bands formed by bonding $p_{x,y}$ orbitals and two π bands, formed by bonding and antibonding hybridized p_z orbitals. The two σ bands, corresponding to bonding orbital of the B sublattice, are of hole-type. The σ holes along the ΓA direction of the Brillouin zone are localized within the B planes and strongly exhibit two-dimensional character. Conversely, the two π bands have three-dimensional character and are

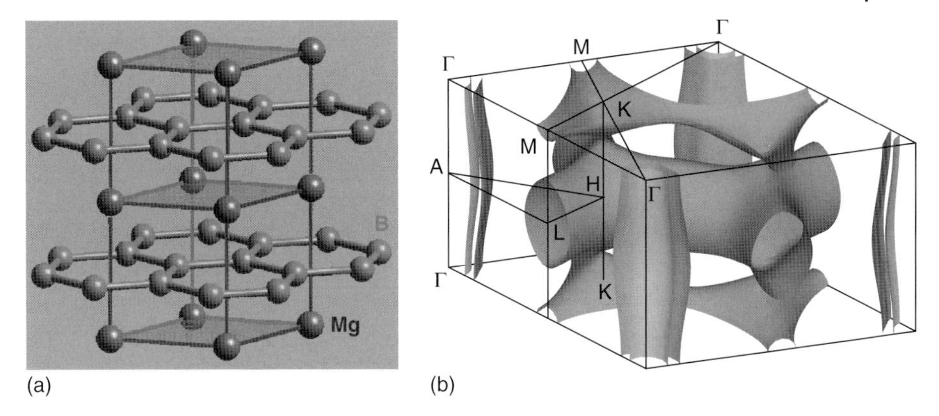

Figure 2.1.2.2 (a) Schematic crystal structure of MgB_2; (b) sketch of MgB_2 Fermi surface. ((b) is reproduced with permission from Ref. [18] (Copyright 2001 by the American Physical Society).)

of hole- and electron-type, respectively. In principle, this band structure could offer the chance of carrying out selective doping in σ and π bands by chemical substitution on the B and Mg site, respectively. The σ and π bands at the Fermi level contribute almost equally to the total density of states. The charge density spatial distribution reflects the fact that the B–B bond is strongly covalent, while the Mg atoms are fully ionized and in the divalent state. The *Fermi surface*, shown in Figure 2.1.2.2b [18], is formed by two coaxial cylindrical sheets around the ΓA line, originating from the σ bands with weak k_z dispersion, and by two tubular networks, an antibonding electron-type and a bonding hole-type sheet, originating from the π bands. Experimental probes of such Fermi surface have been obtained by de Haas van Alphen effect [20].

In order to account for the superconductivity mechanisms in MgB_2, a detailed description of the *phonon spectrum* is crucial [21]. In general, the light elements, such as B, result in higher frequency phonon modes, and, in the case of phonon-mediated conventional superconductors, the Bardeen–Cooper–Schrieffer (BCS) theory [22] predicts that higher frequency phonon modes yield enhanced transition temperatures. In MgB_2, the light B atoms provide indeed most of the weight to the phonon density of states at high energies. On the whole, there are four optical phonon modes, denoted as B_{1g}, E_{2g}, A_{2u}, and E_{1u}, corresponding to in-plane and out-of-plane displacements of either Mg or B atoms. Among these four modes, the two-dimensional E_{2g} mode, corresponding to B in-plane bond stretching, has a characteristic energy of 58–82 meV and is the most relevant, in that it is strongly coupled to planar boron σ bands near the Fermi level. The phonon-mediated scattering between σ and π carriers is described by the coupling matrix, whose elements have been calculated by solving the Eliashberg equations $\lambda_{\sigma\sigma} = 1.017$, $\lambda_{\pi\pi} = 0.448$, $\lambda_{\pi\sigma} = 0.155$, and $\lambda_{\sigma\pi} = 0.213$ [23]. Clearly, among these values, $\lambda_{\sigma\sigma}$ is the largest one.

The change of T_c with isotope mass, $T_c \sim M^{\alpha}$ (α isotope coefficient), the so-called *isotope effect*, is a compelling experiment to assess the phonon-mediated character of the pairing mechanism. Boron isotope effect measured from specific heat has revealed a 1.0 K shift in T_c between $Mg^{11}B_2$ and $Mg^{10}B_2$ samples [24]. The isotope effect coefficients of B and Mg have turned out to be $\alpha_B \sim 0.30$ and $\alpha_{Mg} \sim 0.02$, respectively [25], pointing to B phonons as primarily responsible for the pairing. However, total coefficient $\alpha_B + \alpha_{Mg} \sim 0.32$ is much lower than the value expected for a typical BCS superconductor $\alpha \sim 0.5$, which is indicative of strong coupling.

Clear signatures of *two-gap superconductivity* have been found for the first time from measurements of specific heat under magnet fields [26], where large excess weight near around 0.2 T_c has been ascribed to the presence of an additional smaller gap (see Figure 2.1.2.3a,b). Other experimental measurements

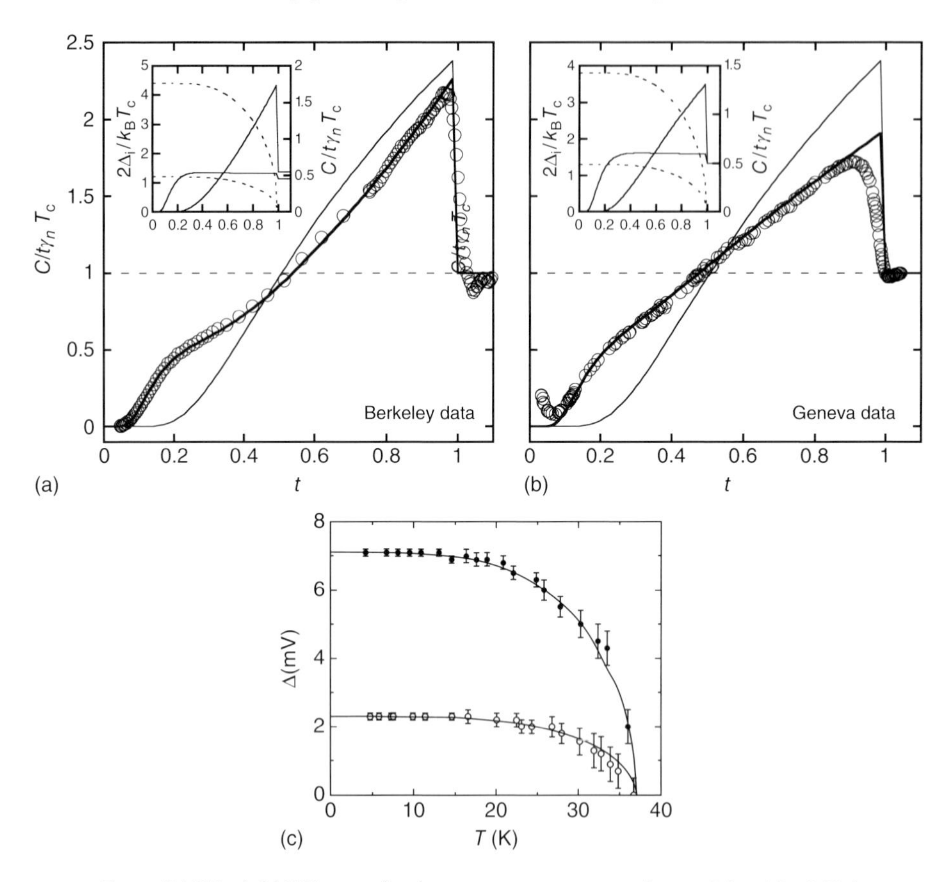

Figure 2.1.2.3 (a,b) BCS-normalized specific heat plotted versus reduced temperature T/T_c, fitted by a two-band model. (Figure is reproduced with permission from Ref. [26] (Copyright 2001 by IOP Publishing).) (c) Temperature dependence of the two gaps extracted from scanning tunneling spectroscopy data and fitted by BCS theory, from a work carried out at the Argonne National laboratory (UChicago Argonne, LLC, U.S. Dept. of Energy under contract No. DE-AC02-06CH11357). (Figure is reproduced with permission from Ref. [30] (Copyright 2002 by the American Physical Society).)

of the two gaps have been obtained from point-contact spectra [27], electronic Raman spectra [28], photoemission spectra [29], scanning tunneling spectra [30], angle-resolved photoemission spectra (ARPES) [31], thermal conductivity curves [32], far infrared optical reflectivity [33], magnetic induction [34], and tunnel junction characteristics [35], all in perfect agreement with the energy distribution of gap values calculated by first principles [36]. The two-gap nature is closely related to the anisotropic multiband nature of MgB_2 and is a consequence of the k dependence of the electron–phonon coupling. Indeed, σ and π carriers induce different s-type nodeless gaps $\Delta_\sigma \sim 7$ meV and $\Delta_\pi \sim 2$ meV, respectively. As both gaps disappear at the same T_c, in terms of reduced values the above Δ_σ and Δ_π correspond to $2\Delta_\sigma/k_B T_c \sim 3.8–4.3$ and $2\Delta_\pi/k_B T_c \sim 1.2–1.3$. The measured temperature dependence of each one of the two gaps follows the BCS prediction [30, 35] (see Figure 2.1.2.3c). From the BCS relationship $2\Delta = 3.53 k_B T_c$, the predicted T_cs related to the two gaps would be 15 and 45 K, but the finite interband coupling yields a single T_c around 40 K. The key point which explains the relatively high T_c and the large gap in MgB_2, as compared to other multiband superconductors, is that interband scattering between σ and π bands is pretty weak ($\lambda_{\pi\sigma}$ and $\lambda_{\sigma\pi}$ are calculated to be five times smaller than $\lambda_{\sigma\sigma}$ [23]), as a consequence of the different parity and almost orthogonal spatial distribution of p_z and $p_{x,y}$ orbitals which form the π and σ bands respectively [37]. Hence, the finite interband scattering yields a single T_c, but its weakness preserves the different character of the bands and determines the T_c value around 40 K, not much depressed as compared to the value associated to the σ gap.

As a further manifestation of the two-gap nature, a collective excitation corresponding to small fluctuations of the phase difference between the two-order parameters, the so-called *Leggett mode* has been predicted [38]. More specifically, the Leggett mode is a longitudinal excitation resulting from equal and opposite displacements of the two superfluids along the direction of the mode's wave vector q. Such effect has been indeed detected in MgB_2 crystals by Raman spectroscopy [39], offering new degrees of freedom for electronic applications.

The *upper critical field in clean MgB_2* is low and anisotropic. Typical low-temperature H_{c2} values are around 3 T for $H^{||c}$ and 16 T for $H^{||ab}$, yielding an anisotropy $\gamma = H_{c2}^{||ab}/H_{c2}^{||c}$ that is around 4 at low temperature and decreases monotonically with increasing temperature up to T_c [40]. This temperature dependence of γ is a consequence of the existence of two different gaps in MgB_2, as the relative contribution of the π gap to superconductivity, which is isotropic, increases with temperature.

Using the relationships $H_{c2}^{||c} = \Phi_0/(2\pi\xi_{ab}^2)$ and $H_{c2}^{||ab} = \Phi_0/(2\pi\xi_{ab}\xi_c)$, with Φ_0 magnetic flux quantum, it can be seen that the in-plane and out-of-plane *coherence lengths* are $\xi_{ab}(0) \sim 10$ nm and $\xi_c(0) \sim 4$ nm. These values are much larger than interatomic spacing and also larger than typical grain boundary thickness, which has important consequences for applications.

Conversely, the *lower critical field* H_{c1} is virtually isotropic, being around 0.1 T for both field directions [41]. The *London penetration depth*, which determines the extent to which a magnetic field penetrates into the superconductor, is isotropic as well [42, 43] and pretty large in value $\lambda \sim 100$ nm. As λ is inversely proportional to the square root of the Cooper pair density, its large value is consistent with the metallic character of the compound.

The *critical current densities measured in as-grown single crystals* are small, due to the weakness of pinning. J_c is below 10^5 A cm^{-2} at low temperatures in self-field [41, 44] and it rapidly decreases with increasing magnetic field, dropping to below 10^3 A cm^{-2} at around 1 T.

In the normal state, MgB_2 is a conventional metal with low-temperature *resistivity* in the range of some tenths of micro-ohm centimeter. Actually, in polycrystalline samples, the measured resistivities are often much larger, up to hundred micro-ohm centimeter, mainly as a consequence of the poor intergrain connectivity. In order to discriminate between intrinsic and extrinsic normal state transport, it has been proposed that the resistivity versus temperature $\rho(T)$ curves must be rescaled by a factor such that for the rescaled resistivity $\Delta\rho = \rho(300\,\mathrm{K}) - \rho(40\,\mathrm{K}) \approx 7.5\,\mu\Omega\,\mathrm{cm}$ [45]. Indeed, this $\Delta\rho$ value is related to the electron–phonon coupling of the π carriers, which dominates the temperature slope of resistivity, and therefore it is an intrinsic parameter, rather independent of the impurity content.

As for other normal state magnetotransport properties, MgB_2 exhibits a positive Hall resistance R_H in the range of few times 10^{-10} m^3 C^{-1}, dominated by hole-type σ bands, and a large longitudinal magnetoresistance up to 130% in clean samples [46, 47]. Indeed, the multiband character magnifies the cyclotron magnetoresistance, as the contributions of carriers of different types add up.

Different *chemical substitutions* have been tried in order to explore the possibility of tuning superconducting properties. A limited number of elements can be substituted in MgB_2, the most effective among these are Al in the Mg site and C in the B site [48]. Substitution of Mg by Al dopes electrons in the system in a rigid band-filling framework and affects mainly the intraband scattering of the π bands. The reduction of the overall hole density suppresses T_c and both the gaps [49]. Substitution of B by C also dopes electrons, thus reducing the hole density and suppressing T_c and both the gaps [49]. However, in this case, the substitution has a large impact on both π and σ bands, driving the system into the dirty limit.

The *effect of impurities* is peculiar for multiband superconductors, where it can originate a great variety of physical phenomena. Moreover, if the scattering with impurities involves carriers of either band having different characteristics, superconducting parameters can be effectively tuned and properties can be significantly improved.

While in single-band s-wave superconductors only magnetic scattering has a pair-breaking effect, thus suppressing T_c, in a two-band s-wave superconductor, interband scattering, even in the case that it is due to nonmagnetic impurities, is predicted to suppresses T_c down to a saturation value [50]. This saturation T_c value is predicted to be around 20 K, where an equivalent one-gap BCS system with isotropic coupling stabilizes. Yet, in experimental data of irradiated MgB_2, T_c, far from saturating, is completely suppressed at rather low levels of disorder (residual resistivity values of the order of 80 μΩ cm, which roughly correspond to a scattering rate of 0.6 eV [51]) and this can be explained by a smearing of the partial density of states caused by intraband scattering mechanisms [52]. Experimental evidence of the merging of the two gaps with increasing disorder, in correspondence of $T_c \approx 11$ K, has been obtained from specific heat measurements in irradiated samples [53] and from point contact spectroscopy measurements in C-doped samples [54].

The effect of disorder is also dramatic on the *upper critical field*. In the seminal work by Gurevich [55], quasiclassic Usadel equations for two-band superconductors in the dirty limit are derived with the account of both intraband and interband scattering by nonmagnetic impurities and the temperature dependence of the upper critical field is obtained. It is shown that the shape of the $H_{c2}(T)$ curve essentially depends on the ratio of the π and σ intraband electron diffusivities, departing significantly from the Werthamer–Helfand–Hohenberg limit $H_{c2}(0) = 0.69T_c\, dH_{c2}/dT_c$ for dirty samples [56]. In particular, in samples where the diffusivity of π carrier is much smaller than that of σ carriers, $H_{c2}(T)$ may exhibit an unusual upward curvature at low temperature, yielding enhanced $H_{c2}(0)$. Remarkably, Gurevich model suggests that $H_{c2}(T)$ curve can be tailored by engineering of the σ and π band scattering. Moreover, this theoretical framework has helped rationalizing experimental data, where $H_{c2}(T)$ curves with very large values and changing curvatures have been observed [57]. Low-temperature $H_{c2}^{||ab}$ values exceeding 60 T have been reached in C-doped MgB_2 thin films [58], which is very close to the paramagnetic limit of 70 T and exceeds the performances of industrial Nb-based superconductors (see Figure 2.1.2.4). H_{c2} has been enhanced from 15 to 30 T by 5% C doping also in single crystals, despite T_c being reduced to 27 K by this doping [4]. H_{c2} values exceeding 30 T have also been measured in C-doped wires [59] and C-nanotube-doped samples [60]. The effectiveness of C doping in enhancing H_{c2} must be partly attributed to the dirty π bands of $Mg(B_{1-x}C_x)_2$, but mainly to structural defects related to C alloying as well as to C-rich amorphous phases at the grain boundaries, which induce strain. Indeed, transmission electron microscopy (TEM) analyses have revealed lattice buckling of the ab planes, possibly due to strain induced by precipitates, which results in strong π scattering due to disturbance of the p_z π orbitals [57]. This multiplicity of factors, including defects, dopants, microstructure, strain, makes C alloying not yet completely under control in view of tailoring H_{c2} behavior at will.

C doping also enhances the *irreversibility field* H_{irr}, which is influenced by the presence of structural defects that behave as effective pinning centers, too [61]. However, the most successful method to improve H_{irr} has been the incorporation of SiC nanoparticles, which generate defects, nano-inclusions, and small grain size, all contributing as pinning centers [62].

As mentioned above, the *critical current density* in clean MgB_2 crystals is rather low, but it can be enhanced by improving H_{c2} as well as vortex pinning, up to a thermodynamic threshold set by the depairing current $j_d = \Phi_0/(3\sqrt{3}\pi\lambda^2\xi\mu_0) \approx 1.3 \times 10^8\ \mathrm{A\,cm^{-2}}$, above which the superconducting order parameter vanishes. Inclusions should be effective pinning centers if their size

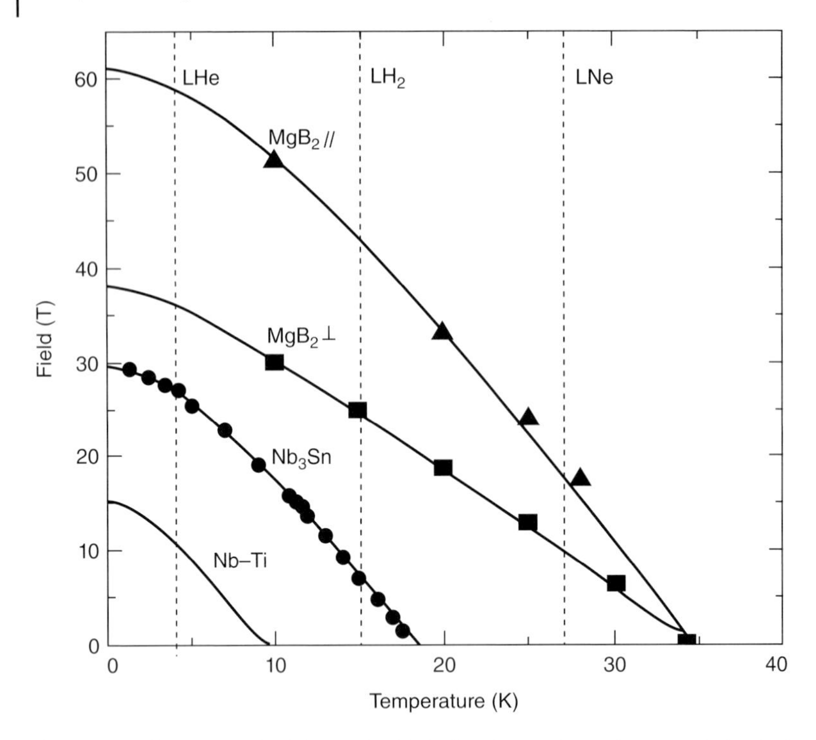

Figure 2.1.2.4 Upper critical fields of a MgB_2 film in parallel and perpendicular field, of a Nb–Ti bulk and of a Nb_3Sn bulk [58]. (Figure is reproduced with permission from Ref. [58] (Copyright 2006 by IEEE).)

matches the coherence length and their spacing matches the vortex distance in the desired field range.

In thin *films*, the J_c values are significantly larger than in single crystals. The highest reported values in self-field and at low temperature are $3-4\times10^7$ A cm^{-2} [63–65], which exceeds 10% of j_d. Strong pinning is provided by grain boundaries between columnar grains formed during the film growth, indeed the highest J_c values are found in clean films with maximum T_c [66]. Among MgB_2 films, there is a general trend of decreasing J_c with decreasing T_c [13], indicating that the defects which reduce T_c are not effective pinning centers. A further confirmation that pinning occurs mainly at the grain boundaries is given by the fit of the experimental pinning force curves $F_p = J_c(H)\times H$, normalized to the maximum value $F_{p\,max}$, plotted versus the reduced field H/H_{irr}. In most cases, the curves are described by the scaling law $F_p/F_{p\,max} \sim (H/H_{irr})^n(1-H/H_{irr})^m$, with exponent values $n = 0.5$ and $m = 2$ associated with grain boundary pinning.

Also in *polycrystalline bulk samples*, grain boundary pinning plays a major role. For example, it has been found that pinning force curves measured in tapes produced by the *ex situ* PIT method and by varying the high energy milling time of the powders perfectly overlap and are described by the scaling law $F_p/F_{p\,max} \sim (H/H_{irr})^{0.5}(1-H/H_{irr})^2$ (see Figure 2.1.2.5a) [67].

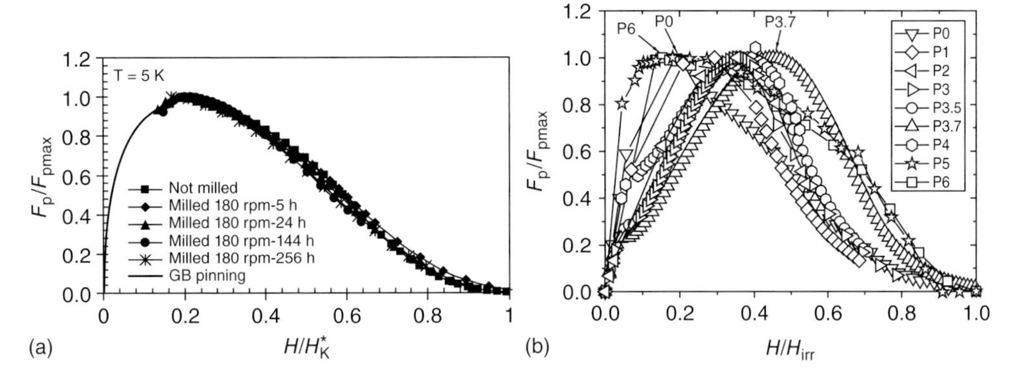

Figure 2.1.2.5 (a) Behavior of the normalized pinning force versus reduced magnetic field as extracted from the magnetic J_c measurements carried out at $T = 5$ K on tapes produced by the *ex situ* PIT method and by varying the high energy milling time of the powders. A comparison with grain boundary pinning model is also shown. (Figure reproduced with permission from Ref. [67] (Copyright 2008 by the American Institute of Physics).) (b) Normalized pinning force as a function of reduced field at 5 K for bulk samples irradiated with increasing thermal neutron doses. The position of the maximum shifts at higher reduced fields with increasing irradiation, indicating that point-like pinning centers are introduced. (Figure reproduced with permission from Ref. [68] (Copyright 2008 by IOP Publishing).)

This behavior is common to many nano-added MgB_2 samples with enhanced J_c and indicates that the dominant effect of doping and nanoparticle addition is to increase H_{c2}, hence making the already present pinning centers more effective rather than introducing new point defects capable of pinning more vortices. On the contrary, nanometer-size defects induced by neutron irradiation have turned out to be effective point-like pinning centers (see Figure 2.1.2.5b) [68, 69]. Several types of dopants and nanoparticles addition have been considered. The most remarkable results have been obtained by adding C in elemental form (microspheres, nanotubes, nanodiamond) or by means of C-containing compounds such as SiC, TiC, B_4C, organic compounds. SiC is by far the most popular dopant, because carbon can be doped into MgB_2 at low temperatures (600 °C), whereas higher processing temperatures are necessary for most of the other carbon sources, leading to grain growth and worse pinning. In SiC-doped wires, the J_c behavior at 20 K and 2 T is few times 10^5 A cm^{-2}, which is comparable to that of the best Ag/Bi-2223 tapes. A dual reaction model has been suggested to explain the improvement of the superconducting properties in SiC-doped MgB_2, based on the J_c dependence on the sintering temperature [62]. The reaction of SiC with Mg at low temperature releases fresh and active carbon, which is easily incorporated into the lattice of MgB_2 at the same temperature. On the other hand, the reaction product Mg_2Si and excess carbon are also high-quality nanosized pinning centers. Thus, the J_c improvements result from a combination of C substitution effects and improved pinning. Also nitrides, borides, silicides, as well as several metal oxides have been added, often obtaining encouraging results

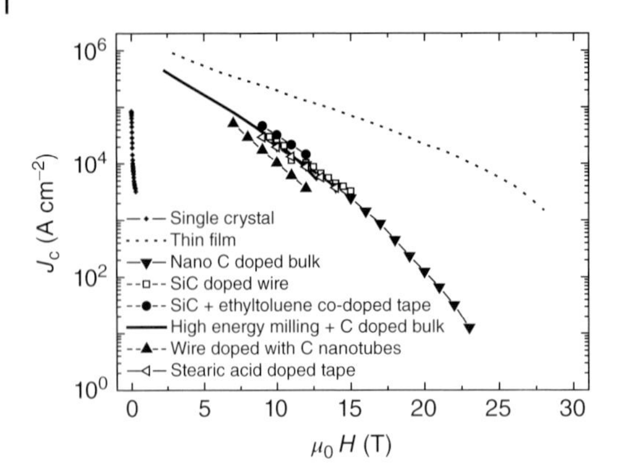

Figure 2.1.2.6 Collection of remarkable literature data of transport J_c as a function of field in different kinds of MgB_2 polycrystals at 4–5 K. (Taken from Refs [62, 73–77].) For comparison also shown are J_c of a thin film [72] and magnetic J_c of a single crystal [44].

(see Table I in Ref. [16]). A collection of some among the best results obtained in MgB_2 polycrystals, with different additions, is shown in Figure 2.1.2.6 [11]. As an alternative route to additions, different strategies have been followed in order to strengthen the grain boundary pinning by decreasing the grain size: among them, the high-energy ball milling of the precursors and/or of the powders [70] and the reduction of sintering temperature to limit the grain growth [71]. Up to now, nanostructuring seems to be the most promising method to increase J_c. This is shown in Figure 2.1.2.6, where extremely large J_c values exceeding 10^4 A cm^{-2} at 22 T are reported for MgB_2 thin film with H//film, with strong grain boundary pinning due to grains of 10–20 nm in size [72].

As reviewed in [13], the highest J_c values among polycrystals, wires, and tapes are up to 10^6 A cm^{-2} at 20 K in self-field [78, 79] and few times 10^4 A cm^{-2} at 5 K in a field of 8 T [62, 73–77]. Hence, the J_c values in polycrystals are more than one order of magnitude lower than in thin films. *Connectivity* seems to be the key issue. Indeed, although it was pointed out soon after the discovery of MgB_2 that clean grain boundaries are in principle no obstacles for super-currents [5], thanks to coherence lengths larger than the typical grain boundary thickness, the connections between the grains remain delicate, since dirty grain boundaries potentially reduce the critical currents. Insulating phases are often present at the grain boundaries, consisting of MgO, boron oxides, or boron carbide. Cracks, porosity, or normal conducting phases can further reduce the cross-section over which super-currents effectively flow. In order to increase the connectivity, a number of strategies have been pursued: improving the density by the application of the external pressure during the synthesis; limiting the presence of impurity phases at grain boundaries by using high-purity elements and avoiding any kind of dopants; limiting the formation of the amorphous MgO layer on the grain surface by performing the synthesis in controlled atmosphere; increasing the grain size in order to reduce the exposed surface of the grains; and texturing the MgB_2 phase by mechanical means or grain orientation under magnetic field.

2.1.2.3 Sample Preparation

In this section, we briefly describe the most used preparation methods for MgB_2 samples, namely bulks, single crystals, and thin films. For a more general description of the experimental techniques, we refer to Chapter 3 and its subsections in this book.

2.1.2.3.1 MgB_2 Phase Diagram and Polycrystals Synthesis

Despite its simple structure and composition, MgB_2 shows high sensitivity of its superconductive, and, more generally, electric and morphological properties with respect to the variations of the synthesis parameters. The powder is usually synthesized starting from the 1 : 2 molar mixture of the two elements, magnesium and boron, in their elemental forms, according to the phase diagram shown in Figure 2.1.2.7 [80]. Alternatively, sources of boron and magnesium are used, such as magnesium hydride [81], which decomposes into magnesium and hydrogen gas around 450 °C. As long as magnesium has a melting point of 650 °C and boiling point of 1090 °C, while boron has a melting point of 2050 °C, MgB_2 synthesis would be very hard if it was necessary to react them at the melting points of both. Luckily, the formation enthalpy is about −41 kJ mol^{-1}, which means that it is highly favored and exothermic [80], to the extent that even the self-propagating high-temperature synthesis technique has been employed to obtain bulk MgB_2 with good superconducting properties [82].

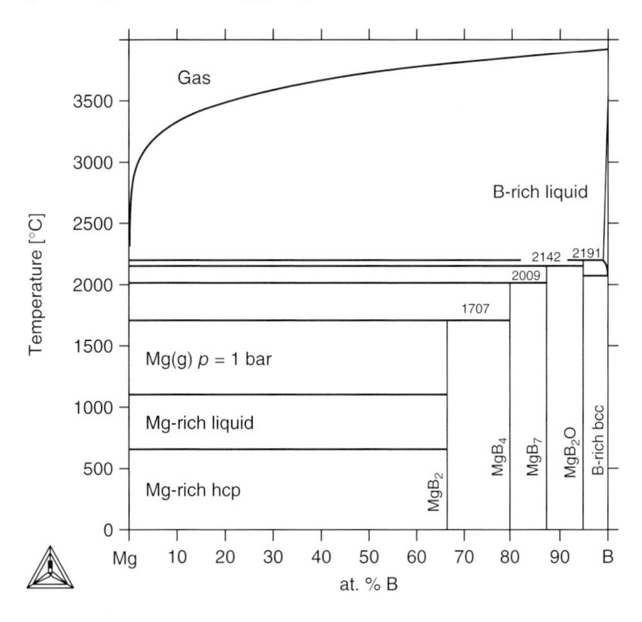

Figure 2.1.2.7 (a–c) Mg–B phase diagram. (Figure is reproduced with permission from Ref. [80] (Copyright 2005 by Elsevier).)

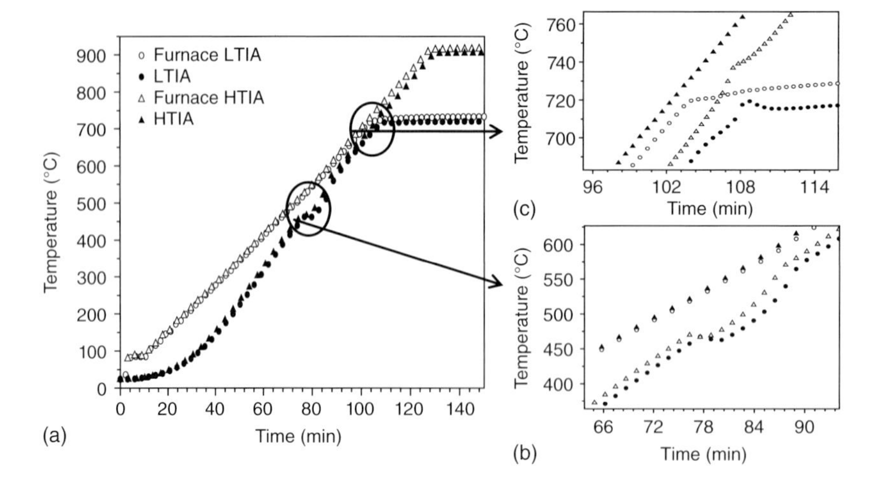

Figure 2.1.2.8 Thermal profile of the MgB_2 synthesis reaction. (a) Temperature profile of the thermal treatments for the low and high synthesis temperature. (b,c) Details of MgH_2 decomposition and of MgB_2 formation. (Figure is reproduced with permission from Ref. [83] (Copyright 2011 by IOP Publishing).)

Boron and magnesium have deeply different physical properties. Indeed, while boron is almost inert at the temperatures of interest, which are well below its melting point, magnesium starts to present evaporation even below its melting point. Owing to the magnesium low melting point and high tendency to evaporate, MgB_2 must be in contact with the correct partial pressure of magnesium, in order to be formed and stabilized. If this is not the case, higher borides are formed, due to variation of existence zone [80]. Thus, the reaction between the two elements gives a quite large spectrum of behaviors along with variations of the temperature of synthesis. From the thermal profile acquired during a reaction in a temperature ramp (see Figure 2.1.2.8) [83], it is visible how the mixture has a general endothermic behavior, which is even more pronounced when, above 650 °C and up to about 750 °C, the magnesium melting absorbs energy. Afterward, the sudden exothermic reaction, triggered by the enhanced intimate contact between the boron powder and the liquefied magnesium, gives a local peak in the temperature profile. This macroscopic manifestation gives the idea that the major part of the synthesis occurs after melting of magnesium.

Nevertheless, thanks to a high-energy X-ray diffraction (HEXRD) analysis performed during a synthesis [83], formation of MgB_2 is discovered at surprisingly low temperatures starting from about 300 °C, which indicates a possible solid–solid reaction at an extremely low temperature. Clearly, although scientifically very interesting, such a low temperature is not easily useful for technical purposes.

In fact, tests of reaction time for a MgB_2 bulk synthesis have shown that at 550 °C the time to reach 80% of completion is as long as 13 h [84]. An even more important fact is witnessed in the plot in Figure 2.1.2.9 [84] reporting heat

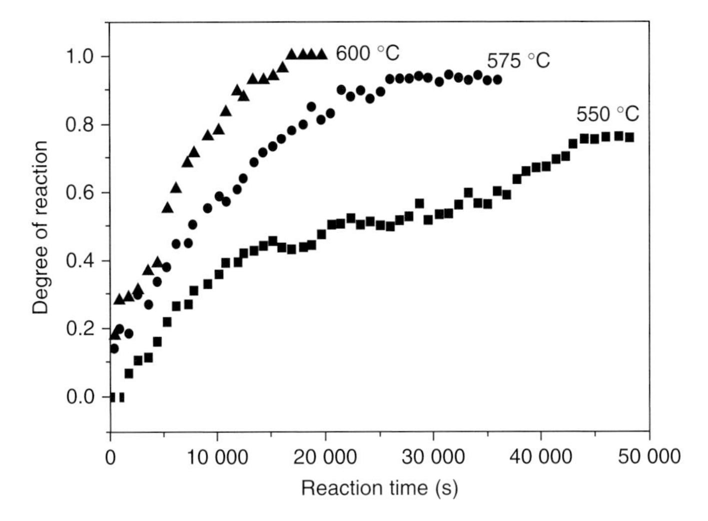

Figure 2.1.2.9 Plots of degree of reaction versus time isothermally annealed Mg–B powders at 550, 575, and 600 °C. (Figure is reproduced with permission from Ref. [84] (Copyright 2011 by Elsevier).)

treatments at various temperatures, analyzed by real-time X-ray diffraction: all the reactions exhibit an asymptotic time behavior, but only above 600 °C the fully 100% completion is reached. Attempts to overcome these hindrances have been made by incorporating small additions (small percentage in weight) of various metals such as copper, silver, or tin. The effect of these additions is to favor the reaction and reduce the reaction time, by forming magnesium alloys with lower melting point, which work as carriers to enhance contact with boron [84].

In summary, the relevant temperature regime for technological applications is the one beyond the melting point of magnesium, specifically from around 650 °C up to about 1000 °C. The dwelling temperature of the heat treatment is of practical interest as well. Indeed, the particle size distribution of the powders can be controlled by the sintering temperature, as the grain growth is promoted by the higher temperature. For example, it has been shown that the particle size distribution is peaked at 0.3 and 1.4 μm for powders sintered at 760 and 910 °C, respectively [83].

The MgB_2 phase reacted in an *in situ* process inside an iron-sheathed tape at various temperatures has been analyzed by different experimental techniques [85]. The cell parameters have turned out to be constant ($c = 3.525$ Å, $a = 3.085$ Å) in the temperature range between 650 and 900 °C. Moreover, a clear trend of decreasing width of the 110 peak of MgB_2 phase with increasing heat treatment temperature has been identified. This effect is related to the in-plane crystallinity and to the stress which varies from one grain to another, thus leading to random deformation of the lattice parameters, causing the peak broadening. This hypothesis has been confirmed also by the strain relieving effect of the heat treatment temperature [85]. The same stress relieving also affects the T_c values, which are enhanced by the cleaning and crystallizing effect of higher thermal treatment temperatures. On the other hand, this mechanism removes the pinning centers, thus affecting the J_c

capability, namely reducing it at higher magnetic fields, but probably enhancing it at lower magnetic fields, as it is suggested by the converging trend of the J_c curves toward lower fields, observed in Figure 5 of Ref. [85].

2.1.2.3.2 MgB_2 Single Crystals

Conventional methods of crystal growth, such as growth from high-temperature solutions at ambient pressure or from a stoichiometric melt, do not work for MgB_2, as it melts noncongruently and, moreover, it often reacts with container materials at high temperatures. The most successful growth method in terms of superconducting properties and crystal size is from a solution in Mg at high pressure. Indeed, the solubility of MgB_2 in Mg is extremely low at temperatures below the boiling temperature of Mg (1107 °C), so that crystals have to be grown at much higher temperature. Owing to the high Mg vapor pressure at these high temperatures [86], processing at high pressures is necessary. MgB_2 single crystals of millimeter size are obtained by high-pressure cubic anvil technique from a precursor containing Mg, B, and BN, via the peritectic decomposition of $MgNB_9$ [86, 87]. MgB_2 single crystals can be also grown at ambient pressure by using a Mg-self-flux method, using MgB_2 powder as the raw material and Mg as the flux, according to a mechanism of liquid-assisted solid-state recrystallization [88]. Finally, slightly smaller submillimeter size MgB_2 crystals can be grown from the vapor phase. In this vapor transport method, the crystals are obtained from a mixture of Mg and B, heated up to 1400 °C in a sealed molybdenum crucible [89].

2.1.2.3.3 MgB_2 Thin Films

Because Mg is volatile and MgB_2 decomposes at high temperature when not under sufficient Mg vapor pressure, the most important requirement for the deposition of MgB_2 films is to provide a high Mg vapor pressure at the elevated deposition temperatures. The earliest successful technique for obtaining high-quality MgB_2 films is to deposit precursor films of B or Mg–B mixture at lower temperatures and anneal them in Mg vapor at high temperatures. Different techniques have been employed for the precursor deposition, such as pulsed laser deposition (PLD) [90], magnetron sputtering [6], electron beam evaporation [91], thermal evaporation [92], chemical vapor deposition [93], and even screen printing [94]. By this two-step process, it is often difficult to eliminate oxygen and carbon contamination, which degrade the film properties. Single step *in situ* depositions by PLD and by molecular beam epitaxy have been also carried out. In these cases, deposition at low temperature circumvents the requirement for high Mg vapor pressure, but results in nonoptimal crystallinity. For depositing the cleanest epitaxial MgB_2 films for various fundamental studies and for applications, hybrid physical–chemical vapor deposition (HPCVD) is by far the most successful. In this technique, the Mg is provided by thermally evaporating bulk Mg pieces (physical vapor deposition) and the B is from the diborane, B_2H_6, precursor gas (chemical vapor deposition). This technique provides a high Mg vapor pressure that satisfies the thermodynamic phase stability condition at the temperature used for the deposition, and uses a reducing hydrogen ambient during the deposition that suppresses the oxidation of Mg. Very clean MgB_2 films with T_c over 40 K, residual resistivity $\rho(T = 40\text{ K})$ lower than 0.1 μΩ cm, and residual resistivity ratio RRR over 80 are fabricated [95]. The T_c value is higher than that in the bulk MgB_2 due to the strain-induced E_{2g} phonon softening that enhances the electron–phonon interaction [96]. Regarding the choice of the substrate, sapphire, SiC, and MgO are the most used, but excellent properties have also been measured in films deposited on $LaAlO_3$, $SrTiO_3$, yttrium-stabilized ZrO_2 [97], and other substrates.

2.1.2.4 Applications of MgB_2

2.1.2.4.1 Wires and Tapes

Immediately after the discovery of MgB_2, researchers have been able to fabricate long-length MgB_2 wires using the 50 year experience in the conductor fabrication with low-T_c superconductors and the 20 year experience with high-T_c superconductors. From the fabrication point of view, the main advantages in using MgB_2 are related to the low synthesis temperature and to the possibility of improving the grain connectivity and the superconducting properties with short heat treatments. Wires of several kilometer length fabricated in a single batch are commercially available [98, 99]. The main methods to produce wires and tapes are three: the PIT, the diffusion method, and the coating technique. For a more general description of these methods, we refer to Section 3.4 in this book.

PIT: *In Situ* and *Ex Situ* Techniques The PIT is the most popular method to fabricate MgB_2 wires. A metallic tube is filled in air or in inert atmosphere with prereacted MgB_2 (*ex situ* technique) or with Mg and B precursors (*in situ* technique). The tube is then cold worked through rolling or drawing machines up to the shape of a wire or a tape, which must be strong enough to sustain the brittle component, but ductile enough to let the cold working be possible. Generally Ni, Fe, and Ti alloys are preferably used because they favor high packing density of the powders, thus allowing better grain connectivity. Several heat treatments are performed at intermediate steps during the cold working to recover the mechanical properties and release structural stress of the metallic matrix. A final heat treatment is performed for the sintering of the MgB_2 powders (*ex situ*) or to form the MgB_2 phase (*in situ*).

The PIT technique is used not only for the fabrication of single-filament wires but also for multifilamentary wires which exhibit larger uniaxial and bend strains, as well as higher thermal stability. Complex structures can be realized by restacking bundles of wires with superconducting core and additional stabilizing components such as copper or aluminum. Figure 2.1.2.10 shows typical cross-sections of an *in situ* wire from Hypertech Research Inc. [100] and two *ex situ* wires and an *ex situ* tape from Columbus Superconductors S.p.A. [99]. *Ex situ* and *in situ* wires are currently able to carry critical current densities exceeding 2000 A cm^{-2} at 20 K and in a magnetic field of 1 T, over long lengths.

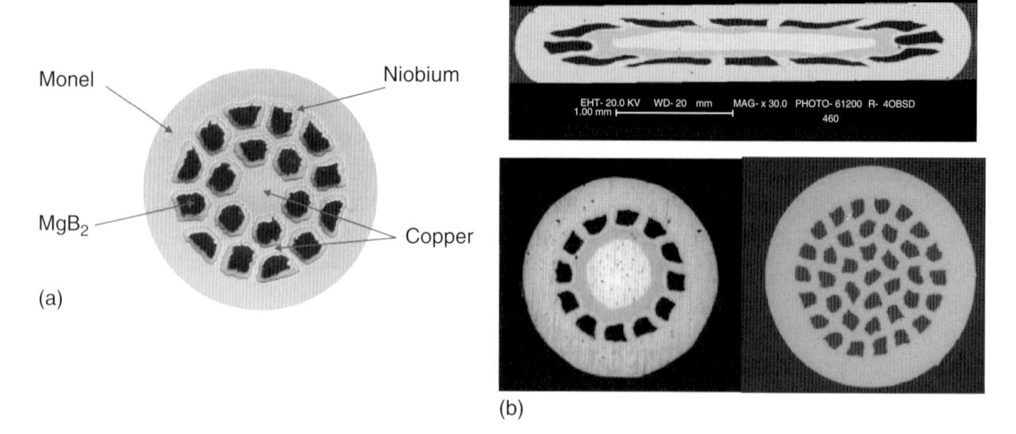

Figure 2.1.2.10 Typical cross-sections of a wire processed with the *in situ* technique from Hypertech Research Inc. [100] (a) and of a tape and two wires processed by the *ex situ* technique from Columbus Superconductors S.p.A. [99] (b). ((a) is reproduced with permission from Ref. [100] (Copyright 2007 by Elsevier).)

Diffusion Technique An interesting variant of the *in situ* method to produce MgB_2 wires is the "internal Mg diffusion" (IMD) process or "Mg-reactive liquid infiltration" (Mg-RLI) process [101]. The conductors processed this way are sometimes called *2G MgB_2 wire conductors* and are leading to a new level of performance for MgB_2 wires. The technique has been developed by Giunchi *et al.* [102] and it consists of the cold working of a B-filled tube with a Mg rod embedded axially in it. After a final heat treatment, the Mg diffusion leaves a hollow, but produces a dense MgB_2 layer structure with excellent longitudinal and transverse connectivity (see Figure 2.1.2.11). In this case, the transport properties are described by a "layer J_c" whose typical value is about $10^5\ \mathrm{A\ cm^{-2}}$ at 4.2 K and 10 T.

Figure 2.1.2.12, taken from Ref. [103], shows the $J_c(H)$ characteristics of a typical PIT-processed MgB_2 wire at 4.2 and 20 K in comparison with those of practical

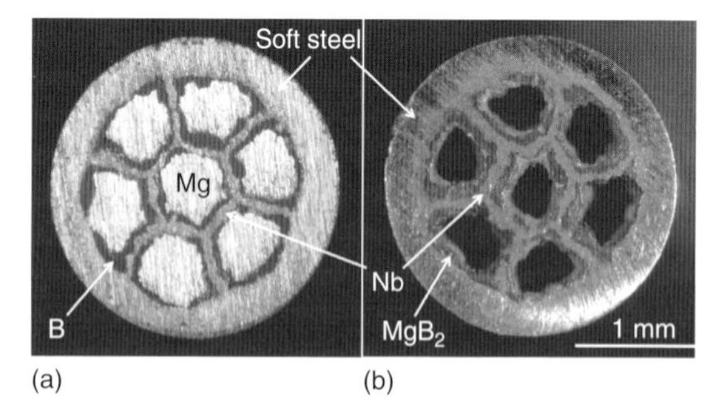

Figure 2.1.2.11 Cross-section of a seven-filament wire fabricated by the Mg diffusion method. (a) Precursor wire; (b) annealed wire. (Figure is reproduced with permission from Ref. [102] (Copyright 2003 by Institute of Physics).)

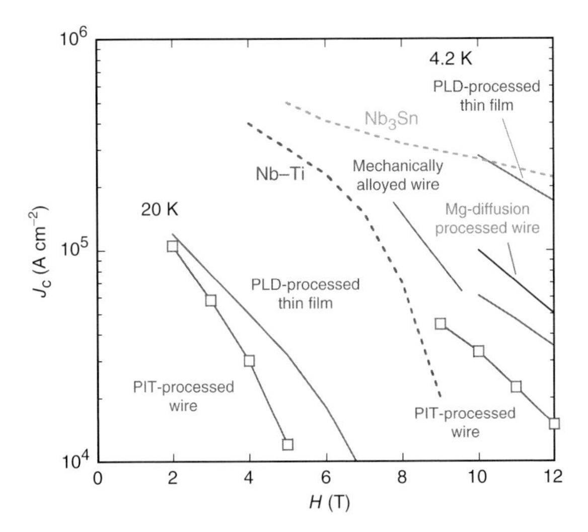

Figure 2.1.2.12 $J_c(H)$ characteristics of a PIT-processed MgB_2 wire in magnetic field at 4.2 and 20 K. For comparison, the values for a mechanically alloyed MgB_2 wire [104], a Mg-diffusion-processed MgB_2 wire [105], a PLD-processed MgB_2 thin film [72], a Nb–Ti wire, and a Nb_3Sn wire are also shown. (Figure is reproduced with permission from Ref. [103] (Copyright 2012 by Japan Society of Applied Physics).)

Nb–Ti and Nb_3Sn wires, a mechanically alloyed PIT-processed MgB_2 wire [104], a Mg-diffusion-processed wire [105], and a PLD-processed MgB_2 thin film [72]. The J_c of MgB_2 wires is approaching Nb–Ti values at high fields >8 T, but it is still well below the values of Nb_3Sn. The temperature range where PIT-processed MgB_2 wires are most competitive is around 20 K, where a practical level J_c value exceeding 10^5 A cm^{-2} at 2 T is achieved.

Coating Techniques The observation of very large H_{c2} and J_c in HPCVD-deposited films as compared to corresponding values obtained in bulk samples has suggested the idea of considering MgB_2-coated conductors as an alternative to PIT wires and tapes. Polycrystalline-coated conductor fibers have been deposited by HPCVD on SiC fibers. In carbon-alloyed fibers, very high upper critical fields $H_{c2} = 55$ T and irreversibility fields $H_{irr} = 40$ T have been obtained [106]. Critical current densities above 10^7 A cm^{-2} in self-field and above 10^5 A cm^{-2} at 3 T have been obtained on MgB_2 deposited on textured Cu (001) tapes, with and without SiC additions, and even larger values have been measured on MgB_2 deposited on polycrystalline Hastelloy tapes [107]. Remarkably, no sign of J_c degradation with increasing thickness up to 2 μm have been detected in these samples, which is very important in view of increasing the filling factor of coated conductor wires or tapes.

The versatility of the coated conductor route is further demonstrated by the deposition of pure films on other different metallic substrates such as stainless steel, Fe, Nb, and Cu [108–111]. All such films show superconducting transitions similar to those of films on SiC or sapphire. Amorphous 50 μm thick MgB_2 films have been deposited on large areas (up to 400 cm^2) of flexible polyamide Kapton-E foils by a two-step method [112]. These films exhibit T_c values around 29 K and impressive mechanical resilience.

The above results indicate that the coated conductor technology is worth further consideration, even if yet far behind the PIT technology.

2.1.2.4.2 **Electronic Applications**

Electronics applications of MgB_2 are of high relevance. In the following, we briefly mention some aspects of this topic, but we refer to Sections 3.3 and Chapter 9 in this book, as well as to reviews in Refs [14, 113] for a more extensive and detailed overview. The primary electronic application of MgB_2 films is the fabrication of Josephson junctions and SQUIDs. Trilayer junctions made of one MgB_2 electrode, a tunneling barrier, and another electrode being a low-temperature superconductor have shown characteristics with typical features related to σ and π gaps [35]. However, the advantage of MgB_2-based Josephson junctions and circuits lies in their T_c of 40 K, which allows operation above 20 K as compared to Nb-based superconductors, which have to be cooled to 4.2 K. In order to exploit this advantage in terms of cost and power consumption, devices fully made of MgB_2 must be considered. AC Josephson effect and Shapiro steps have been observed in superconductor–normal conductor–superconductor (SNS) junctions obtained by patterning MgB_2/TiB_2 bilayers, exhibiting proximity coupling between the two MgB_2 electrodes through the normal metal TiB_2 [114]. SQUIDs have been fabricated from single-layer MgB_2 films by patterning weak-link nanobridges of 70 nm width, 150 nm height, and 150 nm length [115]. Highly uniform junction arrays have been obtained by ion implantation of single-layer MgB_2 films and have exhibited giant Shapiro steps [116]. Beyond the above-mentioned results, MgB_2/barrier/MgB_2-stacked trilayer junctions have represented a technological breakthrough with respect to devices fabricated onto single-layer MgB_2 films. Among these systems, it is also worth mentioning the new method to fabricate MgB_2/graphene/MgB_2 junctions presented in [117]. All these MgB_2/barrier/MgB_2-stacked trilayer junctions have shown clear Fraunhofer patterns. Unfortunately, in most cases, the critical current has been measured to vanish around 20–25 K, due to degradation of MgB_2 at the interfaces [118–122]. Instead, remarkably, $MgB_2/MgO/MgB_2$ Josephson tunnel junctions deposited by HPCVD have exhibited nonvanishing J_c up to 40 K and low temperature values up to 275 kA cm^{-2} [123]. Their $I-V$ characteristics and critical current modulations by applied magnetic field are displayed in Figure 2.1.2.13. The $I_C R_N$ product (I_C is the junction critical current and R_N the normal state resistance; the $I_C R_N$ product determines the signal frequency and switching time at any given working temperature) of these junctions at 4.2 K is in the range 2.1–3.1 mV, about half of the predicted value.

The above-mentioned results and the energy gap in MgB_2 (the smaller one ~2.2 meV) larger than that of Nb (~1.5 meV) imply a limiting clock speed

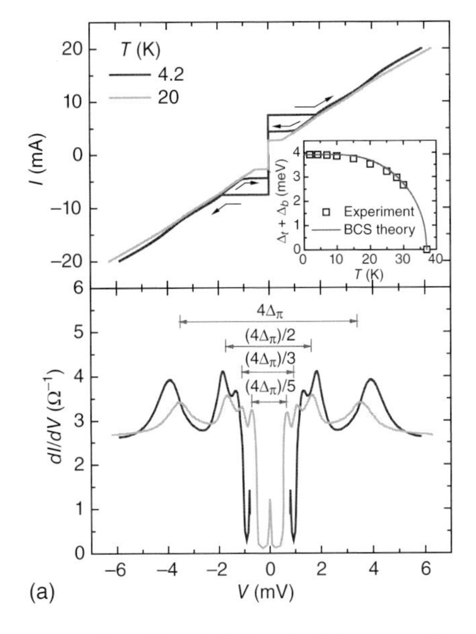

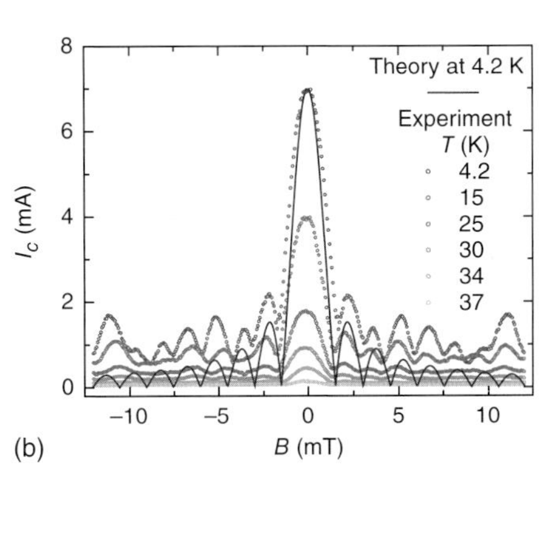

Figure 2.1.2.13 (a) *I*–*V* and *dI*/*dV* versus *V* characteristics of a MgB_2/MgO/MgB_2 junction at 4.2 and 20 K; the two *dI*/*dV* peaks at around ±4 mV correspond to the sum of the π gaps of the top and bottom MgB_2 electrodes. (b) Periodic modulation of I_c versus field resembling the expected Fraunhofer pattern. (Figures are reproduced with permission from Ref. [123] (Copyright 2010 by the American Institute of Physics).)

for MgB_2 digital circuits of above 1 THz, very promising for ultrahigh speed superconducting electronics at temperatures over 20 K.

Application of MgB_2 films in THz electronics, as THz mixers, is highly relevant. Indeed, the high critical temperature and the short electron–phonon interaction time make the MgB_2 very attractive for the fabrication of hot electron bolometer mixers, exhibiting larger gain bandwidth than NbN-based mixers and low noise [124].

Among other potential electronic applications of MgB_2, radio frequency (RF) cavities must be mentioned. Nb is the material currently used for superconducting RF cavities; however, improvement of the RF cavity performance demands new superconductors with higher T_c, higher RF critical field, and higher thermal conductivity. As a key requirement for a higher T_c superconductor to replace Nb, it should have a low surface resistance R_s to allow a high Q. Low R_s is achieved in materials with low residual resistivity ρ_0 and large energy gap, which make MgB_2 a promising candidate. Indeed, in MgB_2 one order of magnitude less RF surface resistance has been measured with respect to Nb at 4 K, with little power dependence [125].

2.1.2.5 **Summary and Outlook**

MgB_2 has a considerable potential for current transport applications, as it can be fabricated in form of round and multifilament wires, which are flexible in use and capable of being used in single-strand or multistrand cables for small and large magnets (see Chapter 4 in this book). The main advantages of MgB_2 are the low cost and the higher T_c of 40 K versus 9 K of Nb–Ti and 18 K of Nb_3Sn. This large T_c is particularly favorable not only for operating temperatures around $T_c/2 \approx 20$ K but also with cryocoolers operating around 4 K, as it allows a larger temperature margin with respect to low-T_c superconductors. However, there is still an edge of improvement in order for MgB_2 to become really competitive with other technical superconductors in high field magnets. In particular, its upper critical field and critical current density should exceed those of Nb–Ti and Nb_3Sn, with the further requirement of cheap fabrication technology. The requirements on H_{c2} have already been achieved at least in films. The results on J_c are approaching Nb–Ti values (around few times 10^5 A cm^{-2} at 8 T), but they are still well below the values of Nb_3Sn and Bi-2212. Connectivity is the main issue in bulk polycrystals, indeed critical current densities as high as 4 $\times$ 10^7 A cm^{-2}, which is 10% of the depairing current limit, have been obtained only in well-connected MgB_2 films deposited by HPCVD. Unfortunately, C alloying, which has proven to be highly beneficial in enhancing H_{c2}, invariably degrades the connectivity and eventually suppresses J_c for high C concentrations. Hence, C alloying process needs further improvement to become fully controlled. New techniques to decrease porosity and mitigate the so-called *cross-sectional deficiency* in PIT wires and tapes are needed. Finally, J_c at high fields must be further improved in order to compete with Nb_3Sn at liquid helium temperature.

At the moment, MgB_2 seems to be mostly competitive in the fabrication of persistent magnets for NMR and magnetic resonance imaging (MRI) (see Section 4.5 in this book), as the technology for superconducting joints of low-T_c superconductors is invaluable, while MgB_2 joints have been fabricated [126, 127]. The competitiveness of MgB_2 must include not only costs themselves but also ease of operation of cryocooled systems and more efficient cryogenics operations well above 4 K. An economic analysis of magnets has calculated that MgB_2 magnets are least expensive for fields <1 T, but the expenses increase sharply above 1.5 T [58]. This again shows the need for and the benefits of further enhancement of J_c in MgB_2.

Finally, regarding electronic applications, thanks to the high quality of MgB_2 thin films, recent advances have been made in the realization and understanding of MgB_2 tunnel junctions and SQUIDs, where multiband transport properties, such as the existence of two energy gaps, phonon spectra, and anisotropy, have been exploited. MgB_2 Josephson junction technology with low critical current spreads is available and reliable SQUIDs and low-noise magnetometers operating close to the T_c of MgB_2 have been realized.

References

1. Jones, M.E. *et al.* (1954) *J. Am. Chem. Soc.*, **76**, 1434.
2. Nagamatsu, J. *et al.* (2001) *Nature*, **410**, 63.
3. Canfield, P.C. *et al.* (2003) *Phys. Today*, **56**, 34.
4. Muranaka, T. *et al.* (2011) *Z. Kristallogr.*, **226**, 385.

5. Larbalestier, D.C. *et al.* (2001) *Nature*, **410**, 186.
6. Eom, C.B. *et al.* (2001) *Nature*, **411**, 558.
7. Bugoslavsky, Y. (2001) *Nature*, **411**, 561.
8. Jin, S. *et al.* (2001) *Nature*, **411**, 563.
9. Buzea, C. *et al.* (2001) *Supercond. Sci. Technol.*, **14**, R115.
10. Xi, X.X. (2008) *Rep. Prog. Phys.*, **71**, 116501.
11. Putti, M. *et al.* (2012) *MRS Bull.*, **36**, 608.
12. Dahm, T. (2005) in *Frontiers in Superconducting Materials* (ed. A.V. Narlikar), Springer, Berlin, 983–1009.
13. Eisterer, M. (2007) *Supercond. Sci. Technol.*, **20**, R47.
14. Xi, X.X. (2009) *Supercond. Sci. Technol.*, **22**, 043001.
15. Vinod, K. *et al* (2007) *Supercond. Sci. Technol.*, **20**, R1.
16. Collings, E.W. *et al.* (2008) *Supercond. Sci. Technol.*, **21**, 103001.
17. Braccini, V. *et al.* (2007) *Physica C*, **456**, 209.
18. Kortus, J. *et al.* (2001) *Phys. Rev. Lett.*, **86**, 4656.
19. Satta, G. *et al.* (2001) *Phys. Rev. B*, **64**, 104507.
20. Carrington, A. *et al.* (2007) *Physica C*, **456**, 92.
21. Kong, Y. *et al.* (2001) *Phys. Rev. B*, **64**, 020501(R).
22. Bardeen, J. *et al.* (1957) *Phys. Rev.*, **108**, 1175.
23. Golubov, A.A. *et al.* (2002) *J. Phys. Condens. Matter*, **14**, 1353.
24. Bud'ko, S.L. *et al.* (2001) *Phys. Rev. Lett.*, **86**, 1877.
25. Hinks, D.G. *et al.* (2001) *Nature*, **411**, 457.
26. Bouquet, F. *et al.* (2001) *Europhys. Lett.*, **56**, 856.
27. Szabó, P. *et al.* (2001) *Phys. Rev. Lett.*, **87**, 137005.
28. Chen, X.K. *et al.* (2001) *Phys. Rev. Lett.*, **87**, 157002.
29. Tsuda, S. *et al.* (2001) *Phys. Rev. Lett.*, **87**, 177006.
30. Iavarone, M. *et al.* (2002) *Phys. Rev. Lett.*, **89**, 187002.
31. Souma, S. *et al.* (2003) *Nature*, **423**, 65.
32. Putti, M. *et al.* (2003) *Phys. Rev. B*, **67**, 064505.
33. Ortolani, M. *et al.* (2008) *Phys. Rev. B*, **77**, 100507.
34. Manzano, F. *et al.* (2002) *Phys. Rev. Lett.*, **88**, 047002.
35. Chen, K. *et al.* (2008) *Appl. Phys. Lett.*, **93**, 012502.
36. Choi, H.J. *et al.* (2002) *Nature*, **418**, 758.
37. Mazin, I.I. *et al.* (2002) *Phys. Rev. Lett.*, **89**, 107002.
38. Anishchanka, A. (2007) *Phys. Rev. B*, **76**, 104504.
39. Blumberg, G. *et al.* (2007) *Phys. Rev. Lett.*, **99**, 227002.
40. Welp, U. *et al.* (2003) *Physica C*, **385**, 154.
41. Lyard, L. *et al.* (2004) *Phys. Rev. Lett.*, **92**, 057001.
42. Niedermayer, C. *et al.* (2002) *Phys. Rev. B*, **65**, 094512.
43. Cubitt, R. *et al.* (2003) *Phys. Rev. Lett.*, **90**, 157002.
44. Lee, S. *et al.* (2003) *Supercond. Sci. Technol.*, **16**, 213.
45. Rowell, J.M. *et al.* (2003) *Appl. Phys. Lett.*, **83**, 102.
46. Li, Q. *et al.* (2006) *Phys. Rev. Lett.*, **96**, 167003.
47. Pallecchi, I. *et al* (2005) *Phys. Rev. B*, **72**, 184512.
48. Kortus, J. *et al.* (2005) *Phys. Rev. Lett.*, **94**, 027002.
49. Gonnelli, R.S. *et al.* (2007) *Physica C*, **456**, 134.
50. Golubov, A. *et al.* (1997) *Phys. Rev. B*, **55**, 15146.
51. Putti, M. *et al.* (2008) *Supercond. Sci. Technol.*, **21**, 043001.
52. Brotto, P. *et al.* (2010) *Phys. Rev. B*, **82**, 134512.
53. Putti, M. *et al.* (2007) *Phys. Rev. Lett.*, **96**, 077003.
54. Gonnelli, R.S. *et al.* (2005) *Phys. Rev. B*, **71**, 060503.
55. Gurevich, A. (2003) *Phys. Rev. B*, **67**, 184515.
56. Werthamer, N.R. *et al.* (1966) *Phys. Rev.*, **147**, 295.
57. Braccini, V. *et al.* (2005) *Phys. Rev. B*, **71**, 012504.
58. Iwasa, Y. *et al.* (2006) *IEEE Trans. Appl. Supercond.*, **16**, 1457.

59. Wilke, R.H.T. *et al.* (2004) *Phys. Rev. Lett.*, **92**, 217003.
60. Serquis, A. *et al.* (2007) *Supercond. Sci. Technol.*, **20**, L12.
61. Ohmichi, E. (2007) *Physica C*, **456**, 117.
62. Dou, S.X. *et al* (2007) *Phys. Rev. Lett.*, **98**, 097002.
63. Kim, H.J. *et al.* (2001) *Phys. Rev. Lett.*, **87**, 087002.
64. Xu, S.Y. *et al.* (2003) *Phys. Rev. B*, **68**, 224501.
65. Zeng, X.H. *et al.* (2003) *Appl. Phys. Lett.*, **82**, 2097.
66. Kitaguchi, H. *et al.* (2004) *Appl. Phys. Lett.*, **85**, 2842.
67. Malagoli, A. *et al.* (2008) *J. Appl. Phys.*, **104**, 103908.
68. Martinelli, A. *et al.* (2008) *Supercond. Sci. Technol.*, **21**, 012001.
69. Pallecchi, I. *et al.* (2005) *Phys. Rev. B*, **71**, 212507.
70. Senkowicz, B.J. *et al.* (2008) *Supercond. Sci. Technol.*, **21**, 035009.
71. Romano, G. *et al.* (2009) *IEEE Trans. Appl. Supercond.*, **19**, 2706.
72. Matsumoto, A. *et al.* (2008) *Appl. Phys Express*, **1**, 021702.
73. Yamada, H. *et al.* (2007) *Supercond. Sci. Technol.*, **20**, L30.
74. Senkowicz, B.J. *et al.* (2005) *Appl. Phys. Lett.*, **86**, 202502.
75. Ma, Y. *et al.* (2007) *Supercond. Sci. Technol.*, **20**, L5.
76. Kim, J.H. *et al.* (2006) *Appl. Phys. Lett.*, **89**, 122510.
77. Gao, Z. *et al.* (2007) *Supercond. Sci. Technol.*, **20**, 485.
78. Xu, X. *et al.* (2006) *Supercond. Sci. Technol.*, **19**, L47.
79. Shields, T.C. *et al.* (2002) *Supercond. Sci. Technol.*, **15**, 202.
80. Balducci, G. *et al.* (2005) *J. Phys. Chem. Solid*, **66**, 292.
81. Nakane, T. *et al.* (2006) *Physica C*, **445–448**, 784.
82. Przybylski, K. *et al.* (2003) *Physica C*, **387**, 148.
83. Vignolo, M. *et al.* (2011) *Supercond. Sci. Technol.*, **24**, 065014.
84. Ma, Z.Q. *et al.* (2011) *Mater. Chem. Phys.*, **126**, 114.
85. Kim, J.H. *et al.* (2007) *Supercond. Sci. Technol.*, **20**, 448.
86. Karpinski, J. *et al.* (2003) *Physica C*, **385**, 42.
87. Karpinski, J. *et al.* (2003) *Supercond. Sci. Technol.*, **16**, 221.
88. Du, W. *et al.* (2004) *J. Cryst. Growth*, **268**, 123.
89. Xu, M. *et al.* (2001) *Appl. Phys. Lett.*, **79**, 2779.
90. Ferdeghini, C. *et al.* (2001) *Supercond. Sci. Technol.*, **14**, 952.
91. Paranthaman, M. *et al.* (2001) *Appl. Phys. Lett.*, **78**, 3669.
92. Plecenik, A. *et al.* (2001) *Physica C*, **363**, 224.
93. Wang, S.F. *et al.* (2003) *Chin. Phys. Lett.*, **20**, 1356.
94. Kühberger, M. *et al.* (2004) *Supercond. Sci. Technol.*, **17**, 764.
95. Xi, X.X. *et al* (2007) *Physica C*, **456**, 22.
96. Pogrebnyakov, A.V. *et al* (2004) *Phys. Rev. Lett.*, **93**, 147006.
97. Zhuang, C. *et al.* (2009) *Supercond. Sci. Technol.*, **22**, 025002.
98. Hyper Tech Research, Inc *http://www.hypertechresearch.com* (accessed 19 May 2014).
99. Columbus Superconductors *http://www.columbussuperconductors.com* (accessed 19 May 2014).
100. Tomsic, M. *et al.* (2007) *Physica C*, **456**, 203.
101. Li, G.Z. *et al.* (2012) *Supercond. Sci. Technol.*, **25**, 115023.
102. Giunchi, G. *et al* (2003) *Supercond. Sci. Technol.*, **16**, 285.
103. Kumakura, H. (2012) *Jpn. J. Appl. Phys.*, **51**, 010003.
104. Häßler, W. *et al.* (2008) *Supercond. Sci. Technol.*, **21**, 062001.
105. Togano, K. *et al.* (2009) *Supercond. Sci. Technol.*, **22**, 015003.
106. Ferrando, V. *et al.* (2005) *Appl. Phys. Lett.*, **87**, 252509.
107. Ranot, M. *et al.* (2012) *Curr. Appl Phys.*, **12**, 353.
108. Abe, H. *et al.* (2004) *Appl. Phys. Lett.*, **85**, 6197.
109. Zhuang, C. *et al.* (2007) *Supercond. Sci. Technol.*, **20**, 287.
110. Auinger, M. *et al.* (2008) *Supercond. Sci. Technol.*, **21**, 015006.
111. Chen, L.P. *et al.* (2007) *Chin. Phys. Lett.*, **24**, 2074.

112. Kúš, P. *et al.* (2002) *Appl. Phys. Lett.*, **81**, 2199.
113. Brinkman, A. *et al.* (2007) *Physica C*, **456**, 188.
114. Chen, K. *et al.* (2006) *Appl. Phys. Lett.*, **88**, 222511.
115. Mijatovic, D. *et al* (2005) *Appl. Phys. Lett.*, **87**, 192505.
116. Cybart, S.A. *et al.* (2006) *Appl. Phys. Lett.*, **88**, 012509.
117. Elsabawy, K.M. (2011) *RSC Adv.*, **1**, 964.
118. Singh, R.K. *et al.* (2006) *Appl. Phys. Lett.*, **89**, 042512.
119. Shimakage, H. *et al.* (2005) *Appl. Phys. Lett.*, **86**, 072512.
120. Ueda, K. *et al.* (2005) *Appl. Phys. Lett.*, **86**, 172502.
121. Kim, T.H. *et al.* (2006) *J. Appl. Phys.*, **100**, 113904.
122. Costachea, M.V. (2010) *Appl. Phys. Lett.*, **96**, 082508.
123. Chen, K. *et al.* (2010) *Appl. Phys. Lett.*, **96**, 042506.
124. Bevilacqua, S. *et al.* (2012) *Appl. Phys. Lett.*, **100**, 033504.
125. Tajima, T. *et al.* (2007) *IEEE Trans. Appl. Supercond.*, **17**, 1330.
126. Nardelli, D. *et al.* (2010) *IEEE Trans. Appl. Supercond.*, **20**, 1998.
127. Yao, W. *et al.* (2009) *IEEE Trans. Appl. Supercond.*, **19**, 2261.

2.2
High-Temperature Superconductors

2.2.1
Cuprate High-Temperature Superconductors
Roland Hott and Thomas Wolf

2.2.1.1 Introduction
Cuprate high-temperature superconductors ("HTSs") have played an outstanding role in the scientific and technological development of superconductors (SCs). Except for semiconductors, no other class of materials has been investigated so thoroughly by thousands and thousands of researchers worldwide. The plethora of preparational degrees of freedom, the inherent tendency toward inhomogeneities and defects, did not allow easy progress in the preparation of these materials. The very short SC coherence lengths of the order of the dimensions of the crystallographic unit cell were on one hand a high hurdle. On the other hand, this SC link of nanoscale microstructure and macroscopic transport properties provided minute monitoring of remnant obstructive defects which could then be tackled by further materials optimization. For cuprate HTS, many material problems have been solved or at least thoroughly discussed which had not been even realized as a problem for other SCs before.

2.2.1.2 Structural Aspects
The structural element of HTS compounds related to the location of mobile charge carriers are stacks of a certain number $n = 1, 2, 3$, and so on, of CuO_2 layers "glued" on top of each other by means of intermediate Ca layers (see Figure 2.2.1.1b) [1–4]. Counterpart of these "*active blocks*" of $(CuO_2/Ca/)_{n-1}CuO_2$ stacks are "*charge reservoir blocks*" $EO/(AO_x)_m/EO$ with $m = 1, 2$ monolayers of a quite arbitrary oxide AO_x (A = Bi [2], Pb [3], Tl [2], Hg [2], Au [5], Cu [2], Ca [6], B [3], Al [3], Ga [3]; see Table 2.2.1.1) "wrapped" on each side by a monolayer of alkaline earth oxide EO with E = Ba, Sr (see Figure 2.2.1.1b). The HTS structure results from alternating stacking of these two block units. The choice of BaO or SrO as "wrapping" layer is not arbitrary but depends on the involved AO_x since it has to provide a good spatial adjustment of the CuO_2 to the AO_x layers.

The general chemical formula $A_mE_2Ca_{n-1}Cu_nO_{2n+m+2+y}$ (see Figure 2.2.1.1b) is conveniently abbreviated as A-$m2(n-1)n$ [4] (e.g., $Bi_2Sr_2Ca_2Cu_3O_{10}$ or Bi-2223), neglecting the indication of the alkaline earth element (see Table 2.2.1.1). The family of all $n = 1, 2, 3$, and so on, representatives with common AO_x are often referred to as *A-HTS*, for example, Bi-HTS. The most prominent compound $YBa_2Cu_3O_7$ (see Figure 2.2.1.1a), the first HTS discovered with a critical temperature T_c above the boiling point of liquid nitrogen [35], is traditionally abbreviated as "YBCO" or "Y-123" ($Y_1Ba_2Cu_3O_{7-\delta}$). It also fits into the general HTS classification scheme as a modification of Cu-1212, where Ca is completely substituted by Y. This substitution introduces extra negative charge in the CuO_2 layers due to the higher valence of Y (+3) compared to Ca (+2).

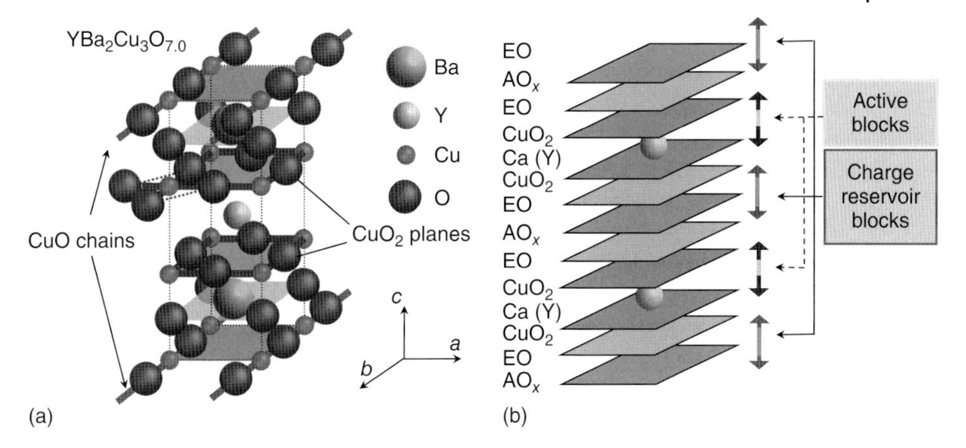

Figure 2.2.1.1 (a) Crystal structure of $YBa_2Cu_3O_7$ ("YBCO"). The presence of the CuO chains introduces an orthorhombic distortion of the unit cell ($a = 0.382$ nm, $b = 0.389$ nm, $c = 1.167$ nm [7]). (b) General structure of a cuprate HTS A-$m2(n-1)n$ ($A_mE_2Ca_{n-1}Cu_nO_{2n+m+2+y}$) for $m = 1$. For $m = 0$ or 2, the missing (additional) AO_x layer per unit cell leads to a ($a/2$, $b/2$, 0) "side step" of the unit cells adjoining in c-axis direction.

The HTS compounds $REBa_2Cu_3O_{7-\delta}$ ("RBCO," "RE-123"), where RE can be La [2] or any other rare earth element [24] except for Ce or Tb [36], can be regarded as a generalization of this substitution scheme. The lanthanide contraction of the RE ions provides here an experimental handle on the distance between the two CuO_2 layers of the active block of the doped Cu-1212 compound [21, 37, 38]. $Y_1Ba_2Cu_4O_8$ ("Y-124") [39] is the $m = 2$ counterpart Cu-2212 of YBCO.

The "214" HTS compounds $E_2Cu_1O_4$ (see Table 2.2.1.1), for example, $La_{2-x}Sr_xCuO_4$ ("LSCO") or $Nd_{2-x}Ce_xCuO_4$ ("NCCO") are a bit exotic in this ordering scheme but may also be represented here as "0201" with $m = 0$, $n = 1$, and $E_2 = La_{2-x}Sr_x$ and $E_2 = Nd_{2-x}Ce_x$, respectively.

Further interesting chemical modifications of the basic HTS compositions are the introduction of fluorine [40, 41] or chlorine [42] as more electronegative substituents of oxygen. For Hg-1223, $T_c = 135$ K can thus be raised to 138 K [43], the highest T_c reported by now under normal pressure conditions (164 K at 30 GPa [44]).

2.2.1.3 **Metallurgical Aspects**

Within 27 years since their discovery [45], cuprate HTS samples have greatly improved toward high materials quality and can in fact nowadays be prepared in a remarkably reproducible way. The enormous worldwide efforts for this achievement may be estimated from the plain number of more than 100 000 articles which meanwhile have been published on cuprate high-T_c superconductivity, strongly outnumbering the only about 15 000 publications [46–47] that appeared within a whole century on the remaining superconducting materials.

Table 2.2.1.1 Classification and reported optimized T_c values of cuprate HTS compounds.

HTS family	Stoichiometry	Notation	Compounds	Highest T_c (K)
Bi-HTS	$Bi_mSr_2Ca_{n-1}Cu_nO_{2n+m+2}$	Bi-$m2(n-1)n$, BSCCO	Bi-1212	102 [8]
	$m = 1, 2$		Bi-2201	34 [9]
	$n = 1, 2, 3, \ldots$		Bi-2212	96 [10]
			Bi-2223	110 [2]
			Bi-2234	110 [11]
Pb-HTS	$Pb_mSr_2Ca_{n-1}Cu_nO_{2n+m+2}$	Pb-$m2(n-1)n$	Pb-1212	70 [12]
			Pb-1223	122 [13]
Tl-HTS	$Tl_mBa_2Ca_{n-1}Cu_nO_{2n+m+2}$	Tl-$m2(n-1)n$, TBCCO	Tl-1201	50 [2]
	$m = 1, 2$		Tl-1212	82 [2]
	$n = 1, 2, 3, \ldots$		Tl-1223	133 [14]
			Tl-1234	127 [15]
			Tl-2201	90 [2]
			Tl-2212	110 [2]
			Tl-2223	128 [16]
			Tl-2234	119 [17]
Hg-HTS	$Hg_mBa_2Ca_{n-1}Cu_nO_{2n+m+2}$	Hg-$m2(n-1)n$, HBCCO	Hg-1201	97 [2]
	$m = 1, 2$		Hg-1212	128 [2]
	$n = 1, 2, 3, \ldots$		Hg-1223	135 [18]
			Hg-1234	127 [18]
			Hg-1245	110 [18]
			Hg-1256	107 [18]
			Hg-2212	44 [19]
			Hg-2223	45 [20]
			Hg-2234	114 [20]
Au-HTS	$Au_mBa_2Ca_{n-1}Cu_nO_{2n+m+2}$	Au-$m2(n-1)n$	Au-1212	82 [6]
123-HTS	$REBa_2Cu_3O_{7-\delta}$	RE-123, RBCO	Y-123, YBCO	92 [21]
	RE = Y, La, Pr, Nd, Sm, Eu, Gd, Tb, Dy, Ho, Er, Tm, Yb, Lu		Nd-123, NBCO	96 [21]
			Gd-123	94 [22]
			Er-123	92 [23]
			Yb-123	89 [24]
Cu-HTS	$Cu_mBa_2Ca_{n-1}Cu_nO_{2n+m+2}$	Cu-$m2(n-1)n$	Cu-1223	60 [2]
	$m = 1, 2$		Cu-1234	117 [25]
	$n = 1, 2, 3, \ldots$		Cu-2223	67 [2]
			Cu-2234	113 [2]
			Cu-2245	< 110 [2]
Ru-HTS	$RuSr_2GdCu_2O_8$	Ru-1212	Ru-1212	72 [26]
B-HTS	$B_mSr_2Ca_{n-1}Cu_nO_{2n+m+2}$	B-$m2(n-1)n$	B-1223	75 [27]
			B-1234	110 [27]
			B-1245	85 [27]

Table 2.2.1.1 (*Continued*)

HTS family	Stoichiometry	Notation	Compounds	Highest T_c (K)
214-HTS	E_2CuO_4	LSCO "0201"	$La_{2-x}Sr_xCuO_4$	51 [28] 25
			Sr_2CuO_4	(75)[29]
		Electron-doped HTS	$La_{2-x}Ce_xCuO_4$	28 [30]
		PCCO	$Pr_{2-x}Ce_xCuO_4$	24 [31]
		NCCO	$Nd_{2-x}Ce_xCuO_4$	24 [31]
			$Sm_{2-x}Ce_xCuO_4$	22 [32]
			$Eu_{2-x}Ce_xCuO_4$	23 [32]
	$Ba_2Ca_{n-1}Cu_nO_{2n+2}$	"02(n − 1)n"	"0212"	90 [33]
			"0223"	120 [33]
			"0234"	105 [33]
			"0245"	90 [33]
Infinite-layer HTS	$ECuO_2$	*Electron-doped I. L.*	$Sr_{1-x}La_xCuO_2$	43 [34]

Nevertheless, the HTS materials quality level is still far from the standards of classical SCs. The reason is the larger number of at least four chemical elements from which the various cuprate HTS phases have to be formed: The majority of the classical SC compounds are made of only two elements. Each element contributes an additional degree of freedom to the preparation route toward new compounds. For the metallurgist, this translates roughly into one order of magnitude more elaborate exploratory efforts. A new ternary cuprate HTS compound requires therefore typically as much metallurgical optimization work as 100 binary compounds.

With respect to the cation stoichiometry, for the RE-123 phase as a particularly well-investigated HTS example, the RE/Ba ratio represents the second chemical degree of freedom which may also deviate from the stoichiometric value 1/2. Y-123 (YBCO) represents here a remarkable exception. However, RE-123 compounds where the Y ions are replaced by larger RE ions like Gd or Nd exhibit a pronounced homogeneity range. For Nd-123, the Nd/Ba ratio encountered in such solid solutions may vary from 0.49 to 2, with the oxygen content being still an almost fully independent chemical variable. The Cu/(EA = Ba or Sr) ratio as chemical degree of freedom of all "hole-doped" HTS compounds (see Table 2.2.1.1) has until now not been studied in great detail. For the 123 phase as a particularly complicated HTS example with respect to the Cu content featuring two CuO_2 layers and an additional CuO chain structure in the unit cell, the present assumption is that the Cu/Ba ratio sticks to the stoichiometric value 3/2. For samples prepared close to their peritectic temperatures or at very low temperatures close to the boundary of the stability field of the 123 phase, this may no longer be the case.

Chemical purity of the starting material is still a topical issue for a reproducible preparation of cuprate HTS. The use of chemicals with a purity of 99.99% and better is mandatory but still not sufficient. The frequently used Ba source material $BaCO_3$ is usually not completely reacted and may thus lead to the incorporation of carbonate ions into the HTS cuprate phase. During the preparation procedure, the formation of even a small amount of liquid has to be carefully avoided since this may corrode the substrate or crucible or may solidify at grain boundaries (see Figure 2.2.1.5), thus impeding the grain-to-grain supercurrent transfer. This corrosion process is particularly harmful for crystal growth experiments where the complete melt encounters the crucible wall. BaO/CuO melts are highly corrosive and attack all conventional types of crucible materials, forming new solid phases. Some of these reaction products have turned out to be well suited as materials for corrosion-resistant crucibles for the preparation of cuprate HTS. $BaZrO_3$ is here the best-known example [23]; $BaHfO_3$ and $BaSnO_3$ are promising candidates. The preparation of an HTS layer structure based on $(CuO_2/Ca/)_{n-1}CuO_2$ stacks with a well-defined number n of CuO_2 layers (see Figure 2.2.1.1) introduces another challenge: As the formation enthalpy of a compound, for example, with a single CuO_2 layer ($n=1$), differs only little from that of the ($n=2$) compound with two adjacent CuO_2 layers, these materials tend to form polytypes [39].

2.2.1.4 Structure and T_c

Experimentally, for all HTS families A-$m2(n-1)n$, the optimized T_c is found to increase from $n=1$ to $n=3, 4$ and to decrease again for higher n (see Table 2.2.1.1). It is still unclear whether this T_c maximum is an intrinsic HTS property since the synthesis of higher n members of the HTS families turns out to be more and more complicated [48–51]. In particular, there is at present no preparation technique that allows to adjust here a sufficiently high oxygen content or to provide otherwise sufficient electronic doping that would allow to clarify the possible range of T_c optimization for higher n HTS A-$m2(n-1)n$ [48, 51]. Hence, the question is still open if such an optimized T_c may eventually continue to increase toward higher n, possibly up to $T_c \sim 200$ K [52].

Another well-investigated experimental T_c trend is the slight increase of the optimized T_c values of the RE-123 HTS with increasing distance between the two CuO_2 layers in the active block. It has been explained in terms of a higher effective charge transfer to the CuO_2 layers [21, 24]. For RE-123 HTS with larger RE ions (La, Pr, Nd), sufficient oxygenation with respect to T_c optimization becomes increasingly difficult. As a further complication, these larger RE ions are comparable in size with the Ba ions which favor cation disorder with respect to the RE and Ba lattice sites [53] with the consequence of substantial T_c degradation [54]. This disorder effect, oxygen deficiency, and/or impurities had been suggested as reasons for the nonappearance or quick disappearance of superconductivity in Pr-123 [55] and Tb-123 as the only non-SC members of the 123-HTS family. Pr-123 samples with $T_c \sim 80$ K at normal pressure [56] and 105 K at 9.3 GPa [57] as measured immediately after preparation lost their SC properties within a few days. The nature of this SC Pr-123 phase is still unclear. It has been suggested that it consisted of Ba-rich Pr-123 which is unstable at room temperature with respect to spinodal decomposition.

In Bi-HTS, cation disorder at the Sr lattice site is inherent and strongly affects the value of T_c. In Bi-2201, partial substitution of Sr by RE = La, Pr, Nd, Sm, Eu, Gd, and Bi was shown to result in a monotonic decrease of T_c with increasing ionic radius mismatch [10]. For Bi-2212, partial substitution of Ca by Y results in a T_c optimum of 96 K at 8% Y doping, apparently due to a trade-off between the respective disorder effect and charge doping [10].

In 214-HTS without BaO or SrO "wrapping" layers around the CuO_2 layers (see Table 2.2.1.1), T_c seems to be particularly sensitive with respect to oxygen disorder [58]. The electron-doped 214-HTS such as $Nd_{2-x}Ce_xCuO_4$ ("NCCO") have here the additional complication of a different oxygen sublattice where oxygen ions on interstitial lattice positions in the $Nd_{2-x}Ce_xO_2$ layer (which are yet for hole-doped 214-HTS the regular oxygen positions in the non-CuO_2 layer!) tend to suppress T_c [59].

With respect to the T_c dependence of the A-$m2(n-1)n$ HTS families on the cation A of the charge reservoir blocks, there is an increase moving in the periodic table from Bi to Hg (see Table 2.2.1.1). However, continuing to Au, the reported T_c is already substantially lower. This T_c trend seems to be related to the chemical nature of the A–O bonds in the AO_x layers [19].

For constant doping, buckling of the CuO_2 layer is observed to decrease T_c [60]. Such deviations from the simple tetragonal crystal structure are found in most of the HTS compounds as a chemical consequence of the enforced AO_x layer arrangement in the cuprate HTS structure. The record T_c values for single CuO_2 layer HTS compounds of Tl-2201 ($T_c = 90$ K) [61, 62] and Hg-1201 ($T_c = 90$ K) [63] with simple tetragonal crystal structure indicate that undistorted flat CuO_2 layers provide optimum superconductivity.

The SC critical temperature values T_c reported in the preceding paragraphs refer to the maximum values obtained for individually optimized doping either by variation of the oxygen content or by suitable substitution of cations. The following scenario applies to *hole-doping* [64] as well as to *electron-doping* of all cuprate HTS (see Figure 2.2.1.2) [65]: The undoped compounds are antiferromagnetic insulators up to a critical temperature T_N well above 300 K, with alternating spin orientations of the hole states that are localized around the Cu atoms in the CuO_2 layers. Adding charge carriers by doping relaxes the restrictions of spin alignment due to the interaction of these additional spin-1/2-particles with the spin lattice. T_N decreases and the insulator turns into a "bad metal." At low temperature, however, the electric transport shows a dramatic change within a small doping range from an insulating to a SC behavior [66]. For $La_{2-x}Sr_xCuO_4$, this happens at a critical hole concentration $x = 0.05$ in the CuO_2 planes (see Figure 2.2.1.1). On stronger doping, superconductivity can be observed up to an increasingly higher critical temperature T_c until the maximum T_c is achieved for "optimal doping" ($x \approx 0.16$ for $La_{2-x}Sr_xCuO_4$). On further doping, T_c decreases

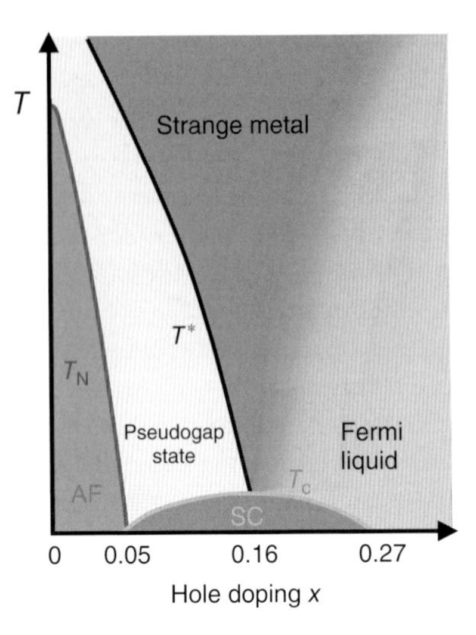

Figure 2.2.1.2 Schematic HTS temperature-doping phase diagram with the interplay of antiferromagnetism ("AF") and superconductivity ("SC") [64, 67–69].

again until finally ($x \geq 0.27$ for $La_{2-x}Sr_xCuO_4$) only normal conducting behavior is observed.

2.2.1.5 **Superconductive Coupling**

The rationale that the phenomenon of superconductivity in HTS can be conceptually reduced to the physics of the CuO_2 layers [70] has evolved to a more and more two-dimensional view in terms of CuO_2 *planes*. The superconductive coupling between these planes within a given $(CuO_2/Ca/)_{n-1}CuO_2$ stack ("interplane coupling") is much weaker than the intraplane coupling, but still much stronger than the superconductive coupling between the $(CuO_2/Ca/)_{n-1}CuO_2$ stacks which can be described as Josephson coupling (see Figure 2.2.1.3).

The charge reservoir blocks $EO/(AO_x)_m/EO$ play in this idealized theoretical picture only a passive role, providing the doping charge as well as the "storage space" for extra oxygen ions and cations introduced by additional doping.

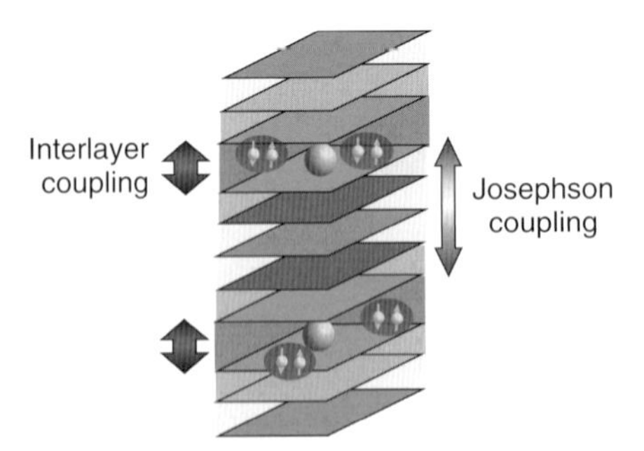

Figure 2.2.1.3 Hierarchy of the superconductive coupling in cuprate HTS.

However, the huge pressure dependence of T_c [44] in combination with the large quantitative variation of this effect for the various A-$m2(n-1)n$ HTS families points to a more active role where the cations change not only their valency but also their transmission behavior for the interstack tunneling of Cooper pairs [71].

This becomes most evident for the Cu-HTS family, in particular for YBCO or the RE-123 HTS where the AO_x layer is formed by 1-d CuO chain structures (see Figure 2.2.1.1). There is experimental evidence that these CuO chains become SC, probably via proximity effect. The intercalation of superconductive CuO chain layers in between the $CuO_2/Ca/CuO_2$ bilayer stacks is most likely the origin of the strong Josephson coupling between these bilayer stacks. This explains the remarkable reduction of the superconductive anisotropy in c-axis direction compared to all other HTS families. Moreover, in contrast to the usually isotropic SC behavior within the $a-b$-plane, the CuO chains seem to introduce a substantially higher SC gap in b-direction compared to the a-direction [72]. This particular SC anisotropy renders YBCO and the RE-123 HTS exceptional among the cuprate HTS.

HTS are extreme type-II SCs [73] with $\lambda > 100$ nm and $\xi \sim 1$ nm. The quasi-two-dimensional nature of superconductivity in HTS leads to a pronounced anisotropy of the SC properties with much higher supercurrents along the CuO_2 planes than in the perpendicular direction [74, 75] and a corresponding anisotropy of λ, for example, $\lambda_{ab} = 750$ nm and $\lambda_c = 150$ nm, in YBCO [76] (the indices refer to the respective orientation of the magnetic field). Material imperfections of the dimension of the coherence length which are required as pinning centers preventing the flux flow of magnetic vortices are easily encountered in HTS due to their small coherence lengths, for example, for optimally doped YBCO $\xi_{ab} = 1.6$ nm, $\xi_c = 0.3$ nm for $T \rightarrow 0$ K [77] which are already comparable to the lattice parameters (YBCO: $a = 0.382$ nm, $b = 0.389$ nm, $c = 1.167$ nm [7]). The high T_c in combination with the small value of coherence volume $(\xi_{ab})^2\xi_c \sim 1\ \text{nm}^3$ allows large thermally induced magnetic fluctuations in the SC phase at temperature close to T_c, an effect which could completely neglected in classical SCs [78]. Moreover, since technical SC materials consist of a network of connected grains, already small imperfections at the grain boundaries with spatial extensions of the order of the coherence length lead to a substantial weakening of the SC connection of the grains and thus to a "weak-link behavior" of the transport properties. Obviously, this effect has to be avoided in technical conductor materials [79]. On the other hand, it has also been widely exploited for the fabrication of HTS Josephson junctions [80].

The low ξ_c, that is, the weak superconductive coupling between the $(CuO_2/Ca/)_{n-1}CuO_2$ stacks may lead for c-axis transport to an intrinsic Josephson effect within the unit cell even for perfect single-crystalline materials [81]. If the thickness of the charge reservoir blocks $EO/(AO_x)_m/EO$ in between these stacks is larger than ξ_c, vortices are here no longer well-defined due to the low Cooper-pair density (see Figure 2.2.1.4). This leads to a quasi-disintegration of the vortices into stacks of "*pancake vortices*" which are much more flexible entities than the continuous quasi-rigid vortex lines in conventional SCs and require therefore individual pinning centers. The extent of this quasi-disintegration is

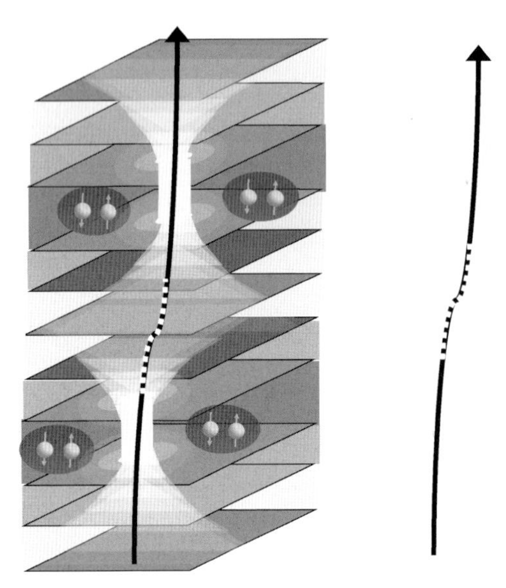

Figure 2.2.1.4 Quasi-disintegration of magnetic vortex lines into "pancake" vortices due to weak SC interlayer coupling and magnetic field overlap of neighboring vortices [83].

different for the various HTS compounds since ξ_c is on the order of the thickness of a single-oxide layers, for example, $d_{\mathrm{TlO}x} = 0.2$ nm for the Tl-HTS [76]. Hence, the number of layers in the charge reservoir blocks $\mathrm{EO/(AO}_x)_m\mathrm{/EO}$ makes a significant difference with respect to the pinning properties and thus to their supercurrents in magnetic fields. This is one of the reasons why YBCO ("Cu-1212") has a higher supercurrent capability in magnetic fields than the Bi-HTS Bi-2212 and Bi-2223 which for manufacturing reasons have been for a long time the most prominent HTS conductor materials. In addition, in the Cu-HTS family, the AO_x layer is formed by CuO chains (see Figure 2.2.1.1) which apparently become SC via proximity effect. This leads here to the smallest superconductive anisotropy among all HTS families [82].

The effects described in the preceding two paragraphs combine to reduce the *irreversibility field* $B_{\mathrm{irr}}[T]$, the tolerable limit for magnetic fields with respect to SC transport, in cuprate HTS substantially below the thermodynamical critical field $B_{\mathrm{c2}}[T]$, a distinction which was more or less only of academic interest in the case of classical SC.

Beside these intrinsic obstacles for the transport of supercurrent in single-crystalline HTS materials, there are additional hurdles since HTS material is not a homogeneous continuum but rather a network of linked grains (see Figure 2.2.1.5). The process of crystal growth is such that all material that cannot be fitted into the lattice structure of the growing grains is pushed aside the growth front with the consequence that, in the end, all remnants of secondary phases and impurities are concentrated at the boundaries in between the grains. Such barriers impede the current transport even if they consist of only a few

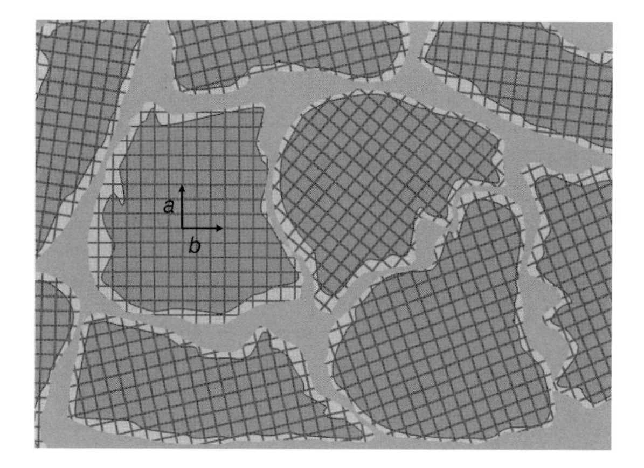

Figure 2.2.1.5 Schematics of the HTS microstructure: Differently oriented single crystal grains are separated by regions filled with secondary phase relicts from the melt growth. In addition, oxygen depletion and thus T_c reduction may occur at grain boundaries.

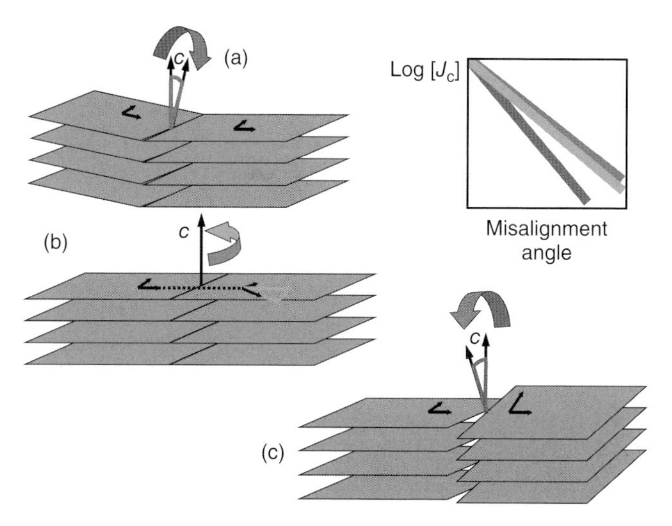

Figure 2.2.1.6 Basic grain boundary geometries and experimentally observed J_c reduction $J_c \sim e^{\alpha/\alpha_0}$ as function of the misalignment angle α: $\alpha_0 \approx 5°$ for (a) and (b), $\alpha_0 \approx 3°$ for (c) independent of temperature [80a].

atomic layers and have to be avoided by careful control of the growth process, in particular of the composition of the offered material.

Another obstacle for supercurrents in HTS (which is detrimental for transport currents, but also enables the fabrication of HTS Josephson junctions) is misalignment of the grains: Exponential degradation of the supercurrent transport is observed as a function of the misalignment angle (see Figure 2.2.1.6).

One of the reasons for this behavior is the d-symmetry of the SC order parameter (see Figure 2.2.1.7) [84]: Cuprate HTS have been established as a textbook example of a d-wave symmetric SC order parameter which can be

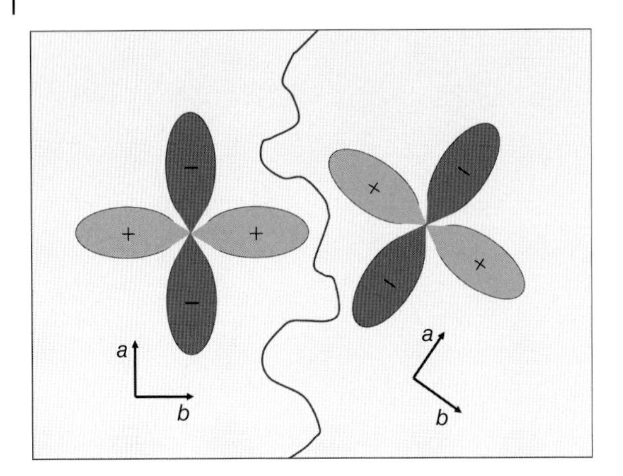

Figure 2.2.1.7 Schematics of a boundary between HTS grains. Misorientation of the SC d-wave order parameter leads to partial cancellation of the supercurrents modified by the faceting of the grain boundaries.

observed directly by means of particularly designed "Superconducting QUantum Interference (Device)" ("SQUID") circuits [84–89].

However, the J_c reduction as a function of the misalignment angle α turns out to be much larger than what is expected from d-wave symmetry alone [81a, 91, 92]. This extra J_c degradation as well as the change of the current–voltage characteristics of the transport behavior [92] were believed to arise from structural defects such as dislocations [93] and deviations from stoichiometry, in particular the loss of oxygen at the grain surfaces [94] and the concomitant local degradation of the SC properties due to the decreased doping (see Figure 2.2.1.2) [95]. A recent microscopic modeling identified the buildup of charge inhomogeneities as the dominant mechanism for the suppression of the supercurrent [96].

Owing to all of these limitations, practical application of cuprate HTS materials has turned out to be restricted to perfectly aligned single-crystalline materials such as epitaxial films and well-textured bulk material without *weak-link behavior*, the drastic reduction of critical currents already in low magnetic fields resulting from the effects described above.

Bi-HTS *(oxide) powder in tube* ("*(O)PIT*") tape conductors seem to be the only exception since they constantly paved their way from the first short samples in 1989 [97] to present large cable projects [98, 99]. The basic idea behind this wire preparation technique can be seen from the solution of the problem how to knot a cigarette: Wrap it in aluminum foil and then go ahead with your mechanical deformation! For Bi-HTS powder, this principle works with Ag tubes as well: Bi-2212 or Bi-2223 powder or respective precursor powder is filled in Ag tubes which are subject to several mechanical deformation steps of drawing and rolling and intermediate annealing for the development of the SC Bi-HTS phase. The oxygen permeability of Ag allows for sufficient subsequent oxygenation. The two neighboring BiO layers in the atomic Bi-2212 or Bi-2223 structure are only weakly bound and lead to graphite-like mechanical properties which allow an easy sliding or splitting of the grains along these layers (that is the chemical reason why for YBCO and Tl-1223 with only a single intermediate oxide layer (O)PIT wire fabrication did not meet with success). The resulting plate-like Bi-HTS grains in the filaments become aligned during the mechanical deformation steps of drawing and rolling within a few degrees. This process has been optimized and results nowadays in conductor material with SC currents that are sufficiently high for cable applications. Unavoidably, microcracks occur in the Bi-HTS filaments, but are fortunately short-circuited by the Ag matrix which stays in close contact to Bi-HTS filaments. Nevertheless, this results in the inclusion of short resistive current paths which are reflected in the current–voltage characteristics that do not allow persistent current operation. The cost of Ag as the only possible tube material is an additional handicap. YBCO-coated metal bands with epitaxial single-crystalline YBCO grain alignment offer the prospect of overcoming both problems within foreseeable future, but are nowadays due to the required complicated fabrication process still quite expensive as well. Anyhow, even though not all cuprate HTS dreams have come true, the broad commercial interest in applications enabled by these two conductor materials will now definitively establish at least one of these cuprate HTS within the next decade as new important technical conductor material.

References

1. Chu, C.W. (1999) *J. Supercond.*, **12**, 85.
2. Chu, C.W. (1997) *IEEE Trans. Appl. Supercond.*, **7**, 80.
3. Yamauchi, H., Karppinen, M., and Tanaka, S. (1996) *Physica C*, **263**, 146.
4. Yamauchi, H. and Karppinen, M. (2000) *Supercond. Sci. Technol.*, **13**, R33.
5. Bordet, P. *et al.* (1997) *Physica C*, **276**, 237.
6. Wu, N.L. *et al.* (1999) *Physica C*, **315**, 227.
7. Harshman, D.R. and Mills, A.P. Jr., (1992) *Phys. Rev. B*, **45**, 10684.
8. Zoller, P. *et al.* (1995) *Z. Phys. B*, **96**, 505.
9. Feng, D.L. *et al.* (2002) *Phys. Rev. Lett.*, **88**, 10700.
10. Eisaki, H. *et al.* (2004) *Phys. Rev. B*, **69**, 064512.
11. Lösch, S. *et al.* (1991) *Physica C*, **177**, 271.
12. Yamauchi, H. *et al.* (1995) *Jpn. J. Appl. Phys.*, **34**, L349.
13. Tamura, T., Adachi, S., Wu, X.-J., Tatsuki, T., and Tanabe, K. (1997) *Physica C*, **277**, 1.
14. Iyo, A. *et al.* (2001) *Supercond. Sci. Technol.*, **14**, 504.
15. Iyo, A. *et al.* (2001) *Physica C*, **357–360**, 324.
16. Tristan Jover, D. *et al.* (1996) *Phys. Rev. B*, **54**, 10175.
17. Chen, Z.Y. *et al.* (1993) *Supercond. Sci. Technol.*, **6**, 261.
18. Antipov, E.V. *et al.* (2002) *Supercond. Sci. Technol.*, **15**, R31.
19. Acha, C. *et al.* (1997) *Solid State Commun.*, **102**, 1.
20. Tatsuki, T. *et al.* (1996) *Jpn. J. Appl. Phys.*, **35**, L205.
21. Williams, G.V.M. and Tallon, J.L. (1996) *Physica C*, **258**, 41.
22. Stangl, E., Proyer, S., Borz, M., Hellebrand, B., and Bäuerle, D. (1996) *Physica C*, **256**, 245.
23. (a) Erb, A., Walker, E., and Flükiger, R. (1995) *Physica C*, **245**, 245; (b) Erb, A., Walker, E., Genoud, J.-Y., and Flükiger, R. (1997) *Physica C*, **282–287**, 89.

24. Lin, J.G. *et al.* (1995) *Phys. Rev. B*, **51**, 12900.
25. Ihara, H. (2001) *Physica C*, **364–365**, 289.
26. (a) Klamut, P.W. *et al.* (2001) *Physica C*, **364–365**, 313; (b) Lorenz, B. *et al.* (2001) *Physica C*, **363**, 251.
27. Kawashima, T., Matsui, Y., and Takayama-Muromachi, E. (1995) *Physica C*, **254**, 131.
28. Bozovic, I. *et al.* (2002) *Phys. Rev. Lett.*, **89**, 107001.
29. Karimoto, S., Yamamoto, H., Greibe, T., and Naito, M. (2001) *Jpn. J. Appl. Phys.*, **40**, L127.
30. Naito, M. and Hepp, M. (2000) *Jpn. J. Appl. Phys.*, **39**, L485.
31. Alff, L. *et al.* (1999) *Phys. Rev. Lett.*, **83**, 2644.
32. Jin, C.Q. *et al.* (1993) *Appl. Phys. Lett.*, **62**, 3037.
33. Iyo, A., Tanaka, Y., Tokumoto, M., and Ihara, H. (2001) *Physica C*, **366**, 43.
34. Jung, C.U. *et al.* (2001) *Physica C*, **364–365**, 225.
35. Wu, M.K. *et al.* (1987) *Phys. Rev. Lett.*, **58**, 908.
36. Chu, J.W., Feng, H.H., Sun, Y.Y., Matsuishi, K., Xiong, Q., and Chu, C.W. (1992) in *HTS Materials, Bulk Processing and Bulk Applications - Proceedings of the 1992 TCSUH Workshop* (eds C.W. Chu, W.K. Chu, P.H. Hor, and K. Salama), World Scientific, Singapore, p. 53; TCSUH preprint 92:043.
37. Guillaume, M. *et al.* (1994) *J. Phys. Condens. Matter*, **6**, 7963.
38. Guillaume, M. *et al.* (1993) *Z. Phys. B*, **90**, 13.
39. (a) Williams, R.K. *et al.* (1994) *J. Appl. Phys.*, **76**, 3673; (b) Zhang, W. and Osamura, K. (1992) *Physica C*, **190**, 396; (c) Voronin, G.F. and Degterov, S.A. (1991) *Physica C*, **176**, 387; (d) Karpinski, J., Rusiecki, S., Bucher, B., Kaldis, E., and Jilek, E. (1989) *Physica C*, **161**, 618; (e) Karpinski, J., Rusiecki, S., Bucher, B., Kaldis, E., and Jilek, E. (1989) *Physica C*, **160**, 449; (f) Karpinski, J., Kaldis, E., Jilek, E., Rusiecki, S., and Bucher, B. (1988) *Nature*, **336**, 660.
40. Al-Mamouri, M., Edwards, P.P., Greaves, C., and Slaski, M. (1994) *Nature*, **369**, 382.
41. Kawashima, T., Matsui, Y., and Takayama-Muromachi, E. (1996) *Physica C*, **257**, 313.
42. Hiroi, Z., Kobayashi, N., and Takano, M. (1994) *Nature*, **371**, 139.
43. Putilin, S.N. *et al.* (2001) *Physica C*, **338**, 52.
44. Gao, L. *et al.* (1994) *Phys. Rev. B*, **50**, 4260.
45. Bednorz, J.G. and Müller, K.A. (1986) *Z. Phys. B*, **64**, 189.
46. Roberts, B.W. (1976) *J. Phys. Chem. Ref. Data*, **5**, 581.
47. Roberts, B.W. (1978) *Properties of Selected Superconductive Materials 1978 Supplement*, NBS Technical Note 983, National Bureau of Standards.
48. Majewski, P. (2000) *J. Mater. Res.*, **15**, 854.
49. Siegal, M.P. *et al.* (1997) *J. Mater. Res.*, **12**, 2825.
50. Capponi, J.J. *et al.* (1994) *Physica C*, **235–240**, 146.
51. Ito, T., Suematsu, H., Karppinen, M., and Yamauchi, H. (1998) *Physica C*, **308**, 198.
52. Leggett, J. (1999) *Phys. Rev. Lett.*, **83**, 392.
53. Kuroda, K. *et al.* (1997) *Physica C*, **275**, 311.
54. Attfield, J.P., Kharlanov, A.L., and McAllister, J.A. (1998) *Nature*, **394**, 157.
55. (a) Pieper, M.W., Wiekhorst, F., and Wolf, T. (2000) *Phys. Rev. B*, **62**, 1392; (b) Markwardsen, A.J. *et al.* (1998) *J. Magn. Magn. Mater.*, **177–181**, 502.
56. (a) Oka, K., Zou, Z., and Ye, J. (1998) *Physica C*, **300**, 200; (b) Zou, Z., Ye, J., Oka, K., and Nishihara, Y. (1998) *Phys. Rev. Lett.*, **80**, 1074.
57. Ye, J. *et al.* (1998) *Phys. Rev. B*, **58**, 619.
58. Lee, Y.S. *et al.* (2004) *Phys. Rev. B*, **69**, 020502.
59. Mang, P.K. *et al.* (2004) *Phys. Rev. B*, **70**, 094507.
60. Chmaissem, O. *et al.* (1999) *Nature*, **397**, 45.
61. Izumi, F. *et al.* (1992) *Physica C*, **193**, 426.

62. Ren, Z.F., Wang, J.H., and Miller, D.J. (1996) *Appl. Phys. Lett.*, **69**, 1798.
63. Yamato, A., Hu, W.Z., Izumi, F., and Tajima, S. (2001) *Physica C*, **351**, 329.
64. Orenstein, J. and Millis, A.J. (2000) *Science*, **288**, 468.
65. Onose, Y., Taguchi, Y., Ishizaka, K., and Tokura, Y. (2001) *Phys. Rev. Lett.*, **87**, 217001.
66. Ando, Y. *et al.* (2001) *Phys. Rev. Lett.*, **87**, 017001.
67. Tallon, J.L. and Loram, J.W. (2001) *Physica C*, **349**, 53.
68. Varma, C.M. (1997) *Phys. Rev. B*, **55**, 14554.
69. Alff, L. *et al.* (2003) *Nature*, **422**, 698.
70. Anderson, P.W. (1987) *Science*, **235**, 1196.
71. Geballe, T.H. and Moyzhes, B.Y. (2000) *Physica C*, **341–348**, 1821.
72. Lu, D.H. *et al.* (2001) *Phys. Rev. Lett.*, **86**, 4370.
73. Brandt, E.H. (2001) *Phys. Rev. B*, **64**, 024505.
74. Grasso, G. and Flükiger, R. (1997) *Supercond. Sci. Technol.*, **10**, 223.
75. (a) Sato, J. *et al.* (2001) *Physica C*, **357–360**, 1111; (b) Okada, M. (2000) *Supercond. Sci. Technol.*, **13**, 29.
76. Saxena, A.K. (2010, 2012) *High-Temperature Superconductors*, Springer-Verlag, Berlin, Heidelberg.
77. Plakida, N.P. (1995) *High-Temperature Superconductivity*, Springer-Verlag, Berlin, Heidelberg.
78. Fossheim, K. and Sudbo, A. (2004) *Superconductivity – Physics and Applications*, John Wiley & Sons, Ltd.
79. Larbalestier, D., Gurevich, A., Feldmann, D.M., and Polyanskii, A. (2001) *Nature*, **414**, 368.
80. (a) Hilgenkamp, H. and Mannhart, J. (2002) *Rev. Mod. Phys.*, **74**, 485; (b) Gross, R. *et al.* (1997) *IEEE Trans. Appl. Supercond.*, 7, 2929.
81. Kleiner, R., Steinmeyer, F., Kunkel, G., and Müller, P. (1992) *Phys. Rev. Lett.*, **68**, 2394.
82. (a) Tanaka, K. *et al.* (2000) *Physica B*, **284–288**, 1081; (b) Watanabe, T. *et al.* (2000) *Physica B*, **284–288**, 1075.
83. Brandt, E.H. (1995) *Rep. Prog. Phys.*, **58**, 1465; condmat/9506003.
84. (a) Tsuei, C.C. and Kirtley, J.R. (2000) *Rev. Mod. Phys.*, **72**, 969; (b) Tsuei, C.C. and Kirtley, J.R. (2002) *Physica C*, **367**, 1.
85. Wollman, D.A. *et al.* (1995) *Phys. Rev. Lett.*, **74**, 797.
86. Bulaevskii, L.N., Kuzii, V.V., and Sobyanin, A.A. (1977) *JETP Lett.*, **25**, 290.
87. Geshkenbein, V.B., Larkin, A.I., and Barone, A. (1987) *Phys. Rev. B*, **36**, 235.
88. Chesca, B. *et al.* (2002) *Phys. Rev. Lett.*, **88**, 177003.
89. Hilgenkamp, H. *et al.* (2003) *Nature*, **422**, 50.
90. Hilgenkamp, H., Mannhart, J., and Mayer, B. (1996) *Phys. Rev. B*, **53**, 14586.
91. Nilsson, P.A. *et al.* (1994) *J. Appl. Phys.*, **75**, 7972.
92. Betouras, J. and Joynt, R. (1995) *Physica C*, **250**, 256.
93. Alarco, J. and Olsson, E. (1995) *Phys. Rev. B*, **52**, 13625.
94. Agassi, D., Christen, D.K., and Pennycook, S.J. (2002) *Appl. Phys. Lett.*, **81**, 2803.
95. Hilgenkamp, H. and Mannhart, J. (1998) *Appl. Phys. Lett.*, **73**, 265.
96. Graser, S. *et al.* (2010) *Nat. Phys.*, **6**, 609.
97. Heine, K., Tenbrink, J., and Thöner, M. (1989) *Appl. Phys. Lett.*, **55**, 2441.
98. Honjo, S. *et al.* (2011) *IEEE Trans. Appl. Supercond.*, **21**, 967.
99. Yamada, Y., Mogi, M., and Sato, K. (2007) *SEI Tech. Rev.*, **65**, 51.

2.2.2
Iron-Based Superconductors: Materials Aspects for Applications

Ilaria Pallecchi and Marina Putti

2.2.2.1 Introduction

In this section, we extend the general aspects of superconductivity described so far, focusing on the special case of Fe-based superconductors (FeSCs). Chemistry is under sufficient control to allow for reasonable comparisons of experimental data without major concerns about sample quality variations and for analysis of general and universal properties. This is not an easy task because diverse, yet reproducible, properties, especially in the superconducting state, are observed: the intrinsic nature of these materials and the near degeneracy of different ground states seem to result in a strong sensitivity to many real and unavoidable perturbations.

In the following, an introductory review on FeSCs is presented, with main focus on the most relevant properties for applications. In Section 2.2.2.2, some distinctive aspects of FeSCs are briefly summarized, such as crystalline structure, magnetic order, electron bands, and Fermi surfaces. In Section 2.2.2.3, we give some information on the preparation of different kinds of samples such as polycrystals, single crystals, and thin films. In Section 2.2.2.4, we analyze some basic superconducting properties such as critical temperature, upper critical field, and critical current density. In Section 2.2.2.5, we describe how the critical current density has been improved by introducing pinning centers. In Section 2.2.2.6, we tackle the issue of grain boundaries (GBs), which may obstruct the superconducting current flow in wires and tapes, thus severely limiting applications. In the following two sections, we review the achievements obtained so far in the fabrication of FeSC wires, tapes, and coated conductors. In Section 2.2.2.9, we present some results obtained in view of FeSC electronic applications, such as Josephson junctions and superconducting quantum interference devices (SQUIDs). Finally, we draw some conclusions about the state of the art and the perspectives of FeSCs for applications.

2.2.2.2 General Aspects of Fe-Based Superconductors

Since the discovery of superconductivity at 26 K in fluorine-doped LaFeAsO by the group of Hideo Hosono [1], the maximum transition temperature T_c obtained to date in compounds of the same family has peaked at around 55 K by replacing La by other rare earth elements [2]. Given the antagonistic relationship between superconductivity and magnetism, the discovery of these FeSCs has been quite unexpected and has triggered fundamental and experimental studies worldwide. Four main iron-based superconducting families with distinctive *crystallographic structures* have been identified so far: the “1111” family with chemical composition LnFeAsO (Ln, lanthanides), the “122” family with chemical composition AFe_2As_2 (A, alkaline earth metal), the “111” family represented by LiFeAs (or another alkali metal in the place of Li), and the “11” family with chemical composition FeCh (Ch, chalcogen ion). Among these, the 1111, 122, and 11

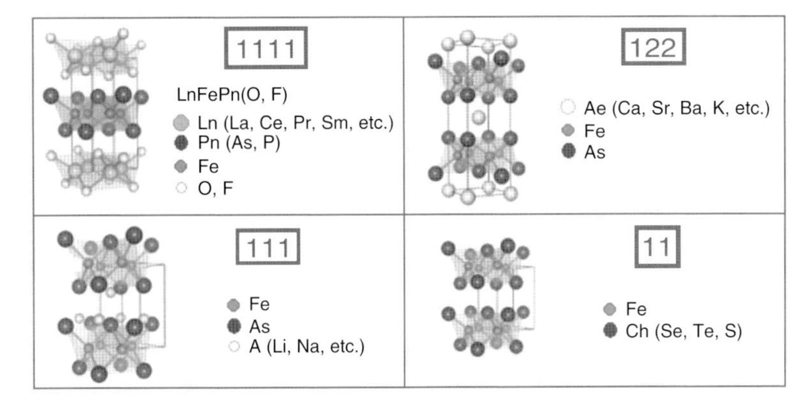

Figure 2.2.2.1 Schematic representation of the crystal structures of the FeSCs.

are the most studied for bulk superconductivity applications. The corresponding crystal structures are sketched in Figure 2.2.2.1. All of these crystal structures share square lattices of iron atoms with tetrahedrally coordinated bonds to either phosphorus, arsenic, selenium, or tellurium anions that are staggered above and below the iron lattice. These slabs are either simply stacked together, as in FeSe, or are separated by spacer layers using alkali (e.g., Li), alkaline earth (e.g., Ba), and rare earth oxide/fluoride (e.g., LaO or SrF). The geometry of the $FeAs_4$ tetrahedra plays a crucial role in determining the electronic and magnetic properties of these systems. For instance, the As–Fe–As tetrahedral bond angles seem to play a crucial role in optimizing T_c. This fact offers plenty of opportunities for tuning the superconducting properties by external control parameters such as strain, applied pressure, or chemical substitution.

Long-range *magnetic order* also shares a similar pattern in all of the FeSC parent compound systems, with an arrangement consisting of spins ferromagnetically arranged along one chain of nearest neighbors within the iron lattice plane, and antiferromagnetically arranged along the other direction.

Common features are easily identified also in the *electronic band structures* of Fe-SCs. The dominant contribution to the electronic density of states at the Fermi level derives from metallic bonding of the iron d-electron orbitals, which form a Fermi surface of at least four quasi two dimensional electron and hole cylinders. These consist of two hole pockets at the Brillouin zone center and two electron pockets at $(0;\pm\pi)$ and $(\pm\pi;0)$ in the tetragonal unit cell. A fifth hole band may be also present at $(0;\pm\pi)$, depending on structural and compositional details. A typical Fermi surface is shown in Figure 2.2.2.2.

The issue of single-gap or multi-gap superconductivity in FeSCs is still open and experimental evidences thereof are reviewed in [4].

It is widely believed that a *spin density wave* (SDW) instability arises from the nesting of two Fermi surface pockets by a large $Q = (\pi;\pi)$ vector that is commensurate with the structure (see Figure 2.2.2.2). This vector corresponds to the magnetic ordering vector measured throughout the FeAs-based parent compounds as well as that for magnetic fluctuations in the related superconducting

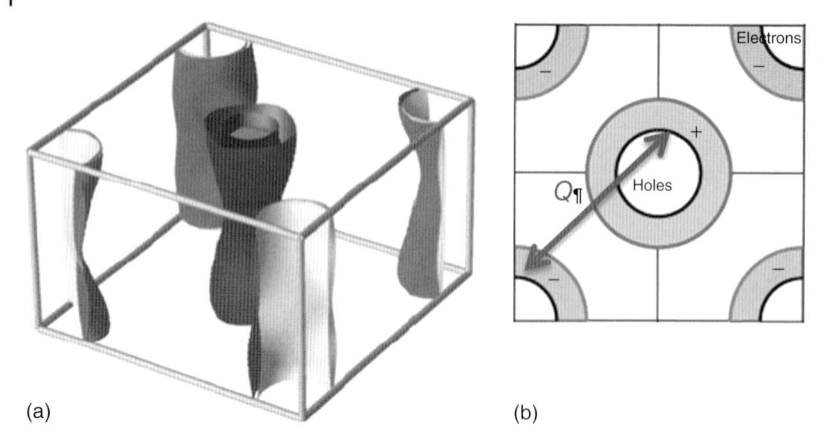

Figure 2.2.2.2 (a) Fermi surface of Co-doped $BaFe_2As_2$, representative of the Fermi surfaces of most FeSCs. (b) Schematic of the two-dimensional (k_x-k_y) projection of the Brillouin zone of FeSCs, with multiple bands reduced to single hole and electron pockets. The spin-mediated mechanism assumes an exchange of antiferromagnetic fluctuations between the hole and electron pockets connected by the antiferromagnetic wave-vector $Q = (\pi;\pi)$. The resulting multiband pairing gap symmetry, drawn as shaded regions on hole and electron pockets, is shown for an s± structure with isotropic gaps on each band. See Ref. [3].

compounds. Experimental evidence for (π;π) Fermi surface nesting across most of FeSC compounds has been found (in fact, a notable exception is $A_xFe_{2-y}Se_2$, where A is an alkali element [5]).

Since the earliest stages of research on these materials, the unconventional nature of the *pairing mechanism* has been pointed out, so that a comparison with the high-T_c cuprates comes out naturally. Although the mediator of pairing is yet unidentified both in FeSCs and in cuprates, it is widely believed that it should be attributed to magnetic spin fluctuations: in all cases, magnetism must be suppressed by either pressure or doping, before optimal bulk-phase superconductivity appears. The *pairing symmetry* is likely to be different in the two classes of materials: d-wave symmetry of the order parameter is widely accepted in cuprates, while most of the experimental evidence and the theoretical predictions indicate s-wave symmetry in FeSCs, with a sign change in the phase of the order parameters in different sheets of the Fermi surface (s± symmetry, pictured in Figure 2.2.2.2). However, the final answer in this respect has yet to be given.

The *phase diagrams* of the main FeSC families, plotted in Figure 2.2.2.3 as temperature versus doping or chemical substitution, are reminiscent of those of high-T_c cuprates, characterized by the competition between magnetic and superconducting orders. In all cases, the SDW order observed in the parent compounds below a characteristic temperature T_N disappears at doping levels of one to small percentage. A superconducting ground state appears with increasing doping, below a dome-shaped region of the phase diagram, with maximum superconducting T_c at optimal doping. In some cases, the SDW and

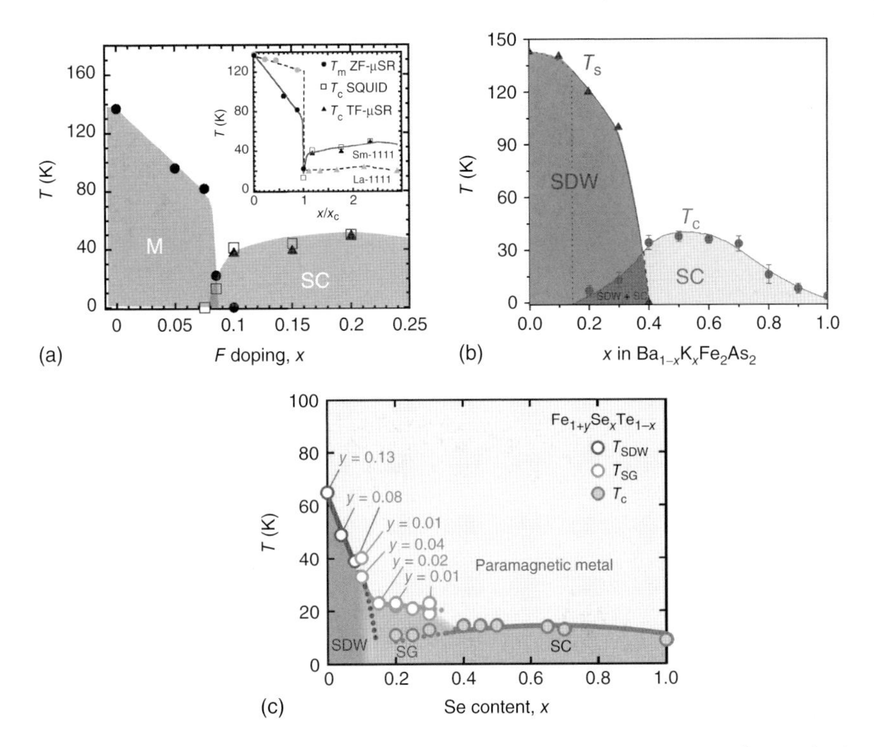

Figure 2.2.2.3 Phase diagrams representative of (a) 1111, (b) 122, and (c) 11 families. (Figures are reproduced with permission from Ref. [6] (Copyright 2009 by the American Physical Society), from Ref. [7] (Copyright 2009 by IOP Publishing), and from Ref. [8] (Copyright 2010 by The Physical Society of Japan).)

superconducting phases coexist in some portions of the phase diagram, with phase separation at nanoscale level.

For details of the above general description of FeSCs, see for example the reviews in Refs [3, 9].

2.2.2.3 Material Preparation

In the following, we briefly mention some peculiar aspects of FeSC preparation. Different well-established synthesis methods have been used for the synthesis of polycrystalline and single crystalline FeSCs. For the preparation of *polycrystalline samples*, solid-state reaction and high-pressure synthesis methods are used. The latter, that is, the *high-pressure method*, is more efficient than the former, that is ambient pressure (solid state) method for the synthesis of gas-releasing compounds such as $LnFeAsO_{1-x}F_x$ at high temperatures [10]. Severe fluorine loss observed in the common vacuum quartz tube seal method can thus be avoided, so that the high-pressure technique may drastically improve the T_c. Moreover, in the case of polycrystalline samples, high-pressure leads to strongly sintered samples with better intergrain connections [11]. Last but not least, the high-pressure anvil technique is relatively safe because the sample is confined in a closed container supported by anvils, while in the ampoule technique, explosions of ampoules may lead to a contamination of the laboratory with poisonous compounds.

Large single crystals of high quality are of fundamental importance to determine the intrinsic properties of the FeSCs and allow crucial experiments aimed to decipher the pairing mechanism. Large and pure crystals grow in fluxes of binary FeAs, which melts at around 1000 °C. This so-called "self-flux" method is especially useful for transition metal (mostly cobalt) doping of 122 compounds [12] because crystals can grow in $Fe_{1-x}Co_xAs$ melts of the desired and constant composition. *Metal flux methods* are less practicable for oxygen-containing 1111-type superconductors and rather tiny crystals are obtained from salt fluxes under high-pressure conditions; however, recently, millimeter-size 1111 crystals have been grown from NaAs and KAs fluxes at high pressure [13].

Successful fabrication of *epitaxial thin films* is essential to explore the anisotropic electromagnetic properties and for the development of electronic

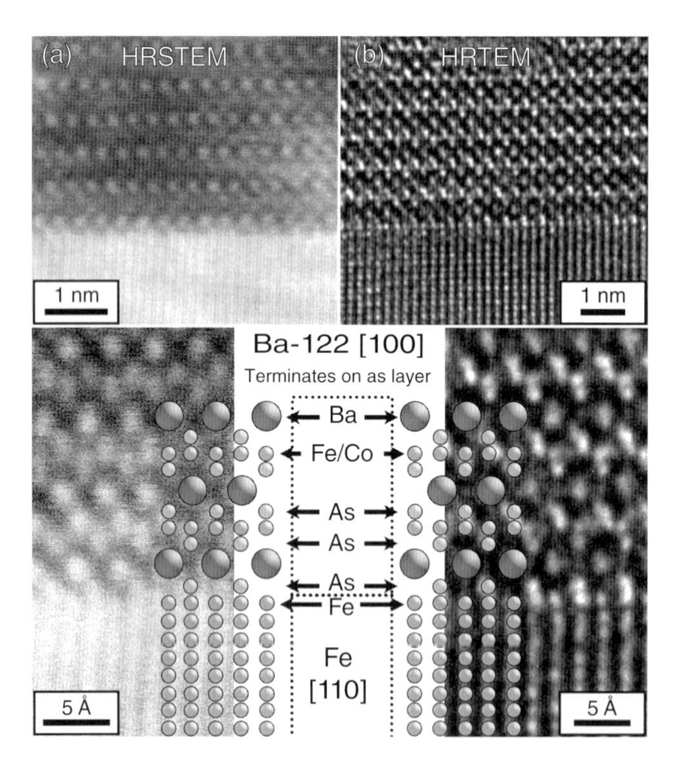

Figure 2.2.2.4 $Ba(Fe_{1-x}Co_x)_2As_2$ thin films grown on a Fe layer. (a) High-resolution scanning transmission electron microscope (HRSTEM) image of the $Fe/Ba(Fe_{1-x}Co_x)_2As_2$ interface. Directly below, a higher-resolution image of a neighboring region is shown. (b) High-resolution transmission electron microscope (HRTEM) image of the $Fe/Ba(Fe_{1-x}Co_x)_2As_2$ interface whose enlargement is shown below. The interface is highly coherent and bonding takes place directly on the FeAs sublattice. (Figure reproduced with permission from Ref. [15] (Copyright 2009 by the American Institute of Physics).)

devices, such as Josephson devices and SQUIDs. Moreover, thin film growth allows investigating interface effects, inducing selective strain on the structure, exploring the effect of pinning mechanisms, as well as designing artificial systems which can exhibit different properties with respect to bulk materials (see Section 3.2 of this volume).

FeSC thin films initially have not been easy to grow because of the volatile nature of As, but at present, epitaxial films, mainly of the 122 and 11 phases, are available. High-quality epitaxial films of the 122 phase grown by pulsed laser ablation (pulsed lased deposition PLD) have shown excellent structural and superconducting properties [14–16], among them the present record value of critical current density is 5.5 MA cm^{-2} at 4.2 K in zero field [14]. 122 thin films have been grown on suitable intermetallic buffer layers on various types of oxide substrates, thus improving the superconducting properties [14]. Epitaxial Fe/$Ba(Fe_{1-x}Co_x)_2As_2$ bilayers of excellent structural quality are shown in Figure 2.2.2.4.

Also epitaxial films of the 11 phase have been grown by pulsed laser ablation [17]. These films exhibit T_cs, which vary with film thickness and reach values around 21 K, much larger than the bulk T_c (16 K). Indeed, in the 11 phase T_c can be widely tuned by applied pressure and the T_c enhancement has been attributed to the strain related to the growth mode. Growth on different substrates has been tested to explore the effect of strain [18]. Moreover, superconductivity has been found to appear in oxygen-doped FeTe films [19], and, most remarkably, superconductivity with enhanced T_c has been found in FeSe monolayers grown on $SrTiO_3$ [20].

The deposition of 1111 films appears to be more difficult because doping is produced by F and O, which are both volatile and hardly controlled in the final films. For PLD growth of 1111 films, both *in situ* and *ex situ* methods have been tried to overcome the problem of F incorporation. Film properties comparable to those of bulk samples have been obtained by molecular beam epitaxy (MBE) growth [21].

2.2.2.4 Superconducting Properties

2.2.2.4.1 Critical Temperature T_c

The first intrinsic property which makes a superconductor interesting for applications is the T_c. Appealing values as large as 55 K have been obtained for the 1111 phase in $SmFeAs(O_{0.9}F_{0.1})$ [2], but noteworthy values have been obtained also for the other phases, namely T_c up to 38 K for the 122 phase in $(Ba_{0.6}K_{0.4})Fe_2As_2$, up to 18 K for the 111 phase in LiFeAs, and up to 16 K in the 11 phase in $Fe(Se_{0.5}Te_{0.5})$ [22].

Moreover, a new family of layered FeSCs with perovskite-type oxide blocking layers has recently been discovered [23]. Since the perovskite-type layers are flexible in terms of chemical composition and crystal structure, there are wide opportunities to create new compounds by modifying the perovskite-type layers. In fact, many compounds, such as $(Fe_2As_2)(Sr_4M_2O_6)$ with M = Sc, Cr, V, and (Mg,Ti), have been reported. The second bracket in each chemical formula represents the local chemical composition of the perovskite-type oxide layer. Moreover, the number of sheets in the perovskite-type layer can be controlled by the nominal composition and the synthesis conditions. Among these systems, T_c up to around 40 K has been found [23a,b, 24]. These new compounds indicate that considerable opportunities for new superconductors still exist within layered iron pnictides.

2.2.2.4.2 **Critical Fields and Characteristic Lengths**

Since the earliest stages of the research on FeSCs, it has been apparent that they are characterized by very high upper critical field values (H_{c2}), shown in Figure 2.2.2.5. The first magnetotransport measurements in high fields have been carried out on the *1111 phase* [10c, 25]. The shape of the 1111 resistive transition is significantly broadened by the magnetic field [10c, 25], but to a smaller extent than in the high-T_c cuprates. The extracted H_{c2} curves exhibit a distinctive upward curvature, reminiscent of the MgB_2 behavior [26], which is a signature of the multiband character. $\mu_0 H_{c2}$ values of up to 60 T and $\mu_0 H_{c2}$-slopes close to T_c, typically around $-10\,T\,K^{-1}$, have been measured [10c]. The critical field anisotropy γ_H, defined as $\gamma_H = H_{c2}^{\|ab}/H_{c2}^{\perp ab}$, is up to 9, close to T_c [10c] and its slight temperature dependence is again indicative of multiband behavior. On the other hand, as a consequence of the broadening of the transition, the irreversibility field H_{irr} (defined as the field at which the resistivity becomes zero) is much smaller than H_{c2} (defined at the onset of the transition). This feature, which is an obvious drawback for applications, is very common in high-T_c cuprates and strongly related to their anisotropic nature, the nearly two-dimensional character of these compounds and a signature of the weakness of flux pinning and/or the significance of thermal fluctuations.

A strikingly different H_{c2} behavior has been observed in the *122 phase*. The in-field 122 resistive transition exhibits no broadening [29], much like the behavior of low-T_c superconductors [30], where the effect of the magnetic field is an almost

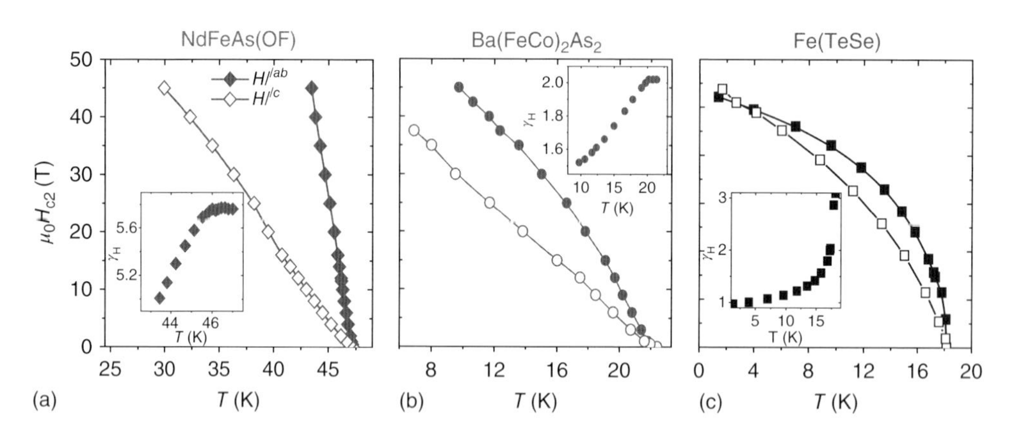

Figure 2.2.2.5 Temperature dependences of $H_{c2}^{\|ab}$ (full symbols) and $H_{c2}^{\perp ab}$ (open symbols) for NdFeAs(OF) (a), $Ba(FeCo)_2As_2$ (b), and Fe(SeTe) (c), representative of the 1111, 122, and 11 families, respectively. In the insets, the H_{c2} anisotropy versus temperature curves are plotted. (The data are taken from Refs [27, 28].)

rigid shift of the transition to lower temperatures. Consequently, the H_{irr} curve closely follows the H_{c2} curve. The H_{c2} anisotropy is 1.5–2 close to T_c and rapidly approaches unity with decreasing temperature [31, 32]. On the whole, the H_{c2} values of this family are smaller than those of the 1111 family, with $\mu_0 H_{c2}$ slopes close to T_c around $-5\,\mathrm{T\,K^{-1}}$ [29, 31, 32].

The magnetoresistance behavior of the *11 family* is midway between the fan-shaped one of the 1111 phase and the rigidly shifted one of the 122 phase [33, 34]. The $\mu_0 H_{c2}$ slopes close to T_c are the largest among FeSCs, ranging from -10 to $-30\,\mathrm{T\,K^{-1}}$ [33–36] and reaching a value of $-500\,\mathrm{T\,K^{-1}}$ in thin films [27]. The H_{c2} curves, after this step-like increase, bend and exhibit anomalous downward curvatures for both field directions. The H_{c2} anisotropy quickly decreases to unity with decreasing temperature and even becomes smaller than unity at the lowest temperatures [36].

The anomalous downward curvature and anisotropy behavior approaching unity at low temperature, well evident in the 11 and 122 families for which the whole phase diagrams are accessible, are consequences of the so-called paramagnetic limit. This effect occurs when H_{c2} is extremely high and the Zeeman splitting exceeds the superconducting energy gap – which represents the bonding energy of Cooper pairs – thus breaking the Cooper pairs.

In Figure 2.2.2.6, the upper critical fields of some FeSCs samples are compared to those of some other superconductors. $H_{c2}(0)$ for $FeSe_{1-x}Te_x$ is almost twice as high as for Nb_3Sn, despite their same $T_c = 18\,\mathrm{K}$. Yet the paramagnetic limited $H_{c2}(0)$ of $FeSe_{1-x}Te_x$ is lower than the $H_{c2}(0)$ of $PbMo_6S_8$, which has $T_c = 14\,\mathrm{K}$. Moreover, the different H_{c2} values found in FeSCs also indicate that H_{c2} does not simply scale with T_c, since the multiband effects and the interplay of orbital and

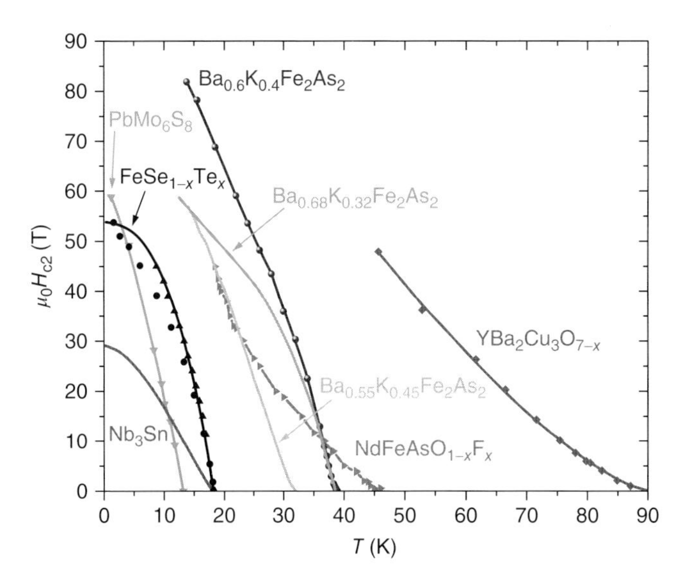

Figure 2.2.2.6 Upper critical field of some FeSCs and other superconductors. (Figure reproduced with permission from Ref. [27], Copyright 2011 by the American Physical Society.)

Table 2.2.2.1 Important superconducting state properties of some FeSC compounds representative of 1111, 122, and 11 FeSC families [33, 39–42], YBCO [37, 43–47], BSCCO [38, 47–50], and MgB_2 [51–54].

	NdFeAs(O,F)	$Ba(FeCo)_2As_2$	Fe(Se,Te)	YBCO	BSCCO	MgB_2
T_c (50% R_n) (K)	47.4	22.0	15	93	85–90	39
$-\mu_0 dH_{c2}^{\perp ab}/dT$ (T K^{-1})	2.1	2.5	14	0.7–1	0.7–6	0.05
$-\mu_0 dH_{c2}^{\parallel ab}/dT$ (T K^{-1})	10	4.9	26	3-4	20–45	0.3
γ_H	5	2	2–3	4–14	50–60	3–5
$\mu_0 H_{c1}^{\parallel c}$ (mT)	15	40	4.5	20	3.7	100
ξ_{ab} (nm)	2.1	2.9	1.5	2.1–2.3	2.7–3.2	10
ξ_c (nm)	0.6	1.5	0.6	0.5–0.6	0.4–0.5	2
λ_{ab} (nm)	200	200	490	160	180	50–100
Ginzburg number, Gi	$4\cdot10^{-4}$	$2\cdot10^{-5}$	$1\cdot10^{-3}$	$>10^{-3}$	$10^{-3}-10^{-2}$	$<10^{-5}$

paramagnetic pair breaking (the former determined by the condition that the normal cores of the vortices overlap in space and the latter by the condition that the Zeemen splitting exceeds the superconducting gap) can result in more complex and interesting behaviors.

As can be seen in Table 2.2.2.1, FeSCs show the largest H_{c2} slopes close to T_c in comparison with all other high-T_c cuprates. This enables high field applications even at high operation temperatures just below T_c. The *anisotropy* of the critical fields γ_H is crucial for applications as well, especially in magnet fabrication. On the whole, the anisotropy of FeSCs is smaller than that of the $YBa_2Cu_3O_{7-\delta}$ (YBCO) family, where values from 4 to 14 have been reported [37] and much smaller than that of the $Bi_2Sr_2CaCu_2O_x$ (BSCCO) family, where it is typically in the range 50–60 [38]. This is a very important point because the anisotropic, nearly two-dimensional nature of high-T_c cuprates is the main reason for the weakness of the pinning and the significance of thermal fluctuations, which cause the broadening of the transition.

The different behavior of the broadening of the transitions in these superconductors is understood if the *coherence lengths* ξ are considered. Rough estimates of ξ obtained from the $\mu_0 H_{c2}$ slopes at T_c are reported in Table 2.2.2.1 (0.6, 1.5, and 0.6 nm along the direction perpendicular to Fe planes, for the 1111, 122, and 11 families, respectively). These values must be compared with the distance d between the Fe–As planes, in order to establish the two-dimensional or three-dimensional character of superconductivity and consequently the degree of dissipation in a magnetic field. The values of d are around 0.86, 0.65, and 0.6 nm in the 1111, 122, and 11 families, respectively. Therefore, the 122 family appears to be the most promising for applications in this respect, with the largest ξ/d ratio among the FeSCs. On the other hand, the 1111 compounds have exhibited two-dimensional superconductivity ($\xi < d$), measured both by the broadening of the resistive transition in high fields [55] and by a dissipation peak in the out-of-plane critical current when the magnetic field is exactly aligned parallel to the FeAs planes [56].

The *London penetration depth* λ is a characteristic length which determines the extent to which a magnetic field penetrates into the superconductor. λ is inversely proportional to the square root of the Cooper pair density and thus it is larger in compounds with low carrier density. The FeSC λ values along ab planes in the low temperature limit, λ_{ab}, reported in Table 2.2.2.1, are midway between the values measured in MgB_2 and in high-T_c cuprates.

The inverse squared London penetration depth λ^{-2} determines the *lower critical field* H_{c1}, which is relevant for electronic applications. H_{c1} values of FeSCs are much smaller than H_{c1} of MgB_2 but comparable or larger than H_{c1} of copper oxides, as reported in Table 2.2.2.1.

Broadening of the transition in zero field is caused by *thermal fluctuations*, which can be parameterized by the Ginzburg number $Gi = (\pi\lambda_{ab}^2 k_B T_c \mu_0 / 2\xi_c \Phi_0^2)^2$, where k_B the Boltzmann constant and Φ_0 the magnetic flux quantum. Typical values of $Gi \approx 4{\cdot}10^{-4}$, $2{\cdot}10^{-5}$, $1{\cdot}10^{-3}$ for the 1111, 122, and 11 families [33] should be compared to Gi larger than 10^{-3} for $YBa_2Cu_3O_{7-\delta}$ and lower than 10^{-5} for MgB_2. Again, the 122 family seems to be the most promising for applications, as its slightly lower T_c and lower anisotropy yield reduced thermal fluctuations.

The usual route to improve the upper critical field of superconductors is to drive the material into the dirty limit by introducing point defects that limit the mean free path ℓ. In the dirty limit, obtained when $\ell/\xi < 1$, H_{c2} is inversely proportional to ℓ. The above-mentioned small values of the coherence length in the nanometer range for the FeSCs indicate that high densities of point defects may be necessary to achieve the dirty limit $\ell/\xi < 1$, with consequently a likely suppression of T_c as well. Moreover, it is possible that the anisotropy of the superconducting properties can be tuned by the introduction of disorder or by chemical substitutions, as is the case for the multiband superconductor MgB_2 [26b, 57].

The investigation of the *effect of disorder* is another fundamental point, the T_c suppression by impurity scattering being related to the pairing symmetry. Indeed, the bizarre fully gapped sign-reversing s-wave state (s± pairing symmetry), which is expected for FeSCs, could give rise to a new and unexpected phenomenology, that is, an odd dependence on impurities that should be properly investigated. Calculations predict that s± pairing is very fragile against nonmagnetic impurities [58], whereas it is preserved by inter-band magnetic scattering, whose effect is to restore the pairing, thus hardly affecting T_c [59]. Preliminary results show that FeSCs exhibit a high tolerance to magnetic impurities [60]. This would really be a unique case, given that magnetic impurity scattering is strongly detrimental to superconductivity both in conventional superconductors (s-wave pairing) and in high-T_c cuprates (d-wave pairing).

2.2.2.4.3 **Critical Current Density J_c**

Measurements of the critical current density J_c in single crystal samples of FeSCs have revealed high and quite isotropic critical current densities (i.e., weakly varying along crystal directions, $J_c^{ab} \approx J_c^c$), which is a very promising combination for applications. Moreover, J_cs show only a weak decrease on application of a magnetic field at low temperatures, similarly to observations in $YBa_2Cu_3O_{7-\delta}$ [37b, 61]. Such a behavior is consistent with the nanometer-scale coherence lengths, the exceptionally high H_{c2} values, and the pinning associated with atomic-scale defects, resulting from chemical doping or nanometer-scale local modulations of the order parameter [62]. For the 1111 family, a high in-plane J_c of $2{\cdot}10^6$ A cm^{-2} at 5 K in a $SmFeAsO_{1-x}F_x$ crystal, almost field-independent up to 7 T at 5 K, has been reported [63]. Many single crystal results have been reported for the 122 system, since larger crystals are easily grown. Significant fishtail peak effects and large current-carrying capability up to $5{\cdot}10^6$ A cm^{-2} at 4.2 K have been found in a $Ba_{0.6}K_{0.4}Fe_2As_2$ single crystal [64]. Fishtail effect and currents in the range 10^5 A cm^{-2} at low temperature have been also reported for the 122 family in Refs [29, 65]. As for the 11 system, J_c of tellurium-doped $FeTe_{0.61}Se_{0.39}$ crystals with $T_c \approx 14$ K exceeding 10^5 A cm^{-2} at low temperatures has been reported [66].

J_c can be further enhanced by introducing effective pinning centers, as described in Section 2.2.2.5.

J_c values measured on FeSC single crystals are reported in Table 2.2.2.2 in comparison with other technical superconductors. The single crystal properties are excellent; indeed, J_c is high and rather *field independent* as in the high-T_c cuprates, whereas the J_c *anisotropy*, defined as J_c^{ab}/J_c^{c}, is around 2 in all FeSC families [33, 67, 68]. The evaluation of the J_c anisotropy, measured either along the *ab*-planes or along the *c*-axis, is a crucial one and it can be extracted by a complex analysis of magnetization measurements on single crystals, by varying the field direction. However, also reliable and model-independent results have been obtained by direct transport J_c measurements in the two main crystallographic directions

Table 2.2.2.2 Indicative values of J_c for various superconductors.

		1111	122	11	YBCO	BSCCO family	MgB_2	A15
Single crystals	$J_c^{(ab)}$ (A cm^{-2}) by inductive measurement	$2{\cdot}10^6$ [63, 69]	$5{\cdot}10^6$ [64] $3{\cdot}10^5$ [29, 65]	10^5 [66]	$3{\cdot}10^6$ [61]	$2.5{\cdot}10^6$ [71] $5{\cdot}10^6$ [72]	10^5	—
	$J_c^{(ab)}/J_c^{(c)}$	2.5 [69]	2 [67]	—	30 [70] 10–50 [73]	—	~1–2	—
	J_c (1 T)/J_c ($H = 0$)	0.97 [69]	0.5 [29]	0.3 [66]	0.8 [61]	—	~0.01	0.9
	J_c (5 T)/J_c ($H = 0$)	0.8 [69]	0.5 [29]	0.3 [66]	0.5 [61]	0.92 [71]	0	0.7
Tapes or wires	J_c (A cm^{-2}) by transport measurement	4600 [74]	10^5 [75]	10^3 [76]	$7{\cdot}0^6$ [77, 78]	$8{\cdot}10^4$ [72]	10^6	10^6

All data refer to 4–5 K. For the 1111 phase, we have considered SmFeAs(O,F) (Sm-1111) samples.
Source: Data for MgB_2 are from Ref. [79] and data for A15 are from Ref. [80]. For YBCO and BSCCO, data obtained on textured tapes are considered.

by fabricating micro-bridges on Sm-1111 single crystals through focused ion beam (FIB) processing [69]. As seen in Table 2.2.2.2, the J_c anisotropy in FeSCs is comparable to the anisotropy in MgB_2 and much lower than the values of up to 30–50 found in the cuprates [70]. This is a very significant result because it could imply that, contrary to the high-T_c cuprates, the FeSCs may not require very high degrees of texturing in the process of fabrication of wires or tapes. Actually, the low J_c anisotropy is only a prerequisite for the current transport in polycrystalline materials. Indeed, the second requirement is the transparency of the GBs to current flow. This is a key issue and it will be discussed in Section 2.2.2.6.

2.2.2.5 **Critical Current Pinning**

Here, we extend the general concepts of pinning in superconductors, described in Section 1.2.3 of this volume, to the case of FeSCs. Thin films are often used to investigate the critical current density and its dependence on the magnetic field, applied either parallel or perpendicular to the crystalline *ab* planes. In general, in layered superconductors, $J_c(H^{\|ab})$ is larger than $J_c(H^{\|c})$ for two main reasons. The first reason is that the upper critical field along the *ab* planes is larger than that along the *c* axis ($H_{c2}^{ab} > H_{c2}^{c}$), due to the anisotropy of the effective masses ($m_{eff}^{ab} < m_{eff}^{c}$), the second reason is that magnetic flux lines are more effectively pinned when their normal cores, of diameter $\sim\xi_c$, lie within adjacent superconducting planes, that is, parallel to Fe planes in the case of FeSCs. The latter effect is known as *intrinsic pinning*. Given that for such applications as in superconducting magnets, it is highly desirable to have superconducting windings made of a material whose J_c is weakly dependent on the direction of the applied field, much research is aimed to introduce evenly spaced pinning centers which could be effective in enhancing $J_c(H^{\|c})$ up to the value of $J_c(H^{\|ab})$ (correlated pinning centers). Thin films are ideal systems to explore the effect of introducing pinning mechanisms effective either for $H^{\|ab}$ or for $H^{\|c}$, in view of obtaining the lowest possible anisotropy, defined as the ratio $J_c(H^{\|ab})/J_c(H^{\|c})$.

In 1111 films, J_c values above $10^6\,A\,cm^{-2}$ have been measured [81]. The J_c angular dependence follows the anisotropic Ginzburg–Landau scaling, also suggesting a role of intrinsic pinning for field parallel to the *ab* planes ($H^{\|ab}$) and no role of correlated pinning for field parallel to the *c* axis ($H^{\|c}$) [82]. In 122 films, vertically aligned, self-assembled $BaFeO_2$ nanorods of diameter comparable to the coherence length have been introduced without suppressing T_c [68]. These nanorods act as strong *c*-axis correlated pins which enhance $J_c(H^{\|c})$ above $J_c(H^{\|ab})$, inverting the intrinsic material anisotropy. The resulting pinning force turns out to be better than in optimized Nb_3Sn at 4.2 K. In addition, multilayer artificial structures in the same systems have been proven to be effective in enhancing $J_c(H^{\|ab})$ as well, yielding a virtually flat angular dependence of J_c, highly desirable for applications, low temperature self-field J_c values up to $10^6\,A\,cm^{-2}$, and J_c values still larger than $10^5\,A\,cm^{-2}$ at fields of 20 T [83, 84]. The J_c of Fe(Te,Se) films deposited onto $LaAlO_3$ substrates reaches $5\cdot10^5\,A\,cm^{-2}$ in self-field and remains above $10^4\,A\,cm^{-2}$ up to 35 T [85]. The major pinning mechanism has been identified as point pinning, likely due to inhomogeneous distribution of Se and Te ions. However, in some cases, $J_c(H^{\|c})$ larger than $J_c(H^{\|ab})$ suggests also the presence of correlated pinning along the *c*-axis [86].

Introduction of pinning centers has been attempted in single crystals as well. For example, point defect introduced by proton irradiation in 122 crystals has enhanced J_c by a factor 2.6 [87]. Also, irradiation with Au ions [88] and neutrons [89] has demonstrated that pinning can be further increased by introducing defects without affecting T_c.

2.2.2.6 Grain Boundaries

One of the essential requirements for large-scale application of superconductivity is the capability to carry currents over long lengths, that is, the superconducting currents need to flow in polycrystalline materials across many GB regions. The "magical" ability of Cooper pairs to cross non-superconducting regions or areas with depressed order parameter is due to their long coherence length, of the order of several tens of nanometers in conventional superconductors. The discovery of the cuprates with high T_c, high H_{c2}, enhanced anisotropy and, consequently, low coherence lengths (1–2 nm in the *ab*-plane and much lower along the *c*-axis) has shown the limits of this capability and the "issue of GBs" has clearly emerged for the first time. In the past decades, enormous efforts have been made to grow textured superconducting tapes with small-angle GBs, but this approach is not yet commercially viable for most applications. Very recently, a theoretical understanding of why the currents in the high-T_c cuprates are so sensitive to the GB mismatch has been proposed [90].

A very different story is that of MgB_2, discovered <10 years ago, with non-evident GB problem [91]. Indeed, MgB_2 has already been successfully used for fabricating magnets for commercial magnetic resonance imaging (MRI) systems.

These considerations highlight the importance of exploring the nature of GBs in novel superconducting materials.

Regarding FeSCs, as shown in Table 2.2.2.1, the large H_{c2} values imply coherence lengths varying between 1 and 3 nm among the families. These values are reminiscent of those in the cuprates, but the reduced J_c anisotropy of FeSCs compared to the cuprates (see Table 2.2.2.2) allows current flow along the *c*-axis, thus making the requirement of texturing less stringent.

Early studies of the critical current density of 1111 polycrystals have emphasized the *strong granularity* of these compounds, which limits the global J_c flow across all the sample to very low values [33, 92]. Two distinct scales of current flow have been found in polycrystalline Sm- and Nd-iron-oxypnictides using magneto-optical imaging and studying the field dependence of the remnant magnetization [93]. Low-temperature laser scanning microscopy (LTLSM) and scanning electron microscopy (SEM) observations on polycrystalline Sm-1111 samples [94] have emphasized cracks and wetting Fe–As phase at the GBs as the main mechanisms of current blocking in polycrystalline materials. However, in the FeSCs, even with substantial blocking by such GB phases, the intergranular current densities appear to be more than one order of magnitude larger at 4 K than in early results on randomly oriented polycrystalline cuprates [95].

Up to date, the difficulty in obtaining fully dense single-phase polycrystalline materials seems to be one of the main reasons for the granular behavior of bulk materials. However, considering the strong similarities between high-T_c cuprates and FeSCs, the existence of intrinsic mechanisms limiting the transmission of current in misaligned GB needs to be carefully considered [96].

A pioneering experiment has investigated the nature of GB weak links in $Ba(Fe_{1-x}Co_x)_2As_2$ *films grown on bicrystals* with different misorientation angles Θ. It has been shown there that the critical current is substantially depressed for Θ larger than 3°. However, in stronger magnetic fields, the J_c reduction due to GBs seems to become of less importance compared with the intrinsic magnetically induced J_c reduction [97]. This J_c suppression with Θ is reminiscent of the behavior of $YBa_2Cu_3O_{7-\delta}$ [98], even if a close comparison suggests that in FeSCs the suppression is less severe than in the high-T_c cuprates: as shown in Figure 2.2.2.7 for Θ from 0° to 24°, J_c decreases by one order of magnitude, while in $YBa_2Cu_3O_{7-\delta}$, by two orders of magnitude. Another similar experiment on $Ba(Fe_{1-x}Co_x)_2As_2$ films grown on bicrystals, whose results are also reported in Figure 2.2.2.7, has been carried out by Katase and coworkers [99]. They have found a slightly larger critical angle $\Theta \approx 9°$ and they have concluded that Josephson-coupled GBs show superconductor–metal–superconductor rather than superconductor–insulator–superconductor behavior, a result which points to higher angle GBs being metallic rather than insulating, as in the cuprates. On the whole, the studies of Refs [97, 99], directly compared in Figure 2.2.2.7, are pretty consistent, both showing a similar exponential falloff of J_c with increasing misorientation angle above a certain critical value.

Very recent results [100] indicate that the misorientation angle is not the only parameter that determines whether or not GBs are transparent to the

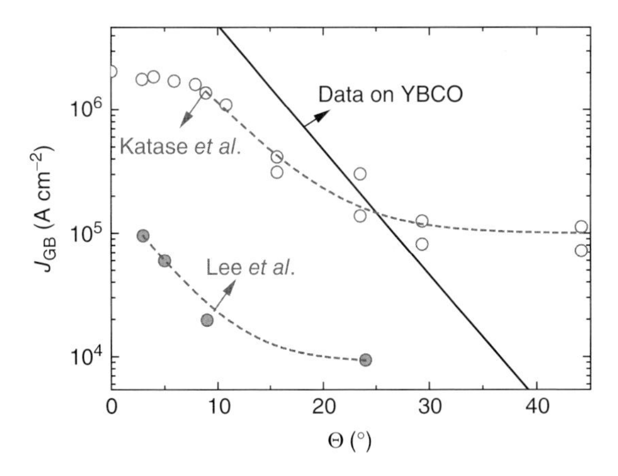

Figure 2.2.2.7 Dependence of the critical current density across GBs. J_{gb} at 12 K and 0.5 T as a function of the misorientation angle Θ measured on $Ba(Fe_{1-x}Co_x)_2As_2$ films grown on bicrystals with different misorientation angles. (Data are taken from Refs [97, 99].) The dotted lines are exponential fits. The continuous line shows the behavior of YBCO GBs according to Ref. [98].

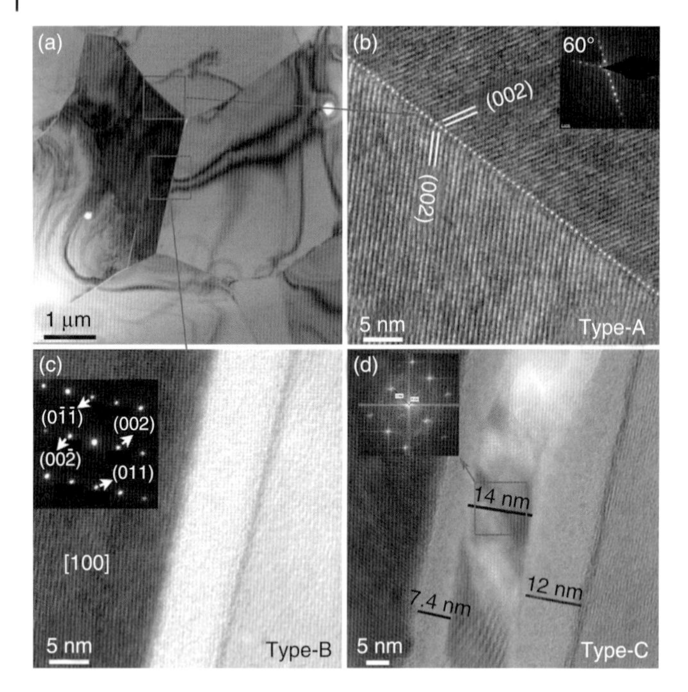

Figure 2.2.2.8 (a–d) Transmission electron microscopy (TEM) images of different types of GBs in $Sr_{0.6}K_{0.4}Fe_2As_2$ polycrystals. (Figure reproduced with permission from Ref. [101], Copyright 2011 by the American Institute of Physics.)

supercurrent, but also that the orientation of the field with respect to the GB also has to be taken into account. Indeed, it has been found that the intergrain J_c degrades only when a significant portion of the vortex lies in the GB, whereas, when the vortex lies obliquely across the GB, there is little or no suppression of the intergrain J_c.

The reliability of these preliminary results needs to be carefully checked to rule out possible extrinsic effects, such as diffusion of oxygen from the substrate, which requires the investigation of ideal GBs. As shown in Figure 2.2.2.8, three distinctive types of GB have been identified in 122 polycrystals [101], namely clean GBs, GBs containing amorphous layer with a width of ~10 nm, and GBs containing amorphous–crystallite–amorphous trilayers ~30 nm in width. The latter two types of GBs are larger than the coherence length and thus result in a current blocking effect; however, improvements in the sample preparation may heal this problem, at least partially. Similar amorphous layers around individual grains have been reported in 1111 polycrystals [100].

2.2.2.7 Wires and Tapes

Soon after the discovery of superconductivity in FeSCs, wires and tapes of 122, 11, and 1111 compounds have started being fabricated by the powder-in-tube (PIT) method. This processing method, described in detail in Section 3.4.1 of this volume, allows to overcome the problem of mechanical hardness and brittleness, which hinders plastic deformation. Stoichiometric amounts of precursor powders are packed into a metal tube, sealed in Ar atmosphere, and successively deformed into wires and heat treated. The synthesis of the superconducting compounds occurs either before filling the metal tube (*ex situ* PIT) or within the final wire (*in situ* PIT). Wires prepared by *ex situ* PIT generally have higher J_c than those prepared by *in situ* PIT [102], as the former process offers more options of optimizing the powder reaction, even by multiple steps.

The optimization of the preparation protocol has required the exploration of different ingredients, namely choice of the best sheath material, addition techniques, heat treatment optimization, *ex situ* processing, and texturing technique. A complete review on this topic is found in [103]. The main focus is improving the global J_c, depressed by cracks, low density, GB-wetting secondary phases, and phase inhomogeneities, which all cause local suppression of the order parameter at GBs.

Most of the work on FeSC wires and tapes has been carried out on the *122* family. This family is less anisotropic, it does not suffer from significant broadening of superconducting transition by thermal fluctuations, it has fairly large T_c up to 38 K, close to that of MgB_2, and it is more favorable for the wire fabrication in that it requires a lower annealing temperature ($T_{an} = 850\,°C$) and no oxygen as compared to the 1111 family. Indeed, the best transport critical current values among FeSC wires and tapes, up to $10^4 - 10^5$ A cm^{-2}, have been obtained with the 122 family so far.

Regarding the choice of the metal sheath, it must not react with the FeSC filling, which would not only shrink the superconducting cross-section but also affect the stoichiometry and microstructure of the FeSC filling, thus degrading the wire properties. Ag, as compared to Ta, Nb, and Fe, has turned out to be the best sheath material for 122 wires and tapes and it has been used since, even in combination with an outer Fe sheath to reduce costs and improve mechanical strength.

Chemical addition is a well-established route to improve grain crystallization, add pinning centers, and enhance the metallic character of GB secondary phases, thus promoting intergrain coupling. In cuprates, for example, the beneficial role of Ca addition in doping the GB phases has been evidenced [90], while C addition in MgB_2 significantly enhances the irreversibility field [104]. Ag substitution is a natural candidate, in that it hardly enters the FeSC lattice, thus affecting only the extrinsic properties. Indeed, 10–20% Ag substitution significantly improves intergrain connectivity and reduces porosity of 122 tapes, resulting in a three-time increase of the global J_c up to high fields [102, 105]. Pb addition, on the other hand, promotes grain growth and improves grain connectivity for contents up to 10%, yielding an enhancement of the global J_c in the low-field regime [105]. Co-doping of both Ag and Pb thus results in a J_c improvement in the whole field range. Sn addition has a similar beneficial effect in 122 [106, 107], as shown in Figure 2.2.2.9.

Nonetheless, with respect to the role of chemical addition in promoting grain growth, it must be remarked that grain growth is not necessarily the right way to pursue. Indeed, it has been shown that untextured polycrystalline $(Ba_{0.6}K_{0.4})Fe_2As_2$ bulks and round wires with high GB density, that is, small

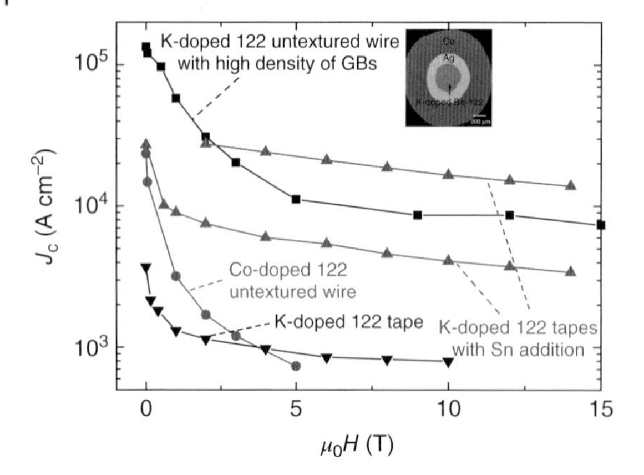

Figure 2.2.2.9 Transport critical current densities of K-doped 122 textured tapes with and without chemical additions and prepared with different heating treatments [106, 107], of a K-doped 122 untextured wire with high density of GBs [75], and of a Co-doped untextured wire [75] as a function of applied magnetic field at 4.2 K.

grains, prepared by *ex situ* PIT, have transport critical current densities well over 10^5 A cm^{-2} in self-field at 4.2 K, larger than the best 122 wires and tapes (see Figure 2.2.2.9) [75]. The enhanced grain connectivity is ascribed to their much improved phase purity and the effectiveness of high density of GBs as pinning centers is explained in terms of low anisotropy and consequent enhanced vortex stiffness of 122 compounds.

The effect of heat treatment is related to chemical addition so that there is no optimal sintering temperature valid in general to yield the best superconducting properties and minimize secondary phases and porosity. For 122 samples, the optimal sintering temperature is 850 °C [108], while for Ag-added samples it is 900 °C.

Texturing methods have also given very good results in achieving high transport J_c values above 10^4 A cm^{-2} [106, 108, 109], surviving up to fields as high as 14 T [107]. Indeed, at high fields, textured tapes perform even better than untextured wires (see Figure 2.2.2.9) [107]. In any case, texturing may not be as crucial as in high-T_c cuprates, whose anisotropy J_c^{ab}/J_c^c is larger than that of FeSCs. The beneficial effect of texturing has led to the idea of preparing FeSC-coated conductors, described in Section 2.2.2.8.

Less work has been carried out in the fabrication of *1111* wires and tapes due to the difficulty in controlling O and F stoichiometry during heat treatments at high temperatures. The commonly used sintering temperature for 1111 wires is 1200 °C, but sintering at 850–900 °C has yielded similarly high T_c and J_c up to 1300 A cm^{-2} [110]. Also in the case of 1111 wires and tapes, Ag has been found to be the best sheath.

Wires and tapes of the *11* family are appealing for applications as well, despite their lower T_c and J_c, as they do not contain toxic elements. However, several difficulties have been encountered do to chemical reactions between the inner

FeSCs and outer sheath during thermal treatments and difficulty in close-packing the powder inside the tube. For this reason, Fe sheath have turned out to be the best choice, as it allows a diffusion process, where Fe-free precursors are sealed inside Fe tubes and the final 11 phase is formed by supply of Fe from the sheath, upon thermal treatment [111]. This diffusion process has yielded a significant improvement in the transport J_c, allowing to obtain J_c values up to 10^3 A cm^{-2} in FeSe wires [76].

Regarding the thermal treatment optimization, it was found that the superconducting properties of FeSe wires improve with increasing annealing temperature up to 1000 °C, where the phase formation is complete [112]. It has been also found that combined melting and annealing processes allow to obtain homogeneous and denser Fe(Te,Se) samples, characterized by large and well-interconnected grains [113]. The resulting samples exhibit enhanced global critical current density, reaching about 10^3 A cm^{-2} (see Figure 2.2.2.10).

The effect of chemical addition in iron chalcogenides has not been thoroughly explored yet, also due to the lower versatility of these compounds in accepting chemical substitutions. Moreover, no clear evidence of insulating secondary phases at the GBs, where carrier depletion occurs, has been identified yet in the 11 family [113], hence it is possible that this family behaves differently from other FeSCs upon chemical addition.

In summary, the best transport critical current values in FeSCs wires and tapes have been obtained so far with the 122 family, namely up to 10^4-10^5 A cm^{-2}. Moreover, in 122 wires, the J_c field dependence is quite flat, with a decrease of one order of magnitude at field well above 10 T, with respect to the self-field value. For the 1111 family, the transport J_c values found in wires and tapes prepared by *ex situ* PIT reach 4600 A cm^{-2} [74] and the field dependence of J_c is steeper as compared to 122 wires and tapes. Wires and tapes of the 11 compounds obtained by *in situ* PIT exhibit the lowest transport J_c values up to 10^3 A cm^{-2} [76, 114], but

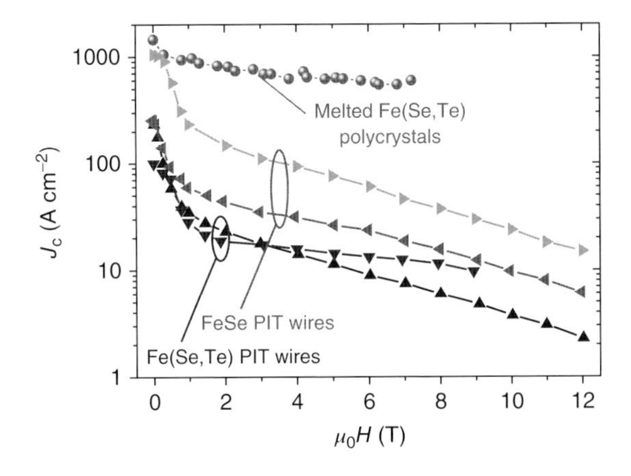

Figure 2.2.2.10 Transport J_c of 11 polycrystals and wires. (Taken from Refs [113–115].)

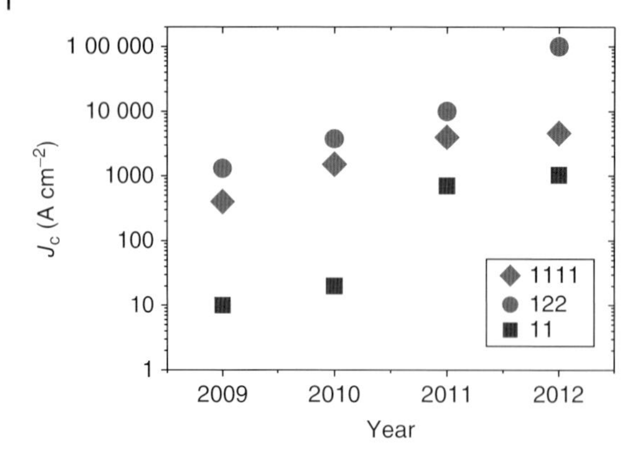

Figure 2.2.2.11 Transport J_c values obtained in FeSCs polycrystals, wires, and tapes versus the year of publication. (The data are taken for the 11 family from Ref. [115] (2009), our own datum obtained in polycrystals (2010), Ref. [116] (2011), and Refs [76, 113] (2012); for the 1111 family from Ref. [94] (2009), Ref. [110] (2010), Ref. [117] (2011), and Ref. [74] (2012); for the 122 family from Ref. [118] (2009), Ref. [119] (2010), Ref. [106, 105] (2011), and Ref. [75] (2012).)

they have the advantages of containing no toxic arsenic and having the simplest crystal structure.

From the state-of-the art results, it can be envisaged that FeSC (122) wires and tapes have good possibilities for magnet applications at 20–30 K, where the niobium-based superconductors cannot play a role owing to their lower T_cs and MgB_2 J_c is rapidly suppressed by the applied field (see Chapter 4 of this volume for a detailed presentation on the state-of-the-art superconducting magnets). Moreover, the steady improvements of J_c values achieved in wires and tapes based on 11, 122, and 1111 FeSCs in the last years, shown in Figure 2.2.2.11, does not seem to be approaching a saturation limit, suggesting the existence of a considerable edge of improvement yet achievable with these materials.

2.2.2.8 Coated Conductors

The general aspects of the technology of coated conductor are presented in Section 3.4.2 of this volume. The evidence of weak-linked behavior in 122 thin films grown onto bicrystals has suggested the idea of applying this technology to FeSCs as well, that is, depositing FeSC films on textured metal substrates with buffer layers, by the technique successfully developed for second-generation cuprate wires.

The coated-conductor templates are manufactured in two steps. First, a Y_2O_3 layer is made on unpolished Hastelloy by sequential solution deposition to reduce the roughness of the tape surface, then a biaxially textured MgO layer is deposited on top by ion beam assisted deposition (IBAD). 122 films have been grown by this method [15, 120]. In-plane misorientation of 3–5° has been measured and most importantly J_c values 10^5-10^6 A cm^{-2}. This route has turned out to be encouraging for the 11 family as well. Fe(Se,Te) thin films deposited on IBAD-MgO-buffered Hastelloy substrates have been able to carry transport critical current of $2{\cdot}10^5$ A cm^{-2} at low temperature and self-field, still as high as 10^4 A cm^{-2} at a field of 25 T [121].

High production cost and low production rates of coated conductor technology must be taken into account when assessing the application potential of this field as compared to that of the PIT technology. However, the elaborate oxide buffer structure, partially designed to protect the metal template from oxidation for cuprates wires may not be needed at all for Fe(Se,Te) wires deposited in vacuum. Growing a Fe(Se,Te) coating directly on textured metal tapes may be possible, thus greatly simplifying the synthesis procedure and reducing production costs.

2.2.2.9 Electronic Applications

The most prominent electronic applications of superconducting materials are certainly Josephson junctions and SQUIDs (see Section 3.3 and Chapters 7–10 of this volume). The Josephson effect is particularly relevant to FeSCs, whose layered nature, particularly in the case of the 1111 family, consists of superconducting sheets weakly coupled through insulating or normal conducting layers. This structure behaves as a periodic stack of Josephson junctions with atomic dimensions along the crystallographic *c*-axis, exhibiting the so-called intrinsic Josephson effect [122, 123, 56]. Yet, wider perspectives are opened by the investigation of Josephson effect in FeSC-based artificial structures, which are on one hand a helpful tool for testing theoretical models about Cooper pair coupling and symmetry of the order parameter and on the other hand allow direct assessment of the electronic application potential as ultra-sensitive magnetic sensors, photon detection in a wide frequency range, voltage standards, tunable frequency sources, latching logic, and digital circuits.

The characteristics of FeSCs such as anisotropy, pairing symmetry, high penetration depth, and small coherence length are crucial parameters in determining the Josephson behavior. However, the key parameter is the superconducting gap, which in the case of an ideal symmetric Josephson junction directly scales with the $I_C R_N$ product (I_C is the junction critical current and R_N the normal state resistance; the $I_C R_N$ product determines the signal frequency and switching time at any given working temperature). Whereas Josephson experiments in high-T_c cuprates have been mostly accounted for in terms of one single gap and d-wave symmetry, the presence of two gaps has been evidenced in the case of MgB_2. As for FeSCs, theoreticians have discussed models with two to five bands with different gaps. In the simplest case, two different gaps for electron-type and hole-type carriers are assumed. Unluckily, different experiments such as scanning tunneling spectroscopy and point contact Andreed reflection have yielded different results about the number and magnitude of the gaps. Most results indicate one gap for the 122 and 11 families and one or two gaps for the 1111 family, with values around 2 meV for the 11, ranging from 2 to 10 meV for the 122, ranging from 1 to 15 meV for the 1111. Relevant literature is reviewed in [124, 4, 125]. The multi-gap nature and the *k*-dependence of the gaps complicates the interpretation of tunneling experimental data, for example, the peak position in $I-V$ curves is in general different from the s-wave case. As for the gap symmetry of FeSC, designs suitable to test pairing symmetry in FeSCs have been proposed [126]. There exists strong support in favor of the s±-wave symmetry, which implies the possibility of "gapless superconductivity" [127], nodes in the gap function, influence of orientation and transparency of interfaces, Josephson currents of opposite directions for electron-type and hole-type carriers, which may cancel out at certain values of applied bias voltage. New kinds of Josephson architectures based on the s±-wave symmetry of FeSCs have been proposed, exhibiting π-phase shift behavior [128, 129]. The unconventional nature of pairing in FeSCs also has consequences for Josephson junctions, as it yields peculiar temperature dependence of the $I_C R_N$ product.

The preparation of polycrystalline, single crystal, and thin film samples is still being optimized, but quality is already at a sufficient level to study the Josephson effects. Single crystalline $Ca_{10}(Pt_4As_8)(Fe_{1.8}Pt_{0.2}As_2)_5$ FeSC nanowhiskers with widths down to hundreds of nanometers can offer new opportunities for device applications as well [130]. Hybrid junctions made of a FeSC electrode and a conventional s-wave counter-electrode have been fabricated using 122 compounds

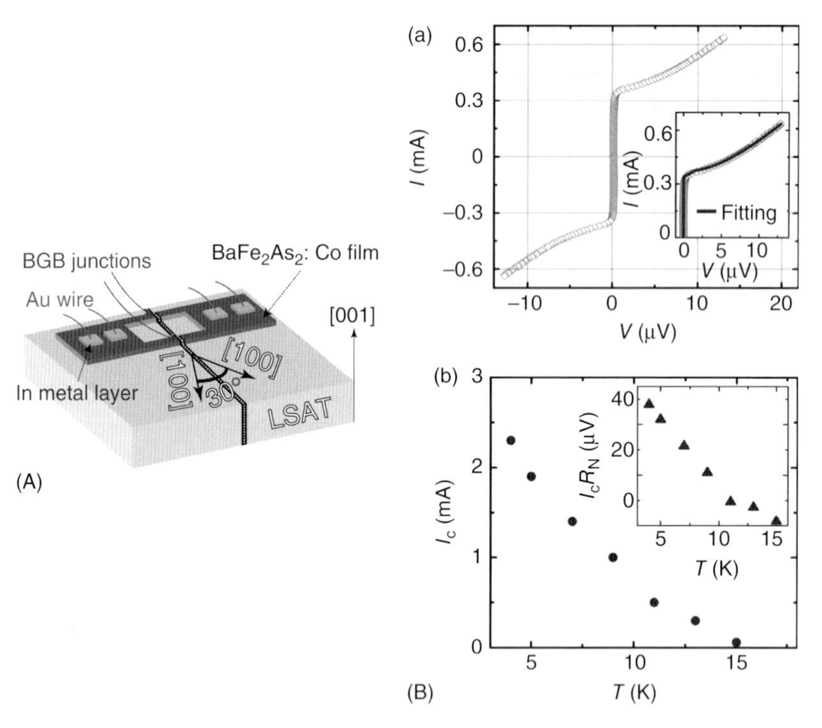

Figure 2.2.2.12 (A) Schematic layout of a bicrystal SQUID fabricated from a Co-doped 122 film. (B) (a) $I-V$ characteristics of the same dc SQUID at $T = 13$ K. The inset shows a comparison to the resistively shunted junction (RSJ) model with noise. (b) Temperature dependence of I_C and $I_C R_N$. (Figures are reproduced with permission from Ref. [136], Copyright 2010 by IOP Publishing.)

[131–133], while in the case of the 1111 the surface quality is still to be improved, especially in thin film technology. All FeSC junctions have been fabricated as well, using different doped 122 single crystals [134] or 122 thin films deposited onto bicrystalline substrates [97, 135]. With such kind of junction, a dc SQUID has been realized [136], providing voltage-flux modulations of 1.4 μV at 14 K and intrinsic flux noise of $7.8 \cdot 10^{-5}$ $Hz^{-1/2}$ above 20 Hz (see Figure 2.2.2.12). Reviews of literature results on FeSC Josephson junctions are found in [124, 137, 138]. In all cases, clear Shapiro steps and Fraunhofer-like magnetic field dependence of the critical Josephson current have been observed. Unluckily, the $I_C R_N$ product values are rather low, in the tens of microvolts range, but experiments suggest that few millivolts values are achievable.

Further studies on improvement of the junction properties via modification of the grain-boundary properties or the use of an artificial barrier are required to realize practical electronic devices. Once some problems are overcome, such as improved surface quality especially along the crystallographic *c* direction, reproducibility, chemical stability, and homogeneity, the all-pnictide junctions may be promising in replacing MgB_2 and cuprate junctions at high working temperatures. They could also provide the advantage of quite high values of H_{c1}, useful for working in unshielded environments.

2.2.2.10 **Summary**

Soon after the discovery of superconductivity in FeSCs, their very high upper critical fields, low anisotropy, and large J_c values, which are only weakly reduced by magnetic fields at low temperatures, have suggested considerable potential in large-scale applications, particularly at low temperature and high fields. Among the different families, the 122 compounds appear to be the most promising, as they are the least anisotropic, have fairly large T_c up to 38 K, close to that of MgB_2, and exhibit the largest values of critical current density. However, 122 compounds contain toxic As and reactive alkaline earth metals, which may be a problem for large-scale fabrication processes. In this respect, also the 1111 compounds present problems, as they contain volatile F and O, whose stoichiometry is hardly controlled. The 11 compounds have lower T_c up to 16 K, but they contain no toxic or volatile elements.

Thanks to the coherence length values of few nanometers, FeSCs are particularly sensitive to the inclusion of nanoparticles as pinning centers to enhance the critical current density. For example, self-assembled $BaFeO_2$ nanorods have proven to enhance the pinning force in 122 films above that of optimized Nb_3Sn at 4.2 K. Critical current densities up to 5.5 MA cm^{-2} at 4.2 K in zero field have been measured in 122 films [14]. Furthermore, inclusion of nanometer scale disorder has proven to suppress the T_c only very weakly [88, 89], suggesting a significant edge of improvement of the pinning mechanism yet achievable.

Wires and tapes of all FeSCs families have been fabricated by the PIT method. The quite isotropic character of J_c with respect to the crystalline direction in all FeSC families suggests that the texturing may not be as stringent as in cuprates. This could mean that cables in the form of untextured wires may provide the required performances, without the need of complex texturing processes as for the fabrication of coated conductors. Unluckily, an exponential decay of the current as a function of the misalignment angle between adjacent grains has been measured in 122 films grown onto bicrystal substrates, similar but not as strong as it has been observed in high-T_c cuprates. However, the performances of FeSCs wires, tapes, and coated conductors are steadily improving in time, indicating that their intrinsic limits are not yet being approached. This positive trend suggests that the intergrain current in real conductors may behave better than that in epitaxially grown bicrystals. Hence, it can be envisaged that FeSC (122) wires and tapes have good possibilities for magnet applications at 20–30 K, where the niobium-based superconductors cannot play a role owing to their lower T_cs and MgB_2 J_c is rapidly suppressed by the applied field.

FeSCs have also been considered for electronic applications such as Josephson junctions and SQUIDs, working at high temperatures. Also in this field, the 122 family is the most promising, thanks to the high structural and morphological quality of films.

References

1. Kamihara, K. *et al.* (2008) *J. Am. Chem. Soc.*, **130**, 3296.
2. Ren, Z.A. *et al.* (2008) *Chin. Phys. Lett.*, **25**, 2215.
3. Paglione, J. (2010) *Nat. Phys.*, **6**, 645.
4. Daghero, D. *et al.* (2010) *Supercond. Sci. Technol.*, **23**, 043001.
5. Dagotto, E. (2013) *Rev. Mod. Phys.*, **85**, 849.
6. Sanna, S. *et al.* (2009) *Phys. Rev. B*, **80**, 052503.
7. Chen, H. *et al.* (2009) *Europhys. Lett.*, **85**, 17006.
8. Katayama, N. *et al.* (2010) *J. Phys. Soc. Jpn.*, **79**, 113702.
9. Wang, N.L., Hideo Hosono, H., and Dai, P. (eds) (2012) *Iron-Based Superconductors: Materials, Properties and Mechanisms*, Pan Stanford Publishing, ISBN-10: 9814303224, ISBN-13: 978-9814303224.
10. (a) Lu, W. *et al.* (2008) *Solid State Commun.*, **148**, 168; (b) Yang, J. *et al.* (2009) *Supercond. Sci. Technol.*, **22**, 025004; (c) Jaroszynski, J. *et al.* (2008) *Phys. Rev. B*, **78**, 064511.
11. Yamamoto, A. *et al.* (2011) *Supercond. Sci. Technol.*, **24**, 045010.
12. Sefat, A.S. *et al.* (2008) *Phys. Rev. Lett.*, **101**, 117004.
13. Zhigadlo, N.D. *et al.* (2012) *Phys. Rev. B*, **86**, 214509.
14. Lee, S. *et al.* (2010) *Nat. Mater.*, **9**, 397.
15. Iida, K. *et al.* (2009) *Appl. Phys. Lett.*, **95**, 192501.
16. Thersleff, T. *et al.* (2010) *Appl. Phys. Lett.*, **97**, 022506.
17. Bellingeri, E. *et al.* (2010) *Appl. Phys. Lett.*, **96**, 102512.
18. Hanawa, M. *et al.* (2012) *Jpn. J. Appl. Phys.*, **51**, 010104.
19. Si, W. *et al.* (2010) *Phys. Rev. B*, **81**, 092506.
20. Wang, Q.Y. *et al.* (2012) *Chin. Phys. Lett.*, **29**, 037402.
21. Kawaguchi, T. *et al.* (2011) *Appl. Phys. Express*, **4**, 063102.
22. Aswathy, P.M. *et al.* (2010) *Supercond. Sci. Technol.*, **23**, 073001.
23. (a) Ogino, H. *et al.* (2010) *Appl. Phys. Lett.*, **97**, 072506; (b) Ogino, H. *et al.* (2010) *Appl. Phys. Express*, **3**, 063103; (c) Kawaguchi, N. *et al.* (2010) *Appl. Phys. Express*, **3**, 063102.
24. (a) Zhu, X. *et al.* (2009) *Phys. Rev. B*, **79**, 220512(R); (b) Ogino, H. *et al.* (2010) *Supercond. Sci. Technol.*, **23**, 115005.
25. Hunte, F. *et al.* (2008) *Nature*, **453**, 903.

26. (a) Gurevich, A. (2003) *Phys. Rev. B*, **67**, 184515; (b) Braccini, V. *et al.* (2005) *Phys. Rev. B*, **71**, 012504.
27. Tarantini, C. *et al.* (2011) *Phys. Rev. B*, **84**, 184522.
28. Pallecchi, I. *et al.* (2012) *Physica C*, **482**, 68.
29. Yamamoto, A. *et al.* (2009) *Appl. Phys. Lett.*, **94**, 062511.
30. Godeke, A. *et al.* (2005) *J. Appl. Phys.*, **97**, 093909.
31. Baily, S.A. *et al.* (2009) *Phys. Rev. Lett.*, **102**, 117004.
32. Yuan, H.Q. *et al.* (2009) *Nature*, **457**, 565.
33. Putti, M. *et al.* (2010) *Supercond. Sci. Technol.*, **23**, 034003.
34. Lei, H. *et al.* (2010) *Phys. Rev. B*, **81**, 094518.
35. Kida, T. *et al.* (2009) *J. Phys. Soc. Jpn.*, **78**, 113701.
36. Fang, M. *et al.* (2010) *Phys. Rev. B*, **81**, 020509(R).
37. (a) Orlando, T.P. *et al.* (1987) *Phys. Rev. B*, **36**, 2394; (b) Nguyen, P.P. *et al.* (1993) *Phys. Rev. B*, **48**, 1148; (c) Moodera, J.S. *et al.* (1988) *Phys. Rev. B*, **37**, 619.
38. (a) Palstra, T.T.M. *et al.* (1988) *Phys. Rev. B*, **38**, 5102; (b) Naughton, M.J. *et al.* (1988) *Phys. Rev. B*, **38**, 9280.
39. Li, G. *et al.* (2008) *Phys. Rev. Lett.*, **101**, 107004.
40. (a) Martin, C. *et al.* (2009) *Phys. Rev. Lett.*, **102**, 247002; (b) Drew, A.J. *et al* (2008) *Phys. Rev. Lett.*, **101**, 097010.
41. Bendele, M. *et al.* (2010) *Phys. Rev. B*, **81**, 224520.
42. (a) Kacmarcik, J. *et al.* (2009) *Phys. Rev. B*, **80**, 014515; (b) Pribulova, Z. *et al.* (2009) *Phys. Rev. B*, **79**, 020508(R).
43. Tajima, Y. *et al.* (1988) *Phys. Rev. B*, **37**, 7956.
44. Zimmermann, P. *et al.* (1995) *Phys. Rev. B*, **52**, 541.
45. Welp, U. *et al.* (1989) *Phys. Rev. Lett.*, **62**, 1908.
46. Basov, D.N. *et al.* (1995) *Phys. Rev. Lett.*, **74**, 598.
47. Tochihara, S. *et al.* (1999) *J. Appl. Phys.*, **85**, 8299.
48. Waldmann, O. *et al.* (1996) *Phys. Rev. B*, **53**, 11825.
49. Cardwell, D.A. and Ginley, D.S. (2003) *Handbook of Superconducting Materials*, IOP Publishing.
50. Yethiraj, M. *et al.* (1994) *J. Appl. Phys.*, **76**, 6784.
51. Putti, M. *et al.* (2008) *Supercond. Sci. Technol.*, **21**, 043001.
52. Zehetmayer, M. *et al.* (2002) *Phys. Rev. B*, **66**, 052505.
53. (a) Niedermayer, C. *et al.* (2002) *Phys. Rev. B*, **65**, 094512; (b) Serventi, S. *et al.* (2004) *Phys. Rev. Lett.*, **93**, 217003.
54. Lyard, L. *et al.* (2004) *Phys. Rev. Lett.*, **92**, 057001.
55. Pallecchi, I. *et al.* (2009) *Phys. Rev. B*, **79**, 104515.
56. Moll, P.J.W. *et al.* (2013) *Nat. Mater.*, **12**, 134.
57. Gurevich, A. *et al.* (2003) *Supercond. Sci. Technol.*, **17**, 278.
58. Onari, S. *et al.* (2009) *Phys. Rev. Lett.*, **103**, 177001.
59. Li, J. *et al.* (2009) *Europhys. Lett.*, **88**, 17009.
60. Tarantini, C. *et al.* (2010) *Phys. Rev. Lett.*, **104**, 087002.
61. Lan, M.D. *et al.* (1991) *Phys. Rev. B*, **44**, 233.
62. Kalisky, B. *et al.* (2010) *Phys. Rev. B*, **81**, 184513.
63. Zhigadlo, N.D. *et al.* (2008) *J. Phys. Condens. Matter*, **20**, 342202.
64. Yang, H. *et al.* (2008) *Appl. Phys. Lett.*, **93**, 142506.
65. Prozorov, R. *et al.* (2008) *Phys. Rev. B*, **78**, 224506.
66. Taen, T. *et al.* (2009) *Phys. Rev. B*, **80**, 092502.
67. Tanatar, M.A. *et al.* (2009) *Phys. Rev. B*, **79**, 094507.
68. Tarantini, C. *et al.* (2010) *Appl. Phys. Lett.*, **96**, 142510.
69. Moll, P.J.W. *et al.* (2010) *Nat. Mater.*, **9**, 628.
70. Santhanam *et al.* (1987) *Sci. News*, **131**, 308.
71. Brandstätter, G. *et al.* (1997) *Phys. Rev. B*, **55**, 11693.
72. Tönies, S. *et al.* (2002) *J. Appl. Phys.*, **92**, 2628.
73. Cabtree, G.W. *et al.* (1987) *Phys. Rev. B*, **36**, 4021.

74. Wang, C.L. *et al.* (2012) *Supercond. Sci. Technol.*, **25**, 035013.
75. Weiss, J.D. *et al.* (2012) *Nat. Mater.*, **11**, 682.
76. Ozaki, T. *et al.* (2012) *J. Appl. Phys.*, **111**, 112620.
77. Foltyn, S.R. *et al.* (2005) *Appl. Phys. Lett.*, **87**, 162505.
78. Sanchez, A. *et al.* (2010) *Appl. Phys. Lett.*, **96**, 072510.
79. Eisterer, M. (2007) *Supercond. Sci. Technol.*, **20**, R47–R73.
80. Harasyn, D.E. *et al.* (1975) *J. Appl. Phys.*, **46**, 2232.
81. Ueda, S. *et al.* (2011) *Appl. Phys. Lett.*, **99**, 232505.
82. Kidszun, M. *et al.* (2011) *Phys. Rev. Lett.*, **106**, 137001.
83. Tarantini, C. *et al.* (2012) *Phys. Rev. B*, **86**, 214504.
84. Lee, S. *et al.* (2013) *Nat. Mater.*, **12**, 392.
85. Li, Q. *et al.* (2011) *Rep. Prog. Phys.*, **74**, 124510.
86. Bellingeri, E. *et al.* (2012) *Appl. Phys. Lett.*, **100**, 082601.
87. Haberkorn, N. *et al.* (2012) *Phys. Rev. B*, **85**, 014522.
88. Nakajima, Y. *et al.* (2009) *Phys. Rev. B*, **80**, 012510.
89. Eisterer, M. *et al.* (2009) *Supercond. Sci. Technol.*, **22**, 095011.
90. Graser, S. *et al.* (2010) *Nat. Phys.*, **6**, 609.
91. Larbalestier, D.C. *et al.* (2001) *Nature*, **410**, 186.
92. Yamamoto, A. *et al.* (2008) *Appl. Phys. Lett.*, **92**, 252501.
93. Yamamoto, A. *et al.* (2008) *Supercond. Sci. Technol.*, **21**, 095008.
94. Kametani, F. *et al.* (2009) *Appl. Phys. Lett.*, **95**, 142502.
95. (a) Larbalestier, D.C. *et al.* (1987) *J. Appl. Phys.*, **62**, 3308; (b) Seuntjens, J.M. *et al.* (1990) *J. Appl. Phys.*, **67**, 2007.
96. Haindl, S. *et al.* (2010) *Phys. Rev. Lett.*, **104**, 077001.
97. Lee, S. *et al.* (2009) *Appl. Phys. Lett.*, **95**, 212505.
98. Hilgenkamp, H. *et al.* (2002) *Rev. Mod. Phys.*, **74**, 485.
99. Katase, T. *et al.* (2011) *Nat. Commun.*, **2**, 409.
100. Durrell, J.H. *et al.* (2011) *Rep. Prog. Phys.*, **74**, 124511.
101. Wang, L. *et al.* (2011) *Appl. Phys. Lett.*, **98**, 222504.
102. Togano, K. *et al.* (2011) *Appl. Phys. Express*, **4**, 043101.
103. Ma, Y. (2012) *Supercond. Sci. Technol.*, **25**, 113001.
104. Ferrando, V. *et al.* (2005) *Appl. Phys. Lett.*, **87**, 252509.
105. Ma, Y. *et al.* (2011) *IEEE Trans. Appl. Supercond.*, **21**, 2878.
106. Gao, Z. *et al.* (2011) *Appl. Phys. Lett.*, **99**, 242506.
107. Gao, Z.S. *et al.* (2012) *Sci. Rep.*, **2**, 998.
108. Ma, Y. *et al.* (2013) *Supercond. Sci. Technol.*, **26**, 035011.
109. Togano, K. *et al.* (2013) *Supercond. Sci. Technol.*, **26**, 065003.
110. Wang, L. *et al.* (2010) *Supercond. Sci. Technol.*, **23**, 075005.
111. Gao, Z. *et al.* (2011) *Supercond. Sci. Technol.*, **24**, 065022.
112. Izawa, H. *et al.* (2012) *Jpn. J. Appl. Phys.*, **51**, 010101.
113. Palenzona, A. *et al.* (2012) *Supercond. Sci. Technol.*, **25**, 115018.
114. Ozaki, T. *et al.* (2011) *Supercond. Sci. Technol.*, **24**, 105002.
115. Mizuguchi, Y. *et al.* (2009) *Appl. Phys. Express*, **2**, 083004.
116. Ding, Q.P. *et al.* (2011) *Supercond. Sci. Technol.*, **24**, 075025.
117. Fujioka, M. *et al.* (2011) *Appl. Phys. Express*, **4**, 063102.
118. Wang, L. *et al.* (2010) *Physica C*, **470**, 183.
119. Qi, Y. *et al.* (2010) *Supercond. Sci. Technol.*, **23**, 055009.
120. Katase, T. *et al.* (2011) *Appl. Phys. Lett.*, **98**, 242510.
121. Si, W. *et al.* (2011) *Appl. Phys. Lett.*, **98**, 262509.
122. Kashiwaya, H. *et al.* (2010) *Appl. Phys. Lett.*, **96**, 202504.
123. Nakamura, H. *et al.* (2009) *J. Phys. Soc. Jpn.*, **78**, 123712.
124. Seidel, P. (2011) *Supercond. Sci. Technol.*, **24**, 043001.
125. Daghero, D. *et al.* (2011) *Rep. Prog. Phys.*, **74**, 124509.
126. Golubov, A.A. *et al.* (2013) *Appl. Phys. Lett.*, **102**, 032601.

127. Glatz, A. *et al.* (2010) *Phys. Rev. B*, **82**, 012507.
128. Tsai, W.F. *et al.* (2009) *Phys. Rev. B*, **80**, 012511.
129. Chen, W.Q. *et al.* (2011) *Phys. Rev. B*, **83**, 212501.
130. Li, J. *et al.* (2012) *J. Am. Chem. Soc.*, **134**, 4068.
131. Zhang, X.H. *et al.* (2009) *Phys. Rev. Lett.*, **102**, 147002.
132. Zhou, Y.R. *et al.* (2009) arXiv:0812.3295
133. Schmidt, S. *et al.* (2010) *Appl. Phys. Lett.*, **97**, 172504.
134. Zhang, X.H. *et al.* (2009) *Appl. Phys. Lett.*, **95**, 062510.
135. Katase, T. *et al.* (2010) *Appl. Phys. Lett.*, **96**, 142507.
136. Katase, T. *et al.* (2010) *Supercond. Sci. Technol.*, **23**, 082001.
137. Tanabe, K. *et al.* (2012) *Jpn. J. Appl. Phys.*, **51**, 010005.
138. Hiramatsu, H. *et al.* (2012) *J. Phys. Soc. Jpn.*, **81**, 011011.

3
Technology, Preparation, and Characterization

3.1
Bulk Materials

3.1.1
Preparation of Bulk and Textured Superconductors
Frank N. Werfel

3.1.1.1 Introduction

Melt processed and textured bulk superconductors have a high potential for electric and magnetic applications such as, magnetic bearings and trains (see Section 4.8), separators, couplings, electric motors and generators (Section 5.5), and superconducting permanent magnets (PMs).

Three families determine the high-temperature superconductors (HTSs). All three high-tech materials exist in form of bulk and thin film geometry. The $RE_1Ba_2Cu_3O_y$ compounds abbreviated as RE123 (rare earth RE = Y, Sm, Gd, Nd, Dy, etc.) have a critical temperature T_c of about 92–95 K and are widely investigated since the discovery in 1986–1987. Parallel, the BiSCCO family with compounds $(Bi,Pb)_2Sr_2Ca_2Cu_3O_y$ (Bi2223) and $Bi_2Ca_2Cu_2Cu_1O_y$ (Bi2221) possess a high critical temperature of ~110 and ~90 K, respectively. Finally, in 2001, the discovery of superconductivity of MgB_2 was unexpected and it has a simple hexagonal structure with a critical temperature of up to $T_c = 39$ K. MgB_2 is a metallic superconductor in contrast to REBCO (rare earth RE) and BiSCCO ceramics families (see Chapter 2.1.2).

While the microstructure of the first two families shows anisotropic behavior and consists of platelets with the *c*-axis perpendicular to the longitudinal plane, bulk textured MgB_2 material of nearly full density exhibit a *c*-axis alignment of the hexagonal MgB_2 grains parallel to a pressure direction, obtainable, for example, by hot deformation of stoichiometric MgB_2 pellets. Independent of the geometry and composition for application, all three phases require a highly textured microstructure to enable the supercurrent flow without weak-links. The microstructure and its texture is the link between the bulk and conductor development. While for thin

Applied Superconductivity: Handbook on Devices and Applications, First Edition.
Edited by Paul Seidel.
© 2015 Wiley-VCH Verlag GmbH & Co. KGaA. Published 2015 by Wiley-VCH Verlag GmbH & Co. KGaA.

film, a two-dimensional texture is easier to obtain, bulk texture requires usually more effort.

Compared to REBCO, the BiSCCO compounds play a less important role if bulk application is concerned. Nevertheless, texturing of BiSCCO is a material issue; it should be briefly mentioned here. The alignment of the microstructure of the Bi compounds depends on various processing steps like precursor powder composition combined with mechanical and heat treatment. Thereby, it seems to give evidence that the necessary alignment is partially or completely achieved before the transformation from Bi2212 to Bi2223 occurs. The texture mechanism of 2212 seems to play a prominent role of 2223 phase.

Bi2223 is used mainly for wires and tapes and are made by the so-called *powder-in-tube* techniques. The method is determined of at least two heat treatment processes separated by an intermediate rolling step. At present, a few industrial companies produce kilometer-long Bi2223 wires for superconducting cable and motor projects.

A key issue in enhancing the BiSCCO texture seems to play the addition of silver in form of substrates or fine dispersed particles. Correspondingly, the oxide powder-in-tube (OPIT) technique uses Ag or Ag alloys tubes, sealed, and then drawn or extruded into round wire fabrication. In addition, in the Bi2223 material, it is in general difficult to align the grains in a perfect way because the formation mechanism of the 2223 superconducting phase is a complex process. The performance of the Bi2223 wires is determined by the density and homogeneity of the microstructure. As the result of the thermomechanical processing steps porosity and cracks limit the critical current transport. By applying isostatic pressure during the thermal processes, one suggests to eliminate the deficits getting a dense core and fewer microcracks. The Bi2223 wire fabrication is described in Section 3.4.

The bulk fabrication of Bi2212 is a sequence of partial melting and subsequent solidification under slow cooling. The process enables a highly oriented grain microstructure if silver metal has been preferentially used in the preparation.

Since 2001, when magnesium diboride (MgB_2) was first reported to have a transition temperature of up to 39 K, conductor development has progressed. In between, MgB_2 superconductor wire in kilometer-long piece-lengths has been presented, for example, in coil form. Now that the wire is available commercially, work has started on not only demonstrating potential MgB_2 applications in superconducting devices: magnetic resonance imaging, fault current limiters, transformers, motors, generators but also utilizing MgB_2 bulk magnetic levitation, and superconducting PMs.

The present chapter provides a survey of melt processing technology of REBCO bulk superconductors, as the most prominent and, in view of applications, the most promising group of superconductors.

We will describe and investigate the fabrication routes, the major technical advancements in high-T_c bulk processing, the obtained bulk properties, and its characterization.

3.1.1.2 Melt Processed REBCO

3.1.1.2.1 Process Steps

Soon after the discovery of HTSs in 1986–1987, the first YBCO ($Y_1Ba_2Cu_3O_y$) samples were mixed in known composition, sintered, and finally slowly cooled down. The pellets had a random grain nucleation without any preferred orientation of the crystalline structure. Nevertheless, the Y123 material showed superconducting properties and could levitate small PMs.

In the pioneer years of HTS development, the YBCO samples have been grown under simplified thermal and pressure conditions. The samples had a polycrystalline structure with a grain size of a few millimeters. Levitating a PM stably above a superconductor cooled in liquid nitrogen (LN_2) was able to demonstrate superconductivity in thousands of presentations. The levitation pictures presented in journals and newspapers became a symbol for the new superconductivity and caused hype in expectations in the scientific community.

The new material promised a wide field of electric and magnetic applications. Soon after analyzing first material properties conducted by many scientific groups at universities and institutes, it was becoming clear that a high critical current density and a large current loop diameter are the essential requirements for future magnetic applications of the bulk-type superconductors.

Among bulk growth technique, two basic methods with respect to the thermal distribution became popular: the temperature gradient growth and the isothermal growth. With the gradient method, polycrystalline melt textured YBCO rings, rods, and plates are easily fabricated. The method is a promising concept for modular upscaling of bulk current transport and magnetic components if the weak-link behavior of grain boundaries (GBs) can be reduced. The fabrication was based on Bridgman-like directional solidification with an applied spatial and time-dependent thermal gradient. The method requires a strict control of the temperature distribution to maintain a virtual seed and nucleation at the point of lowest temperature. This point is usually given with the mechanical support at the bottom of the pressed sample.

The drawbacks of the method were the limited obtained texture, the sensitivity to temperature variations, unwanted parasitic grain growth, and material inhomogeneities caused by large temperature gradients. These factors were difficult to control and led to the strategy of isothermal solidification assuring a better melt texturing of the bulk.

The advantage of the isothermal solidification process is a greater material homogeneity and improved electric and magnetic properties. The challenge of the isothermal fabrication was to search for a mechanism of controlling and preventing the spontaneous nucleation and multicrystal growth. These requirements can be fulfilled by a practical top seeding technique using corresponding small single crystal seeds with a higher melting point compared to the sample material. With successful top seeding developed in the recent 10 years, we have seen an enormous progress in the fabrication and application of bulk high-T_c superconductors.

Experimentally, all texturing methods are performed in precisely temperature-controlled furnaces having a control of the furnace atmosphere as well. The powder compacted pellet is brought to a high temperature and then slowly cooled down so that crystals can nucleate and grow. To control the nucleation and the growth direction in a sample, a seeding technique with small crystals on top has become the best technique.

With successful seeding and the accompanied epitaxial growth, the bulk RE123 becomes well textured in a preferred direction having no or well-linked boundaries. The aim of texturation is to prevent so-called *weak-links*, which can limit drastically the supercurrent in the sample. Misfits, segregations, or strongly disoriented (large-angle) GBs are weak-links for the flowing supercurrent.

Parallel to the melt texturing technique, the critical current density J_c has been improved over the last two decades by nearly two orders of magnitude. Furthermore, with larger-scale grain growth (>50 mm) and multiseeding technique, a preindustrial batch-type fabrication of high-quality bulk superconductors is becoming available. However, large-scale and high-quality single domain growth is still a key issue in the bulk YBCO fabrication. In the following, the individual steps of melt and texturing bulk superconductors are explained.

The first step in the REBCO material preparation route is the mixing of the raw powders, typically in a 99. 9–99.99% purity. For melt texturing, some groups are using pressed pellets made of presintered and calcined precursor phases. Another way is to use mixed powders of RE, barium- and copper oxides. Careful mixing of the raw powders Y_2O_3, $BaCO_3$, and CuO in the desired chemical ratio is a prerequisite of the preparation process. For in-house fabrication of RE(123) and RE(211), the precursors are calcined in a furnace at temperatures of about 940 °C. Chemically, the loss of ignition (LOI) of the raw powders at 900–1000 °C should be <1%. For processing $YBa_2Cu_3O_y$ (Y123) powder, the chemical reaction is given by

$$Y_2O_3 + 4BaCO_3 + 6CuO \rightarrow 2YBa_2Cu_3O_y + 4CO_2\uparrow \qquad (T = 940–950\,°C)$$

The $CaCO_3$ impurity content in raw powder $BaCO_3$ should have a concentration of ≤200 ppm. Other possible impurities like chlorine and sulfur usually occur in minor concentrations. For the other raw powders RE_2O_3 and CuO, the quality should be in a equivalent high quality of 99.95–99.99% purity. The calcination process is performed within 30–36 h and repeated typically twice. Air or oxygen ventilation in the furnace chamber ensures a total calcination reaction. According to the phase diagram, the temperature during the Y123 calcination should be lower than 960 °C to obtain a phase pure Y123 compound. The reacted superconducting powder in a next step is fine-grained milled in mortar grinders up to grain size value of $d_{50} = 5–8\,\mu m$. Alternately, superconducting powders can be purchased from a few chemical companies.

The synthesis of the "green phase" Y_2BaCuO_5 (Y211) follows a similar calcination step at a temperature of 940–950 °C

$$Y_2O_3 + BaCO_3 + CuO \rightarrow Y_2BaCuO_5 + CO_2\uparrow$$

Fabrication of other precursor powders follows equivalent chemical calcination routes. For the liquid phase L(035), this chemical reaction is

$$3BaCO_3 + 5CuO \rightarrow Ba_3Cu_5O_x + 3CO_2 \uparrow \quad \text{processed at } 860\,°C, 30\,h$$

The so-called "*blue phase*" $Y_2Cu_2O_5$ is obtained by

$$Y_2O_3 + 2CuO \rightarrow Y_2Cu_2O_5 \quad \text{processed at } 960\,°C, 30\,h$$

The resultant precursor powder grain size determines the density of the compacted "green body." In general, for melt texturing processing, the properties of the precursor powders, that is, the chemical *and* physical properties determine the growing conditions in the following fabrication steps.

Before texture steps are attempted and can be performed, some thermodynamical considerations are helpful to understand the following experimental processing and texturing steps. The explanation is given for the Y123 superconductor; the other RE123 materials follow a similar thermodynamics.

3.1.1.2.2 Melt Processing Thermodynamics

The principal steps of the melt and texture process are shown in Figure 3.1.1.1. From the thermodynamics, the stability field of the superconducting Y123 phase depends on the composition, the temperature, and the oxygen partial pressure [1]. At a partial oxygen pressure $p(O_2) > 150$ Pa, Y123 decomposes in a peritectic reaction into the two components Y_2BaCuO_5 (Y211) and the melt phase L (BaCuO) accompanied by an oxygen loss in the melt

$$YBa_2Cu_3O_y \rightarrow aY_2BaCuO_5 + bL(BaCuO) + cO_2$$

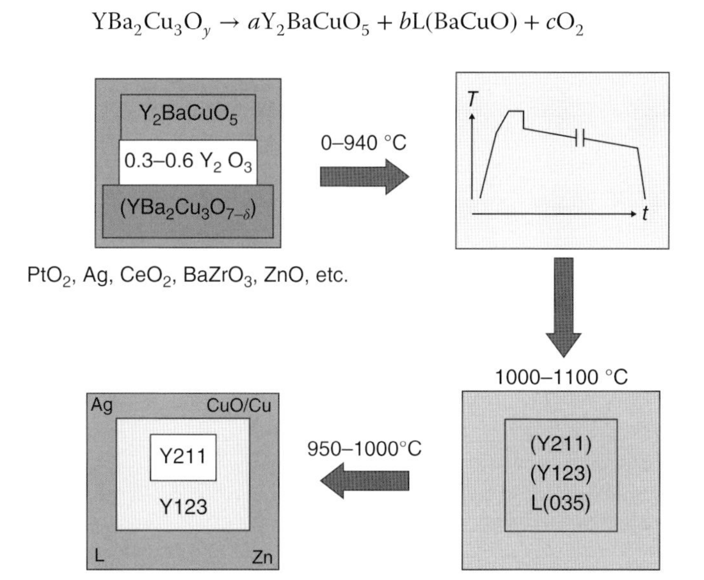

Figure 3.1.1.1 YBCO melt processing steps with mixing phases and dopants at appropriate temperatures.

Studying the Y123 phase diagram, it is found that the basic melt composition consists of the two principal phases Y123 + Y211, but is restricted to a rather narrow primary crystallization field. Thereby, the overheating and growing temperature and the accompanied specific surface area determine the peritectic phase transition. The decomposition temperature T_p of Y123 depends on the partial oxygen pressure in the furnace. The T_p temperature changes from $T_p = 1015\,°C$ at normal pressure $p(O_2) = 0.21 \times 10^5$ Pa to $T_p = 950\,°C$ at reduced pressure $p(O_2) = 150$ Pa.

By applying the temperature profile given in Figure 3.1.1.1 schematically, careful studies of the peritectic transformation have shown the limiting factors of the melt texturing process. During the melt process and growth of Y123, the melt is continuously depleted in $YO_{1.5}$ (yttrium) and oxygen. Because oxygen is usually sufficiently available from the environment in the furnace chamber, the diffusion of yttrium through the melt into the growing solidification front is the limiting factor. In the consequence, only one part of the total Y211 contribution and liquid L phase ingot is finally reacted to Y123, leaving small unreacted 211 phase particles in the 123 phase (Figure 3.1.1.1). Analyzing the experimental results more deeply, two principal diffusion processes seem to influence the single domain 123 growth. First, yttrium- , barium-, and copper cations diffuse within the melt phase to the 123 solidification front proportional to the square of the undercooling temperature $(\Delta T)^2$. Secondly, the oxygen diffusion depends on then vapor–solid interface which is usually the surface from air to the sample. Both processes have to be considered and controlled carefully to enable 123 crystal solidification.

The 211 particles entrapped in the 123 crystal play an important role by the ability to allow effective pinning of magnetic flux. This pinning effect has been widely investigated, and it has been found that especially the interface between 123 and 211 with its microstructure in a dimension of the coherence length of Y123 superconductor ($\xi = 3–5$ nm) is able to pin magnetic flux effectively. This mechanism has been utilized by promoting enhanced pinning due to an excess of 211 (422) phase in the starting powder composition of all REBCO compounds.

YBCO melt processing can be beneficially prepared by adding Y_2O_3 in excess, instead of the usually applied admixture of Y211 [2]. In this process, Y211 grains of small size are formed in a preheating solid state reaction and the crystallization can take place already under equilibrium conditions in a temperature window of $940\,°C < T < 1015\,°C$. The broad temperature interval 940–1015 °C allows less sensitive growing conditions instead of the univariant peritectic solidification temperature in the conventional process

$$\text{Standard process}: \quad Y_{1.8+x}Ba_{2.4}Cu_{3.4}O_y \rightarrow Y_1Ba_2Cu_3O_y + 0.4Y_2BaCuO_5 + L$$

$$Y_2O_3 \text{ excess [2]}: \quad Y_{1.48-1.50}Ba_2Cu_3O_y \rightarrow Y123 + 0.39 \times Y211 + 0.18 \times CuO + L$$

For a given composition, the precursor powders are mixed and calcined in typical stoichiometric compositions of $Y_{1.8}Ba_{2.4}Cu_{3.4}O_y$ by varying the Y211 between 0.3 and 0.5 mol Y211 per 1 mol Y123. The exact 211 percentage is a parameter of variation between the different laboratories in terms of melt processing stability conditions. Because of the effective pinning performance of the 211 phase, several techniques have been investigated to inhibit the coarsening of the 211 phase during the melt process. Addition of either PtO_2 or CeO_2 is commonly applied for this purpose. Both dopants can prevent the so-called *Ostwald ripening mechanism*, after that a growing and coarsening of the 211 particles at high temperatures and longer holding time reduces the pinning efficiency of the non-superconducting inclusions. Larger samples require a longer high temperature step to obtain a homogeneous temperature distribution throughout the sintered body. For refining Y211, a standard doping procedure contains 0.3 wt% Pt, 3–10 wt% Ag, and 0.4–1 wt% CeO_2. With the chemical additions, one attempts to refine the 211 phase and create small precipitates of third and higher order phases to improve magnetic flux pinning. In view of large J_c, however, one has to pay attention to further intrinsic parameters: the macroscopic material density with the grain or sub-GBs, possible further precipitates like Cu/CuO at the 211/123 interface, microcracks, and structure defects. All of these features may influence the mechanical and electromagnetic properties of bulk superconductors.

3.1.1.2.3 **Powder Compacting**

Mechanical compacting of precursor REBCO powders is less carefully studied compared to the powder chemistry. Usually, the powder mixtures are pressed into the desired shape by one-axial or a cold-isostatic pressing (CIP) compacting process. The quality of the pressed green body is essential for the heat transfer during the growing process and determines the final macroscopic material density and the weak-link behavior of the GBs. In addition, homogeneity and high density of the compacted precursor powder is an ultimate prerequisite for growing large grains. Using CIP method, the typical compacting pressure is 1.5–2 kbar. One-axial pressing is used for compacting simple geometrical bodies like pellets. For extended and more complicated green bodies, such as tubes, hollow cylinders, rods, and plates, the compacted powder density may vary substantially at surfaces, in corners, and so on, and cause cracks. In case of very fine precursor powders of less than about 3 μm (d_{50} value) a pre-granulation step and a reduction of the pressure to 0.5–1 kbar may contribute to improved results. Cold-isostatic compacted pellets up to 400 mm diameter and rings up to 200 mm outer diameter are possible. One-matrix pressing at 1 kbar pressure gives a relative powder density of up to 60% of the theoretical density dependent on the grain size distribution. With thermal sintering at temperatures of 940–950 °C, a relative density of up 90% is achieved, for example, for the fabrication of RE123 targets.

After the melt processing step, the samples typically shrink further to a relative material density of 93–95% (theoretical density of Y123: 6.38 g cm^{-3}).

3.1.1.2.4 **Texture Process**

In the first decade of high-T_c superconductivity, several practical methods to fabricate RE123 bulks have been developed [3]. Melt texturing procedure was performed in furnaces operating with *thermal gradient methods* allowing a wide variety of temperature spatial gradients and time-dependent profiles. Figure 3.1.1.2 illustrates two growing arrangements of linear-like vertical and quasi in-plane

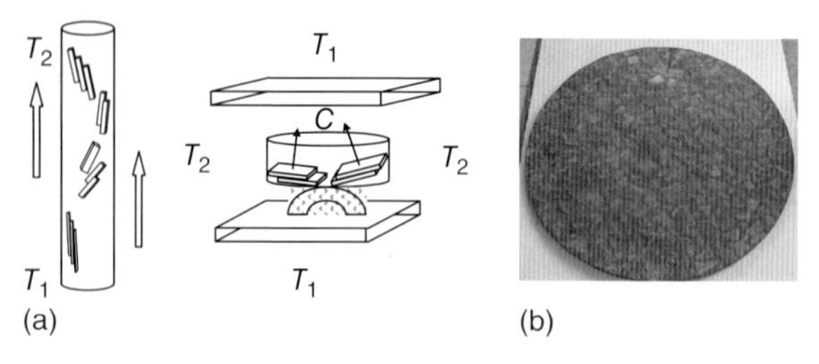

Figure 3.1.1.2 Spatial thermal gradient growth of rods and cylindrical YBCO samples (a). YBCO melt processed plate (270 mm, 7 mm thick) (b) [4]. (Reproduced with permission of ATZ.)

thermal gradients applied for rods and standard cylinder texturing, respectively. The seedless temperature gradient method requires a strict control of the spatial and time-dependent temperature distribution to maintain a virtual seed function and nucleation at a point of lowest temperature [4]. Usually, the sample is located between two Al_2O_3 plates adequately subcooled by the bottom plate of the furnace. While the supporting point of the sample provides a local undercooling and hence, acts as a virtual seed. On top of the sample, unwanted spontaneous nucleation and multiple grain growth may still appear. Achieving a pure single domain sample was for a long time more of an attempt than a scientific technology.

Nevertheless, many samples were grown with a random nucleation and multiple grain structure. After a melting stage around 1060 °C for 1 h holding time, depending on the size and weight, the sample is cooled-down fast to the peritectic solidification temperature T_p, and then slowly ramping-down between 1020 and 970 °C. Without active seeding, a spontaneous nucleation will take place in the melt. In order to prevent an extensive and random nucleation, a local undercooling seeding technique can be performed. In Figure 3.1.1.2b, the cylindrical YBCO plate has a diameter of 270 mm, and is only 7 mm thick. Owing to the extreme aspect ratio between height and diameter, a large percentage of the grains (5–15 mm size) are *a*- and *b*-axis in-plane oriented with random distribution azimuthally.

The technique of punctual undercooling has been used for smaller samples up 30 mm size, giving a nearly single grain structure with a slight turn-off in the *c*-axis orientation (Figure 3.1.1.2a). Finally, rod-like samples obtained with a sufficient connectivity of the grains by a strong thermal gradient along the sample could be obtained. Such kind of bulk samples were fabricated within the first years after the HTS discovery. Not surprisingly, the pellets possess an excellent mechanical stability, and after 20 years they show strong magnetic levitation.

With better understanding of the thermodynamics and the HTS material properties, it was becoming clear that isothermal growth process provides better material opportunities. The active seeding on top of the samples could suppress unwanted and random nucleation, provide higher J_c material, and especially important, allow fabricating in batch processes with growing 30–50 samples simultaneously.

To achieve these goals, various *top seed melt growth* (*TSMG*) methods have been developed [5–9]. Among this fabrication, single grains or domains with dimensions up to more than Ø 150 mm × 20 mm are prepared. Sm123 and Nd123 single crystals are generally used as seed materials, which have higher melting points than YBCO and comparable lattice constants [10, 11]. Parallel Sm123- or Nd123-evaporated thin film seeds on MgO (100) are successfully applied for single grain processing [12, 13]. The seed crystal is placed and aligned on top surface of the pressed YBCO sample with *c*-axis perpendicular to the sample surface (Figure 3.1.1.4b). Dependent on the material under that is growing the small seed on top (3 mm × 3 mm), have to be selected such that during the high temperature step the seed is not degraded or partially melted. If its melting point is high enough relative to the material of the sample, the seed could be prepared and placed on top before the texture procedure is started. This is the so-called *cold seeding* method. The cold seeding procedure provides an easy and convenient orientation and alignment of the seed. The alignment is especially important in case of accurate placement of several seeds on larger samples (multiseeding) [14]. The exact position and orientation of the seeds have to be controlled carefully to allow a perfect epitaxial crystallization. Multiple seeds, like in Figure 3.1.1.3, each generate an individual crystal. Neighbor crystals are linked by GBs which are usually barriers for the supercurrent. In some cases, the GBs are homogeneous and dense, allowing the supercurrent to pass the barrier.

More complicated is the *hot seeding* technique during the melt process. If the difference of the melting temperatures of the seed and the sample are not sufficiently large enough, hot seeding is the preferred texture method. To avoid a possible partial decomposition and interdiffusion, the seed is placed on the hot surface after the process approaches or passes the (much lower) peritectic temperature T_p of the sample material. The hot seeding technique requires a special designed furnace and proficient skills in reproducible handling.

Under the melting process, the seed produces a preferential nucleation site on top of the sample. From there, the crystallization front corresponding to the

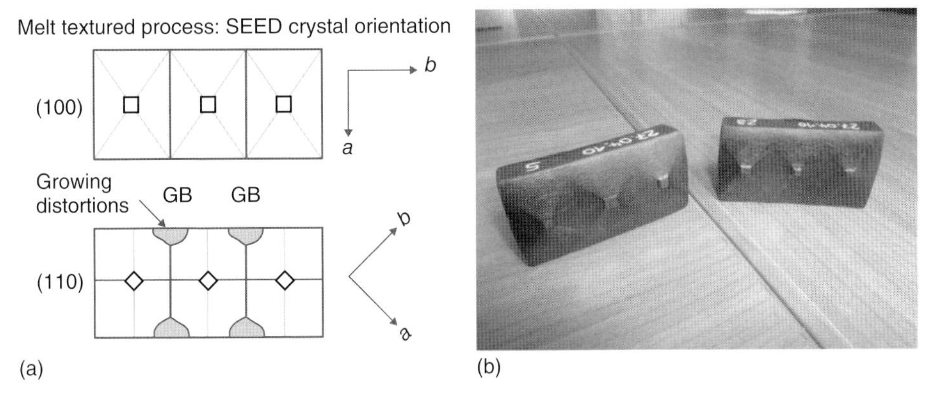

Figure 3.1.1.3 Seed orientation in threefold seeded YBCO (a) triple Sm123 seeded YBCO samples (b) [14]. (Reproduced with permission of ATZ.)

peritectic reaction moves outward, transforming the melted sample into a single grain material. It should be mentioned, however, that top seeding alone does not guarantee perfect single grain samples, especially if the sample size approaches large diameters above 50 mm. One important parameter is the undercooling window controlled by the temperature program which influences both the growing rate as well as the unwanted parasitic domain nucleation. According to a theoretical model by Cima *et al.* [15], the melt growth velocity of YBCO is governed by the yttrium diffusion from the 211 phase to the growth front. Therefore, a large undercooling difference $\Delta T > 15\,K$, relative to the nominal peritectic temperature, supports a higher growth speed while the probability of occurrence of random unwanted nucleation is increased simultaneously. In contrast, a small undercooling temperature $\Delta T = 3-5\,K$ delays the occurrence of random nucleation but causes a low growth rate implying a longer holding time to obtain larger domains. Longer processing routes of more than 100 h cause other complications, such as the 211 coarsening and the loss of the liquid phase L(035).

As mentioned above, the occurrence of GBs in multiseeding procedure is more complicated and has been investigated by different research groups [16, 17]. On the other hand, more than 1000 pieces of three-seed YBCO bulk samples in a typical as-grown geometry of 67 mm × 34 mm × 14 mm has been fabricated for magnetic train application. Beneficially, with the help of misalignment in the nominal $a-b$ growing distance geometry ($a = 35$ mm, $b = 22$ mm), the three-seed bulk shown in Figure 3.1.1.3 is forced, to squeeze and press the crystal growing fronts against each other along the sample length to reduce the weak-link behavior of the two GBs. Thereby, in contrast to the top seeding technique of a single grain cylindrical sample, the exact lateral orientation of the square planar 3 mm × 3 mm Sm123 seeds is an important parameter for the final superconducting properties of the GBs. The [100] alignment of the seed crystals orientation to each other on top of the green body determines the quality of the total multiseed sample. The performance of the GBs of carrying intergrain current depends on the epitaxial connection of the two neighboring crystals and domains. Seed orientation in [110] direction in Figure 3.1.1.3 after our observation very often causes parasitic grains or growth distortion in the corners of the neighbor crystals. This observation is different from samarium barium copper oxide (SmBCO) thin film experiments on YBCO where [110]/[110] GBs formation was found to be clean without trapping liquid [18].

3.1.1.2.5 **Single Grain Fabrication**

With increasing demand of high-quality bulk material for magnetic application, it is necessary to fabricate samples free of GBs. The GBs act as a weak-link in high-T_c superconductors and reduce the intrinsic J_c values substantially. In addition, because of the strong anisotropic behavior of electromagnetic parameters, it is important to align the crystal c-axis direction orthogonally to the sample surface. Large-scale single domain fabrication requires further experimental conditions as the selection of the supporting material during the melt procedure. The

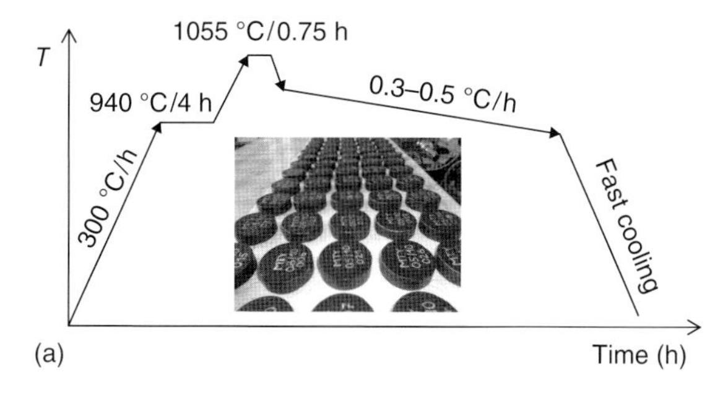

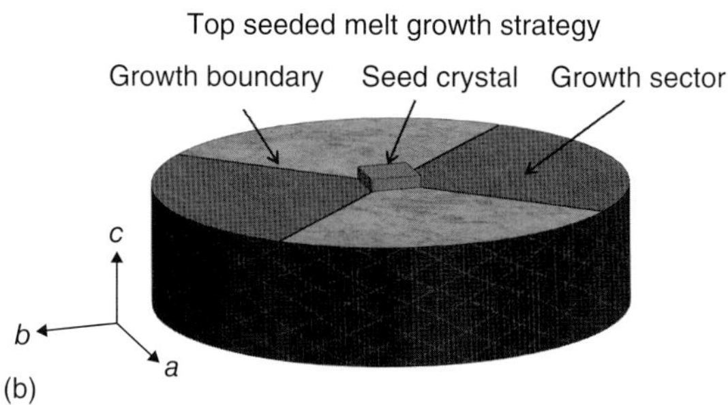

Figure 3.1.1.4 YBCO bulk processing: temperature pattern of TSMG YBCO fabrication (a) [19] (Reproduced with permission of ATZ); top seeded melt growth strategy (b).

supporting material should prevent any additional nucleation from the sample bottom and, simultaneously, reduce the loss of the liquid phase during the process. Typically, RE211, MgO, or a corundum plate is utilized for mechanical support.

In the following, we will review large-scale HTS bulk production in an engineering level serving the increased HTS bulk demand for magnetic applications.

Figure 3.1.1.4a displays the typical growing procedure and temperature profile applied by Adelwitz Technologiezentrum GmbH (ATZ) for single and multiple grain fabrication [19]. The melt texture process follows a temperature route: heating up to about 940 °C for densification and prestep reaction, further heating to 1040–1100 °C, 0.5–1 h dwell time, fast cooling to 980–990 °C and recrystallization with a ramp-down of 0.5–1 K h^{-1} to 940 °C, cooling with 50 K h^{-1}, or furnace cooling to room temperature. In last step, oxygen annealing at 600–350 °C for 200 h is performed. With the top seeding technique, high-quality superconducting magnetic material in blocks of circular or rectangular shapes up to 60 mm can be fabricated in a corresponding chamber furnace. The TSMG strategy for single grain fabrication is demonstrated in Figure 3.1.1.4b. Except the position of the seed crystal, the specimens show a typical top pattern consisting of the four growth sectors and the growth boundaries. Larger samples or superconductor bulks with circular shape (rings, tubes, and segments) could not be grown as single crystalline material. Therefore, the bulks must be mechanically milled into the desired shape, then assembled to the component design, glued together, and fine-machined up to the precise shape.

One question which is left, related to the TSMG technique, is directed to the reproducibility and magnetic quality, which can be achieved in a large-scale production. In most magnetic applications, the textured bulk superconductors have to be machined and assembled, for example, for ring-type bearing stators. Hence, using seeding technique in a *first step*, superconducting bulks exhibit peak values of the trapped field of 1.0 – 1.2 T at 77 K and 1.3 T excitation field, indicating a single domain structure without significant macrocracks. However, by assembling the individual grains, the resulting magnetic pressure of the assembly is limited due to the unavoidable existence of (nonsuperconducting) connections between bulks. For larger applications, the average magnetization determined over the total assembly area is more important than the peak values. To raise the material quality for larger magnetic applications, it is proposed to fabricate, in a *second step*, large-sized high-performance melt textured bulks with multiseed domain structure to increase the average trapped field value. In Figure 3.1.1.5, fabricated high-T_c TSMG YBCO bulks of single and multiple grain structure are shown [20]. For larger superconducting magnetic stator rings or tubes, the use of multiple seed samples reduces the machining and assembling effort and the production costs.

In a *third step*, a technique of multiple seeding of rings and of curved sample with near radial *c*-axis alignment relative to the surface is performed. The technology was combined and fine-tuned with the growing procedure close to the final net shape and size. Rings and cylinders are formed into the desired shape by a CIP compacting process (1 – 1.5 kbar) with a relative powder density between 60% and 70%. The CIP equipment allows pressing rings up to 250 mm diameter in one monolithic piece. The rings are *c*-axis aligned and cold seeded by SmBCO seed crystals on the inner surface and grown under the above temperature profile. Owing to an appropriate seeding and growing, the complete inner surface is nicely covered by 123 crystals with a nearly perfect radial *c*-axis orientation. The obtained material quality is unexpectedly good, and saves machining and assembling effort.

Large-scale application of bulk HTS requires a reproducible production process and a fine-tuned material and growing logistics. TSMG is a well-understood melt texturing process, studied and improved in the last decade in many HTS groups. In addition, light rare earth (LRE = Sm, Gd, Nd) materials are attracting the interest,

Figure 3.1.1.5 YBCO bulks fabricated by top seeded melt growth (TSMG) [20]. (Reproduced with permission of ATZ.)

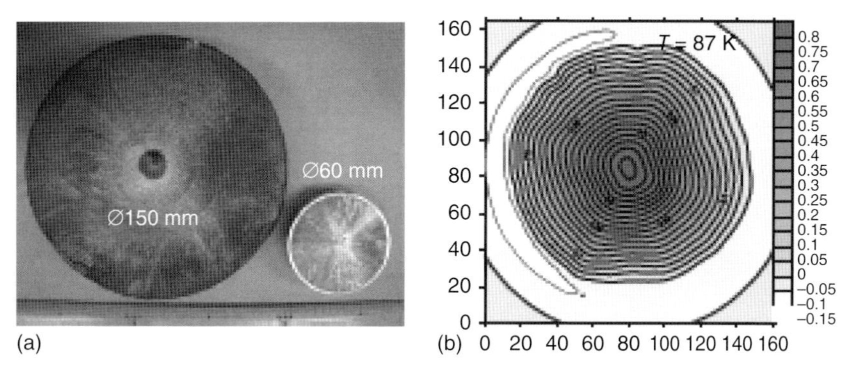

Figure 3.1.1.6 (a) Ø 150 mm and Ø 60 mm single domain GdBCO bulk with a homogeneous flux distribution; (b) trapped flux 0.8 T@87 K [21].

partly by using generic thin film seeds (Sm123, Nd123) and processing in reduced and normal oxygen atmosphere [5, 6]. Especially, gadolinium barium copper oxide (GdBCO) bulk samples prepared by cold seeding seem to have higher J_c values at 77 K compared to YBCO. Bulk YBCO and GdBCO with large domain size up to 150 mm diameter and high intradomain critical current density have been developed by quench melt growth (QMG) method under reduced oxygen atmosphere by Nippon Steel Corporation (NSC) as it is demonstrated in Figure 3.1.1.6 [21].

REBCO single grain bulk fabrication with extreme dimensions above 80–100 mm diameter requires a material-controlled peritectic temperature gradient to prevent parasitic nucleation and unwanted grain growth. The technique utilizes different precursor mixtures of RE components such as $Y_{1-x}Dy_xBaCuO$ ($x = 0.1-0.3$) whereby the different powders are compacted and separated in a *cylindrical shell structure* from the inner to the outer diameter of the green body. A composition gradient is used in two or three concentric shells around the seed center to vary the peritectic temperature from shell to shell by a few degrees. Thereby, the powder mixture of the outer shell possesses the lowest peritectic formation temperature compared to the material in the center. As an example for a two-shell structure, around the central seed a composition of $Y_{1-x}Dy_xBa_2Cu_3O_{6.5}$ ($x = 0.2$) powder is pressed into an annular shell geometry while the following shell consists of pure $Y_1Ba_2Cu_3O_{6.5}$ powder having a lower peritectic decomposition temperature. Because of the difference in the peritectic temperatures of Y123 ($T_p = 1015\,°C$) and Dy123 ($T_p = 1023\,°C$), the undercooling effect causes a continuous movement of the growth front in the melt over large sample distances and helps to prevent unwanted parasitic grain growth in the exterior sample regions.

The same effect can be generated by the addition of Ag_2O which decreases the formation temperature substantially (10 wt% Ag drops T_p by nearly 8 °C) without influencing the material composition.

It should be noted, the fabrication effort and time for such large bulks is substantial, causing material and fabrication costs of several thousands of dollars per large sample.

3.1.1.2.6 Mechanical Properties

For application, it is important to increase the mechanical strength of the REBCO material. The tensile strength of YBCO at 77 K is between 20 and 30 MPa. REBCO bulks above 60 mm cylinder size show sensitive mechanical properties and tend to break under high forces and stresses. The improvement of mechanical properties is therefore highly desirable. The addition of Ag_2O improves the microscopic stability and tensile strength of REBCO bulks. Resin impregnation of the bulks and reinforcement by a surrounding bandage either of metal (Al alloy, stainless steel, and titanium) or of glass and carbon fiber give further stability, and with applying pre-tension it compensates the tensile stress acting on the bulk during the magnetization process.

The magnetization and cool-down process of RE123 bulks is extremely sensitive at large magnetic fields. The resulting stress is a Lorentz force between the trapped field B_0 and circulating current loop $A_c = B_0/\mu_0$ in the magnitude σ $[\mathrm{N\,cm^{-2}}] = A_c \times B$. The Lorentz force can cause distortions in the crystalline and domain structure. Rigorously spoken, the maximum trapped field is not limited by the magnetic properties of the material, rather by the produced internal magnetic force. This force passes the maximum tensile strength of 30 MPa already at about 3 T. Therefore, higher fields often cause material fracturing accompanied by drops in the magnetization curve.

In order to prevent mechanical damage, the samples are armed under prepressure by a steel or carbon fiber bandage. Further on, the tensile strength of bulk YBCO superconductors can be improved by epoxy resin impregnation and wrapping with a carbon fiber fabric [22]. The epoxy is able to penetrate from the surface along microcracks and can fill microstructural defects up to a few millimeter depths. Owing to the stabilizing technology, the internal stress during magnetization from 7 to 0 T at 65 K was reduced from 150 to 40 MPa. Epoxy resin impregnation enhances the mechanical strength of YBCO by a factor of 2.5 and with 60–80 MPa at 77 K, the material strength approaches the properties required for most industrial applications. The same effect of surface stabilization can be obtained by copper surface plating and additional heat treatment. The copper diffuses into the surface and fills the microholes with a beneficial stabilizing effect.

Crystal defects in well-oriented REBCO pellets may occur during oxygenation and thermochemical heat treatments. Standard oxygenation is performed at a temperature of 400–500 °C during at least 150–200 h in 1 bar pure oxygen atmosphere. Surprisingly, the oxygenation is successful even for large samples because the inherent microcracks enable the transport of oxygen atoms into deeper regions of the sample by diffusion and deposit the atoms into the oxygen sublattice sites (CuO chains).

In connection with the formation of crystal defects during oxygenation, Diko *et al.* [23, 24] have investigated the cracking behavior of YBCO bulk. They have shown postgrowth treatments to influence weak-links and modify the effectiveness of pinning centers. By eliminating oxygenation-caused cracks with high-pressure oxygen treatment, up to three times increased critical current density J_c has been demonstrated.

3.1.1.2.7 Doping Strategy

Besides the macroscopic sample size, the bulk applicability is determined by pinning performance of the magnetic flux in magnetic vortices (Shubnikov phase). The nonsuperconducting RE211 phase, preferred in a nanoscale size, provides the basic pinning background. Compared to low-temperature superconductor (LTS), the new HTS possesses a relatively weak intrinsic pinning. This observation has led to wide spectrum of doping concepts, ranging from columnar defects generated by irradiation (in a thickness of a few millimeters), substitution of ions in the HTS atomic structure, and specifically doping by addition of secondary phases.

As already mentioned, a doping content of 0.3 wt% PtO_2 and/or 0.6 wt% CeO_2 is effective to influence the microstructure to refine the Y211 particles. Because of increasing costs, PtO_2 doping is applied more rarely to refine Y_2BCO_5 precursor powder. It is replaced by CeO_2 doping, which seems to have an equivalent beneficial effect on the particle size.

To achieve large-area melt textured YBCO bulks, multiseeding growing technology has been favored and developed in the last years by a few groups. Because the superconducting and magnetic properties can strongly vary with the position on a sample, the performance of the whole sample is important when applications are concerned.

A successful type of chemical pinning by Cu ion substitution is associated with the spatial scatter of the superconducting transition temperature. YBCO melt textured bulks with the substitution of Cu by the neighbor Zn or other transition metal atoms like Fe or Ni on Cu sites in the CuO_2 planes is effective in pinning. It follows the mechanism of local suppression of superconductivity with an increase of the critical current density, especially at fields of 1–3 T. The J_c (B) curves show the typical peak effect. It seems now it is generally accepted that the same peak effect is obtained for solid solutions of binary or ternary compounds such as (Nd, Eu, Gd)BaCuO with composition variations in the matrix providing regions with weaker superconducting properties or causing disorder in the oxygen sublattice. Which pinning mechanism is most effective is still a matter of investigation. Numerous RE_1/RE_2BCO solid solutions and Cu-site Zn doping of a few mole parts per million demonstrate the progress in J_c enhancement [25]. The Zn doping effect is shown in Figure 3.1.1.7 demonstrating the improved trapped flux behavior. The doping concentration M_x in $YBa_2Cu_{3-x}M_xO_y$ is usually small and between $x = 0.001$ and $x = 0.006$. The size of the pinning centers need to correlate with the coherence length of the superconductor ($\zeta = 3$ nm for Y123 at 77 K).

The effective refinement of second nonsuperconducting phases RE211 or RE422 on the critical current density has been shown in by Muralidhar *et al.* [26]. Extremely fine 211 powders in nanometer scale are available due to ball milling treatment technique. By adding Gd211 of 70 nm size and 10 nm NbO_3 particles on a mixed NEG (Nd, Eu, Gd)–Ba–Cu–O matrix system J_c values of

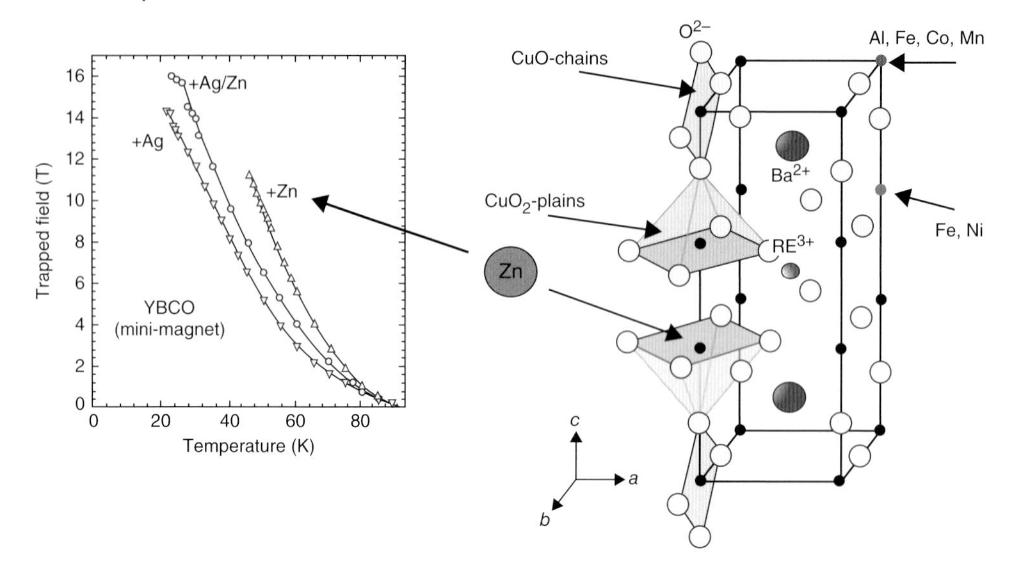

Figure 3.1.1.7 Dilute Zn doping of melt textured YBCO increase the maximum trapped field [25]; doping strategy of RE123 compounds.

925 kA cm^{-2} at 65 K and 640 kA cm^{-2} at 77 K could be obtained. Even for 90 K the critical current density was in a level of 100 kA cm^{-2}.

Large domains, however, very often do not reproduce the J_c values of smaller samples. Consequently, the maximum trapped magnetic fields in melt textured RE123 of 1.5 T at 77 K and up to 17 T between two 30 mm samples at 29 K are measured at small pellets [27]. The YBCO samples in that experiment were resin impregnated and wrapped with carbon fiber fabrics to achieve a higher mechanical strength and a pre-tension. The 17 T trapped field value is the highest obtained with bulk superconductors to date.

At moderate LN_2 temperatures, the measured maximum trapped flux of a superconducting magnet is J_c-limited, but with 5–6 T it approaches the maximum tensile strength at lower temperatures [5]. The research activities clearly show a large potential for improving intrinsic microstructure properties, for example, improving the mechanical stability by Ag_2O contribution or resin impregnation.

3.1.1.3 Characterization

3.1.1.3.1 Electromagnetic Force

Levitation force measurement for HTS bulk characterization has a long tradition. The method is simple and easy to perform but displays its limitation as the HTS bulk material has better electric and magnetic properties. Today, the measurement of the trapped field distribution after field cooled excitation seems a more adequate and reliable parameter of the magnetic bulk performance. Magnetic domain size, its orientation with respect to excitation field, and critical current density determine the quality of the fabricated material. The integral levitation force and the trapped field profile are commonly accepted to test the magnetic properties of the material.

Physically, the magnetic force between a magnet and a superconductor is a function of the size of the flowing shielding current loop and the height of the critical current J_c. The force exerted by a magnet on a superconductor is given by the gradient of the volume integral

$$\boldsymbol{F} = -\text{grad} \int (\boldsymbol{M} \times \boldsymbol{B}) dV$$

where $\boldsymbol{M}$ is magnetic moment of the superconductor and $\boldsymbol{B}$ is the magnetic flux density produced by the PM. The maximum levitation pressure is $P_{\text{max}} = B_r^2/2\mu_0$, if the critical current density is infinity ($\mu \to 0$). For the presently available high-energy PMs (NdFeB, SmCo) with a surface induction B_r ~1.0 T, the pressure can reach about 4 bar.

Experimentally, the repulsive force between an YBCO bulk superconductor and a PM is usually measured in the vertical direction at 77 K. A SmCo or NdFeB PM is kept at a large distance from HTS during cooling down; this situation is called *zero field cooling* (*zfc*). After reaching the superconducting state, the magnet is moved slowly down in a position 1 or 0.5 mm above the HTS surface and is reversed upwards in the original magnet position. Simultaneously, during movement, the force is measured as a function of the distance. Figure 3.1.1.8 gives a typical force versus distance measurement. The maximum levitation force between a 0.4 T SmCo PM of 25 mm × 15 mm size ($d \times h$) and a Ø 40 mm melt textured YBCO cylinder is 88 N at 0.5 mm distance. For comparability, levitation measurements in Europe are usually performed with a standardized SmCo PM Ø 25 mm × 15 mm having a surface induction of about 0.4 T (Figure 3.1.1.8). The levitation curve in Figure 3.1.1.8 is manifold instructive: the forward and backward force curves follow an exponential law, are not identical, and show a hysteretic loop. The area between the curves is equivalent to the hysteretic loss. The maximum repulsive force of 88 N gives a force density of about 18 N cm^{-2} for the applied SmCo standard magnet.

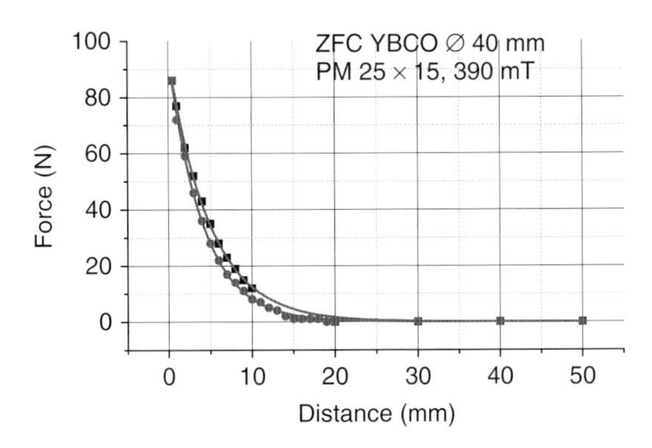

Figure 3.1.1.8 Zero field cooled force versus distance measurement; YBCO sample Ø 45 mm; SmCo PM Ø 25 mm × 15 mm (ATZ measurement).

Much higher levitation forces are achievable with large-scale magnetic devices as it is demonstrated with a HTS magnetic platform in Figure 3.1.1.9. The platform consists of a closed LN_2 vacuum cryostat containing the YBCO bulks in ring form. A corresponding PM ring with a diameter of about 0.6 m after field-cooling (fc) can levitate forces of up to 1000 N at 10–20 mm distance. The platform is a prominent exhibition feature for demonstrating superconductivity to a broad and interested

Figure 3.1.1.9 Demonstration of man-loading levitation platform (1000 N) [19]. (Reproduced with permission of ATZ.)

audience. The picture in Figure 3.1.1.9 is taken from an exhibition in 2011 in Paris honoring the 100 years anniversary of the discovery of superconductivity by Heike Kamerlingh Onnes in 1911.

Levitation measurements are sensitive to the coercitive force of the used PM, to magnetic distance variations, and displacement speeds. Therefore, the estimation of the trapped field distribution seems a more adequate and reliable parameter of magnetic quality of the HTS bulk material. The results of a corresponding round robin test with a 15% level spread give confidence for the reliability of trapped field measurements even between different laboratories and conditions [28].

3.1.1.3.2 **Magnetization and Field Mapping Technique of Bulk Superconductors**

Bulk superconductors are magnetized usually by three methods: fc, zfc, and pulsed field magnetization (pfm). In general, the exciting magnetic flux density B_{exc} has to be larger than the maximum trapped magnetic flux in the bulk B_T^{max}.

If the HTS sample is field cooled in the presence of an external magnetic field created by electromagnetic coil or PM, part of the field will be trapped in the superconductor. The single domain samples are oriented with the a-, b-planes perpendicular to the direction of the external magnetic field. The material is characterized usually at 77 K by field mapping. Thereby, the trapped magnetic field is scanned stepwise on the sample surface using a miniaturized Hall sensor. Higher fields can be trapped by reducing the temperature. For this, the external field is applied above T_c, for example, 100 K, and the temperature is reduced to the measuring point. For trapped fields above 5 T, cooling-down and heating-up have to ramp at low rates to avoid thermally induced flux jumps which can cause material deterioration.

Pulse field excitation is another technique to deposit a magnetic flux in the HTS. Using iteratively magnetizing pulsed-field operation with reduced amplitude (IMRA method) a maximum magnetic field of about 3 T at 38 K could be trapped [29].

The measurement of the trapped field distribution after field cooled excitation seems a more objective and reliable parameter of the magnetic bulk performance. Our magnetization of the superconductors is obtained by a conventional copper magnet with iron pole shoes corresponding to Figure 3.1.1.10. The magnetic excitation field is applied parallel to the c-axis, generating the persistent supercurrents in the $a-b$ planes. Maximum flux density at 40 mm pole shoe diameter is about 1.5 T, while for larger bulk excitation with Ø 150 mm the excitation filed is reduced to $B_{exc.} = 0.8-1.0$ T. The disadvantage of a Weiss-magnet excitation with Fe pole shoes is the strong magnetic attraction of the magnetized HTS bulk at the iron shoes. Part of the trapped flux is lost by removing the bulk from the magnet.

Higher magnetic field excitation up to 12 T is obtained by measurements with superconducting NbTi coils. Experimentally, a large warm bore is useful to deposit a corresponding cryostat in the bore during the magnetization.

After excitation, the resulting trapped magnetic flux distribution is measured using automatized scanning Hall equipment. The external field is reduced to zero and the sample removed from the magnet. The REBCO sample shows then a

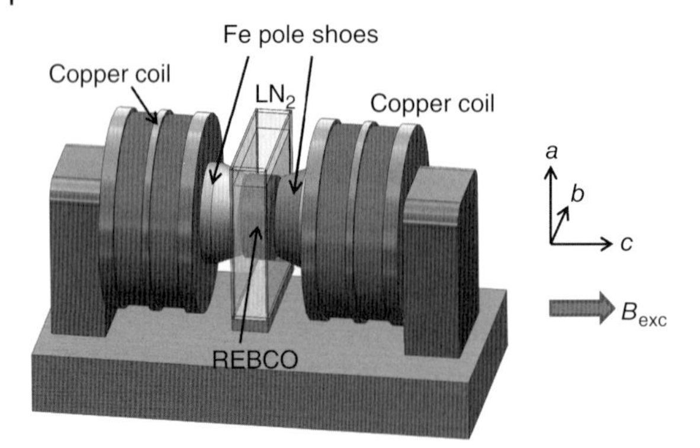

Figure 3.1.1.10 HTS bulk excitation with conventional Weiss magnet ($B_E^{max} = 1.5$ T) [14]. (Reproduced with permission of ATZ.)

smooth reduction of the trapped field due to flux flow and flux creep effects on a logarithmic time scale. At LN_2 temperature, the flux reduction due to thermal relaxation in the first 15–20 min is about 6–8%, at lower temperatures this value is reduced to a small percentage. The bulk superconductor is then fixed in a thermal insulation container immersed to LN_2. The field mapping of the sample is performed using the finger-type Hall sensor which is stepwise scanned in the x-, y-plane 0.5–1 mm above the sample top surface. The step width is 1 mm and a typical measurement of a 40 mm × 40 mm sample takes 1 h. Most of the scanning measurements are performed at LN_2 temperatures of 77 K to get a magnetic flux density distribution as a primary bulk qualification.

High-T_c bulk superconductors are capable of trapping magnetic fields permanently and becoming superconducting magnets. The principle schematics about simple understanding of trapped flux density in a bulk superconductor are shown in Figure 3.1.1.11. The trapping performance is proportional to the critical current density J_c of the material. J_c is the most important parameter of all superconductors. In the successfully applied BEAN approach, one assumes a simplified model

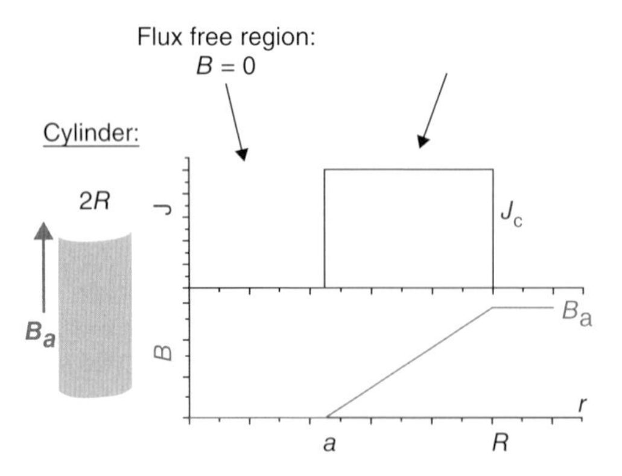

Figure 3.1.1.11 BEAN model: principle of magnetization and critical current density in bulk superconductor.

for the current distribution. Thereby, in this critical state model, the sum of the quantized vortex currents is substituted by a macroscopic screening current density that the sample is capable to carry. In this way, the material can be assumed to have the magnetic permeability of free space (μ_0), and a macroscopic screening current density J_c. J_c has a nonlinear relation with the induced electric field E in the material. The BEAN model considers that this macroscopic current density has a constant value equal to J_c. Inside the superconductor, the electromagnetic properties can be represented by a nonlinear (E–J) relation.

Measurements of the critical current density are performed inductively using corresponding magnetometers. Thereby, the induced magnetic moment loops were measured on small specimens cut from the parent grains using a superconducting quantum interference device (SQUID) magnetometer (see Chapter 9). The J_c value parallel to the sample c-axis can be calculated using the following equation derived from the BEAN model:

$$J_c = \frac{20(\Delta M)}{a(1 - a/3b)}$$

whereby M is the hysteresis in the volume magnetization, a and b are the cross-sectional dimensions of the sample perpendicular to the applied field, and $a < b$.

Following the Maxwell equation, the rotation of the internal magnetic flux density determines the critical current,

$$\nabla \times \boldsymbol{B} = \mu_0 \boldsymbol{J}_c$$

In one dimension, the above equation is reduced to

$$dB_x/dx = m_0 J_c^{\,y} \text{ in rectangular coordinates, respectively}$$
$$dB_z/dr = \mu_0 J_c^{\,\theta} \text{ in cylindrical coordinates}$$

The maximum trapped field flux density in z direction $B_z^{\max}$ of an infinite long cylindrical sample with a diameter of $2R$ is then given by the relation

$$B_z^{\max} = \mu_0 J_c^{\,\theta} R$$

After this equation, the maximum trapped field depends on the critical current density J_c and the diameter $D = 2R$ of the superconducting domain. In practice, the value is reduced by geometrical and demagnetization effects which lower the trapped flux by about 20% relative to applied magnetic flux density B_{exc}.

As an example, in case of radial symmetric geometry, one has to consider $dB/dr = \mu_0 J_c$, integrated in the Bean model gives $B^* = \mu_0 J_c R$ for the maximum trapped magnetic flux B^*. Assuming a critical current $J_c = 10^4\ \mathrm{A\,cm^{-2}}$ and a grain diameter $2R = 40$ mm, it gives a trapped field value $B^* = 1.2$ T.

Analyzing the above equations, a better bulk performance is given by increasing the critical current density J_c as well as by the length scale over which the currents flow, that is, the grain size. Both factors determine the field trapping ability, which is improved in the last decade routinely to maximum values of 1.2 T at 77 K for YBCO.

3.1.1.3.3 Trapped Field Magnetic Flux Density

According to the above relations, bulk performance is given by increasing the critical current density J_c and the grain size. If the superconducting current flows throughout a single domain material without distortions, large magnetic moments can be produced. In the last 5 years, the maximum trapped field B_{max} of selected samples has been increased continuously to values close 1.5 T at 77 K, passing the maximum field of high energy PMs NdFeB and SmCo. Other RE123 material with extremely fine-grained 211 phases, such as Dy123 or Gd123, can trap up to 3 T at 77 K [30].

Although numerous RE123 compounds (RE = Y, Nd, Sm, Dy, Gd) in the basic composition $RE_1Ba_2Cu_3O_{7-\delta}$ were synthesized with excellent superconducting properties and T_c temperatures between 90 and 95 K, for most application $Y_1Ba_2Cu_3O_{7-\delta}$ (Y123) is base material for large-scale magnetic application. In addition, economical arguments cause low cost, and recycling considerations which were important especially in large-scale application, where often tens of kilograms of valuable melt textured bulks, are necessary.

In Figure 3.1.1.12, we demonstrate a magnetic flux density distribution of a Ø 46 mm ATZ YBCO bulk specimen. The measured distribution is instructive in multiple regards. Although the maximum excitation flux was limited to 1.45 T using the copper magnet in Figure 3.1.1.10, the measured trapped flux density in the center of the sample is ~1.2 T at 0.5 mm distance. The flux density exhibits further a concentric distribution with a fourfold symmetry corresponding to the four crystal sector growing boundaries from the center to the surrounding diameter. Another effect of the trapped field distribution is the slope of the peak. It changes from the center to the boundary indicating different J_c values. This behavior displays higher critical current density values of J_c near the bulk center and reduced critical supercurrents in the distance. Larger J_c at higher fields correspond to the often observed peak effect at mixed valence $RE_{1-x}RE_xBaCuO$ superconductors.

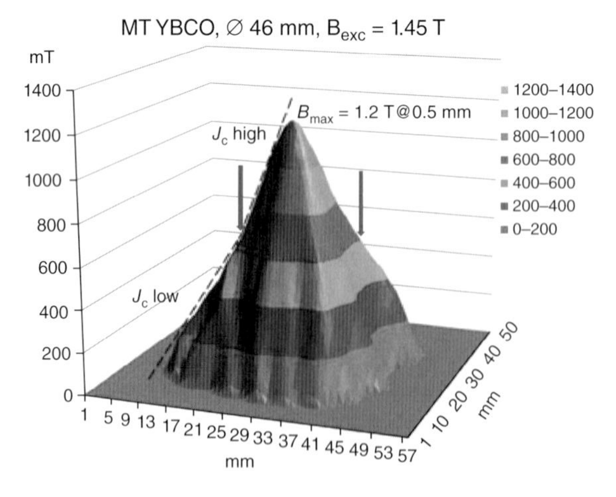

Figure 3.1.1.12 Trapped flux density of a 45 mm melt textured YBCO sample [19]. (Reproduced with permission of ATZ.)

3.1.1.3.4 Multiseeded Bulk Characterization

High resolution trapped field measurements can explain the GB behavior in multiseed samples. Figure 3.1.1.13 shows the trapped flux density distributions of a multiseeded bulk and the combined bulks after cutting at a measurement gap of 1 mm [31]. The shape of the distribution indicates a threefold domain structure without significant macrocracks. For the levitation pressure, the average magnetization is more important than the peak values. To raise the material quality for magnetic bearing purpose, it was common to select single domain bulk samples and assemble them to the desired shape. Alternately, large-sized high performance melt textured bulks with multiseeded domain structure increases the average trapped field value. The common factor is that the field distribution in each of the three grains is characterized by its own single domain peak shape. Between the peaks, the GBs reduce the current transport capability according to the BEAN model. The as-grown sample with the SmBCO seeds in Figure 3.1.1.13a show a nonvanishing trapped field distribution between the three peaks with peak values of 0.6–0.7 T and an integral flux $\Phi = 0.548$ mWb. After sample cutting along the GBs in Figure 3.1.1.13b, the peak values are smaller by about 0.25 T and the integral flux was reduced by more than 40%. The scanning Hall results of multigrain bulks give evidence of linked components allowing a supercurrent to flow across the GBs in the multiseed bulk. While the *intragrain* current determines the three individual magnetic peaks, an additional *intergrain* current can pass the GBs and contribute a substantial part to the total trapped magnetic flux density integrated over the bulk. Figure 3.1.1.13 displays the model of the intra- and intergrain currents flowing in multiseed samples. From Figure 3.1.1.13a, a trapped field

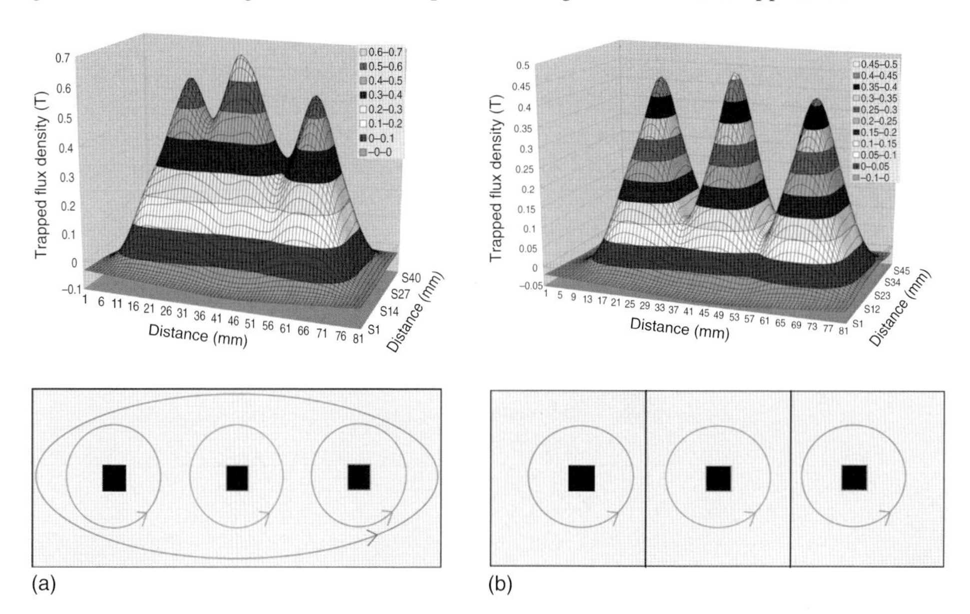

Figure 3.1.1.13 Trapped field distribution of as-grown three-seed YBCO bulk (a); and after cutting into separate crystals (b) [31]. (Reproduced with permission of ATZ.)

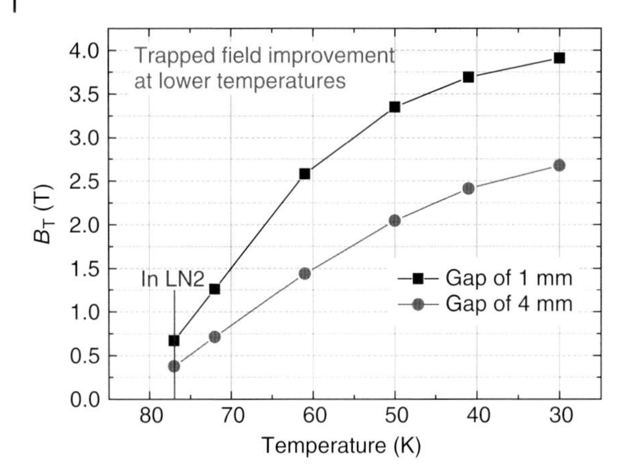

Figure 3.1.1.14 Maximum trapped field values of three-seed YBCO bulk at lower temperatures [31]. (Reproduced with permission of ATZ.)

around 0.3–0.5 T was observed in the connection areas between two seeds. In Figure 3.1.1.13b, the GB contribution is disappeared due to sample cutting. This could be the direct evidence to show that the connected or coupled GBs exist inside the multiseeded bulk, implying some supercurrent flowing across the GBs. This current makes an additional contribution to the trapped magnetic flux in the whole bulk area, besides the intragrain current circulating in each grain. The intergrain current is especially beneficial for large-scale applications.

Figure 3.1.1.14 reproduces the trapped flux improvement of the same sample as in Figure 3.1.1.13. Owing to the higher critical current density at lower temperatures than LN_2, the trapped flux increases continuously. At 30 K, the field cooled multiseed sample in a 5 T excitation coil is capable to trap a peak values of about 4 T [31].

3.1.1.3.5 Comparison of the REBCO Bulk Materials

All bulk applications require significant reduction of the material and fabrication cost. This request is sometimes diametric to the observation of increased RE powder costs within the last years. Single grain bulks of YBCO or nowadays GdBCO up to 150 mm diameter were fabricated by top seeded QMG processes using compositional gradient technology [32]. The fabrication effort and time for growing such large bulks is substantial and approaches often a value of several thousand dollars per large sample. In addition, under machining and cutting procedure, the probability of damage may appear.

In the following, we will briefly compare the various REBCO melt processed bulk materials in view of their trapped field properties. In Table 3.1.1.1, the properties of melt textured YBCO in the ATZ composition and fabrication is summarized. In Table 3.1.1.2, the corresponding peritectic temperatures of several RE123 compounds without and with Ag_2O are presented. Except YBCO as a standard material, in the recent years, significant advances has been made in growing single grain LRE cuprates LRE-123 with LRE = Gd, Eu, Sm, Nd. In contrast to YBCO,

Table 3.1.1.1 Properties of melt textured YBCO.

Parameter	Value
Transition temperature, T_c	90–92 K (–181 °C)
Specific resistance, ρ_{300}	0.6–0.7 m Ω cm
Density Y123 (theoretical), η	6.38 g cm^{-3}
Specific heat capacity, k_{300}	0.18 J K^{-1}g^{-1}
Coefficient heat transfer, λ	4 W m^{-1} K^{-1} (c), 10 W m^{-1} K^{-1} (a, b)
Lin. expansion coefficient, κ	5–12 × 10^{-6} K^{-1}
Tensile strength, σ	25–30 MPa
Critical current density, J_c	10^4 – 10^5 A cm^{-2} (77 K, 0 T) 10^6 A cm^{-2} (4.2 K, 0 T)

Table 3.1.1.2 Peritectic temperatures of RE123 compounds.

RE123	T_p (°C)	T_p (10 wt% Ag) (°C)
Nd123	1086	1061
Sm123	1061	1039
Gd123	1028	1020
Y123	1015	996

most of the LRE-123 pellets are fabricated under reduced oxygen atmosphere. Two principal problems are accompanied with the LRE-123 textured single grain processing: first, the need and availability of appropriate seed crystals with comparable lattice constants. Both SmBCO and NdBCO seeds may have to small distances in the melting point relative to the precursor.

Secondly, another difficulty of processing the LRE-123 melt textured bulks is the observation that a LRE element easily substitutes a Ba site in the Ba sublattice which forms a solid solution (ss) of the type $LRE_{1+x}Ba_{2-x}Cu_3O_y$. This substitution causes a lowering of the carrier density because the divalent ion Ba is replaced by a trivalent LRE ion. An increased substitution depresses the critical temperature T_c seriously. To overcome the solid solution formation, it could be advantageously shown that the melt process is performed in reduced oxygen atmosphere [12]. Growing under reduced O_2 suppress the substitution and improve the superconducting behavior (T_c, J_c) of the LRE-123 compounds. From the investigation of the LRE properties, the oxygen-controlled quench melt-growth (OQMG) process has been developed.

In connection with the fabrication process of LRE-123 pellets the addition of Ag_2O has a twofold beneficial effect; it lowers the peritectic temperature T_p and eliminates macrocracks. With 10 wt% Ag_2O, the mechanical properties of LRE-123 could be stabilized and withstand the mechanical stress caused by higher trapped fields. The peritectic temperatures of some LRE-precursors without and with 10 wt% Ag_2O together YBCO in Table 3.1.1.2 demonstrate the influence of Ag contribution on the peritectic temperature.

Finally, a brief overview about continuous and engaged development of the quality of REBCO single grain processing is given. Although excellent results of many scientific groups have been published within the last 10 years it is impossible to include here all top results.

YBCO YBCO is most successfully processed and investigated bulk material. The maximum trapped fields for YBCO were leveling off at about 1.2–1.3 T at 77 K at samples size of about 40–50 mm. At 51.5 K, a maximum value of 8.5 T between two YBCO samples, each 24 mm in diameter, was reported [33]. Later, the Gruss *et al.* [3] of the Dresden group magnetized 30 mm YBCO bulks and measured between the pellets a trapped field of 16.4 T [3]. In 2003, Tomita and Murakami [27] under the conditions of extremely mechanical (carbon fiber bandage) and thermal stabilization magnetized very slowly two 26 mm large YBCO bulks and measured a maximum static trapped field of 17.24 T at 29 K between the bulks. Parallel showed the YBCO bulk experiment at 46 K at saturated maximum trapped field of 9.5 T. These data are the highest trapped magnetic field values obtained with REBCO bulk to date.

NdBCO As mentioned the $Nd_{1+x}Ba_{2-x}Cu_3O_y$ phase forms a solid solution (ss) because Nd^{3+} can substitute Ba^{2+}site in the Ba sublattice. After Ref. [34] the subsolidus phase diagram under normal pressure conditions and $T = 890\,°C$ the homogeneity region extends from $x = 0.04$ to 0.6, whereas a slightly lower temperature of 885 °C and a reduced oxygen partial pressure $p(O_2) = 100$ Pa the existence field with $x = 0–0.15$ is confined. From this, for $x > 0.15$ a phase composition Nd123 is not stable under these conditions, but Nd123 with $x = 0$ can synthesized at 100 Pa oxygen partial pressure successfully.

The fabrication of melt textured Nd123 is determined by a number of preparation problems, like the above mentioned search for seed crystals having a melt temperature far enough of the precursor, the RE/Ba substitution controlled by the oxygen partial pressure and the existence of macrocracks, preventing homogeneous large grain growth. Most of the Nd123 experiments use 10–20 wt% Ag_2O to decrease the peritectic temperature (see Table 3.1.1.2) and improve the mechanical stability. The trapped field performance of NdBCO is comparable to that of SmBCO, showing 1.23 T at 77 K and 7.0 T at 42 K [35].

SmBCO The first HTS material having a slightly higher $T_c = 94–95$ K compared to Y123 is SmBCO, which is processed by replacing yttrium by the LRE element Sm. Again SmBCO shows under atmospheric pressure the Sm/Ba substitution which seems responsible for the enhanced pinning at an external magnetic field of a few tesla (peak effect). This peak effect is interpreted as caused by a local fluctuation of the superconducting properties due to the Sm/Ba substitution. After that, the Sm ion on the Ba site is believed as a source of field induced pinning. According to the substitution effect, the critical current density shows a J_c peak in the applied field at 1 – 3 T.

SmBCO samples with Ag addition for improvement the mechanical properties show trapped fields as high as 1.7 T at 77 K at hot seeding [36]. This value is the highest trapped flux of RE123 measured in LN_2. With a thin film cold-seeding technique in a 36 mm SmBCO pellet a trapped field value of 1.52 T is reported [12].

In a round robin test, two resin reinforced SmBCO pellets have been investigated by four European laboratories. The trapped flux measurements under different experimental conditions exhibit absolute values between 1.0 and 1.2 T between the laboratories [28]. The 15% value scattering at the applied two reference samples gives some indication about the strong influence of the experimental measuring conditions. Also the field trapping ability of both samples showed some deterioration over the 16 month period investigation time.

GdBCO GdBCO has a high upper critical field B_{c2} exceeding 30 T and is a promising superconductor for magnetic applications. Sm123/MgO thin film seed under reduced oxygen partial pressure is capable to provide the epitaxial growth of GdBCO single grains. Using careful adjustment of the temperature profile and fine Gd_2BaCuO_5 (Gd211) particles with a size smaller than 1 μm increased J_c values could be obtained [37]. Corresponding several 32 mm GdBCO samples showed a trapped flux of up to 1.5 T at 77 K, but with 20 wt% Ag_2O even more than 2.0 T at LN_2. Batch processed GdBCO pellets with Nd/MgO cold seeding at normal pressure under economical considerations were successfully processed by Muralidhar *et al.* [13]. The averaged trapped field of the 24 mm pellets at 77 K was between 0.8 and 0.9 T in the peak maximum. A similar result (B_z^{max} = 0.88 – 0.96 T) was obtained for 30 mm samples using thin film seed and reduced oxygen partial pressure in [38]. This is the first demonstration that a thin film works as a seed in a reduced oxygen atmosphere as well.

Although a small contamination of Mg from the film substrate was observed the overall quality of the samples was comparable to, or even better than the samples prepared by hot-seeding. The results at GdBCO are promising with respect to the simplification of the production process of bulk superconductors.

References

1. Krabbes, G., Schaetzle, P., Bieger, W., Wiesner, U., Stöver, G., Wu, M., Strasser, T., Köhler, A., Litzkendorf, D., Fischer, K., and Görnert, P. (1995) *Physica C*, **244**, 145.
2. Krabbes, G., Bieger, W., Schaetzle, P., and Wiesner, U. (1998) *Supercond. Sci. Technol.*, **11**, 144.
3. Gruss, S., Fuchs, G., Krabbes, G., Verges, P., Stoever, G., Mueller, K.-H., Fink, J., and Schultz, L. (2001) *Appl. Phys. Lett.*, **79**, 3131.
4. Werfel, F.N., Floegel-Delor, U., Rothfeld, R., Wippich, D., and Riedel, T. (2001) *Physica C*, **357 – 360**, 843.
5. Morita, M., Sawamura, M., Takabayashi, S., Kimura, K., Teshima, H., Tanaka, M., Miyamoto, K., and Hashimoto, M. (1994) *Physica C*, **253-240**, 209.

6. Nariki, S., Sakai, N., Kita, M., Fujikura, M., Murakami, M., and Hirabayashi, I. (2006) *Supercond. Sci. Technol.*, **19**, S500.
7. Sakai, N., Kita, M., Nariki, S., Muralidhar, M., Inoue, K., Hirabayashi, I., and Murakami, M. (2006) *Physica C*, **445–448**, 339.
8. Murakami, M. (2007) *Int. J. Appl. Ceram. Technol.*, **4**, 225.
9. Morita, H., Hirano, H., and Teshima, H. (2006) *Nippon Steel Tech. Rep.*, **93**, 18–23.
10. Shi, Y., Babu, N.H., and Cardwell, D.A. (2005) *Supercond. Sci. Technol.*, **18**, L13–L16.
11. Babu, N.H., Jackson, K.P., Denise, A.R., Shi, Y.H., Mancini, C., Durell, J.H., and Cardwell, D.A. (2010) *Supercond. Sci. Technol.*, **25**, 075012.
12. Oda, M., Yao, X., Yoshida, Y., and Ikuta, H. (2009) *Supercond. Sci. Technol.*, **22**, 075012.
13. Muralidhar, M., Tomita, M., Suzuki, K., Jirsa, M., Fukumoto, Y., and Ishihara, A. (2010) *Supercond. Sci. Technol.*, **23**, 045033.
14. Werfel, F., Floegel-Delor, U., Riedel, T., Wippich, D., Goebel, B., Rothfeld, R., and Schirrmeister, P. (2012) *Superconductivity: Recent Developments and New Production Technologies*, Chapter 9, Nova Science Publishers, ISBN: 978-1-62257-137-6.
15. Cima, M.J., Merton, C., Flemings, M., Figueredo, M.A., Nakade, M., Ishii, H., Brody, H.D., and Haggerty, J.S. (1992) *J. Appl. Phys.*, **72**, 179–190.
16. Sawamura, M., Morita, M., and Hirano, H. (2002) *Physica C*, **378-381**, 617–621.
17. Babu, N.H., Whithnell, T.D., Iida, K., and Cardwell, D.A. (2007) *IEEE Trans. Appl. Supercond.*, **17** (2), 2949–2952.
18. Li, T.Y., Wang, C.-L., Sun, L.J., Yan, S.B., Cheng, L., Yao, X., Xiong, J., Toa, B.W., Feng, J.Q., Yu, X.Y., Li, C.S., and Cardwell, D.A. (2010) *J. Appl. Phys.*, **108**, 023914.
19. Werfel, F.N., Floegel-Delor, U., Riedel, T., Goebel, B., Rothfeld, R., Schirrmeister, P., and Wippich, D. (2013) *Physica C*, **484**, 6–11.
20. Werfel, F.N., Floegel-Delor, U., Rothfeld, R., Riedel, T., Goebel, G., Wippich, D., and Schirrmeister, P. (2012) *Supercond. Sci. Technol.*, **25**, 014007.
21. Li, B., Zhou, D., Xu, K., Hara, S., Tsuzuki, K., Miki, M., Felder, B., Deng, Z., and Izumi, M. (2013) *Physica C*, **482**, 50–57.
22. Tomita, M., Murakami, M., and Katagiri, K. (2002) *Physica C*, **378-381**, 783–787.
23. Diko, P., Antal, V., Zmorayova, K., Seficikova, M., Chaud, X., Kovac, J., Yao, X., Chen, I., Eisterer, M., and Weber, H.M. (2010) *Supercond. Sci. Technol.*, **23**, 124003.
24. Diko, P., Krcunoska, S., Ceniga, L., Bierlich, J., Zeisberger, M., and Gawalek, W. (2005) *Supercond. Sci. Technol.*, **18**, 1400.
25. Fuchs, G., Gruss, S., Verges, P., Krabbes, G., Mueller, K.-H., Fink, J., and Schult, L. (2002) *Physica C*, **372-376**, 1131–1133.
26. Muralidhar, M., Sakai, N., Jirsa, M., Murakami, M., and Hirabayashi, L. (2008) *Appl. Phys. Lett.*, **92**, 162512.
27. Tomita, M. and Murakami, M. (2003) *Nature*, **421**, 517.
28. Cardwell, D.A., Murakami, M., Zeisberger, M., Gawalek, W., Gonzalez-Arrabal, R., Eisterer, M., Weber, H.W., Fuchs, G., Krabbes, G., Leenders, A., Freyhardt, H.C., and Babu, N. (2005) *Supercond. Sci. Technol.*, **18**, 173.
29. Oka, T., Hirose, Y., Kanayama, H., Kikuchi, H., Ogawa, J., Fukui, S., Sato, T., and Yamaguchi, M. (2008) *J. Phys. Conf. Ser.*, **97**, 012102.
30. Nariki, S., Sakai, N., and Murakami, M. (2005) *Supercond. Sci. Technol.*, **18**, S126.
31. Deng, Z., Izumi, M., Miki, M., Felder, B., Tsuzuki, K., Hara, S., Uetake, T., Floegel-Delor, U., and Werfel, F.N. (2012) *IEEE Trans. Appl. Supercond.*, **22** (2), 6800110.
32. Morita, M., Hirano, H., and Teshima, H. (2006) *Nippon Steel Tech. Rep.*, **93**, 18–23.
33. Fuchs, G., Krabbes, G., Schaetzle, P., Gruß, S., Stoye, P., Staiger, T., Mueller, K.H., Fink, J., and Schultz, L. (1997) *Appl. Lett.*, **70**, 117.

34. Yoo, S.I., Murakami, M., Sakai, N., Higuchi, T., and Tanaka, S. (1994) *Jpn. J. Appl. Phys.*, **33**, L1000–L1003.
35. Ikuta, H., Hosokawa, T., Yoshikawa, M., and Mizutani, U. (2000) *Supercond. Sci. Technol.*, **13**, 1559.
36. Ikuta, H., Yamada, T., Yoshikawa, M., Yanagi, Y., Itoh, Y., Oka, T., and Mizutani, U. (1998) *Supercond. Sci. Technol.*, **11**, 1345.
37. Nariki, S., Sakai, N., and Murakami, M. (2002) *Physica C*, **378-381**, 631.
38. Oda, M. and Ikuta, H. (2007) *Physica C*, **460–462**, 301.

3.1.2
Single crystal growth of the high temperature superconducting cuprates

Andreas Erb

3.1.2.1 General Problems in the Crystal Growth of the High T_c Cuprate Superconductors

The main problems with the growth of high-quality single crystals are the incongruent melting behavior of basically all the high T_c compounds and the very reactive melts which tend to corrode all the usual crucible materials used in standard crystal growth experiments.

The incongruent melting behavior hampers crystals of the high T_c compounds to be grown out of stoichiometric melts. Rather flux methods are required where either a surplus of one of the components, for example, CuO, or some other low-melting materials, are used as a solvent for the compounds to be grown. The use of a surplus of one of the components is often referred to as *self-flux method*, and it has the advantage that there are no other elements involved which can change composition and doping of the obtained crystals. The use of other solvents like molten salts, NaCl/KCl being among the most commonly used ones in crystal growth, is always problematic due to the possible incorporation of the flux elements in the crystals. Nevertheless, this method has successfully been used in the case of Y-124 crystals [1].

The other big issue of crystal growth of the cuprates is that their melts are very aggressive and tend to react with the commonly used refractory materials like Al_2O_3, Y-stabilized ZrO_2 or even with noble metals like Pt, Au. The corrosion of the crucibles has two different effects: First, it alters the melt composition during the growth process, leading to uncontrollable and ill-defined growth conditions. Second, the corrosion leads to a pollution of the melt with constituents of the container material, and, in turn, to a pollution of the crystals. Depending on the used crucible material, the pollution and additional doping of the obtained crystals can be high enough to drastically affect the superconducting properties. In the extreme, superconductivity is even destroyed by the pollutant.

During the last years, sophisticated methods for the growth of high T_c cuprates have been developed to overcome the problems one has to face when cuprate superconductors are grown as single crystals.

3.1.2.2 $YBa_2Cu_3O_{7-\delta}$, $YBa_2Cu_4O_8$, and $REBa_2Cu_3O_{7-\delta}$ (RE, Rare Earth Element)

$YBa_2Cu_3O_{7-\delta}$ and $YBa_2Cu_4O_8$ are somewhat exceptional compounds amongst the high T_c cuprates since their ratio of the metal atoms is fixed to 1 : 2 : 3 or 1 : 2 : 4, and they are therefore often referred to as *line phases*. This property of being a line phase is lost when $REBa_2Cu_3O_{7-\delta}$ crystals are grown. This is due to the fact that for light RE atoms, a certain solubility of the RE atom on a Ba site of the 123 structure is present, resulting in crystals with a composition of $RE_{1+x}Ba_{2-x}Cu_3O_{7-\delta}$.

While for $YBa_2Cu_3O_{7-\delta}$ all the doping from the antiferromagnetic insulator $YBa_2Cu_3O_6$ to the slightly overdoped superconductor $YBa_2Cu_3O_7$ is done by varying the oxygen content, the solubility of the lighter RE atoms on the Ba sites leads

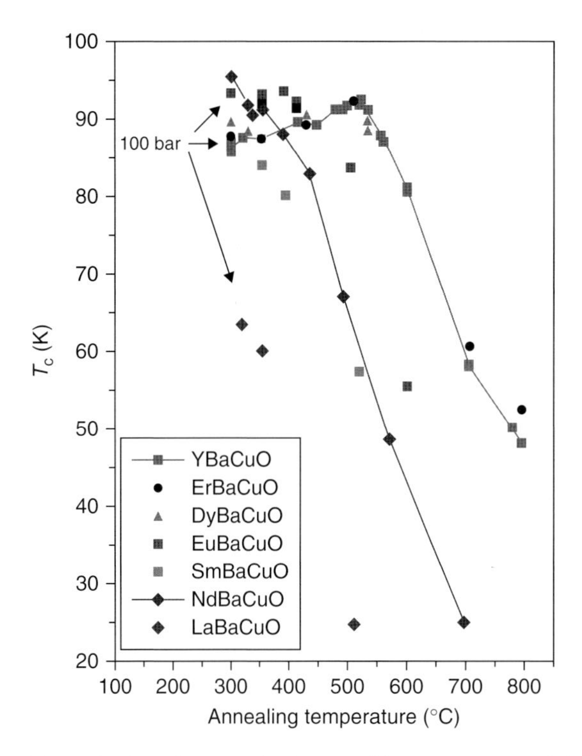

Figure 3.1.2.1 Superconducting transition temperature of the 123 single crystals as a function of the oxygenation temperature. (Adopted from the habilitation thesis of Erb, University Geneva, Switzerland [3].)

to an additional co-doping effect which, besides other effects, alters the superconducting transition temperature as a function of oxygen content. Figure 3.1.2.1 shows this effect for different RE-123 compounds as a function of annealing temperature, which is equivalent to the oxygen content. It should be noted here that the solubility of the RE atom on the Ba site can be influenced by the oxygen content during crystal growth in a way that the solubility can be suppressed at least partially by growing the crystals under low oxygen partial pressure [2].

The main problem in the case of the growth of the $RE_{1+x}Ba_{2-x}Cu_3O_{7-\delta}$ was, however, the chemically very aggressive melt used for the solution growth of these compounds.

$YBa_2Cu_3O_{7-\delta}$ decomposes in air at around 1030 °C in a peritectic reaction into Y_2BaCuO_5 plus a liquid. Thus, it sets the upper temperature limit for the growth procedure to this temperature.

In the phase diagram of the relevant Y_2O_3–BaO–CuO system, a low melting mixture of the phases has been found [4] at a composition of BaO/CuO with an atomic ratio of 30/70 and a melting point of 910 °C in air. This low melting composition, which is also in close proximity to the phase field of $YBa_2Cu_3O_{7-\delta}$, was therefore used as flux or solvent in the standard crystal growth experiments of $YBa_2Cu_3O_{7-\delta}$. However, the solubility of the 123 compounds in this flux is rather low, and only about 10 wt% of $YBa_2Cu_3O_{7-\delta}$ can be dissolved in this flux. This makes the whole crystal growth procedure very inefficient since, at most, only 10% of the invested material can be crystallized as a single crystal. What makes the situation even worse is the chemical attack of virtually all the commercially available crucible materials like refractory ceramics, for example, Al_2O_3, Y-stabilized ZrO_2, or noble metals by the very reactive melt. This chemical attack not only led to unstable growth conditions, since the composition is altered with time due to crucible corrosion, but also the crystals contained considerable amounts of impurities (Al, Au, Pt) depending on the used container material. During the process of learning how to grow crystals of the 123 compound, the use of Y-stabilized ZrO_2 was favored by most experimentalists, since the main pollutant Zr did not incorporate into the crystals but rather formed a solid corrosion product $BaZrO_3$. However, the corrosion process kept altering the composition of the melt, leading to uncontrollable growth conditions and limited success in crystal preparation. Moreover, the crucible corrosion deteriorated the already unsatisfactory efficiency due to the low solubility of the 123 compound in its solvent by the loss of an important amount of the batch due to the formation of corrosion products. For this reason, the use of very high purity starting materials for crystal growth was also relatively seldom applied because of the enormous costs for such experiments.

The solution for this problem came with the invention of an adapted crucible material, namely $BaZrO_3$ [5, 6]. $BaZrO_3$ is found as the main corrosion product in growth experiments of the 123 compounds. Since it is solid in the temperature range where crystal growth of the 123 compounds is performed and since has virtually no solubility in the melt used to grow 123 compounds, it is, of course, an ideal container material. The formation of crucibles out of this compound is possible because $BaZrO_3$, unlike ZrO_2, does not undergo crystallographic phase transitions up to its high melting point of above 2600 °C.

With the availability of $BaZrO_3$ crucibles, the crystal growth of $YBa_2Cu_3O_{7-\delta}$ was immediately brought to a reproducibility not given before, and the growth conditions were now well controllable. Moreover now, much more effective crystallization allowed the use of high purity starting ingredients for the crystal growth procedure, resulting in a crystal purity of 99.995 at.% [3, 6]. Figure 3.1.2.2 shows a picture of freestanding $YBa_2Cu_3O_{7-\delta}$ single crystals in a $BaZrO_3$ crucible after the solvent has been decanted. In marked contrast with the situation one experienced when Y-stabilized ZrO_2 crucibles were used, the crystals were only slightly attached to the crucible walls and could be easily separated from the container.

The crystals obtained in that way showed supreme quality, and their increased purity resulted in the possibility to study many physical properties in more detail, especially in the physics of vortex pinning and vortex phase diagrams and in the transport measurements. The method of crystal growth from a high temperature solution using $BaZrO_3$ crucibles is nowadays state of the art for the 123 compounds [5–7].

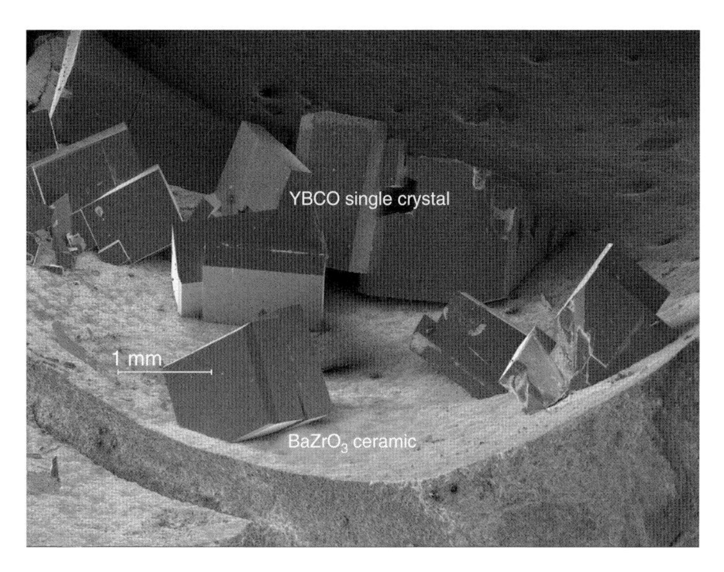

Figure 3.1.2.2 Single crystals of the high-temperature superconductor $YBa_2Cu_3O_{7-\delta}$ inside a $BaZrO_3$ crucible [3].

3.1.2.3 The 214-Compounds $La_{2-x}Sr_xCuO_4$, $Nd_{2-x}Ce_xCuO_4$, and $Pr_{2-x}Ce_xCuO_4$

Besides the 123 compounds, which are the only stoichiometric compounds or line phases in the cuprates, the so-called 214 compounds are the second best systems in terms of crystal perfection. The underdoped mother compounds of these systems are the antiferromagnetic insulators La_2CuO_4, Nd_2CuO_4, and Pr_2CuO_4 which, upon doping with either Sr in the case of La_2CuO_4 or Ce in the case of Nd_2CuO_4 and Pr_2CuO_4, become superconducting solid solution crystals. One can actually state that next to a line phase like 123, the second-best well-defined crystal is a solid solution crystal with homogeneous doping of the whole crystal.

Also in this case, the use of ordinary refractory materials led to ill-defined growth conditions and polluted samples of the 214 compounds. Moreover, typically for a solid solution crystal is that the dopant does not normally incorporate into the crystal at the same ration as it is present in the melt but usually to a smaller extent. The ratio between the concentration of the dopant in the growing crystal and the concentration of the dopant in the melt is called the *segregation coefficient*. For this reason, crystals grown in a crucible must have a concentration gradient of the dopant over their volume. Here, the solution for both problems arising from both the aggressive melt as well as the doping issue came with the development of a container-free crystal growth method using mirror furnaces and the so-called traveling solvent floating zone (TSFZ) technique. Figure 3.1.2.3 shows such a mirror furnace in which the four elliptically shaped mirrors produce a fine focus in the center of the furnace. In that way, only a small zone of the material is then molten and the molten zone can be transferred over the length of the sample by moving the mirror system vertically. Since also the 214 compounds do not congruently melt, the molten zone must consist of a solvent saturated

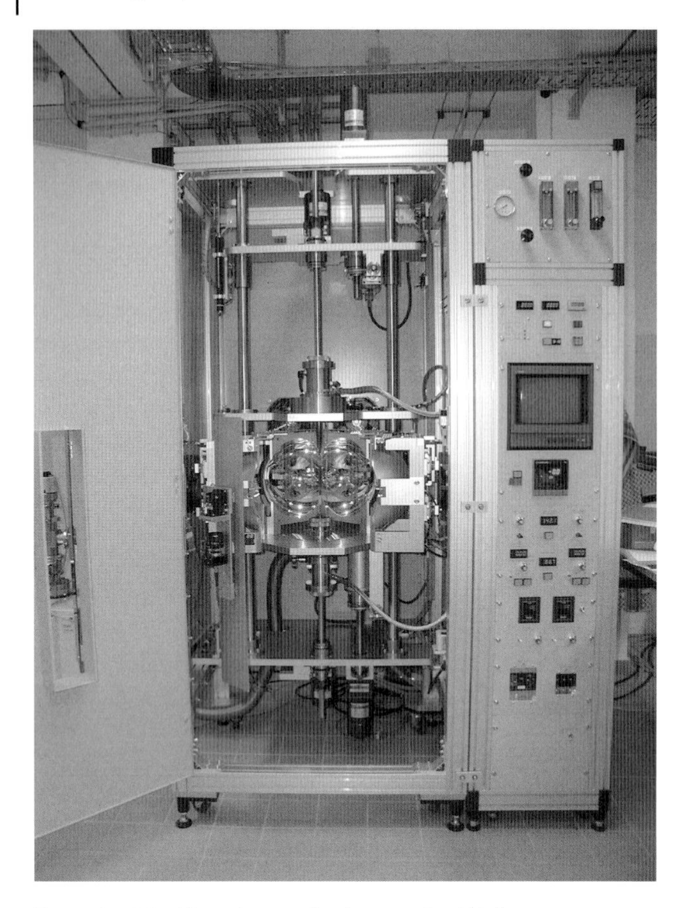

Figure 3.1.2.3 Mirror furnace for the growth of high-temperature superconductors using the traveling solvent floating zone technique (TSFZ).

with the 214 phase, namely a surplus of 80 at.% of CuO, is used as a solvent. The solvent only acts as a transport medium and it is only held by surface tension (Figure 3.1.2.4). By moving the mirror stage upwards, the feed material of the 214 compound is dissolved at the top and crystallizes at the bottom. Even without using a seed crystal, the growth anisotropy of the 214 compounds leads to a grain selection in a way that the crystal direction with the fastest growth velocity is normally aligned along the direction of the mirror movement, while the slow growing *c*-direction normally points out perpendicular to the growth direction and it can even form large growth facets. Figure 3.1.2.5 shows a crystal of $La_{1.85}Sr_{0.15}CuO_4$ grown with a mirror furnace.

Even though the feed and seed rods are counter rotated for better mixing of the molten zone and the relatively steep temperature gradients favor convection and thus facilitate material transport to the growing interface, the whole crystal growth process using a solvent is also diffusion controlled directly in front of the growth interface and the maximum growth velocities are only as low as

Figure 3.1.2.4 Molten zone inside a mirror furnace during the crystal growth of a $Pr_{2-x}Ce_xCuO_4$ single crystal. The molten zone in the middle consists of a CuO, which is saturated with the desired $Pr_{2-x}Ce_xCuO_4$, and only acts a transport medium during growth. Growth direction in that case is along the <110> direction.

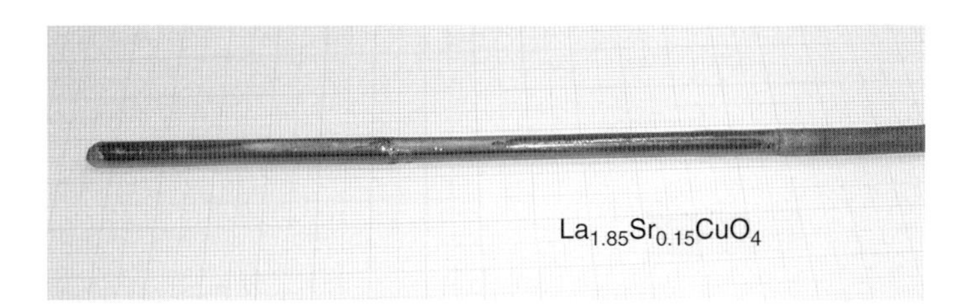

Figure 3.1.2.5 Single crystal of $La_{1.85}Sr_{0.15}CuO_4$ grown in a mirror furnace. The last about 6 cm on the left end of the sample are formed by a single crystallite.

$\sim$0.5 mm h^{-1}. This, however, is a relatively large growth velocity for the cuprates. Thus, big crystals of several grams in weight can be grown by this method. This overcomes both basic problems of crystal growth of the high-temperature superconductors since there is no crucible involved, and the method facilitates the growth crystals with a homogeneous doping.

For $La_{2-x}Sr_xCuO_4$, the growth experiments have led to very big (several grams) and homogeneous single crystals of this compound. The crystals with optimal doping have a transition temperature of $T_c = 37.5$ K with a transition width of <1 K. One can grow series in the whole doping range up to concentrations of 0.3 for the Sr doping. Thus, it is possible to probe the whole doping range from the antiferromagnetic mother compound La_2CuO_4 over the whole superconducting

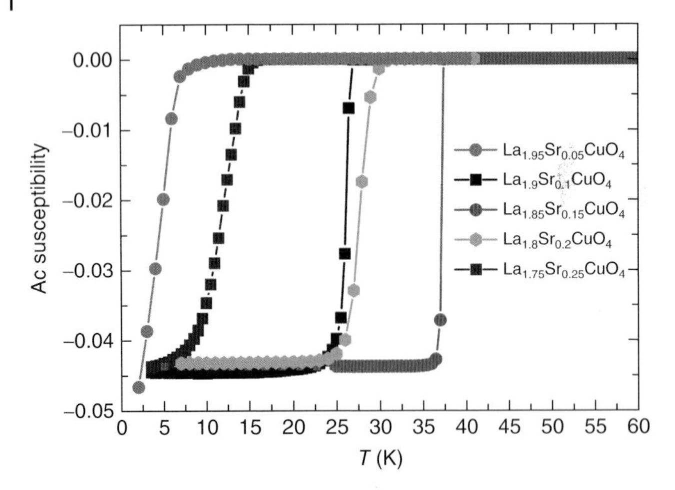

Figure 3.1.2.6 Normalized ac susceptibility plotted versus temperature of $La_{2-x}Sr_xCuO_4$ single crystals with different doping level.

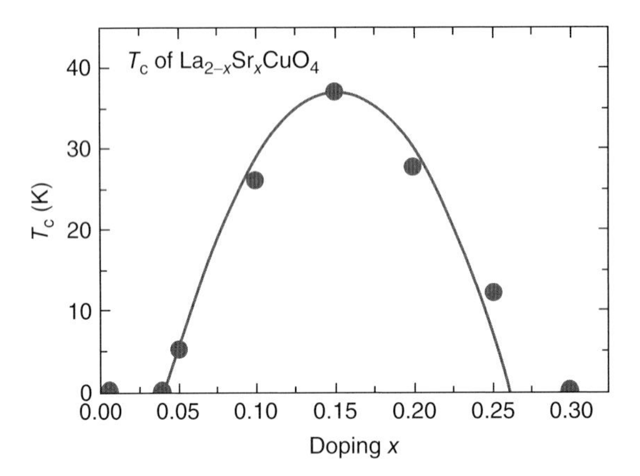

Figure 3.1.2.7 Transition temperatures of $La_{2-x}Sr_xCuO_4$ versus doping for single crystals with different doping level.

range up to a concentration of 0.3, where superconductivity vanishes again at a concentration of around 0.27 (see Figures 3.1.2.6 and 3.1.2.7).

Single crystals of the electron-doped compounds $Nd_{2-x}Ce_xCuO_4$ and $Pr_{2-x}Ce_xCuO_4$ have successfully been grown by the TSFZ method even though the crystal growth of these compounds is more complicated and challenging. The stability of the growth conditions were found to strongly depend on the oxygen partial pressure and doping for these compounds. After optimization of the growth conditions, we reproducibly obtain large single crystals of these compounds. For optimally doped samples of $Nd_{2-x}Ce_xCuO_4$ and $Pr_{2-x}Ce_xCuO_4$, the maximum transition temperatures are $T_c = 23.5$ and 25.5 K, respectively, which are the highest T_c values reported so far [8].

In both cases, the transition width was only about 1 K. Crystals of the whole series of $Nd_{2-x}Ce_xCuO_4$ and $Pr_{2-x}Ce_xCuO_4$ compounds from the antiferromagnet up to the solubility limit of cerium, which was found to be $x = 0.18$ and 0.15, respectively, have been grown. The crystals have also been proven to be absolutely stable over several years without any sign of decomposition. Both the chemical stability and the good superconducting properties were achieved by an optimization of the conditions for the critical annealing process, which is necessary to remove the interstitial oxygen from the crystals after the growth process (Figures 3.1.2.8 and 3.1.2.9).

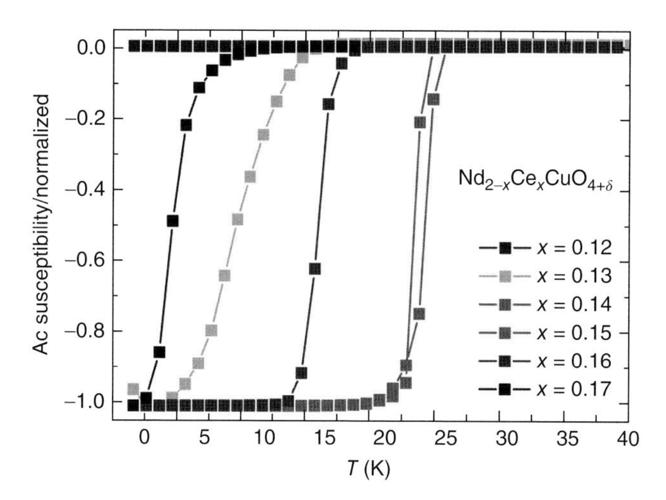

Figure 3.1.2.8 Normalized ac susceptibility plotted versus temperature of $Nd_{2-x}Ce_xCuO_4$ crystals with different doping level.

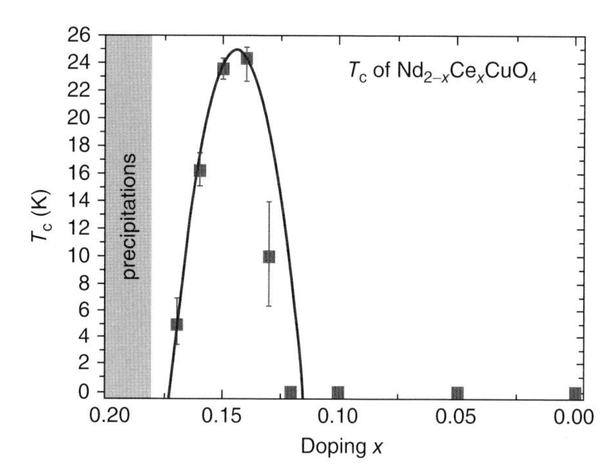

Figure 3.1.2.9 Transition temperatures versus doping for $Nd_{2-x}Ce_xCuO_4$ single crystals with different doping level.

Extensive studies of the electrical transport properties have been performed on single crystals of the $Nd_{2-x}Ce_xCuO_4$ series, and the crystals could prove their high quality by the observation of Shubnikov–de Haas oscillations [9].

3.1.2.4 Conclusions

Since the early beginning of the crystal growth of high-temperature superconductors, there were always two problems that had to be solved before crystal growth of these compounds became a controllable and clean process: the chemical reactivity of the aggressive melts and the incongruent melting behavior of the systems under study. In the cases of the 123 and 124 compounds, the invention of the adapted crucible material $BaZrO_3$ allowed the controllable growth of very high purity single crystals, which in turn led to new observations in their physical properties. Even though only a few groups worldwide have mastered the growth using this technique, it can nowadays be considered state of the art.

As to the other high-temperature superconductors, only the 214 compounds, both on the electron- and the hole-doped side of the phase diagram, can be grown in a similar controllable manner. Since the 214 compounds are solid solution crystals, bigger samples with homogeneous doping cannot be obtained via crucible growth techniques but have to be grown in different manners. Here, the development of the TSFZ technique for the oxide superconductors paved the way toward an equally controllable crystal growth.

For all the other compounds of the high-temperature superconductors, no such highly developed technique exists, partially due the high vapor pressure of, for instance, the Hg and Tl compound, partially because of peculiarities of the phase diagram like in the case of the Bi-compounds.

With the compounds $YBa_2Cu_3O_{7-\delta}$, $Nd_{2-x}Ce_xCuO_4$, and $Pr_{2-x}Ce_xCuO_4$, however, a set of samples of extremely good quality is available to probe the whole doping range of the phase diagram of the cuprates on the electron- and hole-doped side.

References

1. Songa, Y.T., Penga, J.B., Wanga, X., Suna, G.L., and Lina, C.T. (2007) *J. Cryst. Growth*, **300**, 263–266.
2. Wolf, T., Bornarel, A.-C., Küpfer, H., Meier-Hirmer, R., and Obst, B. (1997) *Phys. Rev. B*, **56**, 6308.
3. Erb, A. (1999) Habilitation thesis. University Geneva, Switzerland, available online at: *www.lauecamera.com* (accessed 15 May 2014).
4. Erb, A., Biernath, T., and Müller-Vogt, G. (1993) *J. Cryst. Growth*, **132**, 389.
5. Erb, A., Walker, E., and Fluekiger, R. (1995) *Physica C*, **245**, 245.
6. Erb, A., Walker, E., and Flükiger, R. (1996) *Physica C*, **258**, 9.
7. Ruixing, L., Bonn, D.A., and Hardy, W.N. (1998) *Physica C*, **304**, 105.
8. Lambacher, M., Helm, T., Kartsovnik, M., and Erb, A. (2010) *Eur. Phys. J. Spec. Top.*, **188**, 61.
9. Helm, T., Kartsovnik, M.V., Sheikin, I., Bartkowiak, M., Wolff-Fabris, F., Bittner, N., Biberacher, W., Lambacher, M., Erb, A., Wosnitza, J., and Gross, R. (2010) *Phys. Rev. Lett.*, **105**, 247002.

3.1.3
Properties of Bulk Materials

Günter Fuchs, Gernot Krabbes, and Wolf-Rüdiger Canders

3.1.3.1 Irreversibility Fields of Bulk High-T_c Superconductors

Applications of bulk high-T_c superconductors (HTSs) are limited to magnetic fields below the irreversibility field B_{irr}. In the field range $B_{irr} < \mu_0 H < B_{c2}$, with B_{c2} as the upper critical field, the effect of thermal fluctuations on the vortex lattice becomes so strong that currents cannot flow without losses although the superconductor is not yet in the normal state. Because the irreversibility field B_{irr} sets an upper limit for trapped fields in bulk superconductors, the knowledge of the irreversibility line $B_{irr}(T)$ is a basic prerequisite for applications of high-temperature superconductors.

In Figure 3.1.3.1, irreversibility lines of bulk $REBa_2Cu_3O_{7-x}$ (RE-123, RE = Y or light rare earths LREs = Nd, Sm, Gd) are compared [1, 2]. The bulk Y-123 (YBCO) compound which is free from RE-Ba solid solution can be relatively easily prepared at air and is widely used as trapped field magnet. However, YBCO has only a rather modest irreversibility field of ∼5 T at 77 K for fields $H||c$ applied along the c-axis. In contrast, higher irreversibility fields can be realized in LRE-123 compounds as shown in Figure 3.1.3.1. The highest irreversibility line was achieved for bulk Gd-123 [2].

The better performance of the bulk LRE-123 compounds is due to the ability of the LRE ions to occupy Ba sites and vice versa, forming, in this way, nanometer-sized clusters of solid solutions with Ba which act as strong pinning

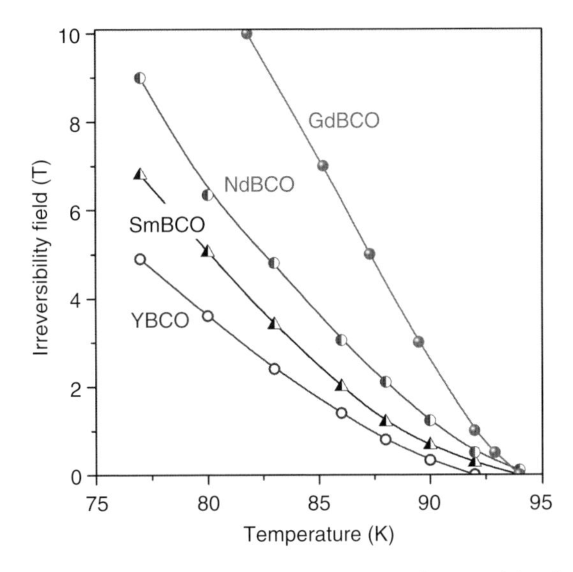

Figure 3.1.3.1 Temperature dependence of irreversibility field for bulk RE-123 samples (RE = Y, Sm, Nd, and Gd). (Data for YBCO, Sm-123, and Nd-123 were taken from Ref. [1] Nature Publishing Group, for Gd-123 from Ref. [2] Elsevier.)

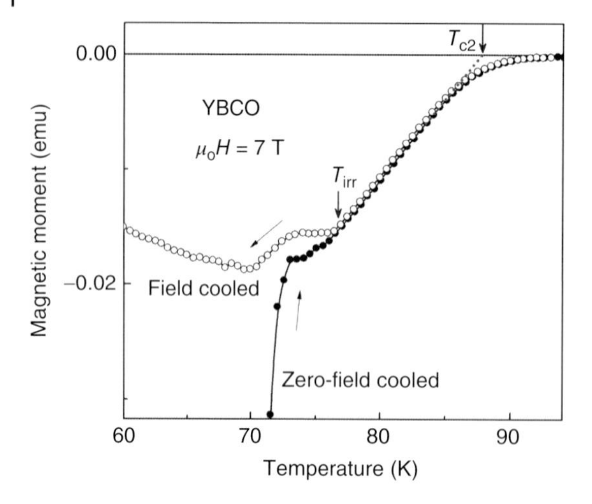

Figure 3.1.3.2 Temperature dependence of magnetization of an YBCO sample at 7 T applied along *c*-axis after zero-field cooling and field cooling.

centers. However, the fabrication of single-domain bulk LRE-123 compounds is more complex than that of bulk YBCO, mainly due to the difficulty of controlling the solid solution formation.

The irreversibility field can be determined from magnetization versus temperature data at constant external field as shown in Figure 3.1.3.2 for a bulk YBCO sample. By measuring $M(T)$ data, both after cooling the superconductor at zero field (zero-field cooling) and by cooling the superconductor at applied field (field cooling), two regions of irreversible and reversible behavior become visible. The irreversibility field is defined at the transition from irreversible to reversible behavior of the magnetization (see Figure 3.1.3.2).

3.1.3.2 **Vortex Matter Phase Diagram of Bulk YBCO in an Extended Field Range up to 40 T**

The range of applications of bulk HTSs can be extended to higher magnetic fields by lowering the temperature using the higher critical current density at low temperatures. This has been demonstrated for trapped field magnets using bulk YBCO for which extremely large trapped fields of 16 T [3] and 17 T [4] have been achieved at temperatures of 24 and 29 K, respectively. This was the motivation to study the vortex matter phase diagram of bulk YBCO, including the irreversibility line $B_{irr}(T)$ and the upper critical field $B_{c2}(T)$ in a wide range of magnetic fields using pulsed fields [5].

In Figure 3.1.3.3, resistive transition curves of a small bulk YBCO sample are shown for magnetic fields up to 50 T applied parallel to the *c*-axis of the melt-textured sample. These measurements were performed in a pulsed field facility in which the duration of the field pulse was 10 ms. The irreversibility field was determined at the onset of resistance using an electric field criterion of 1 μV cm^{-1}. This

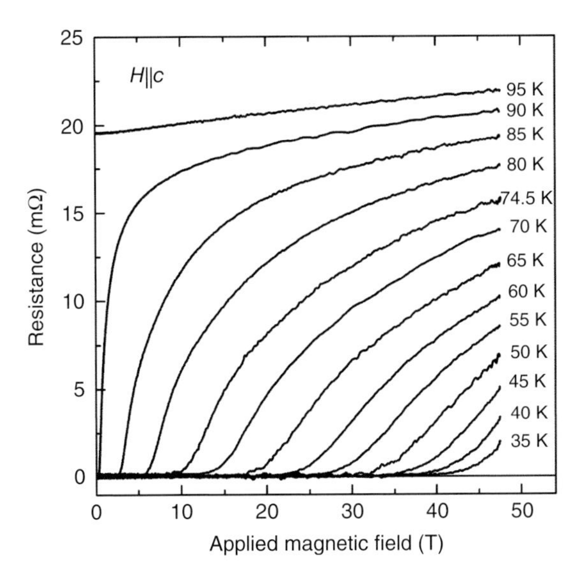

Figure 3.1.3.3 *R*(*H*) transition curves of a bulk YBCO sample for *H*||*c* at different temperatures. The measurements were performed in pulsed magnetic fields up to 50 T.

definition was found to be consistent with the irreversibility field derived from magnetization data (see Figure 3.1.3.2).

In Figure 3.1.3.4, $B_{irr}(T)$ data for this YBCO sample are shown which were obtained by different techniques (ac susceptibility and resistance measured in static field, magnetization and resistance measured in pulsed field). A good agreement was found between B_{irr} data obtained in pulsed and static field. The

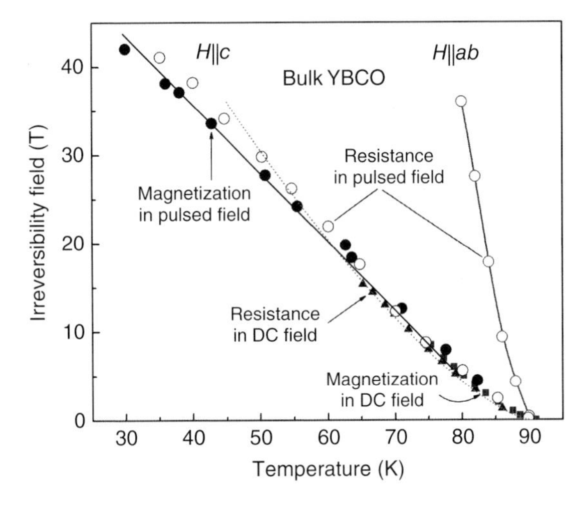

Figure 3.1.3.4 Irreversibility lines of bulk YBCO sample for both field directions obtained from different measurements as indicated in the figure. Dotted line: $B_{irr}(T)$ according to Eq. (3.1.3.1) with $B_{irr}(0) = 95$ T, $T_c = 90$ K, and $n = 1.4$. (Taken from Ref. [5] Wiley.)

dotted line in Figure 3.1.3.4 corresponds to the expression

$$B_{irr}(T) = B_{irr}(0)\left(1 - \frac{T}{T_c}\right)^n \qquad (3.1.3.1)$$

with B_{irr} (0) = 95 T, T_c = 90 K, and n = 1.4 which describes the experimental data at fields up to 25 T. A similar temperature dependence of the irreversibility field is expected both for thermally activated flux creep (n = 1.5) [6] and for the vortex glass–vortex liquid transition (n = 1.3) [7]. Not understood is the almost linear $B_{irr}(T)$ dependence between about 10 and 45 T. By extrapolating the linear part of $B_{irr}(T)$ in Figure 3.1.3.4 (black line) to T = 0, $B_{irr}(0)$ = 70 T is estimated.

The vortex matter phase diagram of bulk YBCO is shown in Figure 3.1.3.5 in the range of magnetic fields up to 50 T for the two field directions. For magnetic fields below B_{irr}, the flux line lattice is plastically deformed by strong pinning and is called a *disordered vortex glass.* As mentioned above, the critical current density disappears in the field range between $B_{irr}(T)$ and $B_{c2}(T)$ due to strong thermal fluctuations.

The upper critical field data plotted in Figure 3.1.3.5 were determined from the reversible part of $M(T)$ data (see Figure 3.1.3.2) showing a linear $M_{rev}(T)$ dependence in accord with the *Ginzburg–Landau* theory. Neglecting temperatures near T_c which are dominated by fluctuations, the linear part of the reversible magnetization in Figure 3.1.3.2 was extrapolated to M = 0. For M = 0, the normal state was reached at the temperature T_{c2} which is related to an upper critical field B_{c2} = 7 T in that case.

From the $B_{c2}(T)$ data in Figure 3.1.3.5, an anisotropy of $B_{c2}^{ab}/B_{c2}^{c} = (5.3 \pm 0.1)$ is obtained for the bulk YBCO sample which was found to be temperature independent in the investigated temperature range above 84 K. A similar anisotropy ($B_{c2}^{ab}/B_{c2}^{c} = 5.5$) was reported for an YBCO single crystal [8].

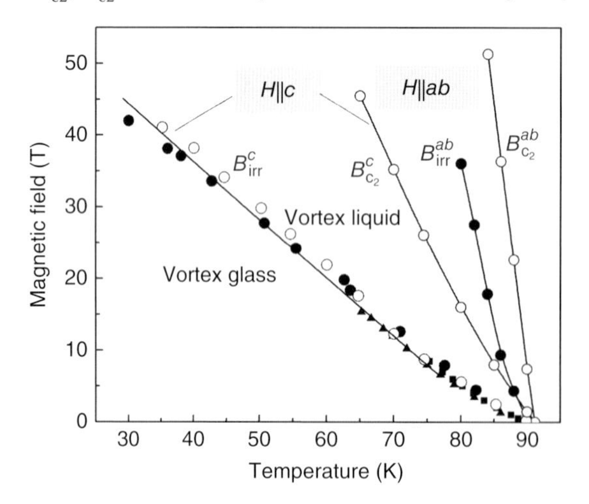

Figure 3.1.3.5 Vortex matter phase diagram of bulk YBCO for magnetic fields up to 50 T, including irreversibility lines $B_{irr}(T)$ and upper critical fields $B_{c2}(T)$ for both field directions. (Taken from Ref. [5] Wiley.)

For applications, most important is the irreversibility line $B_{irr}(T)$ for fields parallel to the c-axis. The irreversibility field parallel to the c-axis increases from about 7 T at $T = 77$ K up to about 42 T at $T = 30$ K. Thus, bulk YBCO can be applied in a wide field range, especially for temperatures below 77 K.

3.1.3.3 Critical Current Density

The critical current density j_c of bulk HTSs is usually determined from contact-free magnetization measurements. In contrast, transport measurements are difficult to realize for bulk superconductors, because one has to overcome the problem to prepare current contacts for high-transport currents of more than 100 A required already at 77 K. Magnetization measurements are easy to perform, especially for applied magnetic fields directed along the c-axis. In this case which is most important for applications, supercurrents are induced in the ab-plane and the size of the current loops is given by the sample geometry. The critical current density $j_c(H,T)$ can be derived from magnetization loops, taking the sample geometry into account [5].

Typical $j_c(H)$ curves of a melt-textured YBCO sample obtained from magnetization measurements in the temperature range between 77 and 59 K are shown in Figure 3.1.3.6 for magnetic fields applied along the c-axis. At 77 K, the critical current density is limited by the irreversibility field of about 7 T. For temperature $T \leq 70$ K, a broad plateau of j_c is observed above the maximum of j_c at zero magnetic field. This field-independent j_c, which is typical for bulk YBCO at these temperatures, points to single vortex pinning, that is, each flux line is pinned individually [9].

The critical current density in bulk YBCO can be substantially improved by enriching the starting composition of the precursor powder with Y_2O_3 rather than Y-211 [10, 11] or by the refinement of the Y-211 precipitations [12] (see

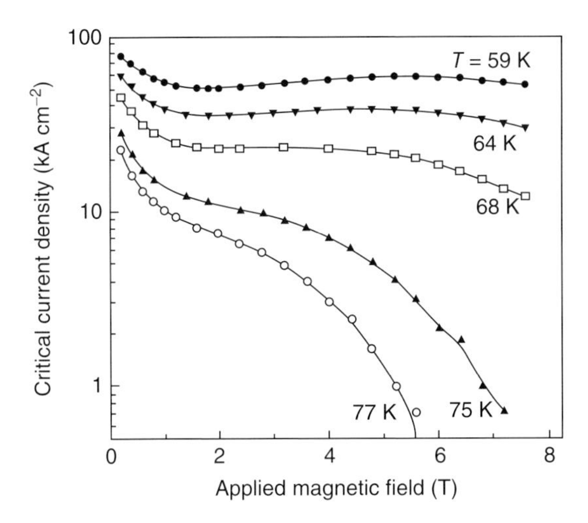

Figure 3.1.3.6 Field dependence of the critical current density of a standard bulk YBCO sample with Y-211 precipitations for $H||c$ at temperatures between 77 und 59 K.

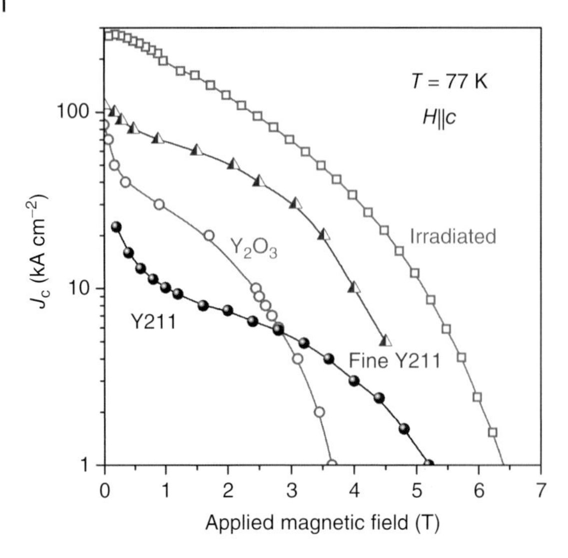

Figure 3.1.3.7 $j_c(B)$ data for bulk YBCO samples with improved flux pinning for $H||c$ at 77 K, using Y_2O_3 [11] IOP Publishing rather than Y-211, refined Y-211 precipitations [12] IOP Publishing, and irradiation-induced defects [13] IEEE.

Figure 3.1.3.7). For instance, high j_c values up to 110 kA cm^{-2} at 77 K in self-field were obtained by the use of very fine Y-211 particles with diameters of 110 nm [12]. Even higher critical current densities up to 275 kA cm^{-2} (at 77 K in low fields) were reported for bulk YBCO irradiated with high-energy U^{238} ions [13]. Flux pinning is significantly improved by discontinuous columnar defects aligned along the *c*-axis. The data shown in Figure 3.1.3.7 are related to columnar defects with a diameter of ~6 nm.

For bulk Gd-123, the addition of ultrafine ball-milled Y-211 powders was found to be very effective in enhancing the critical current density. In bulk (Gd,Y)-123, high j_c values up to 380 kA cm^{-2} (at 77 K in self-field) were reported by adding Y-211 particles with sizes of 50 nm [14]. High trapped fields of 2 T were achieved in a bulk (Gd,Y)-123 sample 33 mm in diameter containing such ultrafine Y-211 particle. The evolution from low critical current densities $j_c(B)$ for samples containing large Y-211 particles to the above mentioned very high critical current densities for the sample with 50 nm sized Y-211 particles is shown in Figure 3.1.3.8.

Bulk $(Nd_{0.33}Eu_{0.33}Gd_{0.33})$-123 compounds, abbreviated NEG-123, constitute a special subclass of LRE-123 compounds with LREs. To improve the pinning properties of NEG-123, various LRE-211 precipitates have been tested. The best pinning performance was found for Gd-211 precipitates. By adding Gd-122 particles (of ~70 nm in size) to mixed ternary NEG-123, a critical current density of 260 kA cm^{-2} was reported at 77 K in self-field [15] (see Figure 3.1.3.9). A careful microchemical analysis of the nano-sized particles obtained by ZrO_2 ball milling of the initial Gd-122 powder revealed the existence of ZrBaCuO and (NEG,Zr)BaCuO particles instead of Gd-122. In addition, it was found that these ball-milled particles with sizes of ~70 nm not only survived the melt-texturing process, but even

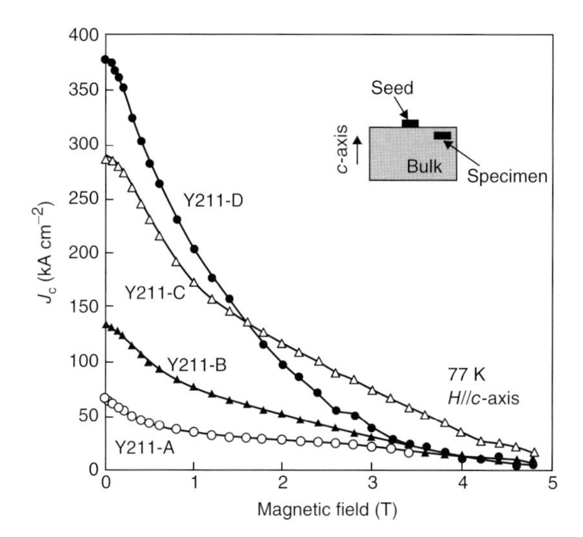

Figure 3.1.3.8 $j_c(B, 77\,\mathrm{K})$ data for bulk (Gd,Y)-123 samples with ball-milled Y-211 particles of different size: 1360 nm (Y-211-A), 110 nm (Y-211-B), 70 nm (Y-211-C), and 50 nm (Y-211-D). (Taken from Ref. [14] Elsevier.)

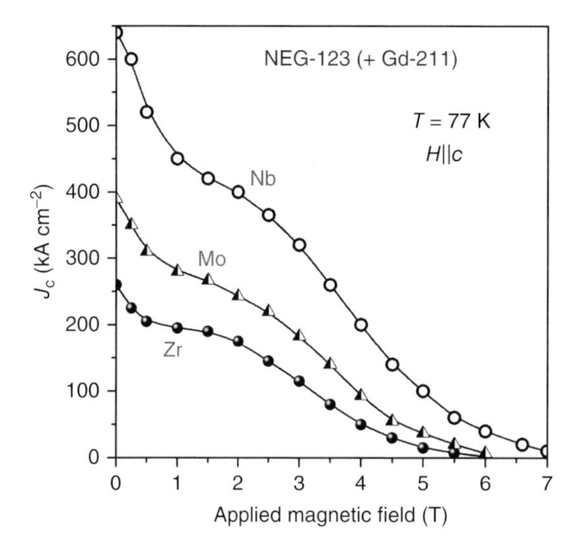

Figure 3.1.3.9 Comparison of $j_c(B)$ data at 77 K for bulk NEG-123 + Gd-122 composite samples doped with Zr, Mo, and Nb. (Taken from Refs. [15] AIP Publishing and [16] Elsevier.)

their size reduced down to 20–50 nm during melt-texturing. This effect of Zr on the particle size was attributed to chemical inertia of Zr in the superconducting matrix [16].

Motivated by this result, NEG-123 + Gd-211 composites were doped by tiny amounts of nanometer-sized Nb and Mo (following Zr in the periodic table). This doping was observed to create extremely small Nb- or Mo-rich NEG-BaCuO nanoparticles with sizes of ~10 nm after melt-texturing. Record flux pinning with critical current densities of 390 and 640 kA cm^{-2} at 77 K in self-field was reported for doping by Mo and Nb nanoparticles, respectively [16] (see Figure 3.1.3.9). These bulk NEG-123 compounds are highly promising for applications as trapped field magnets operating at 77 K and even at higher temperatures.

3.1.3.4 Flux Creep in Bulk YBCO

3.1.3.4.1 Flux Creep in HTS

Thermally activated flux motion is much stronger in HTSs than in low-T_c superconductors, which is due to the low activation energy U for flux creep relative to the thermal energy kT at temperatures in the range of T_c. The ratio $U/k_B T_c$ is ~2 for HTS only, but ~100 in conventional superconductors. Therefore, flux creep becomes strong in HTS.

Flux creep in conventional superconductors is characterized by a logarithmic time dependence of the magnetization and a linear current dependence of the activation energy, and has been described by the flux creep model proposed by *Anderson* and *Kim* [17, 18]. Deviations from the logarithmic time dependence of the magnetization observed for HTS can be explained within the model of collective creep [19] by a nonlinear $U(j)$ relation, assuming weak random pinning and taking the elastic properties of the flux line lattice into account. In this model, the time dependence of the persistent current is described by

$$j(t) = \frac{j_{co}}{[1 + (\mu kT/U_o)\ln(t/t_o)]^{\frac{1}{\mu}}} \quad (3.1.3.2)$$

The parameter μ in Eq. (3.1.3.2) is not only field and temperature dependent, but also depends on the size of the moving flux bundles interacting with weak pinning sites. The pinning potential U_o can be derived from relaxation data obtained at different temperatures following a procedure which has been proposed by Maley *et al.* [20]. The activation energy for flux creep is written as

$$U = kT\left(C - \ln\left|\frac{dM}{dt}\right|\right) \quad (3.1.3.3)$$

and plotted against $M_{irr} \propto j$ using $kT\ln|dM/dt|$ from relaxation data at different temperatures. Eq. (3.1.3.3) is field dependent and the constant parameter C has to be chosen so that all isotherms of the $U(j)$ plot fall on one smooth curve.

As example, the current dependence of the activation energy $U(j)$ of a melt-textured YBCO sample with RuO_2 additions containing a periodic array of nanoscale twin boundaries is shown in Figure 3.1.3.10 [21, 22]. The activation energy $U(j)$ exhibits a crossover from a power law with $\mu = 1$ at high currents ($T \leq 30$ K) to a power law with $\mu = 0.55$ at low currents ($T \geq 30$ K). The high-current data with $\mu = 1$ can be described by the Bose-glass model [23] for correlated pinning arising in this case by the arrangement of twin boundaries along the c-axis. The stronger vortex creep for $\mu = 0.55$ at low currents indicates weaker pinning due to interstitial vortices within the variable-range hopping regime.

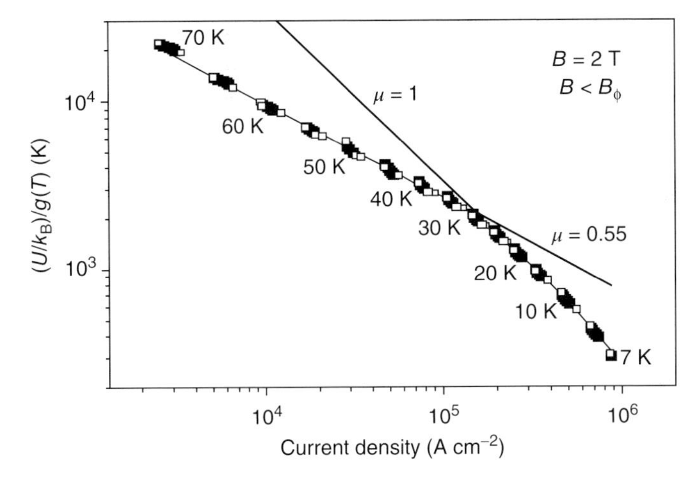

Figure 3.1.3.10 Temperature and current dependence of the activation energy for a bulk YBCO sample with nanoscale twin boundaries along the *c*-axis in applied fields of 2 T for *H*||*c*. (Reproduced from Ref. [22] with permission of IOP publishing.)

The collective flux creep model explains many relaxation data of HTS samples. However, it should be noted that the collective creep can be destroyed if strong pinning centers are introduced into the superconductor. For a small density of strong pins, plastic creep of weakly pinned flux lines along strongly pinned flux lines occurs. Plastic vortex creep associated with the motion of dislocations in the vortex lattice has been observed in YBCO single crystals [24].

For applications of bulk HTS, the normalized flux creep rate $S \equiv 1/M\ \mathrm{d}M/\mathrm{dln}(t)$ is of interest. Typical data for the temperature dependence of the flux creep rate for YBCO are shown in Figure 3.1.3.11 ([5], see also Ref. [25]). A special feature is the temperature-independent plateau of the flux creep rate at intermediate temperatures which cannot be understood within the standard *Anderson–Kim* model for flux creep predicting $S = -kT/U_o$. Within the collective creep model [19], one obtains

$$S = \frac{kT}{U_o + \mu kT \ln t/t_o} \tag{3.1.3.4}$$

According to Eq. (3.1.3.4), the flux creep rate approaches for increasing temperature the limit

$$S = \frac{1}{\mu \ln t/t_o} \tag{3.1.3.5}$$

and thus becomes independent of temperature. Another prediction of this formula is that S decreases with time, which is in agreement with long-time relaxation data. Experimental data for the height of the plateau of S in YBCO samples are in the range of several percent. From Eq. (3.1.3.5), one obtains $S \approx 0.048$ and 0.043 for typical times of flux creep experiments of $t = 10^3$ s and $t = 10^4$ s, respectively, assuming $\mu = 1$ and $t_o = 10^{-6}$ s [26].

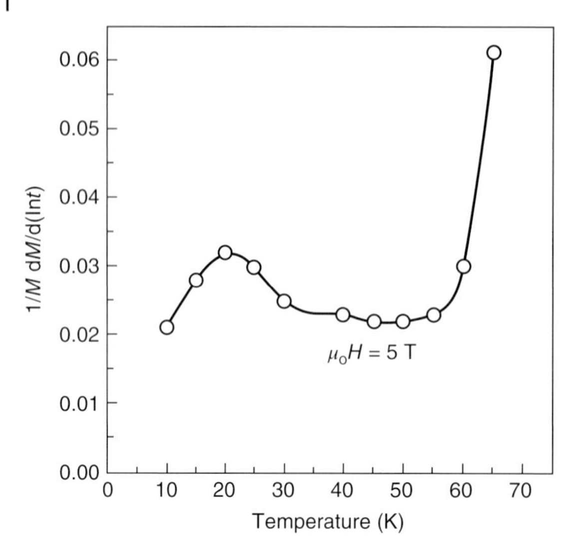

Figure 3.1.3.11 Temperature dependence of normalized flux creep rate $S \equiv 1/M\ dM/d\ln(t)$ for a bulk YBCO sample measured at a magnetic field of 5 T for $H||c$. (Taken from Ref. [5] Wiley.)

3.1.3.4.2 **Reduction of Flux Creep**

Thermally activated flux motion tends to reduce the field gradient and thus the maximum trapped field B_0 in the superconductor or the levitation force of superconducting magnetic bearings. At 77 K, typical relaxation rates of $d(B/B_0)/d(\ln t/t_0) \approx -0.05$ were reported for melt-textured YBCO [27], that is, the trapped field B_0 reduces by about 5% per time decade which is in accordance with the above-mentioned estimation. Hence, flux creep seems to be a serious problem for superconducting permanent magnets, and it is highly desired to control and to reduce the magnetic relaxation.

A very effective experimental procedure to reduce the magnetic relaxation, proposed by Beasley *et al.* [28], is to lower the temperature after establishing the critical state in the superconductor. After reducing the initial temperature T_0 by ΔT, the current density $j = j_c(T_0)$ of the supercurrents becomes less than the critical current density $j_c(T_0 - \Delta T)$ at the lower temperature $T = T_0 - \Delta T$. In this subcritical state, the flux creep rate at the lower temperature is expected to be attenuated exponentially in ΔT [28]. Therefore, flux creep is strongly reduced. Using this procedure, a dramatic suppression of the magnetic relaxation has been achieved in bulk YBCO [29] as shown in Figure 3.1.3.12. The sample, 5 mm in diameter, was magnetized in a superconducting magnet. In order to ensure that the sample was in the critical state, a complete hysteresis loop was passed through starting from $H = 0$ via $\mu_0H = 8$ T, $\mu_0H = -8$ T to $\mu_0H = 1$ T, before the time dependence of magnetization was measured. In order to suppress the flux creep, the sample was cooled from 77 to 65 K shortly after the start of the relaxation measurement at 77 K. Afterwards, no flux creep effect was detectable

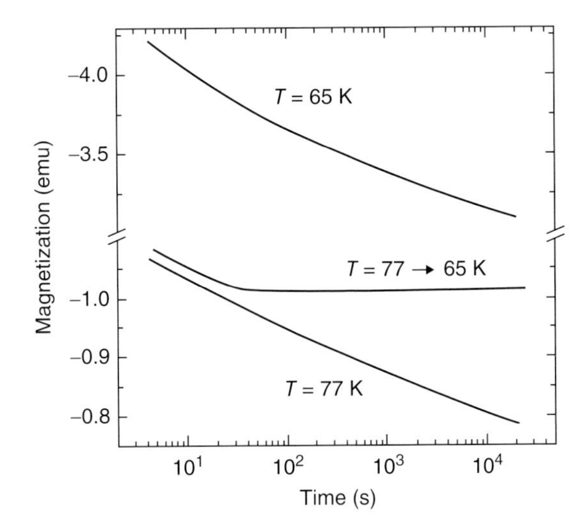

Figure 3.1.3.12 Time dependence of magnetization for a bulk YBCO sample measured at 77 and 65 K. Flux creep was suppressed by cooling the sample from 77 to 65 K shortly after the relaxation measurement was started. (Taken from Ref. [29] Elsevier.)

with the magnetometer. For comparison, relaxation data obtained for $T = 77$ and 65 K are shown in Figure 3.1.3.12, too.

The reduction of the relaxation rate has been also reported for an YBCO mini-magnet consisting of two bulk YBCO disks [27]. A maximum trapped field of $B_o = 4.3$ T has been obtained in this magnet at 65 K. The creep rate at 65 K was found to reduce by factors of 6, 197, or ≥1000 due to lowering the temperature by $\Delta T = 2$, 4, and 6 K, respectively [27].

3.1.3.5 Selected Properties of Bulk YBCO

3.1.3.5.1 Mechanical Properties

HTSs are brittle materials with poor mechanical properties. Typically, the material fractures at a critical strain which is due to the propagation of microcracks which are present in the superconductor. This significantly affects applications of these superconductors. In bulk superconductors, large tensile strains are generated by applying magnetic fields which is due to the large field gradient arising in the critical state within the superconductor. The pinning-induced tensile stress in a fully magnetized cylindrical superconductor increases, according to $\sigma = 2B_o^2/\mu_o$, with the square of the maximum trapped field B_o on its surface. Therefore, it is not surprising that the mechanical strength is very important with respect to the highest trapped fields achievable in these superconductors.

The fracture strength of a brittle material is not an invariant quantity, but rather controlled by the microcracks within the material. The relation between fracture strength σ_{max}, fracture toughness K_{IC}, and the critical length a of microcracks in a brittle material is given by

$$K_{IC} \sim \sigma_{max}(\pi a)^{0.5} \tag{3.1.3.6}$$

These microcracks set a limit for the fracture strength of the material because they start to propagate when the external tensile stress reaches the fracture toughness of the material.

The fracture toughness (K_{IC}) of bulk YBCO which is usually determined by the indentation method has been found to decrease from 1.3 to 0.4 MPa$\sqrt{m}$ by decreasing the temperature from 300 to 40 K [30]. It has been also observed that K_{IC} of bulk YBCO at 300 K decreases with increasing number of thermal cycles between 300 and 77 K. After 50 cycles, a reduction of K_{IC} by about 20% was observed. At the same time, the number of cracks per indentation raised [31]. Therefore, it was concluded that thermal cycling essentially contributes to the generation of microcracks and to the evolution of macrocracks.

Using $K_{IC} \approx 1.0$ MPa$\sqrt{m}$ and $\sigma_{max} \approx 30$ MPa from cracking experiments, one estimates from Eq. (3.1.3.6), $a \approx 350$ μm for the size of the largest microcrack which seems to be reasonable. However, this parameter is difficult to verify experimentally. Instead, the spacing of microcracks is used to characterize the bulk material.

Improved values of the fracture toughness of YBCO at 300 K have been reported by the addition of silver [32] which increased from $K_{IC} \approx 1.9$ to 2.4 MPa$\sqrt{m}$ when 10 mol% Ag was added to the precursor mixture. This improved fracture strength was found to be related to a larger spacing of microcracks in the YBCO/Ag composite material. In particular, the spacing of the $a-b$ microcracks increased from 1.8 to 3.7 μm and the spacing of the c-cracks increased from 34.2 to 133.5 μm when 12 wt% Ag was added [33]. Improved mechanical properties of bulk YBCO have been also reported by resin impregnation [34]. Measuring the tensile strength of bulk YBCO with and without resin impregnation at 300 K, the tensile strength was found to increase from 12 to 18 MPa after resin impregnation. An additional improvement of the tensile strength to 29 MPa has been reported by wrapping the bulk with carbon fiber fabric prior to resin impregnation [34].

Data for the elastic properties of bulk YBCO, as Young modulus E, shear modulus G, and the bulk modulus K, are collected in Table 3.1.3.1.

3.1.3.5.2 Thermodynamic and Thermal Properties

The knowledge of specific heat and thermal conductivity data of HTS is essential to understand the response of the superconductor of heat release due to variations

Table 3.1.3.1 Elastic properties of bulk YBCO.

Young's modulus (E) (GPa)	T (K)	Shear modulus (G) (GPa)	T (K)	Bulk modulus (K) (GPa)	T (K)
92.4	288	38.5	291	70.8	295
93	200	38.9	207	72.5	200
94	150	39.5	139	74.4	100
95.1	87	39.9	79	75.3	20
95.8	11	40.2	15		

Data for Young's and shear modulus were taken from Ref. [35], data for the bulk modulus from Ref. [36].

of the applied magnetic field. Examples are the local heating of the superconductor if it is activated by pulsed fields and the phenomenon of flux jumps occurring at low temperatures. The consequence is, in both cases, a considerable limitation of the trapped field in the superconductor below the value expected from the critical current density at the given temperature.

The typical microcracks in bulk YBCO are mainly caused by the anisotropy of the thermal expansion of the YBCO lattice. Large stresses arising in bulk YBCO on cooling are responsible for the developing of microcracks.

Data of the linear thermal expansion coefficient for melt-textured YBCO and for YBCO single crystals have been measured using a high-resolution capacitance dilatometer [37]. As shown in Figure 3.1.3.13, the thermal expansion coefficient is considerably larger along the *c*-axis than along the *a*- or *b*-axis. Therefore, by cooling the superconductor from 300 to 77 K, the *c*-axis stronger contracts than *a*- and *b*-axes. The resulting stress puts *c*-axis grain boundaries under tension which explains the sensitivity of the bulk superconductor against cracking.

In order to avoid cracking of YBCO permanent magnets during the magnetizing procedure, the unavoidable pinning-induced tensile stress was compensated by reinforcing the YBCO magnet with a metal casing generating a compressive stress in the superconductor after cooling from 300 to 77 K [3]. Steel tubes have turned out to be a good solution. Austenitic steel has, over a temperature range of 50–300 K, an average coefficient of thermal expansion of $\alpha_{av} \sim 10^{-5}\,K^{-1}$, which is comparable with the α value for YBCO in the *c* direction, but significantly larger than its value in the *ab*-plane. Furthermore, austenitic steel is nonmagnetic and has, with 194 GPa, a *Young*'s modulus which is large enough to withstand strong mechanical forces. By using steel tubes, cracking of YBCO permanent magnets

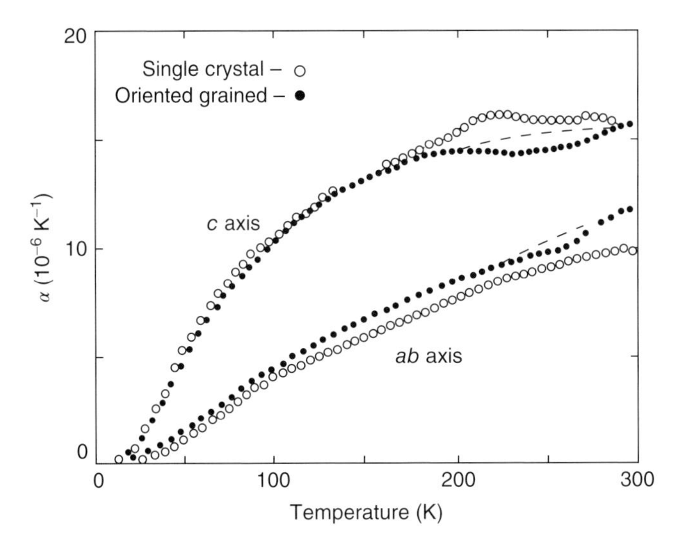

Figure 3.1.3.13 Temperature dependence of the linear thermal expansion coefficient for melt-textured YBCO (solid circles) and for an YBCO single crystal (open circles) along the *c*-axis and in the *ab*-plane. (Taken from Ref. [37] APS.)

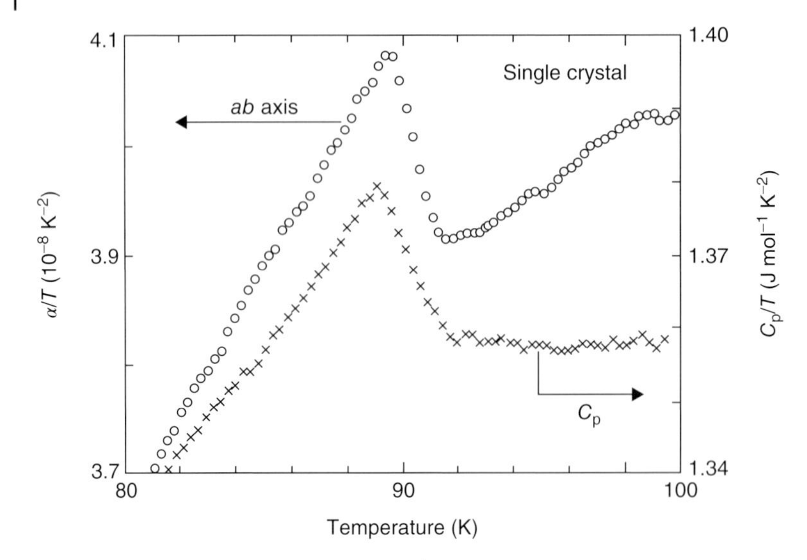

Figure 3.1.3.14 Jump of the thermal expansion coefficient and of the specific heat of a YBCO single crystal at T_c. (Taken from Ref. [37] APS.)

was avoided up to the highest trapped fields of 16 T measured at 24 K in the gap between two YBCO magnets [3].

The thermal expansion coefficient exhibits a jump at T_c which is related to the second- order transition into the normal state (see Figure 3.1.3.14). This jump $\Delta\alpha$ is closely related to the specific heat jump Δc at T_c, according to the *Ehrenfest* relationship

$$\Delta\alpha = \frac{\Delta c}{T_c}\frac{dT_c}{dp} \tag{3.1.3.7}$$

with dT_c/dp as the pressure dependence of T_c. From Eq. (3.1.3.7), values of $dT_c/dp_{ab} \sim 0.036$ and 0.089 K kbar^{-1} have been estimated for the pressure dependence of T_c of bulk YBCO and YBCO single crystals, respectively [37].

For several applications of bulk HTSs, the unusual thermal transport properties of HTS play a decisive role. For the application as bulk magnets at 77 K, the large heat conduction within the *ab*-plane of bulk YBCO is essential to prevent the exudation of the trapped flux due to temperature rise. For the application as power current leads for superconducting coils working at 4.2 K, however, the low thermal conductivity of bulk YBCO at helium temperatures is indispensable.

The temperature dependence of the thermal conductivity κ in the *ab*-plane of YBCO is characterized by a rapid increase of κ below T_c and a peak of κ at approximately $T_c/2$ [38, 39], as shown in Figure 3.1.3.15 for melt-textured YBCO. To understand $\kappa(T)$, one has to take into account the phonon contribution κ_{ph} and the electron contribution κ_{el} to the total thermal conductivity κ. It is well known that at temperatures above T_c, $\kappa(T)$ in HTS is dominated by κ_{ph}. Below T_c, electrons condensed into Cooper pairs cannot carry entropy and thus do not contribute to the thermal conductivity. However, the scattering of

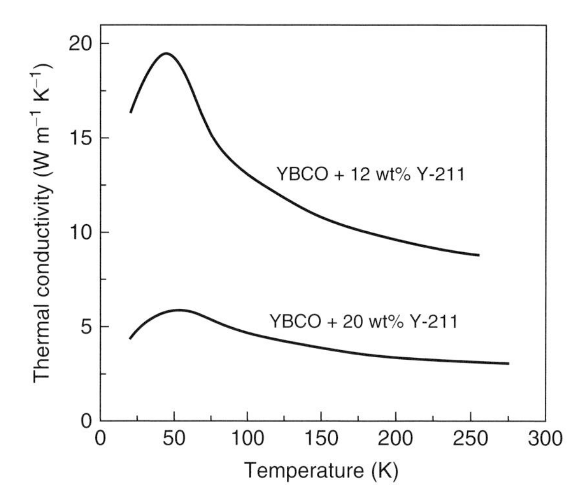

Figure 3.1.3.15 Thermal conductivity versus temperature in the *ab*-plane for two melt-textured $YBa_2Cu_3O_{6.96}$ samples with 12 wt% Y-211 (upper curve) and 20 wt% Y-211 (lower curve). After Ref. [39] APS.

normal electrons in the superconducting state, so-called quasiparticles, has been observed to decrease rapidly with decreasing temperature in YBCO. It is now widely accepted that the rapid increase of κ below T_c and the peak of κ is mostly due to the contribution of quasiparticles located in the CuO_2 planes [38, 40].

The absolute value of κ^* at the peak temperature strongly depends on the Y-211 content as it is demonstrated in Figure 3.1.3.15. κ^* is significantly higher for the YBCO sample with the lower Y-211 content and its value approaches that of good YBCO single crystals which is in the range of $\kappa^* \sim 20-28\,\mathrm{W\,(m{\cdot}K)^{-1}}$. The reason for the lower thermal conductivity of the YBCO sample with the larger Y-211 content appears to be the scattering from the large number of Y-211 grain boundaries [39].

References

1. Hari Babu, N., Shi, Y., Iida, K., and Cardwell, D.A. (2005) *Nat. Mater.*, **4**, 476.
2. Naito, T., Furuta, D., Arayashiki, T., Fujishiro, H., Yanagi, Y., and Itoh, Y. (2012) *Physics Procedia*, **27**, 168.
3. Gruss, S., Fuchs, G., Krabbes, G., Verges, P., Stöver, G., Müller, K.-H., Fink, J., and Schultz, L. (2001) *Appl. Phys. Lett.*, **79**, 3131.
4. Tomita, M. and Murakami, M. (2003) *Nature*, **421**, 517.
5. Krabbes, G., Fuchs, G., Canders, W.-R., May, H., and Palka, R. (2006) *High Temperature Superconductor Bulk Materials Fundamentals–Processing–Properties Control–Application Aspects*, Wiley-VCH Verlag GmbH, Weinheim.
6. Yeshurun, Y. and Malozemoff, A.P. (1988) *Phys. Rev. Lett.*, **60**, 20202.
7. Fisher, D.S., Fisher, M.P.A., and Huse, D.A. (1991) *Phys. Rev. B*, **43**, 130.
8. Welp, U., Kwok, W.K., Crabtree, G.W., Vandervoort, K.G., and Liu, J.Z. (1989) *Phys. Rev. Lett.*, **62**, 1908.
9. Larkin, A.I. and Ovchinnikov, Y.N. (1979) *J. Low Temp. Phys.*, **34**, 409.

10. Krabbes, G., Bieger, W., Schätzle, P., and Wiesner, U. (1998) *Supercond. Sci. Technol.*, **11**, 144.
11. Cardwell, D.A., Yeoh, W.K., Pathak, S.K., Shi, Y.-H., Dennis, A.R., Hari Babu, N., and Iida, K. (2010) in *Advances in Cryogenic Engineering: Transmission Cryogenic Engineering Material Conference -ICMC*, vol. **56** (ed. U. Balachandran), American Institute of Physics, p. 397–406.
12. Nariki, S., Sakai, N., Murakami, M., and Hirabayashi, I. (2004) *Supercond. Sci. Technol.*, **17**, S30.
13. Fuchs, G., Nenkov, K., Krabbes, G., Shlyk, L., Weinstein, R., Gandini, A., Sawh, R., Mayes, B., and Parks, D. (2008) *J. Phys.: Conf. Ser.*, **97**, 012080.
14. Nariki, S., Sakai, N., Murakami, M., and Hirabayashi, I. (2005) *IEEE Trans. Appl. Supercond.*, **15**, 3110.
15. Muralidhar, M., Sakai, N., Jirsa, M., Murakami, M., Koshizuka, N., and Hirabayashi, I. (2005) *Physica C*, **426-431**, 777.
16. Muralidhar, M., Sakai, N., Jirsa, M., Murakami, M., and Hirabayashi, I. (2008) *Appl. Phys. Lett.*, **92**, 162512.
17. Kim, Y.B., Hempstead, C.F., and Strnad, A.R. (1963) *Phys. Rev.*, **131**, 2486.
18. Anderson, P.W. (1962) *Phys. Rev. Lett.*, **9**, 308.
19. Feigel'man, M.V., Geshkenbein, V.B., Larkin, A.I., and Vinokur, V.M. (1989) *Phys. Rev. Lett.*, **63**, 2303.
20. Maley, M.P., Willis, J.O., Lessure, H., and McHenry, M.E. (1990) *Phys. Rev. B*, **42**, 2639.
21. Shlyk, L., Krabbes, G., Fuchs, G., Mickel, C., Rellinghaus, B., and Nenkov, K. (2006) *Appl. Phys. Lett.*, **88**, 062509.
22. Shlyk, L., Krabbes, G., Fuchs, G., Mickel, C., Rellinghaus, B., and Nenkov, K. (2006) *Supercond. Sci. Technol.*, **19**, S472.
23. Nelson, D.R. and Vinokur, V.M. (1992) *Phys. Rev. Lett.*, **68**, 2398.
24. Abulafia, Y., Shaulov, A., Wolfus, Y., Prozorov, R., Burlachkov, I., and Yeshurun, Y. (1996) *Phys. Rev. Lett.*, **77**, 1596.
25. Yeshurun, Y., Malozemoff, A.P., and Shaulov, A. (1996) *Rev. Mod. Phys.*, **68**, 911.
26. Blatter, G., Feigel'man, M.V., Geshkenbein, V.B., Larkin, A.I., and Vinokur, V.M. (1994) *Rev. Mod. Phys.*, **66**, 1125.
27. Weinstein, R., Liu, J., Ren, Y., Sawh, R., Parks, D., Foster, C., and Obot, V. (1996) in *Proceedings 10th Anniversary HTS Workshop on Physics, Materials and Applications* (eds B. Batlogg, C.W. Chu, W.K. Chu, D.U. Gubser, and K.A. Müller), World Scientific, Sinapore, p. 625.
28. Beasley, M.R., Labusch, R., and Webb, W.W. (1969) *Phys. Rev.*, **181**, 682.
29. Fuchs, G., Gruss, S., Krabbes, G., Schätzle, P., Fink, J., Müller, K.-H., and Schultz, L. (1998) in *Advances in Superconductivity X*, vol. **2** (eds K. Osamura and I. Hirabayashi), Springer-Verlag, Tokyo, p. 847.
30. Yoshino, Y.X., Iwabuchi, A., Noto, K., Sakai, N., and Murakami, M. (2001) *Physica C*, **357-360**, 796.
31. Leenders, A., Ullrich, M., and Freyhardt, H.C. (1997) *Physica C*, **279**, 173.
32. Schätzle, P., Krabbes, G., Gruss, S., and Fuchs, G. (1999) *IEEE Trans. Appl. Supercond.*, **9**, 2022.
33. Diko, P., Fuchs, G., and Krabbes, G. (2001) *Physica C*, **363**, 60.
34. Tomita, M., Murakami, M., and Yoneda, K. (2002) *Supercond. Sci. Technol.*, **15**, 803.
35. Yusheng, H., Jiong, X., Sheng, J., Ausheng, H., and Jincag, Z. (1990) *Physica B*, **165**, 1283.
36. Cankurtaran, M., Saunders, G.A., and Goretta, K.C. (1994) *Supercond. Sci. Technol.*, **7**, 4.
37. Meingast, C., Blank, B., Bürkle, H., Obst, B., Wolf, T., and Wühl, H. (1990) *Phys. Rev. B*, **41**, 11299.
38. Zeini, B., Freimuth, A., Büchner, B., Gross, R., Kampf, A.P., Kläser, M., and Müller-Vogt, G. (1999) *Phys. Rev. Lett.*, **82**, 2175.
39. Shams, G.A., Cochrane, J.W., and Russel, G.J. (2001) *Physica C*, **363**, 243.
40. Krishana, K., Ong, N.P., Zhang, Y., Yu, Z.A., Gagnon, R., and Taileffer, L. (1999) *Phys. Rev. Lett.*, **82**, 5108.

3.2 Thin Films and Multilayers

3.2.1 Thin Film Deposition

Roger Wördenweber

3.2.1.1 Introduction

The research on *superconducting thin films* represents an ideal example for the strong interconnection between a successful basic research and its gainful economic application. On the one hand, the inspiring prospects of the applications of superconductor electronics triggered and still trigger tremendous improvements and new developments in field of thin films deposition. Prominent examples are the recent developments due to the discovery of the first high-temperature superconductor (HTS) [1] that led to significant efforts in the research and realization of textured and epitaxial HTS films. This effort is motivated largely by the potential applications of thin HTS films in a number of cryoelectronic devices and by the possibility of using epitaxial single or multilayer HTS films to study new physical properties of these unique 2D layered materials. On the other hand, the successful realization of quite a number of superconducting concepts is still hampered by the lack of superconducting films and (structures) multilayers of sufficient performance with respect to their quality, size, reproducible production, or even cost of production. Prominent examples for the latter case are electronics requiring arrays or stacks of HTS Josephson junctions or HTS-coated conductors of large length. Important existing and future applications that require high-quality superconducting films (HTS and conventional superconductors (low-temperature superconductor, LTS)) can be categorized as follows:

Cryoelectronic devices:

1) Josephson Junctions-based devices:
   (a) *SQUIDs* (superconducting quantum interference devices) represent magnetic sensors of highest sensitivity that represent the backbone of many applications ranging from nondestructive materials evaluation (NDE (nondestructive evaluation)/NDT (nondestructive testing)), geological and environmental prospecting, and neurology and medical diagnostics. For instance *magnetoencephalography* utilizes arrays of SQUIDs for mapping brain activity including basic research of perceptual and cognitive brain processes, localizing regions affected by pathology before surgical removal, determining the function of various parts of the brain as well as neurofeedback. In most cases, conventional superconductor SQUIDs (e.g., $Nb/Al/AlO_x/Nb$) are used, up to now HTS SQUIDs ($YBa_2Cu_3O_{7-\delta}$ (YBCO)) are the exception.

(b) Arrays of Josephson junctions are required for:
SQIFs (superconducting quantum interference filters) are highly sensitive magnetometers for absolute magnetic field measurements which consist of arrays of superconducting loops with Josephson junctions [2, 3]. Since the performance of SQIFs hardly depends on the parameter spread of Josephson junctions they are especially attractive for the use of HTS (e.g., YBCO).
The *Primary DC Voltage Standard* used in all major international standard laboratories is based on superconducting Josephson junction arrays incorporating several thousand Josephson junctions produced in Nb- or NbN technology (see below). Attempts of utilizing YBCO technology are underway.
RSFQ (rapid single-flux-quantum) logic represents a promising alternative to complementary metal oxide semiconductor (CMOS) devices of the future. Here, information is stored and processed in superconductor loops in form of magnetic flux quanta operating at extremely high rates of up to several hundred gigahertz. Due to the requirements of the parameter spread, the perspective technology is typically Nb based.
(c) *Terahertz spectroscopy* using Josephson junctions range from single-junction spectrometer (e.g., HTS Hilbert spectrometer [4]) to stacks of Josephson junctions that are provided by nature in strongly anisotropic HTS material (e.g., Bi–Sr–Ca–Cu–O) and can be used as terahertz emitters and terahertz receiver [5].

2) *Superconducting bolometers* are used to study extremely weak radiation in many environmental and astrophysical experiments. The thin film bolometers are typically produced in Nb or NbN technology.
3) *Microwave devices* (especially high-performance filter) made from HTS films (YBCO and $Tl_2Ba_2CaCu_2O_8$ (TBCCO)) are very promising candidates for mobile communication systems, radio astronomy, and meteorology.
4) *Fluxonic devices* represent innovative concepts for analog and digital device components. They are based on the manipulation of flux in specially patterned LTS (a-Pb, Al, and Nb) or HTS (YBCO) films [6].

Power applications:

1) HTS-coated conductors (YBCO) might provide the basis for a large number of future superconducting power applications ranging from superconducting cables, magnets, generators, engines to fault current limiters.

Considering the requirements of the above-listed applications and starting from the well-established systems to the less established systems, the state-of-art of the different superconducting films can be classified by considering the state-of-art for the deposition of superconducting layers and the preparation of one of the most important component for superconducting electronics, the Josephson junction:

1) Single layers of conventional superconductors can be produced via various techniques at high quality. This includes the classic materials like Nb, Pb,

Al, amorphous superconductors like a-Pb, a-Nb-compounds, as well as more complex compounds like NbN (especially used for superconducting bolometers).

2) Single layers of a number of high-T_c material (e.g., YBCO, Bi–Sr–Ca–Cu–O compounds, Tl–Ba–Ca–Cu–O compounds, and MgB_2) can be deposited via various techniques at high quality and are partially also commercially available. In some cases, even double-sided HTS film deposition is possible. For the more exotic HTS materials (e.g., Hg-based cuprates) or the more recently discovered superconductors (e.g., pnictides), suitable deposition technologies still have to be developed or optimized. Moreover, in case of specific application demands (e.g., large area, large thickness, large length, deposition on specific substrates like microwave suitable substrates, or metallic tapes for coated conductors) HTS deposition technology still requires improvements.

3) More complex technologies are required for the production of Josephson junctions. Single junctions can be prepared in LTS and HTS technology:
LTS Josephson junctions are typically prepared in form of trilayers. Two mainstream technologies are well established: (i) the *Nb technology* based on superconductor–insulator–superconductor (*SIS*), superconductor–insulator–normal conductor–insulator–superconductor (*SINIS*), or *SNS* structures with the superconductor layer $S = \mathrm{Nb}$, the isolating layer $I = $ Al-oxide, and the normal layer $N = $ Al, PdAu, Cu, Ti, HfTi, or multilayer structures, for example, HfTi/Nb/HfTi and (ii) the *NbN technology* in two versions: *SIS* (for 10 K) and *S*SISS** (for 7 K) with $S = \mathrm{NbN}$, $S^* = \mathrm{Nb}$, and $I = \mathrm{MgO}$.
HTS Josephson junctions are typically based on *SNS* contacts employing YBCO. Several types of *SNS* structures have been developed. They range from (i) step-edge junctions utilizing the different epitaxial growth directions of a HTS layer at an artificial step in the single-crystalline substrate, (ii) trilayers using noble metal interlayers (e.g., Au), (iii) ramp-type structures with semiconducting oxide interlayers (e.g., $PrBa_2Cu_3O_7$), (iv) grain boundary junctions generated via epitaxial growth on bicrystal substrates, (v) weak links generated by local damage (e.g., via irradiation damage using electrons or ions) to (vi) weak links generated by constriction of the superconductor (e.g., micro- or nano-contacts). All technologies are suitable for the preparation of single or a few Josephson contacts. Nevertheless, they suffer from drawbacks like reproducibility, restriction in the positioning of the junctions (e.g., bicrystal junctions), or robustness (e.g., microbridges).

4) For the preparation of devices with larger numbers of Josephson junctions (e.g., arrays for RSFQ, superconducting Qbits or voltage standards) reproducibility and especially a sufficiently small spread of the junction parameters are required. Up to now this is only granted for the two LTS technology, the *Nb technology* and the *NbN technology*. Foundries for these LTS technologies are well established, for instance for the *Nb technology* at Hypres (US), TRW (US), NEC (J), FLUXONICS (EU), and for the *NbN technology* at TRW (US).

Nowadays extremely high complexities have been achieved in LTS Josephson devices, for example, (i) RSFQ-RAMs consisting of up to 40 000 externally shunted Nb/Al-oxide junctions (NEC (J)), (ii) programmable voltage standards with up to 30 000 Nb/PdAu/Nb junctions (NIST (US)), 70 000 large-area SINIS junctions (PTB (Ger)), or 70 000 SNS junctions (Supracon/IPHT (Ger)), or (iii) A-to-D and D-to-A converters with a complexity of about 2000 junctions (Hypres (US)) with a performance that challenges the best semiconductor converters [7].

In case of HTS, large numbers of Josephson junctions can be obtained in case of (i) bicrystal junctions which allows for a row of junctions at the junction of the two crystals (in analogy tricrystals would offer the possibility to produce two rows of Josephson junctions) and (ii) stacks of Josephson junctions that are naturally generated in highly anisotropic HTS material (e.g., $Bi_2Sr_2CaCu_2O_8$ and $Ta_2Ba_2CaCu_2O_x$). Whereas junction array with several 100 junctions (e.g., 166 junctions in [8]) are feasible for bicrystal junctions, $Bi_2Sr_2CaCu_2O_8$ single crystal consists of stacks of ~670 intrinsic junctions per micrometer thickness. Both technologies are well developed, whereas a reproducible technology for the fabrication of 2D arrays of HTS Josephson junction with small parameter spread is still missing.

In this contribution the deposition of superconducting thin films and film systems is reviewed. Since LTS deposition is well developed and HTS deposition is more demanding, the contribution will concentrate more on HTS issues. Starting with a sketch of the requirements of superconducting material and substrates, brief introductions to the most commonly used deposition techniques, a number of important issues and problems related to a reproducible fabrication of high-quality thin films for basic research and technical applications are addressed.

3.2.1.1.1 **Material Requirements**

The most commonly examined thin film superconductors are for instance the elements Nb (T_c = 9.2 K), Pb (T_c = 7.2 K), Al (T_c = 1.2 K), LTS compounds NbN ($T_c \approx 16$ K), and HTS material with superconducting transition temperatures T_c partially above the temperature of liquid nitrogen, for example, $ReBa_2Cu_3O_{7-\delta}$ (Re = Y or rare earth element, typical transition temperature $T_c = 90-95$ K), $Bi_2Sr_2CaCu_2O_{8-\delta}$ ($T_c = 85-90$ K), $Bi_2Sr_2Ca_2Cu_3O_{10}$ ($T_c = 110-120$ K), $Tl_2Ba_2CaCu_2O_8$ ($T_c = 108-110$ K), $Tl_2Ba_2Ca_2Cu_3O_{10}$ ($T_c = 125-127$ K), $HgBa_2CaCu_2O_{6+\delta}$ ($T_c \leq 128$ K), $HgBa_2Ca_2Cu_3O_8$ ($T_c \approx 134$ K), MgB_2 ($T_c \approx 39$ K), and, lately, the so-called *ferropnictide superconductors* (e.g., $Sm_{0.95}La_{0.05}O_{0.85}F_{0.15}FeAs$ with $T_c = 57.3$ K) [9]. For HTS cryoelectronic devices, thin films are mainly grown from $YBa_2Cu_3O_{7-\delta}$ (YBCO) material. Reasons for this choice are among others the phase stability, high crystalline quality, high flux pinning level, low surface resistance, the possibility of using a single deposition step with *in situ* oxidation, and the absence of poisonous components that are present in Tl-, Hg-, or As-containing HTS films.

In general, the improvement of the superconducting properties (e.g., increase of T_c, B_{c2}, and energy gap ΔE) of HTS compared to LTS has to be paid for by a

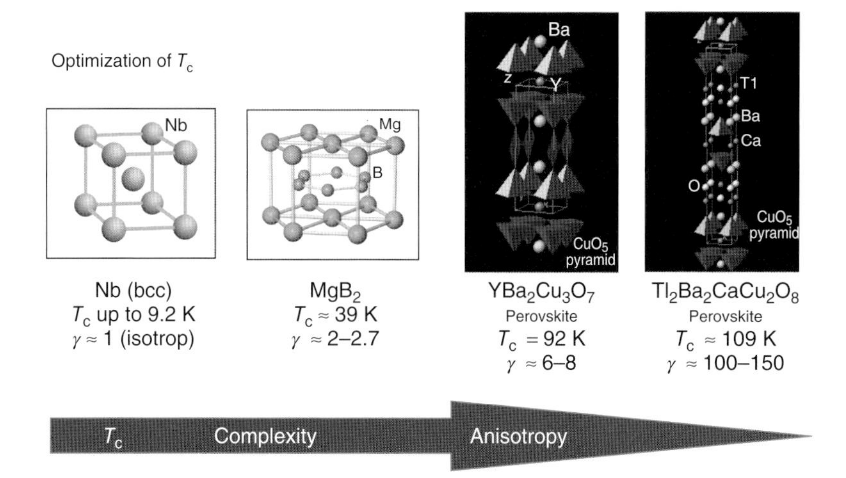

Figure 3.2.1.1 The improvement of the superconducting properties (e.g., increase of T_c and energy gap) is achieved via an enhancement of complexity (structural and stoichiometric) and anisotropy of the superconductor.

significant enhancement of (i) the number of elements, (ii) structural complexity, and (iii) anisotropy (Figure 3.2.1.1). All of these features of the HTS lead to considerable demands on the deposition technology.

First, whereas LTS material like Nb or NbN are generally isotropic superconductors, the quasi two-dimensional (2D) nature of the oxide superconductors related to their layered structure leads to a large anisotropy in nearly all parameters. Anisotropy factors of $\gamma = \xi_{ab}/\xi_c = \lambda_c/\lambda_{ab} \approx 6-8$ in YBCO, $\gamma \approx 55-122$ in Bi compounds, and $\gamma \approx 100-150$ in Tl compounds are reported for the anisotropy between the crystallographic c and $a-b$ directions [10–13]. Furthermore, extremely small coherence lengths are determined for the HTS. For Bi_{2212} values of $\xi_c = 0.02-0.04$ nm and $\xi_{ab} = 2-2.5$ nm, for YBCO $\xi_c = 0.3-0.5$ nm and $\xi_{ab} = 2-3$ nm are measured, which, furthermore, strongly depend on sample quality and oxygen content [14]. Therefore, local variations in the sample properties on the scale of ξ will automatically result in a spatial variation of the superconducting properties. Due to the extremely small coherence length along the crystallographic c-axis, only the current directions along the $a-b$ direction (i.e., along the CuO-planes) can be used in most applications. Since even misalignments in the $a-b$ plane lead to grain boundaries that strongly affect the critical properties (see Figure 3.2.1.2), a perfect biaxial c-axis-orientated epitaxial film growth including the avoidance of antiphase boundaries is necessary for most applications.

Second, due to the large number of elements the superconducting properties of HTS strongly depend upon variations in stoichiometry, for example, in the oxygen concentration. Due to the complex crystallographic structure and the small coherence length, extreme requirements are imposed upon the uniformity and stability of the deposition process.

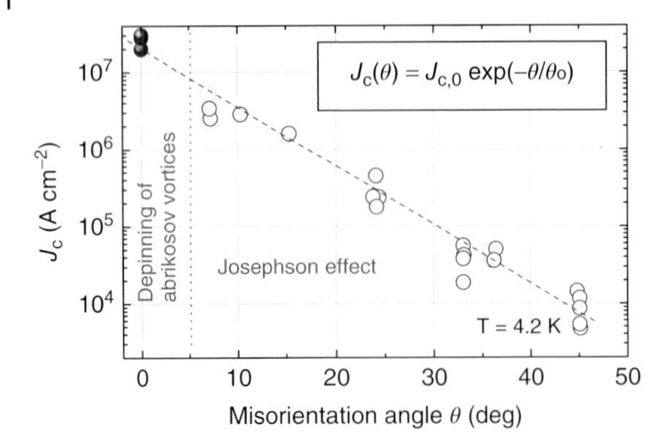

Figure 3.2.1.2 Critical current density in YBCO bicrystals as a function of misorientation angle at 4.2 K [15, 16]. Open symbols represent J_c values of epitaxial films on single crystals.

Finally, in a number of applications HTS films are grown on less ideal carriers, for example, single crystals with large lattice misfit, or, in case of coated conductors, tapes with crystalline buffer layers. In these cases, polycrystalline or textured HTS films are obtained in which the intergranular properties (i.e., grain boundaries) clearly dominate the superconducting properties like the transport current density J_c or microwave surface resistances R_s. The effect the grain boundaries upon the J_c is demonstrated in Figure 3.2.1.2. On the one hand, J_c decreases with increasing misorientation angle. On the other hand, at large angles the grain boundary imposes a Josephson-type behavior that can be utilized for the engineering of Josephson contacts, for example, by deposition of HTS thin films on bicrystalline substrates with perfectly defined grain boundaries [4].

In conclusion, whereas LTS deposition (e.g., *Nb technology* and *NbN technology*) requires high standards with respect to the avoidance of impurities (e.g., small amounts of oxygen contamination lead to strong reduction of T_c in Nb films), the extraordinary properties of HTS material – the large anisotropy, small coherence length, and strong dependence of the superconducting parameters on local modifications – set extremely high requirements for the deposition of HTS thin films which are (i) perfect stoichiometry and structure, (ii) perfect biaxial epitaxial orientation, (iii) avoidance of defects that hamper the superconducting properties, and (iv) avoidance of grain boundaries.

3.2.1.1.2 **Substrate Requirements**

The choice of the substrate material is of vital importance for the performance of a superconducting device. For example, CMOS-compatible cryogenic devices that are based on Si technology, microwave applications require substrates with low microwave losses (typically tan $\delta < 10^{-5}$), whereas coated conductors are deposited on metallic tapes. As a consequence, large lattice misfits or even noncrystalline carriers with complex crystalline buffer systems have to be taken into account (see Figure 3.2.1.3). Especially for HTS films, the impact of the substrate has to be taken into account for the development of a reliable deposition

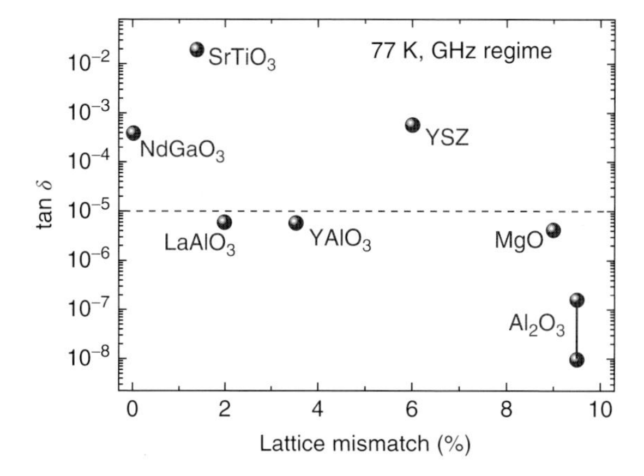

Figure 3.2.1.3 Substrate material for most YBCO microwave applications should exhibit low losses of tan $\delta < 10^{-5}$. Ideal candidates (e.g., MgO and Al_2O_3) suffer from large lattice misfit $(a_{film} - a_{substrate})/a_{film}$, where a represents the in-plane lattice parameter.

technology. In case of epitaxial growth (necessary for most HTS films), the basic requirements for the substrates (including buffer layer) can be summarized as follows:

- reasonable crystallographic lattice match and similar thermal expansion coefficient between film and substrate,
- no chemical interaction at the interface between film and substrate, and
- electronic and physical properties (ranging from conductivity, permittivity, microwave losses, size to robustness) of the substrate suitable for film deposition and for the envisioned application.

In case of HTS deposition, we can distinguish between different classes of substrates:

1) "perfectly compatible" substrates onto which HTS material can be deposited without buffer layer,
2) "less compatible" substrates that require an optional buffer layer, and
3) "noncompatible" substrate materials, which have to be covered with an epitaxial buffer layer prior to the deposition of the HTS film due to large lattice mismatch and/or chemical interaction between substrate and HTS material or due to missing in-plane orientation (e.g., for deposition of biaxial oriented YBCO on polycrystalline substrates), respectively.

Deposition on to compatible substrates is generally easier. Typical candidates for the deposition of YBCO are $LaAlO_3$, $SrTiO_3$, and MgO. Nevertheless, buffer layers can be added, for example, CeO_2 buffer layer on $LaAlO_3$ substrates reduce the probability of a-axis growth. Important properties of different materials that represent suitable substrate materials for HTS deposition (especially YBCO) as well as buffer or interlayer are summarized in Table 3.2.1.1.

Table 3.2.1.1 Properties of crystalline substrate materials suitable for the preparation of YBCO thin films (metallic tapes for coated conductors are not listed).

Substrate materials	Dielectrical properties		Thermal expansion coefficient (10^{-6} °C^{-1})	Melting temperature (°C)	Available substrate size (mm)	Crystalline properties	
	ε	tan δ				Lattice misfit to YBCO (%)	Twinning
$LaAlO_3$	23–24	8×10^{-6} -3×10^{-4}	10–13	2110	Ø100	0.9	Yes
$LaGaO_3$	25		9.7–12.1	1715	Ø40	−1	Yes
$(LaAlO_3)_{0.3}$ $(Sr_2AlTaO_6)_{0.7}$ (LSAT)	22.7	2×10^{-4}	8	1840	Ø50	−0.4	No
MgO	9.6–10	6.2×10^{-6}	12–14	2790	Ø75	−9	No
$NdGaO_3$	23	3×10^{-3}	9–11	1600	Ø50	−0.2	No
$PrGaO_3$	24		7–8	1680	Ø10	−0.2	Yes
Si	16 (3.9 for SiO_2)		2.6	1420	Ø300	0.4	No
$SrTiO_3$	320	$>10^{-2}$	9.4	2080	Ø30–50	−1.4	No
$SrLaAlO_4$	17	8×10^{-4}	8–17*	1650	Ø35	2.7	No
$SrLaGaO_4$	22	5×10^{-5}	10–18*	1520	Ø35	0.4	No
$YAlO_3$	16–20	10^{-5}	2.3–8.7	1870	Ø50	3.8	No
Y : ZrO_2 (YSZ)	25–30	$>6\times10^{-4}$	7–9	2780	Ø100	6	No
$-Al_2O_3$ (sapphire)	9.4–11.6* anisotropic	10^{-8}	5.6–8.4*	2053	Ø200	−(6–11) for *r*-cut −18 for *m*-cut	No
YSZ-buffered *r*-cut sapphire	—	—	—	>2550	Ø200	6	No
CeO_2-buffered *r*-cut sapphire	—	—	—	>2550	Ø200	0.7	No

The lattice misfit $(a_{film} - a_{substrate})/a_{film}$ is normalized with respect to *c*-axis-oriented YBCO using an average in-plane parameter for YBCO, the thermal expansion coefficient of *c*-axis-oriented YBCO is ~$(8.8-11)\times10^{-6}$ K^{-1} [17], and YSZ refers to yttrium stabilized ZrO_2 with ~9 mol% of Y_2O_3. Nonisotropic behavior is marked by an asterisk.

In contrast, the deposition of HTS material onto a number of technically interesting substrate materials (e.g., Si, Al_2O_3, or metal tapes for HTS-coated conductors) requires a previous coating with an adequate buffer layer that enables epitaxial growth (by reducing the lattice mismatch between substrate and HTS material or even providing the crystalline structure for epitaxy in case of coated conductors) and/or presents a barrier against chemical interdiffusion between substrate and HTS material.

As an example, one of the most interesting substrate candidates for microwave application of YBCO is $(1\underline{1}02)$ oriented Al_2O_3 (*r*-cut sapphire), which possesses high crystalline perfection, mechanical strength, and low dielectric permittivity ($\varepsilon \approx 10$) and losses ($\tan \delta$ (77 K, 10 GHz) $\approx 10^{-7} - 10^{-8}$) [18]. However, the complex crystalline structure of sapphire, which is comprised of consecutive planes of oxygen and aluminum hexagons with each third site vacant, provides two planes of rectangular (*m* or $(10\underline{1}0)$ plane) or pseudo-rectangular (*r* or $(1\underline{1}02)$ plane) surface structure. Both planes possess a rather poor lattice match to the rectangular basal plane of the *c*-oriented YBCO structure. Due to smaller mismatch, *r*-cut sapphire is preferentially used for the deposition of YBCO films. The lattice mismatch is accompanied by chemical interaction between YBCO and Al_2O_3 taking place at elevated deposition temperatures [19, 20]. This yields substantial diffusion of Al into the HTS film and formation of an uncontrolled $BaAl_2O_4$ interfacial layer.

Two different approaches have been considered for the choice of buffer layers: (i) In one approach, material is chosen which is similar to YBCO with respect to chemical and structural properties. One of the few promising candidates for such a buffer is the semiconducting perovskite $PrBa_2Cu_3O_7$ (PBCO) [19, 21]. PBCO layers are also good candidates for YBCO multilayers (see, e.g., [22]). The specific resistance can be improved by partial substitution of Cu by Ga without loss of chemical and structural compatibility. Although the Al diffusion into the YBCO film is blocked by the PBCO buffer layer, YBCO films on PBCO/sapphire exhibit higher R_s and lower J_c values than observed for YBCO films of reasonable quality on compatible substrate materials [23]. (ii) In the second approach, oxides with a cubic structure and lattice parameters comparable to the diagonal of both *r*-plane sapphire and (001) plane orthorhombic YBCO are chosen. Among a large number of candidates MgO, yttria-stabilized zirconia (YSZ) (ZrO_2 stabilized with ~9 mol% of Y_2O_3), and CeO_2 are the most attractive candidates. Whereas the lattice parameter of MgO is closer to that of sapphire, those of YSZ and CeO_2 match that of YBCO. CeO_2 has proven to be one of the most effective buffer layers due to its favorable thin film growth characteristics, minimal chemical interaction, and good lattice match with most HTS materials. For instance, (001) CeO_2 buffer layers reduce the lattice mismatch between $(1\underline{1}02)$ sapphire and (001) YBCO and provide a sufficient barrier against diffusion of Al into the YBCO film [24]. (001) CeO_2 itself has a larger lattice mismatch with respect to $(1\underline{1}02)$ sapphire. This can cause insufficient structural perfection of the CeO_2 and YBCO layers despite the very high structural perfection of sapphire substrates. In general, the lattice mismatch between sapphire, buffer, and ceramic film leads to large stress-induced strain in the film. As a consequence, defects will develop and, finally, for films that exceed a critical thickness cracks appear [25, 26]. This will be discussed in Section 3.2. Nevertheless, YBCO films with high quality with respect to morphology, critical properties, and microwave properties can be grown onto technical substrates like sapphire as long as the critical thickness is not surpassed [27].

Another example for the need of buffer layers is given by the HTS-coated conductors. Coated conductors are based on metallic tapes (e.g., hastelloy, NiCrFe, and Ni-alloy). The crystalline structure is generated in a complex oxide buffer system (e.g., $La_2Zr_2O_7$ plus CeO_2 or Nb-doped titanate $Sr_{1-x}(Ca,Ba)_xTi_{1-y} - Nb_yO_3$) onto which the HTS film is deposited. The HTS is normally protected by an additional metallic shunt, typically an Au or Ag layer. Several methods have been developed to obtain biaxially textured buffer layers suitable for high-performance YBCO films. These are ion-beam-assisted deposition (IBAD), rolling-assisted biaxially textured substrate (RABiTS) process, and inclined substrate deposition (ISD). For more details on these different deposition technologies for HTS-coated conductors one should refer to overview articles like [28]. One of the main obstacles to the manufacture of commercial lengths of YBCO wire or tapes has been the phenomenon of weak links, that is, grain boundaries formed by any type of structural misalignment of neighboring YBCO grains are known to form obstacles to current flow. Therefore HTS deposition on metallic tapes require a careful alignment of the grains, low-angle boundaries between superconducting YBCO grains allow more current to flow. For instance, below a critical in-plane misalignment angle of 4°, the critical current density approaches that of YBCO films grown on single crystals (see Figure 3.2.1.2).

3.2.1.2 Deposition Techniques

There exist a large variety of techniques for growing superconducting films and multilayers ranging from physical vapor deposition (PVD) to all chemical deposition processes. In all cases, the actual process of deposition can be divided into three steps that are sketched in Figure 3.2.1.4:

1) The *particle source* can be a stoichiometric compound (e.g., for sputtering and pulsed laser deposition (PLD)), an assembly of different components (e.g., for evaporation), a mixture of organic molecules (e.g., for metal-organic vapour-phase epitaxy (MOCVD)), or a chemical solution (e.g., for chemical solution deposition (CSD) or metal-organiccomposition (MOD)).
2) The *particle transport* is initiated by phonons (evaporation), photons (PLD), or ion bombardment (sputtering) in the case of PVD deposition. In these processes, the initial particle energy at the source, the scatter events between particles and process gas, and the resulting energy of the particles impinging on the substrates play an important role. This will be discussed in detail at the end of section 3.2.1.2.1, Comparison of different PVD techniques. For chemical vapor deposition (CVD), CSD, and MOD, the particle transport is established by evaporation or spray-, spin-, or dip-coating.

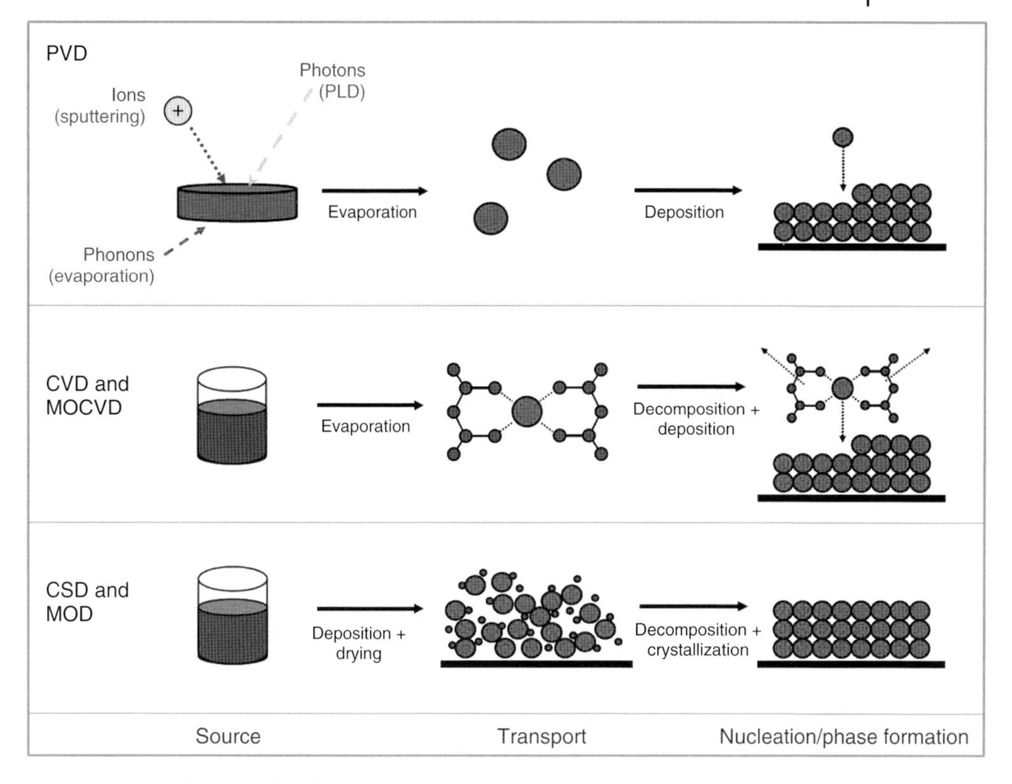

Figure 3.2.1.4 Schematic sketch of the process steps of PVD, (MO)CVD, and CSD/MOD deposition technologies.

3) Finally, the *nucleation, phase formation, and film growth* represent the most important step of the deposition process. During this step, the actual formation of the film takes place which strongly depends on all process parameters (e.g., substrate temperature, process pressure, and density of particles) and, obviously, on the configuration of the oncoming particles (i.e., atoms, (organic) molecules, and chemical solution).

Whereas processes (1) and (2) are reasonably well understood, the actual nucleation, phase formation, and film growth is difficult to analyze, theoretically as well as experimentally. Even simple parameters, like the actual temperature at the surface of the substrate, are difficult to determine. Therefore, this part of the process is object to theoretical models and experimental estimates. Nevertheless, there is a reasonable good understanding on the basic principles of this part of the process.

In the following, different deposition techniques are introduced in detail and the three different deposition steps are illustrated for HTS thin film deposition.

3.2.1.2.1 **PVD Techniques**

PVD involves purely physical processes, it describes all methods to deposit films via condensation of a vaporized form of the material onto a surface. It includes evaporation, PLD, or plasma sputter bombardment (sputter technology) at different regimes of pressure and substrate temperature.

Thermal Co-Evaporation and MBE The thermal evaporation is based on particle sources in form of thermal boats (typically resistively heated), e-guns, or Knudsen cells. As a consequence and in contrast to sputter and PLD technologies, in evaporation or molecular beam epitaxy (MBE)

1) each material is supplied individually from metallic sources and
2) small gas pressures (especially in case of reactive components like oxygen) have to be chosen.

In principle, this approach is most flexible since the stoichiometry can be easily changed. Examination of stoichiometry-dependent effects of the superconducting film or the preparation of artificial superlattices seems to be easier with this method. However, the difficulty lies in the fact that calibration and rate control is rather complex and difficult especially if reactive gas (e.g., oxygen) is involved. Therefore, (i) a very accurate rate control for all components has to be guaranteed and (ii) problems caused by reactive gas and large gas pressures have to be solved.

For example in case of YBCO deposition, the use of an active rate control utilizing (collimated) quartz-crystal monitors or atomic absorption monitors has led to stoichiometry control down to ~1% [29]. Moreover, the oxygen background pressure at the sources should not exceed $\sim 10^{-4}$ mbar during deposition. Therefore, reactive oxygen sources like atomic oxygen, ozone, or NO_2 can be used to increase the oxidation efficiency and allow the formation of the superconducting phase at a lower oxygen pressure. Alternatively, differentially pumped evaporation devices have been developed that provide low pressure at the sources and high pressure at the substrate. This can among others be established by rotation of the substrate in a heater which is partial open-facing the evaporation sources (typically 650 °C at 10^{-5} mbar for YBCO) and partially closed forming an "oxygen pocket" with a 100 times higher oxygen partial pressure [30] (see Figure 3.2.1.5). Thus, in the closed part of the heater the oxygen partial pressure is increased providing the oxygenation of each freshly deposited layer. This method has lately been adopted for the growth of high-quality large-area MgB_2 thin films ($T_c \approx 38-39$ K) using the "heater pocket" for the Mg incorporation [31].

Alternatively, both problems are nicely solved in differentially pumped MBE systems which however have the disadvantage to be very complex and expensive. For instance using an atomic layer-by-layer MBE, a large variety of heterostructures containing Bi- and Dy-based cuprates as well as other complex oxides have been prepared [32], ultrathin films of $La_{2-x}Sr_xCuO_{4+y}$ have been grown [33], and novel superconductors have been designed [34].

Pulsed Laser Deposition Technique and Laser MBE The PLD technology represents a special type of evaporation that has experienced an enormous boost during the development of HTS deposition. Due to its flexibility, it has become an important technique to fabricate layer of novel compounds. It is used for the deposition of

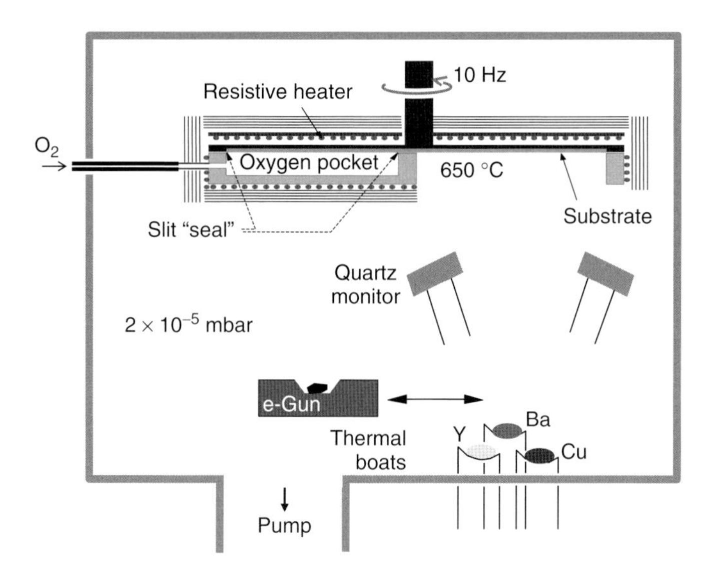

Figure 3.2.1.5 Schematic sketch of a thermal co-evaporation system [30]. An oxidation pocket encloses part of the heater. The substrate is rotated continuously through the pocket, enabling intermittent deposition and oxygenation and providing better uniformity during evaporation.

HTS, ferroelectric, and ferromagnetic oxide materials. Film deposition by PLD is based on the irradiation of a single target by a focused laser beam (e.g., excimer ($\lambda = 308$ and 248 nm) or Nd:YAG (355 nm); energy density 1–3 J cm^{-2} shot^{-1}; frequency of several hertz). The laser beam removes material from the target and this material is transferred to the substrate (Figure 3.2.1.6). During ablation, a luminous cloud (plume) is formed along the normal of the target. Due to the short wavelength, the photons of laser beam interact only with the free electrons of the target material. The subsequent electron–phonon interaction leads to a sudden increase of the local temperature, surface or subsurface vaporization (depending on energy of the laser beam) and an explosive removal of material. Laser-induced thermal evaporation or congruent PLD takes place for lower and high ($>10^7-10^8$ W cm^{-2}) energy densities, respectively. These ablation processes can be explained in various models (see, e.g., [35]):

1) shock wave caused by rapid surface evaporation,
2) subsurface explosion caused by rapid-evaporation-induced cooling of the surface,
3) formation of Knudsen layer due to collision of ejected atoms and
4) superheating of the surface by suppression due to the recoil pressure for evaporated material.

PLD deposition has a number of advantages. In contrast to most other deposition processes, the energy at the target can be controlled independent of the process pressure and gas mixture. Thus, reactive processes can easily be conducted and stoichiometric deposition for high-energy densities, high rates, and

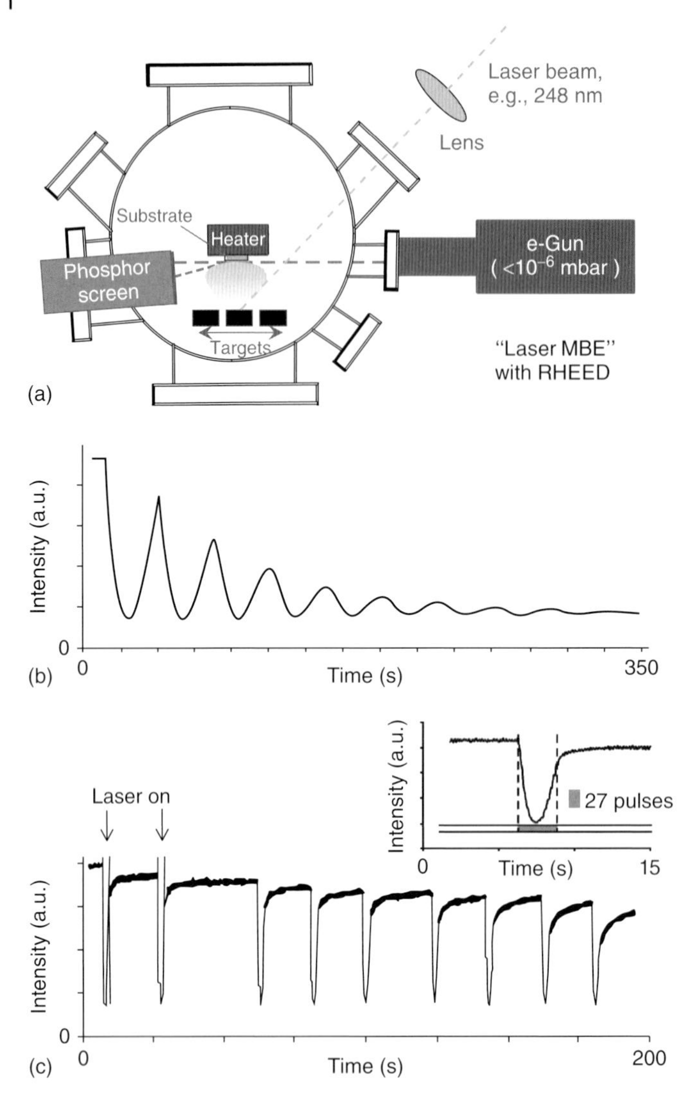

Figure 3.2.1.6 (a) Schematic sketch of a L-MBE with *in situ* RHEED characterization of a continuous deposition (b) and layer-by-layer (c) growth of oxide films according to [36].

high process flexibility are characteristic features of PLD. However, traditionally PLD is limited to small areas (typically <1 cm^2), the pulsed deposition has large impact upon the morphology of the sample and usually droplets of submicrometer size (boulders) are formed on the surface of the deposited film. Via off-axis PLD and rotation of the substrate, films can be deposited on substrates up to 2″ in diameter; utilizing rotation and translation in on-axis configuration HTS films have been deposited even onto 3″ wafers [37]. Much effort has been invested into eliminating the boulders on HTS-film surfaces including optimizing the laser power [38] and masking the ablation plume.

Finally, sophisticated PLD systems have been developed for the deposition of multilayer, complex, or optimized film systems by combining the advantages of PLD and MBE (see for instance [39]). In the so-called *laser-pulsed laser deposition* (L-PLD) (i) a "target carrousel" allows the subsequent deposition of different components or systems and (ii) an *in situ* characterization such as low-energy electron diffraction (LEED) or reflection high-energy electron diffraction (RHEED) enables a perfectly controlled layer-by-layer film growth. Figure 3.2.1.6 shows a L-PLD and the layer-by-layer growth of an oxide film.

Sputter Techniques Sputter deposition has the deserved reputation as being the technique for preparing thin films of alloys and complex materials for industrial application. It is based on a discharge involving free ions and electrons in a gaseous atmosphere (see Figure 3.2.1.7). Three different kinds of discharges can be distinguished: arcs, glow discharges, and dark discharges. They can be distinguished

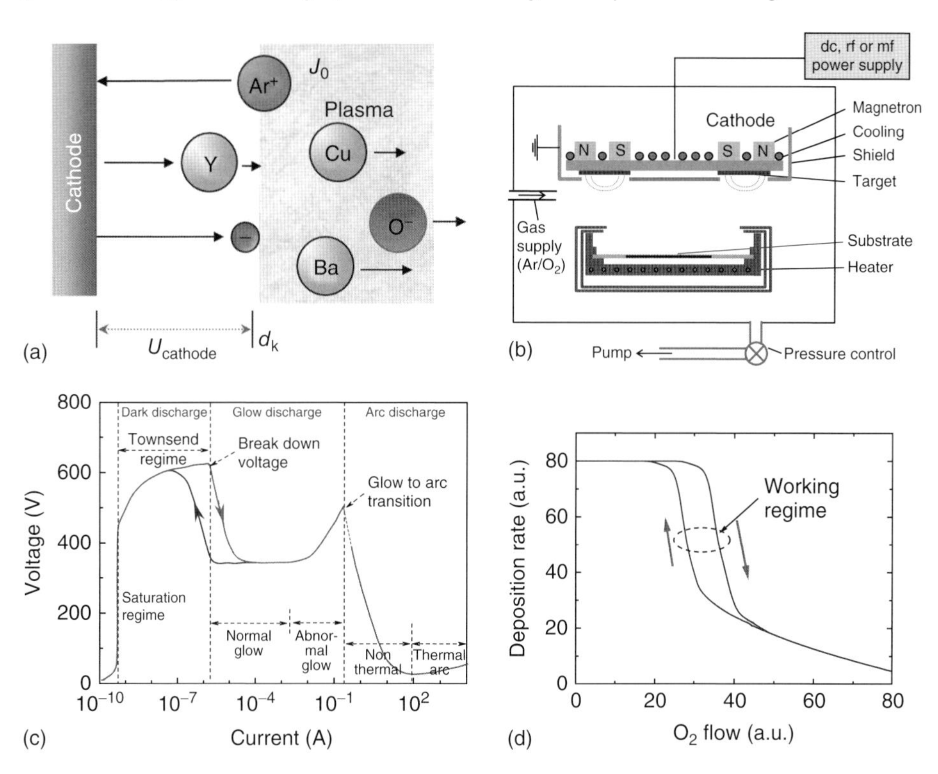

Figure 3.2.1.7 Schematic sketches of the sputter deposition: (a) plasma generation and particle removal at the target, (b) sketch of a magnetron sputter device for HTS deposition, (c) voltage–current characteristic of the discharge, and (d) deposition rate as a function of reactive gas content during the sputter process.

by their luminescence but also by their current–voltage characteristic, the current density, and breakdown voltage (Figure 3.2.1.7c). These main characteristics depend on the geometry of the electrodes and the recipient, the process gas, and the electrode material.

Dark discharge: Under the force of the electrical field, ions and electrons migrate to the electrodes producing a weak electric current. The current increases with increasing voltage in a highly nonlinear way. The region of exponential current increase (Townsend discharge) ends in the electrical breakdown when the ions reaching the cathode start to possess sufficient energy to generate secondary electrons. The breakdown voltage for a particular gas and electrode material depends on the product of the pressure and the distance between the electrodes, it is expressed by Paschen's law [40].

Glow discharge: The regime of glow discharge owes its name to the plasma gas emitting light. In this regime a stable plasma is present. The secondary electrons will generate ions in the plasma that are accelerated toward the target. At the target electrons and particles are removed (Figure 3.2.1.7a). The latter will form the film at the substrate placed opposite to the target (Figure 3.2.1.7b). In the normal glow regime an increase of the energy leads to an increase of the current due to increase of the plasma area. A complete coverage of the target with the plasma is obtained in the regime of abnormal discharge. The abnormal discharge represents the regime of homogeneous sputter deposition where the energy of the particles can be modified by modifying the potential at the cathode. The cathode fall potential increases rapidly with current, and the dark space shrinks.

Arc discharge: The bombardment with ions ultimately heats the cathode causing thermionic emission. Once the cathode is hot enough to emit electrons thermionically, the discharge will change to the arc regime. Arcing will lead to destruction of the target. In spectrochemistry arc discharges are used in spark optical emission spectroscopy (OES) and DC arc spectroscopy.

Sputter deposition is the natural method to deposit LTS material, especially large-scale deposition. It was also tested when HTS materials were first discovered. However, the problem of large particle energies leading to defects and resputtering (see Section 2.1.4) was already recognized in this class of materials some 18 years before the discovery of YBCO for $Ba(PbBi)O_3$ deposition [41] and has been thoroughly investigated later [42]. A method to avoid this problem is given by the thermalization of the energetic species (mainly the negatively charged oxygen that is accelerated in the electric field) either by working at very high pressure (which also provides a particularly rich oxygen environment) or by "off-axis" configuration at a modestly high sputtering pressure which forces all atoms emanating from the target to undergo a few energy-reducing collisions before approaching the substrate.

Several different combinations and techniques using fully oxygenated stoichiometric targets are established for HTS deposition. These are among others (i) off-axis techniques using a cylindrical magnetron [43], (ii) facing magnetrons [44] or a single magnetron [45], (iii) on-axis techniques using nonmagnetron sputtering at very high pressure [4], or (iv) magnetron sputtering at high pressure ([46], see Figure 3.2.1.7b). Furthermore, the cathodes can be charged with DC, radio frequency (rf, 13.56 MHz) or mf (medium frequency ranging from 20 to 200 kHz) power or combination of these [47].

Generally, sputter techniques are highly reproducible. Due to the use of a single stoichiometric target the process is compatible with high oxygen pressure (which is important for HTS deposition), and calibration and rate control are readily obtained. Homogenous composition can be obtained over a rather large area (typically 1–2″ for off-axis sputtering and 4″ and more for on-axis sputtering). However, drawbacks are the relatively small deposition rate (ranging between 0.1 and 20 $\mathrm{nm\,min^{-1}}$ for nonmagnetron sputtering and magnetron sputtering, respectively), the presence of a plasma and high deposition pressures that hamper the use of *in situ* characterization such as LEED or RHEED, and the tendency of an instable plasma in case of reactive sputtering (e.g., for HTS deposition using $Ar-O_2$ process gas, see Figure 3.2.1.7d).

Comparison of the Different PVD Techniques PVD technologies are vacuum technologies. The particles are removed from the source via heating (evaporation and PLD) or ion bombardment (sputtering). The resulting energy of the particles at the substrate is one of the most important parameters that determines the nucleation and growth of resulting film.

At the source the particle energy is described by the Thomson relation $N(E) \propto E/(E+E_B)^3$ [48, 49] and Maxwell–Boltzmann distribution $N(E) \propto E \cdot \exp(-E/kT_\ell)$ for sputtering and evaporation, respectively. The resulting average particle energy is given by ½ of the binding energy E_B for sputtering and the temperature T_ℓ at the source for evaporation, PLD, and CVD. Table 3.2.1.2 displays typical initial (i.e., at the source) particle energies that are characteristic for the different deposition technologies. On the one hand, the smallest energy is associated with CVD followed by evaporation. This is one of the reasons for using these processes for the fabrication of defect-free layers, for example, in semiconductor industry. On the other hand, the largest particle energy is provided in the sputter technologies. Due to the large particle energy, sputter-deposited films have a good mechanical stability and adhesion, but they are usually loaded with

Table 3.2.1.2 Typical energy of particles leaving the "source" for the different PVD technologies.

	Typical particle energy (eV)
Sputtering	6–8
PLD	1–2
evaporation	0.06–0.08
CVD	0.005–0.01

defects. This is of advantage for applications for which the mechanical properties and film adhesion plays a major role, for example, mechanical hardening, optical coating, or CD fabrication. However, in case of structural perfect superconductor films (especially HTS films) defect generation due to large energy represents a problem for sputtering and PLD. A solution of this problem is given by the so-called *thermalization of the particles*, that is, the reduction of the particle energy due to interaction of the particles with the process gas during the particle transport from the source to the substrate. This is one of the reasons why HTS film deposition typically takes place at elevated process pressure.

The transport process and, thus, the thermalization of particles play an important role in high-pressure deposition. It can be simulated via *Monte Carlo* method that describes the statistical energy and angular distributions of particles leaving the source [50]. The initial angular distribution of particles is peaked around the normal to the surface of the source according to $\mathrm{d}N(\theta)/\mathrm{d}\theta \propto \cos^n(\theta)$ with $n = 1$ for evaporation and sputtering, $n = 5-10$ for PLD (explosion-like removal of material in an overcritical liquid state) and θ defining the angle with respect to the normal. In the simulation, the scatter events are described via a quasi-hard-sphere (QHS) model for the interactions between atom particles (sketch Figure 3.2.1.8a). Atoms leaving the target are characterized by their (randomized) (i) initial energy E_0 and (ii) direction of motion. During the transport the atoms move straight over a (randomized) distance δr before colliding with a process gas atom. The motion of the gas atoms is chaotic, that is, position and directions of motion of the colliding atoms are randomized. The scattering process is repeated until the atom reaches either a surface of the chamber or the substrate. The microscopic cross-section σ_{QHS} of the elastic interaction of two particles within the framework of the QHS model depends on the energy E_{C} of their relative motion:

$$\sigma_{\mathrm{QHS}} = \pi \cdot r_{\mathrm{min}}^2(E_{\mathrm{C}}) \tag{3.2.1.1}$$

where r_{min} represents the minimum distance between two atomic particles in case of "head-on" collision, that is, impact parameter $b = 0$ (see Figure 3.2.1.8a). Using Born–Mayer interatomic interaction potential [51] for interatomic distances r, the expression for r_{min} is given by

$$r_{\mathrm{min}} = -b_{\mathrm{BM}}(Z_1, Z_2) \cdot \ln \frac{E_{\mathrm{C}}}{A_{BM}(Z_1 \cdot Z_2)^{\frac{3}{4}}} \tag{3.2.1.2}$$

Here, Z_1 and Z_2 represent the atomic numbers, and A_{BM} and b_{BM} the different constants of the Born–Mayer expression [52, 51]. Accordingly, the mean free path λ_{QHS} of atomic particles in a gas medium is given by

$$\lambda_{\mathrm{QHS}} = \frac{1}{N\pi \cdot r_{\mathrm{min}}^2(E_{\mathrm{C}})} \tag{3.2.1.3}$$

where N is the number density of the gas atoms. The mean free path can be evaluated for any deposition process according to Eq. (3.2.1.3), it can be compared to the simplified standard expression for the mean free path of a particle in atmosphere

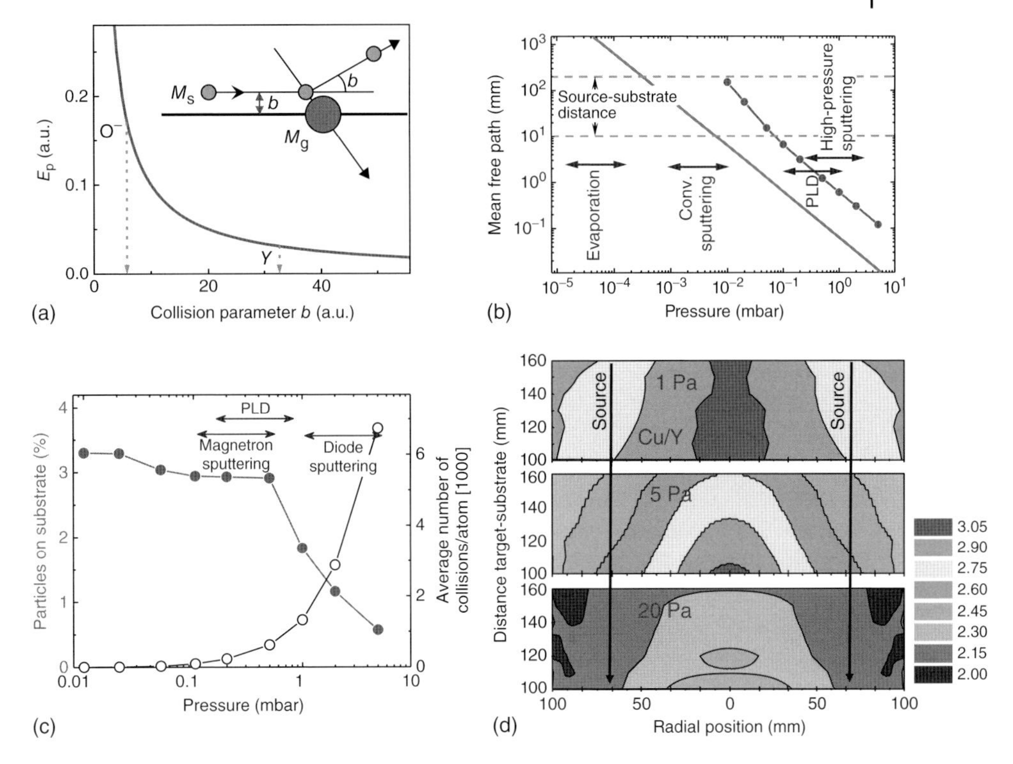

Figure 3.2.1.8 Sketch and data of the simulated particle flux from the source to the substrate according to [50]. In (a) the relation between particle energy and collision parameter b (Eq. (3.2.1.1)) that is used in the QHS model (schematic sketch) is shown. In (b) the simulated values for Cu in $Ar:O_2 = 4:1$ (Eq. (3.2.1.3)) and literature values (Eq. (3.2.1.4)) for the mean free path are compared. In (c) the pressure dependence of the percentage of particles arriving at the substrate and average number of scatter events that these particles undergo during this process is shown. The data are simulated for a realistic Cu sputter process (2″ Cu target, 1 cm^2 substrate, target–substrate distance of 5 cm, and gas mixture of $Ar:O_2 = 1:1$). In (d) the particle flux distribution Cu/Y is given as a function of the position above a YBCO magnetron target. The positions of the magnetrons are marked; the gas mixture is $Ar:O_2 = 4:1$.

and room temperature:

$$\lambda_{\text{lit}} \cong \frac{0.063 \text{ mm}}{p} \tag{3.2.1.4}$$

with p representing the pressure in mbar.

Figure 3.2.1.8b shows the pressure dependence of λ_{lit} and λ_{QHS}. Since the mean free path depends not only on the pressure but also on the energy E_C and the participating elements it is different for different deposition techniques and different elements that are deposited. For instance, for the simulation shown in Figure 3.2.1.8b, λ_{QHS} is slightly larger than λ_{lit}. In general, scatter processes between the particles and the process gas need not be considered for "classical deposition techniques" that work at low pressure. However, for high-pressure processes scatter events have to be considered. The number of scatter events per particle arriving at the substrate increase dramatically in the pressure regime of 1 mbar which is the pressure regime for the deposition of HTS films via sputter and PLD (see Figure 3.2.1.8c). As a result of the scatter events:

- the *particle energy is reduced* (thermalization),
- the *deposition rate is reduced,* and
- the *stoichiometry is modified.*

Whereas the first effect is intended and necessary for the deposition of HTS material, the latter two effects lead to problems for HTS deposition. Furthermore, for potential industrial applications, the deposition rate is of importance, and for large-area deposition (where usually large target–substrate distances are necessary) the deviation of the stoichiometry causes problems (see Figure 3.2.1.8d, [46]). Nevertheless, sputtering and PLD provide the possibility of inducing defects that could be used to engineer mechanical as well as electronic properties [53].

Finally, the different pressure regimes used for the different PVD technologies automatically define the preparation regime in the pressure–temperature phase diagram which is important for the formation of crystalline phases. Especially for the complex phases of HTS, a high mobility of the adsorbed adatoms at the substrate is required. For instance, for YBCO, usually the highest temperature is chosen for which the YBCO phase (1–2–3 phase) is still stable (Figure 3.2.1.9). This typically leads to the tetragonal $YBa_2Cu_3O_y$ phase with y close to 6. The transition to the superconducting orthorhombic phase is performed after deposition (see Figure 3.2.1.9). Activated oxygen or special temperature–pressure process steps are included in order to improve or speed up the uptake of oxygen which is necessary for the phase transition [54].

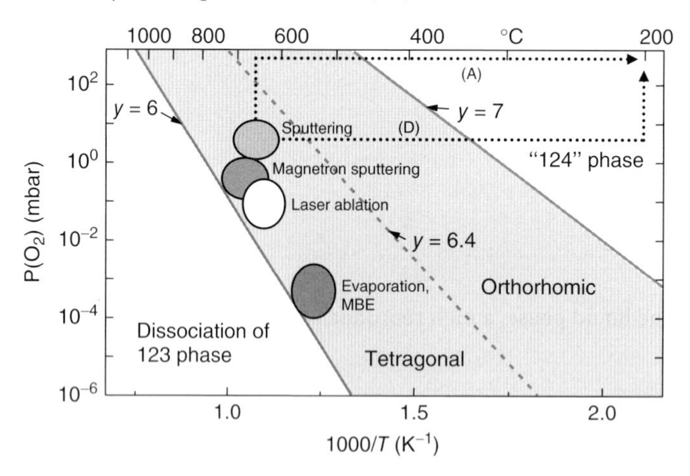

Figure 3.2.1.9 P–T phase diagram for YBCO according to [55, 56]. The 123 phase is stable in the regime indicated by the shaded area, which is divided into the tetragonal and orthorhombic phases. The deposition takes typically place in the tetragonal regime, *in situ* post-deposition treatments are typically following routes (A) or (D), here indicated for high-pressure sputtering.

3.2.1.2.2 CVD Technologies

In general, CVD technologies (CVD, metal-organic Chemical vapour deposition (MOCVD), atomic layer deposition (ALD)) offer the advantages of a good growth control, high throughput, low equipment costs, the capability of coating complex shapes, and ease of scaling-up to manufacturing volumes, all of which are attractive for industrial applications. The disadvantages lie in the small growth rates (typically $< 0.3\ nms^{-1}$) and often the poor reproducibility especially due to the stability of the precursors.

Most LTS material (e.g., Nb and NbN) are deposited by PVD techniques, especially magnetron sputtering. However, since the deposition rates are typically high, the deposition time for ultrathin films, which for instance are required for superconducting bolometers, can be extremely short. Therefore, recently attempts have been made to deposit superconducting NbN thin films via ALD using metal-organic precursors [57]. Since ALD represents a self-limiting CVD process, perfect thickness control can be achieved. However, due to contaminations (e.g., chlorine in case of the use of hydrogen chloride reagent) these NbN films are not suitable for all applications [58].

MOCVD represents a promising technique for the deposition HTS films, for example, for YBCO-coated conductors [59] and even MgB_2 Josephson devices [60]. Attempts to use MOCVD to grow HTS films began soon after the discovery of YBCO in 1987 [61]. However, MOCVD of YBCO, an oxide compound composed of multiple heavy-metal elements, turned out to be quite difficult. The lack of gas-phase or liquid-phase metal-organic precursors for the heavy-metal elements restricted MOCVD of HTS to the use of solid precursors. These solid precursors turned out to be instable in the vapor-phase composition, for example, the commonly used precursors for YBCO deposition (the so-called *thd precursors*: 2,2,6,6,-tetramethyl-3,5-heptanedionate (TMHD) for Y, Ba, and Cu, respectively) exhibit a high melting point and a low vapor pressure. The instability is strongest of the Ba component, as demonstrated for $Ba(thd)_2$ [62, 63, 64, 65, 66]. In order to stabilize the vapor-phase composition of YBCO, alternative precursors or/and novel concepts had to be developed. For example, the successful use of a fluorinated barium precursor ($Ba(tdfnd)_2{\cdot}$tetraglyme, where tdfnd is tetradecafluorononanedione) has been reported that is stable at its operating vaporization temperature of $\sim$145 °C [63]. Using the precursors $Ba(tdfnd)_2{\cdot}$tetraglyme, $Y(thd)_3{\cdot}4t$BuPyNO, and $Cu(tdfnd)_2{\cdot}H_2O$, which are all evaporated from the liquid phase, a high reproducibility of the film composition was reported [66]. However, these fluorinated precursors required the addition of water vapor in the deposition process to avoid the formation of BaF_2, which in turn led to problems associated with the removal of HF. Nagai *et al.* [64] and Yoshida *et al.* [67] reported that high-crystalline-quality MOCVD YBCO films are obtained by liquid-state nonfluorinated sources of $Y(thd)_3{\cdot}4t$BuPyNO, $Ba(thd)_2{\cdot}$2tetraene, and $Cu(thd)_2$. Other adducts that have been used to stabilize the Ba-thd precursor include triglyme and phenanthroline.

Another approach of improved stability is given by the transition from solid-source precursor delivery to liquid-source precursor delivery. These concepts

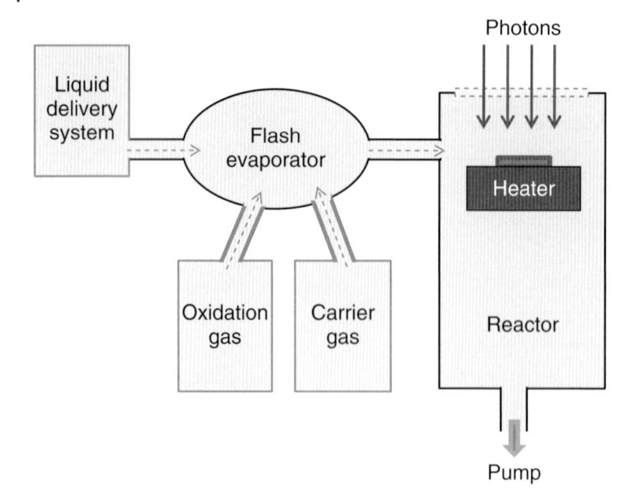

Figure 3.2.1.10 Schematic diagram of photo-assisted MOCVD system with the single liquid precursor delivery technique for the deposition of YBCO [70].

include: (i) liquid-source precursor delivery systems, (ii) direct injection of the precursor in a flash vaporizer [68] or (iii) a two-step evaporation process in a belt-driven system, where the solvent is evaporated in the first step and the precursor is subsequently vaporized in the second step [69].

For example, high-quality YBCO films have been deposited with high growth rates via photo-assisted MOCVD using a single liquid precursor delivery system ([59], Figure 3.2.1.10). The single liquid precursor delivery technique significantly improves the process reproducibility and the photo-irradiation technique increases the growth rate. The reactor is energized by tungsten halogen lamps that also irradiate the surface of substrate. Thus, halogen lamps not only thermally heat the substrate, but also photo-activate the chemical and physical reactions in the MOCVD process. Metal-organic precursors ($Y(TMHD)_3$, $Ba(TMHD)_2$, and $Cu(TMHD)_2$) are nonstoichiometrically dissolved in tetrahydofuran (THF) and injected by a micro-liquid pump into a heated flash evaporator that is maintained at elevated temperature (250 °C). The precursor vapor is then transported to the reactor by high-purity argon carrier gas. Oxygen with a small amount of nitrous oxide (N_2O) is added as oxidizer in the reactor. Growth rates of approx. $3-17\ \mathrm{nm\,s^{-1}}$ and critical currents of $J_c(77\ \mathrm{K}) \approx 4\ \mathrm{MA\,cm^{-2}}$ could be obtained via this technology. Since this deposition technology can easily be scaled up to large size, it is very promising for applications in fields such as coated conductors.

3.2.1.2.3 **CSD Techniques**

CSD of HTS material represents all chemical deposition methods ranging from sol–gel to metal-organic composition (MOD). Since CSD is no vacuum technology, the major advantage of this process lies in the low investment costs and low energy consumption. Among the different *ex situ* techniques, CSD is nowadays one of the most promising candidates for deposition of HTS-coated

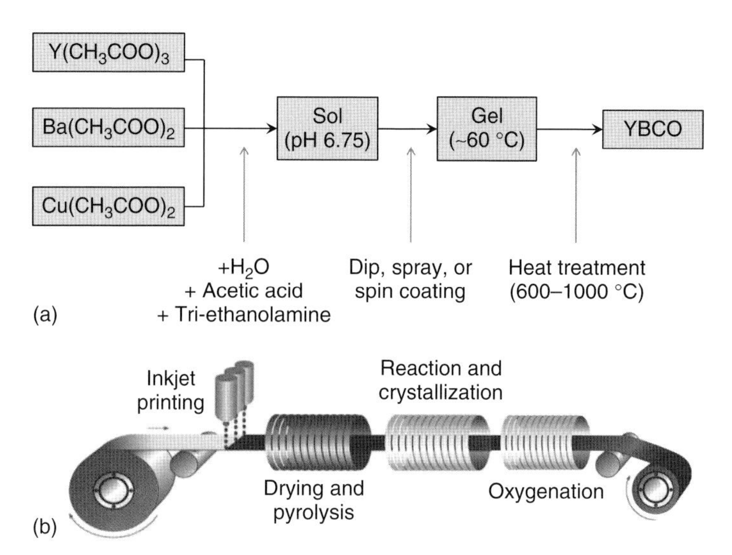

Figure 3.2.1.11 Schematic sketch of a CSD technology according to [78], (a) diagram of a CSD process for YBCO based on acetates, water, acetic acid, and triethanolamine and (b) continuous inkjet-printing deposition for the production of HTS-coated conductors.

conductors including both buffer layer and YBCO [71]. On the one hand, it has been demonstrated that high-quality epitaxial YBCO can be grown by PVD onto MOD buffer layers [72, 73]. On the other hand, it has been shown that high critical currents can be achieved in YBCO thin films grown by the trifluoroacetates (TFAs, ($OCOCF_3$)) route on single crystals or on vacuum-deposited buffer layers [74–77].

Although different chemical solutions can be chosen (e.g., TFA route, fluorine-free route, water, or water-free routes) the basic steps of the CSD processes are fairly similar (Figure 3.2.1.11). Starting with the different educts, the precursor is synthesized. The resulting solution is deposited onto the substrate via spin, dip, or inkjet coating. Depending on the process a consolidation starts due to chemical reaction (sol–gel) or evaporation of the solvent (MOD). Drying steps and pyrolysis (typically at 150–400 °C) lead to an amorphous layer. Finally this layer is crystallized at elevated temperature (typically 600–1000 °C). Nowadays the TFA route is widely recognized as one of the most promising methods for the production of second-generation coated conductors [78].

3.2.1.3 HTS Film Growth and Characterization

Growth of complex films like cuprates has remained at best a complicated matter due to their often rather complex multicomponent crystal structures, which are prone to a variety of defects and growth morphologies, and the often extreme deposition parameters (high process pressure and temperature), which furthermore have to be controlled very accurately. In this section, a brief sketch of the growth mechanisms, in general, and HTS thin film growth, in particular, is given. The general outline of the description is given in Figure 3.2.1.12, detailed

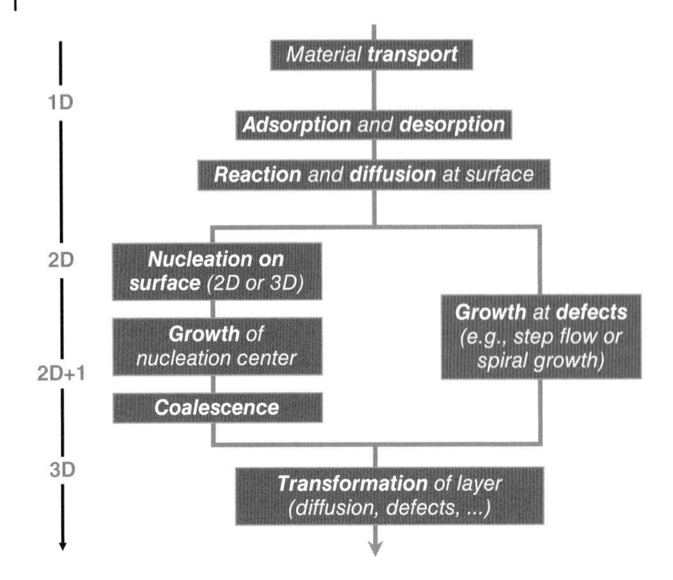

Figure 3.2.1.12 Schematic sketch of the adsorption, nucleation, and thin film growth on ideal substrates (left route) and that dominated by defects of the substrate's surface (right route).

information are given in [79]. The emphasis of the following brief description lies on PVD processes and the specific problems of adsorption of particles, oversaturation, epitaxy, lattice mismatch, defect generation, stress and strain that represent (i) important problems and (ii) options for engineering the properties of the often very complex superconductor films.

3.2.1.3.1 Nucleation and Phase Formation

After transport, particles impinge on the substrate. Chemical adsorption is only possible if the chemical adsorption potential $E_a > 0$, that is, if the system is chemically not inert (Figure 3.2.1.13). The adsorption rate is defined by the number of adatoms per area and time:

$$R = \frac{dn}{dt} \tag{3.2.1.5}$$

and the adatom average migration is given by

$$\lambda = \sqrt{2D\tau} = \sqrt{2}a_o \exp\left(\frac{E_a - E_d}{2kT_s}\right) \tag{3.2.1.6}$$

respectively. $D = D_o \exp(-E_d/kT_s)$ describes the surface diffusion, $a_o = (D_o/\nu)^{1/2}$ is the hopping rate, $\tau = (1/\nu)\exp(E_a/kT_s)$ is the average resident time, $\nu \approx 10^{12}$–10^{14} Hz is the attempt frequency, and T_s, D_o, E_a, and E_d represent substrate temperature, diffusion constant, chemical adsorption potential, and desorption potential, respectively. During growth from melt (e.g., liquid phase epitaxy) or solution the adatom can easily be desorbed from the surface, that is, the residence time is relatively short. During growth from vapor (PVD and CVD techniques) the activation

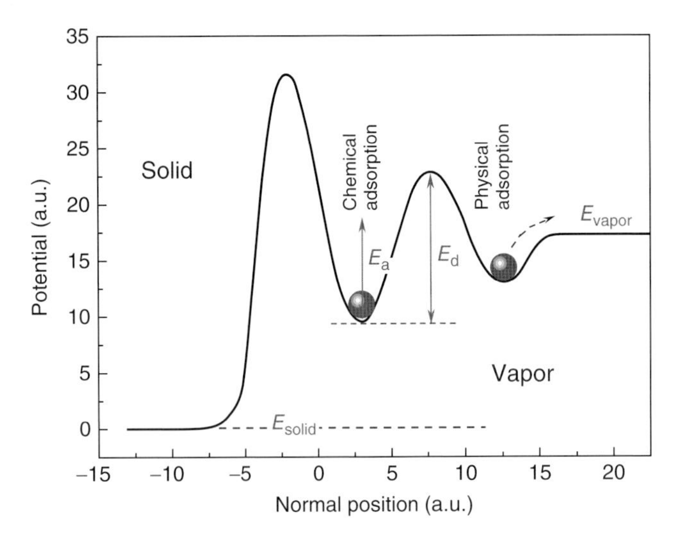

Figure 3.2.1.13 Schematic sketch of the adsorption potentials the impinging atom experiences at the film surface.

energy for desorption is typically larger resulting in a larger migration length and a larger sticking coefficient.

Nevertheless, inserting typical values for E_a, E_d, and D_o for oxides and PVD yields extremely small values for the resident time and migration of $\tau \ll 1$ ps and $\lambda < 1$ nm. Experimentally much larger values of $\tau \approx 0.2-0.5$ s and $\lambda \approx 100-500$ nm are reported [80]. This difference is explained by the particle transport via the gas phase [79]. For this mechanism, a supersaturated gas phase has to be established at the substrate. Desorbed atoms will be readsorbed and, therefore, they can migrate larger distances than theoretically expected (Figure 3.2.1.14). As a consequence supersaturated gas phases are vital for the phase formation

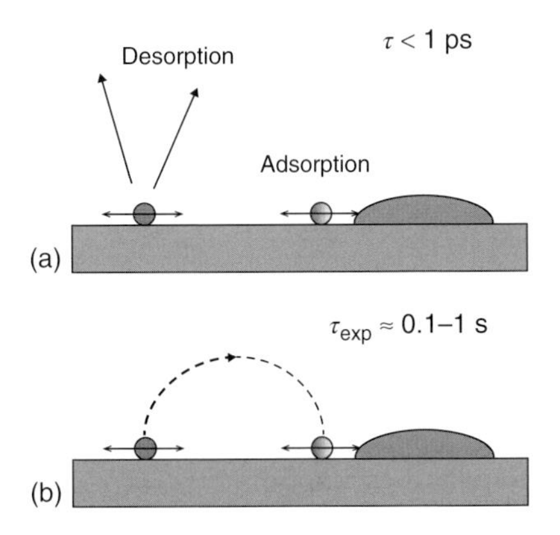

Figure 3.2.1.14 Schematic sketch of the phase formation without (a) and with (b) particle transport via the supersaturated gas phase.

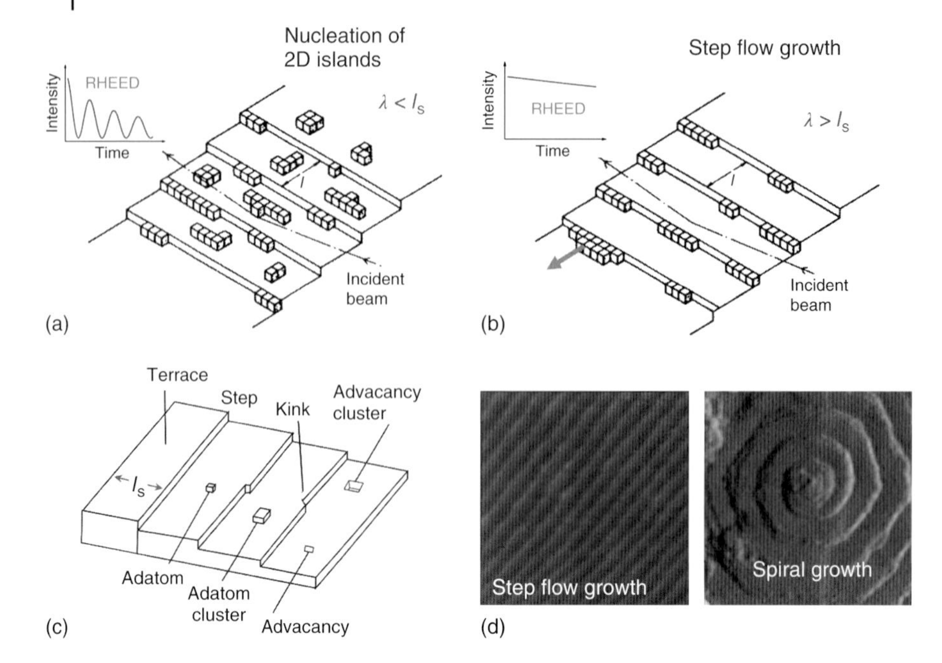

Figure 3.2.1.15 (a) Schematic sketch of the nucleation of 2D islands, (b) step flow growth, (c) topographical structures of "real" single crystalline substrates with a cut slightly tilted to both in-plane crystalline directions, and (d) atomic force microscope (AFM) micrographs of a step flow growth on a vicinal cut $SrTiO_3$ substrate and spiral growth for YBCO films [81].

and deposition. This represents a problem for a number of deposition devices, especially for large-area PVD.

During growth of lattice-matched layers, a layer-by-layer (Frank–van der Merwe) growth mechanism is usually observed. However, even in a layer-by-layer growth the resulting heterojunction interfaces are not perfectly planar. This is especially important in devices (e.g., Josephson junctions) where a monolayer change or defects can strongly affect the electronic properties. Layer-by-layer growth is generally limited by the surface morphology of the underlying substrate (see Figure 3.2.1.15):

1) As long as the surface migration length λ of the oncoming adatoms is clearly smaller than substrate features (e.g., l_s defined by terraces separated by step edges in vicinally cut single-crystal substrates) growth occurs due to the nucleation of 2D islands (Figure 3.2.1.15a).
2) If the migration length is larger than surface terrace size (i.e., $\lambda > l_s$), growth is dominated by attachment of adatoms to the step edges of the terraces. The step edges then propagate at a velocity depending on the step density and growth rate. This limit is called *step-flow growth*. The difference between these two regimes can be observed by RHEED oscillations (Figure 3.2.1.15b).

3) Finally, since "real substrates" are usually cut in a slight vicinal ways in both in-plane directions, kinks are usually present at the steps (Figure 3.2.1.15c). These kinks can be the starting point for spiral growth mode (Figure 3.2.1.15d).

Thus, the growth (heteroepitaxial nucleation and homoepitaxial growth) are strongly determined by the competition between the mobility of the oncoming particles, which can be affected by the preparation method and conditions, and the surface morphology of the substrate, which is among others defined by the cut of the substrate. By modifying the preparation process or the miscut (vicinal) angle of the substrate, the growth mode can be changed from 2D island growth to step flow growth or spiral growth.

3.2.1.3.2 **Heteroepitaxial Growth, Stress, and Defects**

Heteroepitaxial layer growth represents a major technology for advanced thin film electronic – for example, for semiconductor and oxide electronics – and for basic solid-state research. The most fundamental questions in this automatically strained-layer growth nevertheless are:

1) up to what thickness are heteroepitaxial layers stable,
2) which type of misfit defects develops, and
3) what happens upon modifications of the misfit for instance due to cooling of the film?

Generally it is believed that below a critical thickness the strained state is the thermodynamic equilibrium state, and above a critical thickness a strained layer may be metastable or it may relax [82]. Different critical thicknesses might be associated to different types of misfit defects, for example, dislocations, misalignments, or even cracks [25]. Stress in heteroepitaxially grown films results from both an intrinsic and temperature-dependent component. The main reason for the development of intrinsic stress in heteroepitaxially grown films is given by the nominal lattice mismatch

$$\varepsilon_o(T) = \left| \frac{a_{\text{film}}(T) - a_{\text{sub}}(T)}{a_{\text{film}}(T)} \right| \tag{3.2.1.7}$$

at growth condition, where a_{film} and a_{sub} represents the in-plane lattice parameters of the film and substrate, respectively. The thermal contribution is due to the difference in thermal expansion coefficient between the film and the underlying substrate resulting in a temperature-modified strain $\varepsilon_o(T)$ for instance during cooling of the sample [83–85]. Due to these two contributions, the resulting generation of defects in our films is discussed in two steps:

1) generation of lattice misfit dislocations during the growth of the film, for example, HTS films at elevated temperature (typically 700–900 °C for YBCO) and
2) generation of cracks that are most likely generated during the cooling down after deposition.

Ad (1): During deposition, defect-free films grow up to a critical thickness d_c from where on defects develop in the layer. This situation is most generally described initially by van der Merwe [86] and later by Matthews and Blakeslee [87]. In these theories, the line tension of a misfit dislocation of finite length is balanced by the force due to the strain in the layer on the termination of the misfit dislocation. Alternatively and equivalently, the energy of the system with and without misfit dislocation may be considered [88–90]. Although in either approaches several approximations of uncertain effect are made, reasonable values for the critical thickness d_c are obtained that are comparable to experimental values [91] and provided the basis for the discussion of defect development in semiconducting thin films. For instance, the Matthews theory predicts a critical thickness for the development of dislocation lines [87]:

$$d_c = \frac{C}{\varepsilon_{\text{Intrinsic}}} \ln[O(d_c)] \tag{3.2.1.8}$$

where C contains the details of the crystal and the dislocation. Different versions of the term $O(d_c)$ have been discussed by different authors [82, 87, 88, 91], the resulting predictions of critical thickness can vary by at least a factor of 2. For instance the development of misfit dislocation lines has been described with this approach by People and Bean [91]:

$$d_c \cong \left(\frac{b}{4\pi(1+\nu)\varepsilon_{\text{Intrinsic}}}\right)\left(\ln\left(\frac{d_c}{b}\right)+1\right) \tag{3.2.1.9}$$

where b and ν represent the extension of the dislocation line, and Poisson's ratio, respectively.

Typical predictions for the critical thickness describing the development of misfit dislocations in YBCO are given in Figure 3.2.1.16. Inserting reasonable values and taking into account that the theories are estimates of the critical thicknesses within a factor of about 2, for YBCO on $SrTiO_3$ a critical thickness $d_c < 10$ nm is predicted (Figure 3.2.1.16c). Thus, misfit dislocations will be generated already for extremely thin films.

Ad (2): After deposition of the films at elevated temperature (e.g., YBCO at 600–900 °C) the sample is cooled down to room temperature. Due to the mismatch of the thermal expansion coefficient, additional stress is imposed on the layer (see Eq. (3.2.1.7)). Taking into account the presence of misfit dislocations generated during the growth process, the resulting strain imposed on the film can be described by Zaitsev *et al.* [25] and Fitzgerald [89]:

$$\varepsilon_{\text{film}} = \varepsilon_o - \delta = \left|\frac{a_{\text{film}}(T) - a_{\text{substrate}}(T)}{a_{\text{film}}(T)}\right| - \delta \tag{3.2.1.10}$$

where δ describes the release of strain due to the presence of defects or misalignments. While defects and misalignments have developed during growth, cracks are expected to develop during cooling down if is this energetically favorable. According to the theory of fracture of solids, the amount of strain energy that is released per unit length of a 2D crack is given by Zaitsev *et al.* [25] and Cottrell

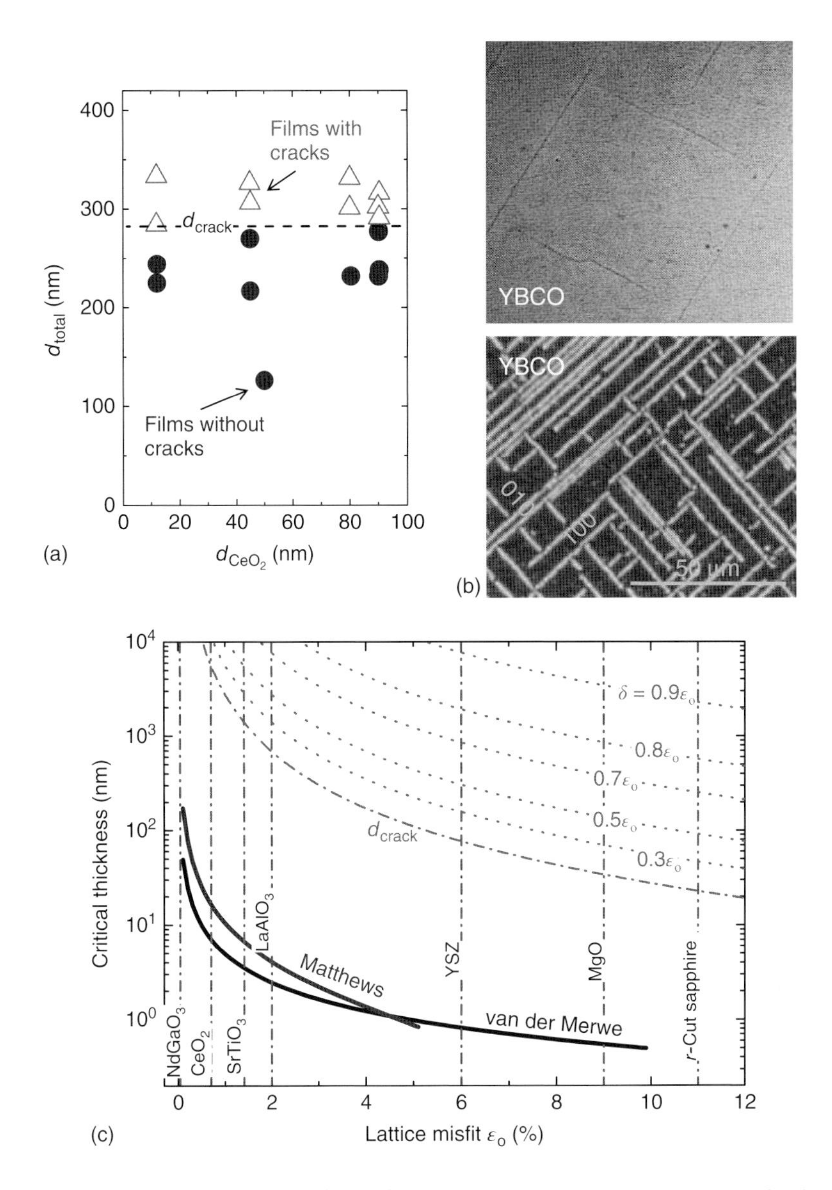

Figure 3.2.1.16 (a) Critical thickness d_{crack} for YBCO/CeO_2 on *r*-cut sapphire (the critical thickness is about 280 nm for buffer layer and YBCO since the mismatch causing the cracks occurs between CeO_2 and sapphire) (b) images of cracks in YBCO on sapphire [92], and (c) theoretical prediction of different critical thicknesses as a function of lattice misfit. In (c), the critical thickness for the generation of misfit dislocations according to van der Merwe [86] and Matthews theory (Eq. (3.2.1.9)) and the critical thickness for crack generation according to Eq. (3.2.1.10) and different values of δ are plotted. The lines indicate the lattice misfit of YBCO with respect to various substrates.

[93]:

$$E_{\text{crack}} = \pi Y \varepsilon^2 \left(\frac{a}{2}\right)^2 \tag{3.2.1.11}$$

where a represents the height of the crack ($a = d$, if the crack is normal to the film surface) and Y is the Young modulus of the material. For example, Young's modulus of bulk YBCO amounts to $Y \approx 300$ [89]. The formation of a crack is energetically favorable, if E_{crack} exceeds the energy E_{surface} (energy per unit crack length) required for the (initial) formation of the two new surfaces of the crack. Therefore, the critical film thickness for crack formation is given for $E_{\text{crack}} = 2E_{\text{surface}}$ resulting in:

$$d_{\text{crack}} = \frac{8E^*_{\text{surface}}}{\pi Y \varepsilon^2} \tag{3.2.1.12}$$

with $E^*_{\text{surface}} = E_{\text{surface}}/d$. The energy of the crack's surfaces is estimated by summation of the binding energies E_{B} of all atoms at these surfaces. Generally binding energies of oxides are of the order of 15 eV. The resulting surface energy for crack along [100], [010], or [001] axes of YBCO results in a surface formation energy $E'_{\text{surface}} \approx 31-34\ \text{J m}^{-2}$ [25, 26]. Figure 3.2.1.16 shows the resulting critical thickness d_{crack} for YBCO and different values of δ (in units of ε_0). The theoretical values of d_{crack} are comparable with experimental observation for epitaxial HTS layers in general. Ideal, defect-free films show the smallest critical thickness. For a misfit of $\sim$10–11% (e.g., YBCO on sapphire) crack-free layers are limited by $d_{\text{crack}} < 30$ nm. With increasing δ the critical thickness increases.

In conclusion, of this section:

1) All heteroepitaxially grown films possess misfit defects that are generated during their growth unless they are extremely thin ($d < d_{\text{c}}$). As a result, $\delta > 0$ and the critical thickness for crack formation is enhanced with respect to the critical thickness of defect-free films.
2) The critical thickness for crack formation d_{crack} is large for small lattice mismatch, it decreases strongly for larger mismatch, for example, YBCO on YSZ, MgO, or sapphire.
3) The critical thickness d_{crack} can be enhanced by artificially inducing adequate defects that take up the stress in the layer. This has been demonstrated for YBCO on sapphire used for microwave applications [53]. The generation of defects (e.g., Y_2O_3 precipitates in YBCO) represents one way to engineer the mechanical and electronic properties of films.

3.2.1.4 **Concluding Remarks**

Superconducting films (LTS and HTS) represent the backbone of today's and future superconducting electronic applications which range from (i) well-established devices for nondestructive material evaluation, communication, geological, and environmental prospecting (e.g., *superconducting bolometers*), primary standards (e.g., *voltage standards*), and neurology and medical diagnostics (e.g., *magnetoencephalography*), (ii) perspective electronic applications like logic devices (e.g., *RSFQ*), terahertz spectroscopy, and fluxonic concepts, to

(iii) future energy applications of HTS-utilizing coated conductors of the second generation. Some of these applications lead to very ambitions requirements for the thin film deposition technology.

In case of conventional LTS deposition, these requirements can be satisfied already today. Nowadays state-or-art technology allows the reproducible and high-quality deposition of among others Nb-, Pb-, Al-, and Nb-based compounds like NbN. Even complex systems like wafers with large arrays of Josephson junctions can be produced via well-established technologies, that is, the *Nb technology* and *NbN technology*.

In contrast, although the situation for HTS deposition is promising and vast progress has been achieved in the past years, there is still a way to go till a reproducible (e.g., for mass production), sufficiently homogeneous (e.g., for large areas or continuous deposition) and cost-efficient (for commercial use) HTS deposition technology is available. On the one hand, it is the task of future research to solve the remaining problems related to HTS deposition. On the other hand, due to the extreme demands of the HTS material, existing deposition technologies have already been strongly improved and novel technologies have been developed. This procedure is not only beneficial for the deposition technology of HTS material but also for the deposition of other complex thin film systems.

Acknowledgment

Valuable support from and discussions with E. Hollmann, J. Schubert, R. Kutzner, M. Bäcker, A. Offenhäuser, and A.G. Zaitsev are greatly acknowledged.

References

1. Bednorz, J.G. and Müller, K.A. (1986) *Z. Phys.*, **B64**, 189.
2. Caputo, P., Oppenlander, J., Haussler, C., Tomes, J., Friesch, A., Trauble, T., and Schopohl, N. (2004) *Appl. Phys. Lett.*, **85**, 1389–1391.
3. Snigirev, O.V., Chukharkin, M.L., Kalabukhov, A.S., Tarasov, M.A., Deleniv, A.A., Mukhanov, O.A., and Winkler, D. (2007) *IEEE Trans. Appl. Supercond.*, **17**, 718.
4. Divin, Y., Poppe, U., Gubankov, V., and Urban, K. (2008) *IEEE Sens. J.*, **8**, 750.
5. Kleiner, R. and Müller, P. (1994) *Phys. Rev. B*, **49**, 1327–1341.
6. Wördenweber, R. (2010) *NanoScience and Technology*, Springer, pp. 25–80, ISBN: 978-3-642-15136-1.
7. ter Brake, H.J.M., Buchholz, F.I., Burnell, G., Claeson, T., Crété, D., Febvre, P., Gerritsma, G.J., Hilgenkamp, H., Humphreys, R., Ivanov, Z., Jutzi, W., Khabipov, M.I., Mannhart, J., Meyer, H.G., Niemeyer, J., Ravex, A., Rogalla, H., Russo, M., Satchell, J., Siegel, M., Töpfer, H., Uhlmann, F.H., Villégier, J.C., Wikborg, E., Winkler, D., and Zorin, A.B. (2006) *Physica C*, **439**, 1–41.
8. Wang, Z., Zhao, X.J., Yue, H.W., Song, F.B., He, M., You, F., Yan, S.L., Klushin, A.M., and Xie, Q.L. (2010) *Supercond. Sci. Technol.*, **23**, 065013.
9. Wei Z., Li H.,·Hong WL., Lv Z., Wu H., Guo X., Ruan K. 2008 *J. Supercond. Novel Magn.* **21** 213
10. Farrell, D.E., Beck, R.G., Booth, M.F., Allen, C.J., Bukowski, E.D., and Ginsberg, D.M. (1990) *Phys. Rev.*, **B42**, 6758.

11. Farrell, D.E., Bonham, S., Foster, J., Chang, Y.C., Jiang, P.Z., Vandervoort, K.G., Lam, D.L., and Kogan, V.G. (1989) *Phys. Rev. Lett.*, **63**, 782.
12. Farrell, D.E., Williams, C.M., Wolf, S.A., Bansal, N.P., and Kogan, V.G. (1988) *Phys. Rev. Lett.*, **61**, 2805.
13. Okuda, K., Kawamata, S., Noguchi, S., Itoh, N., and Kadowaki, K. (1991) *J. Phys. Soc. Jpn.*, **60**, 3226.
14. Ossandon, J.G., Thompson, J.R., Christen, D.K., Sales, B.C., Kerchner, H.R., Thomson, J.O., Sun, Y.R., Lay, K.W., and Tkaczyk, J.E. (1992) *Phys. Rev.*, **B45**, 12534.
15. Dimos, D., Chaudari, P., and Mannhardt, J. (1990) *Phys. Rev. B*, **41**, 4038.
16. Hilgenkamp, H. and Mannhart, J. (1998) *Appl. Phys. Lett.*, **73**, 265.
17. Chaix-Pluchery, O., Chenevier, B., and Robles, J.J. (2005) *Appl. Phys. Lett.*, **86**, 251911.
18. Braginsky, V.B., Ilchenko, V.S., and Bagdassarov, K.S. (1987) *Phys. Lett.*, **120A**, 300.
19. Gao, J., Klopman, B.B.G., Aarnink, W.A.M., Reitsma, A.E., Geritsma, G.J., and Rogalla, H. (1992) *J. Appl. Phys.*, **71**, 2333.
20. Davidenko, K., Oktyabrsky, S., Tokarchuk, D., Michaltsov, A., and Ivanov, A. (1992) *Mater. Sci. Eng., B*, **15**, 25.
21. Jia, C.I., Kabius, B., Soltner, H., Poppe, U., and Urban, K. (1990) *Physica C*, **172**, 81.
22. Triscone, J. and Fischer, Ø. (1997) *Rep. Prog. Phys.*, **60**, 1673.
23. Hammond, R.B., Heyshipton, G.L., and Matthaei, G.L. (1993) IEEE Spectrum 30, 34.
24. Zaitsev, A.G., Wördenweber, R., Ockenfuss, G., Kutzner, R., Königs, T., Zuccaro, C., and Klein, N. (1997) *IEEE Trans. Appl. Supercond.*, **7**, 1482.
25. Zaitsev, A.G., Ockenfuss, G., and Wördenweber, R. (1997) in *Applied Superconductivity*, Institute of Physics Conference Series, vol. **158** (eds H. Rogalla and D. Blank), Institute of Physics, Bristol, p. 25.
26. Hollmann, E., Schubert, J., Kutzner, R., and Wördenweber, R. (2009) *J. Appl. Phys.*, **105**, 114104.
27. Lahl, P. and Wördenweber, R. (2005) *J. Appl. Phys.*, **97**, 113911.
28. Paranthaman, M.P. and Izumi, T. (2004) *MRS Bull.*, **29**, 533.
29. Matijasevic, V., Lu, Z., Kaplan, T., and Huang, C. (1997) *Inst. Phys. Conf. Ser.*, **158**, 189.
30. Utz, B., Semerad, R., Bauer, M., Prusseit, W., Berberich, P., and Kinder, H. (1997) *IEEE Trans. Appl. Supercond.*, **7**, 1272.
31. Moeckly, B.H. and Ruby, W.S. (2006) *Supercond. Sci. Technol.*, **19**, L21.
32. Bozovic, I., Eckstein, J.N., and Virshup, G.F. (1994) *Physica C*, **235–240**, 178.
33. Jaccard, Y., Cretton, A., Williams, E.J., Locquet, J.P., Machler, E., Schneider, E., Fischer, Ø., and Martinoli, P. (1994) *Proc. SPIE*, **2158**, 200.
34. Logvenov, G. and Bozovic, I. (2008) *Physica C*, **468**, 100.
35. Morimoto, A. and Shimizu, T. (1995) in *Handbook of Thin Film Process Technology* (eds D.A. Glocker and S.I. Shah), IOP Publishing Ltd, p. A1.5.
36. Koster, G., Rijnders, G.J.H.M., Blank, D.H.A., and Rogalla, H. (1999) *Appl. Phys. Lett.*, **74**, 3729.
37. Lorenz, M., Hochmuth, H., Grundmann, M., Gaganidze, E., and Halbritter, J. (2003) *Solid State Electron.*, **47**, 2183.
38. Dam, B., Rector, J., Chang, M.F., Kars, S., de Groot, D.G., and Griessen, R. (1994) *Appl. Phys. Lett.*, **65**, 1581.
39. Blank, D.H.A., Rijnders, G.J.H.M., Koster, G., and Rogalla, H. (2000) *J. Electroceram.*, **4**, 311.
40. Paschen, F. (1889) *Ann. Phys.*, **273**, 69.
41. Gilbert, L.R., Messier, R., and Krishnaswamy, S.V. (1980) *J. Vac. Sci. Technol.*, **17**, 389.
42. Rossnagel, S.M. and Cuomo, J.J. (1988) *Am. Inst. Phys. Conf. Proc.*, **165**, 106.
43. Liu, X.Z., Li, Y.R., Tao, B.W., Luo, A., and Geerk, J. (2000) *Thin Solid Films*, **371**, 231.
44. Newman, N., Cole, B.F., Garrison, S.M., Char, K., and Taber, R.C. (1991) *IEEE Trans. Magn.*, **27**, 1276.
45. Eom, C.B., Sun, J.Z., Yamamoto, K., Marshall, A.F., Luther, K.E., Geballe, T.H., and Laderman, S.S. (1989) *Appl. Phys. Lett.*, **55**, 595.

46. Wördenweber, R., Hollmann, E., Poltiasev, M., and Neumüller, H.W. (2003) *Supercond. Sci. Technol.*, **16**, 584.
47. Schneider, J., Einfeld, J., Lahl, P., Königs, T., Kutzner, R., and Wördenweber, R. (1997) *Inst. Phys. Conf. Ser.*, **158**, 221.
48. Thompson, M.W. (1968) *Philos. Mag.*, **18**, 377.
49. Thompson, M.W. (1981) *Phys. Rep.*, **69**, 335.
50. Hollmann, E.K., Vol'pyas, V.A., and Wördenweber, R. (2005) *Physica C*, **425**, 101.
51. Born, M. and Mayer, J.E. (1932) *Z. Phys.*, **75**, 1.
52. Vol'pyas, V.A., Gol'man, E.K., and Tsukerman, M.A. (1999) *Tech. Phys.*, **41**, 304.
53. Einfeld, J., Lahl, P., Kutzner, R., Wördenweber, R., and Kästner, G. (2001) *Physica C*, **351**, 103.
54. Ockenfuss, G., Wördenweber, R., Scherer, T.A., Unger, R., and Jutzi, W. (1995) *Physica C*, **243**, 24.
55. Bormann, R. and Nölting, J. (1989) *Appl. Phys. Lett.*, **54**, 2148.
56. Hammond, R.H. and Bormann, R. (1989) *Physica C*, **162-164**, 703.
57. Ziegler, M., Fritzsch, L., Day, J., Linzen, S., Anders, S., Toussaint, J., and Meyer, H.G. (2013) *Supercond. Sci. Technol.*, **26**, 025008.
58. Hinz, J., Bauer, A.J., and Frey, L. (2010) *Semicond. Sci. Technol.*, **25**, 075009.
59. Molodyk, A., Novozhilov, M., Street, S. *et al.* (2011) *IEEE Trans. Appl. Supercond.*, **21**, 3175.
60. Mijatovic, D., Brinkman, A., Veldhuis, D., Hilgenkamp, H., Rogalla, H., Rijnders, G., Blank, D.H.A., Pogrebnyakov, A.V., Redwing, J.M., Xu, S.Y., Li, Q., and Xi, X.X. (2005) *Appl. Phys. Lett.*, **87**, 192505.
61. Yamane, H., Masumoto, H., Hirai, T., Iwasaki, H., Watanabe, K., Kobayashi, N., Muto, Y., and Kurosawa, H. (1988) *Appl. Phys. Lett.*, **53**, 1548.
62. Becht, M. (1996) *Appl. Supercond.*, **4**, 465.
63. Richards, B.C., Cook, S.L., Pinch, D.L., Andrews, G.W., Lengeling, G., Schulte, B., Jürgensen, H., Shen, Y.Q., Vase, P., Freltoft, T., Spee, A., Linden, J.L., Hitchman, M.L., Shamlian, S.H., and Brown, A. (1995) *Physica C*, **252**, 229.
64. Nagai, H., Yoshida, Y., Ito, Y., Taniguchi, S., Hirabayashi, I., Matsunami, N., and Takai, Y. (1997) *Supercond. Sci. Technol.*, **10**, 213.
65. Burtman, V., Schieber, M., Brodsky, I., Hermon, H., and Yaroslavsky, Y. (1996) *J. Cryst. Growth*, **166**, 832.
66. Busch, H., Fink, A., Müller, A., and Samwer, K. (1993) *Supercond. Sci. Technol.*, **6**, 42.
67. Yoshida, Y., Ito, Y., Hirabayashi, I., Nagai, H., and Takai, Y. (1996) *Appl. Phys. Lett.*, **69**, 845.
68. Onabe, K., Akata, H., Higashiyama, K., Nagaya, S., and Saitoh, T. (2001) *IEEE Trans. Appl. Supercond.*, **11**, 3150.
69. Stadel, O., Schmidt, J., Wahl, G., Jimenez, C., Weiss, F., Krellmann, M., Selbmann, D., Markov, N.V., Samoylenkov, S.V., Gorbenko, O.Y., and Kaul, A.R. (2000) *Physica C*, **341–348**, 2477.
70. Zeng, J., Chou, P., Zhang, X., Tang, Z.J., and Ignatiev, A. (2002) *Physica C*, **377**, 235.
71. Pomar, A., Cavallaro, A., Coll, M., Gàzquez, J., Palau, A., Sandiumenge, F., Puig, T., Obradors, X., and Freyhardt, H.C. (2006) *Supercond. Sci. Technol.*, **19**, L1.
72. Rupich, M.W., Palm, W., Zhang, W., Siegal, E., Annavarapu, S., Fritzemeier, L., Teplitsky, M.D., Thieme, C., and Paranthaman, M. (1999) *IEEE Trans. Appl. Supercond.*, **9**, 1527.
73. Jarzina, H., Sievers, S., Jooss, C., Freyhardt, H.C., Lobinger, P., and Roesky, H.W. (2005) *Supercond. Sci. Technol.*, **18**, 260.
74. Araki, T. and Hirabayashi, I. (2003) *Supercond. Sci. Technol.*, **16**, R71.
75. Castaño, O., Cavallaro, A., Palau, A., González, J.C., Rossell, M., Puig, T., Sandiumenge, F., Mestres, N., Piñol, S., Pomar, A., and Obradors, X. (2003) *Supercond. Sci. Technol.*, **16**, 45.
76. Rupich, M.W., Verebelyi, D.T., Zhang, W., Kodenkandath, T., and Li, X.P. (2004) *MRS Bull.*, **29**, 572.
77. Obradors, X., Puig, T., Pomar, A., Sandiumenge, F., Piñol, S., Mestres, N., Castaño, O., Coll, M., Cavallaro,

A., Palau, A., Gázquez, J., González, J.C., Gutiérrez, J., Romà, N., Ricart, S., Moretó, J.M., Rossell, M.D., and van Tendeloo, G. (2004) *Supercond. Sci. Technol.*, **17**, 1055.
78. Mosiadz, M., Juda, K.L., Vandaele, K., Hopkins, S.C., Patel, A., Glowacki, B.A., van Driessche, I., Soloducho, J., Falter, M., and Bäcker, M. (2012) *Physics Procedia*, **36**, 1450.
79. Wördenweber, R. (1999) *Supercond. Sci. Technol.*, **12**, R86.
80. Koster, G., Rijnders, G.J.H.M., Blank, D.H.A., and Rogalla, H. (1998) *Adv. Laser Ablation Mater.*, **526**, 33.
81. Matijasevic, V.C., Ilge, B., Stäuble-Pümpin, B., Rietveld, G., Tuinstra, F., and Mooij, J.E. (1996) *Phys. Rev. Lett.*, **76**, 4765.
82. Downes, J.R., Dunstan, D.J., and Faux, D.A. (1994) *Semicond. Sci. Technol.*, **9**, 1265.
83. Ohring, M. (1991) *The Materials Science of Thin Films*, Academic Press, New York.
84. Hoffman, R.W. (1976) *Thin Solid Films*, **34**, 185.
85. Taylor, T.R., Hansen, P.J., Acikel, B., Pervez, N., York, R.A., Streiffer, S.K., and Speck, J.S. (2002) *Appl. Phys. Lett.*, **80**, 1978.
86. van der Merwe, J.H. (1962) *J. Appl. Phys.*, **34**, 123.
87. Matthews, J.W. and Blakeslee, A.E. (1974) *J. Cryst. Growth*, **27**, 118.
88. Hu, S.M. (1991) *J. Appl. Phys.*, **69**, 7901.
89. Fitzgerald, E.A. (1991) *Mater. Sci. Rep.*, 7, 87.
90. Freund, L.B. (1990) *J. Mech. Phys. Solids*, **38**, 657.
91. People, R. and Bean, J.C. (1985) *Appl. Phys. Lett.*, **47**, 322.
92. Kästner, G., Hesse, D., Lorenz, M., Scholz, R., Zakharov, N.D., and Kopperschmidt, P. (1995) *Phys. Status Solidi A*, **150**, 381.
93. Cottrell, A.H. (1975) in *Physics of Metals. 2. Defects* (ed. P.B. Hirsch), Cambridge University Press, Cambridge, London, New York, Melbourne, p. 247–280.

3.3
Josephson Junctions and Circuits

3.3.1
LTS Josephson Junctions and Circuits

Hans-Georg Meyer, Ludwig Fritzsch, Solveig Anders, Matthias Schmelz, Jürgen Kunert, and Gregor Oelsner

3.3.1.1 Introduction

Since the discovery of the Josephson effect in 1962, the application of superconductivity to electronics has been a challenging field of work for both physicists and engineers. Josephson tunnel junctions are based on tunneling of Cooper pairs, the superconducting charge carriers, and allow to make use of the Josephson effect technically. Such Josephson junctions have become the fundamental building blocks for any superconductor electronic circuit, similar to transistors in semiconductor electronics.

The most important junction fabrication process is based on Nb–Al/AlO_x–Nb trilayers and was invented in 1983 [1]. Since then it has become the principal technology for all superconductor electronics cooled to liquid helium temperatures (4.2 K). Digital electronics achieved relevance when the rapid single-flux quantum (RSFQ) logic, first reported in 1985 [2], became popular in the late 1980s. Public funding in the 1990s, in particular in Japan and in the United States, has enabled rapid progress in this field, and as a result, several RSFQ IC foundries were established worldwide. In Japan, there are fabrication facilities at ISTEC/SRL; in the United States at HYPRES and NIST; and in Europe at IPHT [3] and PTB [4], both in Germany, at VTT in Finland [5, 6], and at CNR in Italy. Linewidth reduction of the fabrication process and phase engineering are currently the major fields of research for Josephson junctions.

Since the early 1960s, physicists and engineers have identified quite a lot of application fields that benefit from the application of superconductor electronics, among them high-precision magnetometry, radiation detection and spectroscopy, precision metrology, mixed-signal and digital circuitry for communication and computing, and so forth. Compared to the other superconductor circuits and sensors, the digital RSFQ logic requires fabrication processes with the highest complexity. Currently, the niobium-based Nb–Al/AlO_x–Nb junction technology is the only candidate for very large-scale integration (VLSI) superconductor digital electronics circuits.

Most standard Josephson junctions are superconductor–insulator–superconductor (SIS) tunnel junctions. The current–voltage characteristic of a pure (underdamped) SIS junction shows a hysteretic behavior. In various applications, for example, circuits for superconducting quantum interference device (SQUID) or RSFQ logic, a nonhysteretic characteristic of the SIS junction is required. Such a characteristic can be realized with a shunted (damped) SIS junction that consists of a junction in parallel with a resistor (shunt). Another possibility to avoid hysteresis is the integration of a resistive barrier into the Josephson junction itself. Such junction types are superconductor–normal conductor–superconductor (SNS), superconductor–normal conductor–isolator–superconductor (SNIS), and superconductor–insulator–normal conductor–insulator–superconductor (SINIS) junctions.

After the prediction of the Josephson effect – the Cooper-pair tunneling in an SIS system – in 1962 [7], intensive research in Josephson junction technology for digital computing applications started in the middle of the 1960s. The weak superconductor Pb as well as Pb alloys were used as electrode materials and the isolator consisted of the natural oxide of Pb [8–11]. However, because of bad long-time stability and a high initial failure rate, this technology was not suited for high-level integration [12]. Next, refractory superconductors like Nb and NbN and their natural oxides were successfully tested in conjunction with Pb and Pb alloys as counter electrodes [13–15]. These junctions are characterized by good tunneling parameters but high specific capacitances because of the relatively high dielectric constant of the NbO barrier material compared to the formerly used PbO. For high-speed digital circuit applications, this was a drawback. The development of all-refractory Josephson junctions became a success with the change from native oxide barriers to artificial ones. This prevents oxygen diffusion into the counter electrodes, which would form materials with lower T_C at the interface. In the beginning of the 1980s, the all-refractory Nb–Al/AlO$_x$–Nb process was developed [1] and became in conjunction with the selective niobium anodization process (SNAP) [16] the standard technology for preparing thin-film Josephson junctions. It is widely used until now in different variations for analog and digital applications of low-temperature superconductor (LTS) Josephson junctions. In particular, the application of complex digital superconducting electronic circuits in commercial products requires a stable technological process for the fabrication of Josephson junction integrated circuits with high yield and low parameter spread. Currently, the only established process that allows the fabrication of digital circuits with up to tens of thousands of Josephson junctions is based on the all-refractory Nb–Al/AlO$_x$–Nb planar thin-film technology.

Cooling is a major issue for the application of superconducting devices. Therefore, other materials were considered that allow higher operation temperatures. Josephson junctions of the type NbN–MgO–NbN with operation temperatures in the range of 10 K use MgO as a barrier material with NbN electrodes. Such junctions were developed for applications like digital integrated circuits [17–21]. For a more extended historical review, see, for example, Ref. [22].

Table 3.3.1.1 is an overview of different junction types with the most established material pairings and their main fields of applications.

Pushed by the expectations for high operation temperatures, all-NbN technologies [20, 23–25] were developed for the preparation of Josephson junctions and complex superconducting logic circuits. There are no substantial differences between the NbN and Nb processes concerning the types of Josephson junctions (SIS, SNS), the structure sizes, or the number of layers in complex circuits. So far, the parameter spreads and the functional complexity of all-NbN integrated

Table 3.3.1.1 Selected Josephson junction types and their main fields of application.

Junction type	Material	Application
SIS	Nb–Al/AlO$_x$–Nb	RSFQ, SQUID, voltage standard, STJ radiation sensors
	NbN–MgO–NbN	Mixer, RSFQ
	Al–AlO$_x$–Al	Qubit
SNS	Nb–Nb$_x$Si$_y$–Nb	Voltage standard, RSFQ
SINIS	Nb–AlO$_x$–Al–AlO$_x$–Nb	Voltage standard
SINS	Nb–Al–AlO$_x$–Nb	Voltage standard

circuit technologies have not reached the quality and yield level of the established Nb technologies. Partly, this is caused by the fabrication process of the so-called artificial-type tunneling barriers [26] that are prepared by direct deposition of the barrier material (e.g., MgO, AlN, NbN). Any deviation in barrier film thickness or homogeneity strongly influences the wafer-to-wafer or chip-to-chip spread of the junction current density. Presently, high-quality NbN tunnel junctions are used for high-frequency detector applications. Because of their high gap frequency of 1.4 THz [27, 28], these junctions are particularly suited for SIS mixers in the sub-millimeter wavelength range.

3.3.1.2 Junction Characterization

Figure 3.3.1.1 shows a typical hysteretic V–I characteristic of an underdamped Nb–Al/AlO$_x$–Nb junction as an example of a SIS Josephson junction. Parameters that are typically used to characterize Josephson junctions are indicated.

- I_C: Critical current of the junction. Measured I_C values often are influenced by measurement conditions like flux trapping or noise. The unaffected value should be $I_C = \pi V_G / 4R_N$ [29].

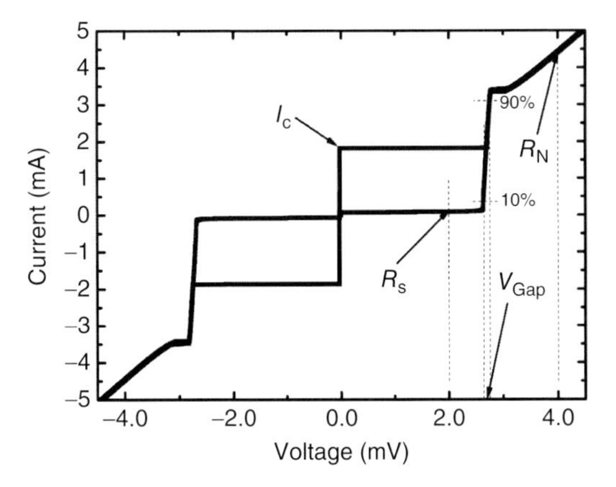

Figure 3.3.1.1 V–I characteristic of an underdamped Nb–Al/AlO$_x$–Nb Josephson junction. The junction area is 100 μm^2. (Permission received from H.-G. Meyer, IPHT Jena, Germany.)

- R_N: Normal state resistance, caused by tunneling of unpaired electrons. By common agreement, for Nb–Al/AlO_x–Nb junctions, it is measured at $V = 4$ mV.
- $I_C R_N$: The product of the critical current I_C and the normal state resistance R_N reflects the material and the superconducting properties of the electrodes. According to the model of Ambegaokar and Baratoff [29], the following equation is valid for direct tunneling in SIS junctions at $T = 0$ K: $I_C R_N = (\pi/2e) \times \Delta_0$. Δ_0 is the superconducting gap energy. This expression is valid in good approximation for temperatures up to $T_C/2$.
- R_S: Subgap resistance. For Nb–Al/AlO_x–Nb junctions, it is measured at $V = 2$ mV by common agreement.
- V_M: V_M is defined as the product of the critical current I_C and the subgap resistance R_S. It characterizes the leakage properties of the junction barrier. It is an essential parameter to evaluate the junction preparation process. By using the often mentioned R_S/R_N ratio, V_M can be derived as $I_C R_N \times (R_S/R_N)$.
- V_G: The gap voltage V_G is the sum of the energy gaps of the junction electrodes and characterizes the quality of the electrode material. It is strongly influenced by proximity effects at the barrier–electrode interface. The BCS theory [30] provides the relationship $eV_G = 2\Delta_0\ (T = 0) = 3.5\ kT_C$.
- C_S: The specific capacitance of the junction depends mainly on the dielectric constant of the barrier material.

Some of these parameters have characteristic theoretical values determined by the ideal superconducting properties of the selected electrode and barrier materials (Table 3.3.1.2). But these theoretical values are in general not realized. Desired circuit parameters like the critical current density or junction size as well as the process technology result in a mutual dependence of the parameters. Such interactions have been experimentally investigated for Nb–Al/AlO_x–Nb Josephson junctions in, for example, Refs. [31–36].

3.3.1.3 Nb–Al/AlO_x–Nb Junction Technology

3.3.1.3.1 General Aspects

Because of its high yield and low parameter spread, the Nb–Al/AlO_x–Nb technology is the most widespread technology for the fabrication of large-scale

Table 3.3.1.2 Typical parameters of SIS Josephson junctions.

Material	Theory		Experimental results				
	V_G (mV)	$I_C R_N$ (mV)	V_G (mV)	V_M (mV)	$I_C R_N$ (mV)	C_S (μF cm^{-2})	References
Nb–Al/AlO_x–Nb (T_C(Nb) = 9.2 K)	2.78	2.18	2.85	40–70	1.68	6	[33]
			2.9	50–70	1.8	4.5	[37]
NbN–MgO–NbN (T_C(NbN) = 15 K)	4.53	3.55	5.1	45	3.2	8	[19]

Theoretical values of V_G and $I_C R_N$ according to Ref. [29, 30], respectively.

integrated LTS superconducting circuits. Therefore, technological aspects of Josephson junction preparation will in this text be restricted to the Nb–Al/AlO_x–Nb system. This does not mean a loss of generality, because there are no principal differences in the application of general thin-film processing steps like lithography, deposition and patterning for other electrode, and barrier material combinations.

Basically, the Nb–Al/AlO_x–Nb standard technology utilizes a sandwich structure of successively deposited and patterned superconducting Nb, normal conductor, and insulating layers. The deposition processes have to be optimized for aspects like minimum film stress, surface morphology, and step coverage [38–40].

The Nb layers are used to wire the Josephson junctions and to shield them from magnetic fields. The number of stacked superconducting layers differs, depending on the complexity of the circuit. At least three Nb layers are used in the basic RSFQ circuit technology. A process for single-flux quantum (SFQ) circuits with up to nine Nb layers [41] has been announced by the Japanese institute ISTEC. Resistors are formed with normal metal layers. Different normal conductors or alloys are in use according to the requirements on the working temperature of the circuit, the sheet resistance, and the process compatibility (see Table 3.3.1.3).

The Josephson junctions are integrated in this process as the so-called trilayer, which is a sequence of the layers Nb–Al/AlO_x–Nb deposited without vacuum interruption, in order to obtain clean interfaces. The tunneling barrier is formed by AlO_x. The AlO_x is generated *in situ* during the trilayer deposition process by oxygen exposure of the freshly deposited Al.

Figure 3.3.1.2 shows a state-of-the-art ultra high vacuum (UHV) sputter system suited for the deposition of metallic and dielectric layers including trilayer stacks with *in situ* oxidation. Substrates up to 150 mm size can be coated with thickness homogeneities of ±2%. Substrate temperatures are controlled by He backside cooling in the range of 0–300 °C.

The critical current density j_C of the Josephson junction is a major issue for both fabrication and application. This parameter depends exponentially on the thickness of the AlO_x tunneling barrier and is set during the fabrication process by the oxygen exposition (the product of oxygen partial pressure and oxidation time) [48] (see Figure 3.3.1.3).

The deposited Al film is about 10 nm thick and the oxide barrier on its surface is quite thin, typically in the range of 1–2 nm. The Al/AlO_x is too an artificial

Table 3.3.1.3 Resistive film materials.

Material	T_C (K)	Resistivity (μΩ cm)	Sheet resistance (Ω sq^{-1})	References
Pd	—	6.6 at 4.2 K	1 at 66 nm	[42]
Au(53 wt%)/Pd	—	35.9 at 4 K	1–20 at 388–15 nm	[43, 44]
Ti	0.39	30 at 4.2 K	2 at 150 nm	[20]
Mo	0.92	—	1 at 100 nm, 1.5 at 70 nm	[45]
Zr	0.61	—	6.4 at 100 nm	[45]

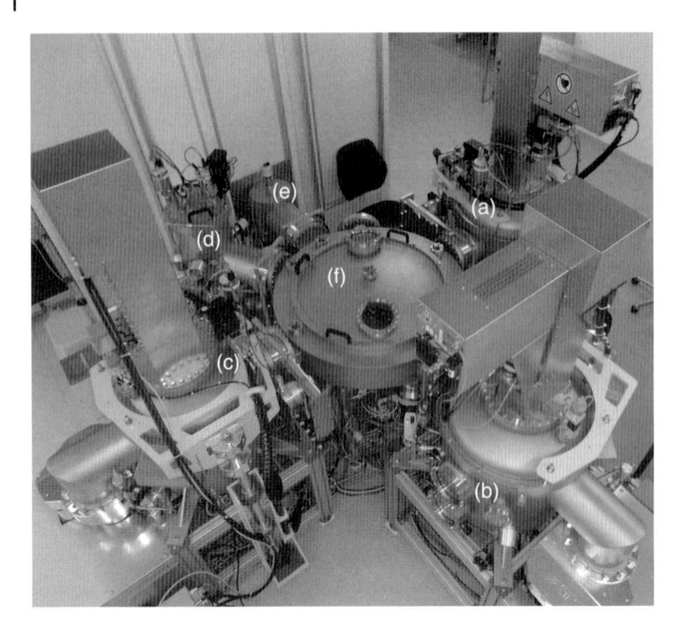

Figure 3.3.1.2 UHV sputter cluster for LTS Josephson circuit technology at IPHT Jena [46]. (a–c) Sputter deposition chambers, each with up to four 3″ magnetron sources in confocal arrangement. (d) Oxidation chamber. (e) Load lock chamber. (f) Handler for substrate transfer between the chambers. (Courtesy Bestec GmbH [47].)

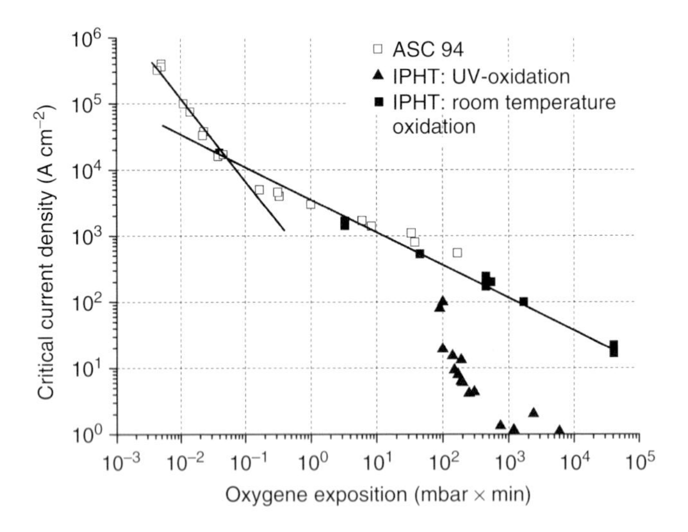

Figure 3.3.1.3 Critical current density as a function of oxygen exposition. Ultraviolet light-assisted oxidation decreases the process time for very low values of the critical current density (black triangles) compared to standard room-temperature oxidation (black squares) [49]. Open squares: room-temperature oxidation [48]. (Reprinted from Ref. [50], Copyright (2010), with permission from Elsevier.)

barrier with a native grown oxide and combines the benefits of an artificial barrier with a well-controllable barrier thickness. With this process, the j_C can be well controlled in a wide range despite of its exponential dependence on the oxide thickness. The residual unoxidized Al has a marginal, but in most cases negligible, influence on the gap voltage. The Al film has the additional most welcome effect that it levels out the surface roughness of the underlying Nb film [51]. The strong increase of the current density for small oxygen exposure (corresponding to j_C larger than 20 kA cm^{-2}) is probably caused by the onset of incomplete oxide coverage of the barrier area. This causes a reduced process reproducibility for extremely high current densities. On the other end of the scale, it is difficult to realize current densities less than 50 A cm^{-2} at room-temperature oxidation conditions because of the self-restricting oxidation process of the aluminum oxide barrier [52]. Increased substrate temperatures or very long oxidation times [53] yield current densities lower than 10 A cm^{-2}; however, problems with the process reproducibility arise. By double oxidation [54] – an *in situ* deposition of a second, extremely thin Al layer onto a normal Al/AlO$_x$ barrier and the following complete oxidation of this second layer – thick oxide barriers can be produced. The drawbacks of this method are the ambitious demands on thickness reproducibility and thickness homogeneity of this second Al film to realize an acceptable chip-to-chip and wafer-to-wafer spread of the current density. Another method to fabricate barriers with a small j_C is the creation of highly reactive atomic oxygen by the application of UV light during standard room-temperature oxidation [49], thereby accelerating the oxide growth. Current densities down to 1 A cm^{-2} can be realized with realistic oxidation time (see Figure 3.3.1.2). Because of the strong chamber surface degassing caused by the UV radiation, a good preconditioning of the oxidation chamber is necessary to obtain reproducible current density values.

The critical current density of the Josephson junctions is an important parameter for applications ranging from single junctions to complex logical RSFQ circuits. Table 3.3.1.4 shows some application fields of LTS Josephson junctions and the corresponding critical current density ranges.

The integration level of superconductor electronic circuitry is governed by scaling rules [55] shown in Figure 3.3.1.4. The clock rate of complex circuits can be

Table 3.3.1.4 Application fields and typical critical current densities of LTS Josephson junctions.

Application field	Typical critical current density (A cm^{-2})
SQUID	100–2 000
STJ	50–500
Voltage standard (SIS)	10–20
Programmable voltage standard (SNS)	2 500
Programmable voltage standard (SINIS)	100–500
Qubit	200–400
SFQ logic circuits	1 000–100 000

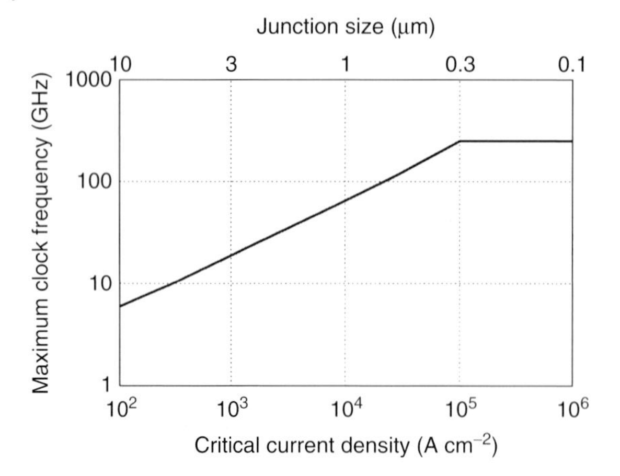

Figure 3.3.1.4 Shrink path for superconductor electronics based on Nb–Al/AlO$_x$–Nb Josephson junctions. By scaling the junction size down to 0.3 μm, the clock speed of integrated circuits can be increased well above 100 GHz. (Reprinted from Ref. [50], Copyright (2010), with permission from Elsevier.)

increased with the square root of the critical current density while the junction size has to be reduced by the reciprocal of the same square root. Apart from the fact that the thermal budget would not permit highly integrated semiconductor circuits to operate at several tens of gigahertz clock rate, it should be noted here that superconducting circuits can achieve such clock rates using a lithography resolution of a few micrometer, thus with a considerably less challenging lithography than semiconductor circuits. A 100 kA cm^{-2} sub-micrometer Josephson junction process would give access to complex digital circuitries with clock rates far above 100 GHz.

After the first successful preparation of high-quality all-refractory Nb–Al/AlO$_x$–Nb Josephson junctions with the so-called SNAP process [16], the technological development aimed to optimize this technology with regard to reproducibility, yield, and parameter spread necessary for large-scale integration (LSI) circuitry. In the traditional SNAP process, the junction area is defined by anodic oxidation of the trilayer top Nb and forms in conjunction with the SiO isolating layer a square contact window for wiring the junction. One main drawback of these window-type junctions is the large parasitic capacitance in parallel to the junction, caused by the wiring overlap. Another drawback is that during the anodization process, niobium oxide forms to a certain degree underneath the resist mask that defines the junction area. This effect can cause open junctions if the junction size is in the sub-micrometer range.

By the so-called cross-type design [56, 57], the junction area is defined by two crossing lines of trilayer and wiring. This way the parasitic capacitance is minimized because any unnecessary overlap of the wiring line with the base Nb of the trilayer is avoided. Additionally, compared to resist squares, the junction area definition by two crossing resist lines is less sensitive to influences of resist development and etching. The cross-type principle seems to be the best choice to realize the small junction sizes necessary to satisfy the demands for high clock frequencies in LSI Josephson circuits (Figure 3.3.1.4).

With increasing complexity of the circuits, more wiring and shielding layers are required, causing stronger height differences in the surface topography of the layer stack. This, in turn, gives rise to problems with edge coverage in the deposition processes and focus depth in high-resolution lithography. Following the developments in the semiconductor industry, isolation layer planarization technologies are applied to overcome these difficulties [58–60].

3.3.1.3.2 Basic Processes of the Nb–Al/AlO$_x$–Nb Technology

Window-Type Process As an example for a window-type technology, the European FLUXONICS Foundry process [50] for RSFQ circuits with a critical current density of 1 kAcm^{-2} is described. It is an SIS Nb–Al/AlO$_x$–Nb trilayer process with externally shunted Josephson junctions and comprises 12 photomask steps. The key parameters are:

- Critical current density j_C of the Josephson junctions: 1 kA cm^{-2}
- Minimum area of a Josephson junction: 12.5 μm^2
- Sheet resistance of the Mo resistor layer: 1 Ω sq^{-1}
- Minimum feature size: 2 μm
- Overlap layer to layer: 2.5 μm

The film stack consists of five metal layers: three superconducting Nb layers and two resistive layers (Mo, Au). The metal layers are isolated by the combination of NbO$_x$ (anodically oxidized Nb) and thermally evaporated silicon monoxide films. The Nb–Al/AlO$_x$–Nb trilayer is deposited without interruption of the vacuum; the aluminum oxidation is performed at room temperature in pure oxygen atmosphere. The resist structures for pattern definition are prepared by contact lithography using a mask aligner. Fluorine-based reactive-ion etching and liftoff processes are used for film patterning. Table 3.3.1.5 lists film materials and thicknesses. The whole layer stack is schematically shown in Figure 3.3.1.5. In Figure 3.3.1.6, a SEM (scanning electron microscope) micrograph of the cross-section of a shunted SIS junction with ground plane is depicted. The sample was prepared by focused ion beam (FIB) etching.

Technologies for Sub-Micrometer-Sized Josephson Junctions Several applications benefit from Josephson junctions with a side length below 1 μm. For example, fast RSFQ logic requires high critical current densities j_C. At the same time, the critical current I_C must be kept at the design value to avoid thermal fluctuations [55]. At a current density 100 kA cm^{-2}, the junction area has to be in the range of 0.1 μm^2 compared to 10 μm^2 at 1 kA cm^{-2}. Another example where smaller junctions increase the performance of the device is the measurement of small magnetic fields with SQUIDs. Here, the energy resolution is improved by the reduced capacitance of small junctions. First, the intrinsic capacitance of the junction itself is smaller. Also, we will see below that the fabrication technologies for

Table 3.3.1.5 Functionality and thickness of the different layers of the FLUXONICS Foundry 1 kA cm^{-2} RSFQ 1D process.

Name	Function	Material	Thickness (nm)
M0	Ground plane	Nb	250
I0A	Isolation	Nb_2O_5	50
I0B	Isolation	SiO	100
I0C	Isolation	SiO	100
M1	Wiring 1	Nb	250
T1	Trilayer	$Nb-Al/AlO_x-Nb$	60/12/30
I1A	Isolation	Nb_2O_5	70
I1B	Isolation	SiO	180
R1	Shunt	Mo	80
I2	Isolation	SiO	170
M2	Wiring 2	Nb	350
R2	Bond	Au	45

Reprinted from Ref. [50], Copyright (2010), with permission from Elsevier.

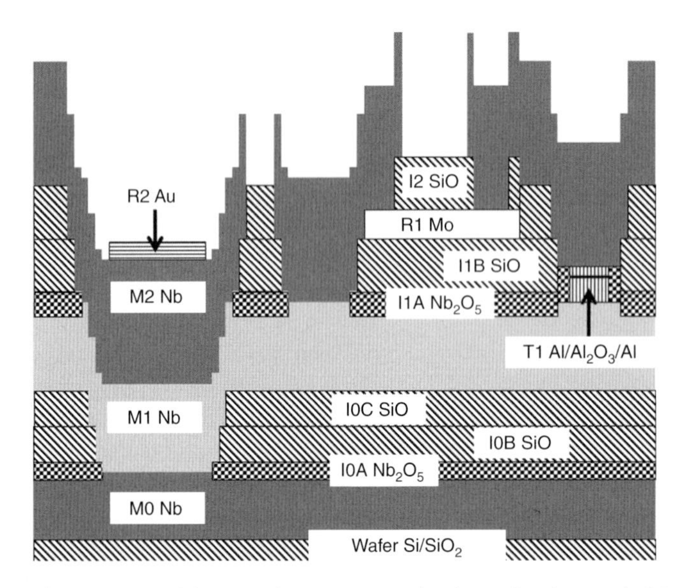

Figure 3.3.1.5 Schematical cross-section of a shunted and grounded Josephson junction. (Reprinted from Ref. [50], Copyright (2010), with permission from Elsevier.)

sub-micrometer junctions can reduce significantly the parasitic capacitance from the surroundings of the junction.

With the window technology described above, junction areas smaller than a few square micrometer cannot be fabricated reliably. The definition of the junction size by anodization is not accurate on a sub-micrometer scale because the electrolyte creeps between the photoresist and the trilayer to various degrees and can even completely lift off the resist mask, thus prohibiting the formation of a junction.

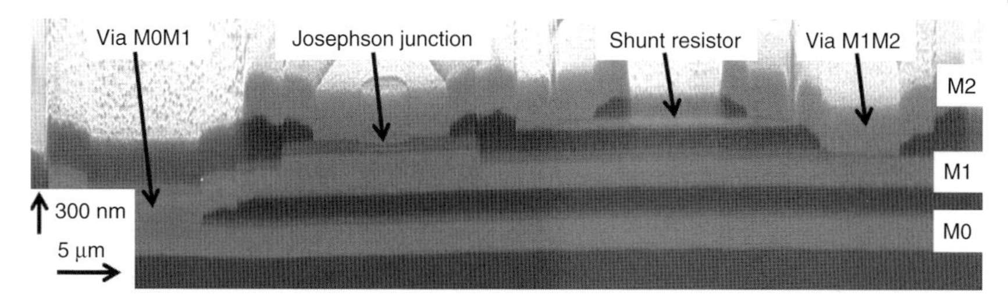

Figure 3.3.1.6 Cross-section of a shunted SIS junction with ground plane. For the SEM image, the sample was prepared by FIB etching. (Reprinted from Ref. [50], Copyright (2010), with permission from Elsevier.)

A variation of the window technology may be used to fabricate sub-micrometer-sized junctions [61]. The junctions were anodized with a hard mask of SiO_2, thus avoiding the softening of a resist mask in the electrolyte.

An advancement of the technologies where the junction area is defined by the anodization of the counter electrode is to define it by the etching of the counter electrode or the complete trilayer. This method allows a more accurate definition of the junction size. However, care must be taken to isolate the base Nb properly so that shorts to the counter electrode are avoided. Several methods have been devised for the isolation. Often, the junction sides are anodized. This anodization is to be distinguished from the anodization used in the window technologies. Still, creeping of the electrolyte between photoresist and trilayer has to be avoided. This may be achieved in various ways. For example, a sacrificial Nb layer can be deposited on top of the junction. It has to be removed after the anodization [62]. A hard mask of either SiO_2 [63] or SiO [64] may be used for anodization. Sidewall anodization is frequently combined with SiO_x or SiO deposition for improved passivation [58, 64–69]. Also, SiO or SiO_2 isolation without Nb anodization has been used [70, 71]. In addition to the deposited SiO_2, the junction sidewall may be passivated in an O_2 plasma [72], by spin-on glass [73], or by a thin layer of Nb_2O_5 and Al_2O_3 [74].

For the definition of the junction by etching, two geometries can be used. The trilayer is either etched in the junction shape (usually a square) [58, 62, 63, 72], or etched as stripe, and after a further lithography step, the wiring is etched as a perpendicular stripe [57, 63, 67, 69]. In the second etching step, the top Nb is removed as well, so that the junction is shaped as square or rectangle. The latter process has the advantage that it does not use small resist dots, where corners may shrink, but resist lines that are more robust.

In the following, an example of this process [65, 69] will be described in detail. It is similar to [57], except that there the first stripe is etched only to the barrier, so that the base electrode remains intact. If the base electrode is removed as well, the parasitic capacitance is significantly reduced.

In [69], the process starts by sputtering the trilayer on a silicon wafer (Figure 3.3.1.7a). The trilayer is etched by reactive-ion etching and sputter etching in the form of a stripe (Figure 3.3.1.7b). With the same resist mask, the sidewalls

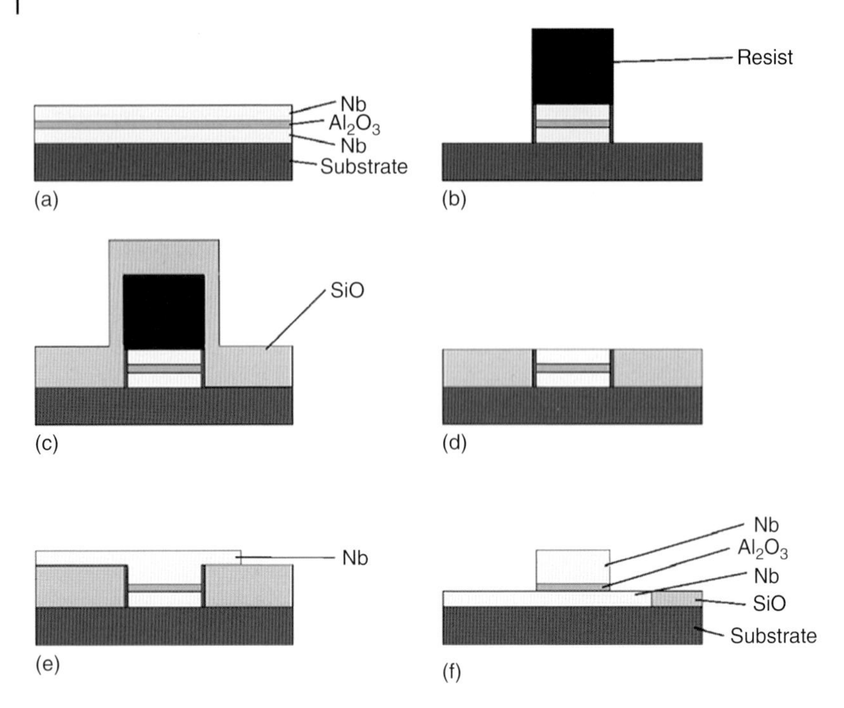

Figure 3.3.1.7 (a–f) Fabrication process as described in the text. Dimensions are not to scale. (Adapted from Ref. [69]. © IOP Publishing. Reproduced by permission of IOP Publishing. All rights reserved.)

of the trilayer are anodized. Still with the same resist mask, SiO is deposited by evaporation (Figure 3.3.1.7c). The SiO serves as added isolation of the base Nb and, after liftoff, planarizes the stripe (Figure 3.3.1.7d). Now the Nb counter electrode is deposited and reactive-ion etched as a stripe perpendicular to the trilayer stripe (Figure 3.3.1.7e). In this step, etching is continued until the Al barrier of the junction is reached. Thereby the junction definition is completed. The view in Figure 3.3.1.7f is rotated in-plane by 90°. By comparing Figure 3.3.1.7e,f, it becomes clear that the trilayer is etched to the barrier on two sides, while on the other two sides, it is etched to the substrate. Figure 3.3.1.8 shows an example of such a cross-type Josephson junction. For lithography, a 5× i-line stepper has been used. Other exposure options include contact lithography and, for deep sub-micrometer-sized junctions, electron beam lithography.

Technologies for Nanometer-Sized Junctions The fabrication of junctions with linear dimensions in the nanometer range is necessary for quantum bits (qubits) and for the observation of Coulomb blockade phenomena, for example, in superconducting single-electron transistors where a small capacitance is needed.

Shadow evaporation [75] of an $Al/AlO_x/Al$ trilayer is the established method for producing small Josephson junctions. From the viewpoint of possible applications, Nb-based junctions are preferred because the superconducting gap of

Figure 3.3.1.8 A cross-type Josephson junction. The horizontal stripe is the etched trilayer and the vertical stripe is the counter electrode. Residues from the resist are visible. The trench around the trilayer indicates that the planarization is not perfect. (Permission received from H.-G. Meyer, IPHT Jena, Germany.)

Nb is about eight times that of Al. However, the frequently used shadow evaporation with a resist mask cannot successfully be applied to Nb-based junctions because the thermal load on the suspended bridge mask is too high for commonly used electron beam lithography resists. Therefore, other technologies have been developed for Nb-based junctions. For example, a four-layer resist for shadow evaporation [76] or shadow evaporation with a hard mask [77, 78]. For the fabrication of Nb–AlO_x–Nb junctions on the sloped edge of an Nb layer, optical lithography was used [79]. Other varieties of edge-type junctions are described in Refs [80, 81]. In general, developing of the resist is a critical process for the transfer of ultrasmall resist structures. To avoid any resist, a focused ion beam system was used to write the structure with implanted Ga ions directly into the top Nb of the trilayer [82]. The Ga-implanted Nb served as an etch mask. Damage to the barrier was prevented by using a low Ga ion energy. It is also possible to use a process similar to the technologies described in the corresponding chapter in this textbook [59, 78], or to use a cross-type technology [83]. To achieve accurate small structures, the etching of the two stripes was done with a hard mask instead of a resist mask. Before the counter electrode was deposited, the structure was planarized with chemical mechanical planarization (CMP). Also, small junctions can be prepared by focused ion beam etching [84, 85].

Planarization Planarization is a key technology for fabricating high-density superconducting circuits. Devices for state-of-the-art RSFQ applications contain up to nine metallization layers [41], all patterned as their purpose requires. Without planarization, after each metallization layer, the resulting topography affects the continuity of subsequent metallization layers across steps and therefore their current-carrying capability. Also, if the topography is too pronounced, exact lithography is not possible due to resist thickness inhomogeneity and depth of focus issues. Various planarization techniques exist.

After etching of each metallization layer, an isolating layer can be deposited and then removed on the elevated areas by CMP, essentially a polishing process suited for the accurate leveling of surfaces [58, 59]. Figure 3.3.1.9 shows a circuit with several planarized isolation layers. For a decreased polishing rate, neutral slurry can been used, thereby polishing without chemical enhancement [86, 87]. Issues that need to be addressed with CMP or related processes are the thickness uniformity of the insulator after polishing as well as possible damage of the junction barriers by a large mechanical pressure. Also, the polishing rate is higher on small mesas that on large structures, so that dummy structures may have to be introduced to avoid overpolishing of small structures. The pattern dependence problem may also be addressed by the so-called Caldera planarization [88–91].

Planarization can also been achieved by depositing spin-on glass on the metallization structure and then opening vias by reactive-ion etching [73]. For the so-called etch-back planarization, SiO_2 and a resist are deposited on the metal structure. The resist layer planarizes the surface because of its thickness. The surface is then etched until the metal mesas are reached [82, 93].

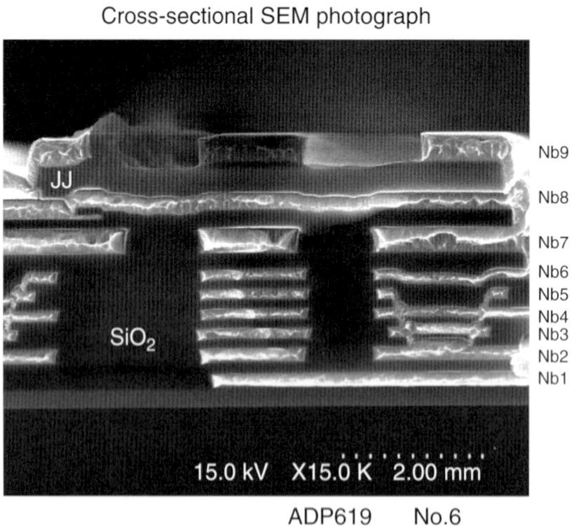

Cross-sectional SEM photograph

JJ
SiO_2
Nb9
Nb8
Nb7
Nb6
Nb5
Nb4
Nb3
Nb2
Nb1
15.0 kV X15.0 K 2.00 mm
ADP619 No.6

Figure 3.3.1.9 Cross-sectional SEM micrograph of a device with CMP-planarized layers. (Reprinted from Ref. [92], Copyright (2009), with permission from Elsevier.)

3.3.1.4 Circuits, Applications, and Resulting Requirements for Josephson Junctions

3.3.1.4.1 Josephson Voltage Standard

With the Josephson voltage standard (JVS) the unit volt is referenced to physical constants. It is a primary standard that needs no recalibration. Such voltage standards are used in many laboratories worldwide for high-precision voltage calibrations. In general, a JVS consists of an array of tens of thousands of Josephson junctions connected in series and fabricated in thin-film technology. A detailed summary of different types of JVS, their applications, and suitable fabrication processes has been compiled by Kohlmann and Behr [94].

For dc applications, JVS are based on underdamped Josephson junctions (SIS junctions) that are driven at 75 GHz. Under microwave irradiation, the junction array exhibits zero-current steps up to 10 V.

The interest in highly precise AC voltages has recently led to programmable JVS arrays based on Josephson junctions with intrinsic damping. Two different junction technologies have been developed: SINIS junctions and SNS junctions. Both types can be adapted for driving frequencies of 15 and 70 GHz. The array is divided into a binary sequence for biasing the subarrays on their zero- or first-order Shapiro step. By fast switching of the bias sources, an AC voltage up to the 10 V level can be sampled by a stepwise approximation of the sine wave. The bandwidth of such programmable JVS arrays is limited to few tens of kilohertz.

Arbitrary waveforms at even higher frequencies can be synthesized by operating the Josephson junctions with a pulse train [95, 96]. A challenge for this type of JVS standards is the generation of practical voltage levels up to 10 V.

Irradiation with 70 GHz generates first-order Shapiro steps at about 145 μV in a single Josephson junction. Thus, tens of thousands of identical junctions are required for a useful output voltage. This is a considerable technological challenge. Especially SINIS junctions ($Nb-AlO_x/Al/AlO_x-Nb$) are very sensitive to process parameter stability due to their extremely thin oxide barriers.

SNS junctions are more robust since their barrier layer is thicker (about 10 nm compared with 1 nm for SINIS junctions). Different materials for the normal conductor in these junctions have been investigated, that is, PdAu [97], $MoSi_2$ [98], and Nb_xSi_{1-x} [99] of which the latter is very promising. The first 10 V SNS arrays with an Nb_xSi_{1-x} barrier were successfully fabricated without junction failures [100].

Another requirement that is common to all kinds of JVS arrays is the uniform supply of all junctions with microwave power. The junctions must be integrated into suitable high-frequency transmission lines. Different types of lines are available. Depending on the required frequency and the level of integration, low-impedance microstrip lines [101], coplanar waveguide transmission lines [102], or coplanar striplines [103] may be used.

3.3.1.4.2 Superconducting Tunnel Junction

Superconducting tunnel junctions (STJs) consist of two superconducting films separated by a thin insulating tunnel barrier. The absorption of photons or particles generates quasiparticles by breaking up Cooper pairs [104]. The increase of the quasiparticle tunneling current can then be measured to determine the energy of the absorbed photon or particle. Due to the small energy gap Δ of superconductors in the order of millielectron volt, a high energy resolution $E/\Delta E$ can be achieved depending on the energy range to be measured. Reported values range from $E/\Delta E$ of about 270 at X-ray energies (22 eV at 5.9 keV [105]) to $E/\Delta E$ of about 12 in the visible spectrum (0.1 eV at 1.24 eV [106]). In addition, the excess charge lifetimes of microseconds allow comparably high count rates of about 10 000 counts/pixel.

In STJs, which have typical lateral dimensions of about $(100 \times 100)\ \mu m^2$, the main focus of attention needs to be placed on limitations or fluctuations in the tunneling process, because as the STJ is a nonthermal detector, the generated quasiparticles have to be measured before they thermalize with the lattice. Therefore, STJs often include a trapping layer, which locally decreases the energy gap due to the proximity effect and leads to a significant improvement in energy resolution [107].

Nowadays, the most common materials for the STJ electrodes are Nb and Ta. However, progress toward a higher resolution requires materials with a lower energy gap.

3.3.1.4.3 SIS Mixer

The active element of SIS quantum mixers are small-area Josephson tunnel junctions. SIS mixers provide coherent millimeter or sub-millimeter wave detection with mixer noise temperatures near the quantum limit $T_N = hf/2k_B$ [108]. Like all heterodyne sensors, their nonlinear characteristic is used to mix the signal from a local oscillator, f_{LO}, with the incoming signal f_{Sig}. The intermediate frequency, typically in the range of 1.5–4 GHz, is amplified by a broadband amplifier. From the resulting spectrum, information on the signal can be obtained.

For an SIS mixer, it is most important that the subgap characteristic of the Josephson junction be undisturbed. The junction capacitance places another limitation on the maximum detection frequency, since large capacitances short the circuit for high-frequency signals.

The energy gap of the superconducting electrodes sets the fundamental limit on the detection frequency. For Nb-based junctions, frequencies up to about 700 GHz can be detected. For even higher frequencies, there are recent developments toward materials with higher gap energy, like NbN and NbTiN, with the prospect to extend the operational range of SIS mixers up to 1.2 THz [109].

3.3.1.4.4 SQUID

SQUIDs belong to the most sensitive detectors for magnetic flux. SQUIDs are formed by a superconducting loop interrupted by one (for radio frequency (rf) SQUIDs [110]) or two (for DC SQUIDs [111]) Josephson junctions. In general, SQUIDs convert magnetic flux Φ (or other physical quantities which can be transformed into magnetic flux) into a voltage across the SQUID. For detailed information on the working principle, the reader is referred to the corresponding chapter in this textbook. SQUIDs cover a variety of application scenarios and reach sensitivities of about 1 fT/Hz$^{1/2}$ as magnetometers [112], 40 fT (mHz$^{1/2}$)$^{-1}$ as gradiometers [113], and about ≈ 10 fA Hz$^{-1/2}$ as current sensors [114, 115]. Furthermore, they are used, for example, as displacement sensors [116], susceptometers [117], or rf amplifiers [118].

The superior sensitivity of SQUIDs is determined by the inductance of the superconducting loop, the operation temperature, and the capacitance of the Josephson junction [119]. A low junction capacitance can be obtained by small physical dimensions and a suitable surrounding area of the junction. This leads to an enhancement of the voltage swing and an improvement in energy resolution. Compared to digital applications, the critical current of the Josephson junctions in SQUIDs is close to the thermal noise limit and amounts typically to about 5–50 μA. For typical DC SQUIDs operated at 4.2 K, j_c is in the range of 0.1–2 kA cm^{-2}, depending on the Josephson junction area.

3.3.1.4.5 **Qubit**

Solid-state qubits are formed by superconducting structures with Josephson junctions as nonlinear elements. Their controllability and good scalability attract interest not only for quantum information processing [120–122] but also for the fundamental study of light–matter interaction [123–126]. Since the main interest is basic research, only small numbers of samples (however with well-defined properties) are needed and only few attempts have been made for wafer-scale fabrication [127].

For the design of superconducting qubits, the relationship between Josephson coupling energy $E_J = I_C\Phi_0/2e$ and charging energy $E_C = 2e^2/C$ of the fabricated Josephson junction is one of the most important parameters [128]. Its value ranges from about 0.1 for so-called charge qubits [129] to about 100 for flux qubits [130] and defines whether charge or phase is a well-defined quantum variable. A feasible Josephson junction size can be estimated from these numbers. With a capacitance of 50 fF μm^{-2} for conventional tunnel junctions and a reasonable current density of 200 A cm^{-2}, the junction size lies between about 0.01 μm^2 for charge qubits and 0.4 μm^2 for flux qubits. The values scale with $j^{-0.5}$ so that at higher current densities, smaller Josephson junction sizes are required. For the fabrication of reproducible nanometer-sized junctions, the shadow evaporating technique is used.

For their use in quantum information processing, the coherence times of the qubits need to be improved. Therefore, recent studies involve the influence of $1/f$ fluctuations of the critical current [131], the morphology of Al-based Josephson junctions [132], and the analysis of two-level fluctuators in the barrier [133].

3.3.1.4.6 **Mixed-Signal Circuit**

The current trend for superconducting radiation detectors is toward an increased number of pixels, for example, for focal plane arrays. For such arrays, it is necessary to keep the technical complexity of wiring between room temperature and the cryogenic stage at an acceptable level. In addition, the thermal load due to the number of wires has to be kept small. Hence, there is a demand for low-power integrated readout circuits that can be operated close to the sensors. The current development toward mixed-signal applications, meaning the combination of, for example, superconductive radiation detectors with digital electronics such as RSFQ [134, 135], may offer this possibility. Similarly, the combination of digital SQUIDs and RSFQ techniques seems very attractive. Here, bandwidth limitations resulting from the delay due to signal propagation may be pushed toward higher frequencies.

These applications require an integrated fabrication technology of analog sensors with critical currents of the Josephson junctions close to the thermal limit and deterministic digital circuits with critical currents of about 100–500 μA. Further issues are the desired unshielded operation of RSFQ circuits and the influence of bias resistor and bias current distribution on the analog sensor performance.

3.3.1.4.7 RSFQ Digital Electronics

RSFQ electronics [2, 136] is a superconductor digital electronics clocked with tens of gigahertz. Its power consumption is very low. The circuit technology has reached a mature level in Japan [91, 137, 138] and in the United States [139–142]. General requirements for state-of-the-art circuits are an increased number of wiring layers, reduced feature sizes, and an increased current density of the Josephson junctions, together with high repeatability and high yield. An improved energy efficiency of superconductor electronics, necessary for high-performance computing systems, is the focus of current R&D programs [143–147]. In Europe [50], the activities concentrate on advanced multichannel sensor applications that require the integration of complex mixed-signal systems. An example for such an application is the combination of an analog sensor with an RSFQ circuit to read out superconductor single-photon nanowire detectors [134, 135].

References

1. Gurvitch, M., Washington, M.A., and Huggins, H.A. (1983) High quality refractory Josephson tunnel junctions utilizing thin aluminum layers. *Appl. Phys. Lett.*, **42**, 472–474.
2. Likharev, K.K., Mukhanov, O.A., and Semenov, V.K. (1985) *Resistive Single Flux Quantum Logic for the Josephson-Junction Technology, SQUID'85*, de Gruyter, pp. 1103–1108.
3. Kunert, J., Brandel, O., Linzen, S., Wetzstein, O., Toepfer, H., Ortlepp, T., and Meyer, H.-G. (2013) Recent developments in superconductor digital electronics technology at FLUXONICS foundry. *IEEE Trans. Appl. Supercond.*, **23** (5), 1101707.
4. Khabibov, M., Dolata, R., Buchholz, F.-I., and Niemeyer, J. (1995) Experimental investigation of RSFQ logic circuits realized by a Nb/Al2O3-Al/Nb technology process. ISEC '95 September 18–21, 1995, Nagoya, Japan.
5. Grönberg, L., Hassel, J., Helistö, P., and Yliammi, M. (2007) Fabrication process for RSFQ/Qubit systems. *IEEE Trans. Appl. Supercond.*, **17** (2), 952–954.
6. Castellano, G., Grönberg, L. *et al.* (2006) Characterization of a fabrication process for the integration of superconducting qubits and rapid-single-flux quantum circuits. *Supercond. Sci. Technol.*, **19**, 860–864.

7. Josephson, B.D. (1962) Possible new effects in superconductive tunneling. *Phys. Lett.*, **1**, 251–253.
8. Greiner, J.H., Kirchmer, C.J., Klepner, S.P., Lahiri, S.K., Warneche, A.J., Basavaiah, S., Yen, E.T., Baker, J.M., Brosious, P.R., Huang, H.C.W., Murakami, M., and Ames, I. (1980) Fabrication process for Josephson integrated circuits. *IBM J. Res. Dev.*, **24**, 195–205.
9. Lahiri, S.K. (1976) Metallurgical considerations with respect to electrodes and interconnection lines for Josephson tunneling circuits. *J. Vac. Sci. Technol.*, **13** (1), 148–151.
10. Suzuki, H., Imamura, T., and Hasuo, S. (1985) Fine-grained Au-Pb-Bi counterelectrodes for lead alloy Josephson junctions. *J. Appl. Phys.*, **57** (7), 2656–2658.
11. Imamura, T., Hasuo, S., and Yamaoka, T. (1985) Improvement of critical current uniformity in lead-alloy Josephson junctions. *J. Appl. Phys.*, **58** (6), 2280–2284.
12. Hoko, H., Yoshida, A., Tamura, H., Imamura, T, and Hasuo, S. (1997) Material dependence of initial failure rates of Josephson junctions. Extract Abstract Internatioanl Conference on Solid State Devices and Materials, Tokyo, Japan, 986, p. 447.
13. Kosaka, S., Shinoki, F., Takada, S., and Hayakawa, H. (1981) Fabrication of NbN/Pb Josephson tunnel junctions with a novel integration method. *IEEE Trans. Magn.*, MAG-17, 314–317.
14. Hikita, M., Takei, K., Iwata, T., and Igarashi, M. (1982) Fabrication of high quality NbN/Pb josephson tunnel junctions with plasma oxidized barriers. *Jpn. J. Appl. Phys.*, **21**, L724–L726.
15. Greiner, J.H., Basavaiah, S., and Ames, I. (1973) Fabrication of experimental Josephson tunneling circuits. *J. Vac. Sci. Technol.*, **11** (1), 81–84.
16. Kroger, H., Smith, L.N., and Jillie, D.W. (1981) Selective Niobium anodization process for fabricating Josephson tunnel junctions. *Appl. Phys. Lett.*, **39** (3), 280–282.
17. Aoyagi, M., Nakagawa, H., Kurosawa, I., and Takada, S. (1992) NbN/MgO/NbN Josephson junctions for integrated circuits; Jpn. *J. Appl. Phys.*, **31** (6A), 1778–1783.
18. Aoyagi, M., Nakagawa, H., Kurosawa, I., and Takada, S. (1992) SubmicronNbN/MgO/NbNJosephson tunnel junctions and their application to the logic circuit. *IEEE Trans. Appl. Supercond.*, **2** (3), 183–186.
19. Shoji, A., Aoyagi, M., Kosaka, S., Shinoki, E., and Hayakawa, H. (1985) Niobium nitride Josephson tunnel junctions with magnesium oxide barriers. *Appl. Phys. Lett.*, **46** (11), 1098–1100.
20. Radparvar, M., Berry, M.J., Drake, R.E., Faris, S.M., Whitley, S.R., and Yu, L.S. (1987) Fabrication and performance of all NbN Josephson tunnel junction circuits. *IEEE Trans. Magn.*, **23**, 1480–1483.
21. Hasuo, S. (1992) High quality niobium Josephson junction for ultrafast computers. *Thin Solid Films*, **216**, 21–27.
22. Misugi, T. and Shibatomi, A. (eds) (1993) *Compound and Josephson High-Speed Devices*, Chapter 8, Plenum Press, New York and London.
23. Aoyagi, M., Nakagawa, H., Kurosawa, I., and Takada, S. (1991) Josephson LSI fabrication technology using NbN/MgO/NbN tunnel junctions. *IEEE Trans. Magn.*, MAG-27 (2), 3180–3183.
24. Kerber, G.L., Abelson, L.A., Elmadjian, R.N., Hanaya, G., and Ladizinsky, E.G. (1997) An improved NbN integrated circuit process featuring thick NbN groundplane and lower parasitic circuit inductances. *IEEE Trans. Appl. Supercond.*, **7** (2), 2638–2643.
25. Terai, H. and Wang, Z. (2001) 9 K Operation of RSFQ logic cells fabricated by NbN integrated circuit technology. *IEEE Trans. Appl. Supercond*, **11** (1), 80–83.
26. Ruggiero, S. (1990) Artificial tunnel barriers, in *Superconducting Devices* (eds S.T. Ruggiero and D.A. Rudman), Academic Press, Inc., San Diego, CA.
27. Wang, Z., Uzawa, Y., and Kawakami, A. (1997) High current density NbN/AlN/NbN tunnel junctions for submillimeter wave SIS mixers.

IEEE Trans. Appl. Supercond., **7** (2), 2797–2800.

28. Plathner, B., Schicke, M., Lehnert, T., and Gundlach, K.H. (1997) NbN-MgO-NbN junctions prepared at room-temperature quartz substrates for quasiparticle mixers. *IEEE Trans. Appl. Supercond.*, **7** (2), 2603–2606.
29. Ambegaokar, V. and Baratoff, A. (1963) Tunneling between superconductors. *Phys. Rev. Lett.*, **10**, 486.
30. Bardeen, J., Cooper, L.N., and Schrieffer, J.R. (1957) Theory of superconductivity. *Phys. Rev.*, **108**, 1175.
31. Laquaniti, V., Maggi, S., Monticone, E., and Steni, R. (1996) Properties of rf-sputtered Nb/Al-AlOx/Nb Josephson SNAP junctions. *IEEE Trans. Appl. Supercond.*, **6** (1), 24–31.
32. Goodchild, M.S., Barber, Z.H., and Blamire, M.G. (1996) Conductance and leakage in superconducting tunnel junctions. *J. Vac. Sci. Technol.*, **A14**, 2427–2432.
33. Nakayama, A., Nagashima, H., Shimada, J., and Okabe, Y. (1995) Effects of aluminum overlayer thickness on characteristics of niobium tunnel junctions fabricated by DC magnetron sputtering. *IEEE Trans. Appl. Supercond.*, **5** (2), 2299–2302.
34. Lehnert, T., Schuster, K., and Gundlach, K.H. (1994) Gap voltage of Nb-Al/AlOx-Nb tunnel junctions. *Appl. Phys. Lett.*, **65** (1), 112–114.
35. Bhushan, M. and Macedo, E.M. (1991) Nb/AlOx/Nb trilayer process for the fabrication of submicron Josephson junctions and low.noise dc SQUIDs. *Appl. Phys. Lett.*, **58** (12), 1323–1325.
36. Du, J., Charles, A.D.M., Petersson, K.D., and Preston, E.W. (2007) Influence of Nb film surface morphology on the sub.gap leakage characteristics of Nb/AlOx-al/Nb Josephson junctions. *Supercond. Sci. Technol.*, **20**, S350–S355.
37. Miller, R.E., Mallison, W.H., Delin, K.A., and Macedo, E.M. (1993) Niobium trilayer Josephson tunnel junctions with ultrahigh critical current densities. *Appl. Phys. Lett.*, **63** (10), 1423–1425.
38. Wu, C.T. (1979) Intrinsic stress of magnetron-sputtered niobium films. *Thin Solid Films*, **64**, 103–110.
39. Imamura, T., Shiota, T., and Hasuo, S. (1992) Fabrication of high quality Nb/AlOx-Al/Nb Josephson junctions: I-sputtered Nb Films for junction electrodes. *IEEE Trans. Appl. Supercond.*, **2** (1), 1–14.
40. Imamura, T., Ohara, S., and Hasuo, S. (1991) Bias-sputtered Nb for reliable wirings in Josephson circuits. *IEEE Trans. Magn.*, **27** (2), 3176–3179.
41. Tanabe, K. and Hidaka, M. (2007) Recent progress in SFQ device technology in Japan. *IEEE Trans. Appl. Supercond.*, **17** (2), 494–499.
42. Cantor, R. and Hall, J. (2005) Six-layer process for the fabrication of Nb/Al-AlOX/Nb Josephson junction devices. *IEEE Trans. Appl. Supercond.*, **15**, 82–85.
43. Sandstrom, R.L., Kleinsasser, A.W., Gallagher, W.G., and Raider, S.I. (1987) Josephson integrated circuit process for scientific applications. *IEEE Trans. Magn.*, **23**, 1484–1488.
44. Sauvageau, J.E., Burroughs, C.J., Booi, P.A.A., Cromar, M.W., Benz, S.P., and Koch, J.A. (1995) Superconducting integrated circuit fabrication with low temperature ECR-based PECVD SiO2 dielectric films. *IEEE Trans. Appl. Supercond.*, **5** (2), 2303–2309.
45. Shiota, T., Ohara, S., Imamura, T., and Hasuo, S. (1993) High-resistivityy Zr resistors with Ti barrier layer for Nb Josephson circuits. *IEEE Trans Appl. Supercond.*, **3** (3), 3049–3053.
46. Ipht-Jena *www.ipht-jena.de* (accessed 20 May 2014).
47. BESTEC GmbH *www.bestec.de* (accessed 19 June 2014).
48. Mallison, W.H., Miller, R.E., and Kleinsasser, A.W. (1995) Effect of growth conditions on the electrical properties of Nb/Al-oxide/Nb tunnel junctions. *IEEE Trans. Appl. Supercond.*, **5** (2), 2330–2333.
49. Fritzsch, L., Köhler, H.-J., Thrum, F., Wende, G., and Meyer, H.-G. (1997) *Preparation and RBS Measurements of Nb/Al-AlOX/Nb Josephson Junctions*

With Very Low Critical Current Density, Institute of Physics Conference Series No. 158, IOP Publishing Ltd, pp. 491–494.

50. Anders, S. *et al.* (2010) European roadmap on superconductive electronics – status and perspectives. *Physica C*, **470**, 2079–2126.
51. Kominami, S., Yamada, H., Miyamoto, N., and Takagi, K. (1993) Effects of underlayer roughness on Nb/AlOX/Nb junction characteristics. *IEEE Trans. Appl. Supercond.*, **3** (1), 2182–2186.
52. Kleinsasser, A.W., Miller, R.E., and Mallison, W.H. (1995) Dependence of critcal current density on oxygen exposure in Nb-AlOX-Nb tunnel junctions. *IEEE Trans. Appl. Supercond.*, **5**, 26–30.
53. Müller, F., Kohlmann, J., Hebrank, F.X., Weimann, T., Wolf, H., and Niemeyer, J. (1995) Performance of josephson array systems related to fabrication techniques and design. *IEEE Trans. Appl. Supercond.*, **5** (2), 2903–2906.
54. Murduck, J.M., Porter, J., Dozier, W., Sandell, R., Burch, J., Bulman, J., Dang, C., Lee, L., Chan, H., Simon, R.W., and Silver, A.H. (1989) Niobium trilayer process for superconducting circuits. *IEEE Trans. Magn.*, **25** (2), 1139–1142.
55. Kadin, A.M., Mancini, C.A., Feldman, M.J., and Brock, D.K. (2001) Can RSFQ logic circuits be scaled to deep submicron junctions? *IEEE Trans. Appl. Supercond.*, **11** (1), 1050–1055.
56. Aoyagi, M., Shoji, A., Kosaka, S., Shinoki, F., and Hayakawa, H. (1986) A 1um cross line junction process. *Adv. Cryog. Eng. Mater.*, **32**, 557–563.
57. Dang, H. and Radparvar, M. (1991) A process for fabricating submicron all-refractory Josephson tunnel junctions. *IEEE Trans. Magn.*, **27** (2), 3157–3160.
58. Ketchen, M.B., Pearson, D., Kleinsasser, A.W., Hu, C.-K., Smyth, M., Logan, J., Stawiasz, K., Baran, E., Jaso, M., Ross, T., Pedrillo, K., Manny, M., Basavaiah, S., Brodsky, S., Kaplan, S.B., Gallagher, W.J., and Bushan, M. (1991) Sub-um, planarized, Nb-Alox-Nb Josephson process for 125 mm wafers developed in partnership with Si technology. *Appl. Phys. Lett.*, **59** (20), 2609–2611.
59. Bao, Z., Bhushan, M., Han, S., and Lukens, J.E. (1995) Fabrication of high quality, deep-submikron Nb/AlOx/Nb Josephson junctions using chemical mechanical polishing. *IEEE Trans. Appl. Supercond.*, **5** (2), 2731–2734.
60. Numata, H., Nagasawa, S., Koike, M., and Tahara, S. (1995) Fabrication technology for a high. Density Josephson LSI using an electron cyclotron resonance etching technique and a bias-sputtering planarization. 5th International Superconductor Electronics Conference (ISEC '95), Nagoya, Japan, September 18-21, 1995.
61. Imamura, T. and Hasuo, S. (1989) Effects of intrinsic stress on submicrometer Nb/AlOx/Nb Josephson junctions. *IEEE Trans. Magn.*, **25**, 1119.
62. Huq, S.E., Blamire, M.G., Evetts, J.E., Husko, D.G., and Ahmed, H. (1991) Fabrication of sub-micron whole-wafer SIS tunnel junctions for millimeter wave mixers. *IEEE Trans. Magn.*, **27**, 3161.
63. Dolata, R., Weimann, T., Scherer, H.-J., and Niemeyer, J. (1999) Sub μm Nb/AlOx/Nb Josephson junctions fabricated by anodization techniques. *IEEE Trans. Appl. Supercond.*, **9**, 3255.
64. Fritzsch, L., Elsner, H., Schubert, M., and Meyer, H.-G. (1999) SNS and SIS Josephson junctions with dimensions down to the submicron region prepared by a unified technology. *Supercond. Sci. Technol.*, **12**, 880.
65. Voss, M., Karpov, A., and Gundlach, K.H. (1993) Submicron Nb-Al-oxide-Nb junctions for frequency mixers. *Supercond. Sci. Technol.*, **6**, 373.
66. Dierichs, M.M.T., Panhuyzen, R.A., Noningh, C.E., de Boer, M.J., and Klapwijk, T.M. (1993) Submicron Niobium junctions for submillimeter-wave mixers using optical lithography. *Appl. Phys. Lett.*, **62**, 774.
67. Maier, D., Rothermel, H., Gundlach, K.H., and Zimmermann, R. (1996) Submicron Nb - Al/Al oxide – Nb tunnel junctions sandwiched between Al films. *Physica C*, **268**, 26.
68. Meng, X., Zheng, L., Wong, A., and Van Duzer, T. (2001) Micron and submicron Nb/Al-AlOx/Nb tunnel

junctions with high critical current densities. *IEEE Trans. Appl. Supercond.*, **11**, 365.

69. Anders, S., Schmelz, M., Fritzsch, L., Stolz, L., Zakosarenko, V., Schoenau, T., and Meyer, H.-G. (2007) Submicrometer-sized, cross-type Nb-AlOx-Nb tunnel junctions with low parasitic capacitance. *Supercond. Sci. Technol.*, **22**, 064012.
70. Bloemhof, E.E. (1994) Production of high quality Nb/Al-oxide/Nb SIS trilayers for submillimeter wave mixers. *Int. J. Infrared Millimetre Waves*, **15**, 2031.
71. Chen, W., Patel, V., and Lukens, J.E. (2004) Fabrication of high-quality Josephson junctions for quantum computation using a self-aligned process. *Microelectron. Eng.*, **73-74**, 767.
72. Worsham, A.H., Prober, D.E., Kang, J.H., Przybysz, J.X., and Rocks, M.J. (1991) High-quality sub-micron Nb trilayer tunnel junctions for a 100 GHz SIS receiver. *IEEE Trans. Magn.*, **27**, 3165.
73. Pavolotsky, A., Weimann, T., Scherer, H., Niemeyer, J., Zorin, A., and Krupenin, V. (1999) Novel method for fabricating deep submicron Nb/AlOx/Nb tunnel junctions based on spin-on glass planarization. *IEEE Trans. Appl. Supercond.*, **9**, 3251.
74. Meng, X. and Van, T. (2003) Duzer; Light-anodization process for high-jc micron and submicron superconductiong junction and integrated circuit fabrication. *IEEE Trans. Appl. Supercond.*, **13**, 91.
75. Dolan, G.J. (1977) Offset masks for lift-off photoprocessing. *Appl. Phys. Lett.*, **31**, 337.
76. Harada, Y., Haviland, D.B., Delsing, P., Chen, C.D., and Claeson, T. (1994) Fabriction and measurement of a Nb based superconducting single electron transistor. *Appl. Phys. Lett.*, **65**, 636.
77. Ono, R.H., Sauvageau, J.E., Jain, A.K., Schwartz, D.B., Springer, K.T., and Lukens, J.E. (1985) Suspended metal mask techniques in Josephson junction fabrication. *J. Vac. Sci. Technol.*, **B 3** (1), 282.
78. Dolata, R., Scherer, H., Zorin, A.B., and Niemeyer, J. (2003) Single electron transistors with Nb/AlOx/Nb junctions. *J. Vac. Sci. Technol.*, **B21** (2), 775.
79. Martinis, J.M. and Ono, R.H. (1990) Fabrication of untrasmall Nb-AlOx-Nb Josephson tunnel junctions. *Appl. Phys. Lett.*, **57**, 629.
80. Bluethner, K., Goetz, M., Haedicke, A., Krech, W., Wagner, T., Muehlig, M., Fuchs, H.-J., Huebner, U., Schelle, D., Kley, E.-B., and Fritzsch, L. (1997) Single electron transistors based on Al/AlOx/Al and Nb/AlOx/Nb tunnel junctions. *IEEE Trans. Appl. Supercond.*, **7**, 3009.
81. Vdovichev, S.N., Klimov, A.Y., Nozdrin, Y.N., and Rogov, V.V. (2004) Edge-type Josephson junctions with silicon nitride spacer. *Tech. Phys. Lett.*, **30**, 374.
82. Akaike, H., Watanabe, T., Nagai, N., Fujimaki, A., and Hayakawa, H. (1995) Fabrication of submicron Nb/Al-AlOx/Nb tunnel junctions using focused ion beam implanted Nb patterning (FINP) technique. *IEEE Trans. Appl. Supercond.*, **5**, 2310.
83. Born, D., Wagner, T., Krech, T., Huebner, U., and Fritzsch, L. (2001) Fabrication of ultrasmall tunnel junctions by electron beam direct-writing. *IEEE Trans. Appl. Supercond.*, **11**, 373.
84. Watanabe, M., Nakamura, Y., and Tsai, J.-S. (2004) Circuit with small-capacitance high-quality Nb junctions. *Appl. Phys. Lett.*, **84**, 410.
85. Fretto, M., Enrico, E., De Leo, N., Boarino, L., Rocci, R., and Lacquaniti, V. (2013) Nano SNIS junctions fabricated by 3D FIB sculpting for application to digital electronics. *IEEE Trans. Appl. Supercond.*, **23**, 1101104.
86. Numata, H. and Tahara, S. (2001) Fabrication technology for Nb integrated circuits. *IEICE Trans. Electron.*, **E84**, 2.
87. Terai, H., Kawakami, A., and Wang, Z. (2002) Sub-micron NbN/AlN/NbN tunnel junction with high critical current density. *Physica C*, **38**, 372–376.
88. Hinode, K., Nagasawa, S., Sugita, M., Satoh, T., Akaike, H., Kitagawa, Y., and Hidaka, M. (2003) New Nb multi-layer fabrication process forlarge-scale SFQ circuits. *IEICE Trans. Electron.*, **E86-C**, 2511.

89. Nagasawa, S., Hinode, K., Sugita, M., Satoh, T., Akaike, H., Kitagawa, Y., and Hidaka, M. (2003) Planarized multilayer fabrication technology for LTS large-scale SFQ circuits. *Supercond. Sci. Technol.*, **16**, 1483.
90. Nagasawa, S., Hinode, K., Satoh, T., Akaike, H., Kitagawa, Y., and Hidaka, M. (2004) Development of advanced Nb process for SFQ circuits. *Physica C*, **1429**, 412–414.
91. Satoh, T., Hinode, K., Akaike, K., Nagasawa, S., Kitagawa, Y., and Hidaka, M. (2005) Fabrication process of planarized multi-layer Nb integrated ciecuits. *IEEE Trans. Appl. Supercond.*, **15**, 78.
92. Nagasawa, S., Satoh, T., Hinode, K., Kitagawa, Y., Hidaka, M., Akaike, H., Fujimaki, A., Takagi, K., Takagi, N., and Yoshikawa, N. (2009) New Nb multi-layer fabrication process for large-scale SFQ circuits. *Physica C*, **469**, 1578–1584.
93. Numata, H., Nagasawa, S., Koike, M., and Tahara, S. (1996) Fabrication technology for a high-density Josephson LSI using an electron cyclotron resonance etching technique and a bias-sputtering planarization. *Supercond. Sci. Technol.*, **9**, A42.
94. Kohlmann, J. and Behr, R. (2011) in *Superconductivity – Theory and Applications* (ed. A.M. Luiz), InTech, pp. 239–260, ISBN: 978-953-307-151-0.
95. Benz, S.P., Hamilton, C.A., Burroughs, C.J., Harvey, T.E., Christian, L.A., and Przybysz, J.X. (1998) Pulse-driven Josephson digital/analog converter. *IEEE Trans. Appl. Supercond.*, **8** (42).
96. Kieler, O.F., Iuzzolino, R., and Kohlmann, J. (2009) Sub-μm SNS Josephson junction arrays for the Josephson arbitrary waveform sysnthesizer. *IEEE Trans. Appl. Supercond.*, **19**, 230.
97. Benz, S.P., Hamilton, C.A., Burroughs, C.J., Harvey, T.E., and Christian, L.A. (1997) Stable 1 V programmable voltage standard. *Appl. Phys. Lett.*, **71**, 1866.
98. Chong, Y., Burroughs, C.J., Dresselhaus, P.D., Hadacek, N., Yamamori, H., and Benz, S.P. (2005) Practical high-resolution programmable Josephson voltage standards using double- and triple-stacked MoSi2 barrier junctions. *IEEE Trans. Appl. Supercond.*, **15**, 461.
99. Baek, B., Dresselhaus, P.D., and Benz, S.P. (2006) Co-sputtered amorphous NbxSi1-x barriers for Josephson-junction circuits. *IEEE Trans. Appl. Supercond.*, **16**, 1966.
100. Mueller, F., Behr, R., Weiman, T., Palafox, L., Olaya, D., Dresselhaus, P.D., and Benz, S.P. (2009) 1 V and 10 V SNS programmable voltage standards for 70 GHz. *IEEE Trans. Appl. Supercond.*, **19**, 981.
101. Niemeyer, J., Hinken, J.H., and Kautz, R.L. (1984) Microwave-induced constant-voltage steps at one volt from a series array of Josephson junctions. *Appl. Phys. Lett.*, **45**, 478.
102. Benz, S.P. (1995) Superconductor – normal- superconductor junctions for programmable voltage standards. *Appl. Phys. Lett.*, **67**, 2714.
103. Schubert, M., May, T., Wende, G., Fritzsch, L., and Meyer, H.-G. (2001) Coplanar strips for Josephson voltage standard circuits. *Appl. Phys. Lett.*, **79**, 1009.
104. Peacock, A., Verhoeve, P., Rando, N., van Dordrecht, A., Taylor, B.G., Erd, C., Perryman, M.A.C., Venn, R., Howlett, J., Goldie, D.J., Lumley, J., and Wallis, M. (1996) Single optical photon detection with a superconducting tunnel junction. *Nature*, **381**, 135–137.
105. Verhoeve, P., Rando, N., Peacock, A., Van Dordrecht, A., Taylor, B.G., and Goldie, D.J. (1998) High-resolution x-ray spectra measured using tantalum superconducting tunnel junctions. *Appl. Phys. Lett.*, **72** (25), 3359–3361.
106. Peacock, A. (1999) *Physica B*, **263-264**, 595.
107. Booth, N.E. (1987) Quasiparticle trapping and the quasiparticle multiplier. *Appl. Phys. Lett.*, **50**, 293.
108. Feldman, M.J. (1987) Quantum noise in the quantum theory of mixing. *IEEE Trans. Magn.*, **MAG-23**, 1054.
109. Karpov, A., Miller, D., Zmuidzinas, J., Stern, J., Bumble, B., and LeDuc, H. (2002) Low noise SIS mixer for the frequency above 1 THz, in *Proceedings Far–IR, Sub–MM, and MM Detector*

Technology Workshop (eds J. Wolf and J. Davidson).

110. Zimmermann, J.E., Thiene, P., and Harding, J.T. (1970) Design and operation of stable rf-biased superconducting point-contact quantum devices. *J. Appl. Phys.*, **41**, 1572.
111. Jaklevic, R.C., Lambe, J., Silver, A.H., and Mercereau, J.E. (1964) Quantum Interference effects in Josephson tunneling. *Phys. Rev. Lett.*, **12**, 159.
112. (a) Drung, D., Knappe, S., and Koch, H. (1995) Theory for the multiloop DC superconducting quantum interference device magnetometer and experimental verification. *J. Appl. Phys.*, **77**, 4088–4098; (b) Schmelz, M., Stolz, R., Zakosarenko, V., Schönau, T., Anders, S., Fritzsch, L., Mück, M., and Meyer, H.-G. (2011) Field-stable SQUID magnetometer with sub-fT Hz $-$ $^{1/2}$ resolution based on sub-micrometer cross-type Josephson tunnel junctions. *Supercond. Sci. Technol.*, **24**, 065009.
113. Stolz, R., Zakosarenko, V., Fritzsch, L., Oukhanski, N., and Meyer, H.-G. (2001) Long baseline thin film SQUID gradiometers. *IEEE Trans. Appl. Supercond.*, **11**, 1257–1260.
114. Gay, F., Piquemal, F., and Genvès, G. (2000) Ultra low noise amplifier based on a cryogenic current comparator. *Rev. Sci. Instrum.*, **71**, 4592–5.
115. Zakosarenko, V., Schmelz, M., Stolz, R., Schönau, T., Fritzsch, L., Anders, S., and Meyer, H.G. (2012) Femtoammeter on the base of SQUIDwith thin-film flux transformer. *Supercond. Sci. Technol.*, **25** (9), 095014.
116. Gottardi, L., de Waard, A., Usenko, O., Frossati, G., Podt, M., Flokstra, J., Bassan, M., Fafone, V., Minenkov, Y., and Rocchi, A. (2007) Sensitivity of the spherical gravitational wave detector MiniGRAIL operating at 5 K. *Phys. Rev. D*, **76** (10), 102005.
117. Ketchen, M.B., Kopley, T., and Ling, H. (1984) Minature SQUID susceptometer. *Appl. Phys. Lett.*, **44**, 1008–1010.
118. Hilbert, C. and Clarke, J. (1985) dc SQUIDs as radiofrequency amplifiers. *J. Low Temp. Phys.*, **61**, 263–280.
119. (a) Tesche, C.D. and Clarke, J. (1977) dc SQUID: noise and optimization. *J. Low Temp. Phys.*, **29**, 301–31; (b) Clarke, J. (1996) in *SQUID fundamentals SQUIDs: Fundamentals, Fabrication and Applications*, NATO ASI Series E: Applied Sciences, vol. **329** (ed. H. Weinstock), Kluwer Academic Publishers, Dordrecht, pp. 1–62.
120. Izmalkov, A., Grajcar, M., van der Ploeg, S.H.W., Hübner, U., Il'ichev, E., Meyer, H.-G., and Zagoskin, A.M. (2006) Measurement of the ground-state flux diagram of three coupled qubits as a first step towards the demonstration of adiabatic quantum computation. *Europhys. Lett.*, **76** (3), 533.
121. DiCarlo, L., Chow, J.M., Gambetta, J.M., Bishop, L.S., Johnson, B.R., Schuster, D.I., Majer, J., Blais, A., Frunzio, L., Girvin, S.M., and Schoelkopf, R.J. (2009) Demonstration of two-qubit algorithms with a superconducting quantum processor. *Nature*, **460**, 240.
122. Fedorov, A., Steffen, L., Baur, M., da Silva, M.P., and Wallraff, A. (2012) Implementation of a toffoli gate with superconducting circuits. *Nature*, **481**, 170.
123. Wallraff, A., Schuster, D.I., Blais, A., Frunzio, L., Huang, R.-S., Majer, J., Kumar, S., Girvin, S.M., and Schoelkopf, R.J. (2004) Circuit quantum electrodynamics: coherent coupling of a single photon to a Cooper pair box. *Nature*, **431**, 162.
124. Astafiev, O., Inomata, K., Niskanen, A.O., Yamamoto, T., Pashkin, Y.A., Nakamura, Y., and Tsai, J.S. (2007) Single artificial-atom lasing. *Nature*, **449**, 588.
125. Astafiev, O., Zagoskin, A.M., Abdumalikov, A.A. Jr.,, Pashkin, Y.A., Yamamoto, T., Inomata, K., Nakamura, Y., and Tsai, J.S. (2010) Resonance fluorescence of a single artificial atom. *Science*, **327**, 840.
126. Oelsner, G., Macha, P., Astafiev, O.V., Il'ichev, E., Grajcar, M., Hübner, U., Ivanov, B.I., Neilinger, P., and Meyer, H.-G. (2013) Superconductivity (cond-mat.supr-con); quantum physics (quant-ph). *Phys. Rev. Lett.*, **110**, 053602.

127. Prunila, M., Meschke, M., Gunnarsson, D., Enouz-Vedrenne, S., Kivioja, J.M., and Pekola, J.P. (2010) Ex-situ tunnel junction process technique characterized by coulomb blockade thermometry. *J. Vac. Sci. Technol. B*, **28**, 1026.
128. Clarke, J. and Wilhelm, F.K. (2008) Superconducting quantum bits. *Nature*, **453**, 1031.
129. Nakamura, Y., Pashkin, Y.A., and Tsai, J.S. (1999) *Nature*, **449**, 786.
130. Mooij, J.E., Orlando, T.P., Levitov, L., Tian, L., van der Wal, C.H., and Lloyd, S. (1999) Josephson persistent current qubit. *Science*, **285**, 1036.
131. Van Harlingen, D.J., Robertson, T.L., Plourde, B.L.T., Reichardt, P.A., Crane, T.A., and Clarke, J. (2004) Decoherence in Josephson-junction qubits due to critical-current fluctuations. *Phys. Rev. B*, **70**, 064517.
132. Roddatis, V.V., Hübner, U., Ivanov, B.I., Il'ichev, E., Meyer, H.-G., Koval'chuck, M.V., and Vasiliev, A.L. (2011) The morphology of Al-based submicron Josephson junction. *J. Appl. Phys.*, **110**, 123903.
133. Grabovskij, G.J., Peichel, T., Lisenfeld, J., Weiss, G., and Ustinov, A.V. (2012) Strain tuning of individual atomic tunneling systems detected by a superconducting qubit. *Science*, **338**, 232.
134. Terai, H., Miki, S. *et al.* (2010) Demonstration of single-flux-quantum readout operation for superconducting single-photon detectors. *Appl. Phys. Lett.*, **97**, 112510.
135. Ortlepp, T., Hofherr, M. *et al.* (2011) Demonstration of digital readout circuit for superconducting nanowire single photon detector. *Opt. Express*, **19**, 18593.
136. Likharev, K.K. and Semenov, V.K. (1991) RSFQ logic/memory family: a new Josephson-junction technology for sub-terahertz-clock-frequency digital systems. *IEEE Trans. Appl. Supercond.*, **1**, 3–28.
137. Hasuo, S. (2008) Special section on recent progress in superconductive digital electronics. *IEICE Trans.*, E91-C, 251.
138. Hasuo, S. (2012) *Digital Electronics in Japan, 100 Years of Superconductivity*, CRC Press Taylor & Francis Group, pp. 397–407, ISBN 978-1-4398-4946-0.
139. Mukhanov, O.A., Kirichenko, D. *et al.* (2008) Superconductor Digital-RF receiver systems. *IEICE Trans. Electron.*, E91-C, 306.
140. Bedard, F. (2012) *Digital Electronics in the USA, 100 Years of Superconductivity*, CRC Press Taylor & Francis Group, pp. 407–415, ISBN 978-1-4398-4946-0.
141. Tolpygo, S.K., Yohannes, D. *et al.* (2007) 20 kA/cm2 process development for superconducting integrated circuits with 80 GHz clock frequency. *IEEE Trans. Appl. Supercond.*, **17**, 946–951.
142. Brock, D.K., Kadin, A.M. *et al.* (2001) Retargeting RSFQ cells to a submicron fabrication process. *IEEE Trans. Appl. Supercond.*, **11**, 369–372.
143. Mukhanov, O.A. (2011) Energy-efficient single flux quantum technology. *IEEE Trans. Appl. Supercond.*, **21**, 760–769.
144. Volkmann, M.H., Sahu, A. *et al.* (2013) Implementation of energy efficient single flux quantum digital circuits with sub-aJ/bit operation. *Supercond. Sci. Technol.*, **26**, 015002.
145. Ortlepp, T., Wetzstein, O. *et al.* (2011) Reduced power consumption in superconducting electronics. *IEEE Trans. Appl. Supercond.*, **21**, 770–775.
146. Herr, Q.P., Herr, A.Y. *et al.* (2011) Ultra-low-power superconductor logic. *J. Appl. Phys.*, **109**, 103903.
147. Takeuchi, N., Ehara, K. *et al.* (2013) Margin and energy dissipation of adiabatic quantum-flux-parametron logic at finite temperature. *IEEE Trans. Appl. Supercond.*, **23**, 1700304.

3.3.2
HTS Josephson Junctions

Keiichi Tanabe

3.3.2.1 Introduction

The discovery of high-temperature superconducting (HTS) cuprates with a layered structure soon stimulated researches on the development of underdamped or superconductor–insulator–superconductor (SIS)-type Josephson junctions with the gap voltage and the characteristic voltage, I_cR_n (here, I_c and R_n are the critical current and the normal resistance of the junction), much larger than those for Nb-based Josephson junctions. In spite of enormous efforts, such researches have not been successful. This is primarily due to the peculiar physical properties of cuprate superconductors.

Because of the high T_c and the resultant larger energy gap, the coherence length of the HTS cuprate superconductors is much shorter than that for Nb-based superconductors. In particular, the coherence length along the c-axis is extremely short and comparable to the thickness of the multiple CuO_2 plane. Since Josephson junctions reflect superconducting properties of material only within the coherence length from the barrier interface, the shorter coherence length imposes strict control of the superconductor–barrier interface. Cuprate superconductors exhibit superconductivity only when a proper amount of charge carries are doped to the antiferromagnetic insulating parent materials and structural disorder readily induces carrier depletion and degraded superconductivity. Thin films of cuprate superconductors are prepared at a high temperature around 700 °C, which could lead to mutual atomic diffusion at the interface. Thus, it is not easy to fabricate junctions with an ideal superconductor–barrier interface. On the other hand, because of the highly anisotropic physical properties, Josephson junctions are naturally formed along the c-axis in cuprate superconductors [1, 2]. These "intrinsic" junctions in single crystals exhibit properties of an array of underdamped Josephson junctions. However, due to the $d_{x^2-y^2}$ symmetry of the order parameter in these materials, it is difficult to obtain properties of ideal SIS junctions with a sharp current rise at the gap voltage even in this case [2].

In contrast, weak-link type or overdamped Josephson junctions with properties good enough for various electronic applications can be fabricated for cuprate superconductors. Due to the physical properties mentioned above, high-angle grain boundaries (GBs) in cuprate superconductors behave as weak links [3], and various types of GB Josephson junctions have been developed [4]. Junctions with a variety of artificial barrier materials, in particular, those having ramp-edge or edge-type junction geometry were also developed with the aim of application to integrated circuits such as single flux quantum (SFQ) devices [5]. In the following sections, the structures, transport properties and their understandings, and the fabrication methods for various types of HTS Josephson junctions with cuprate superconductors are described, mainly focusing on those actually used for electronic application. The key materials in cuprate superconductors are REBCO ($REBa_2Cu_3O_{7-\delta}$, here RE is Y or rare earth elements) superconductors with T_c values above 90 K. Junctions with other cuprate superconductors, MgB_2, and recently discovered iron-based superconductors are also partially described.

3.3.2.2 **Various Types of Junctions**

The structures of representative HTS Josephson junctions so far developed are schematically illustrated in Figure 3.3.2.1. HTS junctions can be roughly classified into two categories, GB junctions and junctions with an artificially formed barrier layer. Bicrystal junctions can be fabricated simply by epitaxially growing an HTS film on a bicrystal substrate with a straight GB line, and this GB is transcribed into the HTS film. Similar GB junctions can be also fabricated by utilizing a buffer layer technique, and such junctions are called *bi-epitaxial junctions* [6]. Step-edge junctions are another type of GB junctions and are fabricated by depositing an HTS film on a substrate with a step fabricated by an ion milling technique. Though careful fabrication of a step is required, step-edge junctions have an advantage of better flexibility not limited by the GB line of the substrate. Planar-type junctions with a weak link in an HTS thin film can also be fabricated utilizing electron- or ion-beam irradiation, and good Josephson properties have been reported in some cases.

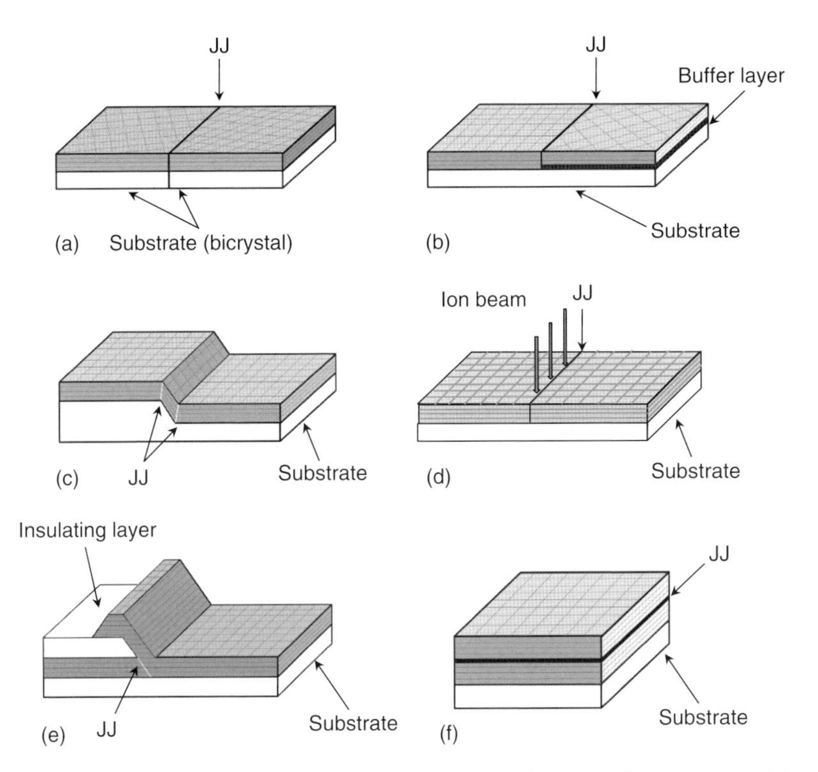

Figure 3.3.2.1 Schematic illustrations of various types of HTS Josephson junctions. (a) Bicrystal junction, (b) bi-epitaxial junction, (c) step-edge junction, (d) planar-type junction (ion-beam damaged), (e) ramp-edge junction, and (f) stacked-type junction.

In HTS junctions with an artificial barrier, ramp-edge or edge-type junction geometry is mostly employed. Fabrication of this type of junctions requires a multilayer technology including deposition of at least two HTS layers and an intermediary insulating layer. The edge of an insulator–HTS bilayer with a slope of typically 30° is fabricated by ion milling. The junctions are fabricated by preparing a barrier layer and subsequent deposition of an upper HTS layer. Cuprate superconductors have large anisotropy in their transport properties and longer coherence length as well as higher critical current density (J_c) along the *ab* plane. In ramp-edge junctions, current flows along the *ab* plane of *c*-axis-oriented HTS layers and limited only by a barrier layer not by anisotropic transport along the *c* axis. Since this type of junctions has two HTS layers, flexible wiring and crossover wiring can be readily realized, which is advantageous in fabricating complicated device structures and integrated circuits. Vertically stacked junctions with a barrier layer similar to Nb-based junction have also been developed. However, it is not easy to implement such junctions in actual devices and circuits because one more HTS layer is required for wiring.

3.3.2.3 Grain-Boundary Junctions

3.3.2.3.1 Bicrystal Junctions

Bicrystal substrate is composed of two halves, at least one of which has been rotated in plane about the *c*-axis or *a*-axis before being fused together in the center. The first systematic study on the transport properties of GBs in HTS films was performed by Dimos and coworkers [3] using bicrystal substrates. They grew YBCO epitaxial films on $SrTiO_3$ bicrystal substrates with three different GB configurations, [001]-tilt, [100]-tilt, and [100]-twist boundaries, as schematically shown in Figure 3.3.2.2, and observed systematic decreases in J_c across the GB with an increase in the misorientation angle θ. More comprehensive reviews on GBs in HTS materials were given by Hilgenkamp and Mannhart [4], as well as Tafuri and Kirtley [7]. Figure 3.3.2.3 shows the misorientation angle dependence of the intergrain J_c at 4.2 K for YBCO thin films on [001]-tilt $SrTiO_3$ bicrystal substrates [4, 8–10]. The J_c decreases almost exponentially with increasing θ, although a substantial spread of J_c is observed because the intergrain J_c is sensitive to the intragrain J_c or the film quality as well as the quality of bicrystal substrates. Similar exponential decay, expressed as

$$J_{c,\mathrm{GB}}(\theta) = J_{c,\mathrm{grain}} \exp\left(-\frac{\theta}{\theta_0}\right)$$

is also observed at 77 K and for [100]-tilt, [100]-twist boundaries, and bicrystal GBs based on other cuprate superconductors such as $Bi_2Sr_2CaCu_2O_{8+x}$ and $HgBa_2CaCu_2O_{6+x}$ [4]. The characteristic angle θ_0 is ~4.2–4.8° for YBCO junctions.

The studies on the transport properties of low-angle GBs also indicated that J_c starts to decrease at a critical angle of ~3–5° and shows an exponential decay at higher angles [4, 10]. I–V characteristics of bicrystal junctions with $\theta < 10°$ are mostly dominated by flow of Abrikosov vortices, while junctions with $\theta >$

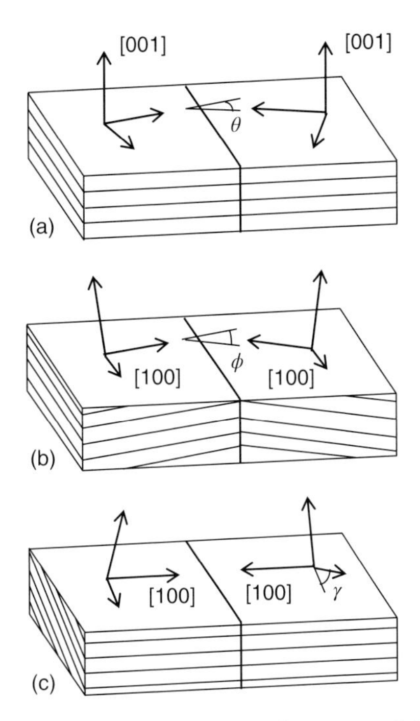

Figure 3.3.2.2 Schematic illustrations of three types of bicrystal grain boundaries. (a) [001]-tilt, (b) [100]-tilt, and (c) [100]-twist boundaries.

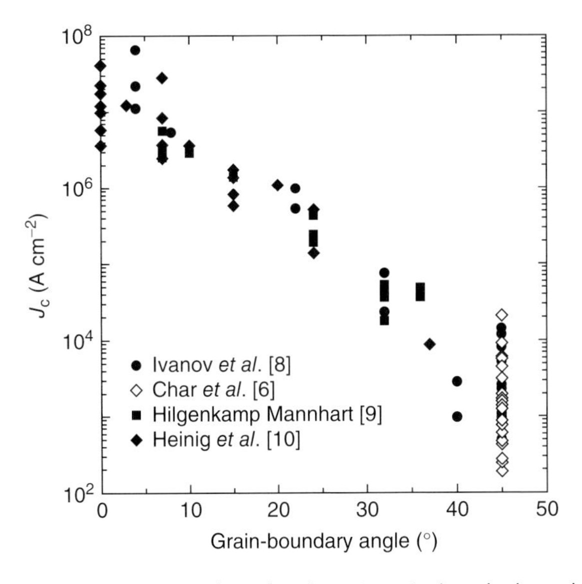

Figure 3.3.2.3 Grain boundary (misorientation) angle dependence of J_c at 4.2 K for YBCO [001]-tilt bicrystal junctions reported by several groups. (Adapted and reproduced from Ref. [4]. Reproduced with permission of American Physical Society (APS))

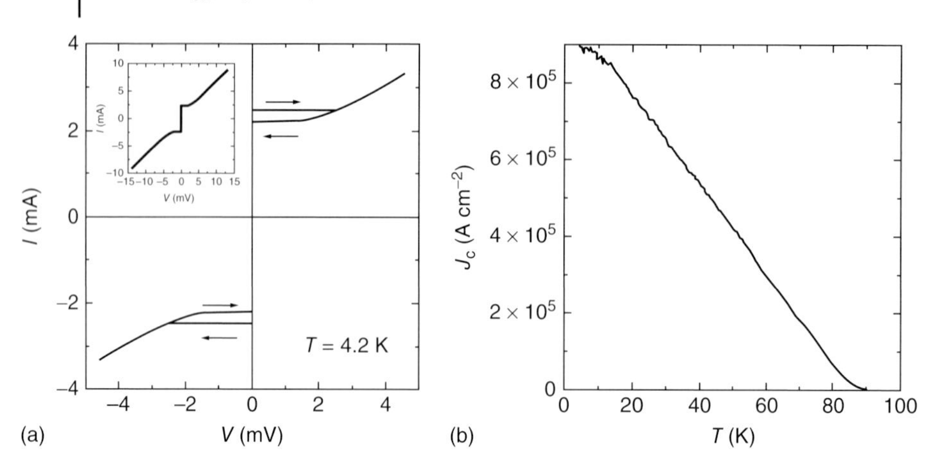

Figure 3.3.2.4 (a) I–V characteristics at 4.2 K and (b) temperature dependence of J_c for a typical YBCO [001]-tilt bicrystal junction with $\theta = 24°$. (Adapted and reproduced from Ref. [4]. Reproduced with permission of American Physical Society (APS))

20° behave as Josephson junctions with low excess current. Typical $I-V$ curve and temperature dependence of critical current (I_c) of such Josephson junctions are shown in Figure 3.3.2.4. The $I-V$ curve exhibits a small hysteresis at low temperatures and is explained by resistively and capacitively shunted junction (RCSJ) model. The hysteresis disappears at temperatures >30–40 K, and the $I-V$ curve can be simply fitted by that expected from resistively shunted junction (RSJ) model. At even higher temperatures near 77 K, rounding of the $I-V$ curve due to thermal fluctuation is observed, and the curve is fitted by the Ambegaokar and Halperin model [11]. To estimate the fraction f of the excess current due to flux flow (FF) in the whole θ range, a phenomenological model has also been proposed [12], that is, $I_{\text{total}} = (1-f)I_{\text{RSJ}} + fI_{\text{FF}}$, where the RSJ current follows $I_{\text{RSJ}} = \left(\left(\frac{V}{R_n}\right)^2 + I_c^2\right)$ and the FF current follows $I_{\text{FF}} = I_S - A\exp\left(-\frac{V}{V_0}\right)$ (R_n is the junction normal-state resistance, and I_S, A, and V_0 are constants).

The critical current shows quasi-T-linear temperature dependence (see Figure 3.3.2.4b), though a quadratic dependence like $(1-T/T_c)^2$ is observed at temperatures very close to T_c in many cases, indicating that the junction basically has an SIS junction character. The specific junction normal resistance AR_n (here, A is a junction area) is ~10^{-9}–10^{-7} Ωcm^2 and increases with increasing θ. The junction-specific capacitance is approximately 10^{-6}–10^{-4} F cm^{-2} [4]. In order to apply bicrystal junctions to electronic devices, the required I_c level is obtained by choosing a proper misorientation angle θ and adjusting the width of the bridge crossing the bicrystal GB. For instance, θ between 24° and 36.8° is typically used for application to superconducting quantum interference devices (SQUIDs) operating at 77 K with the junction width ranging from one to several micrometers. The optimization of [001]-tilt bicrystal junctions for application to SQUIDs with a large voltage modulation and low flux noise has been extensively studied and reported [8, 13, 14]. In addition to $SrTiO_3$, bicrystal substrates made

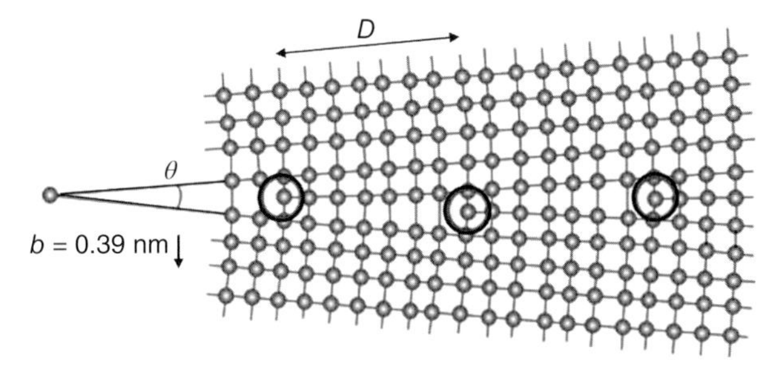

Figure 3.3.2.5 Schematic illustration of [001]-tilt bicrystal grain boundary showing the existence of a regular array of misfit dislocations.

of MgO and (La,Sr)(Al,Ta)O_3 (LSAT) are also commercially available. Though less systematic data have been reported, junction properties similar to those for $SrTiO_3$ bicrystal can be obtained for these substrates.

The microstructure of [001]-tilt GB is characterized by a regular array of misfit dislocations with an interval distance $D = |b|/2\sin(\theta/2)$, where b is the Burgers vector [15] and almost equal to a-, b-axis length, as schematically shown in Figure 3.3.2.5. It has also been pointed out that there exist strain fields around the dislocation cores. Such local disorder and strain around the dislocation core give rise to carrier depletion and thus reduced superconducting order parameter and even insulating characteristics [4, 15]. With increasing θ, the distance D decreases, as seen in the transmission electron microscopy (TEM) pictures of Figure 3.3.2.6 [16]. When the width of the superconducting channel between the regions with significantly reduced order parameters around the dislocation cores becomes comparable to the superconducting coherence length, weak-link behavior starts to manifest itself, which is the origin of the critical angle. At higher angles, the carrier-depleted regions overlap together and form a continuous layer along the

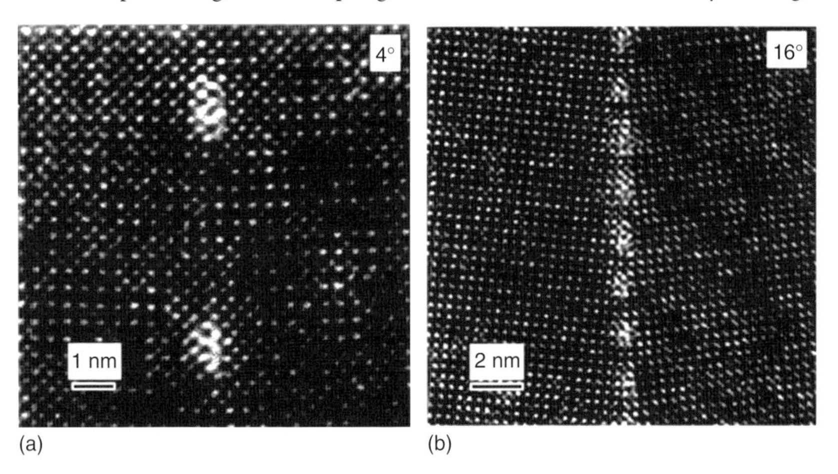

Figure 3.3.2.6 TEM planar view images of (a) 4° and (b) 16° [001]-tilt grain boundaries of YBCO films grown by a liquid phase epitaxy method. (Adapted and reproduced from Ref. [16]. Reproduced with permission of Institute of Physics (IOP))

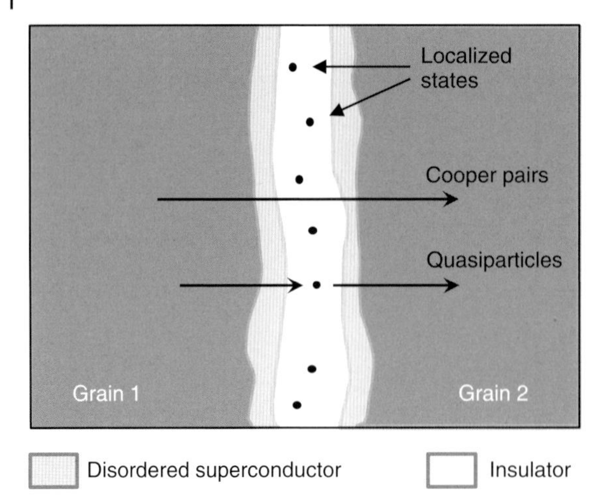

Figure 3.3.2.7 Physical image of a bicrystal grain boundary and current transport across the boundary according to the model proposed by Gross *et al*. (Adapted and reproduced from Ref. [4]. Reproduced with permission of American Physical Society (APS))

GB interface. It has been reported that the width of the non-superconducting regions adjacent to the GB interface, which is estimated by TEM observation and consideration based on bond valence sum, increases linearly with an increase in θ from 11° to 45° [17]. Such a linear increase explains the exponential decay of J_c due to SIS tunneling. The interface charging and band bending model [4, 9] and intrinsically shunted junction models [18, 19] have also been proposed as mechanisms to explain the GB properties. In any case, the high-angle GB is characterized by the insulating layer at the interface and adjacent regions with reduced order parameters, as schematically shown in Figure 3.3.2.7.

The I_cR_n product of Josephson junction is an important figure of merit for device application. The I_cR_n of [001]-tilt YBCO bicrystal junctions tends to increases with decreasing θ, and that for junctions with $\theta = 24° - 30°$ is $\sim 1-2$ mV and 100–300 μV at 4.2 and 77 K, respectively [4]. For higher-angle junctions, $I_cR_n \propto J_c^n$ ($n = 0.3-0.5$) correlation has been reported [4, 19]. Such correlation has also been found in ramp-edge junctions, as will be described later, and this has been attributed to quasiparticle tunneling via localized states in the insulating layer at the interface.

The $d_{x^2-y^2}$ symmetry of the superconducting order parameter for cuprate superconductors does influence the transport properties of bicrystal junctions. The extreme case is the 45° asymmetric junction, as schematically shown in Figure 3.3.2.8a. Since the grain size in YBCO thin films is typically 100 nm, there is significant meandering or faceting of the bicrystal GB, giving rise to a π phase shift at adjacent facets where Josephson current flows in opposite directions. Thus, the bicrystal interface is considered as an array of microscopic "π-junctions," which results in peculiar magnetic field dependence of I_c with a dip at zero field [4]. For symmetrical bicrystal junctions which show normal magnetic field dependence, the effect of d-wave symmetry on the J_c reduction has been estimated, indicating

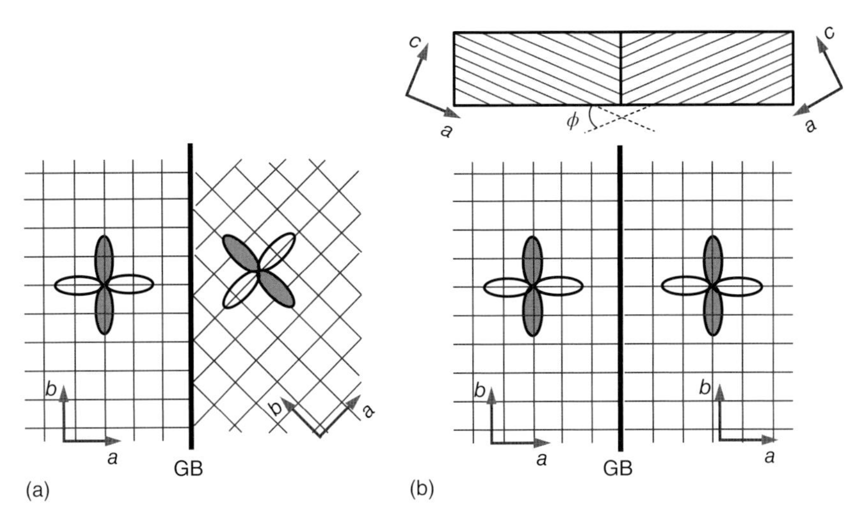

Figure 3.3.2.8 Schematic illustrations of (a) 45° asymmetric [001]-tilt bicrystal junctions and (b) symmetric [100]-tilt bicrystal junction (upper: side view, lower: planar view) functions. Superconducting wave functions with $d_{x^2-y^2}$ symmetry in cuprates are also shown.

that the *d*-wave symmetry can explain only one order of magnitude reduction [20]. Thus, the more than three orders of magnitude J_c reduction depending on θ is mostly attributed to the existence of insulating layer at the interface.

Although the structural and transport properties of [100]-tilt bicrystal YBCO junctions have been less extensively investigated than those for [001]-tilt junctions, substantially higher I_cR_n products of several to 10 mV at 4.2 K and up to 1 mV at 77 K have been reported [21–23]. In the case of [100]-tilt junctions, lobes with the largest amplitude and the same sign of the $d_{x^2-y^2}$ order parameter face at the interface, as schematically shown in Figure 3.3.2.8b. The larger I_cR_n products have been attributed to this situation and possibly less strain near the interface. A similar difference in the I_cR_n product between [100]-tilt and [001]-tilt bicrystal junctions has also been reported for junctions based on $HgBa_2CaCu_2O_{6+x}$ films fabricated by an *ex situ* process [24]. There are two faces of a [100]-tilt bicrystal substrate with a valley-type or mountain-type configuration of [001] planes, and the former type of configuration leads to better morphology of the bicrystal interface without significant overgrowth of grains. It seems difficult, however, to apply [100]-tilt bicrystal junctions to actual devices because both halves of the substrate are so-called vicinal substrates on which the *c*-axis of cuprate superconductor is tilted and the transports in the two orthogonal directions exhibit substantial anisotropy.

3.3.2.3.2 **Step-Edge Junctions**

Step-edge junctions are fabricated by preparing a step structure on the surface of single-crystal substrates such as $SrTiO_3$, $LaAlO_3$, and MgO by an ion milling technique, and growing an HTS film with a thickness smaller than the step height epitaxially over the step structure. Due to the rather simple fabrication process, the

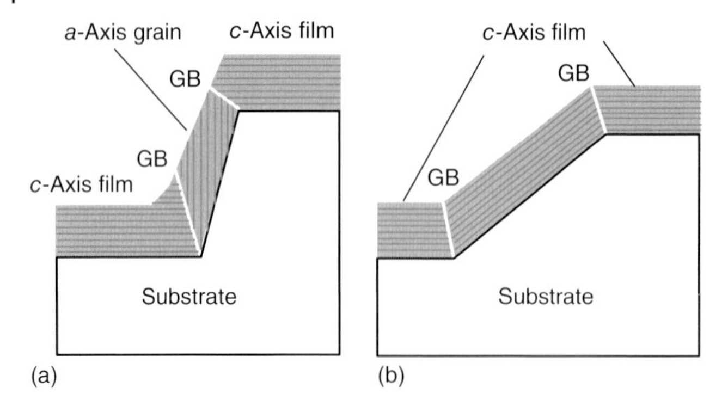

Figure 3.3.2.9 Schematic illustrations of cross-sections for two types of REBCO step-edge junctions on (a) substrates made of perovskite oxides such as $SrTiO_3$ and $LaAlO_3$, and (b) MgO substrates.

fabrication procedure and the properties of step-edge junctions have been extensively studied since early 1990s [25–29]. GBs of the HTS film nucleate from the edges of the step. The microstructure of the junction differs very much depending on the substrate materials, as schematically shown in Figure 3.3.2.9. For the cases of perovskite materials nearly lattice-matched with YBCO, such as $SrTiO_3$ and $LaAlO_3$, a-axis grains nucleate on the slope when the angle of the step is larger than the critical angle of approximately 50° [25, 26], and GBs are formed between this a-axis and c-axis grains on the flat parts. On the other hand, no weak link is formed when the angle is lower than the critical value. For the case of MgO, YBCO grains grow with their a-, b-axis parallel to the slope surface, and GBs are formed near the upper and lower edges of the step [27–29].

Formation of two GBs around the step is not favorable for device application because two junctions with different I_c values are connected in series. However, it was reported that the lower GB disappears when the lower edge of the step is somewhat rounded [30]. Foley *et al.* [31, 32] reported that a step profile with a sharp angle at the top edge and a rounded, smooth curve at the lower edge can be reproducibly obtained by employing Ar ion-beam etching at specific angles to the surface, as schematically shown in Figure 3.3.2.10. First, a step-edge defined by photoresist is patterned on the MgO substrate parallel to the (100) MgO direction. They employed the configuration for etching with ion-beam facing the photoresist edge and aligned typically at $\alpha = 20°$ to the MgO substrate plane in the $x-z$ plane. The MgO substrate is also rotated by $\beta = 10°$ in the $x-y$ plane. After ion-beam etching at an acceleration voltage of typically 500 V and removal of the original photoresist layer, the second etching with the ion-beam normal to the MgO substrate at lower voltages (~300 V) is performed for cleaning prior to film deposition. These procedures provide a step with a step-angle ϕ of ~40° with some scatter. It was also reported that by changing the etching angle α, the step-angle can be varied in a wide range from 0° to 80° [33].

The GBs in step-edge junctions are basically [100]-tilt junctions with the misorientation angle equal to the step-angle ϕ. The junction properties depend on

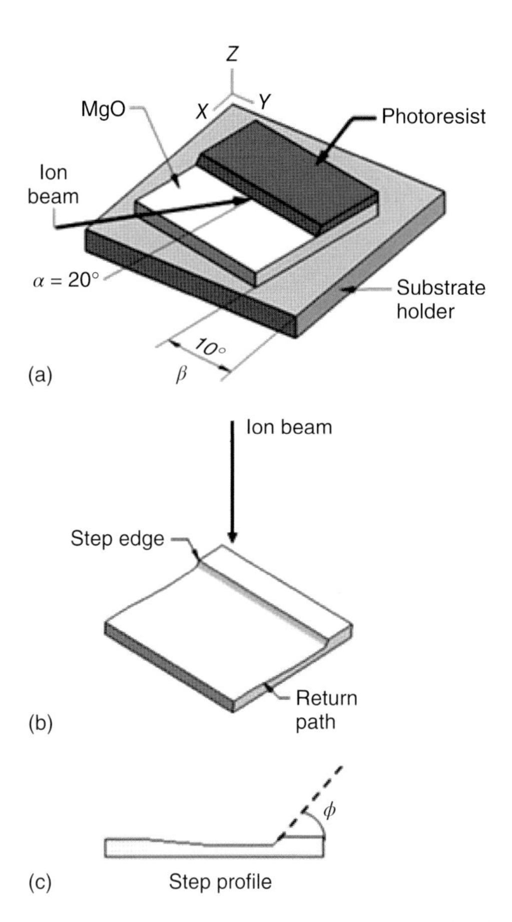

Figure 3.3.2.10 Fabrication process for a step on an MgO substrate generating only one grain boundary junction at the upper edge. (a) Fabrication of a step by ion milling, (b) cleaning by ion-beam irradiation, and (c) cross-sectional profile of the step. (Adapted and reproduced from Ref. [32]. Reproduced with permission of Institute of Physics (IOP))

the film quality, as in the case of bicrystal junctions, and also the ratio of the film thickness t to the step height h. It was reported that the ratio t/h of ~0.7 gave better junctions [34], and t/h of 0.4–0.5 has been reported to be preferable for the case of coevaporated films [35]. The properties of step-edge junctions fabricated by using the controlled step profile described above and high-quality YBCO thin films have recently been reported by Mitchell and Foley [32]. $I-V$ curves for a typical junction shown in Figure 3.3.2.11 exhibit a hysteresis at temperatures lower than 30 K, which is very similar to the case for bicrystal junctions, though excess current with the ratio to I_c of 0–30% is observed at 4.2 K. The MgO steps with the angle ϕ of 33–52° were fabricated by changing the step height from 250 to 550 nm. The junctions with 200 nm-thick YBCO films exhibit an increase in R_n and a decrease in I_c with increasing ϕ, as shown in Figure 3.3.2.12. They obtained I_cR_n products of 3–5 mV and 100–400 μV at 4.2 and 77 K, respectively. These values

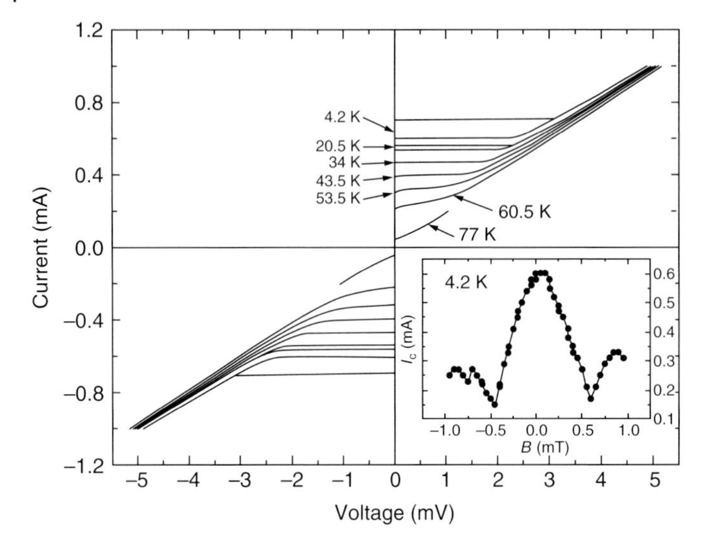

Figure 3.3.2.11 *I–V* characteristics at various temperatures for a typical YBCO step-edge junction on an MgO substrate with a step-angle ϕ of approximately 40°. Inset shows the magnetic field dependence of critical current at 4.2 K. (Adapted and reproduced from Ref. [32]. Reproduced with permission of Institute of Physics (IOP))

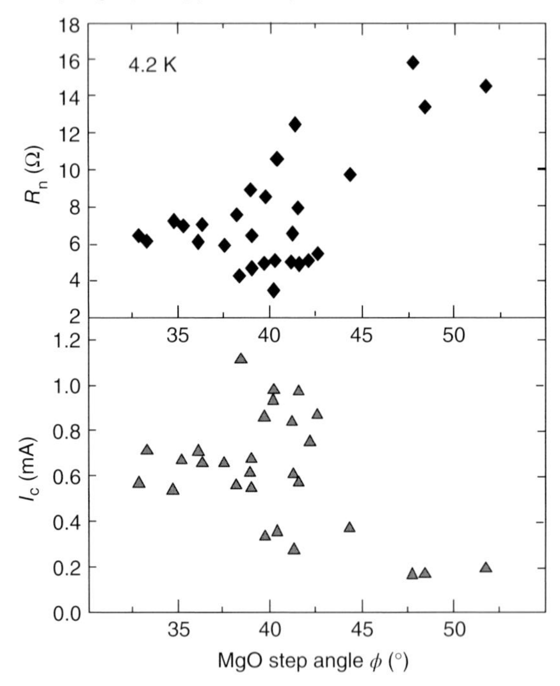

Figure 3.3.2.12 Step-angle dependence of junction normal resistance R_n and critical current I_c for YBCO step-edge junctions on MgO. (Adapted and reproduced from Ref. [32]. Reproduced with permission of Institute of Physics (IOP))

are similar to those reported for [100]-tilt bicrystal junctions [21–23]. A nearly exponential dependence of J_c on ϕ with a characteristic angle of ~15° was also reported.

Because of the high-quality junction properties and flexible layout allowed, step-edge Josephson junctions have been applied to electronic devices such as SQUIDs and detectors for terahertz radiation. It has also been reported that SQUIDs with step-edge junctions exhibit a low flux noise at 77 K of 4.5 $\mu\Phi_0/\mathrm{Hz}^{1/2}$ at 10 Hz [35].

3.3.2.4 Ramp-Edge Junctions

Since the initial study in early 1990s, ramp-edge or edge-type junctions with a variety of materials for a thin-film multilayer and a barrier have been developed [5, 36–39]. Figure 3.3.2.13 schematically illustrates the fabrication process of ramp-edge junctions. First, a REBCO and insulator bilayer with typically 200–400 nm

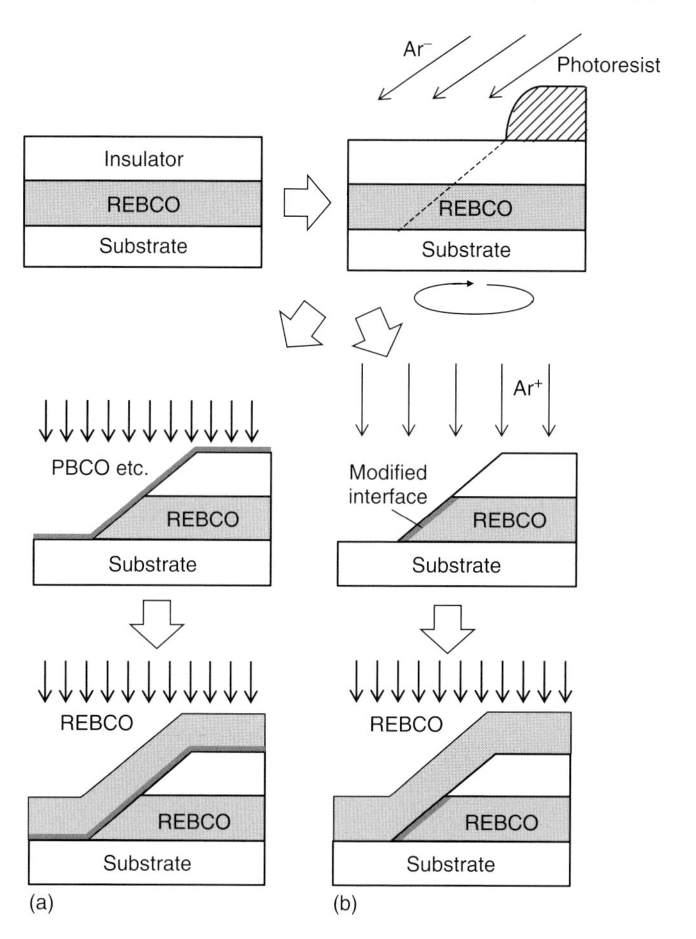

Figure 3.3.2.13 Schematic illustration of fabrication process for ramp-edge junctions with (a) an artificially deposited barrier and (b) an interface-modified barrier.

thickness for each is deposited. As an insulating material, perovskite materials such as $SrTiO_3$ and $SrSnO_3$ or CeO_2 are used. The ramp-edge structure of this bilayer is fabricated by an ion milling technique. To obtain the slope with a uniform angle of typically 20–40° and prevent formation of a sharp edge, which could cause nucleation of a GB, a photoresist pattern with rounded edges made by a reflow process, heating at about 150 °C, is employed. After ion milling and removing the photoresist pattern, a barrier layer and subsequently an upper REBCO layer are formed. The process is completed by pattering the upper REBCO layer and forming a microbridge structure over the ramp edge of the lower bilayer.

As for a barrier layer, there are two types of successful options. One is artificially deposited cuprate materials which have the same structure as REBCO and exhibit no superconductivity or reduced T_c. The other utilizes the surface of the REBCO ramp edge modified by ion bombardment and subsequent annealing at high temperatures including the deposition of the upper REBCO layer. The latter type of junction is called *interface-engineered junction* (IEJ) or interface-modified junction.

Ramp-edge junctions with a deposited barrier made of Co-doped YBCO, $YBa_2(Cu_{1-x}Co_x)_3O_y$ [39], and Ca-doped YBCO, $Y_{1-x}Ca_xBa_2Cu_3O_y$ [40] exhibit RSJ-type $I-V$ characteristics in a temperature range roughly between the T_c of the barrier materials and that of the YBCO thin films. Due to the metallic nature of the barrier materials, the junction-specific resistance AR_n of $10^{-10}-10^{-9}\ \Omega cm^2$ is temperature dependent and much smaller than those for GB junctions, and the temperature dependence of $I_c \propto (1 - T/T_c)^2$ like a superconductor–normal conductor–superconductor (SNS) junction has also been reported. The significant temperature dependence of the junction parameters and the I_cR_n products below 100 μV at 77 K make it difficult to apply these junctions to actual devices.

On the other hand, junctions with a non-superconducting $PrBa_2Cu_3O_y$ (PBCO) barrier, typically 10–30 nm in thickness, exhibit higher specific resistance similar to those for GB junctions [37, 41–44]. Such junctions show RSJ-type $I-V$ characteristics in a wide temperature range and a small hysteresis at temperatures typically below 40 K. As shown in Figure 3.3.2.14, a nearly exponential decay of J_c with increasing barrier thickness is observed, and the decay length, or "normal coherence length," of around 2–4 nm has been reported. The junction conductance, the inverse of junction-specific resistance, also shows a nearly exponential decrease with a substantially longer decay length. This slower decay of the conductance and its peculiar temperature dependence have been attributed to quasiparticle tunneling via localized states in the barrier material [41, 42, 44, 45] as predicted by Glazman and Matveev [46], while Cooper pairs tunnel directly through the barrier. Typical I_cR_n products for ramp-edge junctions with a PBCO barrier are 1–3 and 0.1 mV at 4.2 and 77 K, respectively. Even higher I_cR_n products were reported for junctions with a Ga-doped PBCO barrier [43]. A rather small spread of I_c with a standard deviation σ divided by the average value (1σ spread) <30 % was also reported [42]. Small-scale SFQ circuits and DC SQUIDs operating at 77 K were fabricated using this type of junctions. However, the junction properties such as J_c as a function of the barrier thickness largely depend on the composition, in

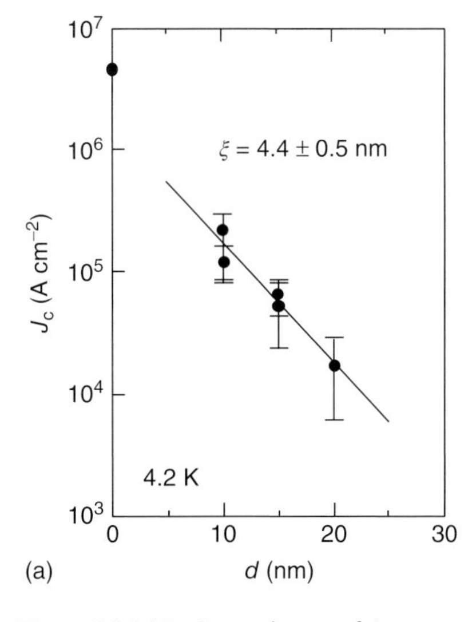

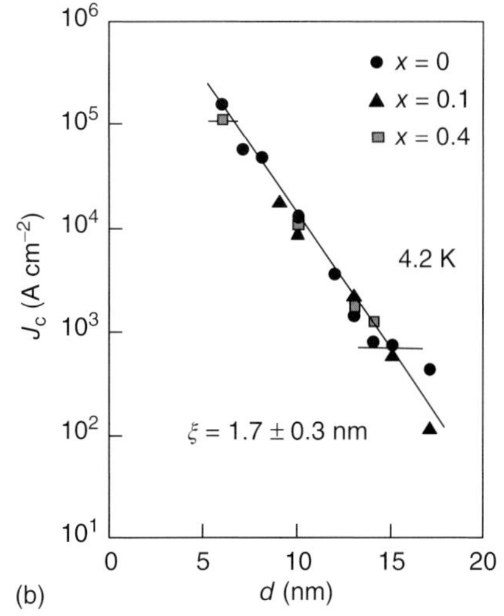

Figure 3.3.2.14 Dependences of J_c on the deposited barrier thickness reported for YBCO ramp-edge junctions with (a) a $PrBa_2Cu_3O_y$ barrier (Adapted and reproduced from Ref. [42]. Reproduced with permission of IEEE) and (b) a $PrBa_2Cu_{3-x}Ga_xO_y$ ($x = 0,0.1,0.4$) barrier (Adapted and reproduced from Ref. [43]. Reproduced with permission of American Institute of Physics (AIP)).

particular oxygen content, of the PBCO layer as well as the quality of the surface of the REBCO ramp edge, possibly leading to insufficient reproducibility.

Interface-engineered or interface-modified junctions [47, 48] attracted much attention because of the smaller I_c spread and excellent reproducibility first reported by Moeckly and Char [47]. To fabricate this type of junctions, after preparing a ramp-edge structure and removing a photoresist pattern, the surface of the ramp structure is subjected to ion bombardment, as schematically shown in Figure 3.3.2.13. Although reverse sputtering in radio frequency (rf) plasma and subsequent vacuum annealing at a high temperature were employed in their original work, later extensive studies have revealed that Ar ion bombardment normal to the substrate surface is favorable to obtain uniform J_c distribution for ramps facing different directions [49, 50]. Typical acceleration voltage and bombardment time are 250–500 V and 1–5 min, respectively. The junction J_c can be controlled by changing these conditions and the deposition temperature for the upper REBCO layer.

IEJs exhibit $I-V$ characteristics similar to those for junctions with a PBCO barrier and GB junctions. Figure 3.3.2.15 shows the relation between I_cR_n products and J_c at 4.2 K. I_cR_n is roughly proportional to $J_c^{1/2}$ for junctions with J_c values lower than 10^4 A cm^{-2}, and tends to saturate to values of 2–3 mV at higher J_c. The formation mechanism of IEJs has been examined by cross-sectional TEM observations [51–53]. After ion bombardment, a few nanometer-thick amorphous layer with a Cu-deficient composition is formed [51], as schematically indicated in the triangular phase diagram of Figure 3.3.2.16. By depositing the

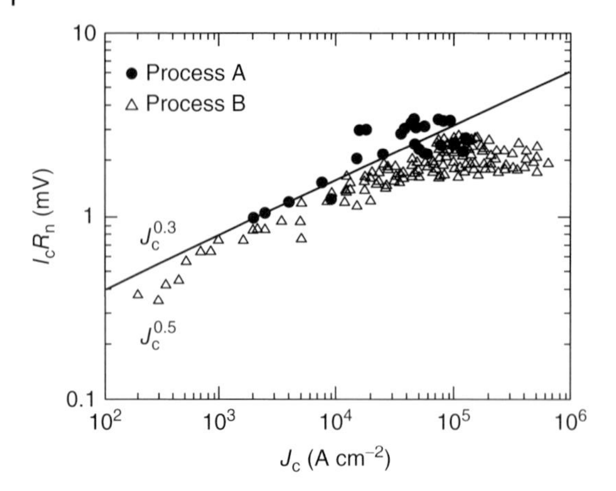

Figure 3.3.2.15 Correlation between I_cR_n products and J_c for ramp-edge junctions with an interface-modified barrier fabricated by ion bombardment (Process B) and deposition of Cu-deficient precursory layer (Process A). (Adapted and reproduced from Ref. [57]. Reproduced with permission of IEEE)

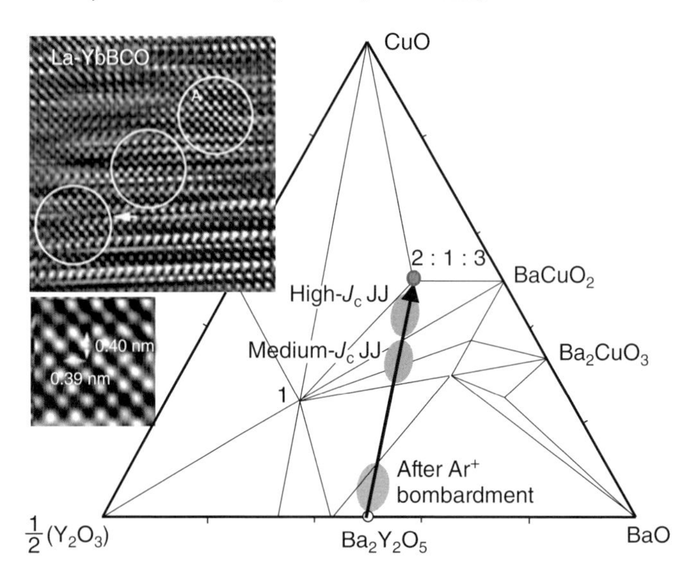

Figure 3.3.2.16 Triangular phase diagram schematically explaining the change of the interface composition through the fabrication process of ramp-edge junctions with an interface-modified barrier. TEM images of the junction interface for a typical junction with a medium J_c of 10^4–10^5 A cm^{-2} and the pseudo-cubic phase observed at the interface are also shown.

upper REBCO layer at a high temperature, this amorphous layer is recrystallized. Simultaneously, atomic diffusion occurs at the junction interface. When the upper layer is deposited at a relatively high temperature to form junctions with J_c well over 10^5 A cm^{-2}, the interface composition is close to that of REBCO, and

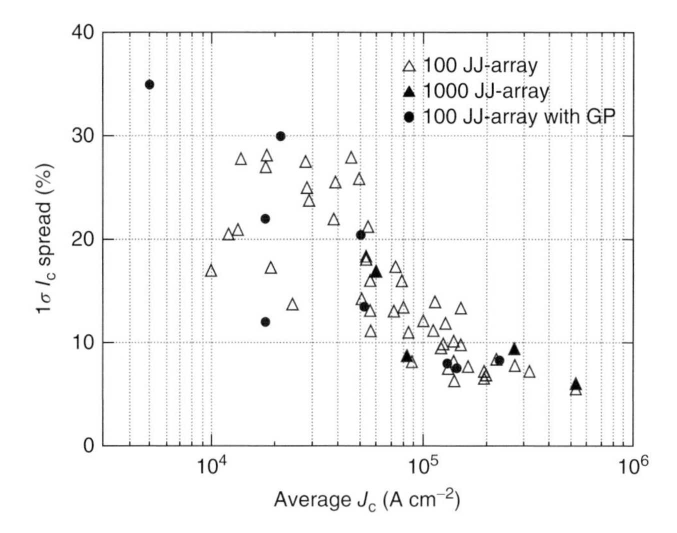

Figure 3.3.2.17 Correlation between 1σ I_c spread and average J_c for series arrays of 100 and 1000 ramp-edge junctions with an interface-modified barrier.

only stacking faults are observed at the interface [53]. For the case of junctions with the upper layer deposited at a lower temperature and J_c of 10^4-10^5 A cm^{-2}, the composition at the interface is more Cu-deficient, and existence of Ba-based pseudo-cubic phases with the lattice parameter of 0.39–0.40 nm is observed [51, 53]. Thus, the junction interface is considered to consist of distributed nanometer-size insulating regions and carrier-depleted REBCO with structural disorder and possibly strain, which is similar to the case of GB junctions. The ramp-edge junctions with an interface-modified barrier show 1σ I_c spreads <10 % [48, 50, 54], which is substantially smaller than those for GB junctions and advantageous for application to integrated circuits such as SFQ circuits. However, the I_c spread depends on the junction J_c, as shown in Figure 3.3.2.17. It has been also pointed out that the I_c spread better correlates with the magnitude of I_c [55]. That is, junctions with larger I_c values exhibit smaller spreads, and this has been explained by the model of the junction interface where the number of distributed microscopic SNS contacts determines the statistical spread of I_c [55].

Ramp-edge junctions with an interface-modified barrier have been used to demonstrate operation of various elementary and some practical SFQ circuits [5]. A cross-sectional scanning electron microscope (SEM) image for an oxide multilayer structure including a superconducting groundplane and ramp-edge junctions for SFQ circuits is shown in Figure 3.3.2.18a. It is necessary to suppress the surface average roughness R_a for the multilayer consisting of four layers typically <2 nm to obtain I_c spreads below 10% [5, 56]. Such a smooth multilayer can be fabricated by employing magnetron sputtering and the combination of proper materials for insulating and superconducting layers. To achieve higher I_cR_n products, slightly modified barrier fabrication process has been also employed [57]. By depositing a thin Cu-deficient precursory layer on the ramp surface

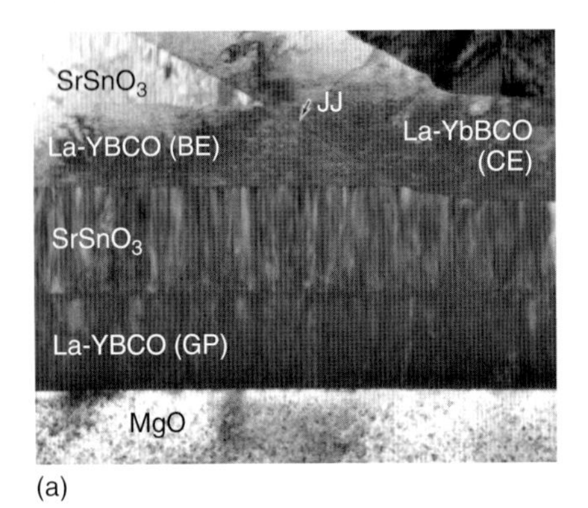

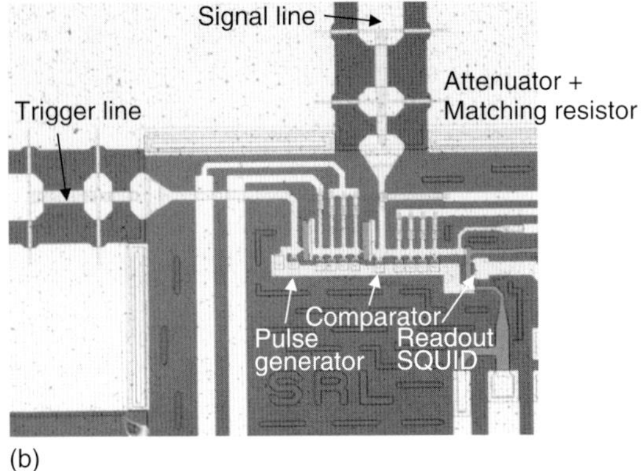

Figure 3.3.2.18 (a) Cross-sectional TEM image for a thin-film multilayer structure with a superconducting groundplane and ramp-edge junctions and (b) optical microscope image of a single-flux-quantum sampler circuit with 15 junctions which was fabricated using the multilayer structure.

instead of Ar ion bombardment, I_cR_n products larger than 1 mV at 40 K which enables clock frequency over 100 GHz have been achieved. A photograph of an SFQ sampler circuit fabricated using such junctions is shown in Figure 3.3.2.18b. Recently, ramp-edge junctions with a modified interface barrier have also been applied to SQUIDs with a multilayer structure [58], and white field noise as low as 10 fT $Hz^{-1/2}$ at 77 K has been reported for a magnetometer with an integrated multi-turn input coil [59]. Very low flux noise $<10^{-5}$ $\Phi_0/Hz^{1/2}$ at 1 Hz has also been confirmed for a gradiometer using this type of junctions [59].

3.3.2.5 Other Types of Junctions

It has been reported that planar-type overdamped Josephson junctions can be fabricated by nanolithography and ion irradiation. For instance, irradiation of 200 keV Ne^+ ion on a 20–100 nm-long region of a YBCO thin film microbridge converted this region to a normal conductor, resulting in Josephson junctions which exhibited RSJ-type $I-V$ characteristics and SNS-junction-like temperature dependence of I_c [60]. The junction T_c and I_c were tuned by changing the length of the damaged weal link and the ion fluence. Although there is not enough data on the reproducibility and controllability, this technique has also been applied to fabrication of series array of ten junctions operating at 77 K with the aim of application to voltage standards and rf devices [61].

As mentioned in the first section, Josephson junctions are naturally formed along the *c*-axis of the crystal of cuprate superconductors. By simply fabricating a mesa structure along the *c*-axis in a single crystal of underdoped or optimally doped $Bi_2Sr_2CaCu_2O_{8+\delta}$ (Bi2212), $I-V$ characteristics with many branches resembling those of a series-connected underdamped Josephson junctions are observed [1, 2]. In this case, the approximately 0.3 nm-thick multiple CuO_2 plane and the 1.2 nm-thick block layer are considered as the superconducting electrode and the barrier layer, respectively. By decreasing the thickness of mesa structure,

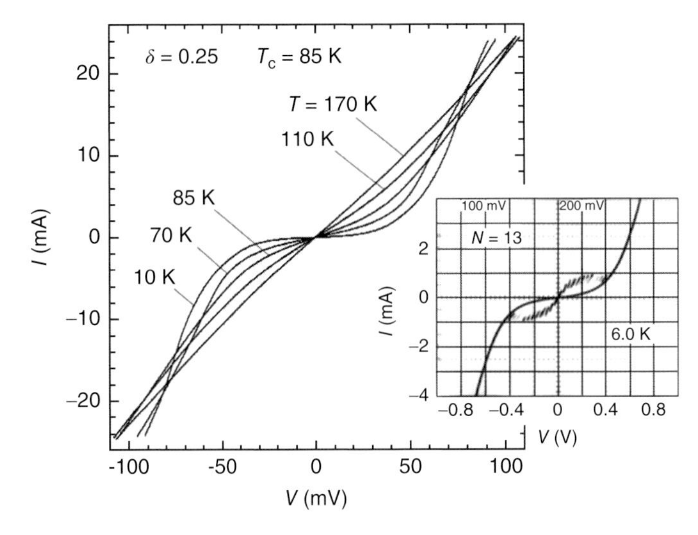

Figure 3.3.2.19 Quasiparticle tunneling characteristics at different temperatures for a mesa structure fabricated in a $Bi_2Sr_2CaCu_2O_{8+\delta}$ ($\delta = 0.25$) single crystal. Inset shows the I–V curve at 6.0 K of the same mesa structure containing a series array of 13 intrinsic junctions. (Adapted and reproduced from Ref. [62]. Reproduced with permission of American Physical Society (APS))

an array of a few junctions or even a single junction can be fabricated. The quasiparticle tunneling branch, however, rises gradually from zero voltage due to $d_{x^2-y^2}$ symmetry of the superconducting gap [2, 62], as shown in Figure 3.3.2.19. Thus, it is difficult to apply intrinsic junctions to "conventional" electronic devices. Recently, the intrinsic junctions have attracted much attention since emission with a rather large power of microwatts range is observed by biasing a mesa structure at a voltage close to the gap voltage [63] (more detail, see Sections 8.5 and 11.1).

3.3.2.6 Summary and Outlook

As described in this chapter, for REBCO HTS materials, overdamped or weak-link-type Josephson junctions with quality high enough for device application can be fabricated by employing bicrystal, step-edge, or ramp-edge junction technologies. These HTS junctions exhibit basically similar $I-V$ characteristics which are explained by the RSJ model at high temperatures, and RCSJ model at temperatures typically below 30–40 K. Their typical I_cR_n products are ~1–3 and 0.1–0.3 mV at 4.2 and 77 K, respectively. These three types of junctions have advantages and disadvantages. The bicrystal junction is fabricated through the simplest process but there is limitation in the layout of junctions by the existence of the bicrystal line. Strict control of edge profile of a substrate step is required to reproducibly fabricate step-edge junctions. However, junctions can be laid out in any place on a substrate. The ramp-edge junction is fabricated through the most complicated process. Once the fabrication conditions for a multilayer and a barrier are optimized, however, junctions with a small I_c spread can be fabricated rather reproducibly.

This type of junction is also compatible with complicated device structures including crossover wiring. These three types of junctions have been actually applied to SQUIDs with low flux noise operating at 77 K and various SQUID-based practical systems. The ramp-edge junctions have also been applied to fabrication of small-scale digital and mixed-signal circuits such as an SFQ sampler. They will be used more in SQUID-based systems and other devices such as detectors in the near future.

In contrast to the success in overdamped Josephson junctions, fabrication of underdamped junctions with ideal SIS junction properties remains a big technological challenge which seems difficult to overcome. This is due to the peculiar physical properties of cuprate superconductors and the rather high substrate temperature required for film growth. Fabrication of SIS junctions based on MgB_2, another HTS material discovered after cuprates, has also been studied by several groups [64, 65]. In spite of less anisotropic physical properties and *s*-wave symmetry of the order parameter, the reported junction $I-V$ curves are still far from the ideal one [65], suggesting that fabrication of a sharp superconductor–barrier interface without degradation of order parameter is still difficult even at a substrate temperature of about 300 °C. As for the recently discovered iron-based superconductors, overdamped Josephson junctions have been fabricated on bicrystal substrates utilizing weak-link properties of high-angle GBs [66–68]. Although their peculiar physical properties in some part similar to cuprates suggest difficulty in fabricating SIS or SNS junctions, some initial study on the junctions with a Au barrier has been also reported [69, 70].

References

1. Kleiner, R., Steinmeyer, F., Kunkel, G. *et al.* (1992) Intrinsic Josephson effects in $Bi_2Sr_2CaCu_2O_8$ single crystals. *Phys. Rev. Lett.*, **68**, 2394–2397.
2. Yurgens, A.A. (2000) Intrinsic Josephson junctions: recent developments. *Supercond. Sci. Technol.*, **13**, R85–R100.
3. Dimos, D., Chaudhari, P., and Mannhart, J. (1990) Superconducting transport properties of grain boundaries in $YBa_2Cu_3O_7$ bicrystals. *Phys. Rev. B*, **41**, 4038–4049.
4. Hilgenkamp, H. and Mannhart, J. (2002) Grain boundaries in high-T_c superconductors. *Rev. Mod. Phys.*, **74**, 485–548.
5. Tanabe, K., Wakana, H., Tsubone, K. *et al.* (2008) Advances in high-T_c single flux quantum device technologies. *IEICE Trans. Electron.*, **E91-C**, 280–292.
6. Char, K., Colclough, M.S., Garrison, S.M. *et al.* (1991) Bi-epitaxial grain boundary junctions in $YBa_2Cu_3O_7$. *Appl. Phys. Lett.*, **59**, 733–735.
7. Tafuri, F. and Kirtley, J.R. (2005) Weak links in high critical temperature superconductors. *Rep. Prog. Phys.*, **68**, 2573–2663.
8. Ivanov, Z.G., Nilsson, P.A., Winkler, D. *et al.* (1991) Weak links and dc SQUIDs on artificial nonsymmetic grain boundaries in $YBa_2Cu_3O_{7-\delta}$. *Appl. Phys. Lett.*, **59**, 3030–3032.
9. Hilgenkamp, H. and Mannhart, J. (1998) Superconducting and normal-state properties of $YBa_2Cu_3O_{7-\delta}$ bicrystal grain boundary junctions in thin films. *Appl. Phys. Lett.*, **73**, 265–267.
10. Heinig, N.F., Redwing, R.D., Nordman, J.E. *et al.* (1999) Strong to weak coupling transition in low misorientation angle thin film $YBa_2Cu_3O_{7-\delta}$ bicrystals. *Phys. Rev. B*, **60**, 1409–1417.

11. Ambegaokar, V. and Halperin, B.I. (1969) Voltage due to thermal noise in the dc Josephson effect. *Phys. Rev. Lett.*, **22**, 1364–1366.
12. Saitoh, K., Ishimaru, Y., Fuke, Y. *et al.* (1997) A model analysis for current-voltage characteristics of superconducting weak links. *Jpn. J. Appl. Phys.*, **36**, L272–L275.
13. Gross, R., Chaudhari, P., Kawasaki, M. *et al.* (1990) Low noise $YBa_2Cu_3O_{7-\delta}$ grain boundary junction dc SQUIDs. *Appl. Phys. Lett.*, **57**, 727–729.
14. Minotani, T., Kawakami, S., Kiss, T. *et al.* (1997) High performance dc superconducting quantum interference device utilizing a bicrystal junction with a 30° misorientation angle. *Jpn. J. Appl. Phys.*, **36**, L1092–L1095.
15. Gurevich, A. and Pashitskii, E.A. (1998) Current transport through low-angle grain boundaries in high-temperature superconductors. *Phys. Rev. B*, **57**, 13878–13893.
16. Wen, J.G., Takagi, T., and Koshizuka, N. (2000) Microstructural studies on YBCO film bicrystals with large single facet grain boundaries grown by liquid phase epitaxy. *Supercond. Sci. Technol.*, **13**, 820–826.
17. Browning, N.D., Buban, J.P., Nellist, P.D. *et al.* (1998) The atomic origins of reduced critical currents at [001] tilt grain boundaries in $YBa_2Cu_3O_{7-\delta}$ thin films. *Physica C*, **294**, 183–193.
18. Halbritter, J. (1992) Pair weakening and tunnel channels at cuprate interfaces. *Phys. Rev. B*, **46**, 14861–14871.
19. Gross, R. (1994) in *Interfaces in High-T_c Superconducting Systems* (eds S.L. Shinde and D.A. Rudman), Springer-Verlag, New York, pp. 176–209.
20. Hilgenkamp, H., Mannhart, J., Mayer, B. *et al.* (1996) Implication of d_{x2-y2} symmetry and faceting for the transport properties of grain boundaries in high-T_c superconductors. *Phys. Rev. B*, **53**, 14586–14593.
21. Poppe, U., Divin, Y.Y., Faley, M.I. *et al.* (2001) Properties of $YBa_2Cu_3O_7$ thin films deposited on substrates and bicrystals with vicinal offcut and realization of high I_cR_n junctions. *IEEE Trans. Appl. Supercond.*, **11**, 3768–3771.
22. Sarnelli, E., Testa, G., Crimaldi, D. *et al.* (2005) A class of high-T_c $YBa_2Cu_3O_{7-x}$ grain boundary junctions with high-I_cR_n products. *Supercond. Sci. Technol.*, **18**, L35–L39.
23. Liatti, M.V., Poppe, U., and Divin, Y.Y. (2006) Low-frequency voltage noise and electrical transport in [100]-tilt $YBa_2Cu_3O_{7-x}$ grain-boundary junctions. *Appl. Phys. Lett.*, **88**, 152504.
24. Ogawa, A., Sugano, T., Wakana, H. *et al.* (2006) Properties of (Hg, Re)$Ba_2CaCu_2O_y$ [100]-tilt grain boundary Josephson junctions. *J. Appl. Phys.*, **99**, 123907.
25. Jia, C.L., Kabius, B., Urban, K. *et al.* (1991) Microstructure of epitaxial $YBa_2Cu_3O_7$ films on step-edge $SrTiO_3$ substrates. *Physica C*, **175**, 545–554.
26. Jia, C.L., Kabius, B., Urban, K. *et al.* (1992) The microstructure of epitaxial $YBa_2Cu_3O_7$ films on steep steps in $LaAlO_3$ substrates. *Physica C*, **196**, 211–226.
27. Edwards, J.A., Satchell, J.S., Chew, N.G. *et al.* (1992) $YBa_2Cu_3O_7$ thin film step junctions on MgO substrates. *Appl. Phys. Lett.*, **60**, 2433–2435.
28. Ramos, J., Ivanov, Z.G., Olsson, E. *et al.* (1993) $YBa_2Cu_3O_{7-\delta}$ Josephson junctions on directionally ion beam etched MgO substrates. *Appl. Phys. Lett.*, **63**, 2141–2143.
29. Tanaka, S., Kado, H., Matsuura, T. *et al.* (1993) Step-edge junction of YBCO thin films on MgO substrates. *IEEE Trans. Appl. Supercond.*, **3**, 2365–2368.
30. Mitsuzuka, T., Yamaguchi, K., Yoshikawa, S. *et al.* (1993) The microstructure and I-V characteristics of step-edge Josephson junctions in $YBa_2Cu_3O_7$ films on MgO (100) steps. *Physica C*, **218**, 229–237.
31. Foley, C.P., Mitchell, E.E., Lam, S.K.H. *et al.* (1999) Fabrication and characterization of YBCO single grain boundary step edge junctions. *IEEE Trans. Appl. Supercond.*, **9**, 4281–4284.
32. Mitchell, E.E. and Foley, C.P. (2010) YBCO step-edge junctions with high I_cR_n. *Supercond. Sci. Technol.*, **23**, 065007.
33. Foley, C.P., Lam, S.K.H., Sankrithyan, B. *et al.* (1997) The effects of step angle

on step edge Josephson junctions on MgO. *IEEE Trans. Appl. Supercond.*, **7**, 3185–3188.

34. Hermann, K., Kunkel, G., Siegel, M. *et al.* (1995) Correlation of $YBa_2Cu_3O_7$ step-edge junction characteristics with microstructure. *J. Appl. Phys.*, **78**, 1131–1139.
35. Du, J. (2003) Optimization and passivation of HTS step-edge Josephson junction rf SQUIDs. *IEEE Trans. Appl. Supercond.*, **13**, 865–868.
36. Gao, J., Aarnink, W.A., Gerritsuma, G.J. *et al.* (1990) Controlled preparation of all high-T_c SNS-type edge junctions and DC SQUIDs. *Physica C*, **171**, 126–130.
37. Gao, J., Boguslavskij, Y., Klopman, B.B.G. *et al.* (1991) Characteristics of advanced $YBa_2Cu_3O_x$/$PrBa_2Cu_3O_x$/$YBa_2Cu_3O_x$ edge type junctions. *Appl. Phys. Lett.*, **59**, 2754–2756.
38. Stolzel, C., Siegel, M., Adrian, G. *et al.* (1993) Transport properties of $YBa_2Cu_3O_x$/$Y_{0.3}Pr_{0.7}Ba_2Cu_3O_x$/$YBa_2Cu_3O_x$ Josephson junctions. *Appl. Phys. Lett.*, **63**, 2970–2972.
39. Mallison, W.H., Berkowitz, S.J., Hirahara, A.S. *et al.* (1996) A multilayer $YBa_2Cu_3O_x$ Josephson junction process for digital circuit applications. *Appl. Phys. Lett.*, **68**, 3808–3810.
40. Antognazza, L., Moeckly, B.H., Geballe, T.H. *et al.* (1995) Properties of high-T_c Josephson junctions with $Y_{0.7}Ca_{0.3}Ba_2Cu_3O_{7-\delta}$ barrier layers. *Phys. Rev. B*, **52**, 4559–4567.
41. Sawada, Y., Terai, H., Fujimaki, A. *et al.* (1995) Transport properties of YBCO/PBCO/YBCO junctions. *IEEE Trans. Appl. Supercond.*, **5**, 2099–2102.
42. Satoh, T., Hidaka, M., and Tahara, S. (1997) Study of *in-situ* prepared high-temperature superconducting edge-type Josephson junctions. *IEEE Trans. Appl. Supercond.*, **7**, 3001–3004.
43. Verhoeven, M.A.J., Gerritsma, G.J., Rogalla, H. *et al.* (1996) Ramp type junction parameter control by Ga doping of $PrBa_2Cu_3O_{7-\delta}$ barriers. *Appl. Phys. Lett.*, **69**, 848–850.
44. Boguslavskij, Y.M., Gao, J., Rijinders, A.J.H.M. *et al.* (1992) Transport processes in $YBa_2Cu_3O_x$/$PrBa_2Cu_3O_x$/$YBa_2Cu_3O_x$ ramp type Josephson junctions. *Physica C*, **194**, 268–276.
45. Yoshida, J. and Nagano, T. (1997) Tunneling and hopping conduction via localized states in thin $PrBa_2Cu_3O_{7-x}$ barriers. *Phys. Rev. B*, **55**, 11860–11871.
46. Glazman, L.I. and Matveev, K.A. (1988) Inelastic tunneling across thin amorphous films. *Sov. Phys. JETP*, **67**, 1276–1282.
47. Moeckly, B.H. and Char, K. (1997) Properties of interface-engineered high T_c Josephson junctions. *Appl. Phys. Lett.*, **71**, 2526–2528.
48. Satoh, T., Hidaka, M., and Tahara, S. (1999) High-temperature superconducting edge-type Josephson junctions with modified interfaces. *IEEE Trans. Appl. Supercond.*, **9**, 3141–3144.
49. Ishimaru, Y., Wu, Y., Horibe, O. *et al.* (2002) Evaluation of fabrication process and barrier structure for interface-modified ramp-edge junctions. *Physica C*, **378-381**, 1327–1333.
50. Soutome, Y., Fukazawa, T., Saitoh, K. *et al.* (2002) HTS surface-modified junctions with integrated ground-planes for SFQ circuits. *IEICE Trans. Electron.*, Gothic one, 759–763.
51. Wen, J.G., Koshizuka, N., Tanaka, S. *et al.* (1999) Atomic structure and composition of the barrier in the modified interface high-T_c Josephson junction studied by transmission electron microscopy. *Appl. Phys. Lett.*, **75**, 2470–2472.
52. Huang, Y., Merkle, K.L., Moeckly, B.H. *et al.* (1999) The effect of microstructure on the electrical properties of YBCO interface-engineered Josephson junctions. *Physica C*, **314**, 36–42.
53. Wu, Y., Ishimaru, Y., Wakana, H. *et al.* (2002) Identification of different phases in barriers of interface-engineered ramp-edge Josephson junctions: formation mechanisms and influences on electrical properties. *J. Appl. Phys.*, **92**, 4571–4577.
54. Suzuki, T., Ishimaru, Y., Horibe, M. *et al.* (2003) Evaluation of fabrication process for interface-modified ramp-edge junctions. *Physica C*, **392-396**, 1378–1381.

55. Yoshida, J., Katsuno, H., Nakayama, K. *et al.* (2004) Current transport and the fluctuation of critical current in high-temperature superconductor interface-engineered Josephson junctions. *Phys. Rev. B*, **70**, 054511.
56. Wakana, H., Adachi, S., Kamitani, A. *et al.* (2005) Improvement in reproducibility of multilayer and junction process for HTS SFQ circuit. *IEEE Trans. Appl. Supercond.*, **15**, 153–156.
57. Wakana, H., Adachi, S., Nakayama, K. *et al.* (2007) Fabrication of ramp-edge junctions with high I_cR_n products by using Cu-poor precursor. *IEEE Trans. Appl. Supercond.*, **17**, 233–236.
58. Wakana, H., Adachi, S., Hata, T. *et al.* (2009) Development of integrated HTS SQUIDs with a multilayer structure and ramp-edge Josephson junctions. *IEEE Trans. Appl. Supercond.*, **19**, 782–785.
59. Adachi, S., Tsukamoto, A., Oshikubo, Y. *et al.* (2011) Fabrication of low-noise HTS-SQUID gradiometers and magnetometers with ramp-edge Josephson junctions. *IEEE Trans. Appl. Supercond.*, **21**, 367–370.
60. Katz, A.S., Sun, A.G., Woods, S.I. *et al.* (1998) Planar thin film $YBa_2Cu_3O_{7-\delta}$ Josephson junctions via nanolithography and ion damage. *Appl. Phys. Lett.*, **72**, 2032–2034.
61. Chen, K., Cybart, S.A., and Dynes, R.C. (2004) Planar thin film $YBa_2Cu_3O_{7-\delta}$ Josephson junction pairs and arrays via nanolithography and ion damage. *Appl. Phys. Lett.*, **85**, 2863–2865.
62. Suzuki, M. and Watanabe, T. (2000) Discriminating the superconducting gap from the pseudogap in $Bi_2Sr_2CaCu_2O_{8+\delta}$ by interlayer tunneling spectroscopy. *Phys. Rev. Lett.*, **85**, 4787–4790.
63. Ozyuzer, L., Koshelev, A., Kurter, C. *et al.* (2007) Emission of coherent THz radiation from superconductors. *Science*, **318**, 1291–1293.
64. Mijatovic, C., Brinkman, A., Oomen, I. *et al.* (2002) Magnesium-diboride ramp-type Josephson junctions. *Appl. Phys. Lett.*, **80**, 2141–2143.
65. Ueda, K., Saito, S., Semba, K. *et al.* (2005) All-MgB_2 Josephson tunnel junctions. *Appl. Phys. Lett.*, **86**, 172502.
66. Katase, T., Ishimaru, Y., Tsukamoto, A. *et al.* (2010) Josephson junction in Co-doped $BaFe_2As_2$ epitaxial film on (La, Sr)(Al, Ta)O_3 bicrystal substrate. *Appl. Phys. Lett.*, **96**, 142507.
67. Katase, T., Ishimaru, Y., Tsukamoto, A. *et al.* (2011) Advantageous grain boundaries in iron pnictide superconductors. *Nat. Commun.*, **2**, 409.
68. Tanabe, K. and Hosono, H. (2012) Frontiers of research on iron-based superconductors toward their applications. *Jpn. J. Appl. Phys.*, **51**, 010005.
69. Schmidt, S., Doring, S., Schmidl, F. *et al.* (2010) $BaFe_{1.8}Co_{0.2}As_2$ thin film hybrid Josephson junctions. *Appl. Phys. Lett.*, **97**, 172504.
70. Seidel, P. (2011) Josephson effects in iron based superconductors. *Supercond. Sci. Technol.*, **24**, 043001.

3.4 Wires and Tapes

3.4.1 Powder-in-Tube Superconducting Wires: Fabrication, Properties, Applications, and Challenges

Tengming Shen, Jianyi Jiang, and Eric Hellstrom

3.4.1.1 Overview of Powder-in-Tube (PIT) Superconducting Wires

3.4.1.1.1 Introduction

Powder-in-tube (PIT) is an important technology for manufacturing many superconductors, including Nb_3Sn, MgB_2, $Bi_2Sr_2CaCu_2O_x$ (Bi-2212), $(Bi,Pb)_2Sr_2Ca_2Cu_3O_x$ (Bi-2223), and the iron-based superconductors (FBSs), into twisted, multifilamentary metal/superconductor composite wires. In its simplest form, PIT wires are made by putting a powder inside a tube, drawing the tube through a set of dies into a wire, and then doing a heat treatment to convert the powder into a superconductor. These wires are the foundation of superconducting magnet technology and superconducting power devices such as lossless underground power cables, superconducting fault-current limiters, transformers, and rotating machines. In this chapter, we discuss the wire manufacturing, conductor design, and heat-treatment processes of major PIT superconducting wires, and describe typical superconducting performance in relation to magnetic field and strain.

PIT superconducting wires are typically round or tape-shaped composite conductors, in which multiple superconductor filaments are embedded in a matrix consisting of a normal metal, such as Cu or Ag. The multifilamentary conductor structure is important for stabilization against heat generated by flux jumping [1]. The first superconducting PIT wires were made by Kunzler *et al.* [2] at Bell labs using Nb tubes filled with crushed Nb and Sn powder that were heat-treated at 900–1100 °C to react the precursors into Nb_3Sn. Many superconductors, such as Nb_3Sn and all the ceramic, high-temperature superconductors (HTSs), are brittle and hard, and are difficult to process into wires. The PIT method provides a convenient and economical fabrication technology that yields versatile wires of various size, filament count and architecture, and aspect ratio. There are six major considerations for designing and developing a PIT superconductor wire suitable for applications, including:

1) *High critical current density*: High critical current density is of paramount importance for applications. For superconducting magnet applications, the wire must carry an engineering current density J_E ($J_E = I_c/A$, where I_c is the critical current of the wire and A is the cross-section of the entire wire) of $>100\ \mathrm{A\ mm^{-2}}$ in strong magnetic fields. This requires the superconducting portion of the wire to have high critical current density J_c ($J_c = I_c/A_s$, where A_s is the cross-section of the superconductor). Over the past decades J_E and J_c of PIT conductors have increased considerably by decreasing porosity in the filaments, tuning the critical transition temperature (T_c) and the irreversibility

field (B_{irr}), and controlling the nanostructures of superconductors to enhance flux pinning.

2) *Mechanical properties*: Superconducting wires must have sufficient strength to withstand conductor fabrication, coil winding, cabling, thermal stresses during cooldown, and the Lorentz force during operation. However, many superconductors are brittle and their superconducting properties are sensitive to strain. For example, in bulk form Nb_3Sn fractures at ~0.2% strain and applying stress can affect its superconducting transition temperature T_c and upper critical field B_{c2} [3]. The mechanical support for brittle superconductors is provided by the normal metal sheath surrounding the superconducting filaments in a PIT wire.
3) *Stabilization and protection*: The normal metal sheath in PIT wires also provides cryogenic stabilization and protection. The sheath must be able to carry large current during a quench without producing enormous joule heating that could overheat the conductor. The sheath material cannot react with the superconductor to either degrade the filament superconductivity or develop an insulating barrier – often an oxide layer – at the superconductor interface that reduces the beneficial current sharing between superconductor and sheath material.
4) *Cabling*: The ability to economically fabricate cables that carry high current is important for large magnet systems. The high-energy physics (HEP) particle colliders use Rutherford cables with operating currents of several kiloamperes (4.3 and 11.8 kA for the Tevatron cable and Large Hadron Collider (LHC) cable, respectively). The Rutherford cable is composed of superconducting strands twisted and transposed on each other. It is much easier to make cables with round wires than flat tapes.
5) *AC loss*: Type II superconductors dissipate energy when flux lines move through them due to time-varying conditions of magnetic field, transport current, or both. Therefore, for AC applications, it is important to design conductor to minimize such AC loss. AC loss in Type II superconductor composite has been treated by Carr [4], Wilson [1], and Amemiya *et al.* [5]. Reducing AC loss requires small diameter filaments, decoupling the filaments by twisting the wire, and using resistive matrix material or resistive barriers to increase the resistance along the path of the coupling current.
6) *Manufacturability*: Ease of manufacturing and use are key considerations for designing and manufacturing a PIT conductor. It is very important to develop conductor with kilometer-long piece lengths with consistent properties.

3.4.1.1.2 General Comments about PIT Wire Manufacture

PIT wires and tapes are made by loading superconductor precursor powders into a metal tube, which is typically sealed and drawn through a series of dies with gradually decreasing diameters into a fine wire. Theses wires are cut and restacked for a series of additional drawing, rolling, and heat-treatment steps, leading to the final multifilamentary PIT wire, which can be cold-rolled into a tape. The final wire can contain hundreds of filaments, each of which has a diameter ranging from 10 to 100 μm.

For fabricating these superconductors, a key challenge has always been engineering the conductors to carry higher I_c in strong magnetic fields and then to incorporate the high performance developed in laboratory samples into large production billets economically. The interplay between final J_c and conductor design, fabrication, and heat treatment is complicated but must be well understood to utilize the full potential of superconductors. In the following sections, we use Bi-2212 as a model system to elucidate key practical aspects of PIT wire design, fabrication, and heat treatments, and we review the state-of-the-art of other major commercialized PIT superconducting wires – Nb_3Sn, MgB_2, and Bi-2223 – and PIT wires made from the new FBSs.

3.4.1.2 Manufacturing, Heat Treatment, and Superconducting Performance of PIT Wires

3.4.1.2.1 $Bi_2Sr_2CaCu_2O_x$ (Bi-2212) Round Wire

Historical Perspectives Bi-2212 was the first HTS material in round wire form that could develop high J_c. The first PIT Bi-2212 round monocore wire was made by Heine and coworkers in 1989 [6, 7] using a partial melt heat treatment. Their wire exhibited a J_c of 150 A mm^{-2} at 4.2 K, 26 T. The first round multifilamentary Bi-2212 wire was made by Motowidlo *et al.* in 1994 [8] and their wires were cabled into Rutherford cables by Scanlan *et al.* in 1999 [9], demonstrating that it was feasible to make high-capacity Bi-2212 cable for high-field accelerator magnets. Significant advances were also made in the 1990s to develop Bi-2212 tapes. The National Institute for Materials Science in Japan achieved J_c up to 5000 A mm^{-2} at 4.2 K, 10 T (field parallel to tape surface) in dip-coated tape by a process called *PAIR* (*pre-annealing-intermediate rolling*) [10].

Beginning in the early 2000s, the focus of Bi-2212 conductor development has been almost exclusively on round wire. At present, commercial Bi-2212 round wires are available from Oxford Superconducting Technology (OST) in New Jersey, Supercon in Massachusetts, and SupraMagnetics in Connecticut, and Nexans SuperConductors GmbH (Nexans) in Germany produces Bi-2212 powder. Figure 3.4.1.1 shows transverse and longitudinal cross-sections of a typical as-drawn Bi-2212 round wire made by OST [11]. In 2005, an OST Bi-2212 round wire measured by Trociewitz *et al.* [12] exhibited J_c over 10^3 A mm^{-2} at 4.2 K in fields up to 45 T. The high J_c in this Bi-2212 round wire, together with the Bi-2212 irreversibility field, which is >100 T at 4.2 K, showed that there is huge head room compared to Nb_3Sn to fabricate very high-field magnets.

Yet its very high-field applications did not quickly follow. In 2009, a U.S. collaboration, called the *Very High Field Superconducting Magnet Collaboration* (*VHFSMC*), was formed between three universities and five national laboratories to evaluate the potential of Bi-2212 for high-field applications. To their dismay, industrial-scale, long-length conductors showed a performance much lower than the 2005 conductor Trociewitz measured. For example, in 2010, 11 different industrial batches of Bi-2212 conductor delivered to VHFSMC showed J_c (4.2 K,

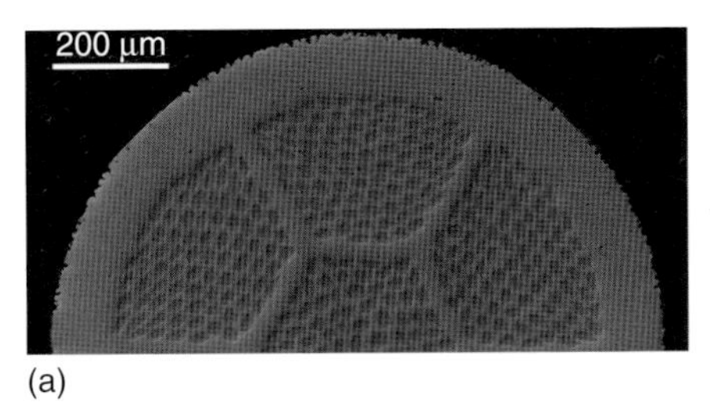

(a)

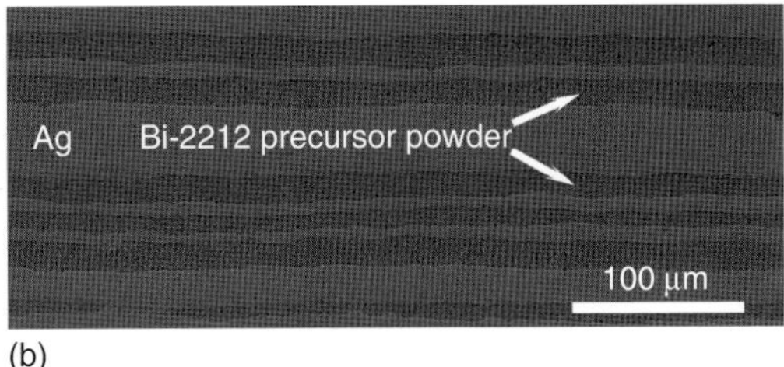

(b)

Figure 3.4.1.1 (a) Half transverse and (b) partial longitudinal cross-sections of a typical as-drawn Bi-2212 round wire with an 85 × 7 architecture. The designation 85 × 7 indicates the wire has a double restack architecture that contains 85 filaments in each of seven bundles for a total of 595 filaments in the wire. (Figure from Ref. [13]. Reproduced with permission from IOP Publishing.)

5 T) that varied from 800 to 2800 A mm^{-2}. Such J_c was obtained in short-length samples that were heat-treated with their ends open. When long-length Bi-2212 was wound into coil and reacted, its J_c further degraded and rarely exceeded 65% of its short-sample J_c. Coils also often show signs of Bi-2212 leakage that results when Bi-2212 liquid leaks out through the sheath material and reacts with insulation and coil former and flanges.

Examining high-temperature microstructures of VHFSMC and 2005 champion J_c conductors led to the discovery that the major critical current limiting mechanism is the large gas bubbles that form during the melt processing from residual porosity in as-drawn wires [13–15]. Removing porosity in short-length wires using cold isostatic pressing (CIP) and swaging quickly improved the J_c (4.2 K, 5 T) of many of VHFSMC wires to 3500 A mm^{-2} [16, 17]. The J_c degradation in long-length conductors was soon found to be caused by wire expansion due to silver creep driven by internal gas pressure [18–21]. The source of the leakage was identified as the creep rupture of Ag alloy sheath [21]. This deleterious effect of internal gases can be controlled by an overpressure (OP) processing, which was used by the National High Magnetic Field Laboratory (NHMFL) to develop $J_E > 700$ A mm^{-2} at 4.2 K, 20 T, which is sufficient to build high-field accelerator magnets [18].

Fabricating and Heat-Treating Bi-2212 Round Wires The quality of the precursor powder packed in the Ag tube is critical to the performance of the final conductor. Important powder variables include powder composition, phase assemblage (i.e., amount of Bi-2212, $Bi_2(Sr,Ca)_2CuO_x$ (Bi-2201), and alkaline earth cuprate (AEC) phases in the powder), particle size distribution, and impurity phases such as Fe oxide, and carbon and hydrogen compounds that can form gaseous species during the heat treatment. Currently the main supplier of Bi-2212 powder is Nexans SuperConductors GmbH, which makes Bi-2212 powder through a melt-casting and jet-milling process. Other methods of making Bi-2212 powder include solid-state reaction, coprecipitation, sol–gel techniques, freeze-drying, spray-drying, and aerosol spray pyrolysis. The Nexans's powder has the nominal composition of $Bi_{2.17}Sr_{1.94}Ca_{0.98}Cu_{2.00}O_x$ [22].

Silver is the only material that can be used for the sheath material for Bi-2212 wires. This is because it does not oxidize during the heat treatment, it is chemically compatible with Bi-2212, and it is permeable to oxygen. However, pure Ag is mechanically weak, so oxide dispersion-strengthened (ODS) Ag alloys such as Ag–Mg and Ag–Al are widely used as the outer sheath for Bi-2212 wires [23, 24]. Studies to further improve ODS Ag alloys are currently being done [25].

Bi-2212 wires can be made with different filament architectures. They can be single stack wires, such as the 1050 filament single stack wire made by Supercon, and double restack wires, such as OST's 85 × 7, 37 × 18, 85 × 18, and 121 × 18 architectures with 15–20 μm diameter filaments [26]. Hitachi developed a round, isotropic conductor, called *ROSAT* conductor (*ROtation Symmetric Arranged Tape-in-tube*), using anisotropic flat tapes. They stacked 18 flat tape conductors on top of one another, then they inserted three stacks of tape into a Ag tube maintaining a 120° rotation between each stack of tapes [27]. SupraMagnetics is developing a random restack conductor [28].

A partial melt processing has been widely used to heat-treat Bi-2212 for high J_c. Figure 3.4.1.2 shows a standard melt-processing schedule used by VHFSMC, which includes a short time (12 min) at the maximum temperature of about 890 °C during which Bi-2212 melts incongruently, forming liquid and crystalline phases. A slow cooling (2.5 °C/h) is used to convert the melt into Bi-2212, the 836 °C/48 h step removes residual liquid phase, and during the final cooling to room temperature the Bi-2212 takes up oxygen, which overdopes the Bi-2212 and improves J_c at 4.2 K [29]. Crystalline phases in the melt change with the overall composition, temperature, and oxygen partial pressure (pO_2). One of the problems with melt processing Bi-2212 in air, which is what was originally done, is large grains of $(Sr,Ca)CuO_2$ (1 : 1 AEC) form in the melt [30]. It reacts slowly with the melt during the reformation of Bi-2212, resulting in large remnant grains of 1 : 1 AEC in the conductor that block current flow. In 1993, Endo and Nishikida [31] found that melt processing in 1 bar pure O_2 replaces the 1 : 1 AEC phase with $(Sr,Ca)_{14}Cu_{24}O_x$ (14 : 24 AEC), which grows less substantially in the melt and reacts faster than 1 : 1 AEC during the reformation of Bi-2212. Currently all Bi-2212 is heat-treated in 1 bar of pure O_2 or in gas mixtures with pO_2 = 1 bar.

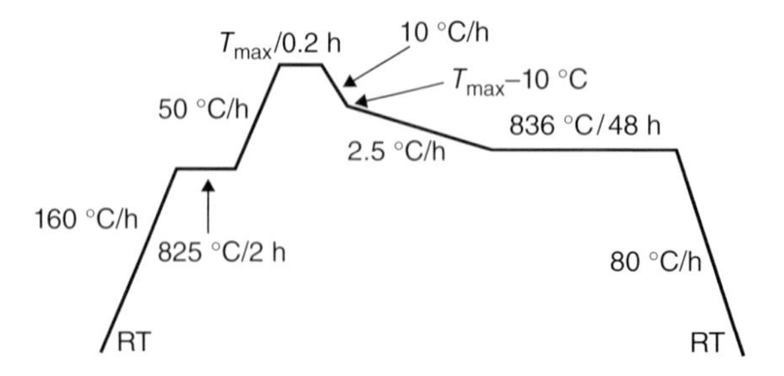

Figure 3.4.1.2 Schematic heat-treatment schedule used for Bi-2212 wire.

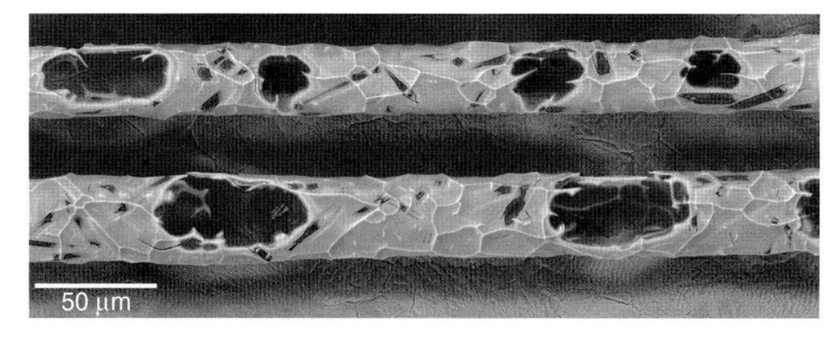

Figure 3.4.1.3 SEM images of two Bi-2212 filaments of a 27 × 7 wire quenched from the melt, showing large gas bubbles (large dark regions). (Figure from Ref. [14]. Reproduced with permission from IOP Publishing.)

Overpressure (OP) Processing as the Route to High J_c in Coil-Length Bi-2212 Round Wires In as-drawn wire, the Bi-2212 powder is only about 70% dense [32], which means that there is about 30% void space in the as-drawn wire filled with gas. In the as-drawn wire, the gas is uniformly distributed in the voids around the grains of powder but when the Bi-2212 powder melts, this uniformly distributed gas agglomerates into filament-sized, gas-filled bubbles, as shown by microstructures of Bi-2212 round wires quenched from the melt by Shen *et al.* [13]. Further studies by Kametani *et al.* [14], Scheuerlein *et al.* [15], Malagoli *et al.* [19, 20], Jiang *et al.* [16, 17], and Shen *et al.* [21] provide firm evidence that these large gas bubbles (Figure 3.4.1.3) are a major current limiting mechanism. A direct implication of this finding is that decreasing the void fraction in fully heat-treated wire will improve J_c in Bi-2212 wires. Jiang *et al.* [17] showed that this is true. They more than doubled J_c by replacing residual air in the filaments by pure oxygen and then using CIP at 2 GPa to compress the Ag sheath and Bi-2212 powder to greatly decrease the filament void fraction. Swaging also doubled J_E of short samples [16].

But later tests on 1 m-long CIPped wires showed cracks and leakage after the heat treatment. Malagoli *et al.* [20] examined this issue and found that J_c decays with increasing distance from the ends of the wire in 1 m, heat treated conductors, which was also reported by Kuroda *et al.* [33]. Malagoli *et al.* [20] also discovered that the diameter increased with increasing distance from the ends of the wire, with longer samples (up to 2.4 m) often exhibiting damage and leakage after the heat treatment. Malagoli *et al.* [20] speculated that the wire expansion was caused by the internal gas pressure. Only oxygen and hydrogen can diffuse through the Ag sheath and all other gases can only be removed by diffusing out the open ends of the wire, which becomes difficult when the length of the conductor is more than a few meters because the time constant of gas diffusion goes up with the square of the conductor length. These other gases would normally be residual N_2 or Ar from when the tube was packed with powder, and CO_2 and H_2O that form during the heat treatment from impurities in the powder. Normally these gases, which are trapped or form in long lengths of wire, cause the wire to expand causing leakage and decreasing J_c. This hypothesis was verified by Shen *et al.* [21], who observed the gases released during melt processing of Bi-2212 and observed the expansion and creep rupture of Ag-sheathed Bi-2212 wires. Power-law creep of Ag explains the behaviors found by Malagoli *et al.* [20] and Shen *et al.* [21] well.

Although CIPping and swaging can increase J_c and J_E in short wires, these techniques are not useful for long wires, because the deleterious gases cannot escape from the ends of very long wires. The solution is to apply sufficient isostatic pressure to the Bi-2212 wire during the heat treatment to oppose the internal gas pressures and swelling and compress the Ag sheath. The applied pressure has to be high enough to overcome the internal gas pressure in the wire and to compress the Ag so the volume of each hole in the Ag wire is decreased until it just matches the volume of Bi-2212 that is in it. This compression is done using OP processing, which is a form of hot isostatic pressing (HIP). This increases the Bi-2212 filament density to approach 100%. A pressure of 10 bar was used in 1997 for processing Bi-2212 tapes [34, 35], which limited the bubble formation in the Bi-2212 tapes. OP processing was jointly developed as a processing technique at the University of Wisconsin–Madison and American Superconductor, Corp. (AMSC), in the early 2000s to heal cracks and close up small residual pores and increase J_c in Bi-2223 wire [36]. In 2012, the NHMFL OP processed Bi-2212 round wires with a total pressure up to 100 bar [18]. Figure 3.4.1.4 shows that 25–100 bar OP processing can increase J_E by up to seven times compared to 1 bar, long-length wire processing. The best short wire (reacted under 100 bar) achieved J_E of 725 A mm^{-2} at 4.2 K, 20 T.

OP processing is used for the entire Bi-2212 heat treatment shown in Figure 3.4.1.2. Ar, which cannot diffuse through the Ag, compresses the wire, whereas oxygen, which can diffuse through the Ag, does not compress the wire.

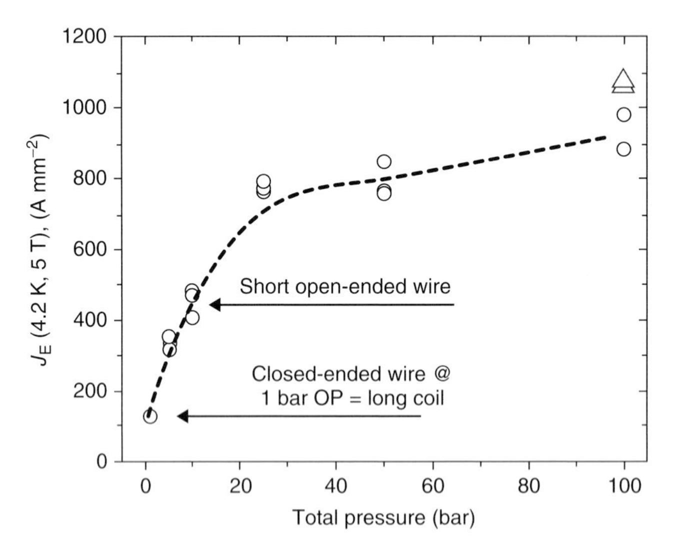

Figure 3.4.1.4 Engineering critical current density J_E (4.2 K, 5 T) as a function of total pressure for overpressure processed wires [18]. (Reproduced with permission from Nature publishing.)

The oxygen partial pressure is set to $pO_2 = 1$ bar for all Bi-2212 OP processing, which fixes the thermodynamic oxygen activity needed to form Bi-2212. In contrast, OP processing is only used in the second heat treatment for Bi-2223 and the $pO_2 = 0.075$ bar for all total OP pressures [36].

Bi-2212 Cables and High-Field Magnet Technology Because Bi-2212 is a brittle ceramic, Bi-2212 magnets are typically built using the Wind-and-React approach. But one of the major challenges is the weak mechanical strength of the wires. The management of the mechanical and thermal stresses with acceptable and reversible critical current degradation is very important. Rutherford-type cables with 18 Bi-2212 strands and a Ni–Cr core were successfully made by Scanlan *et al.* in 1999 [9]. The Ni–Cr core provides high interstrand resistance for reduced coupling losses and a gentler bending of the wires at the edge of the cable. The bare Ni–Cr core contaminated the Bi-2212 wire during the melt processing but this was prevented by coating the Ni–Cr with an oxide layer such as Al_2O_3. Shen *et al.* made a (6 + 1) cable shown in Figure 3.4.1.5, which has a high-strength alloy (Fe–Cr–Al) coated with TiO_2 in its center. The cable was OP-processed under 10 bar and retained 100% of the I_c of the short, open-end strands [37]. This showed that Bi-2212 wire can be made into a strengthened cable and then be OP-processed to attain high J_E for accelerator magnet applications.

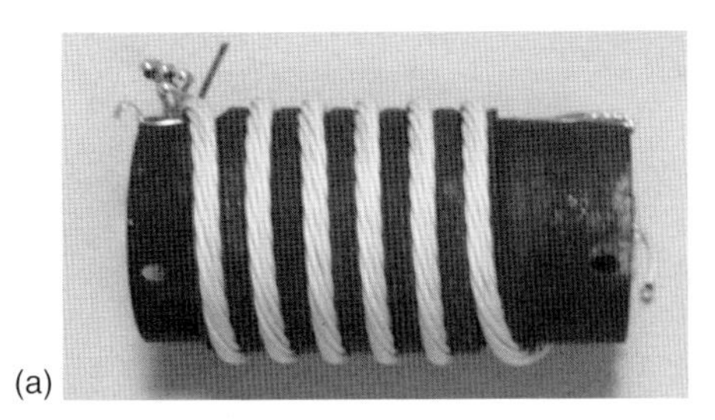
(a)

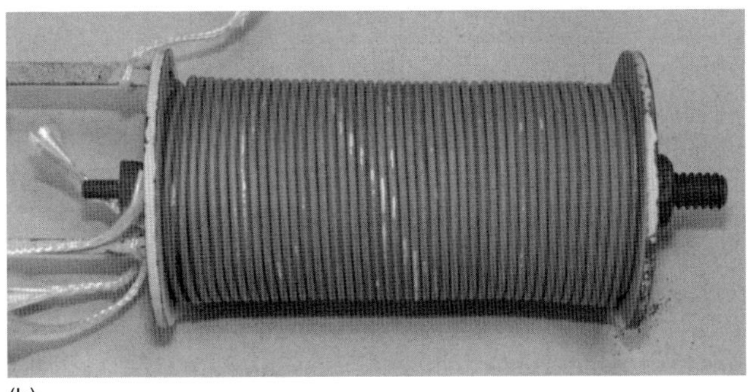
(b)

Figure 3.4.1.5 (a) 1 m long 6 + 1 Bi-2212 cable made by Fermi National Accelerator Laboratory. It was OP processed at 10 bar. (b) Bi-2212 coil made by National High Magnetic Field Laboratory that was OP processed at 10 bar. This coil was insulated with nGimat TiO_2. It generated 2.6 T in a background field of 31.2 T [18, 38].

The development of insulation for Bi-2212 wire is also very challenging. The insulation needs to be applied to the wire before being wound into a coil and survives the winding process and the high-temperature heat treatment (Figure 3.4.1.2). The insulation should neither react with the Bi-2212/Ag-alloy wire nor prevent Bi-2212 from access to oxygen. In addition, the insulation needs high dielectric strength at 4.2 K but it has to be thin to maximize the winding current density. Alumino-silicate fiber braid has been used, but its layer thickness is about 150 μm, which significantly reduces the winding current density, and it reacts with the Ag during the heat treatment. Recently, nGimat LLC and the NHMFL have independently developed TiO_2 coating for Bi-2212 wire [38, 39]. As shown in Figure 3.4.1.5, TiO_2 worked nicely on a test coil. The advantages of the TiO_2 coating are that it is chemically compatible with the Bi-2212 wire, it can be made as thin as 10 μm, and it is porous so it allows oxygen flow during the Bi-2212 heat treatment.

Many test magnets have been successfully made using Bi-2212 tapes and wires, such as a 5 T insert in a 20 T background field [40–42] made from Bi-2212 tapes, and race-track coils made from Rutherford cable [43]. A recent solenoid insert made by NHMFL (Figure 3.4.1.5) generated 2.6 T in a 31.2 T background field [18]. This high-field demonstration coil was wound with 30 m of 1.4 mm diameter Bi-2212 wire and heat-treated at 10 bar in an OP furnace. This OP pressure was high enough to prevent wire expansion but not high enough to compress the wire. For comparison, the previous coil heat-treated at 1 bar total pressure only generated an additional field of 1.1 T [42]. The OP-processed coil was quenched multiple times without damage. It achieved a maximum quench current of 388 A ($J_E = 252$ A mm^{-2} at 33.8 T). These demonstrations provide the new and specific evidence that round, multifilamentary 2212 wire is now ready for magnet application. Indeed, based on the best short-sample OP result in Figure 3.4.1.4, it is estimated that a coil OP-processed at 100 bar, rather than the 10 bar employed for the 33.8 T coil, would have approximately doubled its J_E, allowing a Bi-2212 coil to equal or even exceed the recent world record 35.4 T (also done with a 31.2 T background) REBCO-coated conductor coil [44].

3.4.1.2.2 $(Bi,Pb)_2Sr_2Ca_2Cu_3O_x$ (Bi-2223) Tapes

Bi-2223, whose nominal composition is $(Bi,Pb)_2Sr_2Ca_2Cu_3O_x$, is produced as a PIT tape. It is the workhorse HTS used in many prototype applications because it was the first HTS available in kilometer-long piece lengths from AMSC and Sumitomo Electric Industries, Ltd. (SEI). There was intense work worldwide in the 1990s and early 2000s to develop this conductor, which is processed very differently than Bi-2212. Below we briefly describe the constraints that shape Bi-2223 processing, how these have been managed to make the current conductor, and describe the current Bi-2223 conductor available from SEI. Bi-2223 has a higher T_c (110 K) than Bi-2212 (~75–92 K) and it can be used at 77 K, whereas Bi-2212 can only be used below ~20 K. In contrast to Bi-2212, Bi-2223 cannot be melt-processed because the Bi-2212 phase rather than Bi-2223 forms from the melt on cooling. Instead the Bi-2223 phase is formed *in situ* in the PIT tape at elevated temperature by a liquid-assisted solid-state reaction from precursor powder that is a mixture of phases that has the overall Bi-2223 stoichiometry and includes the Bi-2212 phase. The precursor powder is packed in pure Ag tubes, drawn to hexagonal-shaped wire, cut and restacked in a Ag-alloy sheath, and then drawn into a round wire. The multifilamentary round wire is then rolled to a flat tape. Bi-2223 grains need *c*-axis alignment, which comes from the rolling, to have high J_c. When the round wire is rolled into tape, the micaceous Bi-2212 grains in the precursor powder become aligned with their *c*-axis perpendicular to the rolling direction. The Bi-2223 grains adopt this *c*-axis alignment as they grow during the heat treatments. Just as with Bi-2212, Ag is used for the Bi-2223 sheath material, again because it does not react with the Bi-2223 and it is permeable to oxygen.

Bi-2223 tape is heat-treated in two distinct steps with an intermediate cold rolling step between the two heat treatments. The heat-treatment temperature is selected so a small amount of liquid forms to accelerate the formation of the Bi-2223 phase. Pb is added to Bi-2223 because it improves the formation of the Bi-2223 phase. The length of the first heat treatment is adjusted so only a portion of the powder reacts to form Bi-2223. The tape is cooled to room temperature and receives a small amount of cold rolling. This mainly densifies the Bi-2223 filaments and also helps break up, mix, and densify unreacted powder, helping the reaction go further toward completion. A downside of rolling is that it creates cracks in the filaments that run perpendicular to the rolling direction and, if not removed, block current flow. In the second heat treatment, the remainder of the powder reacts to form Bi-2223 and hopefully the cracks heal.

Like as-drawn Bi-2212 wires, as-rolled Bi-2223 tapes are porous. Yuan *et al.* [45] developed OP processing, which is used during the second heat treatment to apply pressure to the tape to heal the cracks and densify the tape. Their laboratory-size system applied pressures up to ~150 bar with a precise mixture of Ar and oxygen gas. As described above for Bi-2212, during OP processing Ar applies the densifying pressure, and the oxygen partial pressure, which was $pO_2 = 0.075$ bar, at the total working pressure of 150 bar in the OP system, sets the thermodynamic conditions in the powder needed to form Bi-2223. When, AMSC decided to quit manufacturing Bi-2223 in the mid-2000s to focus on yttrium barium copper oxide (YBCO)-coated conductors, SEI became the major company in the world making Bi-2223 conductor. They adopted the OP process, which they call controlled overpressure (CT-OP) sintering and built an industrial-scale OP furnace that is large enough to heat-treat several tens of kilometers of Bi-2223 tape at a time.

Rupich *et al.* [46] wrote an interesting history of the development of Bi-2223 at AMSC several years after AMSC was no longer making Bi-2223. In it they relate fascinating details about Bi-2223 processing that had been closely held information at AMSC when it was manufacturing Bi-2223.

SEI now offers several types of Bi-2223 tape, which they sell under the trademark DI-BSCCO, for different applications [47]. The Ag-sheathed Bi-2223 conductor is weak, which is not a problem for some applications, but other applications require stronger conductor. Stronger conductor is made by laminating the faces of fully heat-treated Bi-2223 tape with a stronger metal. They also make Bi-2223 with a

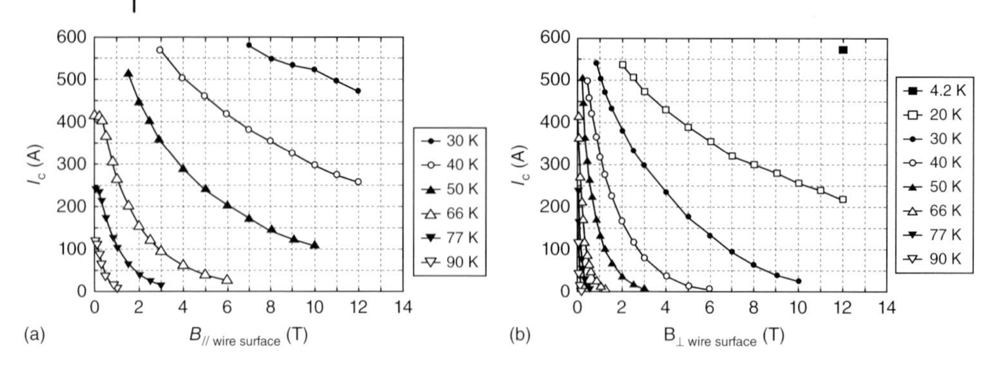

Figure 3.4.1.6 Transport I_c in SEI Bi-2223 tape (4.3 mm × 0.23 mm) as a function of applied magnetic field at different temperatures. $I_c = 240$ A at 77.3 K, self-field. (a) Field is applied parallel to the surface of the tape and (b) field is applied perpendicular to the surface of the tape. (From Ref. [48]. Reproduced with permission from Japan Society of Applied Physics.)

Ag–Au alloy sheath that has low thermal conductivity and is used for magnet current leads. They also offer tape with fully transposed filaments that is made by twisting the round wire while it is being rolled flat. This transposition reduces AC losses.

As mentioned above, Bi-2223 has to be *c*-axis-textured to have high J_c. A consequence of this texturing is that J_c varies with the angle of the applied magnetic field. This is shown in Figure 3.4.1.6, where I_c at a given temperature drops slowest with increasing magnetic field when the field is parallel to the surface of the tape, that is, parallel to the *ab* planes (Figure 3.4.1.6a). Just as in YBCO-coated conductor, the I_c performance of Bi-2223 tape degrades when the field deviates from this favorable orientation.

SEI has increased I_c over the years as shown in Figure 3.4.1.7 for their standard size conductor (~4.3 mm × ~0.23 mm). When they started using OP processing in 2004, they had a step increase in I_c. They can achieve I_c of 273 A (77 K, self-field) in short laboratory samples and will incorporate these R&D modifications in their commercial, kilometer-length conductor. They currently sell kilometer-long piece lengths of tape with $I_c > 200$ A at 77 K, self-field. This corresponds to $J_E = 202$ A mm^{-2} and to a critical current of 465 A cm^{-1} width.

3.4.1.2.3 Nb_3Sn

Nb_3Sn, a brittle intermetallic compound with a T_c of 18 K and a B_{c2} of 23–29 T at 4.2 K, is used for constructing superconducting magnets to generate magnetic fields in the 10–22 T range. Nb_3Sn superconducting magnets are used in HEP accelerators, fusion energy devices such as the international thermonuclear experimental reactor (ITER), nuclear magnetic resonance (NMR) systems for chemical analysis, and standard high-field laboratory magnets. Nb_3Sn wires can be manufactured through various processes, including the bronze process, the internal-tin process, and the PIT process. The first Nb_3Sn monocore PIT wire that showed the possibility of high-field superconductivity was made by Kunzler

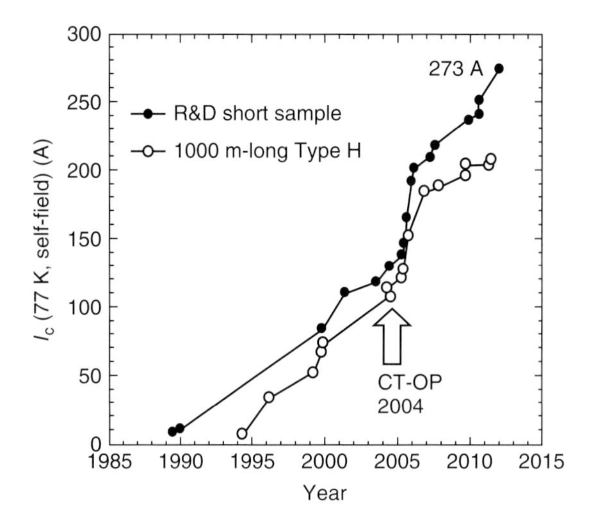

Figure 3.4.1.7 Increase in I_c with time in short samples and commercial, 1 km lengths of SEI Bi-2223 tape. (Figure courtesy of SEI.)

et al. [2] by drawing a mixture of Nb and Sn powders in a Nb tube and reacting at 900–1000 °C. Overall current densities of Kunzler's wires were low. The method was improved by using $NbSn_2$ powder and a small amount of Cu additive in a Nb tube [49], which converted to Nb_3Sn via Nb_6Sn_5 with a ~600 °C heat treatment. This low temperature reaction produces a fine-grain structure and hence strong pinning and good J_c because grain boundaries are the dominant flux pinning centers in Nb_3Sn. Filamentary composite PIT Nb_3Sn wires were made by ShapeMetal Innovation, The Netherlands (presently Bruker-EAS). Industrial PIT wires from Bruker regularly achieve non-Cu J_c in excess of 2500 A mm^{-2} at 4.2 K and 12 T.

Applications of Nb_3Sn in HEP accelerator dipoles and quadrupoles demand an ideal conductor with non-Cu J_c of >3000 A mm^{-2}, RRR (residual resistivity ratio) >100, and effective filament diameter $d_{eff} < 20$ μm [50]. Achieving high J_c with small filaments and high RRR is challenging for any of the wire manufacturing routes. PIT-processed Nb_3Sn wires have demonstrated a combination of very high current density (presently non-Cu J_c is up to 2500 A mm^{-2} at 4.2 K, 12 T) with fine (35 μm diameter), well-separated filaments, and RRR >150. Thus, PIT Nb_3Sn can be used in NMR magnets and are being developed for HEP accelerator applications. However, present manufacturing costs of PIT Nb_3Sn conductors are about two to three times higher than Nb_3Sn made by conventional metal working method, because the production quantity of PIT Nb_3Sn produced is small compared to internal-tin and bronze Nb_3Sn and also because delicate care needs to be given to obtain high-purity precursor powder. For more information about PIT Nb_3Sn, see Godeke [51], where he discusses Nb_3Sn processing and the dependence of Nb_3Sn J_c on A15 composition, grain morphology, and strain state, and

Godeke *et al.* [52], where they give a detailed review of the historical and current development of PIT Nb_3Sn wires.

3.4.1.2.4 MgB_2

MgB_2, which was discovered in 2001, has a T_c of 39 K and a high irreversibility field B_{irr} reaching 30 T at 4.2 K [53] in wires by alloying with carbon, using different additives that are described below. These properties may enable MgB_2 to challenge the dominance of NbTi and Nb_3Sn in superconducting magnets. PIT is used to produce multifilamentary MgB_2 wires and tapes. Polycrystalline MgB_2 PIT round wires are feasible because randomly oriented grain boundaries in MgB_2 are not obstacles to current flow [54]. At 4.2 K, 10 T, J_c and J_E of MgB_2 PIT tapes and wires have reached 1070 A mm^{-2} and 167 A mm^{-2}, respectively [55]. At 20 K, 1 T, J_c and J_E of commercial MgB_2 tapes and wires are 2200 and 590 A mm^{-2}, respectively (S. Brisigotti *et al.*, personal communication, 2013).

Reaction Routes, Strand Design, and Industrial Manufacturing There are two variants of PIT MgB_2 fabrication, *in situ* and *ex situ*. The *in situ* approach uses Mg or MgH_2 and B powder that react to form MgB_2 during a single-step heat treatment (700 °C, 20–40 min in flowing Ar; note that Mg melts at 650 °C). Compared to the *ex situ* route, the lower processing temperature results in smaller grains, leading to stronger pinning because the grain boundaries are the dominant pinning centers in MgB_2. It also helps minimize reactions between the MgB_2 and the sheath, and makes the reaction more compatible with existing insulation and structural materials developed for Nb_3Sn magnets. The low processing temperature also produces poor crystallinity, resulting in high B_{c2}. Another advantage of the *in situ* method is its flexibility to incorporate dopants and additives. A drawback of *in situ* wires is that the wires are porous, with the density of the reacted MgB_2 cores generally being as low as 50% because Mg and B powders cannot be 100% densely packed and because additional porosity develops during the *in situ* reaction.

The *ex situ* approach uses cold-work to crush the hard MgB_2 powder inside iron or nickel tubes. Owing to the prolonged cold-working, the as-formed conductor has a T_c of 30 K. To recover the T_c and to improve the granular connectivity, the wire is heat-treated at high-temperature in Ar. The sintering temperature is often as high as 900 °C; otherwise the connectivity is poor and the critical currents are small. This heat treatment leads to recrystallization, which increases the crystallinity and reduces B_{c2}. In addition, disorder cannot be introduced simply by low-temperature processing in the *ex situ* process. The consequence is that their high-field performance is lower than that of *in situ* wires, thus *ex situ* wires are generally for low-field applications (<1.5 T). Thermally stable defects (i.e., C atoms) are needed to enhance B_{c2} in high-field conductors but they may also reduce T_c. The filaments in *ex situ* wires are significantly denser (70–80%) than *in situ* wires and the density can be increased by HIP [56]. However, the grain coupling of *ex situ* wires is much weaker than that of *in situ* wires.

Strands of MgB_2 wires are currently commercially available in kilometer-long piece length. *In situ* wires are available from Hyper Tech Research, Columbus,

Ohio, and *ex situ* wires are available from Columbus Superconductors Spa, Genoa, Italy. The Hyper Tech wires feature MgB_2 filaments encased in Nb barrier and Cu sheath, which are again encased in a Monel external sheath [57]. The *ex situ* wires from Columbus Superconductors consist of nickel clad, copper stabilizer, and MgB_2, and are available in various configurations, such as sandwich-type tape conductor laminated with single or double oxygen-free high conductivity (OFHC) copper strips and round wire for cable applications (S. Brisigotti *et al.*, personal communication, 2013).

Columbus Superconductor makes their *ex situ* wire by packing prereacted MgB_2 powder in a tube and drawing the tube. This is what we normally think of as the PIT method. Hyper Tech has developed a continuous tube forming and filling (CTFF) process to make a wire with powder inside it. They begin with a coil of flat metal that runs through a set of rollers so the transverse cross-section of the metal has the shape of the letter U. The U-shaped metal passes under a dispenser that drops a precise amount of premixed Mg + B into the U-shaped metal. The U-shaped metal and powder are put through another set of rollers that deform the metal around the powder until it is shaped like the letter O and completely encloses the powder. Although both processes produce a wire that has powder in a tube, they are very different processes from the viewpoint of patents. However, most people refer to wires made with both processes as PIT MgB_2 wire.

J_c in MgB_2 Wires The transport J_c of MgB_2 wires or tapes has been considerably improved over the last decade. Higher J_c at high magnetic fields are associated with increases in B_{c2} and the decrease of the anisotropy through alloying. Substituting C for B is an effective route to introduce disorder and impurity scattering to improve B_{irr} and B_{c2} of MgB_2. C doping reduces the a_0 axis length, T_c, and the crystallinity of MgB_2 but significantly improves its B_{irr} and B_{c2}. Notable additives reported are pure C or C-based compounds, including SiC [58], B_4C [59], carbohydrates such as malic acid [60], and hydrocarbons [61].

Second, J_c increases from improving connectivity by using high-purity starting powders and oxygen-free atmosphere during synthesis to minimize the formation of large non-superconducting precipitates such as MgO, boron oxides, or boron carbide at grain boundaries [62]. Additionally, J_c is improved by precluding excessive porosity using cold or hot densification methods. Notable methods applied to *in situ* wires include cold high-pressure densification [63], an internal-Mg diffusion route in which the B powder surrounds a Mg rod [55], and using MgB_4 and B as the starting powders [64].

3.4.1.2.5 **Iron-Based Superconductors (FBS)**

The first FBS (LaFePO called 1111) was discovered by Hosono's group in 2006 [65]. Other FBS systems that contain As were soon discovered including $BaFe_2As_2$, which is referred to as 122 [66]. Superconductivity was also discovered in other As-containing systems and in the FeSe [67] and $Fe(Se_{1-x}Te_x)$ [68] systems, which are called 11. This short summary focuses mainly on PIT 1111 and 122 wires and tapes, which are studied most and have the highest J_c. For more information, see Putti *et al.* [69] who reviewed the properties of FBS relevant for practical applications and Ma's [70] comprehensive review of FBS wires. We note that the 1111 and 122 systems are usually referred to as *pnictide superconductors*; however, when the 11 systems were discovered the more general term *FBS* was adopted to describe all the systems.

As summarized by Putti *et al.* [69], FBS are potentially interesting for practical applications because T_c is as high as 55 K in 1111, the B_{c2} is >100 T in 122, their anisotropy is as low as 2 in 122, and the intragranular J_c is >10 kA mm^{-2} (4.2 K, self-field) in 122. The early bulk 1111 samples showed two distinct scales of current flow with a low global current density [71] due to poor grain connectivity from observable extrinsic cracks and grain boundary wetting phases, but the intrinsic limit to transport across grain boundaries could not be determined [72]. Bicrystal studies using $Ba(Fe,Co)_2As_2$ showed that grain boundaries block supercurrent with increasing angle, but not as much as in the cuprate superconductors [73, 74].

All the FBS wires have all been made using the PIT technique. Shortly after the FBS were discovered, Ma's group made the first 1111 wires, which had the nominal composition $LaFeAsO_{0.9}F_{0.1}$ [75] then $SmFeAsO_{1-x}F_x$ [76], followed closely by the first 122 wire with composition $(Sr_{0.6}K_{0.4})Fe_2As_2$ [77].

The *in situ* La-1111 wire was made using a hand-ground stoichiometric mixture of La, As, LaF_3, Fe, and Fe_2O_3 powder that was packed in an Fe tube lined with a thin Ti foil whose purpose was to prevent reactions between the Fe sheath and FBS core during the heat treatment. The tube was rotary swaged then drawn. Short sections of wire were cut, sealed in an Fe tube, and then heat-treated at 1140 °C for 40 h. X-ray diffraction (XRD) showed 1111 and impurity phases (LaOF, and LaAs) present after the heat treatment. The J_c calculated from magnetization measurements was ~10^3 A cm^{-2} at 5 K, self-field. Microscopic images of the heat-treated wire showed that the Ti buffer prevented chemical reactions between the Fe sheath and that the FBS core contained a lot of porosity. The Sm-1111 wire was made similarly using a Ta tube rather than the Ti-lined Fe tube. The FBS reacted with the Ta. These first wires had extremely low transport J_c.

The first *in situ* 122 wire [77] was made similarly to the 1111 wire but the hand-ground mixture of the elements was packed in a Nb tube and heat-treated at 850 °C for 35 h. There was a reaction between the FBS and the Nb tube.

These first wires highlight the challenges for making PIT FBS wires, which are forming phase-pure FBS, preventing reaction between the sheath and the FBS, and forming a well-connected, high-density ceramic core. Below we address advances in FBS PIT round wire and flat tape that have led to transport J_c in 122 that is as high as 300 A mm^{-2} at 4.2 K, 10 T.

Figures 3.4.1.8 and 3.4.1.9 show transport J_c values for 122 and 1111 wires and tapes measured at 4.2 K in field. These figures show the increase in J_c in 122 that has been made, and that J_c in 122 is much higher than in 1111. Table 3.4.1.1 gives details for the data in Figure 3.4.1.8.

It is more difficult to form high-purity 1111 than 122, in part because the rare earth (RE) reacts with oxygen from the oxide in the starting powders, Fe_2O_3 in the

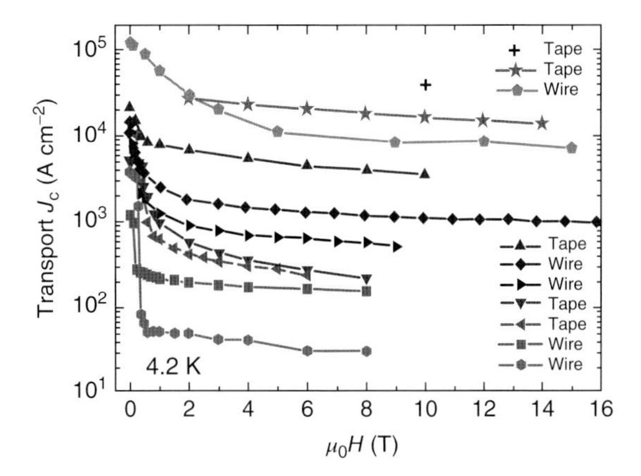

Figure 3.4.1.8 Transport J_c as a function of magnetic field at 4.2 K for 122 wires and tapes. The identification and details for these data are given in Table 3.4.1.1. For the tapes, the magnetic field was applied parallel to the face of the tape. (Figure from Ref. [70]. Reproduced with permission from IOP Publishing.)

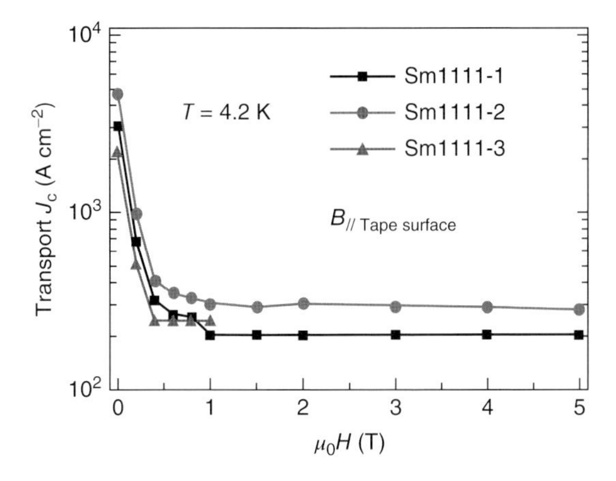

Figure 3.4.1.9 Transport J_c as a function of field for Sm-1111 tapes with nominal composition $SmFeAsO_{0.8}F_{0.2}$. The tapes were made with three different mixtures of starting materials given by Sm1111-1 = $Sm_{3-x}Fe_{1+2x}As_3$, Fe_2O_3, SmF_3; Sm1111-2 = SmAs, FeO, Fe_2As, SmF_3; Sm1111-3 = Sm, As, SmF_3, Fe, Fe_2O_3. (From Ref. [78]. Reproduced with permission from IOP Publishing.)

examples given above, forming RE_2O_3, which is extremely thermodynamically stable. It is difficult to get this RE_2O_3 to react with the arsenides to form 1111 without going to high temperature for long time, which causes F loss [78]. In addition, rare earth oxyfluoride (REOF) compounds can form that are also very thermodynamically stable and require long time at high temperature to form 1111. Wang *et al.* [78] studied starting mixtures and found the highest phase purity in Sm-1111 PIT wire using a mixture of SmAs, FeO, Fe_2As, and SmF_3 with a 900 °C heat treatment.

Table 3.4.1.1 Details about 122 wires and tapes shown in Figure 3.4.1.8 and the legends for Figure 3.4.1.8.

Symbol	Nominal stoichiometry	Type of powder – PIT heat treatment	Texturing	Sheath	Form	References
✚	$Ba_{0.6}K_{0.4}Fe_2As_2$	*Ex situ* rolling + 850 °C/2 h (2×) then pressing + 850 °C/2 h	Flat rolling + uniaxial pressing	Ag	Tape	[79]
—★—	$Sr_{0.6}K_{0.4}Fe_2As_2$ + 5–10 wt% Sn	*Ex situ* 800–950 °C/1–30 min + 600 °C/5 h	Rolling	Fe	Tape	[80]
—⬟—	$Ba_{0.6}K_{0.4}Fe_2As_2$	*Ex situ* (MSR powder) – 600 °C/12 h/192 MPa	None	Ag	Round wire	[81]
—▲—	$Sr_{0.6}K_{0.4}Fe_2As_2$ + 10 wt% Sn	*Ex situ* – 1100 °C/0.5–15 min	Rolling	Fe	Tape	[82]
—◆—	$Ba_{0.6}K_{0.4}Fe_2As_2 + Ag_{0.5}$	*Ex situ* – 850 °C/30 h	None	Ag	Round wire	[83]
—▶—	$Ba_{0.6}K_{0.4}Fe_2As_2$ + 15 wt% Ag	*Ex situ* – 600–900 °C/12–36 h	None	Ag	Square wire	[84]
—▼—	$Sr_{0.6}K_{0.4}Fe_2As_2$	*Ex situ* – 1100 °C/5–15 min	Rolling	Fe	Tape	[85]
—◀—	$Ba_{0.66}K_{0.48}Fe_2As_2 + Ag_{0.5} + Pb_{0.2}$	*Ex situ* – 1100 °C/5 min	Rolling	Fe	Tape	[86]
—■—	$Sr_{0.6}K_{0.4}Fe_2As_2$ + 10 wt% Ag	*Ex situ* – 900 °C/20 h	None	Ag	Round wire	[87]
—⬢—	$Sr_{0.6}K_{0.4}Fe_2As_2$ + 10 wt% Pb	*Ex situ* – 900 °C/20 h	None	Ag	Round wire	[87]

The first 1111 and 122 wires were made using a single, *in situ* heat treatment. The advantage of using a single-step, *in situ* heat treatment is that the sheath helps prevent loss of volatile species making it easier to control the final stoichiometry. However, with a single heat treatment, the sample cannot be reground, repressed, and reheated to increase the extent of reaction. It is not surprising that these early wires contained appreciable amounts of impurity phases. *Ex situ* 122 wires are now made by heat-treating the powder at least once, or using powder to form the 122 phase before it is packed in the tube and made into a wire. Alternately, the 122 phase is made using the mechanically activated, self-sustained reaction (MSR) developed by Weiss *et al.* [81, 88] that forms the 122 phase from the elements using high-energy ball milling. 122 wires made with the very fine-grained MSR powder can be sintered as low as 600 °C in a HIP, yielding high-phase purity, high-density 122 with high J_c in untextured round wire.

Ag, Sn, and Pb have been added to the powder used in 122 wires because empirically it has been observed that they increase J_c [82, 87]. Zhang *et al.* [89] investigated Pb additions in 122 and found that when the Pb was added to the powder immediately before it was packed in the tube its effect on J_c, changed but it is still not clear how these additions actually increase J_c.

The choice of sheath material depends on the temperature used for the PIT heat treatment. Wang *et al.* [90] found that Ag does not react with 122 below ~900 °C so Ag tubes are typically used for 122 and also for 1111 wires that are heat-treated below 900 °C. Fe tubes are used for high temperature reactions (~1100 °C) for both 1111 and 122.

1111 and 122 PIT wires can be rolled into flat tape to increase the core density and to texture the grains to increase J_c. Recently, Togano and coworkers [79, 91] showed that cold-rolling created cracks that ran perpendicular to the rolling direction and blocked current, just as in Bi-2223 tape. They found that when they ended the deformation sequence using uniaxial pressing rather than rolling the cracks ran parallel to the long axis of the tape and did not block the current path. Weiss *et al.* [81] heat-treated their 122 round wire in a HIP, which densified the core but did not align the 122 grains and retained the round wire shape, which is the geometry that magnet designers and builders prefer for single-strand coils and to make Rutherford or 6 + 1 cables.

Significant advances have been made to improve J_c in 122 round wires and tapes. However, to be a practical material, J_c has to be raised to above 10^3 A mm^{-2} in field (>10 T), preferably in round wire. This will require further understanding of what limits current transport across grain boundaries in 122 conductors and designing ways to increase the transport J_c.

3.4.1.3 Strain Sensitivity of PIT Superconductor Wires

After reaction, Nb_3Sn, MgB_2, Bi-2212, and Bi-2223 (BSCCO) are brittle. Nb_3Sn, MgB_2, and BSCCO PIT wires can be visualized as continuous, brittle fibers inside a ductile matrix. The modulus and yield stress of the composite conductors can therefore be estimated using the rule of mixture. However, their I_c exhibits rather complex strain dependence. Figure 3.4.1.10 summarizes two typical $I_c(\varepsilon)$

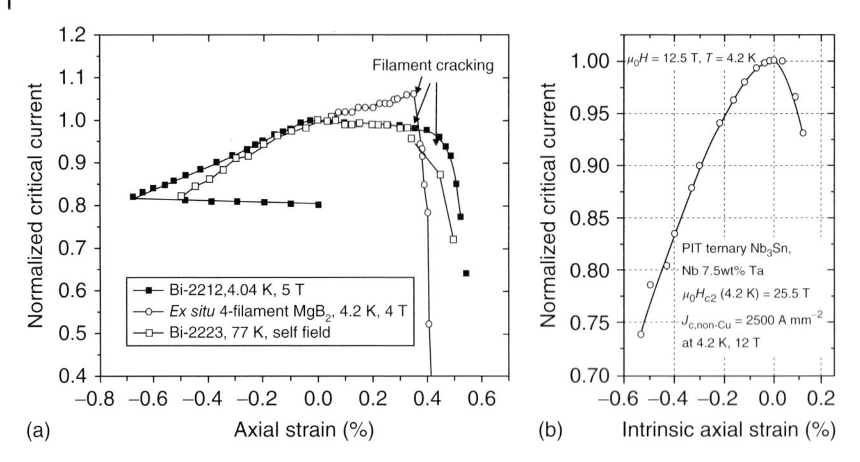

Figure 3.4.1.10 Normalized critical current as a function of axial strain for (a) PIT Bi-2212, Bi-2223, and MgB_2 wires, and (b) PIT Nb_3Sn wires. (Data sources: Bi-2212 [92] (Reproduced with permission from IEEE); Bi-2223 [93] (Reproduced with permission from IEEE); MgB_2 [94] (Reproduced with permission from IOP publishing); Nb_3Sn [52]. Reproduced with permission from Elsevier.)

dependence of superconducting composites under axial loads. Figure 3.4.1.10a shows that MgB_2, Bi-2223, and Bi-2212 have similar $I_c(\varepsilon)$ behavior under axial loads. I_c is virtually insensitive to tensile strain up to a sample-dependent strain limit. When this limit is exceeded, I_c decreases steeply and irreversibly. Applying compressive strain to Ag-sheathed BSCCO conductors and MgB_2 causes a gradual and irreversible decrease of I_c. Magnetic fields have no noticeable effect on this typical $I_c(\varepsilon)$ dependence. This $I_c(\varepsilon)$ dependence can be understood by yielding of the soft matrix followed by cracking of the polycrystalline filaments under tensile stress or buckling of grains under compressive loads.

As Figure 3.4.1.10b shows, Nb_3Sn exhibits a generically different $I_c(\varepsilon)$. I_c reaches a maximum at a wire-dependent tensile strain level, and decreases reversibly when this tensile strain is either released or further increased. The $I_c(\varepsilon)$ decrease upon the release of tensile strain continues smoothly and reversibly when the wire is put under compression. Its self-field critical current is relatively strain-insensitive but, in sufficiently high magnetic field, applied strain has a strong effect. Strain effects in Nb_3Sn can be understood as a shift of $B_{c2}(T)$ while the crystallographic unit cell is stressed.

For commercial Monel/Nb/MgB_2 wires, a reasonable working maximum tensile stress appears in the region of 280 MPa whereas the irreversible tensile strain at which the critical current density starts to degrade varies between 0.37% and 0.5% [57]. For commercial Ag–0.2 wt% Mg/Ag/Bi-2212 (area ratio AgMg : Ag : Bi-2212 = 0.25 : 0.5 : 0.25) round wire, the working maximum is around 120 MPa at 4.2 K whereas the irreversible tensile strain is between 0.3% and 0.45% [95]. Tensile stress tolerance of the commercial Bi-2223 tapes depends on conductor architecture and is often improved with a lamination of reinforcing material such as stainless steels [48].

To first order, the irreversible strain limits of BSCCO and MgB_2 metal composites are determined by the level of the thermal pre-compression of the superconductor filaments [94]. In commercial composites based on brittle superconductors, it is common to place the superconductor under compression. This can be done by using a sheath material whose thermal contraction coefficient is larger than that of superconductor. In the case of BSCCO conductor, the Ag and Ag alloys serve this purpose and the total thermal contraction from 293 to 4 K is −0.413% for Ag, and −0.152% for Bi-2212 (//ab planes). For MgB_2, stainless steels can provide a similar function.

3.4.1.4 Successful Applications Using PIT Wires, Remaining Challenges, and PIT Wires in the Future

PIT wires have many applications. High-capacity HTS current leads made from Ag–Au Bi-2223 have been successfully used in the Tevatron [96] and LHC [97, 98], and are being developed for ITER magnets [99] to minimize heat invasion into the liquid helium bath. Electric power applications represent the largest potential market. A slew of prototype devices, such as power cables, high-efficiency industrial motors, and lightweight ship propulsion systems have been successfully tested using Bi-2223 tapes worldwide. Some prototype projects in the United States are a 36.5 MW HTS ship motor produced by AMSC for the U.S. Navy [100], a 350 m HTS underground power cable connecting two substations in Albany, New York [101], a 138 kV HTS cable system installed in Long Island, NY, transmitting up to 574 MW of electricity, and a 2.25-MVA-rated distribution-level standalone fault current limiter made by AMSC and Siemens [102]. Despite considerable progress, acceptance of HTS in the demanding utility market requires significant further demonstrations of reliability and reduction in cost. The cost of the HTS wire, which is currently the largest initial cost component for HTS devices, should be reduced by further enhancing J_c. In addition, AC loss, which results in significant operating cost for the HTS power systems, should also be reduced [103]. Prototype systems based on Bi-2223 have also been demonstrated in the areas of transportation, information and communication, manufacturing, and medical imaging and analysis [48]. Quasi-DC electric cables cooled by helium gas are being developed using MgB_2 round wires for the LHC magnets near the interaction region where radiation is high [104].

Cryogen-free magnetic resonance imaging (MRI) based on MgB_2 represents another big potential market for PIT wires. Typical MRI magnets running in persistent current mode are often made of wire with a total length of ~10 km with a total resistance $<10^{-10}\ \Omega$. Commercial 0.84 mm MgB_2 short wires carry an $I_c = 200$ A (defined at $1\ \mu V\ cm^{-1}$) and $n = 30$ at 15 K, 2 T. To operate MgB_2 MRI in persistent current mode, the operation current I_{op} needs to be <109 A, which is <55% of I_c. To better use the critical current margin, it is necessary to keep n-value >30 over 10 km, a nontrivial task for manufacturers.

At present, nearly all superconducting magnets are made from NbTi and Nb_3Sn conductors, both of which have matured toward their theoretical limits. Bi-2212,

Bi-2223, and FBS PIT wires present opportunities for making superconducting magnets that generate fields >20 T for NMR systems, advanced materials research, and frontier accelerator technologies. Such systems will represent a quantum leap in technology and will require significant further research and development efforts to improve on many fronts, such as increasing stress tolerance of Bi-2212 and applying OP processing to practical coil systems.

Acknowledgments

We would like to thank our colleagues in the VHFSMC and at our respective institutions. The work at Florida State University was supported by the National Science Foundation under award numbers DMR-1157490, DMR-1306785, and by the State of Florida. Work at Fermilab was supported by the Office of Science at the U.S. Department of Energy (DOE) under contract No. DE-AC02-07CH11359. T.S. was partially supported by the U.S. DOE Early Career Research Program.

References

1. Wilson, M.N. (1983) *Superconducting Magnets*, Oxford University Press, New York.
2. Kunzler, J.E., Buehler, E., Hsu, F.S.L., and Wernick, J.H. (1961) Superconductivity in Nb_3Sn at high current density in a magnetic field of 88 kgauss. *Phys. Rev. Lett.*, **6**, 89–91.
3. Ekin, J. (2006) *Experimental Techniques for Low-Temperature Measurements: Cryostat Design, Material Properties and Superconductor Critical-Current Testing*, Oxford University Press, New York.
4. Carr, W.J. (2001) *AC Loss and Macroscopic Theory of Superconductors*, 2nd edn, Taylor & Francis Group, New York.
5. Amemiya, N., Murasawa, S.-i., Banno, N., and Miyamoto, K. (1998) Numerical modelings of superconducting wires for AC loss calculations. *Physica C: Supercond.*, **310**, 16–29.
6. Heine, K., Tenbrink, J., and Thoner, M. (1989) High-field critical current densities in $Bi_2Sr_2CaCu_2O_{8+x}$/Ag wires. *Appl. Phys. Lett.*, **55**, 2441–2443.
7. Tenbrink, J., Wilhelm, M., Heine, K., and Krauth, H. (1991) Development of high-T_c superconductor wires for magnet applications. *IEEE Trans. Magn.*, **27**, 1239–1246.
8. Motowidlo, L., Galinski, G., Ozeryansky, G., Zhang, W., and Hellstrom, E. (1994) Dependence of critical current density on filament diameter in round multifilament Ag-sheathed $Bi_2Sr_2CaCu_2O_x$ wires processed in O_2. *Appl. Phys. Lett.*, **65**, 2731–2733.
9. Scanlan, R., Dietderich, D., Higley, H., Marken, K., Motowidlo, L., Sokolowski, R., and Hasegawa, T. (1999) Fabrication and test results for Rutherford-type cables made from BSCCO strands. *IEEE Trans. Appl. Supercond.*, **9**, 130–133.
10. Kitaguchi, H., Kumakura, H., Togano, K., Miao, H., Hasegawa, T., and Koizumi, T. (1999) $Bi_2Sr_2CaCu_2O_x$/Ag multilayer tapes with J_c(4.2 K, 10 T) of 500,000 A/cm^2 by using PAIR process. *IEEE Trans. Appl. Supercond.*, **9**, 1794–1799.
11. Miao, H., Marken, K.R., Meinesz, M., Czabaj, B., and Hong, S. (2005) Development of round multifilament Bi-2212/Ag wires for high field magnet applications. *IEEE Trans. Appl. Supercond.*, **15**, 2554–2557.

12. Trociewitz, U., Schwartz, J., Marken, K., Miao, H., Meinesz, M., and Czabaj, B. (2006) Bi2212 superconductors in high field applications. *NHMFL Rep.*, **13**, 31.
13. Shen, T., Jiang, J., Kametani, F., Trociewitz, U., Larbalestier, D., Schwartz, J., and Hellstrom, E. (2010) Filament to filament bridging and its influence on developing high critical current density in multifilamentary $Bi_2Sr_2CaCu_2O_x$ round wires. *Supercond. Sci. Technol.*, **23**, 025009.
14. Kametani, F., Shen, T., Jiang, J., Scheuerlein, C., Malagoli, A., Di Michiel, M., Huang, Y., Miao, H., Parrell, J., and Hellstrom, E. (2011) Bubble formation within filaments of melt-processed Bi2212 wires and its strongly negative effect on the critical current density. *Supercond. Sci. Technol.*, **24**, 075009.
15. Scheuerlein, C., Di Michiel, M., Scheel, M., Jiang, J., Kametani, F., Malagoli, A., Hellstrom, E., and Larbalestier, D. (2011) Void and phase evolution during the processing of Bi-2212 superconducting wires monitored by combined fast synchrotron micro-tomography and x-ray diffraction. *Supercond. Sci. Technol.*, **24**, 115004.
16. Jiang, J., Miao, H., Huang, Y., Hong, S., Parrell, J.A., Scheuerlein, C., Di Michiel, M., Ghosh, A.K., Trociewitz, U.P., and Hellstrom, E.E. (2013) Reduction of gas bubbles and improved critical current density in Bi-2212 round wire by swaging. *IEEE Trans. Appl. Supercond.*, **23**, 6400206.
17. Jiang, J., Starch, W., Hannion, M., Kametani, F., Trociewitz, U., Hellstrom, E., and Larbalestier, D. (2011) Doubled critical current density in Bi-2212 round wires by reduction of the residual bubble density. *Supercond. Sci. Technol.*, **24**, 082001.
18. Larbalestier, D., Jiang, J., Trociewitz, U.P., Kametani, F., Scheuerlein, C., Dalban-Canassy, M., Matras, M., Chen, P., Craig, N.C., Lee, P.J., and Hellstrom, E.E. (2014) Isotropic round-wire multifilament cuprate superconductor for generation of magnetic fields above 30 T. *Nat. Mater.*, **13**, 375.
19. Malagoli, A., Kametani, F., Jiang, J., Trociewitz, U., Hellstrom, E., and Larbalestier, D. (2011) Evidence for long range movement of Bi-2212 within the filament bundle on melting and its significant effect on J_c. *Supercond. Sci. Technol.*, **24**, 075016.
20. Malagoli, A., Lee, P., Ghosh, A., Scheuerlein, C., Di Michiel, M., Jiang, J., Trociewitz, U., Hellstrom, E., and Larbalestier, D. (2013) Evidence for length-dependent wire expansion, filament dedensification and consequent degradation of critical current density in Ag-alloy sheathed Bi-2212 wires. *Supercond. Sci. Technol.*, **26**, 055018.
21. Shen, T., Ghosh, A., Cooley, L., and Jiang, J. (2013) Role of internal gases and creep of Ag in controlling the critical current density of Ag-sheathed $Bi_2Sr_2CaCu_2O_x$ wires. *J. Appl. Phys.*, **113**, 213901.
22. Miao, H., Marken, K., Meinesz, M., Czabaj, B., Hong, S., Rikel, M., and Bock, J. (2006) Studies of precursor composition effect on J_c in Bi-2212/Ag wires and tapes. *AIP Conf. Proc.*, **824**, 673.
23. Marken, K., Miao, H., Meinesz, M., Czabaj, B., and Hong, S. (2003) BSCCO-2212 conductor development at Oxford superconducting technology. *IEEE Trans. Appl. Supercond.*, **13**, 3335–3338.
24. Motowidlo, L., Selvamanickam, V., Galinski, G., Vo, N., Haldar, P., and Sokolowski, R. (2000) Recent progress in high-temperature superconductors at Intermagnetics General Corporation. *Phys. C: Supercond.*, **335**, 44–50.
25. Kajbafvala, A., Nachtrab, W., Lu, X.F., Hunte, F., Liu, X., Cheggour, N., Wong, T., and Schwartz, J. (2012) Dispersion-strengthened silver alumina for sheathing $Bi_2Sr_2CaCu_2O_{8+x}$ multifilamentary wire. *IEEE Trans. Appl. Supercond.*, **22**, 8400210.
26. Miao, H., Meinesz, M., Czabaj, B., Parrell, J., and Hong, S. (2008) Microstructure and J_c improvement in multifilamentary Bi-2212/Ag wires for high field magnet applications. *AIP Conf. Proc.*, **986**, 423.

27. Okada, M., Tanaka, K., Wakuda, T., Ohata, K., Sato, J., Kumakura, H., Kiyoshi, T., Kitaguchi, H., Togano, K., and Wada, H. (1999) A new symmetrical arrangement of tape-shaped multifilaments for Bi-2212/Ag round-shaped wire. *IEEE Trans. Appl. Supercond.*, **9**, 1904–1907.
28. Myers, C., Susner, M., Motowidlo, L., Distin, J., Sumption, M., and Collings, E. (2011) Transport, magnetic, and SEM characterization of a novel design Bi-2212 strand. *IEEE Trans. Appl. Supercond.*, **21**, 2804–2807.
29. Shen, T., Jiang, J., Yamamoto, A., Trociewitz, U., Schwartz, J., Hellstrom, E., and Larbalestier, D. (2009) Development of high critical current density in multifilamentary round-wire $Bi_2Sr_2CaCu_2O_{8+x}$ by strong overdoping. *Appl. Phys. Lett.*, **95**, 152516.
30. Hellstrom, E., Ray, R. II,, and Zhang, W. (1993) Phase development and microstructure in Bi-based 2212 Ag-clad tapes processed at 880, 890, and 905 °C: the Cu-free phase and (Sr, Ca) CuO_2. *Appl. Supercond.*, **1**, 1535–1545.
31. Endo, A. and Nishikida, S. (1993) Effects of heating temperature and atmosphere on critical current density of $Bi_2Sr_2Ca_1Cu_2Ag_{0.8}O_y$ Ag-sheathed tapes. *IEEE Trans. Appl. Supercond.*, **3**, 931–934.
32. Karuna, M., Parrell, J.A., and Larbalestier, D.C. (1995) Study of powder density, Ag: superconductor ratio, and microhardness of BSCCO-2212 Ag-sheathed wires and tapes during wire drawing and rolling. *IEEE Trans. Appl. Supercond.*, **5**, 1279–1282.
33. Kuroda, T., Tanaka, Y., Togano, K., Suga, Y., Sakamoto, T., Abe, Y., Miura, K., Yanagiya, T., and Ishizuka, M. (2001) Critical current densities of AgCu-sheathed Bi-2212 multifilamentary round wires in long length. *Phys. C: Supercond.*, **357**, 1098–1101.
34. Reeves, J.L., Hellstrom, E.E., Irizarry, V., and Lehndorff, B. (1999) Effects of overpressure processing on porosity in Ag-sheathed Bi-2212 multifilamentary tapes with various geometries. *IEEE Trans. Appl. Supercond.*, **9**, 1836–1839.
35. Reeves, J.L., Polak, M., Zhang, W., Hellstrom, E.E., Babcock, S.E., Larbalestier, D.C., Inoue, N., and Okada, M. (1997) Overpressure processing of Ag-sheathed Bi-2212 tapes. *IEEE Trans. Appl. Supercond.*, **7**, 1541–1543.
36. Hellstrom, E., Yuan, Y., Jiang, J., Cai, X., Larbalestier, D., and Huang, Y. (2005) Review of overpressure processing Ag-sheathed $(Bi, Pb)_2Sr_2Ca_2Cu_3O_x$ wire. *Supercond. Sci. Technol.*, **18**, S325.
37. Shen, T., Tollestrup, A., Tompkins, J., and Cooley, L. (2012) High-strength, 20-T class $Bi_2Sr_2CaCu_2O_x$ cables for high-field magnets. Applied Superconductivity Conference, Portland, OR.
38. Chen, P., Trociewitz, U.P., Dalban-Canassy, M., Jiang, J., Hellstrom, E.E., and Larbalestier, D.C. (2013) Performance of titanium oxide-polymer insulation in superconducting coils made of Bi-2212/Ag-alloy round wire. *Supercond. Sci. Technol.*, **26**, 075009.
39. Kandel, H., Lu, J., Jiang, J., Matras, M., Chen, P., Natanette, C., Viouchkov, Y., Trociewitz, U., Hellstrom, E., and Larbalestier, D. (2013). Development of thin ceramic coating in $Bi_2Sr_2CaCu_2O_{8+x}$ (Bi-2212) round wire. 23rd International Conference on Magnet Technology, Boston, MA.
40. Dalban-Canassy, M., Myers, D., Trociewitz, U., Jiang, J., Hellstrom, E., Viouchkov, Y., and Larbalestier, D. (2012) A study of the local variation of the critical current in Ag-alloy clad, round wire $Bi_2Sr_2CaCu_2O_{8+x}$ multi-layer solenoids. *Supercond. Sci. Technol.*, **25**, 115015.
41. Weijers, H., Trociewitz, U., Marken, K., Meinesz, M., Miao, H., and Schwartz, J. (2004) The generation of 25.05 T using a 5.11 T $Bi_2Sr_2CaCu_2O_x$ superconducting insert magnet. *Supercond. Sci. Technol.*, **17**, 636.
42. Weijers, H., Trociewitz, U., Markiewicz, W., Jiang, J., Myers, D., Hellstrom, E., Xu, A., Jaroszynski, J., Noyes, P., and Viouchkov, Y. (2010) High field magnets with HTS conductors. *IEEE Trans. Appl. Supercond.*, **20**, 576–582.

43. Godeke, A., Acosta, P., Cheng, D., Dietderich, D., Mentink, M., Prestemon, S., Sabbi, G., Meinesz, M., Hong, S., and Huang, Y. (2010) Wind-and-react Bi-2212 coil development for accelerator magnets. *Supercond. Sci. Technol.*, **23**, 034022.
44. Trociewitz, U.P., Dalban-Canassy, M., Hannion, M., Hilton, D.K., Jaroszynski, J., Noyes, P., Viouchkov, Y., Weijers, H.W., and Larbalestier, D.C. (2011) 35.4 T field generated using a layer-wound superconducting coil made of $(RE)Ba_2Cu_3O_{7-x}$ (RE = rare earth) coated conductor. *Appl. Phys. Lett.*, **99**, 202506.
45. Yuan, Y., Jiang, J., Cai, X., Larbalestier, D.C., Hellstrom, E.E., Huang, Y., and Parrella, R. (2004) Significantly enhanced critical current density in Ag-sheathed $(Bi, Pb)_2Sr_2Ca_2Cu_3O_x$ composite conductors prepared by overpressure processing in final heat treatment. *Appl. Phys. Lett.*, **84**, 2127.
46. Rupich, M. and Hellstrom, E. (2012) in *100 Years of Superconductivity* (eds H. Rogalla and P. Kes), Boca Raton, FL, CRC Press, pp. 671–688.
47. SUMITOMO ELECTRIC *http://global-sei.com/super/hts_e/index.html* (accessed 30 July 2013).
48. Sato, K.-I., Kobayashi, S.-I., and Nakashima, T. (2012) Present status and future perspective of bismuth-based high-temperature superconducting wires realizing application systems. *Jpn. J. Appl. Phys.*, **51**, 010006.
49. van Beijnen, C. and Elen, J. (1975) Potential fabrication method of superconducting multifilament wires of the A-15 type. *IEEE Trans. Magn.*, **11**, 243–246.
50. Bottura, L. and Godeke, A. (2012) Superconducting materials and conductors: fabrication and limiting parameters. *Rev. Accel Sci. Technol.*, **5**, 25–50.
51. Godeke, A. (2006) A review of the properties of Nb3Sn and their variation with A15 composition, morphology and strain state. *Supercond. Sci. Technol.*, **19**, R68.
52. Godeke, A., Den Ouden, A., Nijhuis, A., and Ten Kate, H. (2008) State of the art powder-in-tube niobium-tin superconductors. *Cryogenics*, **48**, 308–316.
53. Sumption, M., Bhatia, M., Rindfleisch, M., Tomsic, M., Soltanian, S., Dou, S.X., and Collings, E.W. (2005) Large upper critical field and irreversibility field in MgB_2 wires with SiC additions. *Appl. Phys. Lett.*, **86**, 092507.
54. Larbalestier, D., Cooley, L., Rikel, M., Polyanskii, A., Jiang, J., Patnaik, S., Cai, X., Feldmann, D., Gurevich, A., and Squitieri, A. (2001) Strongly linked current flow in polycrystalline forms of the superconductor MgB_2. *Nature*, **410**, 186–189.
55. Li, G., Sumption, M., Susner, M., Yang, Y., Reddy, K., Rindfleisch, M., Tomsic, M., Thong, C., and Collings, E. (2012) The critical current density of advanced internal-Mg-diffusion-processed MgB_2 wires. *Supercond. Sci. Technol.*, **25**, 115023.
56. Serquis, A., Civale, L., Hammon, D., Liao, X., Coulter, J., Zhu, Y., Jaime, M., Peterson, D., Mueller, F., and Nesterenko, V. (2003) Hot isostatic pressing of powder in tube MgB_2 wires. *Appl. Phys. Lett.*, **82**, 2847–2849.
57. Tomsic, M., Rindfleisch, M., Yue, J., McFadden, K., Doll, D., Phillips, J., Sumption, M.D., Bhatia, M., Bohnenstiehl, S., and Collings, E. (2007) Development of magnesium diboride MgB_2 wires and magnets using in situ strand fabrication method. *Physica C: Supercond.*, **456**, 203–208.
58. Dou, S., Soltanian, S., Horvat, J., Wang, X., Zhou, S., Ionescu, M., Liu, H., Munroe, P., and Tomsic, M. (2002) Enhancement of the critical current density and flux pinning of MgB_2 superconductor by nanoparticle SiC doping. *Appl. Phys. Lett.*, **81**, 3419.
59. Yamamoto, A., Shimoyama, J.-i., Ueda, S., Iwayama, I., Horii, S., and Kishio, K. (2005) Effects of B_4C doping on critical current properties of MgB_2 superconductor. *Supercond. Sci. Technol.*, **18**, 1323.
60. Kim, J.H., Zhou, S., Hossain, M.S.A., Pan, A.V., and Dou, S.X. (2006) Carbohydrate doping to enhance electromagnetic properties of MgB_2

superconductors. *Appl. Phys. Lett.*, **89**, 142505.

61. Yamada, H., Hirakawa, M., Kumakura, H., and Kitaguchi, H. (2006) Effect of aromatic hydrocarbon addition on in situ powder-in-tube processed MgB_2 tapes. *Supercond. Sci. Technol.*, **19**, 175.
62. Jiang, J., Senkowicz, B., Larbalestier, D., and Hellstrom, E. (2006) Influence of boron powder purification on the connectivity of bulk MgB_2. *Supercond. Sci. Technol.*, **19**, L33.
63. Flukiger, R., Shahriar Al Hossain, M., Senatore, C., Buta, F., and Rindfleisch, M. (2011) A new generation of in situ MgB_2 wires with improved J_c and B_{irr} values obtained by cold densification (CHPD). *IEEE Trans. Appl. Supercond.*, **21**, 2649–2654.
64. Nardelli, D., Matera, D., Vignolo, M., Bovone, G., Palenzona, A., Siri, A., and Grasso, G. (2013) Large critical current density in MgB_2 wire using MgB_4 as precursor. *Supercond. Sci. Technol.*, **26**, 075010.
65. Kamihara, Y., Hiramatsu, H., Hirano, M., Kawamura, R., Yanagi, H., Kamiya, T., and Hosono, H. (2006) Iron-based layered superconductor: LaOFeP. *J. Am. Chem. Soc.*, **128**, 10012–10013.
66. Rotter, M., Tegel, M., and Johrendt, D. (2008) Superconductivity at 38 K in the iron arsenide $(Ba_{1-x}K_x)\ Fe_2As_2$. *Phys. Rev. Lett.*, **101**, 107006.
67. Hsu, F.-C., Luo, J.-Y., Yeh, K.-W., Chen, T.-K., Huang, T.-W., Wu, P.M., Lee, Y.-C., Huang, Y.-L., Chu, Y.-Y., and Yan, D.-C. (2008) Superconductivity in the PbO-type structure α-FeSe. *Proc. Natl. Acad. Sci.*, **105**, 14262–14264.
68. Yeh, K.-W., Huang, T.-W., Huang, Y.-l., Chen, T.-K., Hsu, F.-C., Wu, P.M., Lee, Y.-C., Chu, Y.-Y., Chen, C.-L., and Luo, J.-Y. (2008) Tellurium substitution effect on superconductivity of the α-phase iron selenide. *Europhys. Lett.*, **84**, 37002.
69. Putti, M., Pallecchi, I., Bellingeri, E., Cimberle, M., Tropeano, M., Ferdeghini, C., Palenzona, A., Tarantini, C., Yamamoto, A., and Jiang, J. (2010) New Fe-based superconductors: properties relevant for applications. *Supercond. Sci. Technol.*, **23**, 034003.
70. Ma, Y. (2012) Progress in wire fabrication of iron-based superconductors. *Supercond. Sci. Technol.*, **25**, 113001.
71. Yamamoto, A., Jiang, J., Kametani, F., Polyanskii, A., Hellstrom, E., Larbalestier, D., Martinelli, A., Palenzona, A., Tropeano, M., and Putti, M. (2011) Evidence for electromagnetic granularity in polycrystalline Sm1111 iron-pnictides with enhanced phase purity. *Supercond. Sci. Technol.*, **24**, 045010.
72. Kametani, F., Polyanskii, A., Yamamoto, A., Jiang, J., Hellstrom, E., Gurevich, A., Larbalestier, D., Ren, Z., Yang, J., and Dong, X. (2009) Combined microstructural and magneto-optical study of current flow in polycrystalline forms of Nd and Sm Fe-oxypnictides. *Supercond. Sci. Technol.*, **22**, 015010.
73. Katase, T., Ishimaru, Y., Tsukamoto, A., Hiramatsu, H., Kamiya, T., Tanabe, K., and Hosono, H. (2011) Advantageous grain boundaries in iron pnictide superconductors. *Nat. Commun.*, **2**, 409.
74. Lee, S., Jiang, J., Weiss, J., Folkman, C., Bark, C., Tarantini, C., Xu, A., Abraimov, D., Polyanskii, A., and Nelson, C. (2009) Weak-link behavior of grain boundaries in superconducting $Ba(Fe_{1-x}Co_x)_2As_2$ bicrystals. *Appl. Phys. Lett.*, **95**, 212505.
75. Gao, Z., Wang, L., Qi, Y., Wang, D., Zhang, X., and Ma, Y. (2008) Preparation of $LaFeAsO_{0.9}F_{0.1}$ wires by the powder-in-tube method. *Supercond. Sci. Technol.*, **21**, 105024.
76. Gao, Z., Wang, L., Qi, Y., Wang, D., Zhang, X., Ma, Y., Yang, H., and Wen, H. (2008) Superconducting properties of granular $SmFeAsO_{1-x}F_x$ wires with $T_c = 52$ K prepared by the powder-in-tube method. *Supercond. Sci. Technol.*, **21**, 112001.
77. Qi, Y., Zhang, X., Gao, Z., Zhang, Z., Wang, L., Wang, D., and Ma, Y. (2009) Superconductivity of powder-in-tube $Sr_{0.6}K_{0.4}Fe_2As_2$ wires. *Physica C: Supercond.*, **469**, 717–720.
78. Wang, C., Yao, C., Zhang, X., Gao, Z., Wang, D., Wang, C., Lin, H., Ma, Y., Awaji, S., and Watanabe, K. (2012) Effect of starting materials on

the superconducting properties of $SmFeAsO_{1-x}F_x$ tapes. *Supercond. Sci. Technol.*, **25**, 035013.

79. Kumakura, H., Ye, S.-J., Gao, Z., Matsumoto, A., Zhang, Y., and Togano, K. (2013) Development of MgB_2 and $(Ba(Sr),K)Fe_2As_2$ wires and tapes. US-Japan Superconductor Workshop, Dayton, OH.

80. Gao, Z., Ma, Y., Yao, C., Zhang, X., Wang, C., Wang, D., Awaji, S., and Watanabe, K. (2013) High critical current density and low anisotropy in textured $Sr_{1-x}K_xFe_2As_2$ tapes for high field applications. *Sci. Rep.*, **2**, 998.

81. Weiss, J., Tarantini, C., Jiang, J., Kametani, F., Polyanskii, A., Larbalestier, D., and Hellstrom, E. (2012) High intergrain critical current density in fine-grain $(Ba_{0.6}K_{0.4})Fe_2As_2$ wires and bulks. *Nat. Mater.*, **11**, 682–685.

82. Gao, Z., Wang, L., Yao, C., Qi, Y., Wang, C., Zhang, X., Wang, D., Wang, C., and Ma, Y. (2011) High transport critical current densities in textured Fe-sheathed $Sr_{1-x}K_xFe_2As_2$ + Sn superconducting tapes. *Appl. Phys. Lett.*, **99**, 242506.

83. Togano, K., Matsumoto, A., and Kumakura, H. (2011) Large transport critical current densities of Ag sheathed $(Ba,K)Fe_2As_2$ + Ag superconducting wires fabricated by an ex-situ powder-in-tube process. *Appl. Phys. Express*, **4**, 043101.

84. Ding, Q.-P., Prombood, T., Tsuchiya, Y., Nakajima, Y., and Tamegai, T. (2012) Superconducting properties and magneto-optical imaging of $Ba_{0.6}K_{0.4}Fe_2As_2$ PIT wires with Ag addition. *Supercond. Sci. Technol.*, **25**, 035019.

85. Wang, L., Qi, Y., Zhang, X., Wang, D., Gao, Z., Wang, C., Yao, C., and Ma, Y. (2011) Textured $Sr_{1-x}K_xFe_2As_2$ superconducting tapes with high critical current density. *Physica C: Supercond.*, **471**, 1689–1691.

86. Yao, C., Wang, C., Zhang, X., Wang, L., Gao, Z., Wang, D., Wang, C., Qi, Y., Ma, Y., and Awaji, S. (2012) Improved transport critical current in Ag and Pb co-doped $Ba_xK_{1-x}Fe_2As_2$ superconducting tapes. *Supercond. Sci. Technol.*, **25**, 035020.

87. Qi, Y., Wang, L., Wang, D., Zhang, Z., Gao, Z., Zhang, X., and Ma, Y. (2010) Transport critical currents in the iron pnictide superconducting wires prepared by the ex situ PIT method. *Supercond. Sci. Technol.*, **23**, 055009.

88. Weiss, J., Jiang, J., Polyanskii, A., and Hellstrom, E. (2013) Mechanochemical synthesis of pnictide compounds and superconducting $Ba_{0.6}K_{0.4}Fe_2As_2$ bulks with high critical current density. *Supercond. Sci. Technol.*, **26**, 074003.

89. Zhang, X., Yao, C., Wang, C., Gao, Z., Wang, D., Wang, C., Lin, H., Qi, Y., Wang, L., and Ma, Y. (2012) Mechanism of enhancement of superconducting properties in a $Ba_{1-x}K_xFe_2As_2$ superconductor by Pb addition. *Supercond. Sci. Technol.*, **25**, 084024.

90. Wang, L., Qi, Y., Gao, Z., Wang, D., Zhang, X., and Ma, Y. (2010) The role of silver addition on the structural and superconducting properties of polycrystalline $Sr_{0.6}K_{0.4}Fe_2As_2$. *Supercond. Sci. Technol.*, **23**, 025027.

91. Togano, K. and Gao, Z. (2013) Large enhancement of transport critical current density of ex-situ PIT $Ag/(Ba,K)Fe_2As_2$ tapes achieved by applying cycles of cold deformation and heat treatment. arXiv preprint arXiv:13043161.

92. Lu, X., Cheggour, N., Stauffer, T., Clickner, C., Goodrich, L., Trociewitz, U., Myers, D., and Holesinger, T. (2011) Electromechanical characterization of Bi-2212 strands. *IEEE Trans. Appl. Supercond.*, **21**, 3086–3089.

93. ten Haken, B., Beuink, A., and ten Kate, H.H. (1997) Small and repetitive axial strain reducing the critical current in BSCCO/Ag superconductors. *IEEE Trans. Appl. Supercond.*, 7, 2034–2037.

94. Kovac, P., Dhalle, M., Melisek, T., van Eck, H., Wessel, W., ten Haken, B., and Husek, I. (2003) Dependence of the critical current in ex situ multi- and mono-filamentary MgB_2/Fe wires on axial tension and compression. *Supercond. Sci. Technol.*, **16**, 600.

95. Cheggour, N., Lu, X., Holesinger, T., Stauffer, T., Jiang, J., and Goodrich, L. (2012) Reversible effect of strain on transport critical current in $Bi_2Sr_2CaCu_2O_{8+x}$ superconducting wires: a modified descriptive strain model. *Supercond. Sci. Technol.*, **25**, 015001.
96. Citver, G., Feher, S., Limon, P., Orris, D., Peterson, T., Sylvester, C., and Tompkins, J. (1999) HTS power lead test results. Proceedings of the 1999 Particle Accelerator Conference (IEEE), pp. 1420–1422.
97. Ballarino, A. (1999) High temperature superconducting current leads for the large hadron collider. *IEEE Trans. Appl. Supercond.*, **9**, 523–526.
98. Ballarino, A. (2008) Large-capacity current leads. *Physica C: Supercond.*, **468**, 2143–2148.
99. Bauer, P., Bi, Y., Cheng, A., Cheng, Y., Devred, A., Ding, K., Huang, X., Liu, C., Lin, X., and Mitchell, N. (2010) Test results of 52/68 kA trial HTS current leads for ITER. *IEEE Trans. Appl. Supercond.*, **20**, 1718–1721.
100. Snitchler, G., Gamble, B., and Kalsi, S.S. (2005) The performance of a 5 MW high temperature superconductor ship propulsion motor. *IEEE Trans. Appl. Supercond.*, **15**, 2206–2209.
101. Weber, C., Reis, C., Dada, A., Masuda, T., and Moscovic, J. (2005) Overview of the underground 34.5 kV HTS power cable program in Albany, NY. *IEEE Trans. Appl. Supercond.*, **15**, 1793–1797.
102. Malozemoff, A., Fleshler, S., Rupich, M., Thieme, C., Li, X., Zhang, W., Otto, A., Maguire, J., Folts, D., and Yuan, J. (2008) Progress in high temperature superconductor coated conductors and their applications. *Supercond. Sci. Technol.*, **21**, 034005.
103. Malozemoff, A.P. (2007) High T_c for the power grid. *Nat. Mater.*, **6**, 617–619.
104. Ballarino, A. (2010) Design of an MgB_2 feeder system to connect groups of superconducting magnets to remote power converters. *Journal of Physics*, Conference Series 234, 032003.

3.4.2
YBCO-Coated Conductors

Mariappan Parans Paranthaman, Tolga Aytug, Liliana Stan, Quanxi Jia, and Claudia Cantoni

3.4.2.1 Introduction

Since the discovery of high-temperature superconductors (HTSs) such as $YBa_2Cu_3O_{7-\delta}$/ $REBa_2Cu_3O_{7-\delta}$ (YBCO or REBCO (rare earth barium copper oxide), where (rare earth) RE = Nd, Sm, Eu, Gd, Dy …), $(Bi,Pb)_2Sr_2Ca_2Cu_3O_{10}$ (BSCCO or Bi2223) in 1987, researchers have demonstrated several ways to produce high-performance flexible wires carrying a very high critical current density at liquid nitrogen temperature. In early 1990s, the U.S. Department of Energy set the target price for the wire to the copper wire cost of $10/kA-m (i.e., a meter of wire carrying a current of 1000 A should cost about $10). The main objective is to demonstrate a HTS wire with a current carrying capacity of over 100 times than that of copper. American Superconductor Corporation (AMSC) in United States and Sumitomo Electric in Japan have been widely recognized as the world leaders in manufacturing the first-generation (1G) HTS wires based on Bi2223 materials. AMSC has demonstrated critical currents, I_c, of over 125 A in piece lengths of several hundred meters or more, and a champion current of 170 A at 77 K and self-field at the standard 4.1 mm width and 210 μm thickness in short lengths [1]. Similarly, Sumitomo Electric has demonstrated an I_c of 176 A at 77 K and self-field with the standard 4.3 mm and 230 μm thickness in 2084 m [2]. However, due to the possible higher cost of 1G wire, the researchers have developed second-generation (2G) wires based on REBCO. One of the main drawbacks to the manufacturing of long lengths of REBCO wire has been the presence of weak-links, that is, grain boundaries formed by the misalignment of REBCO grains form obstacles to current flow. Low-angle boundaries between superconducting REBCO grains are formed by carefully aligning the grains, and thereby it allows more current to flow through them. The schematics of the 1G HTS wires based on Bi2223 superconducting oxide in Ag or Ag alloy powder-in-tube and 2G HTS wires based on REBCO multilayer thin film architecture on flexible Ni-alloy substrate are shown in Figure 3.4.2.1.

3.4.2.2 RABiTS and IBAD Technology

The main goal of the YBCO-coated conductor research is to support Office of Electricity Delivery and Energy Reliability, U.S. Department of Energy strategy, and to identify the key barriers for the development of commercial 2G REBCO-based HTS for electric-power applications including direct current (DC) cables, fault current limiters, and rotatory machines. The current status of the 2G HTS wire is reported later in this chapter. The 2G wire cost is still too high. To further reduce the cost of these 2G tapes, it is essential to increase the throughput of buffer, reduce the number of buffer layers and YBCO deposition rates, maintain high critical current performance (A cm-width^{-1}), while improving the price/performance ($ kA^{-1}m^{-1}) of the tapes. The standard industry practice for characterizing the 2G wire is to divide the current by the width of the wire. With either a 5 μm thick

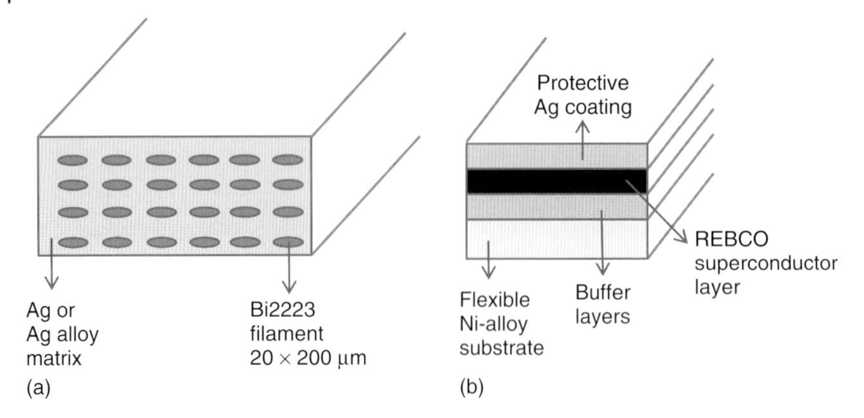

Figure 3.4.2.1 The schematics of both first-generation and second-generation high-temperature superconductor wires.

REBCO layer carrying a J_c of 1 MA cm^{-2} or 1 μm thick REBCO layer carrying a J_c of 5 MA cm^{-2}, the electrical performance would be equal to 500 A cm-width^{-1}. Converting these numbers to industry standard of 0.4 cm wide HTS wire would correspond to 200 A, which is comparable to that of the commercial 1G wire manufactured by Sumitomo Electric. Further increase in thickness, critical current density, or finding a way to incorporate two layers of REBCO (either double-sided coating or joining two REBCO tapes face to face) in single wire architecture would then give performance exceeding 1G, that is, high overall engineering critical current density, J_E at 77 K. The other important advantages of 2G wires over 1G wires are: REBCO has better in-field electrical performances at temperatures ranging from 50 to 77 K; potentially lower cost process due to the need for less silver and yields higher critical currents; higher mechanical strength due to the suitable selection of starting Ni-alloy substrates; and low alternating current (AC) losses by scribing the coated conductor tape into multifilaments.

One of the objectives of the YBCO-coated conductor research in United States is to enable potentially low-cost, high-throughput, high-yield manufacturing processes for substrate/buffer template fabrication, and to gain fundamental understanding of the YBCO-coated conductors. This understanding is critical to the development of a reliable, robust, long-length manufacturing process of 2G wires. Rolling Assisted Biaxially Textured Substrates (RABiTS) and ion-beam assisted deposition (IBAD) templates (as shown in Figure 3.4.2.2) have been developed for fabricating high-performance 2G-coated tapes [1, 3–9]. In standard RABiTS multilayer architecture, up to three buffers comprising of sputtered yttria (Y_2O_3) seed layer (75 nm thick), sputtered yttria-stabilized zirconia (YSZ) barrier layer (75 nm thick), and sputtered ceria (CeO_2) (75 nm thick) are used. On the other hand, the IBAD template technology uses different layer stacks depending on the materials used [10]. Several manufacturers have commercialized globally these technologies and have been producing HTS wires/tapes in kilometer lengths. In standard IBAD-MgO architecture, up to five buffer layers comprising of sputtered alumina (Al_2O_3) barrier layer

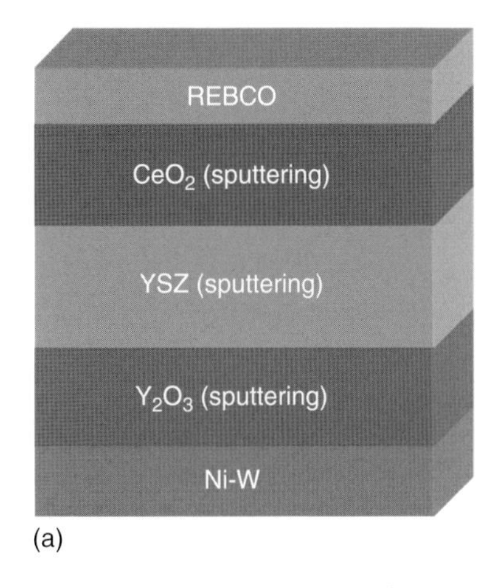

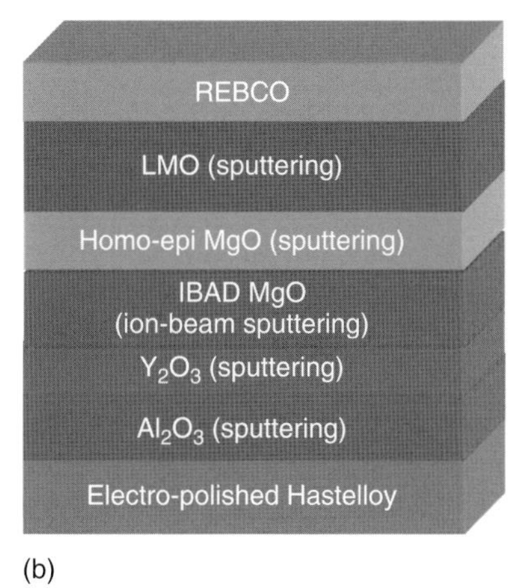

Figure 3.4.2.2 The schematic of both rolling assisted biaxially textured substrates (RABiTS) (a) and ion-beam assisted deposition (IBAD) – Magnesium Oxide (MgO) (b) multilayer architectures developed for second-generation high-temperature superconductor tapes.

(80 nm thick); sputtered yttria (Y_2O_3) nucleation layer (7 nm thick); ion-beam sputtered IBAD-MgO layer (10 nm thick); sputtered homoepitaxial MgO layer (30 nm thick); sputtered $LaMnO_3$ (LMO) cap layer (30 nm thick) are used [1, 6, 11]. Typically, 50 μm thick mechanically polished untextured Hastelloy C-276 substrates are used as the substrate. An electropolishing step is introduced to reduce the roughness of the substrates for the subsequent growth of high-performance buffers and YBCO on top of the substrates. The electropolishing conditions have to be developed for each substrate compositions. It is preferred to eliminate the need for electropolishing steps. In addition, it is also desirable to reduce the number of buffer layers if possible. This is possible by utilizing a multifunctional layer that can combine the tasks of two buffers into one. We have developed previously process conditions to grow LMO buffer cap layers directly on IBAD-MgO templates using sputtering and eliminated the need for homoepitaxial MgO layers by demonstrating the growth of high critical current density, J_c, YBCO films using pulsed laser deposition [12–14]. Several simplified buffer layer architectures such as combining Y_2O_3 barrier and Al_2O_3 nucleation layers into a single layer of Y_2O_3–Al_2O_3 composite (YALO) [10, 15] and use of solution Y_2O_3 [16, 17] and solution $La_2Zr_2O_7$/$Gd_2Zr_2O_7$ [18, 19] as a planarization layer were developed to grow highly aligned IBAD-MgO layers. Recently, we reported our success in developing a chemical solution processed Al_2O_3 planarization layer to reduce the surface roughness of mechanically polished Hastelloy substrates without the need for an electrochemical polishing step while at the same time replacing the sputtered Al_2O_3 barrier layers [20]. We have also demonstrated the growth of highly aligned IBAD-MgO layers as well as high J_c YBCO films on these solution-based Al_2O_3 planarization layers. Here, we describe the approaches for increasing the solution Al_2O_3 thickness per coat and reduce the number of layers required to obtain smooth mechanically polished Hastelloy C-276 substrates.

3.4.2.3 Simplified IBAD MgO Template Based on Chemical Solution Processed Al_2O_3

Mechanically polished Hastelloy C-276 substrates used were 50 μm thick, 12 mm wide, and 100 mm long. Alumina chemical precursor solution was prepared by dissolving aluminum acetylacetonate (Al $(acac)_3$; aluminum 2,4-pentanedionate (Aldrich, 99%)) in glacial acetic acid (Alfa Aesar; 99+ %) solution. Spin coating was used to deposit Al_2O_3 layers on mechanically polished Hastelloy substrates at a spin rate of 2000 rpm for 30 sec. The spin-coated films were then heat-treated at 450 °C for 15 min in flowing Ar gas. About 2 M alumina precursor solutions were used to obtain 20–30 nm thick Al_2O_3 films in a single coat. To obtain thicker Al_2O_3 films, the spin-coating and heat-treatment steps were repeated several times. On chemical-solution-derived Al_2O_3 films, both 7 nm thick Y_2O_3 nucleation layer and 10 nm thick IBAD-MgO layers were deposited using a reel-to-reel ion-beam sputtering system. The details of the IBAD-MgO deposition procedures were published elsewhere [14, 15]. Reflection high-energy electron diffraction (RHEED) images were taken during the IBAD-MgO film growth. Both 30 nm thick homoepi MgO and 30 nm thick LMO layers were subsequently deposited on IBAD-MgO layers using radio frequency (rf) sputtering to complete a buffer stack required for epitaxial growth of REBCO films. atomic force microscopy (AFM) (digital instruments nanoscope III) with a continuous mode operation was used to measure the surface roughness of Al_2O_3 films after each coating. The surface roughness, R_a, was determined from the average R_a values of five different 10 × 10 μm grid area measurements of each sample. Phase and texture of the films were characterized by X-ray diffraction (XRD) (Model XRG3100, Philips, Eindhoven, The Netherlands and Model 4-circle, Picker Corp., Cleveland, OH with Cu K_α radiation). Microstructural analysis was done by cross-sectional transmission electron microscopy (TEM) using a Hitachi HF-3300 TEM/STEM operating at 300 kV and equipped with an energy-dispersive X-ray (EDX) spectrometer. TEM samples were prepared by the focused ion beam (FIB) technique, followed by low voltage ion milling and plasma cleaning.

A typical X-ray $\theta - 2\theta$ scan (as shown in Figure 3.4.2.3) of the chemical solution processed three coats of Al_2O_3 layers deposited on the Hastelloy substrates shows absence of crystalline peaks indicating their nanocrystalline or amorphous nature. AFM images of the surfaces for the uncoated Hastelloy substrate and coated counterparts with one coat, two coats, and three coats of Al_2O_3 layers are shown in Figure 3.4.2.4. The R_a decreased from ~9.5 nm for the starting Hastelloy substrate to 5.15 nm for one coat, 3.99 nm for two coats, and 2.18 nm for three coats of Al_2O_3 layers. The measured surface roughness of electropolished Hastelloy substrates from the 5 × 5 μm grid area is less than 1 nm.

Figure 3.4.2.5 represents the surface roughness, R_a, values versus the number of coats of the Al_2O_3 layers on Hastelloy substrates. Samples generated from two different batches show a similar reduction of R_a with increasing the number of the coats, indicating the uniformity and reproducibility of the chemical solution deposition process. Compared to the previously reported results of films with a R_a value of ~2 nm for eight coats of Al_2O_3 layers, we have demonstrated a similar R_a

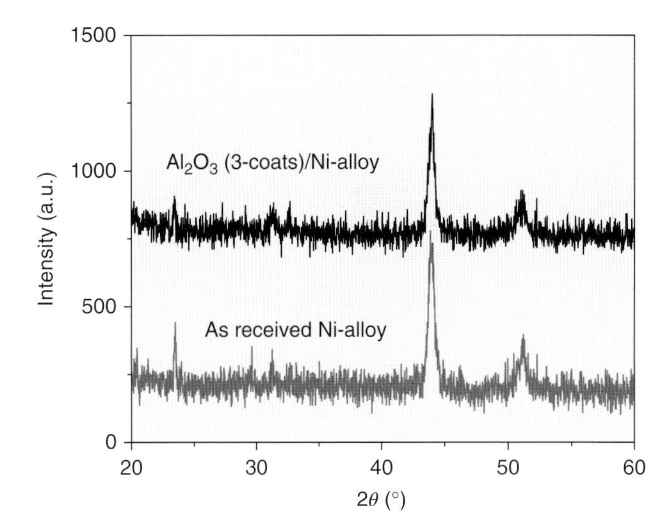

Figure 3.4.2.3 Room temperature X-ray diffraction patterns of both starting as-received Ni-alloy (Hastelloy C-276) substrates and three coats of solution Al_2O_3 deposited on Ni-alloy substrates.

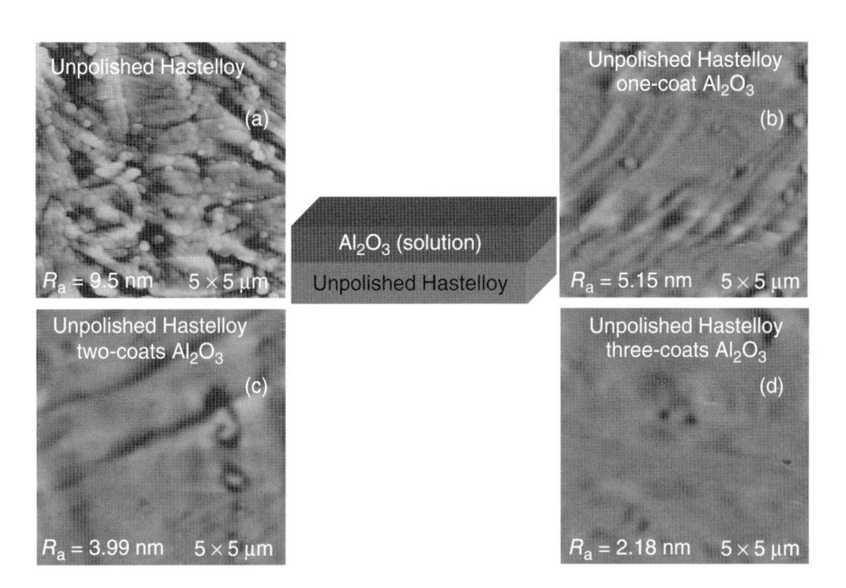

Figure 3.4.2.4 (a) AFM image of a typical mechanically polished Hastelloy C-276 substrate. AFM images of one coat (b), two coats (c), and three coats (d) of solution Al_2O_3 deposited on Hastelloy substrates.

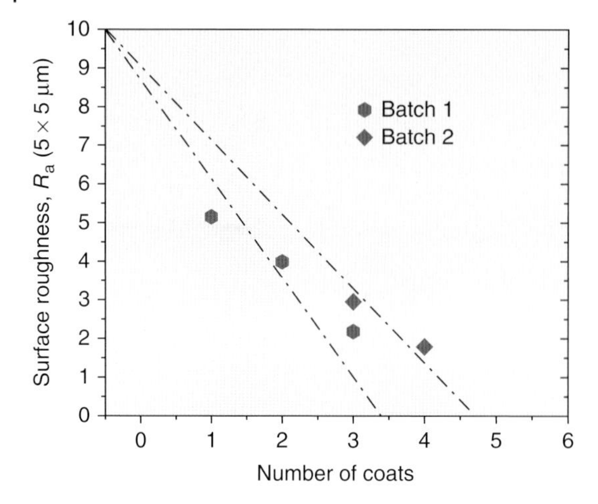

Figure 3.4.2.5 Substrate surface roughness, R_a, as a function of the number of Al_2O_3 layers. R_a data were measured from the 5 × 5 μm grid area on two different batches of samples. The linear dotted line guides the eye.

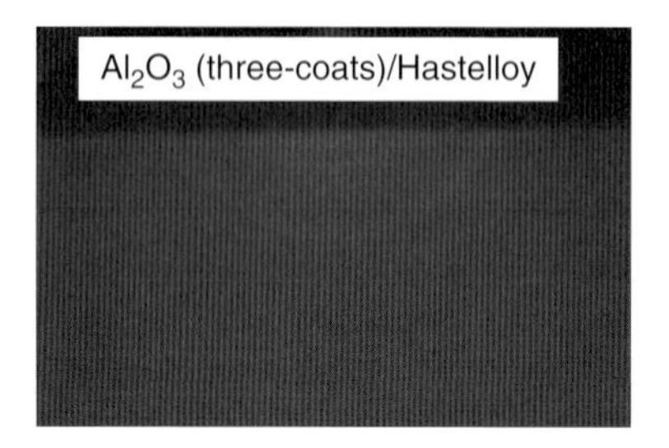

Figure 3.4.2.6 RHEED images of three coats of chemical-solution-deposited Al_2O_3 surface on mechanically polished Hastelloy substrates.

value with four coats of Al_2O_3 by increasing the Al_2O_3 precursor solution concentration. RHEED image of the surface for three-coat solution-deposited Al_2O_3 layer is shown in Figure 3.4.2.6. The presence of a weak ring pattern in the RHEED image further confirms nanocrystalline/amorphous nature for solution-deposited Al_2O_3 layer. The Y_2O_3 nucleation layer followed by IBAD-MgO and homoepi MgO layers were deposited on surface planarized Hastelloy substrates using a reel-to-reel ion-beam sputtering system (as shown in Figure 3.4.2.7). RHEED images of the surfaces for one coat, two coats, three coats, and four coats of solution-deposited Al_2O_3 layer and the subsequent IBAD-MgO layer are shown in Figure 3.4.2.7. The presence of clear spots in the RHEED image for the IBAD-MgO layer indicates the formation of a biaxially textured MgO layer. To complete the full buffer stack for REBCO deposition, epitaxial MgO and LMO cap layers were deposited on top of the IBAD-MgO layer via rf sputtering. A typical X-ray $\theta-2\theta$ scan (as shown in Figure 3.4.2.8) of LMO-buffered IBAD-MgO layers indicates the presence of highly crystalline *c*-axis aligned MgO and LMO layers. The full width at

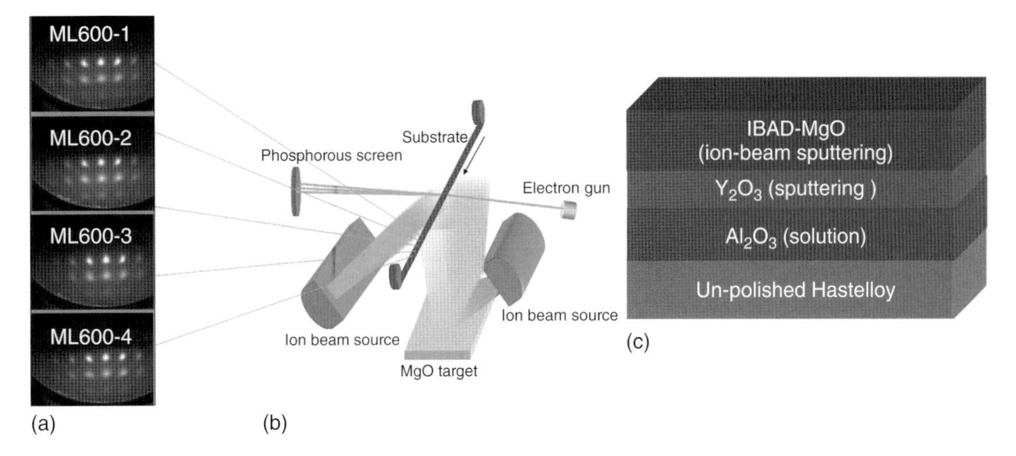

Figure 3.4.2.7 The schematics of the reel-to-reel IBAD-MgO deposition system (b). RHEED images of IBAD-MgO grown on one coat (sample# ML 600-1), two coats (sample# ML 600-2), three coats (sample# ML 600-3), and four coats (sample# ML 600-4) of chemical-solution-deposited Al_2O_3 surface on mechanically polished Hastelloy substrates (a). The schematic of the IBAD-MgO multilayer architecture for these are shown on (c).

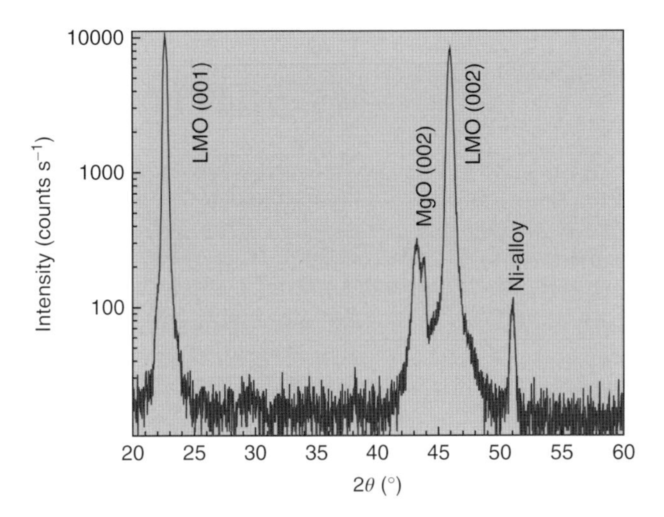

Figure 3.4.2.8 The room temperature X-ray diffraction pattern for LMO-buffered IBAD-MgO templates. The schematic of the IBAD-MgO/LMO multilayer architecture for this film is shown on the right.

half-maximum (FWHM) values for the out-of-plane texture (ω scans) of $LaMnO_3$ (004) are 3.9° and in-plane texture (ϕ scans) of $LaMnO_3$ (110) are 9.5°, respectively. These results indicate that the solution Al_2O_3 planarization layer provides a very good surface for the growth of highly aligned IBAD-MgO and LMO layers. Cross-section TEM image of the LMO-buffered IBAD-MgO templates is shown in Figure 3.4.2.9.

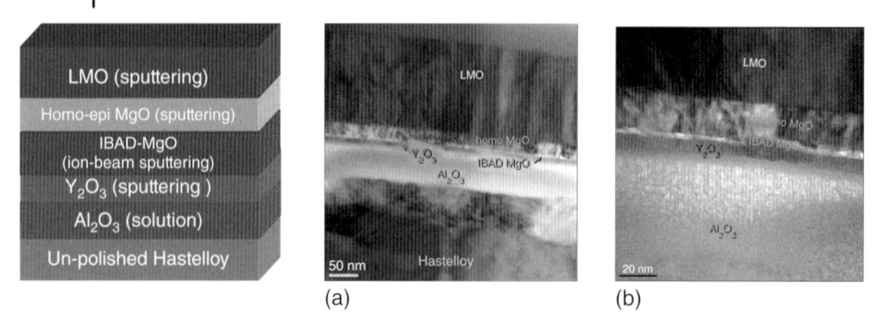

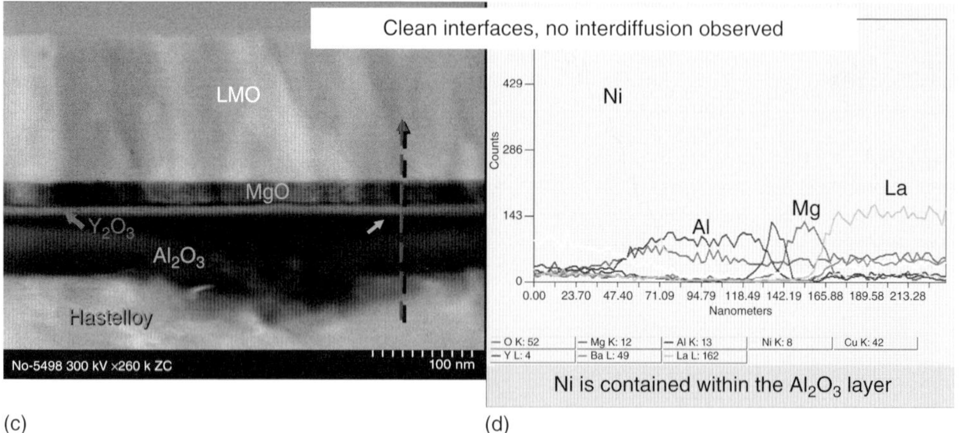

Figure 3.4.2.9 (a–c) The cross-sectional TEM images of the IBAD-MgO/LMO multilayer architecture of the schematics shown on top left. (d) An EDX line profile across the interfaces of the cross-sectional TEM image (c) indicate that Ni, Mn, and Cr are contained within the buffer layers. A dense microstructure for LMO layer and clean interfaces between the IBAD-MgO/homoepitaxial MgO/LMO and Al_2O_3/IBAD-MgO layers are observed.

The enlarged cross-sectional images (Figure 3.4.2.9a–c) show a clean and abrupt interface between the LMO and MgO layers and no signs of interfacial reactions between them. Although the metal alloy substrate oxidation has occurred, this reaction has been confined below the Al_2O_3 layer, which acts as a proper chemical barrier. In fact, the EDX profile scans (Figure 3.4.2.9d) confirmed that the solution Al_2O_3 layer was helpful in preventing diffusion of metal elements (e.g., Ni, Mn, Cr) from the Hastelloy substrate into the LMO layers. These results demonstrate that the solution Al_2O_3 layer is compatible with IBAD-MgO templates, and efforts are being made to fabricate long lengths of dip-coated Al_2O_3 layers and REBCO layers.

The successful demonstration of solution-derived Al_2O_3 planarization layers eliminated the need for both electropolishing of Hastelloy substrates and sputtered Al_2O_3 layers. This result promises a route for producing long lengths of high-performance REBCO-coated conductors using the combination of a

Table 3.4.2.1 Current status of world's second-generation HTS wires [21].

Industry	Template	I_c at 77 K and self-field (A)	Length (m)	Remarks
Fujikura, Japan	IBAD-MgO; REBCO (PLD)	572	816	Demonstrated in February 2011
AMSC, USA	RABiTS; REBCO (TFA-MOD)	466	540	Demonstrated in October 2010
SWCC Showa Cable Systems, Japan	RABiTS; REBCO (TFA-MOD)	310	500	Demonstrated in May 2008
SuperPower/Furakawa, USA/Japan	IBAD-MgO; REBCO (MOCVD)	282	1065	Demonstrated in August 2009
SUNAM, Korea	IBAD-MgO; REBCO (RCE)	275	470	Demonstrated in Oct. 2010

PLD – pulsed laser deposition; REBCO – rare earth barium copper oxide – $REBa_2Cu_3O_{7-\delta}$; TFA-MOD – trifluoro acetate-based metal-organic deposition; MOCVD – metal-organic chemical vapor deposition; RCE - reactive coevaporation.

low-cost chemical-solution-derived Al_2O_3 barrier layers with the well-established IBAD-MgO and $LaMnO_3$ buffer layer technologies.

3.4.2.4 Current Status of 2G HTS Wires

The current status of the world's 2G HTS wires from various groups in United States, Japan, and Korea is reported in Table 3.4.2.1. The typical dimensions of the REBCO-coated conductors are 10 mm (width) × 0.10 mm (thickness). As per Table 3.4.2.1, two templates comprising IBAD-MgO and RABiTS have been developed and superconductivity companies around the world are producing in commercially acceptable 500–1000 m lengths. In addition, four different methods including metal-organic deposition, metal-organic chemical vapor deposition, high-rate pulsed laser deposition, and reactive coevaporation have been used to demonstrate high I_c in long lengths of REBCO-coated conductors. The HTS wires are currently used in the following electric-power applications: SMES (superconducting magnetic energy storage), superconducting power cables, and superconducting transformers; see the corresponding chapters.

3.4.2.5 Future Outlook

Significant advances have been made during the last 8 years toward the manufacturing of coated conductors in United States, Japan, and Korea. Industries around the world are scaling up their manufacturing to meet the demand for various demonstrations. The near-term (1–5 years) HTS device market includes cable, fault current limiters, generators, induction heaters, and current leads. In addition, the mid-term (3–7 years) and long-term (5–10 years) HTS device portfolio includes motors, energy storage devices such as SMES, Maglev, and MRI magnets.

As the technology advances, the wire cost will continue to decrease. The challenge is to improve the price/performance so that price will be competitive with existing alternative technologies. International collaborations are needed to find new sources of funding for continued success with wire development and device demonstrations.

Acknowledgments

This work was sponsored by U.S. Department of Energy, Office of Electricity Delivery, and Energy Reliability – Advanced Conductors and Cables Program. QXJ acknowledges support from the Center for Integrated Nanotechnologies, U.S. Department of Energy, Office of Basic Energy Sciences user facility at Los Alamos National Laboratory. CC and MPP also acknowledge the support by the Materials Sciences and Engineering Division, Office of Basic Energy Sciences, U.S. Department of Energy, and through a user project supported by ORNLs Center for Nanophase Materials Sciences (CNMS), which is sponsored by the Scientific User Facilities Division, Office of Basic Energy Sciences, U.S. Department of Energy.

References

1. Paranthaman, M.P. and Izumi, T. (2004) *MRS Bull.*, **29**, 533–536.
2. Sumitomo Electric *http://global-sei.com/super/hts_e/type_h.html* (accessed 17 May 2014).
3. Goyal, A., Norton, D.P., Budai, J.D., Paranthaman, M., Specht, E.D., Kroeger, D.M., Christen, D.K., He, Q., Saffian, B., List, F.A., Martin, P.M., Klabunde, C.E., Hatfield, E., and Sikka, V. (1996) *Appl. Phys. Lett.*, **69**, 1795–1797.
4. Paranthaman, M.P. (2006) *McGraw-Hill 2006 Yearbook of Science and Technology*, McGraw-Hill Publishers, New York, pp. 319–322.
5. Goyal, A., Paranthaman, M., and Schoop, U. (2004) *MRS Bull.*, **29**, 552–561.
6. Arendt, P.N. and Foltyn, S.R. (2004) *MRS Bull.*, **29**, 543–550.
7. Paranthaman, M., Aytug, T., Christen, D.K., Arendt, P.N., Foltyn, S.R., Groves, J.R., Stan, L., DePaula, R.F., Wang, H., and Holesinger, T.G. (2003) *J Mater. Res.*, **18**, 2055–2059.
8. Selvamanickam, V., Chen, Y., Xiong, X., Xie, Y., Zhang, X., Rar, A., Martchevskii, M., Schmidt, R., Lenseth, K., and Herrin, J. (2008) *Physica C*, **468**, 1504–1509.
9. Shiohara, Y., Yoshizumi, M., Izumi, T., and Yamada, Y. (2007) *Physica C*, **463-465**, 1–6.
10. Stan, L., Chen, Y., Xiong, X., Holesinger, T.G., Maiorov, B., Civale, L., DePaula, R.F., Selvamanickam, V., and Jia, Q.X. (2010) *Supercond. Sci. Technol.*, **23**, 014011.
11. Xiong, X. *et al* (2007) *IEEE Trans. Appl. Supercond.*, **17**, 3375–3377.
12. Stan, L. *et al* (2008) *Supercond. Sci. Technol.*, **21**, 105023.
13. Stan, L. *et al* (2007) *IEEE Trans. Appl. Supercond.*, **17**, 3409–3412.
14. Polat, O., Aytug, T., Paranthaman, M., Kim, K., Zhang, Y., Thompson, J.R., Christen, D.K., Xiong, X., and Selvamanickam, V. (2008) *J. Mater. Res.*, **23**, 3021–3028.
15. Stan, L., Feldmann, D.M., Usov, I.O., Holesinger, T.G., Maiorov, B., Civale, L., DePaula, R.F., Dowden, P.C., and Jia, Q.X. (2009) *IEEE Trans. Appl. Supercond.*, **19**, 3459–3462.
16. Sheehan, C., Jung, Y., Holesinger, T.G., Feldmann, D.M., Edney, C., Ihlefeld, J.F.,

Clem, P.G., and Matias, V. (2011) *Appl. Phys. Lett.*, **98**, 071907.

17. Jung, Y., Sheehan, C., Coulter, J.Y., Matias, V., and Youm, D. (2011) *IEEE Trans. Appl. Supercond.*, **21**, 2953–2956.
18. Paranthaman, M.P., Sathyamurthy, S., Aytug, T., Arent, P.N., Stan, L., and Foltyn S.R. (2009) Chemical solution deposition method of fabricating highly aligned MgO templates. US Patent 7,553,799.
19. Paranthaman, M.P., Sathyamurthy, S., Aytug, T., Arent, P.N., Stan, L., and Foltyn, S.R. (2012) Chemical Solution Deposition Method of Fabricating Highly Aligned MGO Templates. US Patent 8,088,503.
20. Paranthaman, M.P., Aytug, T., Stan, L., Jia, Q., Cantoni, C., and Wee, S.H. (2014) *Supercond. Sci. Technol.*, **27**, 022002. doi: 10.1088/0953-2048/27/2/022002
21. Shiohara, Y., Taneda, T., and Yoshizumi, M. (2012) *Jpn. J. Appl. Phys.*, **51**, 010007. doi: 10.1143/JJAP.51.010007

3.5 Cooling

3.5.1 Fluid Cooling

Luca Bottura and Cesar Luongo

3.5.1.1 Introduction

Practical superconducting materials require active cooling to cryogenic temperatures to attain suitable operating conditions with sufficient operating margin. Cooling of superconducting devices, such as magnets, power transmission cables, current leads, superconducting cavities, or other, can be achieved in a variety of manners that depend on the required operating temperature, the heat load to be removed, the dimensions, the numerous constraints imposed by the design of the device, and, obviously, the fluid chosen as a coolant.

In broad terms, we can classify the cooling methods in the following categories:

- *Direct cooling*: the coolant is in direct contact with the superconductor, and the coolant fills or flows inside channels integrated in the device. The dominant heat transfer mechanism is convection.
- *Fluid-mediated cooling*: the superconducting device is cooled by a fluid loop that mediates heat transport from a cold source (e.g., a cold head of a cryocooler) to a thermal anchor that conducts heat. The dominant heat mechanisms are conduction (in the magnet or device) and convection in the fluid loop.
- *Conduction cooling*: the cold source is directly connected to the device by thermal links, and the heat is removed by pure conduction.

This chapter covers the theory and practice of direct cooling by fluid contact, that is, the first case in the above categories. We include in the discussion the case of cooling channels attached to the device, for example, cooling pipes on the outer surface of a magnet. In this case, although mediated by conduction, the fluid cooling becomes an intimate part of the superconducting device, and cannot be conceptually separated from the device itself. The second and third cases in the above list are treated in Section 3.5.3, and are hence not discussed here.

Even when restricted to the scope of direct cooling, there are many possible arrangements for how the fluid cools the superconductor. Once again, taking a rather simplified approach, we can distinguish between two different modes of direct cooling:

- *Bath cooling*. The superconducting device is immersed in, or permeated by a pool of fluid. The bath, in turn, is kept at the desired temperature by heat exchange or phase separation (evaporation and boiling). The dominant heat transport mechanism in this mode is natural convection in the bath, or, in the case of a bath of superfluid helium, counterflow heat exchange. The bath is in principle stagnant, and only needs refill when evaporation causes a fraction of the bath to be lost;

- *Internal cooling*. The device is cooled by a flow in channels and/or pipes that are built into the device. The flow can be driven by two mechanisms, either caused by buoyancy (natural convection), or driven by pumps or compressors (forced convection). The dominating heat transport mechanism in this mode is convection, the energy transport conveyed by the flow itself. When superfluid helium is used in an internally cooled conductor, there is usually no circulation given the excellent heat removal characteristics of HeII.

A further classification of cooling modes is based on the status of the coolant. Specifically, we can distinguish between using a coolant in the following conditions:

- Saturated liquid, when the fluid is at equilibrium conditions of pressure and temperature along the saturation line. Heat transfer in this condition involves evaporation and/or boiling;
- Single-phase subcooled liquid, when the pressure is below the critical pressure, and the temperature is lower than the corresponding value on the saturation line. A sufficient temperature increase in this state may cause evaporation and/or boiling;
- Single-phase supercritical fluid, when the pressure is above the critical pressure, such that the fluid always remains in a single phase, and no evaporation or boiling can take place.

In principle, all of the above states can be combined with bath or internal cooling. Some combinations are more appropriate and efficient, as discussed in this chapter. For helium, it is further possible to conceive cooling systems operating in superfluid state, either saturated or subcooled. Because of the peculiarities of superfluid helium when compared to normal coolants, it will be discussed separately.

Finally, the fluid used for cooling gives rise to a number of variants in the technology and practical realization of a cooling system for a superconducting device. For devices based on low-temperature superconductors (LTSs), helium is the only fluid of interest. For high-temperature superconductors (HTSs) devices other coolants are possible (liquid or gas): hydrogen (also suitable for the intermediate T_C material MgB_2), neon, nitrogen, or oxygen, in addition to helium. This chapter will concentrate mainly on liquid helium cooling, for which the available database is well established and most varied. Other fluids will be included to the extent that the heat transfer mechanisms and basic nondimensional equations are similar. Table 3.5.1.1 and Figure 3.5.1.1 report the main quantities of interest for the selected cryogens, and their range of application compared to the critical temperature of technical superconductors.

The properties in Table 3.5.1.1 and the considerations above apply to cooling fluids in the liquid state. For HTS devices, and under certain circumstances, gaseous fluids may be used. Helium, neon, and hydrogen have been considered, or used, as gas coolants.

Table 3.5.1.1 Main properties of cryogenic fluids of interest for the cooling of superconducting devices [1, 2].

	Boiling temperature at 1 atm (K)	Latent heat of vaporization (kJ kg^{-1})	Critical temperature (K)	Critical pressure (bar)
Helium	4.2	21	5.2	2.3
Hydrogen	20.3	445	32.9	12.8
Neon	27.1	86	44.4	26.5
Nitrogen	77.3	199	126.3	34.0
Oxygen	90.2	213	154.6	50.4

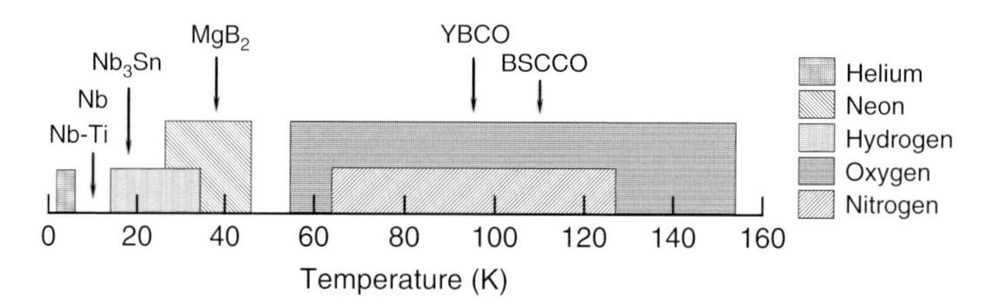

Figure 3.5.1.1 Range of temperature for the use of the various cryogenic fluids considered, as compared to the critical temperature of technical superconductors. (Drawn after B. Baudouy, CEA, Saclay.)

3.5.1.2 Bath Cooling

3.5.1.2.1 Principle

Bath cooling of a superconducting device was the earliest, and possibly the simplest form of cooling of a superconducting device. Example of bath-cooled devices are early high-energy physics detector magnets such as the ANL Bubble Chamber [3] and the CERN Big European Bubble Chamber [4, 5], accelerator magnets such as the large Hadron collider (LHC) [6], and most laboratory-scale magnets that generate the background magnetic field for experiments and material research.

A bath-cooled device looks as schematically represented in Figure 3.5.1.2. The device is submerged in the cryogen, in the liquid state. The heat produced in the device (e.g., AC loss during ramps, resistive heat from the current leads), or entering the cryostat (radiation, supports), is deposited in the bath. We can distinguish here two situations, depending on whether the bath has a free surface, where the liquid and vapor phases are in equilibrium (*saturated* bath), or whether the liquid is in a thermodynamic state sufficiently far from saturation and fills the whole cooling vessel (*subcooled* bath). While in the first case the heat is removed from the bath by evaporation at the free surface, in the second case a heat exchanger plunged in the bath is necessary to provide the cold source.

The advantages of bath cooling are (i) simplicity and modest cost for systems of small dimensions, and (ii) the benefit of a very large heat exchange at the wetted surface, when in nucleate boiling regime, which in the case of coils with low current density can be used to achieve unconditional stability.

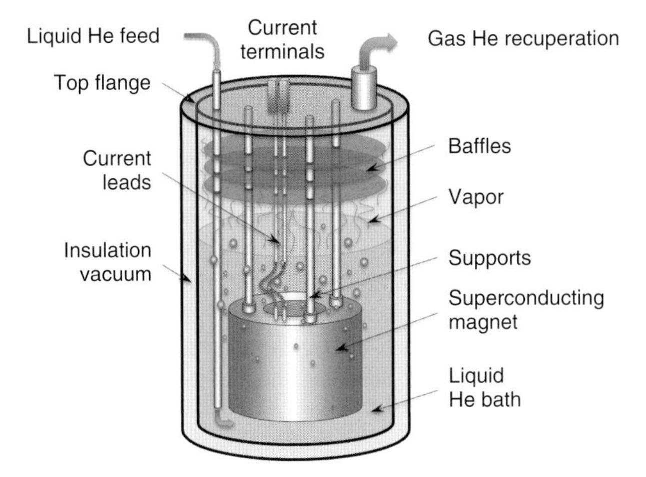

Figure 3.5.1.2 Prototype bath-cooled superconducting device. In this schematic the device (a magnet) is plunged in a bath of boiling cryogen contained in a cryostat.

The disadvantages are that (i) the temperature is fixed, or can be varied only in a modest temperature range, close to the saturation line, (ii) the heat exchange depends on the orientation of the wetted surfaces, and can experience a crisis with a sharp drop if the critical heat flux for nucleate boiling is exceeded, and (iii) a relatively large volume of cryogen is stored in the cryostat, which needs allowances for venting to avoid overpressure in case of sudden heating (e.g., quench or loss of cryostat vacuum insulation), creating a practical limit on magnet size especially if pressure vessel codes need to be met, (iv) high cost associated with large amounts of coolant in the case of helium, and lastly (v) the fact that bath-cooled magnets usually have high-voltage parts exposed to the cryogen, which tends to limit the maximum device voltage to medium values (kilovolt range).

3.5.1.2.2 **Heat Removal in a Bath**

Whichever the configuration, saturated or subcooled, two main *thermal impedances* need to be considered in the design and operation of a bath-cooled device:

- heat transfer from the device itself to the cryogen bath, taking place at the wetted surface of the device, and strongly dependent on the heat flux, the thermodynamic state of the cryogen, and the geometry and nature of the surface of the device;
- bulk heat transport in the bath, from the surface of the device to the surface of the bath (saturated bath) or to the cold source (subcooled bath) and, eventually, to the cryogenic plant.

Of the two, the first, heat transfer at the surface of the device, is the most important, and we will discuss it in detail later. As to the second, the heat transport in the bath, we remark that bulk heat transport mechanisms and regimes in a bath cryostat can be very complex [2], and are outside the scope of this chapter. The discussion below is aimed at providing only a glimpse of the governing phenomena.

Taking a saturated bath as an example, and in much simplified terms, we can distinguish between the behavior under small and large heat fluxes. At small heat flux (a few to a few tens of watts per meter square), the heat does not cause boiling and is transported by natural convection of superheated fluid. The superheated fluid reaches the free surface, evaporates, and the heat is carried by the exhaust vapor. At larger heat fluxes (hundreds of watts per meter square and larger), boiling takes place at the heated surfaces, and heat transfer in the bulk bath is greatly enhanced by bubble formation and phase separation. Vapor bubbles can travel to the surface, or collapse in this process, which promotes turbulence in the bath. The temperature gradient in a saturated bath is small. As an example, if we take a 1 m liquid helium column at 4.2 and 1 bar at the surface, the difference in pressure from the surface of the bath to the bottom of the cryostat is 12 mbar, which corresponds to a difference in the saturation temperature of only 14 mK. Somewhat larger temperature gradients are required to establish natural convection, in the range of few tens of millikelvin. In practice, a saturated helium bath can be regarded as a constant temperature heat sink for the device.

The situation is less favorable in the case of a subcooled bath of normal liquid, where at small heat fluxes the fluid can stratify and sustain temperature gradients of the order of hundreds of millikelvin or more. The exception is helium-II, where the counterflow heat exchange provides large bulk heat removal capacity up to a critical flux, at which point superfluidity breaks down. Temperature gradients of few millikelvin are normally sufficient to remove several tens of kilowatts per meter square over lengths of the order of 1 m from a stagnant bath of superfluid helium at 1.9 K and 1 bar. As for a saturated bath, a bath of superfluid helium can be assimilated to a constant temperature heat sink.

The rate of evaporation of the cryogen $\dot{m}_{\text{evap}}$ from the bath[1] is related to the heating power $\dot{q}$ deposited in the bath by a simple relation:

$$\dot{m}_{\text{evap}} = \frac{\dot{q}}{h_{\text{lv}}} \tag{3.5.1.1}$$

where $\dot{q}$ includes all heat sources in the bath and h_{lv} is the latent heat of vaporization.

Table 3.5.1.1 reports the values of the latent heat of vaporization for the most common cryogens. It is interesting to note that helium, the working fluid for baths at 4.2 K, has a relatively small value of h_{lv}, among the smallest of all liquids (for reference, h_{lv} for helium is two orders of magnitude smaller than for water). The quantity of vapor generated by a given heat load is hence relatively large, and pipework of appropriate dimensions is necessary to recuperate it. As the lowest temperature attainable in the bath is directly related to the minimum pressure at the bath, it is important to reduce the pressure drop along the vapor recovery line. Finally, for reference, a boil-off of 1 g s^{-1}, a relatively large but representative value

1) Note that for subcooled baths, the cold side of the heat exchanger is usually filled with a saturated cryogen, and the heat removed from the heat exchanger is then governed by the same equation.

for a large bath, corresponds to a heat removal of 20 W for helium at 4.2 K and 1 bar, and to 200 W for nitrogen at 77 K and 1 bar.

In summary, provided that the technical aspects of cryostat design, fill, and recovery are properly addressed, a bath of a cryogen is a cold source of approximately constant temperature, up to the capacity of the liquefier. This is why the attention is rather devoted to the other thermal impedances, and especially to the heat transfer from the surface of the device to the bath.

3.5.1.2.3 Heat Transfer from a Solid Surface to a Bath

Steady State Two surface heat transfer regimes are of interest in bath-cooled devices: natural convection and boiling. During steady state, or quasi-steady-state operation, it is expected that the heat load would be small. In that case, if the solid surface temperature is only slightly above that of the coolant, the heat transfer coefficient is determined by natural convection mechanisms.

Natural convection in cryogens (excluding superfluid helium) is no different than for any other fluid, and heat transfer coefficients can be computed using standard correlations among nondimensional numbers of the form [7]:

$$Nu = c(GrPr)^n \tag{3.5.1.2}$$

Nu is the Nusselt number $Nu = hl/k$ in which h is the convective heat transfer coefficient, k is the thermal conductivity of the fluid, and l is a characteristic length. The choice of l can be arbitrary, but it is usually a length along the boundary layer. The Nu number represents the ratio of convective to conductive heat transfer in the fluid and configuration under consideration.

Gr is the Grashof number defined as $Gr = g\beta(T_s - T_o)l^3/\nu^2$ in which g is the gravity constant, β is the volumetric thermal expansion coefficient for the fluid, T_s and T_o are the surface and bulk fluid temperatures, respectively, l is a characteristic length, again chosen as a relevant dimension along the boundary layer and ν is the dynamic viscosity of the fluid. The Grashof number represents the ratio between buoyancy and viscous forces acting on the fluid, and thus holds importance as a determining parameter in the heat transfer coefficient in natural convection flows. Like the Re number for forced flows in pipes for instance, the Gr number determines the transition from a laminar to a turbulent boundary layer. Transition to turbulence occurs for Gr between 10^8 and 10^9 depending on the geometry and is determined experimentally in each case.

Pr is the Prandtl number defined as $Pr = \frac{c_p\mu}{k}$ where c_p is the specific heat, μ is the viscosity, and k the thermal conductivity. The Pr number represents the ratio of viscous diffusion to thermal diffusion. For a fluid at a given pressure and temperature, as is the case for the coolant in a superconducting device, the Pr number can be taken to be a constant (property of the fluid).

In Eq. (3.5.1.2), c is a constant which depends on the fluid, its conditions, and the specific geometry for which the heat transfer coefficient is being evaluated. It is determined experimentally. The exponent n is 1/4 in the laminar regime, and 1/3 in the turbulent regime. In the case of helium, an approximate expression for the natural convection heat transfer coefficient was given in [8]:

$$h = 0.02\Delta T^{0.33} \quad \text{for} \quad \Delta T < 0.2\,\text{K} \tag{3.5.1.3}$$

where ΔT is the temperature difference between the surface and the bulk fluid, or $(T_s - T_o)$ in our notation.

As the heat flux is increased, the solid surface temperature goes above the boiling point, and bubbles begin to form at the surface. A new regime sets in and consideration of two-phase energy exchange at the surface comes into play. Heat transfer in boiling is of practical importance since a superconducting device will be in this regime during any relevant stability or quench event.

Boiling in cryogens, like natural convection, also follows standard correlations and exhibits a similar behavior as any other normal fluid. The different regimes involved in boiling heat transfer are illustrated in Figure 3.5.1.3. The onset of boiling is characterized by a sharp increase in heat transfer coefficient from that in natural convection, the result of a phase change at the solid/fluid interface. Bubbles of vapor form at the surface (around imperfections), and these bubbles of initially superheated vapor, detach from the surface and reach equilibrium with the liquid as they rise to the liquid/vapor interface of the bath. This regime is referred to as *nucleate boiling*, and thanks to the latent heat of vaporization, and the mass transfer associated with bubbles separating from the surface and convecting energy away from it, the effective heat transfer coefficient in nucleate boiling can be quite high when compared to natural convection. There is a limit, however, to the surface temperature difference that can be sustained in nucleate boiling. At certain point, a critical heat flux, q_c, is reached and a boiling crisis ensues.

This boiling crisis marks the transition from nucleate to film boiling. The formation of bubbles is such that they overwhelm the capacity to detach from the

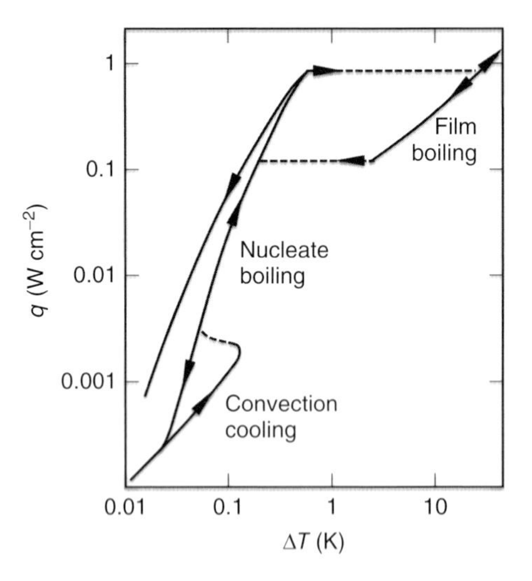

Figure 3.5.1.3 Heat flux during boiling as a function of solid–fluid temperature difference for helium. (Based on original in [7], Plenum Press.)

surface and drift away. The bubbles coalesce to form a continuous vapor film that separates the solid surface from the liquid. Heat is now transferred by conduction through the vapor film to the liquid, a very inefficient mechanism, and therefore the effective heat transfer coefficient in the film regime is an order of magnitude less than in nucleate boiling. The only way to maintain the critical heat flux, or higher, while in film boiling is for the solid surface to assume a significantly higher temperature than the cooling bath.

Also illustrated in Figure 3.5.1.3 is the fact that during a transient, the transition from one regime to the next depends on the sign of the rate of change in heat flux. In other words, the heat flux exhibits hysteresis, transition from nucleate to film boiling occurs at a certain point, but once heat flux is decreasing, the return to nucleation happens later (at lower heat flux). The same occurs for the transition from single-phase natural convection to nucleate boiling. This behavior makes it difficult to develop a universal correlation that accurately predicts heat transfer rates in all regimes.

Since overall boiling behavior is generic to all coolants, a universal expression to approximately derive the heat transfer coefficient during nucleate boiling has been derived [9]. A much-simplified expression based on the universal scaling, and useful for orders-of-magnitude estimates, is given by:

$$h = A\Delta T^{1.5} \quad \text{for} \quad 0.2\,\text{K} < \Delta T < 1\,\text{K} \tag{3.5.1.4}$$

where A is a parameter that ranges from 0.6 [8] to 1.0 [10].

Of practical importance for the design of superconducting devices is the derivation of an expression for the critical heat flux, giving rise to the boiling crisis. This relation can be put in the following form [11]:

$$Ku = 0.16 \tag{3.5.1.5}$$

where Ku is the Kutateladze heat flux parameter defined as:

$$Ku = \frac{q_\text{c}}{h_\text{lv}\rho_\text{v}^{\frac{1}{2}}[\sigma g(\rho_\text{l} - \rho_\text{v})]^{\frac{1}{4}}} \tag{3.5.1.6}$$

where q_c is the critical heat flux, h_lv is the heat of vaporization, ρ is the density, the subscripts l and v refer to the liquid and vapor phases, σ is the surface tension, and g the gravity constant. This correlation works well for all cryogens of interest and has been tested for helium, hydrogen, and nitrogen.

Transient There is special interest to understand and quantify the heat transfer between solid and fluid under transient conditions, particularly during very fast and short heat pulses. This is very relevant to the design for stability in superconducting devices since many of the events of interest, such as flux jumps, mechanical motion, or sudden current redistribution triggered by these events, can occur in the time frame of milliseconds. Indeed, the effective heat transfer during the early times of a transient can be what determines the stability limit in a superconducting device.

During a fast transient there is no initial boundary layer, so effective heat transfer rates to the fluid can briefly be much higher than predicted by the correlations presented above. The formation of the boundary layer is governed by conduction, and the fluid acts as an ideal heat sink. The topic of transient heat transfer in bath cooling has been experimentally studied and reviewed by Schmidt [10], Steward [12], and Giarratano and Frederick [13]. Summary of their findings can be also found in [7, 14]. Typical values of the transient heat transfer coefficient are in excess of the steady-state heat transfer by a factor up to 5, until a maximum energy transfer condition is reached.

3.5.1.3 Internal Cooling

The characteristic of *internally cooled* superconducting devices is that the flow of the coolant is confined to channels that are integral part of the device itself. Recent and major examples of internally cooled magnets are the large fusion magnets of the present generation, all built with cable-in-conduit conductors (CICC) (EAST [15], KSTAR [16], SST1 [17], Wendelstein 7X [18], ITER [19], JT60-SA [20]), or the high-energy physics detector at the LHC (compact muon solenoid (CMS) [21] and a toroidal LHC apparatus (ATLAS) [22]). The class of internally cooled device also comprises applications such as power transmission cables, see the examples in [23, 24], and the overview in [25].

As seen from the variety of the examples above, the flow in internally cooled devices can be either proximal to the superconductor, when the coolant is in direct contact with the superconducting wire or tape, or can cool indirectly the superconductor by conduction, being in contact only to a part of the device. Whichever the configuration, internally cooled devices can be seen conceptually as a cooling loop consisting of a cold source that determines the temperature of the incoming flow, a circulator that provides the pressure head for the flow, and the device to be cooled, where most of the heat load enters the cooling loop. This is schematically represented in Figure 3.5.1.4, a prototype of a simple loop with heat exchange to a bath, a cold pump, and a parallel of cooling pipes into the cooled device. The

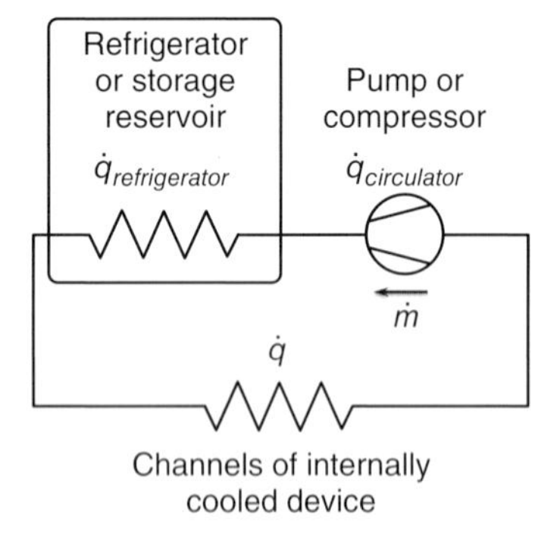

Figure 3.5.1.4 Prototype flow loop for an internally cooled device, showing a cold recirculator providing the flow, a heat exchanger setting the flow temperature, and the load. Also shown the quantities of work and heat entering and exiting the loop.

example is not the only possibility for internally cooled devices, but can be used to point to the main advantages and drawbacks of this configuration.

The advantages of internal cooling were discussed very early in the development of superconducting magnets, and summarized by Arp [26]. In general, internally cooled devices profit from (i) large and easily predictable heat transfer capability, associated with a controlled mass flow, (ii) the absence of the heat transfer crisis which can take place in case of bath cooling when nucleate boiling exceeds critical heat transfer, (iii) compact cryogenics and reduced helium inventory compared to bath cooling, and (iv) compatibility with very high dielectric strength when requesting voltage insulation in the range of tens of kilovolt, which is not practical for bath-cooled devices.

The case of thermo-syphon-driven loops, where the pressure head is provided by gravity and the density difference between the hot and cold part, is limiting advantage (i) in the sense that the range of mass flow available is limited by the density and height of the loop. Using a cooling fluid in the two-phase regime can also limit benefits (i) and (ii), as the operating temperature and mass-flow range are confined to a restricted range of vapor fraction.

The advantages listed above must be contrasted to drawbacks such as: (i) the need of leak tight circuits for the flow, and (ii) the necessity of a circulator to initiate and sustain the desired mass flow. This last is not only a technological challenge, but bears consequences on the thermodynamic efficiency of the overall cooling system.

3.5.1.3.1 Heat Removal from an Internally Cooled Loop

Similarly to the discussion on bath cooling, we identify two main thermal impedances to the heat removal using internal cooling:

- heat transfer from the device to the flow of cryogen, in the cooling channel, mostly dependent on the flow conditions and geometry, as well as the thermodynamic state of the cryogen and the heat flux;
- heat transport along the cooling channel, from the internals of the device to the cold source (e.g., a heat exchanger).

Contrary to bath cooling, however, internally cooled devices can be limited both by the amount of heat that can be transported by the flow, as well as heat transfer at the wetted surface of the cooling pipe.

The power $\dot{q}$ (W) removed by a mass flow $\dot{m}$ (kg s^{-1}) of a fluid at a pressure p, temperature T, and vapor fraction x, of specific enthalpy $h(p,T,x)$ (J kg^{-1}), in stationary state, is given by the product of mass flow and enthalpy difference between inlet and outlet conditions:

$$\dot{q} = \dot{m}[h(p_{\text{out}}, T_{\text{out}}, x_{\text{out}}) - h(p_{\text{in}}, T_{\text{in}}, x_{\text{in}})] \tag{3.5.1.7}$$

where the subscripts indicate the outlet and inlet conditions of the flow. The above expression, albeit simple, contains the complete physics of internally cooled devices. The heat removal capability of the loop can be maximized by providing a large mass flow of a fluid with a large enthalpy difference between the given inlet condition and the desired outlet condition, in turn determined by the operating margins desired in the superconductor.

We focus first on the enthalpy difference. For perfect fluids (or for negligible pressure changes in the flow), in single phase, the enthalpy can be written as follows:

$$h(p, T) = c_p T \tag{3.5.1.8}$$

where c_p is the specific heat of the fluid under constant pressure conditions. Fluids with large specific heat hence provide large heat removal ability under smaller temperature difference, which is an advantage in superconducting device that always profits from increased operating margin. As we did for bath cooling, we can evaluate heat removal capability of helium at 4.2 K and 3 bar (supercritical). Under these conditions, and allowing a *reasonable* temperature increase of 1 K, a flow of $1\,\mathrm{g\,s^{-1}}$ can remove approximately 6 W, that is, a third of what can be removed by evaporation. We can repeat the same evaluation for nitrogen, at 77 K and 3 bar, and we obtain in this case a heat removal capability of 2 W, well below the heat that can be removed from a boiling bath. Note however that in an internally cooled device the total mass flow can be made large by using several cooling channels in parallel. It is not uncommon to reach total cooling flows ranging from several tens to several hundreds of grams per second, thus achieving large heat removal.

As discussed in [27, 28] in the case of helium, the highest heat removal under minimum temperature increase is found when the flow conditions are close to the saturation line or, in the case of supercritical fluid, in the vicinity of the critical point (see Table 3.5.1.1). In this region of operation, where the fluid properties change rapidly with temperature and pressure, the specific heat has a peak, and the enthalpy difference between given inlet and outlet conditions is the highest.

3.5.1.3.2 Mass Flow and Circulator Mechanisms

Internal cooling requires a mechanism to generate a pressure difference between the inlet and outlet of the cooling channels. Several choices are possible, depending on the device dimensions, the heat load, the desired mass flow, and other system-dependent considerations. A first possibility is to choose among:

- natural convection, which is also called a *thermo-syphon loop*. The flow is driven by the weight difference between a *warm* leg and a *cold* leg in the system. The flow in the loop is hence linked to the heat load. To profit from density variations in the two legs, the cryogen is usually at, or close to, saturation conditions, where heat transfer at the device surface is also large. Natural convection leads to a simple design, with no need of active components, minimum maintenance, and high thermodynamic efficiency. The disadvantage is that the circulation requires gravity (i.e., cannot work in an arbitrary orientation), an arrangement of the cooling loop producing the required pressure head (height and reservoirs), and can be affected by instabilities at low or large heat fluxes;
- forced convection, where a pump or a compressor is the component that drives the desired flow. In this case, the flow is in principle independent of the heat

load in the cooling loop, and can be regulated at will, under any orientation and cooling pipe geometry. The control is hence much simplified, at the expense of an additional component (the pump/compressor) that pressurizes the loop, and the need of a heat exchanger.

Specifically to forced convection, the simplest mechanism is to make use of the high-pressure gas coming from the compressor of the refrigerator. The gas is cooled through a series of heat exchangers, and fed to the cooling circuit. The exhaust gas can be recovered at low temperature, which is the most efficient option and is possible if the heat load is small, or needs to be warmed further to be fed back to the refrigerator, which is necessary if the outlet temperature of the cooling loop is too high to reenter the cold end. Overall this is the most efficient way to provide a force flow of cryogen, but has the drawback that the cooling loop is directly coupled to the refrigerator. The overall design of the thermodynamic cycle of the refrigerator depends on the mass flow and heat load of the cooling loop, which may be difficult to optimize over a wide range, and the regulation of dynamic heat loads becomes delicate.

These issues can be solved by using a cold circulator (a pump or compressor) in the cooling loop, where the cold source is a heat exchanger in a bath of liquid generated by a liquefier. In this case, the flow in the cooling loop can be driven and controlled independently of the thermodynamic cycle of the liquefier. If the amount of liquid in the heat exchanger is sufficiently large, the bath also provides a means to smooth dynamic heat load variations. The decoupling of cycle and cooling loop is beneficial for stability and control, but requires a work, which is performed by the cold circulator at the lowest temperature in the system, is dissipated by friction, and needs to be removed at the heat exchanger in addition to the heat load on the cooling circuit. The amount of heat produced is proportional to the work done by the cold circulator and depends on the isentropic efficiency of the circulator $\eta_{\text{circulator}}$ as follows:

$$\dot{q}_{\text{circulator}} = \frac{1}{\eta_{\text{circulator}}} \frac{\Delta p}{\rho} \dot{m} \tag{3.5.1.9}$$

where the mass flow $\dot{m}$ is driven under a pressure head Δp and ρ is the fluid density. This amount of heat can be significant.

When considering circulation mechanisms for internal cooling, it is important to take into account the requirements for cooldown and warm-up of the system, which may impose conditions on the required flow and pressure drop more restricting than normal operation. Finally, in case of an active component driving the flow, care should be taken that the device is protected during off-normal operation conditions such as a quench.

3.5.1.3.3 Heat Transfer in Internal Flows

Steady State As with bath cooling, heat transfer in internal flow of cryogen is similar to other fluids and is described by a well-established theory and correlations once they are properly adjusted for the operating conditions in the vicinity of the saturation line and critical point. Taking the case of supercritical helium, in steady-state conditions, and sufficiently high Reynolds number, defined as $Re = \frac{\rho v D_h}{\nu}$, the heat transfer is dominated by the thickness of the thermal boundary layer and is independent of the orientation of the channel. An accurate forecast of the heat transfer coefficient is possible using a modified Dittus–Boelter correlation in the form proposed in Ref. [29] for helium:

$$Nu = 0.025 Re^{0.8} Pr^{0.4} \left(\frac{T_\mathrm{b}}{T_\mathrm{w}} \right)^{0.716} \tag{3.5.1.10}$$

where the subscript b refers to the fluid bulk temperature, and w indicates the solid wall. In this case the Nusselt number is defined as $Nu = \frac{hD_h}{k}$ where the characteristic length is the hydraulic diameter D_h of the flow. The above correlation takes into account that in the vicinity of the saturation line, and under large heat fluxes, the thermophysical properties of the cryogen (e.g., density, specific heat, conductivity) may change significantly across the boundary layer. Taking typical values for the properties of helium at 4.5 K, the heat transfer coefficient obtained from Eq. (3.5.1.10) for Re in the range of $10^4 – 10^5$ is from 1000 to 10000 W m^{-2} K^{-1}, that is, comparable to the range of nucleate boiling heat transfer. The advantage of force-flow cooling is that the heat transfer does not experience a crisis, and can be reliably reproduced and sustained by acting on the mass flow, at the cost of the pumping work discussed earlier.

In the case of two-phase force-flow cooling, the steady-state heat transfer coefficient is nearly independent of the mass-flow rate and the fraction of vapor in the fluid, and the values are very close to those reported for pool boiling discussed earlier. For force-flow conditions, the heat transfer crisis is reached at values of the Kutateladze heat flux parameter given by [30]:

$$Ku = 0.031 + 0.078(1 - x)^{3.92} \tag{3.5.1.11}$$

where x is the vapor fraction.

Transient Transient heat transfer to a flow of cryogen is a complex phenomenon as it involves changes in the fluid temperature, in the thermal boundary layer, as well as variations of the boundary layer thickness caused by heating-induced flow. In general, transient heat transfer to a forced flow tends to exhibit high heat transfer coefficient, which is beneficial to extract sudden heat depositions from the superconducting device, and thus contribute to its stability. The case of supercritical helium was studied in [31–33] in dedicated measurements on short test sections. The experiments show a peak heat transfer coefficient at early times, below 1 ms, followed by a drop proportional to the inverse of the square root of time, at times in the range of some milliseconds to a few tens of milliseconds. This behavior is consistent with the diffusion of heat in the thermal boundary layer, which can be approximated analytically as follows [33]:

$$h = \left(\frac{\pi k \rho c_\mathrm{p}}{4t} \right)^{\frac{1}{2}} \tag{3.5.1.12}$$

where t is the time measured from the beginning of the transient. The expression above is shown to fit properly the experimental data for times longer than a millisecond and until the thermal boundary layer is fully developed. At earlier times, the values of h are limited by the *Kapitza resistance* [34] at the contact surface, which gives a significant contribution only when the transient heat transfer coefficient is in the order or larger than 10^4 W m^{-2} K^{-1} (see also the later discussion on superfluid helium).

3.5.1.3.4 **Helium Expulsion**

Cryogen venting from an internally cooled device is different from conditions already discussed earlier for a bath, in that specific allowance must be made for the onset and consequences of heating-induced flow. Indeed, in spite of the relatively small cryogen inventory, the helium expulsion from cooling pipes of small diameter and long length under significant heat rate can lead to large values of pressure increase, which can affect the structural integrity of the pipes. Typical events that need to be considered for this analysis are a quench in an internally cooled cable, or loss of vacuum insulation for pipework at the surface of an internally cooled cold mass. The pressure in the pipe depends on the heating rate, its distribution along the pipe, the pipe geometry, and friction factor. To establish values for the pressure increase it is necessary, in principle, to solve the equations for compressible, transient flow, which is a complex task. Codes have been developed to this aim, and various experiments were performed to provide benchmark values for helium. Analysis of experimental data [35] demonstrated that it is possible to estimate the peak pressure in a pipe cooled by helium by a relatively simple formula [36]:

$$p_{\max} = 0.335\left(\frac{fq^2L^3}{D_{\mathrm{h}}}\right)^{0.36} \qquad (3.5.1.13)$$

where L is the total length of the pipe, q is the heating rate per unit helium volume, and f is the fanning friction factor of the flow.

3.5.1.3.5 **HeII Cooling**

At atmospheric pressure, liquid helium below $T_\lambda = 2.17$ K, the so called lambda temperature, undergoes a phase change to what is called *superfluid helium*, or HeII, a quantum fluid. HeII has very different heat transfer characteristics compared to all other normal fluids and thus is treated separately here. Superfluid helium is attractive as a coolant for LTSs not only because it extends the operating regime to higher magnetic fields and current densities (significantly in the case of NbTi), but also because of its excellent heat transfer properties. Disadvantages of relying on superfluid cooling include higher power requirements to remove the same heat load, and the need to have either the coolant, or a heat exchanger space, at subatmospheric pressure.

Superfluid helium is used as coolant in many magnets for a variety of applications, some examples being Tore Supra [37], NHMFL's 45T Hybrid superconducting outsert [38], CERN's LHC [6], and the 12T ISEULT MRI system at CEA-Saclay [39].

For the case of superfluid helium in bath-cooling configuration it should be noted that two regimes exist. For the case of cooling a solid surface that remains below the lambda temperature, heat transfer is determined by two mechanisms that operate in series.

Heat transfer at the solid surface is dominated by what is called *Kapitza resistance* [34]. At the solid/fluid interface there is a temperature discontinuity and the heat transfer through it (acting as a resistance) is given by:

$$\dot{q} = a(T_s^n - T_f^n) \tag{3.5.1.14}$$

$\dot{q}$ is the heat flux, T_s and T_f are the solid wall and fluid temperatures, respectively, n is a constant exponent, and a is a coefficient that depends on the solid material and temperature. Even though the Kapitza resistance is always present in cryogenic fluids, it is only relevant in practice at very low temperatures, that is, for superfluid helium. Values of n are predicted to be 4 by theory, but measured values are ~3, and the constant a is about 400 W m^{-2} K^{-3} for a copper/helium interface.

Heat transfer in the HeII bulk bath, being a nonclassical fluid, does not follow traditional correlations and is controlled but what is called *Gorter–Mellink mechanism*, in which the heat flux in the fluid deviates from the linear Fourier mode and is instead given by:

$$\dot{q} = -k(\nabla T)^{\frac{1}{3}} \tag{3.5.1.15}$$

The practical implication of this nonlinear behavior is that superfluid helium can transfer heat (up to a critical flux) while having extremely low-temperature gradients, that is, HeII can act as a nearly isothermal coolant. The effective heat transfer coefficient in HeII (if a linearization were to be done) is extremely high and it is this property that makes it attractive as a coolant for superconducting devices.

For the case when the solid surface exceeds the lambda temperature, the mechanisms at play become more complex, but the fundamental consideration of heat transfer through the bulk of HeII remains the same as discussed above. In this case, as soon as the temperature of the surface exceeds T_λ, a thin film of HeI forms next to the surface, with a HeI/HeII interface (at T_λ) standing very near the metal surface. Further, within this HeI layer, and if the solid temperature is high enough, there may be yet another interface liquid/vapor as boiling ensues. In all situations, it can be seen that the HeI/HeII interface, which acts as a boundary with its temperature clamped at T_λ, remains infinitesimally close to the solid surface. This is because of the extreme difference in heat transfer rates that can be sustained through normal and superfluid helium, being much higher in the latter. The result is that for analysis and design purposes, the HeII bulk bath can be considered to be removing heat with the metal surface being clamped at T_λ.

For the case of internally cooled conductors using superfluid helium, it is useful to calculate the maximum heat flux that can be sustained in the cooling channel (an actual cooling channel, or the cable interstices in the case of a CICC). Assuming there is a uniform heat input along the HeII channel of magnitude q (W m^{-1}), the channel is of length $2L$, and both ends are kept at constant temperature T_o. Using the Gorter–Mellink expression and treating the channel as a conduction fin, we can derive the temperature difference between the middle of the channel (hot spot) and the clamped-cooled end kept at T_0:

$$\Delta T = \left(\frac{q}{kA}\right)^3 \frac{L^4}{4} \tag{3.5.1.16}$$

where k is the same heat transfer coefficient of Eq. (3.5.1.15) and A is the channel cross-section.

References

1. Frey, H. and Haefer, R.A. (1981) *Tieftemperaturtechnologie*, VDI-Verlag, Düsseldorf.
2. Scurlock (2006) *Low-Loss Storage and Handling of Cryogenic Liquids: The Application of Cryogenic Fluid Dynamics*, Kryos Publications.
3. F. Wittgenstein (1968) The 1.8 Tesla, 4.8 m I.D. bubble chamber magnet. in Brookhaven National Laboratory – BNL-50155(C-55), Proceedings of the 1968 Summer Study On Superconducting Devices and Accelerators, Part III (69, Rec. June), 1968, pp. 765–785.
4. Haebel, E.U. and Wittgenstein, F. (1970) Big European Bubble Chamber (BEBC) magnet progress report. Proceedings of 3rd International Conference on Magnet Technology, DESY, Hamburg, Germany, 1970, pp. 874–895.
5. Wittgenstein, F. (1972) Preliminary test results of BEBC superconducting magnet. Proceedings of 4th International Conference on Magnet Technology, Upton, NY, pp. 295–300.
6. Bruning, O., Collier, P., Lebrun, P., Myers, S., Ostojic, R., Poole, J., and Proudlock, P. (eds) (2004) *LHC Design Report. Part 1. The LHC Main Ring*, CERN-2004-003-V-1, CERN, Geneva.
7. Van Sciver, S. (1986) *Helium Cryogenics*, Plenum Press, New York.
8. Krafft, G. (1986) *Heat Transfer Below 10 K, Cryogenic Engineering*, Academic Press.
9. Kutateladze, S.S. and Borishansky, V.M. (1966) *A Concise Encyclopedia of Heat Transfer*, Pergamon Press, New York.
10. Schmidt, K. (1981) Review of Steady State and Transient Heat Transfer in Pool Boiling Helium I, IIF – IIR Commission A1/2 – Saclay (France) – 1981/6.
11. Kutateladze, S.S. (1948) On the transition to film boiling under natural convection. *Kotloturbostroenie*, **3**, 10–12.
12. Steward, W.G. (1978) Transient helium heat transfer. Phase I – static coolant. *Int. J. Heat Mass Transfer*, **21**, 863–874.
13. Giarratano, P.J. and Frederick, N.V. (1980) Transient pool boiling of liquid helium using a temperature controlled heater surface. *Adv. Cryog. Eng.*, **25**, 455–466.
14. Dresner, L. (1995) *Stability of Superconductors*, Plenum Press, New York.
15. Weng, P.D. and the HT-7U Team (2001) The engineering design of the HT-7U tokamak. *Fusion Eng. Des.*, **58-59**, 827–831.
16. Oh, Y.-K. *et al.* (2009) Commissioning and initial operation of KSTAR superconducting tokamak. *Fusion Eng. Des.*, **84**, 344–350.
17. Saxena, Y.C. and SST-1 Team (2000) Present status of SST-1. *Nucl. Fus.*, **40**, 1069.
18. Riße, K. and for the W7-X Team (2009) Experiences from design and production of Wendelstein 7-X magnets. *Fusion Eng. Des.*, **84** (7–11), 1619–1622.
19. Mitchell, N. *et al.* (2009) Status of the ITER magnets. *Fusion Eng. Des.*, **84** (2–6), 113–121.
20. Tsuchiya, K. *et al.* (2007) Design of the superconducting coil system in JT-60SA. *Fusion Eng. Des.*, **82**, 1519–1525.
21. Herve, A.for the CMS Collaboration (2000) The CMS detector magnet. *IEEE Trans. Appl. Supercond.*, **10**, 389.
22. ten Kate, H.H.J. (2008) The ATLAS superconducting magnet system at

the Large Hadron Collider. *Physica C: Superconductivity*, **468** (15–20), 2137.

23. Watanabe, M. *et al.* (2007) Development of 22.9 kV high-temperature superconducting cable for KEPCO. *Physica C*, **463–465**, 1132–1138.
24. Maguire, J. *et al.* (2009) Development and demonstration of a fault current limiting HTS cable to be installed in the Con Edison grid. *IEEE Trans. Appl. Supercond.*, **19** (3), 1740–1743.
25. EPRI (2012) Superconducting Power Equipment: Technology Watch 2012, Product ID 1024190, December 2012.
26. Arp, V. (1972) Thermodynamics of single-phase one-dimensional fluid flow. *Cryogenics*, **15**, 342–351.
27. Arp, V. (1975) Forced-flow, single-phase, helium cooling systems. *Adv. Cryog. Eng.*, **17**, 285–289.
28. Katheder, H. (1994) Optimum thermo-hydraulic operation regime for cable in conduit superconductors (CICS). *Cryogenics*, **34**, 595–598.
29. Giarratano, P.J., Arp, V.D., and Smith, R.V. (1971) Forced convection heat transfer to supercritical helium. *Cryogenics*, **11**, 385–393.
30. Giarratano, P.J., Hess, R.C., and Jones, M.C. (1974) Forced convection heat transfer to subcritical helium I. *Adv. Cryog. Eng.*, **19**, 404–416.
31. Giarratano, P.J. and Steward, W.G. (1983) Transient forced convection heat transfer to helium during a step in heat flux. *Trans. ASME*, **105**, 350–357.
32. Schmidt, K. (1988) Transient heat transfer into a closed small volume of liquid or supercritical helium. *Cryogenics*, **28**, 585–598.
33. Bloem, W.B. (1986) Transient heat transfer to a forced flow of supercritical helium at 4.2 K. *Cryogenics*, **26**, 300–308.
34. Kapitza, P.L. (1941) The study of heat transfer in helium II. *J. Phys. (USSR)*, **4** (3), 181–210.
35. Miller, J.R., Dresner, L., Lue, J.W., Shen, S.S., and Yeh, H.T. (1980) Pressure rise during the quench of a superconducting magnet using internally cooled conductors. Proceeding of the ICEC-8, Genova, Italy, 1980, pp. 321–329.
36. Dresner, L. (1981) Thermal expulsion of helium from a quenching cable-in-conduit conductor. Proceeding of 9th Symposium on Engineering Problems of Fusion Research, Chicago, IL, 1981, pp. 618–621.
37. Torossian, A. (1993) TF-coil system and experimental results of tore supra. *Fusion Eng. Des.*, **20**, 43–53.
38. Miller, J.R. (2003) The NHMFL 45-T hybrid magnet system: past, present, and future. *IEEE Trans. Appl. Supercond.*, **13** (2), 1385–1390.
39. Schild, T. *et al.* (2008) The Iseult/Inumac whole body 11.7 T MRI magnet design. *IEEE Trans. Appl. Supercond.*, **18** (2), 904–907.

3.5.2 Cryocoolers

Gunter Kaiser and Gunar Schroeder

3.5.2.1 Motivation

3.5.2.1.1 The Principle of "Invisible" Cryogenics

Products and applications which include superconducting elements offer advantages in contrast to conventional ones. Superconducting sensors like superconducting quantum interference device (SQUID) magnetometers and bolometers provide higher sensitivity and better resolution also with low level incoming signals. Superconducting filters offer the possibility to operate with higher pole numbers, providing more telecommunication channels within a certain bandwidth. Due to the fact that the copper losses can be eliminated, superconducting motors, generators, and cables can be made in smaller sizes and with higher efficiencies. In some cases, technical problems only can be addressed with a superconducting solution. This is, for instance, the case in certain types of superconducting fault current limiters.

The disadvantage of all these applications is the operation on temperature levels which are considerably lower than the ambient temperature. In laboratory use, the cryogenic temperatures are typically provided by the application of boiling cryogenics, like liquid helium or nitrogen. For practical applications, this is often no option. In some cases, there is no access to the superconducting application, for instance, when it is operated in remote areas. So there is no practical means for refilling of liquid cryogens. In some other applications, the users like to avoid or even refuse the use of liquid cryogens for their hazards and practical needs.

In order to overcome these problems, closed-cycle cryocooler, operated by use of electricity, is the technical solution. So in some applications, the user even does not recognize that his device is operating by use of cryogenics. There are several requirements for this "invisible cryogenics":

- high efficiency (less contribution to the energy consumption)
- long lifetime (longer than the lifetime of the device to be cooled)
- low noise (low or no contribution of disturbances for application)
- low price (not considerably determining the application price)
- small size (not considerably contributing to the size of the total system)

3.5.2.1.2 Pros and Cons

Advantages The advantages of cryocoolers are already summarized under the aspect of "invisible cryogenics." The main advantage is the consumption of electricity instead of the use of evaporating liquid cryogens. The cryogenic capacity, which is required for the application, is not necessarily to be stored in voluminous vessels. This is, for instance, an advantage for cryogenic high-power applications.

Disadvantages Beside advantages, cryocooler also have certain typical disadvantages, as they have:

- vibrations (mechanical noise)
- electromagnetic interferences (conducted and radiated emissions)
- temperature variations (for regenerative cryocoolers due to the thermodynamic cycle)

In most cases, these disadvantages can be tolerated. In certain cases, however, they must be controlled by additional means for noise reduction.

Mechanical vibrations can be reduced through mechanical solutions by decoupling the vibrating cold head from the application. This is done by flexible thermal interfaces like aluminum or copper foil strips or wire ropes. In this way also geometry variations due to thermal dilatation can be addressed. Furthermore, such thermal resistances and the heat capacity of the application can be used to suppress higher-frequency temperature fluctuations. Electromagnetic interferences are mostly reduced spatially by the cryocooler from the application. There are also possibilities of electronic suppression or elimination by use of a cryogenic storage (latent heat storage) and intermitted cryocooler and application operations.

Dependent on their mechanical operating principle and design, cryocoolers are also more or less sensitive to mechanical loads like bending forces acting on their cold heads or higher level vibrations. In some cases, the more or less "point-like" cold area of a cryocooler can be a disadvantage for cold distribution.

3.5.2.2 Classical Cryocoolers

Since the mid-twentieth century, a lot of cryocooler types have been developed. From all these types, two different electrically driven regenerative cryocoolers are mostly used today, the higher-frequency Stirling and the lower-frequency Gifford–McMahon cryocoolers. The basic design and common principle of these cryocoolers is shown in Figure 3.5.2.1.

The cold head is a parallel configuration of the cycle space and the regenerator. The hot end of the cycle space is directly connected to the pressure wave generator or compressor. During compression and expansion, the displacer increases warm or cold space, respectively, so heat is pumped from cold to warm heat exchanger.

3.5.2.2.1 Stirling Cryocoolers

Principle Stirling cryocoolers consist of a pressure wave generator and a cold head. The pressure wave generator can be connected directly to the cold head, forming a so-called integral cryocooler. In most cases, the pressure wave generator is separated by a hollow tube, called the *split line*. Theses cryocoolers are called *split cryocoolers*. The cold head is equipped with a displacer, which increases the hot space during compression and the cold space during expansion. In most cases, the displacer contains the regenerator. But it is also possible to place the regenerator in a separate space between warm and cold heat exchangers (as depicted in Figure 3.5.2.1). The displacer can be driven by the cryocooler drive, by a separate drive (mostly a linear drive) or by the pressure wave in case of a free linear displacer. The last option with a regenerator-filled displacer is mostly

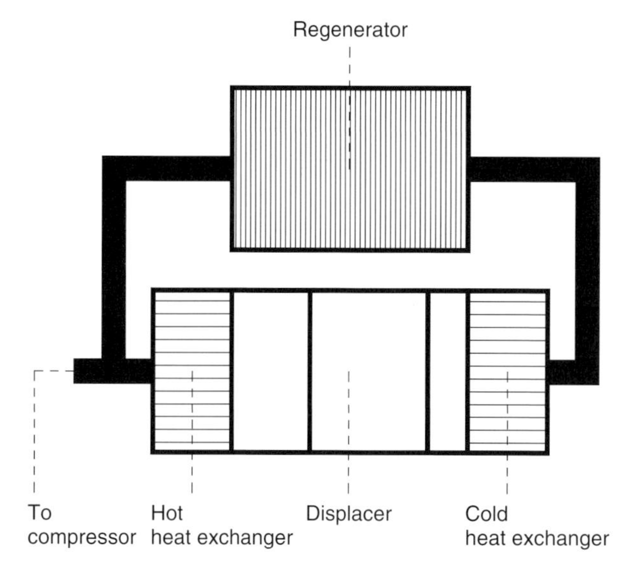

Figure 3.5.2.1 Principle design of Stirling and Gifford–McMahon cryocoolers.

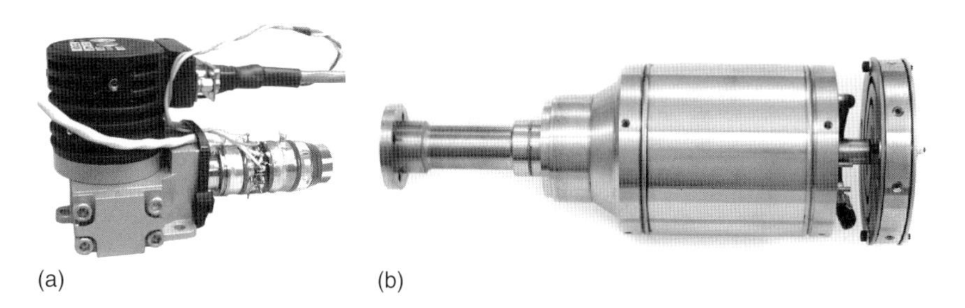

Figure 3.5.2.2 Integral Stirling cryocoolers: (a) RICOR K548[1] and (b) Sunpower Cryotel GT.[2]

used, to keep the design as simple and reliable as possible. Stirling cryocoolers are usually driven with frequencies between 30 and 100 Hz. In most cases, the operating frequency corresponds with that of the electric line (50 or 60 Hz).

Examples Most Stirling cryocoolers are single-stage low-power cryocoolers (<1–15 W at 80 K), developed for military applications. Their zero-load operating temperatures usually reach 30–50 K. Some examples of integral cryocoolers are given in Figure 3.5.2.2. Some split Stirling cryocoolers are shown in Figure 3.5.2.3. There are also some special cryocoolers on the market. For cooling power in the range between 1 and 4 kW, a special Stirling Cryogenics cryocooler

1) *http://www.ricor.com/products/integral-rotary/k548/*
2) *http://www.sunpowerinc.com/cryocoolers/cryotel.php*

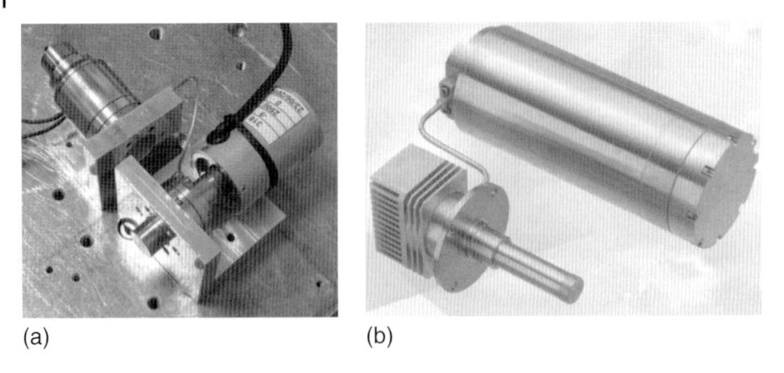

Figure 3.5.2.3 Split Stirling cryocoolers: (a) RICOR K527[3] and (b) AIM SL150.[4]

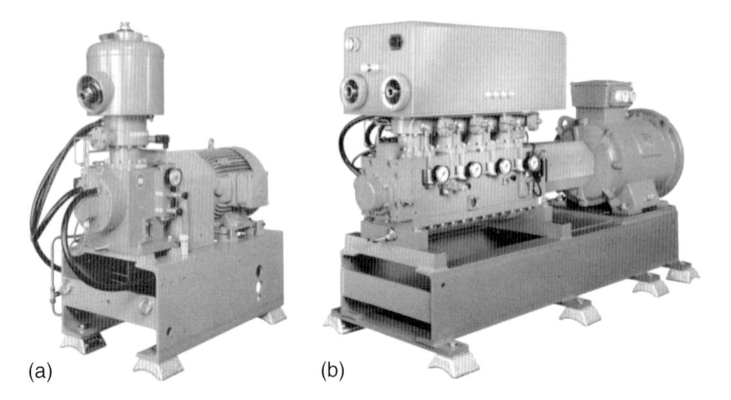

Figure 3.5.2.4 High-power Philips Stirling cryocoolers: (a) 1 kW and (b) 4 kW.[5]

is used (see Figure 3.5.2.4). A special low-temperature Stirling cryocooler was developed for space use working down to 10 K (see Figure 3.5.2.5).

3.5.2.2.2 Gifford–McMahon Cryocoolers

Principle In contrast to Stirling cryocoolers, Gifford–McMahon cryocoolers operate at lower frequencies between 1 and 5 Hz. These cryocoolers are always split-type cryocoolers, consisting of a compressor package and a cold head package. The compressor package contains a hermetic compressor mostly derived from air condition or cooling industry modified to pump helium in a circuit (extra cooling and special seals). It furthermore contains an air or water cooler to reject the heat of compression and a complex oil rejection system in order to avoid oil and oil vapor inside the cold head. The cold head contains the working cylinder with the displacer, the regenerator, and a valve system, mainly designed as a rotary valve. The rotary valve connects the working cylinder with the high- or low-pressure side of the compressor package, respectively. The displacer is

3) *http://www.ricor.com/products/split-linear/k527/*
4) *http://www.aim-ir.com/fileadmin/files/Data_Sheets_Cooler/2011/AIM_Data_Sheet_SL150.pdf*
5) *http://www.stirlingcryogenics.com/products/Cryocoolers/1Stage-Cryocoolers/*

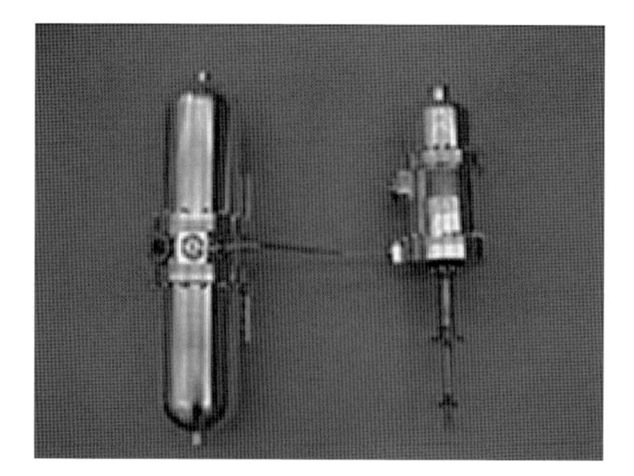

Figure 3.5.2.5 Two-stage Stirling cryocooler for space use.[6)]

driven by an electric motor or additional pneumatics in such a way that the warm space is maximized during high-pressure phase and the cold space during the low-pressure phase. Gifford–McMahon cryocoolers are used for higher cooling powers at 80 K and/or lower temperatures (2–20 K). In the latter case, two-stage configurations of the cold head are used. In the low-temperature range of 20 K, the cold stage is equipped with a lead shot regenerator part. For even lower temperatures of 2–10 K, rare earth shot regenerators are used.

Examples For higher power at 80 K, one-stage configurations as shown in Figure 3.5.2.6 are used. For lower temperatures in the 4 K region, two-stage configurations as shown in Figure 3.5.2.7 are used.

3.5.2.3 Special Types of Cryocoolers

3.5.2.3.1 Pulse Tube Cryocoolers

Principles Pulse tube cryocoolers are, like Gifford–McMahon cryocoolers, also regenerative cryocoolers. Their cold heads, however, has a totally different design and operating principle (see Figure 3.5.2.8).

In contrast to the classical regenerative coolers, the pulse tube cold head consists of a serial configuration of cycle space and regenerator. Cold and hot heat exchangers have changed their positions. So the cold heat exchanger is arranged between pressure wave generator and hot heat exchanger. During compression, the gas inside the pulse tube is compressed and moves towards and partially superheated through the hot heat exchanger. During expansion, the opposite takes place. The returning gas moves towards and partially subcooled through the cold heat exchanger. In this way, heatispumped from the cold to the hot heat exchanger.

In the past three decades, several kinds of pulse tube configurations were developed. They can be distinguished by the design and principle of the additional

6) *http://www.ard.jaxa.jp/eng/research/thermalsg/th-index.html*

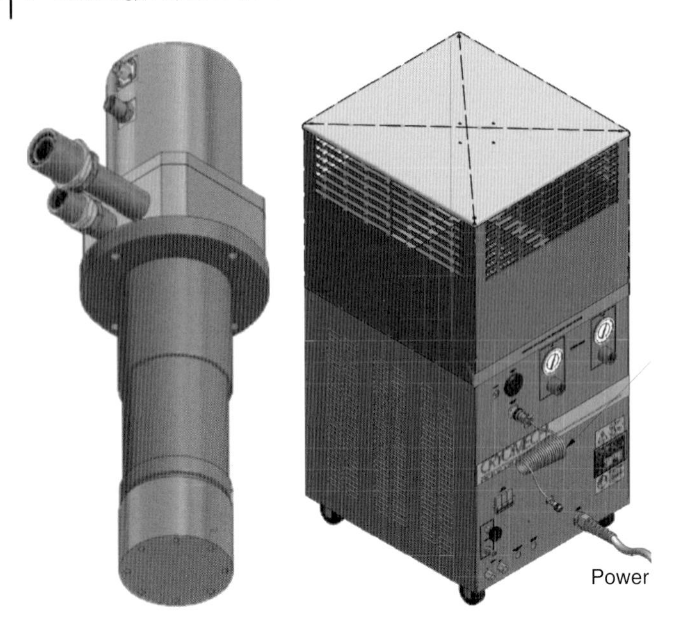

Figure 3.5.2.6 One-stage 80 K Gifford–McMahon cryocooler Cryomech AL 300.[7]

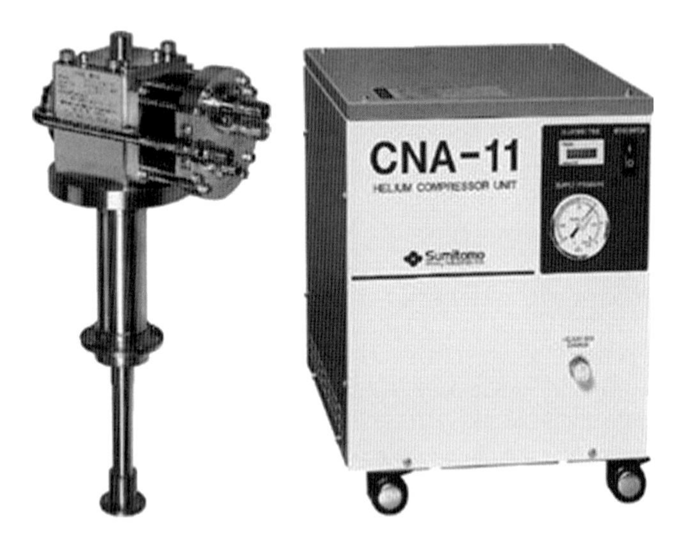

Figure 3.5.2.7 Two-stage 4 K Gifford–McMahon cryocooler Sumitomo CNA 11.[8]

phase shifter. Today, only the orifice pulse tube and the inertance pulse tube are practically relevant (see Figures3.5.2.9 and 3.5.2.10). The orifice with reservoir was introduced by Mikulin in order to increase the phase angle between mass flow and

7) *http://www.cryomech.com/AL300.php*
8) *http://www.shi.co.jp/ejsite/no149/10.html*

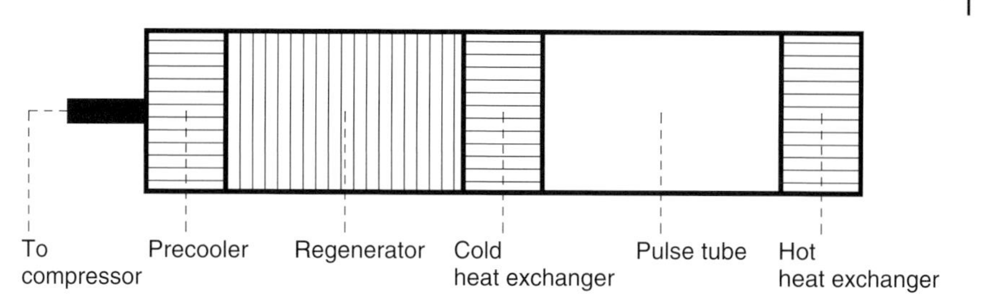

Figure 3.5.2.8 Principle design of pulse tube cryocoolers (basic pulse tube) [1].

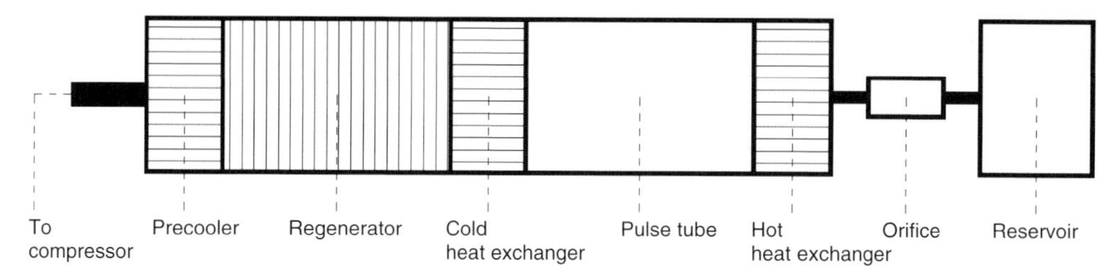

Figure 3.5.2.9 Orifice pulse tube cryocooler [2].

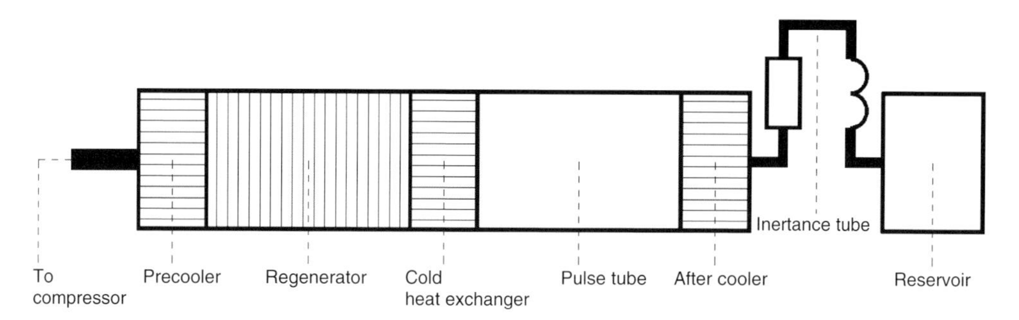

Figure 3.5.2.10 Inertance pulse tube cryocooler.[9]

pressure wave. It acts as a fluidic resistor–capacitor (RC) configuration. The mass flow across the hot heat exchanger can be increased. The inertance tube uses the inertia of the gas contained inside a long tube to increase the phase shift even more. It acts as a fluidic resistor–inductor–capacitor (RLC) configuration.

The pressure wave can be generated by use of a higher-frequency pressure wave generator. These kinds of pulse tube cryocoolers are called *Stirling-type* pulse tube cryocoolers. The other possibility is to operate with a hermetic compressor and a valve system. According to the similar configuration, these kinds of pulse tube cryocoolers are called *Gifford–McMahon-type* pulse tubecryocoolers. There are also configurations with active phase shifters. The alpha-Stirling-like configuration is called *active reservoir* or *active buffer*. There are also configurations which

9) Several publications by Matsubara *et al.* and Kittel *et al.* in 1997.

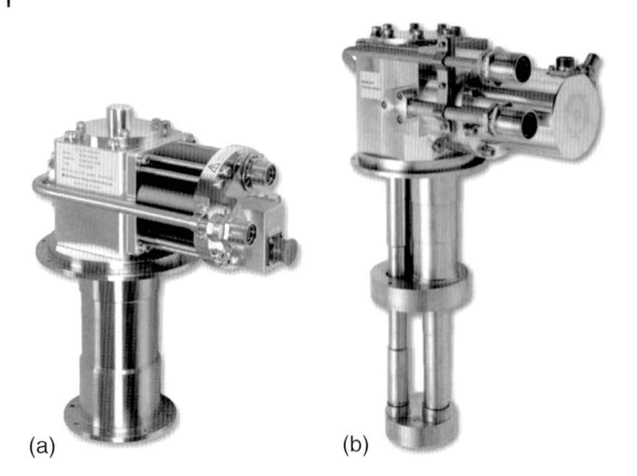

Figure 3.5.2.11 Gifford–McMahon-type pulse tube cold heads: (a) high power[10] and (b) 4 K[11] of SHI (Sumitomo Heavy Industries).

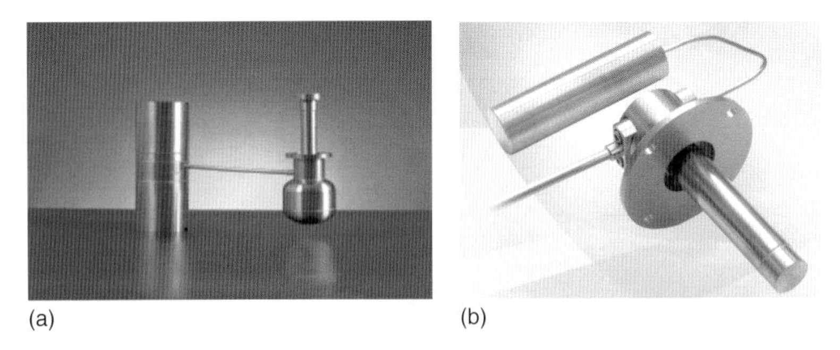

Figure 3.5.2.12 Stirling-type pulse tube cryocoolers: (a) Thales Cryogenics[12] and (b) AIM.[13]

have controlled valves directly at the hot end of the pulse tube. These kinds of pulse tube cryocoolers are called *four-valve.*

The main advantage of pulse tube cryocoolers in comparison to classical regenerative cryocoolers is the absence of moving parts inside the cold head. This fact leads to less design effort, less mechanical vibrations, higher reliability, and longer lifetime of the cold head.

Examples First on the market were Gifford–McMahon-type pulse tube cryocoolers for higher cooling power at 60–80 K or low cooling temperatures (4 K), respectively (see Figure 3.5.2.11).

10) *http://www.shicryogenics.com/products/specialty-cryocoolers/rdk-400b-40k-cryocooler-series*
11) *http://www.thales-cryogenics.com/index.php/coolers/lpt*
12) *http://www.aim-ir.com/fileadmin/files/Data_Sheets_Cooler/2013_neu/2013_AIM_Pulse-Tube.pdf*
13) *http://www.shicryogenics.com/products/pulse-tube-cryocoolers/rp-082b2-4k-pulse-tube-cryocooler-series/*

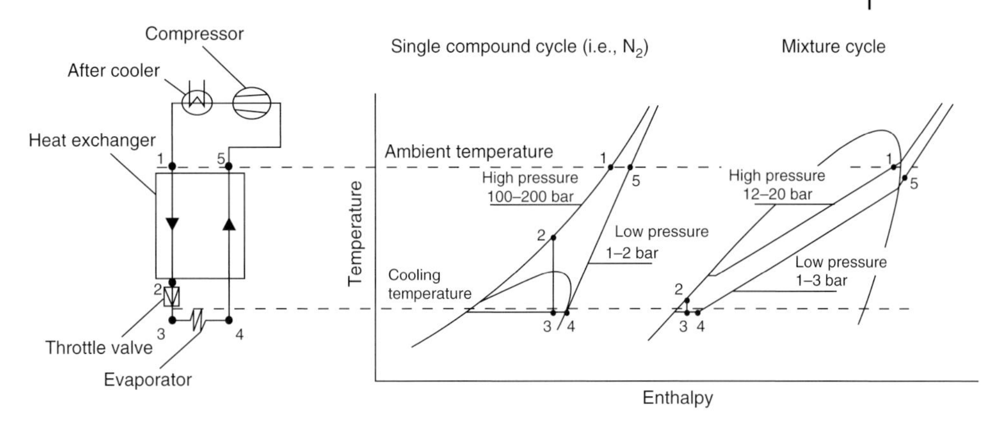

Figure 3.5.2.13 Principle design and cycle diagram of Joule–Thomson cryocoolers [3].

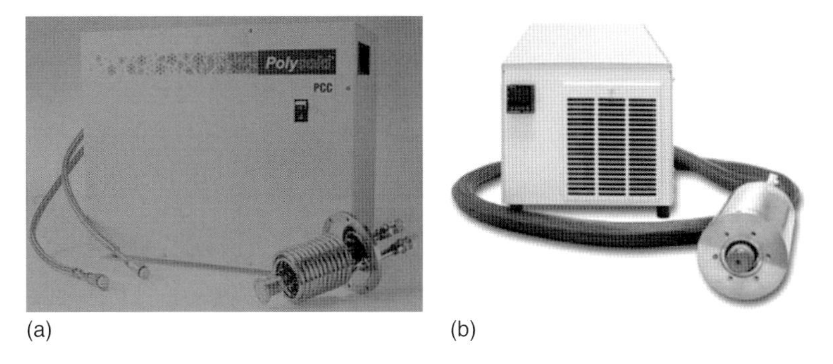

(a) (b)

Figure 3.5.2.14 Low-power mixture Joule–Thomson cryocoolers: (a) CryoTiger[14] and (b) Bio 085.[15]

Stirling-type pulse tube cryocoolers are also now commercialized by several companies. Some examples are shown in Figure 3.5.2.12. Due to their compactness and robustness, they are often used for aerospace or military applications.

3.5.2.3.2 **Mixture Joule–Thomson Cryocoolers**

Principle The principle of a mixture Joule–Thomson cryocooler is comparable to that of a Joule–Thomson cryocooler operating with a single-phase fluid. Principle design and cycle diagram are shown in Figure 3.5.2.13. It consists mainly of a compressor followed by an after cooler, a recuperator, an expansion valve after the high-pressure cold end of the recuperator, and a cold heat exchanger. The low-pressure gas returns through the low-pressure channel of the recuperator to the compressor.

Due to the higher enthalpy of fluid mixtures, consisting of nitrogen and alkanes, it is possible to reduce the operating high-pressure of 12–20 bar. Condensation and evaporation takes place in the high- and low-pressure channels of the

14) *http://www.brooks.com/products/cryopumps-cryochillers/cryochillers/pcc-compact-coolers*
15) *http://www.mmr-tech.com/kleemenko.php*

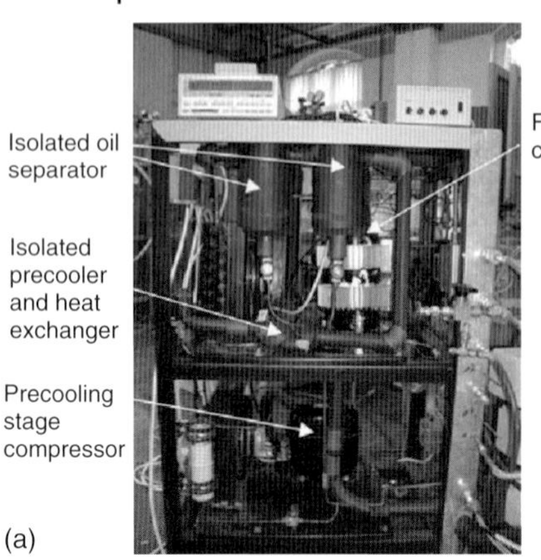

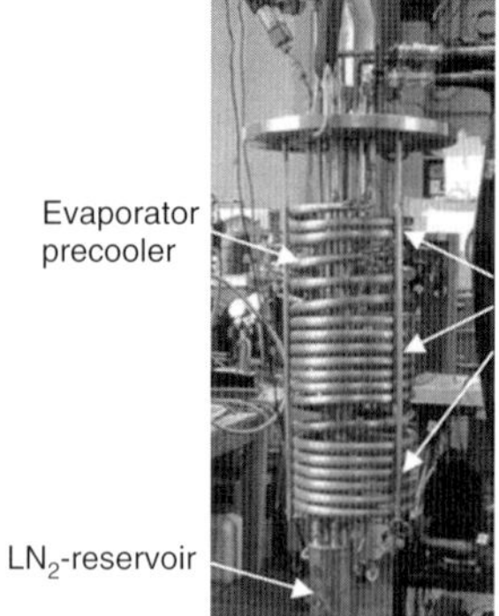

Figure 3.5.2.15 Mixture Joule–Thomson cryocooler for current lead cooling [4] copyright by ILK.

recuperator, respectively. The cooling temperatures of mixture Joule–Thomson cryocoolers have their limit at about 60 K.

The main advantage of such cryocoolers are similar to that of pulse tube cryocoolers, as there are no moving parts in the cold head. A further advantage is the possible scalability from micro-miniaturization up to very large sizes and cryogenic cooling powers from a few milliwatt up to the higher kilowatt range.

Examples There are only just a few mixture Joule–Thomson cryocoolers on the market, which are designed for cooling power in the range of a few watt at temperatures between 70 and 100 K (see Figure 3.5.2.14).

Mixture Joule–Thomson cryocoolers with higher cooling power only exist in the form of laboratory demonstrators. Figure 3.5.2.15 shows the system of Technical University of Dresden, which is designed to cool the current leads of a resistive fault current limiter.

References

1. Gifford, W.E. and Longsworth, R.C. (1963) "Pulse-Tube Refrigeration" ASME paper No. 63-WA-290, presented at the Winter Annual Meeting of the American Society of Mechanical Engineers, Philadelphia, Pennsylvania (November 17–22, 1963), p. 264–268.
2. Mikulin, E.I., Tarasov, A.A, and Shkrebyonock, M.P. (1984) *Advances in Cryogenic Engineering*, vol. 29, Plenum Press, New York, p. 629.
3. Goloubev, D. (2003) Kühlung eines resistiven HTSL-Kurzschlussstrombegrenzers mit einer Gemisch-Joule-Thomson-Kältemaschine. Dissertation an der Technischen Universität Dresden, p. 29.
4. Goloubev, D. (2003) Kühlung eines resistiven HTSL-Kurzschlussstrombegrenzers mit einer Gemisch-Joule-Thomson-Kältemaschine. Dissertation an der Technischen Universität Dresden, p. 59.

3.5.3
"Cryogen-Free" Cooling

Gunter Kaiser and Andreas Kade

3.5.3.1 Motivation and Basic Configuration

3.5.3.1.1 Motivation

Advanced cryogenic applications tend in their realization to more and more distributed structures. These structures need to be cooled according to their needs. Sometimes, the given structures have different cooling power or even temperature requirements. Usually, the cold is generated by a single cooling power source, providing the whole cooling power of the structure at a temperature below the minimum operating temperature of the system. This cooling power source is usually the cold tip of a cryocooler.

Cryocoolers provide the advantage to have an entire cooling system without the need of liquefied cryogens like LN2 or LHe to be evaporated in open systems. Cold is generated by use of an electric energy supply. That is why such systems are usually called *cryogen-free* systems. "Cryogen-fee" does not mean that there are no liquid or gaseous cryogens required in such systems. It only means that these cryogens, if needed, are contained in closed vessels.

"Cryogen-free" systems offer the advantage to provide so-called invisible cryogenics in user applications. They do not confront the system user with the requirements to supply their applications with liquefied cryogens for operation. Sometimes, the user does not even know that he is operating a system with cryogenic subcomponents. Once reached the stationary operating state, the user can use this system like any other system, which does not contain cryogens.

3.5.3.1.2 Basic Configuration

The basic configuration of a "cryogen-free" system consists mainly of three components, the cooling power source (usually a cryocooler), the heat transfer system, and the heat load (the cryogenic application or system). In some cases also thermal interceptors are contained in the system. Usually, they are integrated in order to reduce the heat load of nonoperating defect or redundant parts of the cooling system.

The heat transfer system and the thermal interceptors are the most interesting elements of "cryogen-free systems." They shall be discussed in the following subsections in more detail.

3.5.3.2 Heat Transfer Systems

3.5.3.2.1 Heat Conduction

The simplest way to provide a heat transfer is to use a thermally conducting structure between the cooling power source and the heat load. This arrangement is usually applied for systems, where it is only necessary to provide heat transfer along short distances. Heat conducting interfaces offer the possibility to provide mechanical flexibility in order to enable moving subsystems over short

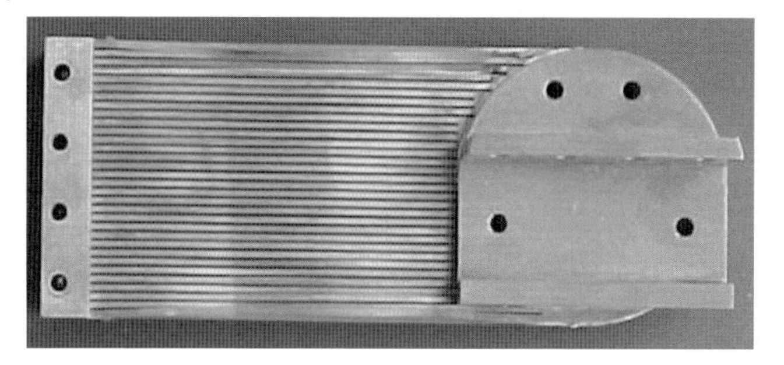

Figure 3.5.3.1 Flexible thermally conducting interface for space cryogenics (ILK Dresden).

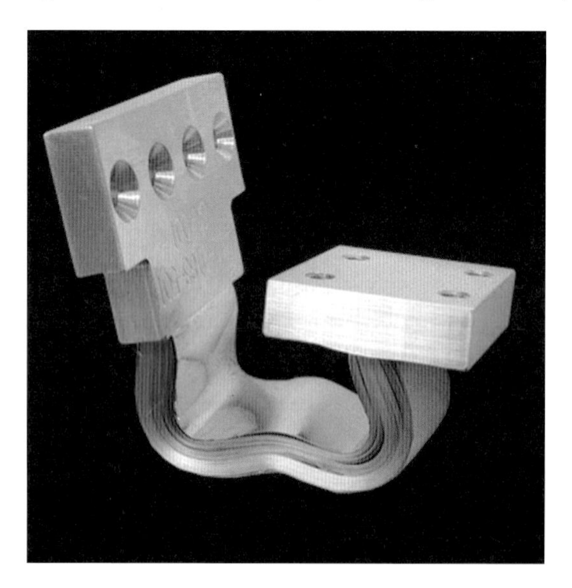

Figure 3.5.3.2 Flexible thermally conducting interface (Thermal Management Technologies[1]).

propagation or relief mechanical stress due to thermal dilatation between system components (see Figures 3.5.3.1 and 3.5.3.2).

That is why also other heat transfer systems are often coupled to flexible heat conduction paths. The disadvantage of heat conduction paths is their limited heat transfer possibility, which leads to larger temperature drops or massive structures, if providing larger cooling power.

3.5.3.2.2 **Thermosiphon**

The use of thermosiphons (see Figure 3.5.3.3) is one way to overcome the limitations of heat conduction. They consist typically of a cooled volume (thermally connected to a cold source), a heated volume (connected to the application), and

1) *http://www.tmtsdl.com/flexibleThermalLinks.html* accessed on 26 June 2013.

Figure 3.5.3.3 Neon-operated thermosiphon in the foreground of a HTS (High-Temperature Superconductor) motor (Siemens[2]).

a fluid transfer structure, consisting of one or more tubes. The heat transfer of a thermosiphon is driven by density gradients due to gravitational force between the cold and warm media.

There are mainly two ways to arrange a thermosiphon, using a single phase (liquid or gas) or using a phase change. Thermosiphons, which use a phase change, offer larger transfer rates of cooling power, due to the larger density difference between the two phases and the higher heat transfer coefficients of evaporation and liquefaction.[3]

Nevertheless, thermosiphons have also certain disadvantages: They need gravitational forces for operation. Therefore, it is not possible to use them in position-independent systems. Furthermore, it is necessary to capture the medium in warm state either/or in a larger vessel at large pressures.

3.5.3.2.3 **Two-Phase Tubes**

A two-phase tube can be used for the media transfer or as evaporator in a thermosiphon structure operating with a phase transition. It offers the advantage to contain the liquid and the gaseous phase in a single tube. It is possible to operate a two-phase tube under direct flow or counter flow condition. The counterflow two-phase tube gives the advantage to save cooling power due to internal recondensation.

Two-phase tubes have the disadvantage to be arranged in a single plain with respect to gravity. The filling state must be controlled or set before operation in order to manage the liquid-to-gas interface level (Figure 3.5.3.4).

2) *http://www.conectus.org/images/HTS_motor_22.jpg* accessed on 26 June 2013.
3) *http://webbook.nist.gov/chemistry/fluid/* accessed on 26 June 2013.

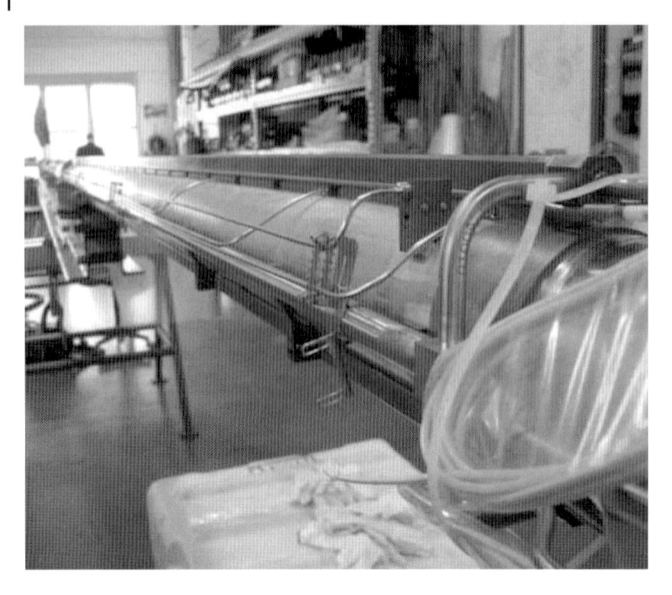

Figure 3.5.3.4 Two-phase tube along both sides of the beam line for for KArlsruhe TRItium Neutrino (KATRIN) experiment (FZ Karlsruhe[4]).

3.5.3.2.4 Heat Pipes

Heat pipes are the most advanced type of phase change heat transfer means. A heat pipe consists of a condenser, an evaporator, and both of a gas path and a liquid path. Therefore, besides the thermal conductivity, the phase transition of media can be used. Cryogenic heat pipes in the 80 K range are typically operated with nitrogen or oxygen. A cryogenic heat pipe made of EADS Astrium is shown in Figure 3.5.3.5.

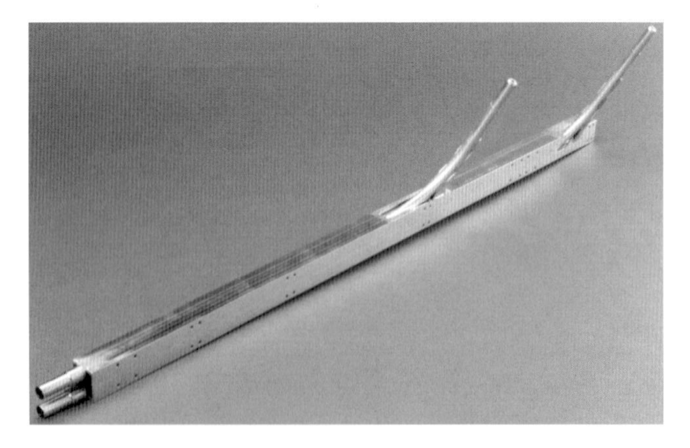

Figure 3.5.3.5 Cryogenic heat pipes from detector modules (EADS Astrium[5]).

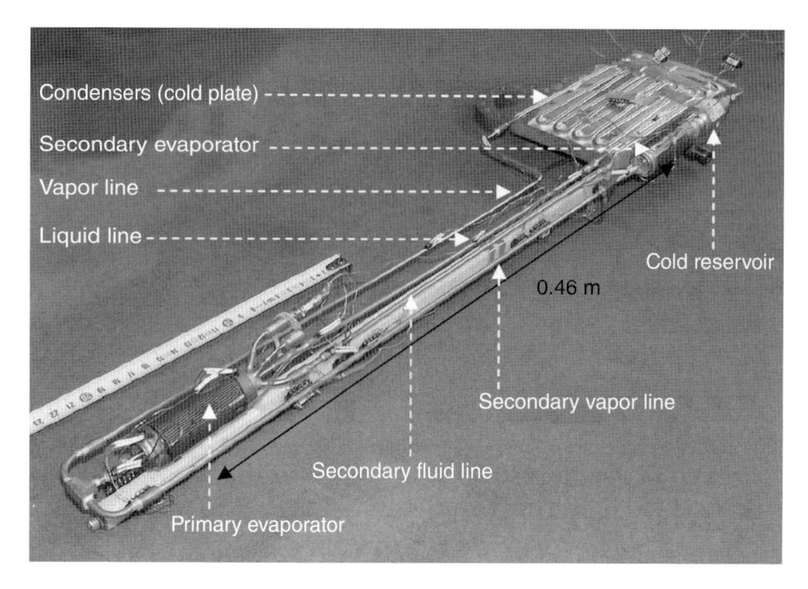

Figure 3.5.3.6 Design of a capillary-pumped loop heat pipe [2].

In cryogenics for large heat fluxes or long distances between heat source and condenser, it is usual to separate the capillary-pumped liquid path from the gas path in order to design the whole system more effectively. Such a separated system is called *capillary-pumped loop* [1] (see Figure 3.5.3.6).

Heat pipes are not very common in cryogenics. This is mainly caused by the facts that, like in the case of thermosiphons, the media must be contained in warm state either/or in a larger vessel at large pressures. There are some uses in space cryogenic systems, because there is no gravity and the need of another mechanism like capillary forces to provide the return of the liquid phase to the evaporator.

3.5.3.2.5 **Circulations**

Cryogenic circulations are used to provide a controlled flow of a media to provide a certain amount of cooling power. They contain a cooling power source, a heat load, two transfer lines, and a circulation pump. Circulations are operated in phase change or single-phase modes. The circulation pump is usually located in the downstream of single-phase systems in order to avoid the transfer of the pumping power to the application. In two-phase systems, the circulation pump is located in the upstream or the liquid phase line to the application.

Single-phase circulations are often used with subcooled, supercritical, or gaseous when a phase change inside the heat exchanger to the application is not

4) *http://www.katrin.kit.edu/img/beamtube.jpg*accessed on 26 June 2013.
5) *http://atmos.caf.dlr.de/projects/scops/sciamachy_book/sciamachy_book_figures_springer/chapter_3/fig_3_13.jpg* accessed on 26 June 2013.

Figure 3.5.3.7 Liquid argon circulation for the Italian ICARUS (Imaging Cosmic And Rare Underground Signals) neutrino project (DH Industries[6)]).

desired or must be avoided. Figure 3.5.3.7 shows an example for a subcooled liquid argon circulation.

According to the needs of the requirements, there are several possibilities to get a circulation pump. In the following pictures, two examples of circulation pumps are shown. Figure 3.5.3.8 shows a liquid cryogen pump, which can be immersed into the cryogenic liquid. It was designed for LH2 and also used for LN2 and LAr already.

Figure 3.5.3.8 Liquid cryogen pump, e.g. 360 l h^{-1} against 7 bar (ILK Dresden).

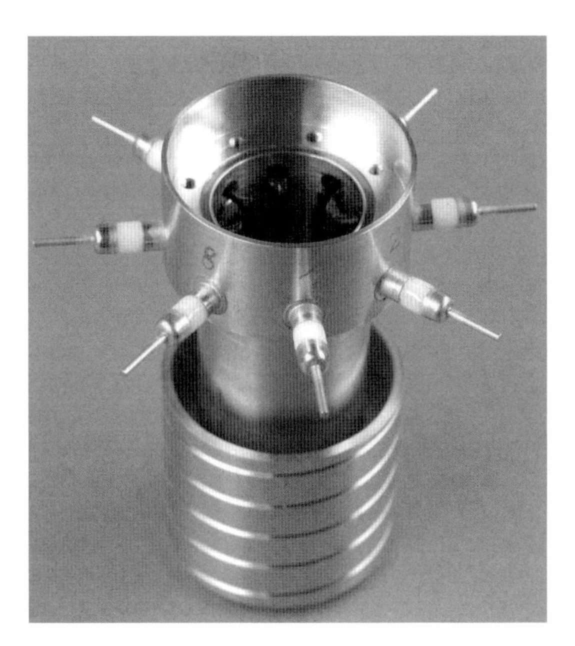

Figure 3.5.3.9 Space thermomechanical pump for LHeII, 0.2 $g\,s^{-1}$ against 0.2 bar (ILK Dresden).

In Figure 3.5.3.9, a thermomechanical pump for a space experiment is shown. The advantage of a thermomechanical pump is that there is no moving mechanical part inside, which could probably fail.

3.5.3.3 Thermal Interceptors

3.5.3.3.1 Mechanically Actuated Switches

In order to set a thermal path "on" or "off," a thermal switch can be used. One of the possibilities is a mechanically actuated thermal switch. It consists mainly of the two thermal contact areas and a mechanical mean to bring them together or apart from each other. Typically a piezoelectric device is used to maintain the mechanical operation because of its very low electric losses. Figure 3.5.3.10 shows such mechanically actuated thermal switch.

3.5.3.3.2 Thermal Dilatation Switches

Another way to provide mechanical operation is the use of thermal dilatation between two thermal contact areas. Such a thermal switch is called *thermal dilatation switch*. It consists mainly of the two contact elements made of two different heat conducting materials which are separated by a small gap. During cooling, the gap becomes smaller and vanishes at a certain onset temperature. An example of a thermal dilatation switch for a space cryogenics application is shown in Figure 3.5.3.11.

6) *http://www.d-h-industries.us/~uploads/img/ica.jpg* accessed on 26 June 2013.

Figure 3.5.3.10 Quad-redundant mechanically actuated thermal switch (Space Dynamics Lab[7]).

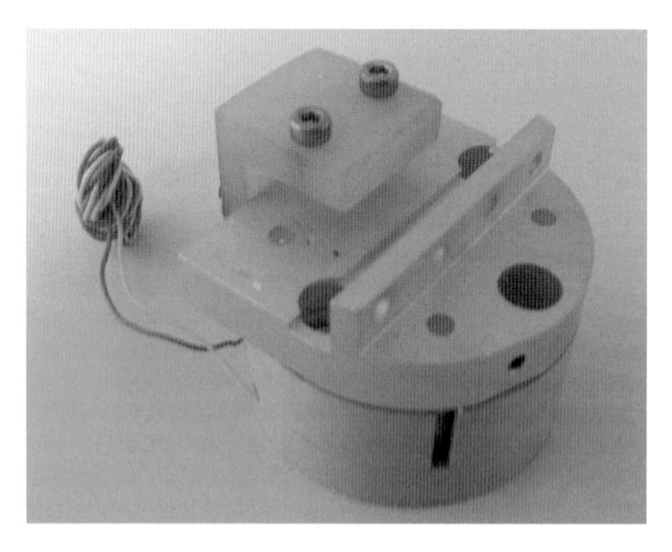

Figure 3.5.3.11 Thermal dilatation switch for a space cryogenics application (ILK Dresden).

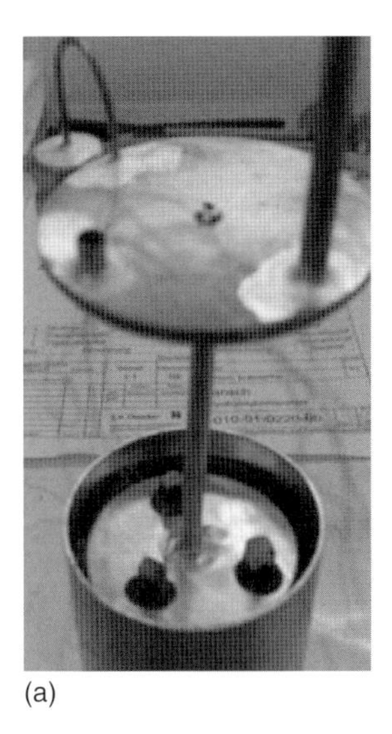
(a)

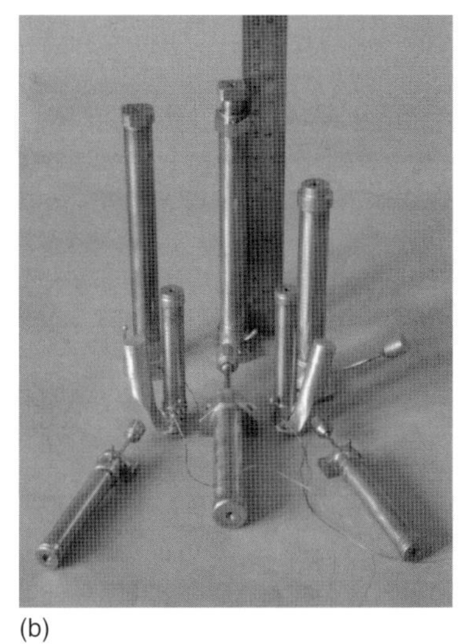
(b)

Figure 3.5.3.12 Examples of gas gap thermal switches for (a) a Millikelvin system (ILK Dresden) and (b) commercial 6 K sorption-pumped thermal switches (Chase Research Cryogenics[8]).

3.5.3.3.3 Gas Gap Switches

Gas gap thermal switches consist also of mainly two elements providing the heat transfer area along a certain gap. This gap can be filled with gas or evaporated. Typically, gas gap thermal switches are pumped and refilled by use of a cryosorption pump. Figure 3.5.3.12 shows an example of a gas gap thermal switch operated by use of a sorption pump used for an ILK Millikelvin system. Figure 3.5.3.12 shows several commercially available sorption-pumped gas gap switches for 6 K range.

References

1. Chandratilleke, R., Hatakeyama, H., and Nakagome, H. (1998) Development of cryogenic loop heat pipes. *Cryogenics*, **38** (3), 263–269.
2. Gully, P., Qing, M., Seyfert, P., Thibault, P., and Guillemet, L. (2008) Nitrogen cryogenic loop heat pipe: results of a first prototype. Cryocoolers 15, pp. 242.

7) *http://www.sdl.usu.edu/programs/jwst-heat-switch* accessed on 26 June 2013.
8) *http://www.chasecryogenics.com/gas-gap.html* accessed on 26 June 2013.

4
Superconducting Magnets

4.1
Bulk Superconducting Magnets for Bearings and Levitation

John R. Hull

4.1.1
Introduction

Levitation, whether an illusion caused by a stage magician, or a permanent magnet (PM) stably levitating over a bulk high-temperature superconductor (HTS), such as the system shown in Figure 4.1.1, often leaves the viewer with a sense of awe. In the case of the PM/HTS system, the levitation also provides a tactile and visual indication of many of the basic phenomena associated with HTSs:

- Zero resistance to DC current flow
- Strong diamagnetic behavior, keeping a change of applied magnetic flux from entering the superconductor
- Ability to trap an applied magnetic field
- Hysteretic magnetization.

As long as the superconductor remains cold, the levitation will persist, due to the zero resistance. From its equilibrium position, if one pushes the PM in Figure 4.1.1 up, down or sideways, or tries to tilt it, a restoring force returns the PM to its initial position, due to field trapping and diamagnetic behavior. The forces between the PM and the HTS can also be highly hysteretic. If one pushes the PM hard enough, its equilibrium position can be changed to almost any orientation or the center of mass of the PM can be moved to a new equilibrium position. If the PM is a cylinder with a relatively symmetric magnetic field, it readily rotates about its axis of symmetry, as indicated by the blurred arc to the left of the star in Figure 4.1.1. Such behavior suggests that the HTS could be used in the construction of a superconducting bearing with minimal rotational loss. In many applications, the advantages of noncontacting surfaces without an active feedback system, ability to operate in a vacuum, and potential for extremely low rotational drag often outweigh the inconvenience of refrigerating the HTS.

Applied Superconductivity: Handbook on Devices and Applications, First Edition.
Edited by Paul Seidel.
© 2015 Wiley-VCH Verlag GmbH & Co. KGaA. Published 2015 by Wiley-VCH Verlag GmbH & Co. KGaA.

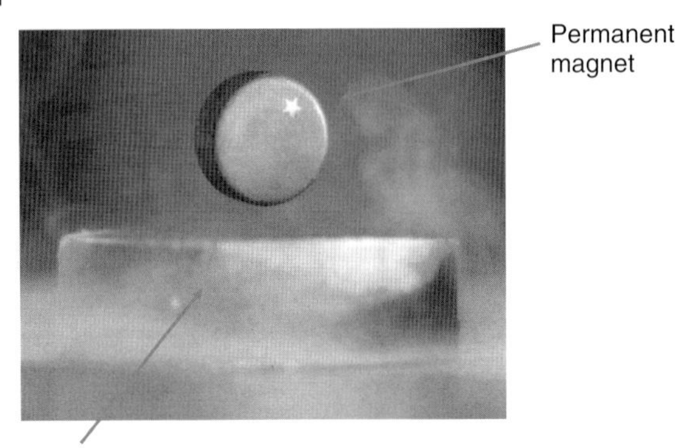

Figure 4.1.1 PM (with star) stably levitated over HTS. The PM levitates with passive stability, rotating freely while maintaining position.

Figure 4.1.2 Examples of large domain, bulk HTS preferred for bearings. The two leftmost HTS are single domain. The ring to the right contains eight domains.

Before discussing the details of how levitation is implemented in a superconducting bearing, several general comments are warranted. First, although superconducting wires and thin films have been investigated for bearing applications, the use of single-domain bulk HTS, such as shown in Figure 4.1.2, has received the most attention and is the subject of this article. In addition, while the ability to achieve strongly stable levitation is due to the technical development of bulk HTS with large engineering current density and large-domain size, the ability to realize low rotational drag in a PM/HTS bearing is a testament to the technical advances over many years to enable a high degree of magnetic homogeneity in PMs. Although bulk HTSs probably do not strictly satisfy most formal definitions of a diamagnetic material (magnetic permeability <1), the HTS often strongly mimics the behavior of a pure diamagnetic material (permeability close to zero), and it is useful to invoke this picture when we strive to understand many of the levitational phenomena. Another fundamental property of superconductors is their tendency to exclude magnetic flux from their interiors. This exclusion of magnetic flux (the Meissner effect) also makes them behave like a diamagnet; however, for type-II superconductors, the forces generated by the Meissner effect are only a small fraction of those generated by flux pinning and are usually ignored in any analysis.

A detailed treatment of HTS bearings is beyond the scope of this article. Fortunately, a number of reviews [1–10] exist that expound on many details of HTS levitation and bearing development. The goal of this article is to highlight general features of the bearing technology and levitational phenomena and discuss selected technical developments that have occurred since the previous reviews.

4.1.2
Understanding Levitation with Bulk Superconductors

4.1.2.1 Simplified Model: Double-Image Dipole

The equilibrium and magnetomechanical stiffness of the levitation process can be understood as a combination of trapped magnetic field and diamagnetic response of the HTS by means of a simple image model. The diamagnetic response is shown schematically in Figure 4.1.3. We consider the PM to be represented by a magnetic dipole. The diamagnetic response of the HTS is such that when a PM approaches the HTS surface, currents in the HTS form the equivalent of a magnetic mirror image of the PM. This diamagnetic image follows the movement of the PM. The height of the image below the HTS surface is equal to the height of the PM above the surface. When the PM moves horizontally, the image also moves horizontally, keeping directly beneath the PM. The diamagnetic response provides vertical repulsive force and vertical stability but does not provide horizontal stability. If the HTS is zero-field cooled (ZFC), that is, cooled without the PM present, and the PM is brought in close proximity to the HTS surface and then released, one often observes the PM moving horizontally off of the HTS – a manifestation of horizontal instability. In terms of the model shown in Figure 4.1.3, this would be a result of the image being constrained by the sidewall of the HTS and not being able to perfectly track the horizontal motion of the PM. Incorporating the HTS sidewall into the model formalism, the position of the diamagnetic image would be closer to the centerline of the HTS than the PM and would provide a horizontal destabilizing force.

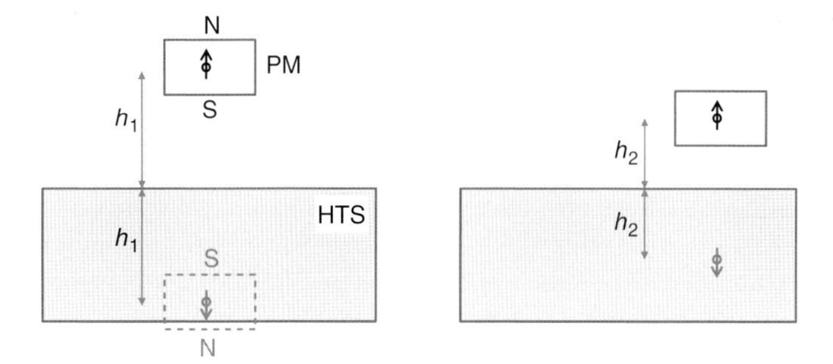

Figure 4.1.3 Diamagnetic response of HTS to a PM, where the PM is modeled as a magnetic dipole (small circle with upward-pointing arrow through it), and the diamagnetic response is modeled by a mirror image PM (dashed rectangle) with a mirror image dipole (downward-pointing arrow) below the surface of the HTS.

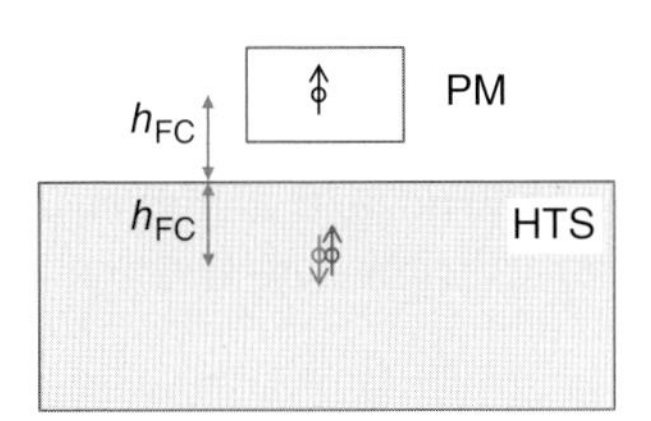

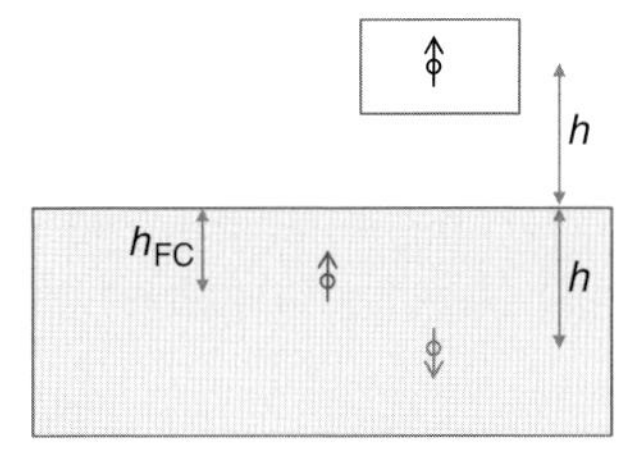

Figure 4.1.4 Trapped flux and diamagnetic response of HTS to a PM, where the PM is modeled as a magnetic dipole (small circle with upward-pointing arrow through it), the diamagnetic response is modeled by an by a downward-pointing dipole below the surface of the HTS and the trapped flux is modeled by an upward-pointing dipole below the surface of the HTS.

In order to reliably attain the stable levitation shown in Figure 4.1.1, the HTS is typically field cooled (FC), that is, the PM is mechanically supported in proximity to the HTS while the HTS is cooled below its superconducting critical temperature T_c to its operating temperature. In this case, the system is modeled by the schematic shown in Figure 4.1.4. When the HTS is FC, both a diamagnetic image (magnetized in the opposite direction as the PM) and a flux-trapped image (magnetized in the same direction as the PM) are formed, both at the diamagnetic image position. The net force on the PM in the FC position is zero. When the PM moves, the diamagnetic mirror image follows the PM, as before, but the flux-trapped image remains in its original position. The flux-trapped image is often termed the *frozen* image. With both images operating as described, there is both vertical and horizontal stability.

Modern PMs have a magnetization slightly greater than 1 T. The following example helps to appreciate the magnitude of the forces generated in an HTS bearing. A PM cylinder magnetized to 1 T along its axis of symmetry and interacting with its magnetic image with a small gap will experience a magnetic pressure of 400 kPa, where the area is that of the circular base of the cylinder.

4.1.2.2 Magnetomechanical Stiffness

To appreciate the stability of the PM/HTS system, one need to only compare it with most other passive magnetic systems which are statically unstable. Most magnetic systems are governed by Earnshaw's theorem, which states that there is no stable, static three-dimensional arrangement of a collection of poles (electric, magnetic, or gravitational) whose magnitudes do not change and which interact via a $1/r^2$ force law [11]. Particles that experience inverse-square law forces obey Laplace's equation, and Earnshaw's theorem is developed from the property of the associated curl- and divergence-free fields that precludes the existence of local, detached, scalar potential maxima or minima, and allows only saddle-type equilibria. In terms of magnetic stiffness (the magnetomechanical analog of a spring constant) K, the theorem for a collection of poles can be written as

$$K_x + K_y + K_z = 0 \tag{4.1.1}$$

where subscripts x, y, and z denote the coordinate directions. For axisymmetric systems, where $K_x = K_y$, the theorem may be rewritten as $2K_x = -K_z$, and the system will be unstable in the direction that exhibits the negative stiffness.

With a PM/HTS system, Earnshaw's theorem does not apply. One may show both experimentally and analytically from the double-image dipole model that in FC, the horizontal stiffness of a vertical dipole is exactly half the FC vertical stiffness (i.e., $2K_{x,\mathrm{FC}} = K_{z,\mathrm{FC}}$) [12]. In the PM/HTS system in FC, all of the stiffnesses are positive.

4.1.2.3 More Advanced Models

The critical-state model [13, 14] is often invoked to explain details of the levitational phenomena, especially hysteretic behavior. It is rooted in a physics that contains the essential features of flux pinning in type-II superconductors. In a critical-state model, one often assumes that at any location in the superconductor, the current is either flowing with critical current density J_c or it is zero. J_c is defined as the critical current density at which the Lorentz force is exactly balanced by the pinning force. J_c may be assumed constant or assigned a dependence on magnetic field. One approach is to represent the electrical behavior of the superconducting material by the power law [15]

$$E = E_0\left(\frac{J}{J_\mathrm{c}}(B)\right)^n \tag{4.1.2}$$

where E is the electric field, B is the magnetic field, $E_0 = 10^{-4}\ \mathrm{V\,m^{-1}}$, and the value of n is determined by the characteristics of the superconducting materials. Maxwell's equations are then evaluated by a numerical solver. A second approach numerically solves the magnetic equations based on a magnetic-energy minimization procedure [16].

4.1.3 Rotational Loss

The low rotational drag of the HTS bearing has important practical ramifications. For example, when used to support a flywheel energy storage device, the self-discharge rate of the HTS bearing, as shown in Figure 4.1.5, is much smaller than for other bearing technologies and is approximately that of a conventional battery. In the case of a mechanical bearing, the loss is dominated by mechanical friction between contacting surfaces. In the case of the active magnetic bearing, the rotor drag is usually negligible, and the loss is dominated by the power loss in the bearing coils and the control power electronics. For the HTS bearing, the rotor drag is also small, and the loss is dominated by the power to the cryocooler that keeps the HTS cold.

A figure of note for the rotational decay of a bearing is the coefficient of friction (COF), defined as the rotational drag force divided by the levitational force (weight of the levitated rotor) [17].

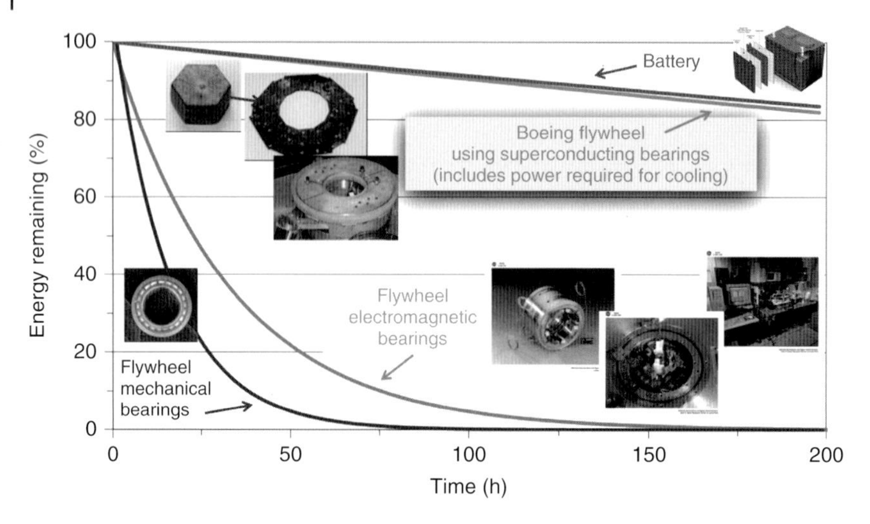

Figure 4.1.5 Notional comparison of self-discharge in a flywheel suspended by different bearing technologies and a conventional electrochemical battery.

$$\mathrm{COF} = -\frac{(2pR_\gamma^2 \mathrm{d}f/\mathrm{d}t)}{(gR_\mathrm{D})} \tag{4.1.3}$$

where R_γ is the radius of gyration of the rotor, f is the rotational frequency, t is the time, g is the acceleration of gravity, and R_D is the mean radius at which the drag force acts. In Eq. (4.1.3), all terms are easily measured or calculated except R_D, the value of which is somewhat ambiguous because, a priori, we do not know the bearing loss mechanism nor how it is distributed over the radius of the bearing. To be definitive, the author always takes R_D to be the outer radius of the bearing part of the rotor.

The COF is useful when comparing the performance of different bearing designs or designs that use various types of HTS, because it helps to normalize the effects of bearing size and differences in moment of inertia (R_γ term) among experimental devices. The COF for a mechanical roller bearing is of the order of 10^{-3}. The COF for an active magnetic bearing is $\sim 10^{-4}$ when parasitic losses for the feedback circuits and power for the electromagnets are factored in. Measured COFs for simple HTS bearings are as low as 10^{-7}.

4.1.3.1 **Hysteresis Loss**

The ease with which a PM disc spins when levitated over an HTS, and the absence of contact between the surfaces, produce the illusion that the rotation is lossless. In reality, small magnetic losses gradually slow the rotation. Usually, the major loss in an HTS bearing or levitational system is hysteretic magnetization. For an HTS with an incomplete penetration of magnetic field, the energy dissipation E_h in a complete hysteresis cycle is

$$E_\mathrm{h} = \frac{A(\Delta B)^3}{J_\mathrm{c}} \tag{4.1.4}$$

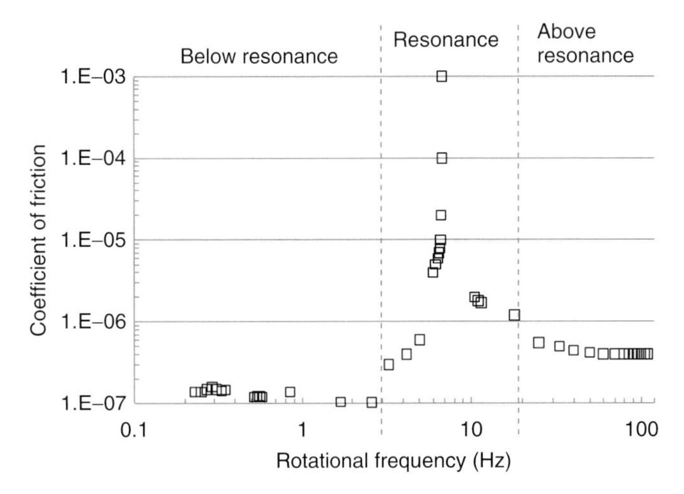

Figure 4.1.6 COF versus frequency for a 25.4 mm diameter, 9.6 mm high PM disc levitated 10 mm above a HTS cylinder.

where A is a geometric factor of order unity and ΔB is the peak-to-peak variation of the magnetic field.

The COF as a function of rotational speed for an experimental PM/HTS bearing is shown in Figure 4.1.6. The speed range is conveniently divided into three separate regions: the bearing resonance and above and below the resonance. At low speed, the bearing rotates about its center of magnetism. The losses in this range are primarily the result of azimuthal inhomogeneities in the magnetization of the PM disc, which produce hysteretic loss in the superconductor. Typically, for PMs with the best homogeneities, the amplitude of the AC component of the magnetic field at a fixed radius above the rotating surface is ~1% of the average field at that radius. Although small, this inhomogeneity is sufficient to cause a detectable decay in rotational rate when the magnet spins in vacuum. Because the loss is hysteretic, the drag force is constant and the COF is independent of frequency to first order.

At rotational rates above the bearing resonance range, the rotation is about the center of mass of the rotor, which is usually offset from the center of magnetism by an eccentricity distance. In this speed range, the AC component of the magnetic field above a position on the HTS is governed by the radial gradient of the PM magnetic field and the eccentricity. The hysteretic drag force is constant, and the COF is again independent of frequency but at a higher level than in the lower speed range.

The bearing resonance occurs when the rotational rate $\omega = 2\pi f$ is equal to the radial vibration frequency of the bearing, that is,

$$\omega = \left(\frac{K_x}{m}\right)^{\frac{1}{2}} \tag{4.1.5}$$

where m is the mass of the rotor and K_x is the radial stiffness. The resonance is characterized by a relatively large radial vibration amplitude, which depends on the eccentricity, and a corresponding large rotational drag.

The stiffness of an HTS bearing is typically smaller than that of other bearing types. A design consequence of the lower stiffness is that the running gaps of the bearing are correspondingly larger. An advantage of the low stiffness is that the resonant frequency is relatively low, and the centrifugal forces are correspondingly more modest when passing through the resonance.

4.1.3.2 High-Speed Loss

At sufficiently high rotational rates, the bearing loss transitions from hysteretic to that of flux-flow resistivity. The COF versus frequency of this effect is shown in Figure 4.1.7. The results are for an experimental rotor 15.4 mm in diameter and 150

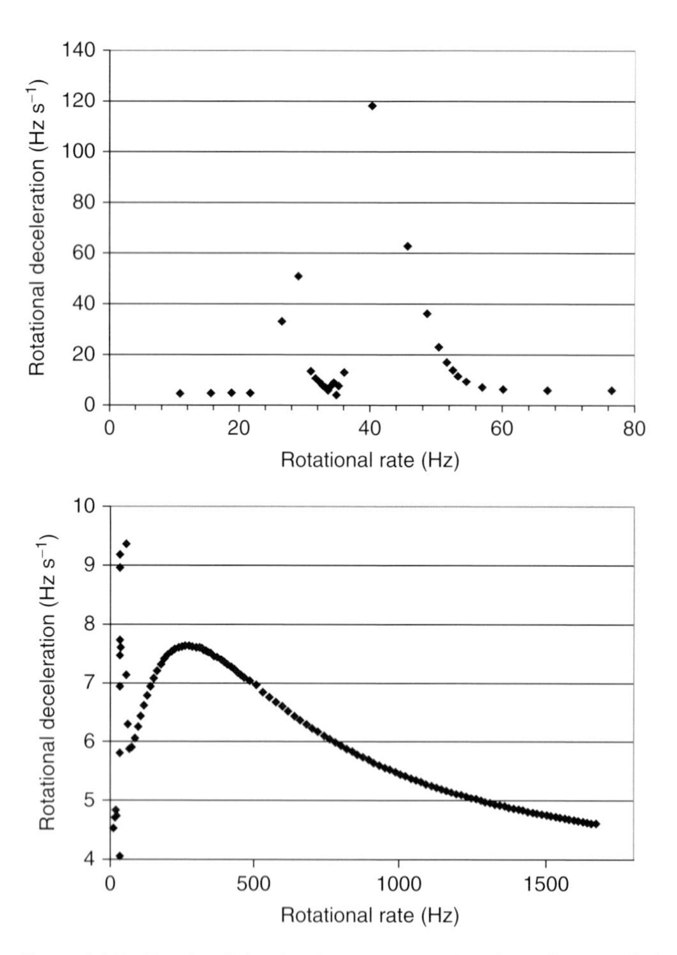

Figure 4.1.7 Rotational deceleration versus rotational rate for a small diameter, large aspect ratio rotor, showing a frequency-dependent drag force at high rotational rates.

mm long, with a PM/HTS bearing at each end. The low-speed data, shown in the upper graph, is characterized by two distinct resonances. This is typical for rotors with a transverse moment of inertia larger than the axial moment of inertia. The lower-speed resonance is a radial vibration mode, and the higher-speed resonance is for a conical vibration mode.

As seen in the lower graph in Figure 4.1.7, at rotational rates above 80 Hz, the loss is no longer independent of frequency but rather shows a drag peak at 274 Hz, after which the drag torque declines with frequency. As discussed in the analysis of this experiment [18], for $f > 160$ Hz, the rotational deceleration is well represented by the term $B + A(f/f_0)/[1 + (f/f_0)^2]$, where B and A are constants and f_0 is the frequency at the drag peak. The form of the frequency-dependent term is analogous for a magnet moving over an electrically conducting guideway. The physics responsible for the term may be interpreted as the circumferential inhomogeneity of the PM interacting with the flux-flow resistivity of the HTS. At higher speeds, the frequency-dependent term is governed by the skin depth and becomes proportional to $1/f^{1/2}$. The frequency-independent term represented by B was interpreted as hysteretic drag associated with the vibrations at the bearing resonance frequencies.

The inhomogeneity of the PMs used in this experiment was rather high. For the more homogeneous PM used in the experiment shown in Figure 4.1.6, the rotational rate at which this frequency-dependent term would show up would be significantly higher than that shown in Figure 4.1.7.

4.1.4 A Rotor Dynamic Issue

A rotor dynamic instability that has plagued some large rotors suspended by HTS bearings and kept these rotors from achieving their full design speed is a sub-synchronous resonance instability [19]. Internal damping, that is, the presence of damping on a rotating object, is a well-known source of sub-synchronous whirl instability in rotor dynamics. Normally, this instability is suppressed by applying damping to the bearing that supports the object. This is problematic for an HTS bearing, because of its inherent low damping, especially at small vibrational amplitudes.

The characteristic of this instability is that its onset appears at some rotational frequency above that of the bearing critical, and that it consists of a forward whirl, with the whirl frequency equal to that of the bearing critical frequency. Typically, the amplitude of this whirl grows with rotational frequency. The physical basis for the sub-synchronous whirl is "internal damping instability" in which hysteretic damping in the rotor material creates a force that adds energy to the forward whirl. On a fiber-composite rotor, internal damping is caused by hysteretic loss in the rotor material when there is a center of mass offset and creates a force that adds energy to the forward whirl. One method to reduce the tendency for sub-synchronous whirl would be to fabricate a rotor that has negligible mechanical imbalance. Unfortunately, this adds considerably to the manufacturing cost of the rotor. A second method would be to increase the size of the HTS bearing, but this would add cost and increased bearing loss.

A way to suppress the sub-synchronous whirl by the use of a secondary damper has recently been demonstrated [19]. The secondary damper mechanically connects the stator of the HTS bearing to the ground foundation. Inserting a secondary damper between the HTS stator and the foundation ground can dramatically increase the onset frequency of the sub-synchronous whirl, effectively eliminating it as a source of instability in large rotor systems. This has been shown both theoretically and experimentally. Theory predicts that the effect of the secondary damping in this regard is especially enhanced if the resonant frequency of the HTS stator with the ground is lower than that of the rotor with the HTS bearing.

4.1.5
Practical Bearing Considerations

In the most common configuration for a basic HTS bearing design, the PM is levitated and free to rotate while the HTS is held stationary. This is mostly due to the convenience of cooling the HTS in boiling liquid nitrogen or by a good thermally conductive contact with a cold source. If the bearing is to operate in vacuum, the only heat transfer mechanism for the levitated component is radiation, which is governed by Stefan's law, and losses are very low at cryogenic temperatures. Because the Curie temperature of commercially available PMs is significantly higher than the critical temperature of HTSs, the configuration in which the PM is levitated exhibits the greatest thermal stability. In situations where the bearing application allows gaseous cooling or circulation of a liquid cryogen, it may be preferable to hold the PM stationary and allow the HTS to levitate and rotate. In non-vacuum maglev applications, it may be possible to carry sufficient cold mass, or stored energy to operate a cryocooler, so that the HTS can be on the levitated component.

The two basic rotating bearing types are thrust and journal. In a thrust bearing, the major load force is in the direction of the rotational axis; in a journal bearing, the load force is perpendicular to the rotational axis. A typical PM/HTS journal bearing configuration could consist of a rotating shaft with PM discs mounted as endcaps, and the shaft assembly supported by HTS bushings that completely or partially surround the PMs at both ends of the shaft assembly.

One may classify a magnetic bearing either as a levitation bearing, which is expected to provide a mostly constant force, or as a stiffness bearing, which provides a stabilizing force against disturbances and for which the average net force is zero. Most practical magnetic bearings are often a combination of levitation and stiffness type. In a magnetic levitation bearing, the thrust bearing has a significant advantage over the journal bearing. To provide a levitation force, the magnetic pressure must be greater at the bottom of the levitated component than it is at the top (assuming that the levitation force is to overcome gravity). The magnetic pressure P is determined by the Maxwell stress tensor as

$$P = \frac{(B_t^2 - B_n^2)}{(2m_0)} \quad (4.1.6)$$

where B is the magnetic field along the plane on which the pressure is to be determined, subscript t denotes the component that is tangential to the plane, and subscript n denotes the component that is normal to the plane, and μ_0 is the permeability of free space. Therefore, a system with a levitation force must experience a different magnetic field at the top than it does at the bottom. In a journal bearing, the rotating component experiences the difference in magnetic field on every part of the perimeter during each rotation; this produces eddy currents and hysteretic losses. In a levitational thrust bearing, the magnetic field remains the same on the top and bottom horizontal surfaces during the rotation, unless there is some azimuthal inhomogeneity in the PM or HTS. The use of nonconducting ferrite PMs would eliminate the problem of eddy currents in journal bearings, but the available magnetization of these materials is significantly lower than that of NdFeB or SmCo, and the levitation pressures obtainable are thus much lower. In a pure stiffness bearing, there is usually no advantage of a thrust over a journal type.

In practical HTS bearings, the low levitational pressure available in the interaction between the PM and the HTS is often augmented by various hybrid schemes in which interactions between pairs of PMs provide the bulk of the levitational force. These methods are sometimes termed *magnetic biasing*. The PM/PM interactions are statically unstable, as Earnshaw's theorem predicts, but the inclusion of a properly designed HTS component in the bearing is sufficient to stabilize the complete bearing. Figure 4.1.8 shows augmentation in the form of an Evershed-type design. Here, the levitated rotor consists of a levitation PM, below which hangs a rigid rod with a smaller stabilizing PM on the bottom. The levitation PM of the rotor experiences an attractive force toward the stator PM immediately above it. This system is statically stable in the radial direction but unstable in the vertical direction. The gap between the levitation PM and the stator PM is adjusted in such a manner that the attractive force between the pair of magnets is just less than 100% of the rotor weight, so that the rotor would tend to fall. The remainder of the weight is provided by the interaction between the stabilizing PM and the HTS; this interaction also supplies sufficient stiffness for vertical stability and some additional radial stability.

As opposed to using a single HTS bearing, the use of the lift bearing reduces the cost of the bearing and reduces the losses in the system by minimizing the amount of HTS required. Larger levitation pressures are possible with PM/PM interaction compared to PM/HTS. Higher pressure in PM/HTS can be obtained by reducing the gap, but this increases the loss. Eddy current losses between rotating PM pairs are small and are removed at room temperature. Hysteretic loss in the HTS is removed at cryogenic temperature and carries multiplicative penalty from Carnot efficiency plus cryocooler efficiency.

Cooling of the HTS component has evolved considerably since the early levitation experiments in which the HTS was bathed in liquid nitrogen in an open container. Today, systems similar to the bearing demonstration device shown in

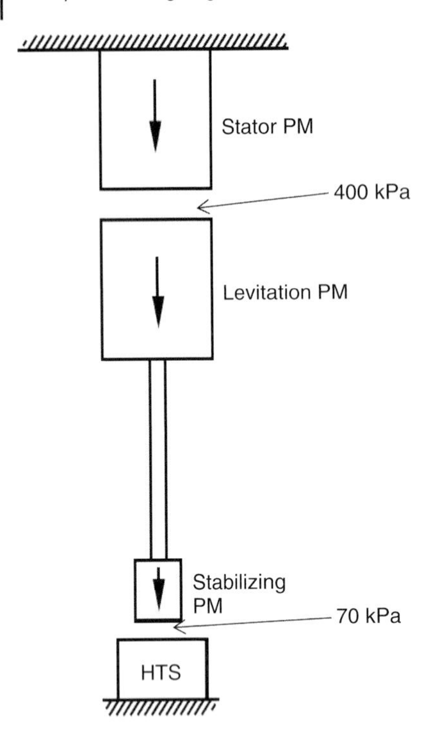

Figure 4.1.8 Lift bearing concept, where PM/PM interaction provides most of the levitational force and PM/HTS interaction provides stability.

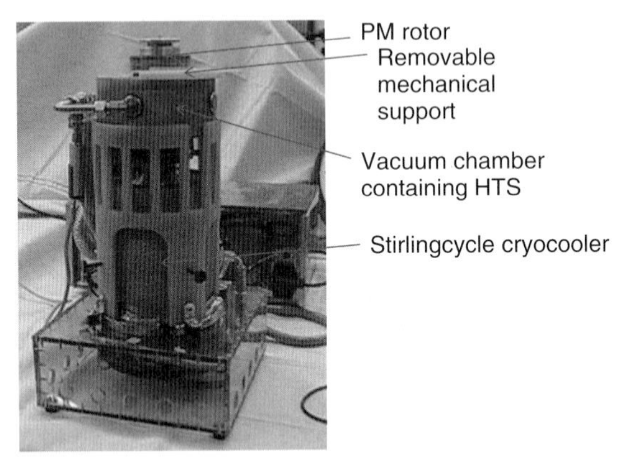

Figure 4.1.9 HTS bearing demonstration model, using cryocooler and conduction cooling to HTS. Shown in the background are a small vacuum pump and power and control for the cryocooler.

Figure 4.1.9 are common. In Figure 4.1.9, several HTS elements are located in a circle inside a G-10 chamber that is kept under vacuum. Inside, the HTS elements are typically covered by several layers of reflective film that reduces radiative heat transport. The HTS elements are thermally connected to the cold head of a cryocooler by thermal conduction through copper. Even though the HTS may be at 40 K, the outside of the vacuum chamber is warm to the touch and does not gather water condensate from the atmosphere.

4.1.6 Applications

HTS bearings have been used and proposed for a number of applications as discussed in the review articles [1–10]. Here, we briefly discuss several selected applications.

The availability of HTS bearings that are so nearly friction free naturally leads to their consideration for flywheel energy storage, and several major projects have investigated this application [10]. Figure 4.1.10 shows the design and pictures of selected components for a 5 kWh, 3 kW flywheel that has been tested as a laboratory prototype at the Boeing Company. The flywheel rotor is suspended totally passively by a PM lift bearing and a lower axial HTS bearing [20]. The lift bearing is a push–pull arrangement that results in a near net zero stiffness and is arranged to be radially stable. The HTS stator consists of yttrrium-barium-copper-oxide bulks in the hexagonal geometry, arranged in a ring and conduction-cooled by a cryocooler. The cryocooler shown for the small bearing demonstration device in Figure 4.1.9 is identical to the cryocooler used for the flywheel apparatus.

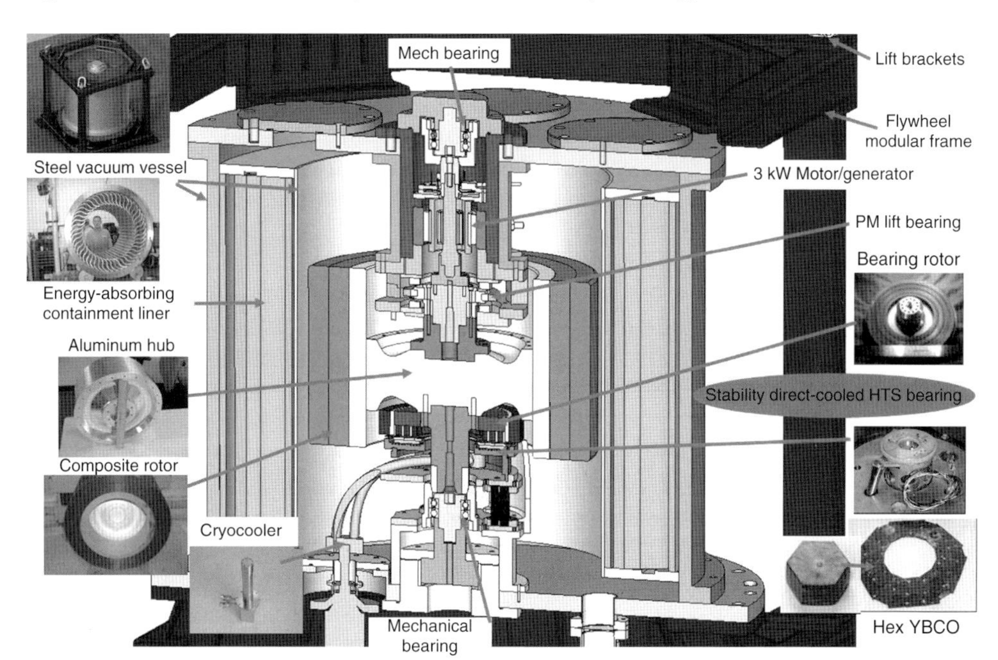

Figure 4.1.10 Design schematic of flywheel energy storage system supported by HTS bearing, showing critical components.

HTS bearings have been used in a sensitive detector to study the polarization of the cosmic background radiation (CMB) [21]. MAXIPOL is a bolometric balloon-borne experiment designed to measure the E-mode polarization of the CMB. The detector uses a rotating half plate that modulates the CMB from a given patch of the sky passing into a polarization analyzer. The challenge in adapting the technique to CMB polarimetry is coupling the rotation mechanism to bolometric detectors, which have high sensitivity to microphonic noise. Standard mechanical rotation mechanisms, such as gears and bearings, have stick–slip friction, which induces vibrations. The vibrations inject mechanical energy in the detectors and cause wires and other components to vibrate, thereby inducing excess electrical noise. The low rotational drag and vibration isolation characteristics of the HTS bearing make it suitable for providing a low-noise environment. Cryogenic cooling is already available in the CMB experiment to detect the long wavelength radiation. The ability to operate in a balloon in the occasionally tempestuous environment at altitudes greater than over 30 km is a testament to the ruggedness of HTS bearings.

Whereas the low damping of the HTS levitation was an obstacle for rotating heavy rotors to high speed because of the sub-synchronous resonance instability, the same low damping is a boon for the scientific and engineering study of non-linear dynamics. This development has been mainly conducted by T. Sugiura and his students [22–25].

References

1. Brandt, E.H. (1990) Rigid levitation and suspension of high-temperature superconductors by magnets. *Am. J. Phys.*, **58**, 43–49.
2. Moon, F.C. (1994) *Superconducting Levitation*, John Wiley & Sons, Inc., New York.
3. Hull, J. (2000) Superconducting bearings. *Supercond. Sci. Technol.*, **13**, R1–R14.
4. Hull, J.R. (2001) in *Engineering Superconductivity* (ed. P.J. Lee), John Wiley & Sons, Inc., New York, pp. 563–568.
5. Hull, J.R. (2002) Levitation, in *Handbook of Superconducting Materials*, Chapter E2.1 (eds D. Cardwell and D. Ginley), Institute of Physics Publishing, Bristol.
6. Hull, J.R. (2002) Superconducting bearings, in *Properties, Processing, and Applications of YBCO and Related Materials*, Chapter 9.4 (eds W. Lo and A.M. Campbell), IEE Books, Stevenage.
7. Ma, K.B., Postrekhin, Y.V., and Chu, W.K. (2003) Superconductor and magnet levitation devices. *Rev. Sci. Instrum.*, **74**, 4989–5017.
8. Hull, J.R. (2004) in *High Temperature Superconductivity 2: Engineering Applications* (ed. A.V. Narlikar), Springer, Berlin, pp. 91–142.
9. Hull, J.R. and Murakami, M. (2004) Applications of bulk high-temperature superconductors. *Proc. IEEE*, **92**, 1705–1718.
10. Werfel, F.N., Floegel-Delor, U., Rothfeld, R., Riedel, T., Goebel, B., Wippich, D., and Schirrmeister, P. (2012) Superconductor bearings, flywheels and transportation. *Supercond. Sci. Technol.*, **25**, 014007.
11. Earnshaw, S. (1842) On the nature of the molecular forces which regulate the constitution of the luminferous ether. *Trans. Camb. Philos. Soc.*, 7, 97–112.
12. Hull, J.R. and Cansiz, A. (1999) Vertical and lateral forces between a permanent magnet and a high-temperature superconductor. *J. Appl. Phys.*, **86**, 6396–6404.
13. Bean, C.P. (1962) Magnetization of hard superconductors. *Phys. Rev. Lett.*, **8**, 250–253.

14. Bean, C.P. (1964) Magnetization of high-field superconductors. *Rev. Mod. Phys.*, **36**, 31–39.
15. Hong, Z., Vanderbemden, P., Pei, R., Jiang, Y., Campbell, A.M., and Coombs, T.A. (2008) The numerical modeling and measurement of demagnetization effect in bulk YBCO superconductors subjected to transverse field. *IEEE Trans. Appl. Supercond.*, **18**, 1561–1564.
16. Goncalves, G.G., Dias, D.H.N., de Andrade, R. Jr., Stephan, R.M., Del-Valle, N., Sanchez, A., Navau, C., and Chen, D.-X. (2011) Experimental and theoretical levitation forces in a superconducting bearing for a real-scale maglev system. *IEEE Trans. Appl. Supercond.*, **21**, 3532–3540.
17. Hull, J.R., Passmore, J.L., Mulcahy, T.M., and Rossing, T.D. (1994) Stable levitation of steel rotors using permanent magnets and high-temperature superconductors. *J. Appl. Phys.*, **76**, 577–580.
18. Hull, J., Strasik, M., Mittleider, J., Gonder, J., Johnson, P., McCrary, K., and McIver, C. (2009) High rotational-rate rotors with high-temperature superconducting bearings. *IEEE Trans. Appl. Supercond.*, **19**, 2078–2082.
19. Hull, J., Strasik, M., Mittleider, J., McIver, C., McCrary, K., Gonder, J., and Johnson, P. (2011) Damping of subsynchronous whirl in rotors with high-temperature superconducting bearings. *IEEE Trans. Appl. Supercond.*, **21**, 1453–1459.
20. Strasik, M., Hull, J.R., Mittleider, J.A., Gonder, J.F., Johnson, P.E., McCrary, K.E., and McIver, C.R. (2010) Overview of Boeing flywheel energy-storage systems with high-temperature superconducting bearings. *Supercond. Sci. Technol.*, **23**, 034021.
21. Johnson, B.R., Collins, J., Abroe, M.E., Ade, P.A.R., Bock, J., Borrill, J., Boscaleri, A., de Bernardis, P., Hanany, S., Jaffe, A.H., Jones, T., Lee, A.T., Levinson, L., Matsumura, T., Rabii, B., Renbarger, T., Richards, P.L., Smoot, G.F., Stompor, R., Tran, H.T., Winant, C.D., Wu, J.H.P., and Zuntz, J. (2007) MAXIPOL: cosmic microwave background polarimetry using a rotating half-wave plate. *Astrophys. J.*, **665**, 42–54.
22. Takazakura, T., Sakaguchi, R., and Sugiura, T. (2013) Experimental study of suppressing a parametric resonance in a HTSC levitation system by a horizontal pendulum. *IEEE Trans. Appl. Supercond.*, **23**, 3600204.
23. Sasaki, M., Takabayashi, T., and Sugiura, T. (2013) Transition between nonlinear oscillations of an elastic body levitated above high-Tc superconducting bulks. *IEEE Trans. Appl. Supercond.*, **23**, 3600604.
24. Amano, R., Kamada, S., and Sugiura, T. (2013) Dynamics of a flexible rotor with circumferentially non-uniform magnetization supported by a superconducting magnetic bearing. *IEEE Trans. Appl. Supercond.*, **23**, 5202104.
25. Yubisui, Y., Kobayashi, S., Amano, R., and Sugiura, T. (2013) Effects of nonlinearity of magnetic force on passing through a critical speed of a rotor with a superconducting bearing. *IEEE Trans. Appl. Supercond.*, **23**, 5202205.

4.2 Fundamentals of Superconducting Magnets

Martin N. Wilson

4.2.1 Windings to Produce Different Field Shapes

In conventional electromagnets, the field shape is determined almost entirely by the profile of the iron yoke. But superconducting magnets are able to produce fields much higher than the saturation of iron, indeed this is their main attraction. Iron is often used in superconducting magnets, for example, to screen the fringe field, but its ability to shape the field is limited because much of it is saturated. Thus, the fields in superconducting magnets are determined mainly by the winding shape, and different shaped windings are needed for different applications.

The simplest coil shape used in most research magnets, NMR spectrometers, MRI scanners, magnetic separators, and so on, is the solenoid. Figure 4.2.1 sketches some typical examples of solenoid windings.

For an infinitely long solenoid winding, the interior field is perfectly uniform and given by:

$$B = \mu_o J_e t \tag{4.2.1}$$

where J_e is the engineering current density, defined as the current crossing unit cross-section of winding, that is, taking account of dilution by added copper, insulation, structural members, and so on. In shorter solenoids, the field is reduced by a factor depending mainly on the ratio of length to inner radius. It is nonuniform, falling as one moves axially away from the center, and rising as one moves radially toward the winding. Split pair coils, as shown in Figure 4.2.1b are used in situations where access is needed along a line perpendicular to the field, and Figure 4.2.1c shows how thicker regions at the ends may be used to produce a more uniform field profile.

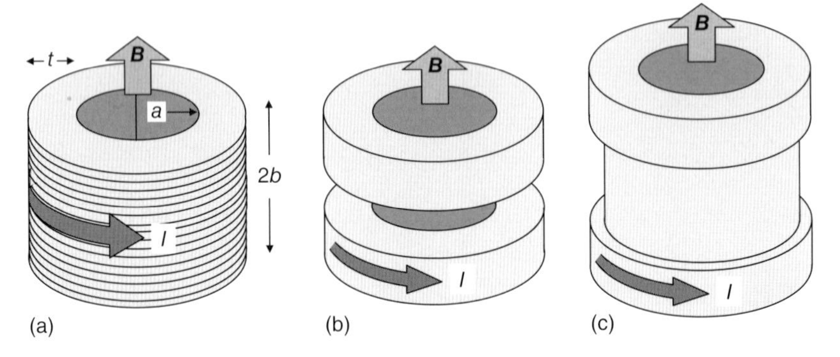

Figure 4.2.1 Solenoid windings, (a) simple solenoid, (b) split pair for transverse access, and (c) with end cheeks to improve field uniformity.

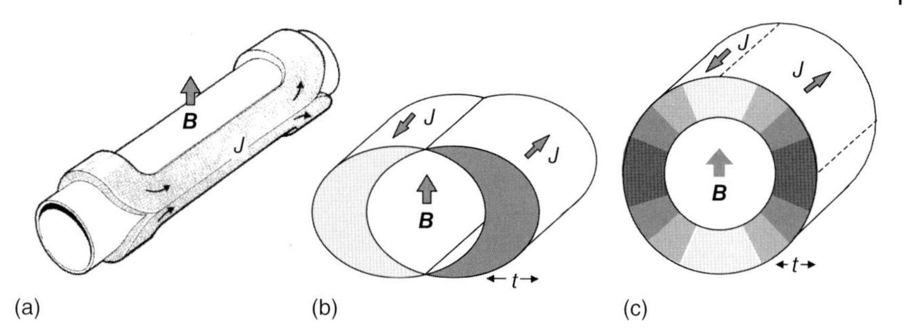

Figure 4.2.2 (a) Saddle winding to produce transverse field, (b) overlapping ellipse cross-section, or (c) current density varying as cos(θ) produce a perfect dipole field.

Accelerator magnets must produce field perpendicular to the motion of the particles, which is usually the long dimension of the magnet, and this demands windings of a "saddle" shape, as sketched in Figure 4.2.2. Winding cross-section must be shaped appropriately to produce the desired field shape, the simplest of which is a uniform or *dipole* field. Figure 4.2.2b,c shows the two simple sections which may be shown analytically to produce a uniform dipole field: (b) overlapping ellipses and (c) an engineering current density which varies as $J_e(\theta) = J_{eo}\cos(\theta)$. All practical windings are derived from these cross-sections, optimized within practical constraints by numerical optimization techniques. The cos(θ) variation of overall current density is obtained by putting spacers into a winding of constant current density. Gradient and higher-order field variations, that is, quadrupole, sextupole, and so on, may be produced by cos(2θ), cos(3θ) variation of the current density.

Finally, Figure 4.2.3 shows a group of circular coils arranged in a toroidal configuration. The key feature of this arrangement is that the field lines close on themselves within the magnet. Thus, it is used in thermonuclear confinement to prevent the hot plasma escaping and also in energy storage magnets to avoid fringe fields.

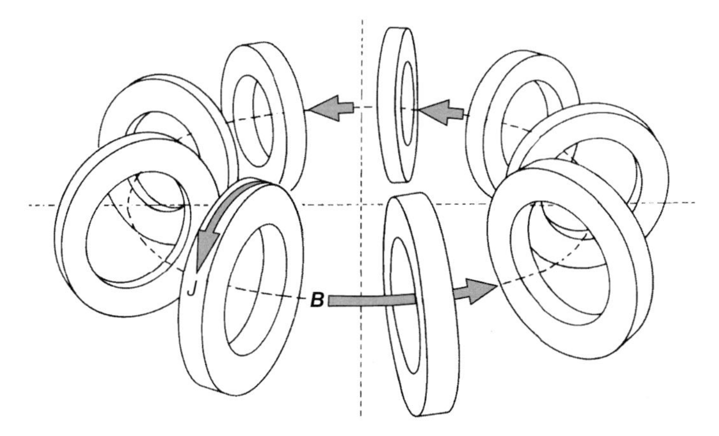

Figure 4.2.3 Toroidal coils, producing a field which closes on itself within the magnet.

4.2.2
Current Supply

Magnets sit at low temperature and power supplies sit at room temperature. Joining the two together introduces a substantial heat leak, often the biggest refrigeration load and a significant cost, particularly when operating at liquid helium temperature. For a given current taken in (and out), we want to minimize the heat input to low temperature. This heat comes from two sources: conduction down from room temperature and ohmic generation along the length of the lead. So what we need is a material with low electrical resistivity and low thermal conductivity. Unfortunately, the conductivity in metals is subject to the Wiedemann–Franz law, which states that:

$$k(\theta)\rho(\theta) = L_o\theta \tag{4.2.2}$$

where L_o is the Lorentz number $= \pi^2 k_B^2/3e^2 = 2.45\times 10^{-8}\ \mathrm{W\,\Omega\,K^{-1}}$ and k_B is Boltzmann's constant. Although Wiedemann–Franz is only accurate for pure metals, the deviations found in alloys are generally in the wrong direction for our purpose. Thus, all metals are similar and our only design variable is the shape of lead: short and fat gives low ohmic power but high thermal conduction, whereas long and thin gives the reverse. It turns out that the optimum shape comes when there is zero temperature gradient at the top of the lead, that is, the only thermal conduction is from ohmic heat generated – nothing comes in from room temperature. For a lead obeying Wiedemann–Franz and connected only to a top temperature θ_2 and bottom temperature θ_1, it may be shown [1] that the optimum (minimum watts per ampere) heat leak is given by:

$$\frac{\dot{Q}_1}{I} = \sqrt{L_o\left(\theta_2^2 - \theta_1^2\right)} \tag{4.2.3}$$

With top and bottom temperatures of 300 and 4 K, respectively, this gives a heat leak per lead of $47\ \mathrm{mW\,A^{-1}}$. Note that the heat leak is linearly related to current – twice the current needs a lead twice as thick which has twice the heat leak. The latent heat of boiling of liquid helium is only $20.7\ \mathrm{kJ\,kg^{-1}}$ or $2.56\ \mathrm{kJ\,l^{-1}}$, so that a pair of leads carrying 100 A will boil off 13 liquid liters per hour – an unacceptably large expense.

The latent heat of helium is exceptionally low, but there is a lot of cold left in the gas boiled off; the enthalpy change $\Delta H = \int C_p(\theta)\,\mathrm{d}\theta$ between 4 and 300 K is $1543\ \mathrm{kJ\,kg^{-1}}$, that is, 74 times as much as the latent heat. So for leads going down into liquid helium, a substantial reduction in heat leak may be achieved by making the lead in the form of a heat exchanger to utilize the cold in the gas which is boiled off. For any lead obeying the Wiedemann–Franz law with a 100% efficient heat transfer between the lead and the upstreaming helium gas, it may be shown [2] that the optimum heat leak is reduced to $1.04\ \mathrm{mW\,A^{-1}}$.

Although the number for optimum heat leak per ampere is universal for all metals, the actual shape of lead depends on the material used, for example, impure copper needs a thicker lead than pure copper. The actual shape must be found by numerical integration of the thermal conductivity with respect to temperature [1, 2]. For example, the ratio of length l to cross-sectional area A of an optimum well-cooled lead in pure copper is given by $l/A = 2.6 \times 10^7/I$, whereas an impure phosphorous deoxidized copper lead has $l/A = 3.5 \times 10^6/I$.

While different purity materials have the same optimum heat leak, there are distinct advantages in using impure materials because they are more stable. At low temperature, the resistivity of very pure metal rises rapidly with temperature, so that any increase in temperature causes higher resistivity and therefore more heating – positive feedback. The resistivity of impure metals varies much less with temperature, so they are more stable. For example, a 10% current excess in a pure copper lead pushes the peak temperature from 300 to 700 K, whereas an impure lead rises to only 310 K.

A useful property of leads obeying the Wiedemann–Franz law is that the voltage drop across any lead at optimum is the same regardless of current. For well-cooled leads going from room temperature to liquid helium, it is 75 mV. This universal number offers a quick and easy way of finding whether the lead is optimum – measuring cryogenic heat leaks via helium boil-off is a notoriously slow and laborious process. It may also be used for safety monitoring: measure the lead voltage and switch off the current if it indicates dangerously high temperature.

High-temperature superconductors (HTS) offer a way around the Wiedemann–Franz law; they have low thermal conductivity, but of course they only have zero resistivity up to their critical temperature. So it is necessary to make a two-part lead, with copper at the top and superconductor below. A common choice of intermediate temperature is the boiling point of liquid nitrogen 77 K. Liquid nitrogen has a higher latent heat than helium and the gas has a lower specific heat, so that the cold in the boiled-off gas is only about the same as the latent heat and it is usually not worth making a heat-exchanging lead. For an uncooled lead equation 4.2.3 shows that the optimum heat leak is 45 mW A^{-1} – not much less than that of helium, but the cost of liquid nitrogen or of 77 K refrigeration is very much less than that of helium. Conduction along the HTS lead down to liquid helium temperature is very small and it is this technology that has made possible the advent of "cryofree" magnet systems cooled by a small cryocooler and requiring no liquid helium.

A warning about the danger of HTS leads under fault conditions; if the temperature of the intermediate junction rises, for example, due to a cryocooler malfunction, the HTS section will become resistive and will start to heat up. Unless the magnet current is reduced in a timely fashion, the lead will overheat and may burn out. We will return to this question in Section 4.2.10.

A unique characteristic of superconducting magnets is that, if their terminals are connected together by a superconducting short circuit, the current will continue to circulate forever. Persistent current operation is routine in magnet systems for NMR spectroscopy, MRI, and general research. Using superconducting joints, the magnet terminals are connected by a persistent current switch which consists of a length of superconducting wire and a heater enclosed in a thermally insulating case. For charging the magnet, the heater is energized so that the wire is resistive and current from the power supply flows into the magnet rather than the switch. When the desired current is reached, the heater is switched off, the switch becomes superconducting, current from the external power supply is reduced to zero, and the magnet current flows instead through the switch. The magnet current then continues to circulate, in principle forever but in practice, because some small resistance remains, with a time constant of centuries. To reduce the cryogenic heat load, the current leads may then be disconnected and removed completely – they will not be needed for a very long time.

4.2.3
Load Lines, Degradation, and Training

A useful aid in designing magnets is to take the usual critical current plot shown in Figure 4.2.4 and add to it the *load line* which relates the highest field produced by the magnet to the current flowing in its windings. Unfortunately, for magnet makers, this peak field always occurs at some point on the superconducting winding. The load line will be straight for an air-cored magnet and curved for an iron-cored magnet, but the argument remains the same. As the current in the winding is increased, the peak field increases along the load line. Thus we expect that the superconductor will revert to its resistive state at the intersection point IP of load line and critical current line, that is, 6 T and 500 A, respectively, for the magnet shown in Figure 4.2.4. When this happens, there is intense ohmic heat generation, the current falls abruptly, and the magnet is said to *quench.*

High-field solenoids are often divided into concentric sections where the critical current is adjusted to match the local peak field seen by that section. The outer low-field sections are wound from thinner wires (with lower critical current) and the inner high-field sections will use thicker wires (with higher critical current), so

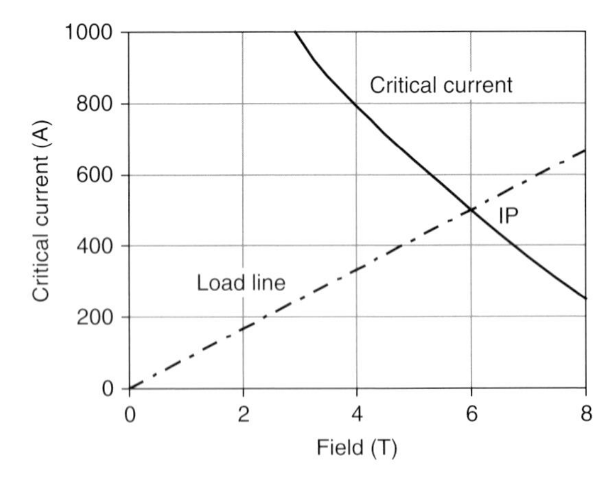

Figure 4.2.4 Critical current for superconductor and load line for the magnet.

the magnet has several local load lines and critical current lines. All sections are connected in series and, with suitable design, it can be arranged that all the local load lines intersect their local critical current lines at the same operating current. In this way, considerable savings in materials cost may be achieved because the superconductor everywhere is being used to its maximum capability.

Magnets made from low-temperature superconductors (LTS), rarely go straight to IP and, in the early days of the technology, their performance fell far short of it. Furthermore, the magnets often exhibit *training*, whereby their performance gets closer to IP with further attempts to increase the current after quenching. The effect becomes more serious with increasing size and it was a serious obstacle to the early utilization of superconducting magnets. Although significant improvements have been made in recent years, degradation and training still present problems for LTS, particularly in large systems and when the performance is pushed toward the highest engineering current density. It seems likely that HTS magnets will not suffer from degradation because specific heats and temperature margins are much greater at higher operating temperatures, but as yet we have little experience of HTS in large magnets. Cures for degradation and training go under the general name of *stabilization*.

4.2.4 Cryogenic Stabilization

It soon became apparent that the performance of magnets was being degraded by energetic disturbances within their windings which were causing local temperature rises as the current, field (and consequent stresses) were increased. The first cure was cryogenic stabilization [3], and it is still the most reliable approach for very large systems. Here, the magnet conductor is a composite bar of superconducting wire embedded in a good normal conductor, usually copper. Figure 4.2.5 shows, as a function of temperature, the ohmic heat generation in such a conductor when it is carrying a constant current I_m. Starting at the base temperature θ_o, there is no heat generation until the temperature reaches θ_g, when the critical

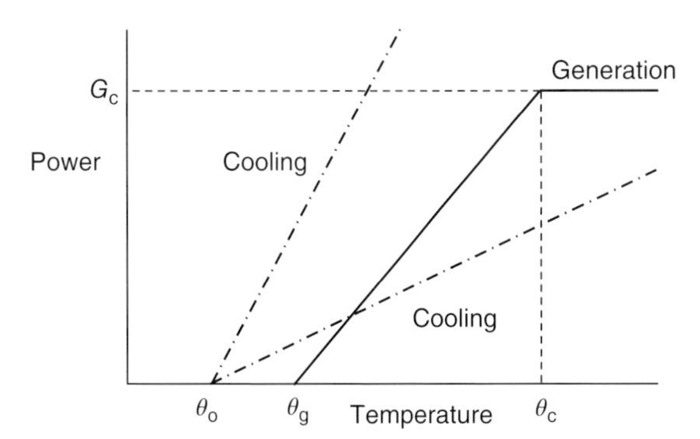

Figure 4.2.5 Power generation in a composite conductor of superconductor and copper, plus lines of cooling to the surroundings.

current I_c of the superconductor equals I_m. Beyond this point, further increases in temperature reduce I_c below I_m and the difference $I_m - I_c$ flows in the copper, where it produces an electric field and generates power. Because the copper and superconductor are connected in parallel, the electric field also exists in the superconductor, which goes into the flux flow state and also generates power. For most superconductors, I_c falls roughly linearly with fall in temperature, so the electric field and hence the power generation rise linearly with temperature until critical temperature θ_c is reached, when all current is in the copper and the generation becomes constant at G_c. If the fraction of superconductor in the conductor cross-section is λ, the resistivity of the copper is ρ, and the magnet current density per unit area of superconductor is J_m, the critical generation per unit volume of conductor is:

$$G_c = \lambda^2 J_m^2 \frac{\rho}{1 - \lambda} \tag{4.2.4}$$

Assuming a linear drop in critical current density from J_{co} at θ_o to zero at θ_c, we may write:

$$\theta_g = \theta_c - (\theta_c - \theta_o)\frac{J_m}{J_{co}} \tag{4.2.5}$$

Also shown in Figure 4.2.5 are two lines representing cooling to the surroundings. If the cooling is good, corresponding to the upper line, it will be greater than the generation at any temperature. So the conductor will always recover from any disturbance which increases its temperature and will return to stable operation at θ_o. When the cooling is not so good, however, as shown by the lower line, there is a region where the heating exceeds the cooling and the conductor will not recover from any disturbance which takes it into this region. Thus, our simplest criterion for stability against disturbances is that the cooling should exceed the generation at all temperatures. If this criterion is obeyed, the conductor can recover from arbitrarily large disturbances.

A common method of cooling is to make channels throughout the winding containing boiling liquid helium at temperature θ_o in direct contact with the copper of the conductor. Boiling heat transfer has the characteristic shape shown in Figure 4.2.6, with a very high heat flux at low-temperature differences in the nucleate boiling region, followed by a lower heat flux at higher temperatures in film boiling, where the surface is effectively insulated by a film of vapor. A further complication is that the energy disturbance in the winding is usually of finite size, so that the resulting resistive zone has a finite length with cold ends. Both these complications can be handled rather elegantly by the equal area theorem [4] which says that the resistive zone will shrink from the cold end if the total area enclosed by the cooling curve exceeds that enclosed by the generation curve. In Figure 4.2.6, the balance point for stability is thus that the two shaded areas are equal. Pictorially, we may think in terms of the excess heating at high-temperature draining along the conductor to low temperature where there is excess cooling.

Providing a system of cooling channels throughout the magnet winding raises many practical problems; for example, the need for liquid to contact the bare

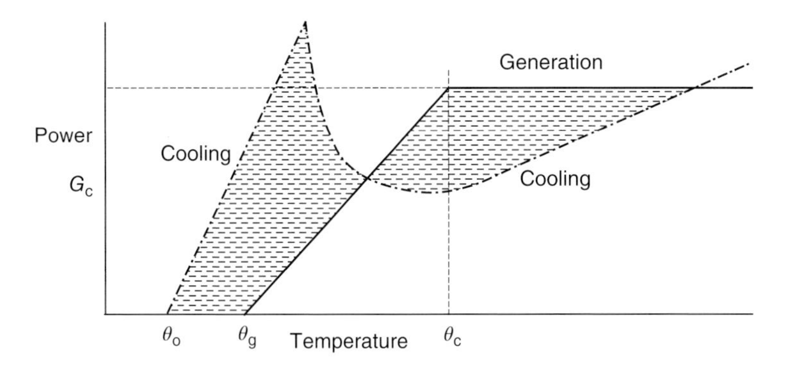

Figure 4.2.6 The equal area criterion for cold end recovery.

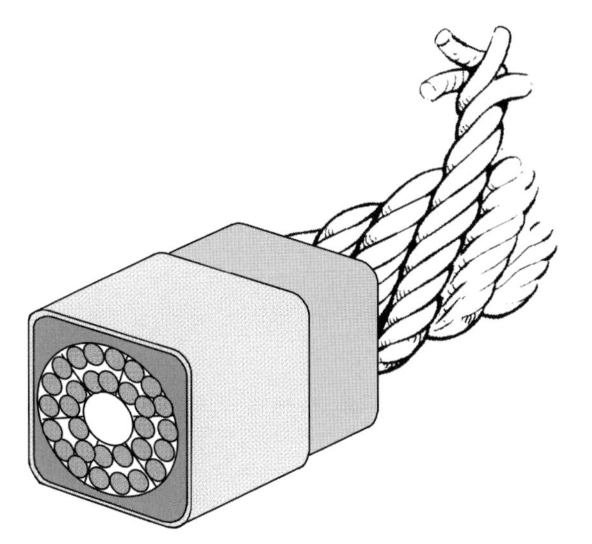

Figure 4.2.7 Cable in conduit conductor CICC.

metal makes the winding vulnerable to shorted turns, and the open structure is mechanically soft against the large electromagnetic forces which occur in high-field magnets. For reasons of this kind, most cryogenically stabilized magnets are now made using cable in conduit conductor CICC. As sketched in Figure 4.2.7, a cable of composite copper and superconducting wires is enclosed within a conduit and helium coolant is pumped through the conduit, flowing in the interstices of the cable. Electrical insulation outside the conduit may be continuous and any residual space between turns may be filled with epoxy resin to produce a very rigid mechanical structure. CICC is now the conductor of choice for large fusion magnets and Section 4.6 discusses their stability in some depth.

Cryogenic stabilization provides a total cure to the problem of degradation and training, enabling large systems to be constructed in the safe and certain knowledge that they will work as designed. It does however have one drawback: the large proportion of winding space occupied by copper and cooling channels, which greatly dilutes the engineering current density. This dilution is too much for many important applications, including MRI and particle accelerators. For these applications to be viable, the superconductor must work reliably at high engineering current density without the dilution of cooling channels or large quantities of copper. In order to achieve this goal, it is necessary to think in a bit more depth about the processes involved.

4.2.5 Mechanical Disturbances and Minimum Quench Energy

It seems that degradation and training are caused by two kinds of energy disturbances occurring in the winding as current and field are ramped up: magnetic instabilities and mechanical movements. Magnetic instabilities, usually known as *flux jumping,* are dealt with in Section 4.2.6 and are a solved problem. Mechanical movements remain a problem to this day, although techniques have been developed to mitigate their worst effects. The root cause of the problem is the very low specific heat of all materials at low temperatures – typically 1/2000 of room temperature. This means that an energy input which at room temperature would raise the temperature by only 1 mK will produce a 2 K rise at low temperature.

Superconducting magnets are not precisely machined structures, they are made by winding a length of wire onto a former – and they do not fit together perfectly. Imagine a region of winding in a field B running at an engineering current density J_e and let it suddenly slip a distance δ under the influence of the electromagnetic force. The work done by the magnetic force per unit volume of conductor is:

$$W = BJ_e\delta \tag{4.2.6}$$

For example, in a field of 5 T with an engineering current density of 500 A mm^{-2}, a slip of only 5 μm does work of 12.5 kJ m^{-3}. If this energy is released as frictional heating, it will heat a typical winding from 4.2 to 6.4 K, probably taking the superconductor into the resistive region and causing the magnet to quench. On a second attempt to energize the magnet, this section of conductor, having already moved to a stable position, may not move again and a higher current may be reached, that is, training. However, it is quite likely that another region of conductor will move and precipitate a quench at current not much higher.

Clearly, it is impossible to wind a superconducting coil such that no turn at any place can move by a few microns and the obvious way to avoid the problem is to impregnate the winding with a material such as epoxy resin which will fill every tiny space and which can then be cured to form a rigid mass. This approach can indeed prevent movement, but unfortunately brings with it a new problem. All the suitable impregnating materials which are electrically insulating are organic compounds with thermal contractions much greater than metals. As a result, when the impregnated winding is cooled down, the metal goes into compression and the resin into tension. Unfortunately, it is also a feature of these compounds that, below their characteristic glass transition temperature, they become brittle. Thus, we have a brittle material under tension. When additional stress is applied by the electromagnetic forces, it is very likely that brittle fracture will occur. When this happens, a large fraction of the strain energy stored in the resin will be released as heat. Because metals are much stiffer than resins, we may take the tensile strain induced in the resin by cooldown to be $\varepsilon = \alpha_r - \alpha_m$, where α_r and α_m are the thermal contractions of resin and metal between room temperature and the magnet operating temperature, respectively. For uniaxial strain, that is, the materials only attached along one axis, the resulting strain energy is:

$$E_1 = \frac{\sigma^2}{2Y} = \frac{Y\varepsilon^2}{2} \tag{4.2.7}$$

where Y is Young's modulus of the resin. Typical numbers for epoxy resin glued to copper give $E_1 = 2.5 \times 10^5\ \mathrm{J\,m^{-3}}$. If all this energy were released as heating during a crack, the temperature of a typical winding would rise from 4.2 to 16 K. For triaxial strain, where the materials are glued together in all directions, the strain energy is even higher:

$$E_3 = \frac{3\sigma^2(1-2\nu)}{2Y} = \frac{3Y\varepsilon^2}{2(1-2\nu)} \tag{4.2.8}$$

where ν is Poisson's ratio. Typical numbers give $2.3 \times 10^6\ \mathrm{J\,m^{-3}}$ and a temperature rise from 4.2 to 25 K.

It is clear that, even if only a fraction of the strain energy finds its way into the conductor after a crack, the magnet will quench, and these simple numbers explain the disappointing performance of early impregnated magnets. Over the years, there has been some progress in using crack resistant resins and other polymers and in reducing their thermal contraction by filling them with inorganic materials such as glass fiber or alumina powder. But the problem remains to this day and most magnets running at high current density still train to a greater or lesser degree.

Magnet construction techniques to minimize degradation and training in high current density windings may be divided into two categories. Solenoids are usually vacuum-impregnated with a crack resistant epoxy resin, but the volume of resin is kept to the absolute minimum. Where possible, gaps in the winding, for example, where the wire rises from one layer to the next, are filled during winding with a low contraction material such as filled epoxy. Larger conductors should be insulated by a glass fiber braid or wrap. Above all, the most important thing is to avoid any significant region of clear resin which, if it cracks, will release sufficient energy to trigger a quench. Accelerator magnet technology has gone one stage further by eliminating clear epoxy resin completely. Any chance of movement is reduced by precompressing the winding a level higher than the maximum anticipated electromagnetic stress. Insulation is usually by a Kapton wrap around the cable and the remaining spaces are filled by a thin (5 μm) layer of polyimide adhesive on the tape; this layer is on the outside face of the wrap, away from the conductor surface. Spacers needed to produce the desired winding shape are made from metal or glass fiber reinforced epoxy.

While careful attention to mechanical design of the coil is very important, the response of the conductor to energy inputs also has an effect. Without cooling channels, if the energy disturbance afflicts a large volume, it will trigger a quench whenever the temperature rises above θ_g in Figure 4.2.5. For disturbances afflicting a small volume, and there are reasons to believe that this is usually the case, the resulting resistive zone is sufficiently small for there to be significant heat conduction out of it, both along the wire and transverse to it through the insulation. We define a minimum propagating zone MPZ as a resistive zone which is just large enough for the heat generation within it to exceed the heat conduction out of it. So resistive zones greater than the MPZ will grow without limit and quench the magnet, but smaller zones will shrink and disappear. The energy to create a MPZ is called the *minimum quench energy* (*MQE*). For any given range of energy disturbances, determined by the mechanical properties of the magnet, it is clear that a conductor with high MQE will perform better than one with low MQE. So an essential adjunct to good mechanical design of the magnet is to choose a conductor with the maximum possible MQE.

Calculating the MQE is complicated; it requires a time-dependent solution of the three-dimensional thermal diffusion equation with highly nonlinear generation terms. It can be done numerically with one of the many finite element solvers now available, but it is difficult to draw general conclusions from specific calculations. Alternatively, the MQE can be measured using a small pulsed heater in good thermal contact with the conductor, which is wound into a small coil with the appropriate thermal environment. Measurement probably gives the most reliable result, but here again it is difficult to generalize for a range of conductor designs.

An approximate analytical treatment which gives some insight of how the MQE depends on each of the constituent parameters is described in [2]. It approximates the winding by a homogeneous anisotropic medium, with no structure of conductor or insulation, just averaged properties in each direction. Solutions are obtained for a range of resistive zones, all of which are in thermal equilibrium. They range from very short zones with a high central temperature to long zones with a lower central temperature, but they are all at the balance point, and if any of them were to be slightly bigger or hotter, they would grow without limit. The energy needed to set up each of these zones is calculated and a minimum is found. This is taken to be the MQE. As shown in [2], all the energies scale from a base energy E_o, defined as:

$$E_o = \frac{\pi}{3} a^2 R_g^3 C_o \theta_o \tag{4.2.9}$$

where C_o is the specific heat per unit volume at base temperature θ_o; R_g and αR_g are the distances along the wire and transverse to it from the center of the zone to the point at which the temperature drops to θ_g. We shall come across the factor α again in the quench velocity calculations of Section 4.2.9; it is defined by:

$$\alpha = \sqrt{\frac{k_t}{k_l}} \tag{4.2.10}$$

where k_t and k_l are the average thermal conductivities transverse to the wire and along it, respectively. The MQE is scaled from E_o by a factor e_t which depends on $\beta = (\theta_g - \theta_o)/\theta_o$ and is found as noted above by numerical minimization. Over the range of interest, this minimum value may be approximated by $e_t = 40\beta^{3/2}$. After some manipulation and using the Weidemann–Franz law, we find:

$$\mathrm{MQE} = e_t E_o = \frac{40\pi^4}{3} \frac{L_o^{\frac{1}{2}} k_t C_o}{J_m^3 \rho^2} (\theta_g - \theta_o)^{\frac{3}{2}} (\theta_c - \theta_g)^{\frac{3}{2}} \tag{4.2.11}$$

To maximize the MQE, we see from Eq. (4.2.11) that a good normal conductor like copper is essential because its resistivity is three to four orders of magnitude less than superconductor at low temperature so, as with cryogenic stabilization, we need a composite conductor. If we increase the fraction of copper in the cross-section, the overall resistivity goes down, improving the MQE. But for a given J_m and critical current density in the superconductor J_c, the temperature margin $\theta_g - \theta_o$ is reduced by increasing the fraction of copper, which means reducing the fraction of superconductor. Thus, there is an optimum mix of superconductor and copper which maximizes the MQE. Of the other factors in Eq. (4.2.11), novel inorganic insulating materials have been proposed as a way of increasing k_t and metals with anomalously high specific heat have been suggested for increasing C_o, but none has been used in practice. Accelerator magnets are made with their windings slightly porous to liquid helium, which has the effect of increasing C_o. Of course, the most powerful driver for increasing MQE would be to reduce the operating current density J_m, but this is exactly the parameter that magnet designers wish to maximize.

Thus, we can calculate (or measure) the MQE to a reasonable accuracy and Eq. (4.2.11) shows what needs to be done to maximize it. Magnets with high MQE should perform better than those with low MQE, but this still does not answer the question of how high the MQE needs to be for any specific magnet. For sure, it is not difficult to calculate the energy released by a particular mechanical event. The problem in a structure as complex a superconducting winding is knowing which event will happen where and how big it will be. Rough estimates have been made, but it would be fair to say that none of them has any real predictive value. However, there is an accumulating body of knowledge about how magnets of a given type, size, current density, field, and so on behave, and from their training performance, one can make an estimate of the size of mechanical disturbances occurring. With this knowledge, one should be able to estimate the effect on performance of changing the conductor type, for example, how it might be improved by using a conductor of higher MQE or how it might be changed by operating at higher or lower temperatures.

4.2.6
Screening Currents and the Critical State Model

The magnet current, as introduced by the power supply, is often called the *transport current*. Unfortunately, there are usually additional currents flowing in the

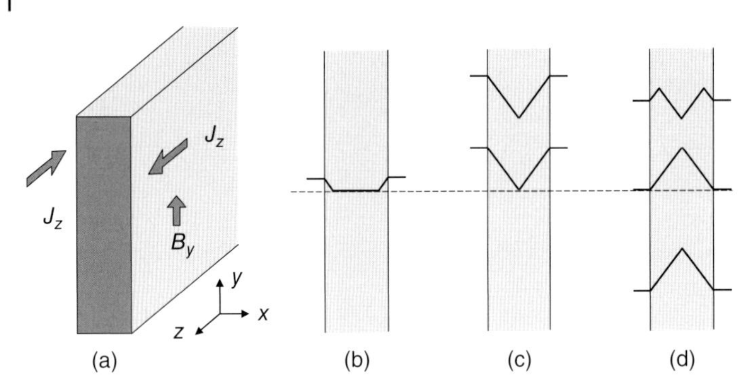

Figure 4.2.8 (a) Superconducting screening currents induced in a slab by changing field, (b) small increase, (c) increase to full penetration and further, and (d) reduction to zero and then negative.

wire and these are invariably a nuisance. They arise whenever the field is changed because this change induces screening currents to flow, very much like eddy currents, but unlike eddy currents, they do not decay with time because there is no resistance. Although wires are invariably round, we usually discuss screening current in terms of an infinite slab, as sketched in Figure 4.2.8, because the one-dimensional analysis is much simpler.

When the magnetic field in Figure 4.2.8 changes, it induces an electric field along the outer surfaces of the slab. The superconductor can only respond to this field by going slightly beyond its critical current into the flux flow resistance region until the resistive voltage is equal to the imposed electric field. Note that this means power dissipation in any superconductor exposed to changing fields; we shall return to this in the next three sections. Once the field stops changing and there is zero electric field, the superconductor relaxes slightly until it is just carrying its critical current density. Writing Maxwell's equation in one dimension, we have:

$$\mathrm{Curl}\,\boldsymbol{B} = \frac{\delta B_y}{\delta x} = \mu_o \boldsymbol{J}_c \tag{4.2.12}$$

So the field decreases linearly from the surface into the interior with a gradient $\mu_o J_c$. When the field is increased further as in Figure 4.2.8c, we reach a point of full penetration where the oppositely directed currents meet at the center. After this, the superconductor can do no more to screen the field and the same pattern of current remains with increasing field. If the field is reduced as in Figure 4.2.8d, the outer layers of current reverse, but the inner regions remain unchanged until the field change penetrates. After a certain field change, all the current has reversed and the current pattern then remains the same the field reduces to zero and then goes negative. This way of envisaging the response of hard irreversible superconductors to magnet field is known as the *Bean London model* and it forms the basis of many calculations about magnetization, loss, stability, and so on. To apply the model, one must remember two things:

- The current density is everywhere $\pm J_c$ or zero.
- Changes come in from outside the material.

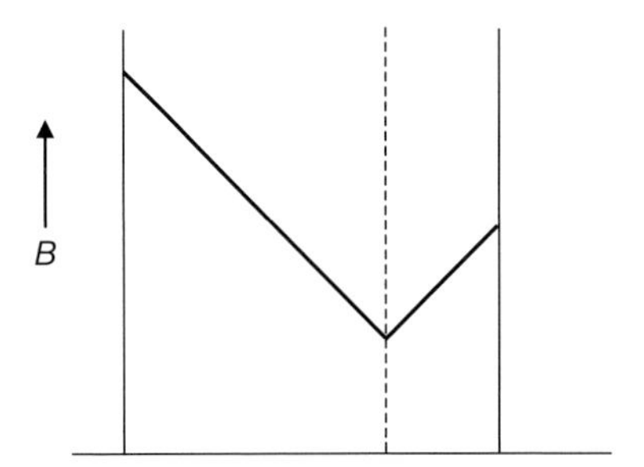

Figure 4.2.9 Transport current added to magnetization currents.

All screening currents are equal and opposite "go and return" (crossing over the wire at each end), so the field patterns are symmetrical. When a unidirectional transport current is added to the induced currents, the above rules still apply but the patterns become unsymmetrical as sketched in Figure 4.2.9. Transport current will be ignored in the sections following because they complicate the picture without changing its nature and also because the problems of magnetization, and so on, are greatest in the low-field region of the magnet where the critical current is very much higher than the transport current.

One-dimensional models are helpful in gaining an understanding of what is going on, and they give a fair approximation of the numerical result. For better accuracy, complete two- or three-dimensional models are needed. They are more complicated and usually need numerical methods to solve, but the physical principles involved are just the same.

4.2.7 Magnetization and Flux Jumping

The persistent currents flowing in Figure 4.2.8 make the superconductor behave like a magnetic material and we may define a magnetization in the usual way as the product of circulating current × area enclosed per unit volume. Figure 4.2.10 illustrates the calculation for a slab above full penetration. We consider a stripe of

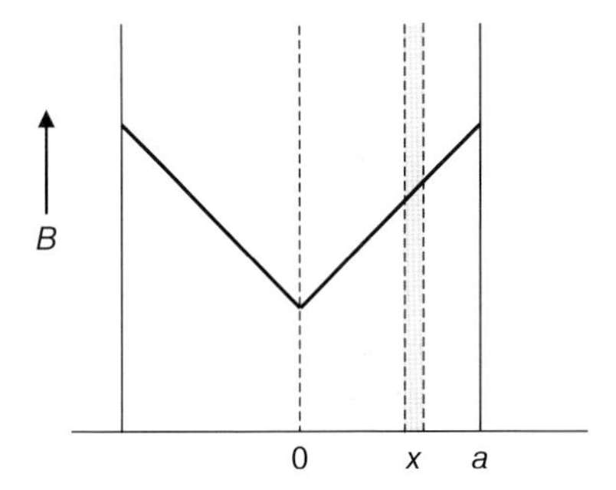

Figure 4.2.10 Integration of screening currents to calculate magnetization.

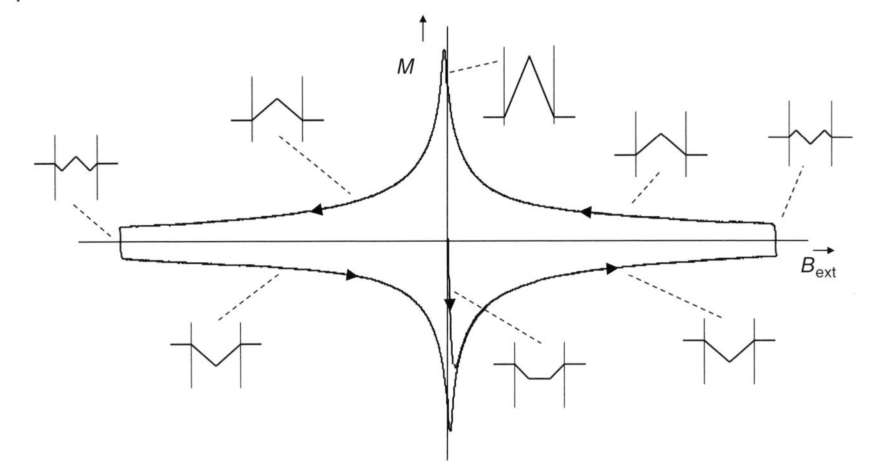

Figure 4.2.11 Measured magnetization loop for a sample of NbTi wire at 4.2 K; surrounding sketches show slab model field profiles at each stage of the loop.

width δx at distance x from the line of symmetry at 0, carrying a current $J_c\delta x$. To obtain the magnetization of the slab, we integrate overall stripes, multiplying by the area enclosed and dividing by the slab volume:

$$M = \frac{1}{a}\int_0^a xJ_c\mathrm{d}x = J_c\frac{a}{2} \tag{4.2.13}$$

A similar, but more complicated, calculation for a round wire gives $M = J_c a\cdot(4/3\pi)$, where a is now the wire radius.

As with any magnetic material, magnetization of a superconducting sample may be measured using a magnetometer. Figure 4.2.11 shows a measured magnetization curve for a sample of NbTi wire at 4.2 K. Starting at the origin, after the sample had been cooled down in zero field, the field is ramped up in a positive direction. Magnetization opposes the change, that is, goes negative, and initially changes at a very rapid rate where, as shown by the surrounding sketch, screening currents have not yet filled the sample. After full penetration, the magnetization changes at a much slower rate, determined only by the reduction in J_c with increasing field. When the field is swept down, magnetization again changes very rapidly as the screening currents reverse, after which the pattern has fully penetrated and magnetization increases slowly as J_c increases with reducing field. Reversing the field produces an exact mirror image and raising the field again retraces the original path.

The important thing to notice about Figure 4.2.11 is that it is irreversible and hysteretic, that is, the magnetization at any given field depends on the history. In common with other hysteresis curves such as the magnetization of iron, there is an energy loss around a cycle equal to the area enclosed by the loop; we shall return to this later. An interesting detail of Figure 4.2.11 is that the low-field peaks are slightly displaced to either side of the magnetization axis. The reason for this is that, in addition to the dominant magnetization coming from irreversible flux pinning and bulk currents, there is still the reversible magnetization coming from the reversible surface "Meissner" currents.

Magnetization creates problems in the design of precision magnets such as those used for NMR spectrometers, particle accelerators, and MRI scanners. The winding behaves like a magnetic material and thereby perturbs the working field shape in a variable fashion depending on the history of energization. Even at zero current, the magnet retains a residual field after its first energization. In common with other problems to be discussed, field errors are dealt with by dividing the superconductor into fine filaments. As shown by Eq. (4.2.13), the magnetization reduces *pro rata* with the radius. The most stringent requirement in this regard comes from accelerator magnets, where the need to minimize field errors at injection demands a filament diameter of 5–10 μm.

Another problem arising from magnetization currents is that they store energy which may be released, rendering the superconductor more susceptible to energy inputs and, in the extreme, causing a catastrophic collapse known as a *flux jump* which will quench the magnet. To see how this happens, consider a "thought experiment" to measure specific heat of the superconductor when it is fully magnetized. We input a quantity of heat and look at the temperature rise. Current density J_c falls with this temperature rise, producing a change in the pattern of screening currents as shown in Figure 4.2.12.

Changing the current pattern causes changes in flux, which produce electric field, drive the superconductor into the flux flow resistive state, and thereby dissipate energy. First, we calculate the flux change (per unit length) within the plane at x:

$$\delta\phi(x) = \int_0^x \Delta B(x)\mathrm{d}x = \int_0^x \mu_o \Delta J_c(a-x)\mathrm{d}x = \mu_o \Delta J_c \left(ax - \frac{x^2}{2}\right) \tag{4.2.14}$$

The power generation per unit volume is $J_c E = J_c \mathrm{d}\phi/\mathrm{d}t$, so the energy dissipated per unit volume is $J_c \Delta\phi$. Thus, the heat per unit volume averaged over the slab is:

$$\Delta Q = \frac{1}{a}\int_0^a \delta q(x)\mathrm{d}x = \frac{1}{a}\int_0^a J_c \mu_o \Delta J_c \left(ax - \frac{x^2}{2}\right)\mathrm{d}x = \mu_o J_c \Delta J_c \frac{a^2}{3} \tag{4.2.15}$$

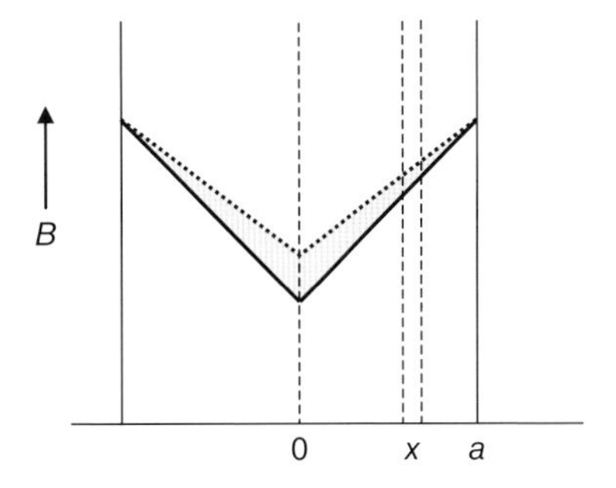

Figure 4.2.12 Change in screening current pattern caused by a temperature rise and consequent fall in J_c.

To a good approximation, we may assume that J_c falls off linearly with temperature:

$$\Delta J_c = -J_c \Delta\theta(\theta_c - \theta_o) \tag{4.2.16}$$

So that, if we supply heat per unit volume ΔQ_s to the slab, the heat balance may be written:

$$\Delta Q_s + \frac{\mu_o J_c^2 a^2}{3(\theta_c - \theta_o)} \Delta\theta = C\Delta\theta \tag{4.2.17}$$

where C is the specific heat per unit volume. From the viewpoint of our thought experimenter who supplies heat ΔQ_s and sees a temperature rise $\Delta\theta$, the effective specific heat C_e is therefore:

$$C_e = \frac{\Delta Q_s}{\Delta\theta} = C - \frac{\mu_o J_c^2 a^2}{3(\theta_c - \theta_o)} \tag{4.2.18}$$

The effective specific heat has been reduced by the energy dissipation when the screening currents change. Thus, the sample has become more susceptible to energy inputs from outside. Catastrophe comes when the right hand side of Eq. (4.2.18) goes to zero, so that an infinitesimally small temperature fluctuation is able to heat the superconductor to its critical temperature. Such events are known as *flux jumps* and they were a strong cause of degradation in the early days of superconducting magnet technology. However, it soon became clear [5] that the problem may be solved by reducing a, that is, making the superconductor into fine filaments. Flux jumping is avoided by reducing a to make the right of Eq. (4.2.18) zero, and this is known as the *adiabatic stability criterion*. For the most reliable magnet performance, however, it is advisable to reduce a even further so that the effective specific heat C_s is not too much smaller than C and the superconductor is not unduly sensitive to the mechanical disturbances discussed in Section 4.2.4. For NbTi, the criterion requires a filament diameter less than ~50 μm.

4.2.8
Filamentary Wires and Cables

The stability criterion calls for filament diameter of $\lesssim$50 μm, but such filaments are fragile and carry a current of only a few amps, so no use for making magnets. Furthermore, Section 4.2.4 shows that the MQE may be enormously increased by combining the superconductor with a good normal conductor such as copper. Practical magnet conductors are therefore made as composite wires of filamentary superconductor in copper. Figure 4.2.13 shows the cross-section of a NbTi/Cu wire, typically drawn to a diameter of ½ mm and routinely used in the manufacture of solenoids for research applications.

In using Eq. (4.2.11) for the MQE of a composite wire, we make the assumption that the superconductor and copper are intimately mixed so that we may take the average thermal and electrical conductivities to be volume average of the constituent materials. However, if the superconducting filaments are too big, the

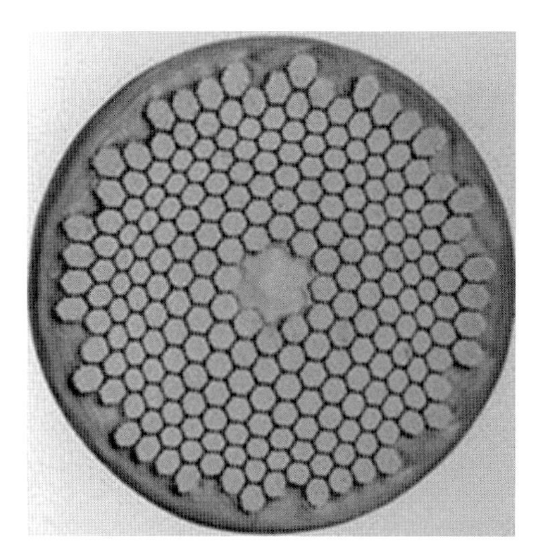

Figure 4.2.13 Cross-section of a filamentary NbTi copper composite wire.

temperature at their center will be greater than that of the rest of the composite and their critical current will be less than that calculated by the homogeneous theory. The criterion on filament radius a for avoiding this situation is known as the *dynamic stability condition* [2] and is given by:

$$a \leq \sqrt{8\frac{\lambda k_s(\theta_c - \theta_o)}{G_c}} \tag{4.2.19}$$

where k_s is the thermal conductivity of the superconductor and other factors are as defined for and by Eq. (4.2.1). Coincidentally, it turns out that, for a typical NbTi copper composite wire with $\lambda \sim 0.4$, the dynamic stability criterion comes with about the same number as the adiabatic stability criterion of Eq. (4.2.18), that is, a filament diameter ~50 μm. However, this is a pure coincidence, adiabatic stability is about heat capacity and dynamic stability is about heat conduction.

The magnetization of wire with 50 μm filaments is too much for applications requiring a precision field shape, such as particle accelerators. For these magnets, much finer filaments are needed, like those in the composite shown in Figure 4.2.14. When this wire is drawn down to a diameter of ½ mm, the filament diameter is 4.2 μm.

The copper matrix improves stability and handing and, as we shall see in Section 4.2.9, is essential for quench protection. However, it does bring one serious problem in that it causes the filaments to be coupled together in changing fields. As shown in Figure 4.2.15a, the desired situation is that each filament behaves independently of its neighbors, with the "go and return" screening currents contained within each filament and crossing over at the ends. In a changing field, however, there is a greater flux linkage between filaments on opposite sides of the wire, so that screening currents are induced to flow unidirectionally in filaments on one side and return in filaments on the other, as in Figure 4.2.15b. When the field

Figure 4.2.14 Filamentary composite wire containing 6264 filaments of NbTi in a copper matrix, which has been etched away in the enlarged view. (Photo courtesy of Bruker EST.)

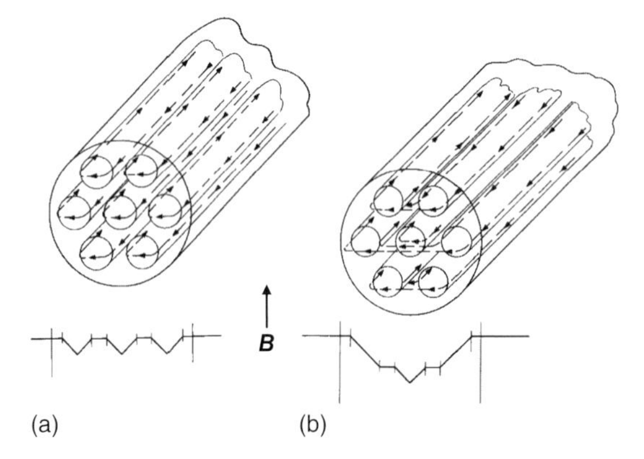

Figure 4.2.15 Screening current in a filament composite wire: (a) uncoupled filaments and (b) filaments coupled via currents crossing the resistive matrix.

change stops, the resistance where the currents cross over the matrix at the ends will eventually force these coupling currents to decay to (a), but in a long thin wire, the crossing resistance is extremely small such that the decay time can be hundreds of years. The filamentary composite behaves essentially like solid wire and advantages of fine subdivision for stability, magnetization, and so on are lost.

Fortunately, the coupling between filaments can be reduced by twisting the wire, as sketched in Figure 4.2.16. Coupling current still flow along the filaments, but they cross over the matrix continuously, not just at the ends, and they flow in a direction parallel to the changing field. It may be shown, see for example [2], that the magnetization coming from coupling currents is given by:

$$M_e = \frac{\boldsymbol{B}}{\rho_t}\left[\frac{p_w}{2\pi}\right]^2 = \frac{2}{\mu_o}\boldsymbol{B}\tau, \quad \text{where} \quad \tau = \frac{\mu_o}{2\rho_t}\left[\frac{p_w}{2\pi}\right]^2 \qquad (4.2.20)$$

where p_w is the twist pitch of the wire.

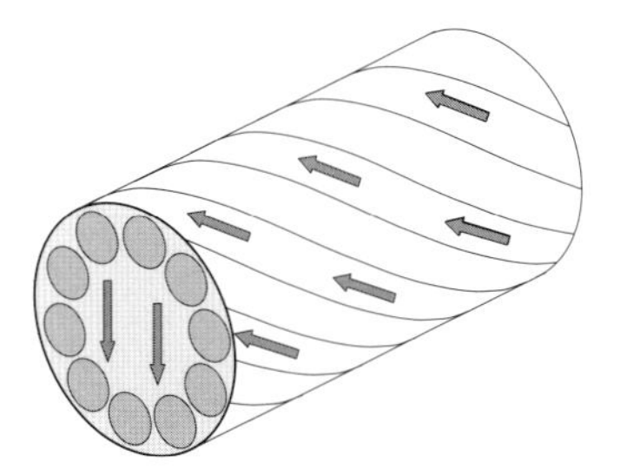

Figure 4.2.16 Coupling currents are reduced by twisting the wire.

The term ρ_t in Eq. (4.2.20) is the effective resistivity for current crossing the matrix perpendicular to the wire axis. Carr [6] has shown how this effective resistivity lies between two extremes, the first corresponding to good contact between filaments and matrix, when the crossing currents can take a shortcut through the filaments:

$$\rho_{tg} = \rho_{Cu}\frac{1-\lambda}{1+\lambda} \tag{4.2.21}$$

where ρ_{Cu} is the copper resistivity and λ is the fraction of superconductor in the wire cross-section. When the contact between filaments and matrix is bad, the currents are constrained to flow in the copper between the filaments, so the transverse resistivity is higher.

$$\rho_{tb} = \rho_{Cu}\frac{1+\lambda}{1-\lambda} \tag{4.2.22}$$

Wires destined for use in rapidly changing fields need to have a high transverse resistivity for low coupling while keeping a low longitudinal resistivity to achieve a high MQE. This is often done by placing a resistive barrier around each filament, but Eq. (4.2.22) defines the maximum resistivity that can be achieved in this way. If higher resistivities are needed, resistive barriers across the matrix may also be produced, but such composites are more difficult to manufacture.

For the superconducting switches mentioned in Section 4.2.2, copper cannot be used in the matrix because it would reduce the "off" resistance too much. So the matrix is made entirely from a resistive alloy such as cupro-nickel. Unfortunately, this greatly reduces the MQE so that the wire is less stable and susceptible to quenching by movement. Great care is therefore needed in the mechanical design and construction of persistent current switches.

Some applications, notably particle accelerators and fusion magnets, require conductors able to carry much higher currents than a single wire. For these applications, many wires must be combined together in a cable. To reduce coupling between the wires in changing fields, they must be twisted, just like the filaments in each wire. The simplest cable is a rope, as sketched in Figure 4.2.17a, but this will not do because the inductance of wires on the outside is less than that of wires inside. As a result, all the current tends to flow in the outer layers, which reach their critical current while wires on the inside are carrying no current at all.

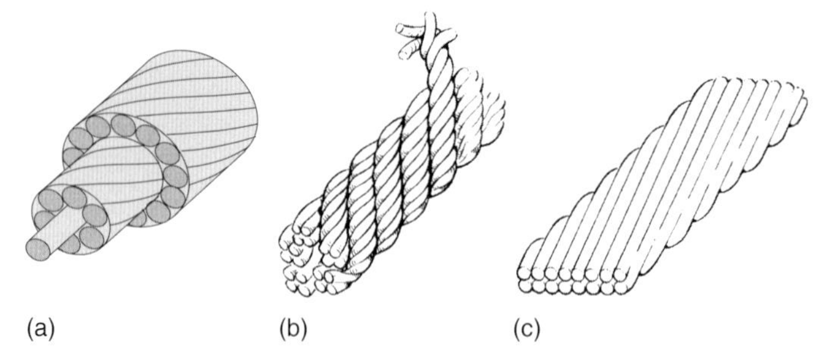

Figure 4.2.17 Different styles of superconducting cable: (a) rope, (b) Litz, and (c) Rutherford.

To get a uniform sharing of current between all wires in the cable, they must be fully transposed, that is, along the length of cable, all wires must change places with each other so that their inductance over a long length of cable is the same. The two types of transposed cable that have been used in superconducting magnets are the Litz cable shown in Figure 4.2.17b and the Rutherford cable shown in Figure 4.2.17c. Cables of the Litz type form the basis of CICC, used in fusion magnets. Rutherford cable has been used in all superconducting accelerator magnets to date.

Although the effect is not well understood, it is a matter of common experience that the degradation and training performance of magnets wound from cable is much improved when there is electrical contact between the strands. So wires in a cable are never insulated, but are allowed to remain in metallic contact with each other. Unfortunately, this electrical contact causes the wires to be coupled together in changing fields. As an example of this effect, we calculate the additional magnetization coming from coupling between wires in a Rutherford cable, using a simplified model with averaged properties. The strongest coupling comes, as sketched in Figure 4.2.18a, when the changing field is perpendicular to the broad face of the cable and induces screening currents to flow in loops like the one shown. Current flowing along a wire at the top face of the cable transfers through the contact resistance to another wire on the bottom face and follows that wire over the edge of the cable so that it emerges onto the top face. It then flows along

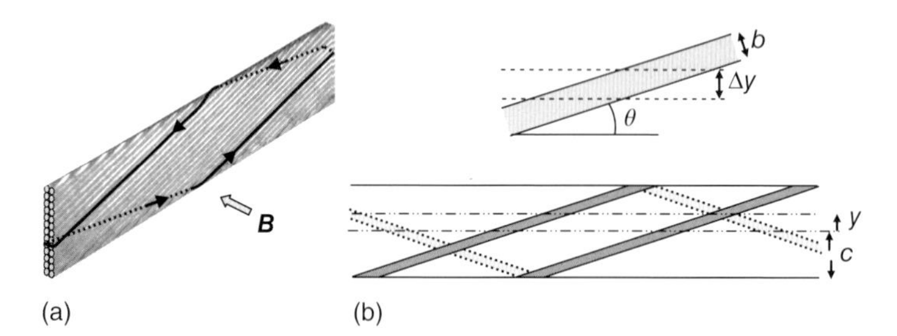

Figure 4.2.18 (a) Coupling current in a Rutherford cable with changing field perpendicular to broad face; dashed lines in (b) show wire on back face, inset shows crossover.

the top face until it reaches a point where it is above the first wire, where it flows down through a second contact resistance to join the first wire again.

To calculate the distribution of induced currents, we calculate the changing flux enclosed by a typical loop shown in Figure 4.2.18b and equate it to the resistive voltage drop at the two crossover contacts. The loop area is given by simple geometry:

$$A_1 = \frac{p(c^2 - y^2)}{2c} \tag{4.2.23}$$

where p is the cable twist pitch and $2c$ is the cable width. The voltage induced around this loop by a changing field $\boldsymbol{B}$ may be equated to the resistive voltage drop of current ΔI leaking from the wire through the two crossover contacts with areas of height Δy as shown in the inset to Figure 4.2.18b.

$$\boldsymbol{B}\frac{p(c^2 - y^2)}{2c} = 2\Delta I\left(\rho_c \frac{\sin\theta}{\Delta y}\right) \tag{4.2.24}$$

where ρ_c is the average resistivity through the cable as measured between the wire centers. Letting $\Delta y \to 0$, we may integrate to get I with the boundary condition that $I = 0$ along the center line of the cable where, by symmetry, the induced currents reverse. The solution, resolved as a current density in the x direction along the cable is:

$$J_x(y) = \frac{\boldsymbol{B}}{b^2\rho_c}\left(\frac{p}{4c}\right)^2\left(c^2y - \frac{y^3}{3}\right) \tag{4.2.25}$$

where b is the thickness of the wires which, for this argument, are assumed to be square so that the cable thickness is $2b$. Finally, by integrating these currents × the area enclosed, we obtain the magnetization.

$$M_{ce} = \frac{\boldsymbol{B}}{\rho_c}\frac{p^2}{60}\left(\frac{c}{b}\right)^2 \tag{4.2.26}$$

It is interesting to note the similarity to Eq. (4.2.20) and see how the magnetization has been increased by the square of the aspect ratio c/b of the cable. Similar expressions may be found for coupling via the contact to adjacent turns and in a changing field which is parallel to the broad face of the cable [7], but these terms are much smaller than Eq. (4.2.26) and usually negligible.

For a more detailed picture of coupling in cables, it is better to use a network model such as CUDI [8]. Network models use the actual resistance per wire crossover R_c which may be related to the averaged resistivity used in Eq. (4.2.26) by:

$$R_c = \rho_c \frac{b}{2pc} N(N-1) \tag{4.2.27}$$

where N is the number of wires in the cables.

Figure 4.2.19 illustrates the different components of magnetization described above. The innermost loop comes from persistent screening current within the superconducting filaments; the amplitude varies with field according to

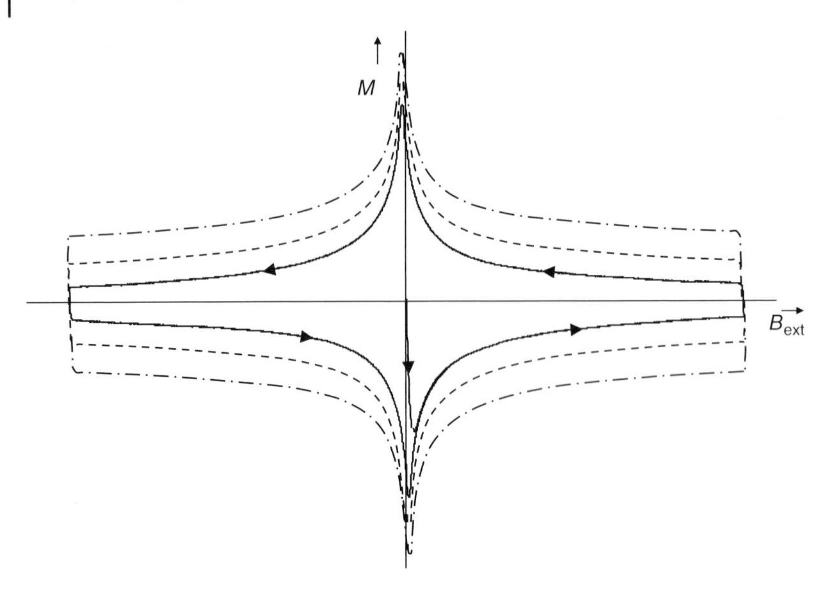

Figure 4.2.19 Showing the different components of magnetization, innermost curve comes from persistent currents within filament, next curve from coupling between filaments within the wire, and outer curve from coupling between wires in the cable.

Eq. (4.2.13). In a changing field, coupling between filaments by current flowing through the wire matrix adds more magnetization to produce the intermediate loop. Note that the additional component of magnetization does not depend on field, only on $\boldsymbol{B}$. Finally, coupling between wires in the cable adds further component to produce the outermost loop. For most situations of practical interest, the coupling currents take up a small fraction of the critical current of the superconductor and, in this situation, the components may be calculated independently and simply added together.

4.2.9 AC Losses

As discussed earlier, changing magnetic fields produce electric fields which drive the superconductor into the flux flow regime and thereby produce losses. To illustrate this, we again consider a simplified slab model, as sketched in Figure 4.2.20. In the strip of width dx, the electric field due to changing flux enclosed is x, giving a loss power of $\boldsymbol{B}xJ_c$. Integrating over the slab thickness, we obtain the average loss power per unit volume:

$$P = \int_0^a \boldsymbol{B}J_c x \mathrm{d}x = \boldsymbol{B}J_c \frac{a}{2} = \boldsymbol{B}M \tag{4.2.28}$$

It turns out that, in most situations of practical interest, where the amplitude of the field change is much greater than the size of the magnetization, the loss per unit volume is $\boldsymbol{B}M$. For example, the loss of coupling currents in a round wire is:

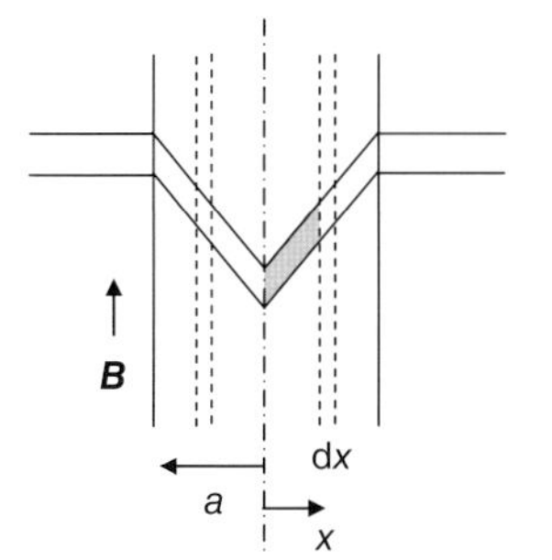

Figure 4.2.20 Profile of changing field in a slab.

$$P_e = \boldsymbol{B}M = \frac{\boldsymbol{B}^2}{\rho_t}\left[\frac{p_w}{2\pi}\right]^2 = \frac{2}{\mu_o}\boldsymbol{B}^2\tau \tag{4.2.29}$$

Note that the coupling loss goes as the square of $\boldsymbol{B}$ just like an eddy current loss. Indeed, coupling currents are eddy currents whose strength has been amplified by the extra flux change enclosed within the superconducting lengths of the circuit. Similarly, the coupling loss in a cable may be calculated from $\boldsymbol{B}$ and Eq. (4.2.26).

If we integrate the power loss $\boldsymbol{B}M$ with respect to time, we find that the total energy dissipated around a hysteresis loop like Figure 4.2.19 is the area enclosed by the loop. In this regard, the irreversible behavior of superconductors is just like other material showing hysteresis, for example, magnetic iron. Regardless of what goes on inside the superconductor, it can be shown that the work done by the external magnetic field around the closed loop is also $\int M\,\mathrm{d}B$.

In situations like that shown in Figure 4.2.9, where a transport current I_m is added to the magnetization currents, the total loss is increased by a factor $(1 + i^2)$, where $i = I_m/I_c$.

Over large field excursions, the magnetization changes as J_c changes with field. A convenient approximation for this dependency is the Kim–Anderson model:

$$J_c(B) = J_o\frac{B_o}{B + B_o} \tag{4.2.30}$$

where J_o and B_o are constants for the particular superconducting material. With this approximation, the loss on ramping between two fields is:

$$Q_{12} = \int_{B_1}^{B_2} M\mathrm{d}B = \frac{2d_f}{3\pi}\int_{B_1}^{B_2} J_c(B)\mathrm{d}B = \frac{2d_f}{3\pi}\int_{B_1}^{B_2}\frac{J_oB_o}{B + B_o}\mathrm{d}B = \frac{2}{3\pi}d_fJ_oB_o\ln\left\{\frac{B_2 + B_o}{B_1 + B_o}\right\} \tag{4.2.31}$$

In all these situations, it may be seen that the loss may be reduced *pro rata* by reducing the filament size. Micron-sized filaments have made it possible for superconductors to work economically in pulsed applications such as accelerators and energy storage systems at ramp times down to ~1 s. Unfortunately, they have not proved economical in magnetic applications at 50/60 Hz with LTS, despite the development of submicron filaments. Perhaps this situation will be changed by the smaller refrigeration needs of HTS, provided they can be made into very fine filaments which are well decoupled in the matrix.

Here we have covered AC losses in the simplest situations, but things can get more complicated, for example, at higher frequencies or where the external field change is not much greater than the magnetization of the sample. For a more complete coverage of the loss mechanism and hysteresis loss in various configurations, the reader is referred to [9], for coupling currents in wires and cables to [10], and for numerical techniques to [11].

4.2.10
Quenching and Protection

Quenching is the most dangerous prospect for superconducting magnets and has been their most common cause of death. As noted earlier, a quench happens when the superconductor reverts to the resistive state and therefore starts to generate ohmic heat, which is then conducted to neighboring conductors, causing them to quench as well. Once it has started, the process is irreversible and proceeds until all the magnet's stored energy has been dissipated as heat and the current has decayed to zero. It is a problem because of the large stored energy and the high operating current density of superconducting magnets, which means a high ohmic power density in the conductor, heating it to a high temperature. The problem is exacerbated by the fact that quenches start at a point and then spread out, so that heating goes on for longest at the start point, which is always the place to suffer the greatest temperature rise.

Of course, nobody intends their magnet to quench but, even if the magnet works well, the cryogenic system or refrigerator may fail. So quenching remains an ever-present possibility and the consequences must always be anticipated and designed for.

The energy density of a magnetic field is $B^2/2\mu_0$, which at 5 T amounts to 10 MJ m^{-3} and at 10 T is 40 MJ m^{-3}. If dissipated in an equal volume of copper winding, these energies would only raise the temperature from 4 to 45 or 73 K, respectively. But high current densities mean that windings can often have a smaller volume than the field they produce. Furthermore, the energy is not uniformly distributed, but concentrated at the quench point, raising the peak temperature still further and sometimes causing to burnout.

As well as temperature rise, the other danger from quenching is excessive voltage. Contrary to common expectation, the voltage across a magnet's terminals during quenching is small, but inside the winding, there can be large voltages, with resistive voltage drop in one part of the winding being opposed by an equal and opposite inductive voltage in another part. During routine operation, the voltage between turns of a magnet when ramping is no more than a few milivolts, but at quench, it can be very much higher and this higher voltage must be designed for. One particular possibility, which has caused some disasters, is the occurrence of a short circuit between adjacent turns, coming perhaps from a tiny metal particle which punches through the conductor insulation. Unless it has a very low resistance, such a short circuit may not be noticed in routine operation, but at quench, there may be large inter-turn voltages which drive large currents through the short, perhaps melting it and starting an arc which is then driven by the stored energy of the magnet. So it is important to guard against inter-turn shorts by good quality control of the insulation, clean conditions in the winding area, and regular electrical measurement as winding proceeds.

However, temperature rise is usually the main problem and will be the main concern here. Firstly, we calculate the temperature rise assuming the current decay profile is known, making a conservative "worst case" assumption of adiabatic conditions. This simple calculation of temperature rise from ohmic heating is made more complicated by the rapid variation of parameters at low temperature, but it may be simplified by grouping the temperature-dependent terms together:

$$\int_0^\infty J_e^2(t)\rho(\theta)\mathrm{d}t = \int_{\theta_o}^{\theta_m} C(\theta)d\theta \quad \text{and so} \quad \int_o^\infty J_e^{\ 2}(t)\mathrm{d}t = \int_{\theta_o}^{\theta_m} \frac{C(\theta)}{\rho(\theta)}\mathrm{d}\theta = U(\theta_m) \tag{4.2.32}$$

where $J_e(t)$ is the engineering current density, $\rho(\theta)$ the resistivity, and $C(\theta)$ the specific heat per unit volume, each factor being averaged over all constituents in the winding cross-section. In practice, the resistivity is dominated by the copper, but all constituents contribute to the specific heat. The function $U(\theta)$ depends only on properties of the winding constituents (Figure 4.2.21 shows a plot for a typical winding (an accelerator dipole)); also shown for comparison is pure copper and stainless steel (which will be needed later for HTS current leads).

So if we know the current decay curve, we can find $\int J_e^2 \mathrm{d}t$ and find the peak temperature by simply reading off the horizontal axis on Figure 4.2.21. With the simplest method of quench protection sketched in Figure 4.2.22, we do know the current decay. In normal operation, the circuit breaker is closed

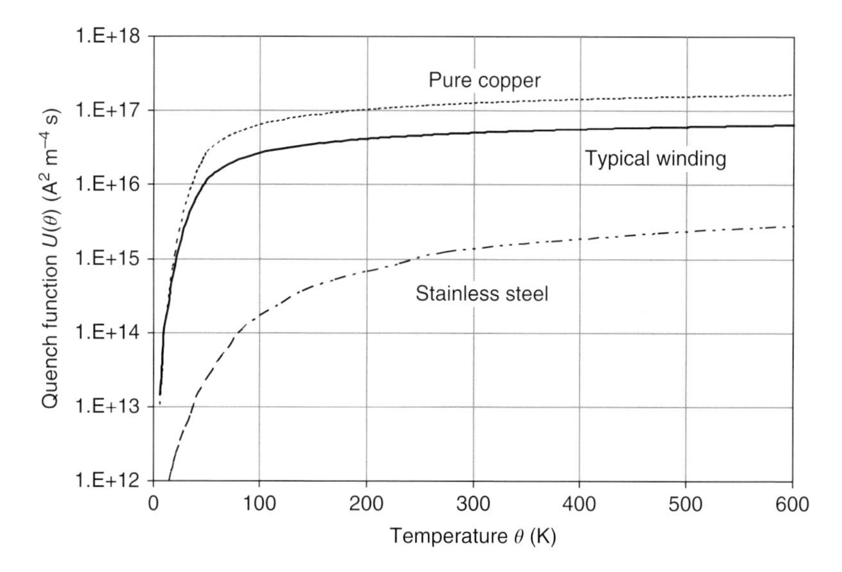

Figure 4.2.21 The quench-heating function $U(\theta)$.

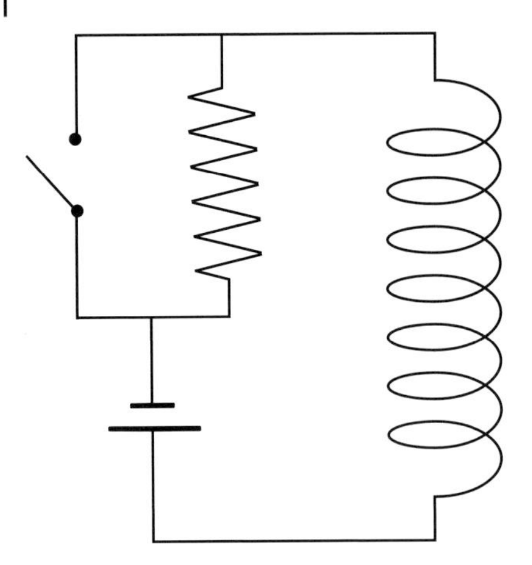

Figure 4.2.22 Quench protection by an external dump resistor.

and current from the power supply flows directly around the coil. If a quench is detected, the breaker is opened and current is forced to flow through the resistor so that it decays according to $I = I_o e^{-t/\tau}$, where $\tau = L/R$. Thus, we have $\int J^2 \mathrm{d}t = J_o^2 \tau/2 = J_o^2 L/2R$ and the maximum temperature follows from $U(\theta)$. For example, the "typical winding" of Figure 4.2.20 was operated at an engineering current density $J_e = 580\ \mathrm{A\,mm^{-2}}$. With a decay time of 0.3 s, we have $U(\theta) = 5 \times 10^{16}\ \mathrm{A^2\,m^{-4}\,s}$, indicating a final quench temperature of 300 K. Note that, because the upper part of the $U(\theta)$ curve is rather flat, the final temperature is very sensitive to τ; for example, with $\tau = 0.4$ s, the final temperature would be 660 K, which would be dangerous for the insulation.

As noted in Section 4.2.2, quench protection is also important for HTS current leads. The HTS conductors themselves, for example, coated YBCO tape, run at very high current density, and would usually burn out if they were to quench, for example, if the cooling at their hot end failed. For this reason, it is recommended that their heat capacity is enhanced by soldering them to a low-conductivity backing material such as stainless steel. The quantity of backing needed is such that it dominates the situation and we may approximately use the $U(\theta)$ curve for pure stainless steel which, as shown in Figure 4.2.21, is much less than a typical winding. For a given decay time, therefore, we must reduce the overall current density. With a 0.3 s decay time, for example, keeping the final temperature down to 300 K demands a reduction of current density to $67\ \mathrm{A\,mm^{-2}}$. This large cross-section of stainless steel will increase the heat leak, but it is the price of safety.

It is not always possible or convenient to use an external circuit breaker and dump resistor, for example, where the magnet is operated in persistent mode with the current leads removed or in a superconducting accelerator where many magnets are connected in series without individual current leads running to room temperature. Furthermore, the circuit breaker is bulky and expensive, so this technique is not used much these days. To ensure safe quenching without the safeguard of an external dump, it is necessary to look more deeply into the quench process.

The quench usually starts at a single point in the winding where perhaps a turn has moved slightly. It then grows via ohmic heat generation and conduction along the wire and transverse to it by conduction through the insulation. By making certain approximations about the interface between superconducting and resistive regions, it may be shown [2] that the resistive zone advances along the wire with a propagation velocity given roughly by:

$$v_l = \frac{J_e}{C}\sqrt{\frac{\rho k}{(\theta_g + \theta_c)/2 - \theta_o}} \tag{4.2.33}$$

where all quantities are averaged over a unit cell of the winding, C is the specific heat per unit volume and θ_g is defined by Eq. (4.2.5). In a direction transverse to the wire, the propagation velocity is αv_l, where α is defined by Eq. (4.2.10). Following a quench, the resistive zone therefore grows in the form of an ellipsoid, with its long axis along the wire. Total resistance in the magnet circuit grows rapidly as the resistive zone expands and also as the temperature within its volume rises, thereby increasing the resistivity. The situation is complicated, with nonlinearities being introduced where the growing ellipsoid hits the coil boundaries, and is best solved numerically using codes such as QUENCH. Alternatively, the full time-dependent thermal diffusion equation may be solved with a nonlinear heat generation term in the superconductor, for example, using the magnet code OPERA.

For small magnets, it is found that internal resistance of the quenched region rises sufficiently rapidly to force a current decay before $U(\theta)$ has become too large. Such magnets are said to be self-protecting and no further action is needed. If the magnet has a large stored energy or the resistive region does not grow quickly enough, the decay time will be too long for self-protection. One cause of slow resistance growth, which will become more important in future, is the slow propagation velocity of HTS conductors. Whereas the longitudinal propagation velocity of LTS conductors is typically in the range 5–20 m s^{-1}, for HTS conductors, it falls to ~5–20 mm s^{-1} so that the resistance grows much more slowly. As it becomes economic to use HTS in large magnet systems, this problem will become very important.

In addition to the external dump resistor, two other techniques are available to protect magnets which are in danger of burnout at quench. They are quench-back heaters and subdivision with diodes. Quench-back heaters, which are used extensively with accelerator magnets, comprise one or more pulsed heaters in close thermal contact with a large fraction of the winding. When a quench is detected electronically, the heaters are pulsed, thereby creating resistive zones in many parts of the winding. This added resistance forces the magnet current to decay more quickly than would otherwise be the case, so that $U(\theta)$ and the maximum temperature are reduced. In the absence of an external dump, the inductive stored energy of the magnet can only be dissipated within the winding; quench-back heaters may be thought of as a way of spreading this energy over larger volume.

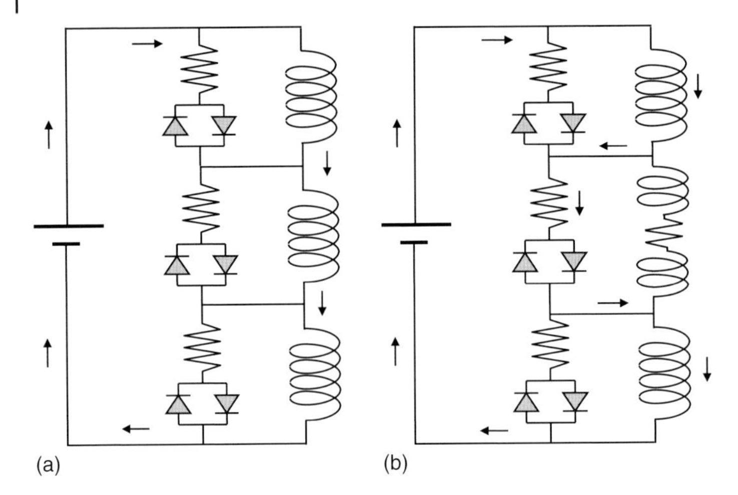

Figure 4.2.23 Quench protection by subdivision; (a) normal operation and (b) at quench, showing how the operating current can bypass the quenched section.

Quench-back heaters and external dump resistors are active protection techniques, that is, they require the quench to be detected electronically and the appropriate action to be triggered. These actions must be available 24/7 with 100% reliability throughout the lifetime of the magnet, requiring the whole paraphernalia of redundant circuitry, uninterruptable power supplies, and so on, plus a structured program of maintenance and testing. All these facilities are readily available in a "big science" laboratory, but may not be available in a smaller institution using, for example, an NMR spectrometer for biology or high-field solenoid for optics research. In this situation, it is desirable to have a passive protection system which is an inherent part of the magnet circuit requiring no external intervention. Subdivision by diodes is such a system. As sketched in Figure 4.2.23, the coil is divided into sections, across each of which is connected a resistor. To prevent power dissipation in the resistors during charging, a diode is connected in series with each resistor and to accommodate either polarity, the diodes are connected "back to back." Note the diode is here being used as a forward breakdown device, with the breakdown voltage being chosen to be higher than the charging voltage. If a quench occurs, the current in that section of the magnet can decay through the shunt resistor, with current from the other sections of the magnet bypassing the quenched section via the resistor. Current decay in the quenched section is driven by the inductance of that section, which is much smaller than that of the whole magnet, so the decay is faster and the temperature rise is lower. Furthermore, the mutual inductance between sections means that a decaying current in the quenched section will drive an increasing current in its neighbors. If this increasing current reaches critical, the neighboring section will quench, thereby spreading the energy around and forcing a faster current decay. Unfortunately, these effects only start to happen sometime after start of the quench when the resistance of the quenched zone has become appreciable. So subdivision is less effective than the active protection techniques where, with a sensitive electronic quench detector, action may be initiated very shortly after the quench occurs. Using an approximate analytical model, it may be shown [2] that subdivision can reduce the decay time by $1/N^{1/2}$, where N is the number of sections. Despite its limitations, protection by subdivision is much used, particularly in commercial solenoid magnets, where the sections can be conveniently arranged to coincide with the concentric sections of different wire size, as mentioned in Section 4.2.3.

One final remark about quenching: it is strongly recommended that the quench calculations be done before testing the magnet – afterwards may just be too late.

References

1. Mercouroff, W. (1963) *Cryogenics*, **3** (3), 171.
2. Wilson, M.N. (1983) *Superconducting Magnets*, Oxford University Press.
3. Stekly, Z.J.J. and Zar, J.L. (1965) *IEEE Trans. Nucl. Sci.*, **12**, 367.
4. Maddock, B.J., James, G.B., and Norris, W.T. (1969) *Cryogenics*, **19**, 261.
5. Wilson, M.N., Walters, C.R., Lewin, J.D., and Smith, P.F. (1970) *Br. J. Appl. Phys. D*, **3**, 1517.
6. Carr, W.J. Jr., (1977) *IEEE Trans. Magn.*, **MAG-13** (1), 192.
7. Wilson, M.N. (2013) Lectures on Superconducting Magnets at JUAS, *https://indico.cern.ch/conferenceTimeTable.py?confId=218284#20130219* (accessed 20 May 2014).
8. Verweij, A.P. (2008) *http://cern-verweij.web.cern.ch/cern-verweij/* (accessed 20 May 2014).
9. Campbell, A.M. (1998) *Handbook of Applied Superconductivity*, Chapter B4.1 and B4.2, Institute of Physics Publishing, Bristol.
10. Duchateau, J.L., Turck, B., and Ciazynski, D. (1998) *Handbook of Applied Superconductivity*, Chapter B4.3, Institute of Physics Publishing, Bristol.
11. Niessen, E.M. and Roovers, A.J.M. (1998) *Handbook of Applied Superconductivity*, Chapter B4.4, Institute of Physics Publishing, Bristol.

4.3
Magnets for Particle Accelerators and Colliders

Luca Bottura and Lucio Rossi

4.3.1
Introduction

Since the pioneering time of Ernest O. Lawrence in Berkeley, the nursery of the first cyclotrons in the 1930s, the quest for higher energy accelerators has driven the demand for larger size and higher field magnets. It is not surprising that superconductivity has become a key technology for modern accelerators and that, in exchange, particle accelerators have passionately fostered superconducting (SC) magnet technology [1]. The timeline of record accelerator energy is shown in the compilation reported in Figure 4.3.1, where we have collected the main accelerators for nuclear and particle physics, highlighting the projects that are based on superconductivity. Many of them feature SC magnets as main technology and performance driver.

In this chapter, we will focus on the SC magnets for high-energy physics (HEP), which is the most important and continuous driver for research and development (R&D) and innovation in applied superconductivity. HEP has been pursuing superconductivity for many large projects, the most recent one being the Large Hadron Collider (LHC) at CERN, and plans are being made for further, even stronger and larger colliders, as shown in the timeline forecast in Figure 4.3.1. We have attempted to cover the most relevant features and issues in SC accelerator magnets, but it is neither possible nor practical to condense half a century of R&D in a short chapter. The reader is directed to the abundant bibliography for further insight and practical design solutions. We wish in particular to refer to the monographs in Refs [2–4] that provide consistent treatment to many of the topics listed in this chapter, as well as to the more recent reviews on SC magnets and conductors for accelerators, Refs [5–7].

4.3.2
Accelerators, Colliders, and Role of Superconducting Magnets

4.3.2.1 Magnet Functions and Type

We can identify the following four main functions of magnets in particle accelerators:

1) To guide a beam of charged particles in a closed orbit or along a beam transfer line: dipole fields, that is, fields with given direction and high uniformity, accomplish this task. Strong dipoles decrease the bending radius and reduce the overall size of the accelerator.
2) To confine the particle in a small space, that is, to focus the beam onto the desired orbit. Quadrupole magnets, where the field grows linearly with the radius from zero on the axis, provide a perfect focusing force. In case of very

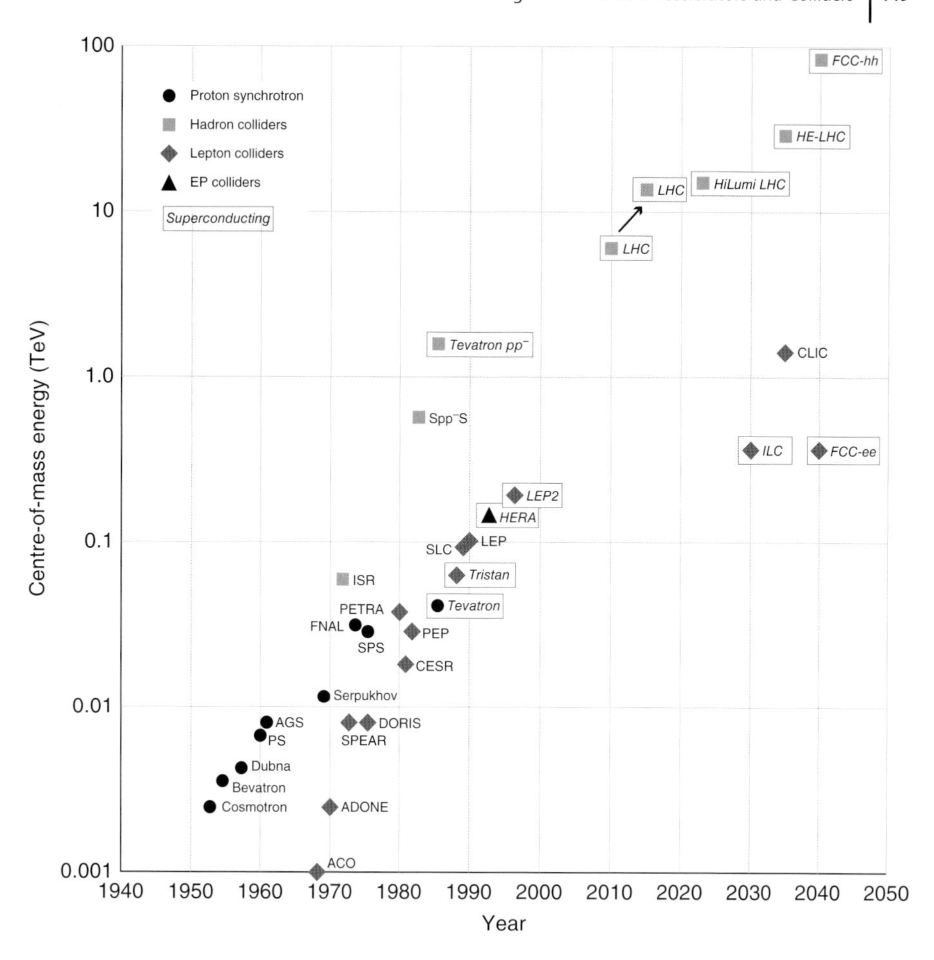

Figure 4.3.1 Energy evolution of main accelerators and colliders for high-energy physics versus time, with superconducting ones in evidence. The ordinate represent the *center-of-mass collision energy* (or the equivalent one for fixed target accelerators). The date of high-energy Large Hadron Collider (HE-LHC), Future Circular Collider hadron-hadron (FCC7-hh), International Linear Collider (ILC), Compact Linear Collider (CLIC), and FCC-ee (electron-positron) are highly hypothetical.

low energy beam (up to a few mega electron volts), solenoids can be used for this function, however rarely of SC type. Strong focusing results in smaller beam dimensions, a clear advantage because of the reduction of the magnet bore and beam chamber.

3) Chromatic aberrations and collective effects can lead to beam instabilities, especially relevant for the intense beams that are common to modern accelerators. High-order multipole magnets, such as *sextupoles* (sometimes also called *hexapoles*), *octupoles*, *decapoles*, and *dodecapoles*, whose characteristic is to provide a focusing strength that depends on the beam radius, are used for the correction of the above effects.

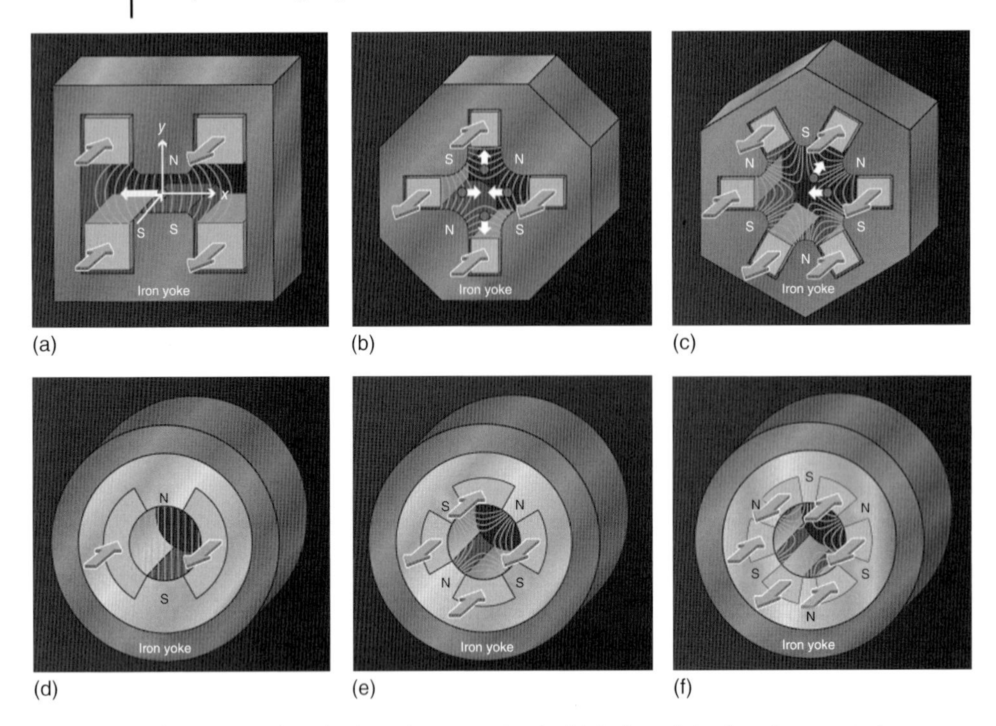

Figure 4.3.2 Iron-dominated magnets (a–c) with indicated the force for an entering positive particle, and superconducting magnets (d–f), from dipoles to sextupoles.

4) To generate intense and brilliant beams of synchrotron light or produce a free electron laser (FEL) light. This is done with undulators and wiggler magnets, employed as insertion devices in a circular synchrotron or an electron linear accelerator (LINAC). This class of magnets also has a limited use as diagnostic or for emittance cooling purposes in accelerators and colliders, but will not be treated further in this chapter.

The schematic cross-section of the various types of magnets described above (excluding wigglers and undulators) is reported in Figure 4.3.2, where a comparison is made between fields generated by magnetizing a precisely shaped iron yoke (iron-dominated electromagnets) and fields generated by a precisely positioned distribution of current (coil-dominated electromagnets). The latter is the main domain of application of SC magnets, in which the coils generate most of the field and the iron yoke is mainly for return flux containment. In fact, SC coils are sometimes used in iron-dominated magnets, when reduction of magnet size and power consumption are important goals. Such class of magnets is called *superferric magnets*. Table 4.3.1 shows a matrix for use of SC magnets for various types of accelerators.

Table 4.3.1 Matrix use of superconducting magnets for accelerators (NC, normal conducting).

	Accelerator type	Guiding field	Focusing field	Stabilizing field	Light emission
Circular accelerators	Cyclotron (hadrons)	Circular split coils 0.3–2 m	NC iron shape	—	—
	Synchrotrons (hadrons)	Dipoles 3–10 m	Quadrupoles 1–3 m	NC	—
	Storage rings (electrons)	NC	NC	NC	Dipoles, undulators, wigglers, 1–5 m
	Colliders (hadrons)	Dipoles 5–15 m	Quadrupoles 1–5 m	Sextu-, octu-deca-, dodecapoles	wigglers 1–3 m
Linacs	Low energy (hadrons)	—	Solenoids	—	—
	High energy (e+e– colliders)	—	Quadrupoles	—	Undulators, wigglers, 1–20 m

4.3.2.2 Transverse Fields

The steering effect of most magnet types described above is given by the force exerted by the field component perpendicular to the main direction of motion of the particles, which is why we define these magnets as generating *transverse fields*, that is, with field direction transverse to the particle entrance and exit axis. Solenoids, where field lines in the magnet body are ideally parallel to the beam axis direction, produce a focusing effect by inducing a particle spiraling motion at the magnet entrance. Although also in this case the field components which are acting are always perpendicular to local particle trajectory direction, the process is different from the dipoles and higher order multipole magnets that are of interest in this chapter. We will neglect solenoids in the following text.

In Figure 4.3.2d–f, we show schematic cross-sections of ideal SC dipoles, quadrupoles, and sextupoles. The circle inscribed in the coils is the space for the beam, which is customarily called the *magnet aperture* or *bore*, and its diameter is a very important characteristic affecting basic design quantities, as discussed later.

Accelerator magnets are slender, small aperture, and long-length objects, to maximize effectiveness and minimize stored energy and cost. We can think of one such magnet as obtained "extruding" the cross-sections shown in Figure 4.3.2 along the desired particle trajectory and for the necessary length. The field is then truly 2D, except for the ends of the magnet. For beam dynamic studies what is important, to the first order, is the integrated effect of the main field component along the beam trajectory, causing the change of position and divergence of a particle after having crossed the entire magnetic element (i.e., the *transfer function* of the beam through the magnet). If we call L the magnetic length, the integrated strength for a dipole of field B_0 is B_0L, which is the magnet bending strength in terms of the angle of deviation of the beam. For quadrupoles, we define an integrated gradient, GL and similarly for higher order magnets. Although approximately equivalent in terms of effect on the beam, magnets of identical integrated strength and different lengths are very different in practice. Indeed, longer lengths imply higher cost and larger infrastructure, while stronger fields bear difficulties for any technology. This is where superconductivity plays its most important role in reducing dimensions by making higher fields accessible.

A static magnetic field exerts a force $\boldsymbol{F}$ on a charged particle given by

$$\boldsymbol{F} = q\boldsymbol{v} \times \boldsymbol{B} \tag{4.3.1}$$

where $\boldsymbol{v}$ is the velocity of the particle and $\boldsymbol{B}$ is the magnetic field induction (or flux density). A uniform, purely transverse field bends the trajectory along a perfect circle, which is the case of a perfect dipole field. By equating the magnetic force to the general expression of the centripetal force, one obtains the following expression for the kinetic energy E of an elementary charged particle (proton or electron)

$$E = \frac{1}{2}\frac{e}{m}(B\rho)^2 \tag{4.3.2}$$

where e/m is the charge to mass ratio, ρ the curvature radius.

Equation (4.3.2) is valid in the nonrelativistic case, the kinetic energy is expressed in electron volts, while all other quantities are in SI units (International System). For ions, the energy per nucleon is the relevant quantity, so the expression becomes $E/A = (1/2)(e/m_{\mathrm{p}})(Z/A)^2(B\rho)^2$ with m_{p} being the proton mass, Z the charge state (for fully stripped ions is the atomic number), and A the mass number, respectively.

In the case of highly relativistic particles, Eq. (4.3.2) has the following simplified limit

$$E = cB\rho \quad \text{where } c \text{ is the speed of light in vacuum} \tag{4.3.3}$$

Which, for ions, reads as $E/A = (Z/A)cB\rho$.

Taking Eq. (4.3.3), the most interesting case for HEP accelerators, a convenient expression in practical units is

$$E \cong 0.3B\rho \tag{4.3.4}$$

where the energy is expressed in tera electron volts, B in tesla, and ρ in kilometer.

4.3.2.3 Dipoles and Relation to Beam Energy

Equation (4.3.4) clearly indicates that in a circular accelerator of length $L = 2\pi\rho$, identical beam energy (the final objective) can be obtained in a compact/high-field, or a large-size/low-field accelerator. The cost trade-off between field level and length is nontrivial. To give a practical example, we can take the case of the proposed superconducting supercollider (SSC), targeting 20 TeV beam energy. Three different designs were considered for the SSC, based on (i) 3 T super-ferric magnets and a length of 190 km, (ii) 5 T SC magnets and 113 km length, and (iii) 6.6 T

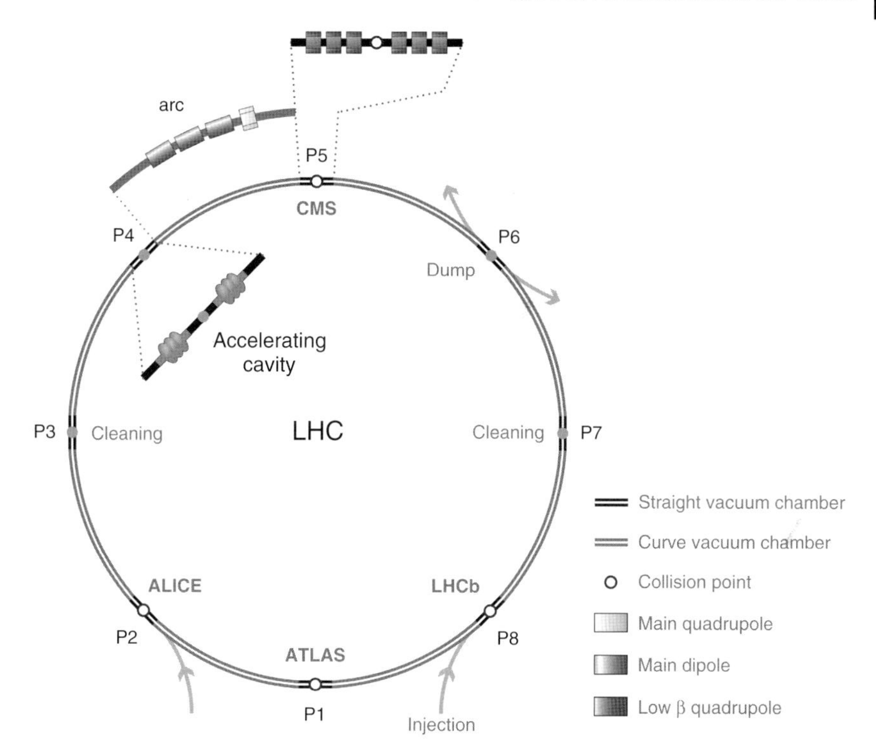

Figure 4.3.3 Collider scheme with main components (not in scale). Dipoles fill about two-third of the ring. There are in total 8 arcs or sectors containing the main dipoles and quadrupoles: P1-P2, P2-3, etc...; there are 4 collision points with low-β quadrupoles: P1, P2, P5, P8; there is one point for superconducting accelerating RF cavities, P4.

SC magnets and 86 km length. The last option was retained, as the medium- and low-field options were evaluated as more costly.

In practice, an accelerator consists of several types of magnets, as well as a wealth of other components as shown schematically in Figure 4.3.3. The dipole filling factor in a modern collider is about 2/3 (i.e., $\rho \approx (2/3)(L/2\pi)$) which is how a 8.3 T field in the 26.7 km tunnel yields the 7 TeV proton kinetic energy of the nominal LHC beam.

4.3.2.4 **Quadrupoles and Focusing**

In a pure quadrupole field, $B = G \cdot r$, G as the module of the gradient, r being the distance from the axis, the force on the beam is radial, linear in r, and its intensity does not depend on the azimuthal position. As shown schematically in Figure 4.3.2, a quadrupole focusing in one plane (e.g., X) is defocusing in the other plane (e.g., Y). However, as shown in Figure 4.3.4, if the polarity of two subsequent quadrupoles is reversed, the total effect is always focusing, in analogy to linear optics. The quadrupole strength also determines the number of oscillation around the axis that a particle does in one revolution, that is, the "tune." Higher tune values imply higher oscillation frequency and stronger focusing. The strength of

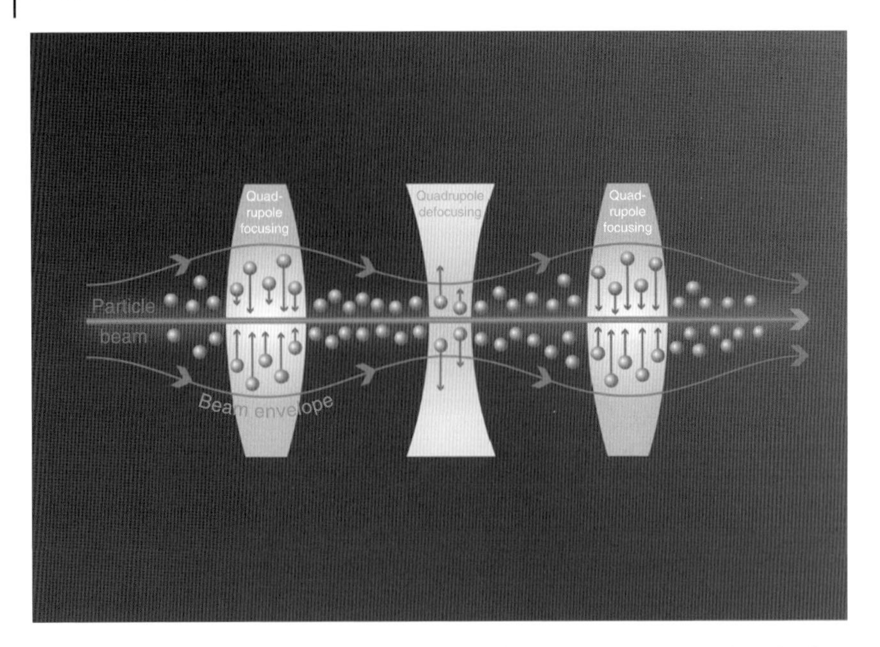

Figure 4.3.4 Beam containment effect of an array of alternating polarity quadrupole. The sketch is only intended to show that focusing forces are always larger (in average) than defocusing forces.

the quadrupole also determines a fundamental quantity in beam physics, namely the "β" function, which relates to the amplitude of the beam oscillation around the nominal orbit. Stronger focusing corresponds to smaller β function, and the cross-section of the beam gets smaller. A small aperture is generally beneficial to reduce the dimensions of the magnets, and the vacuum space allocated to the beam. Finally, for a collider, it is important to make the β function as small as possible at the collision point, which is accomplished by special array of quadrupoles called *low-β triplet*. In practice, β at the collision point is smaller by several orders of magnitude with respect to the arc, to decrease the beam size, increase the density of particles, and increase the collision rate.

4.3.2.5 **Higher Order Multipoles**

Higher order multipole fields act in the same way as quadrupoles by focusing (or defocusing) the beam (depending on their polarity). While the focusing force in quadrupoles grows linearly with the radius r, in higher order multipoles the force is a higher power of the radius. Sextupoles magnets exert a restoring/diverging force whose intensity increases quadratically with r. They correct *chromatic* effects, that is, the fact that beam energy spread generates a spread in oscillation frequency (the tune, see above). Octupoles offer fields strongly nonlinear ($B \propto r^3$), which are used to damp dangerous coherent oscillation mode (Landau damping). In colliders like LHC, the beam undergoes up to 1 billion revolutions: the field errors in the main magnets, dipoles, and quadrupoles (both of regular arcs and of low-β insertions), need to be compensated to order as high as the dodecapole (b_6) to avoid insurgence of long-term beam instabilities driven by resonant terms.

4.3.3 Magnetic Design

4.3.3.1 General

The SC accelerator magnet design is dominated by two – much interlinked – concepts: (i) providing high field with high accuracy; (ii) minimizing the field volume. Since reducing the length is not an option because $E \propto L$, see Eq. (4.3.4), the design consists in minimizing the cross-section, finding the optimal distribution of compact superconductors around the beam pipe separating the ultra-high vacuum from the helium vessel hosting the coils. As we will see, the compactness is a key feature and the dominant factor is the overall current density. In fact, in contrast to classical electromagnets, the field in a SC accelerator magnet is mainly, if not only, produced by the current in the conductor, the magnetization of an iron yoke being negligible for high field. The coil shape is optimized to maximize the bore field and achieve acceptable field quality, as described later. The large forces that are experienced by the coil (several tens to hundreds of tons per meter) cannot be reacted by the winding alone, that has the characteristic shape of a slender racetrack, see Figure 4.3.5, hence the force must be transferred

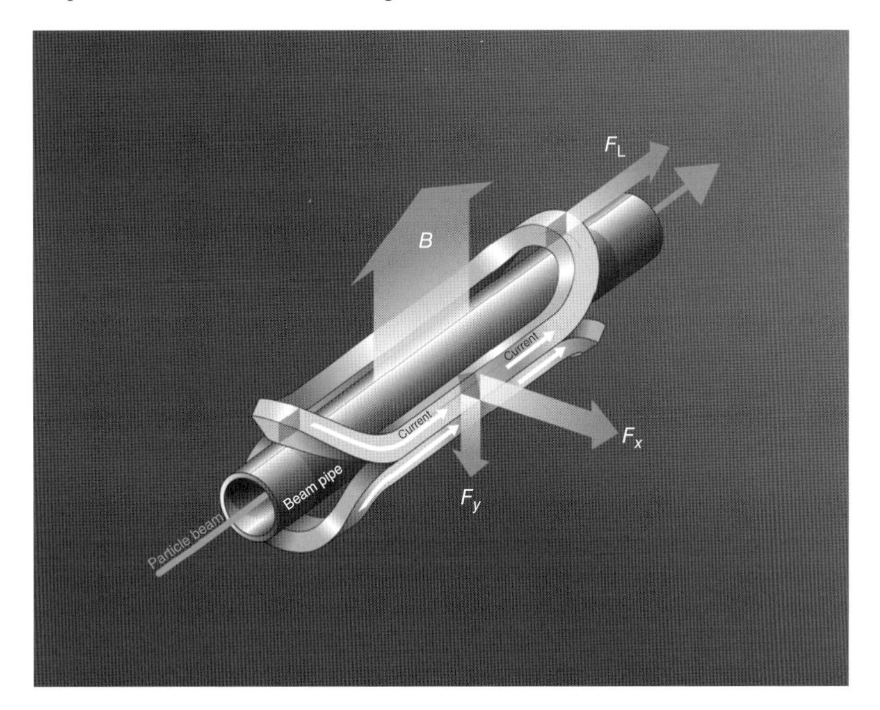

Figure 4.3.5 Schematic of a dipole magnet (only coils and beam pipe) with field and electromagnetic (e.m.) forces exerted on the coils.

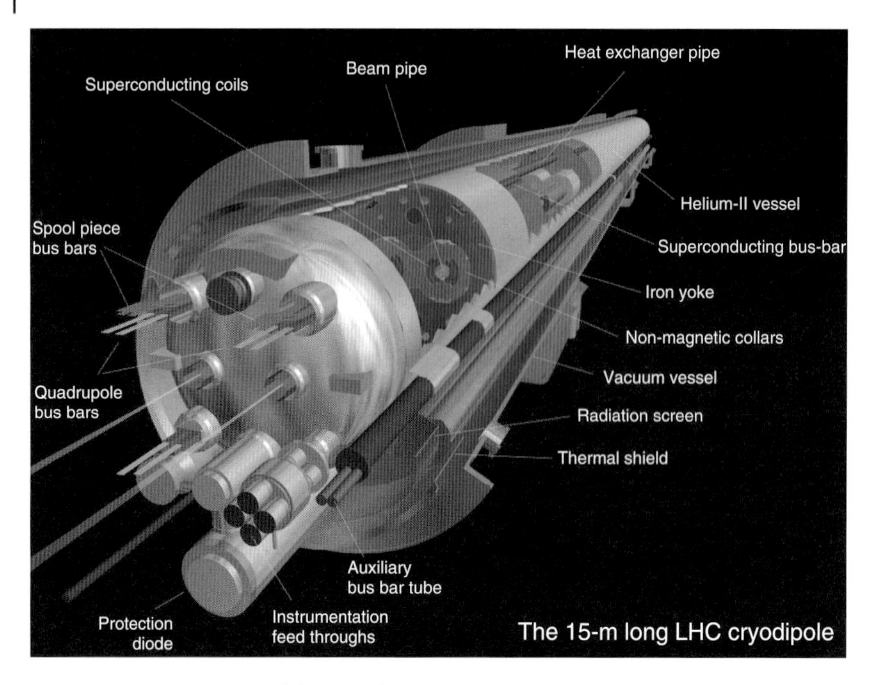

Figure 4.3.6 Artistic view of the 15 m long LHC main dipole in its HeII cryostat (Source: CERN archive).

to a structure that guarantees mechanical stability and rigidity. The iron yoke that surrounds this assembly closes the magnetic circuit, see Figure 4.3.2, shielding the surrounding from stray fields and providing a marginal gain of magnetic field in the bore. In addition, the yoke can have a structural function in reacting or transferring the Lorentz forces from the coil to an external cylinder. Finally, the magnet is enclosed in a cryostat that provides the thermal barrier features necessary to cool the magnet to the operating temperature, which is in the cryogenic range (1.9–4.5 K for accelerators built to date). An implementation of this basic concept is illustrated in Figure 4.3.6 with an artistic view of the LHC dipole.

4.3.3.2 **Current Density**

The requirement to reduce the coil cross-section is translated in a practical demand of high current density in the coil. This is customarily indicated as $J_{\mathrm{overall}} \equiv IN_{\mathrm{turn}}/A_{\mathrm{coil}}$, where the total ampere-turns are divided by the total coil cross-section. A high value of J_{overall} needs the following conditions:

1) A high critical current density in the SC wire. This is usually defined as J_{c}, averaged over the nonstabilizer (non-copper) cross-section, including barriers, residuals from chemical reactions, reinforcements, and so on. It is sometimes more properly called $J_{c\ non\text{-}Cu}$.
2) Minimum stabilizer to nonstabilizer (copper to non-copper) ratio, in either round wire or tape, as required for stability and protection (see later). Usually accelerator magnets require this ratio be in the range 1–2 with a strict

tolerance necessary for the control of the magnetic moment (typical range is ±0.05). Values lower than 1 are dangerous for magnet protection. The current density in a single strand or tape referred to the total cross-section including the stabilizer is the engineering critical current density J_e.

3) Compact cables: strands and tapes typically carry few hundred amperes, need to be assembled in cables capable of several kiloamperes, so that the magnet inductance is reduced, which is helpful for magnet powering (voltage) and protection (dump time). Strand compaction (defined as the conductor cross-section over the total cable cross-section) is around 90% for Nb–Ti, and around 85% for Nb_3Sn, and up to 5% of the original current density is lost due to cabling deformation. As a result, the cable current density ranges from $J_{cable} \sim 0.9J_e$ for Nb–Ti to $J_{cable} \sim 0.8J_e$ for Nb_3Sn.
4) Thin insulation: modern magnets use an insulation thickness of 250 μm or less. Multilayers polyimide tapes are used for Nb–Ti; glass tape (sometimes mica-assisted) are used for Nb_3Sn, followed by epoxy resin impregnation (the actual electrical insulation) after heat treatment. The insulated cable current density is reduced further to typical values $J_{overall} \sim 0.8J_{cable}$.

The above factors explain why all accelerator operating dipoles, and most design for the future projects, work at $J_{overall}$ around 400–500 A mm^{-2}, which corresponds to critical currents J_c in the range of 1200–1800 A mm^{-2}. Larger values would not be useful, as the coil would hit two other limits: maximum stress and protection in case of quench. It is an interesting consequence that once a design attains an overall current density in the above range, a mere change in superconductor properties (e.g., higher J_c) is not sufficient to increase the magnet field. Instead, an increase of the coil thickness, see Figure 4.3.7, Ref. [8], is necessary to extend the field range. Note, however, that thick coils not only increase the cost but also complicate the mechanics and can make the magnet

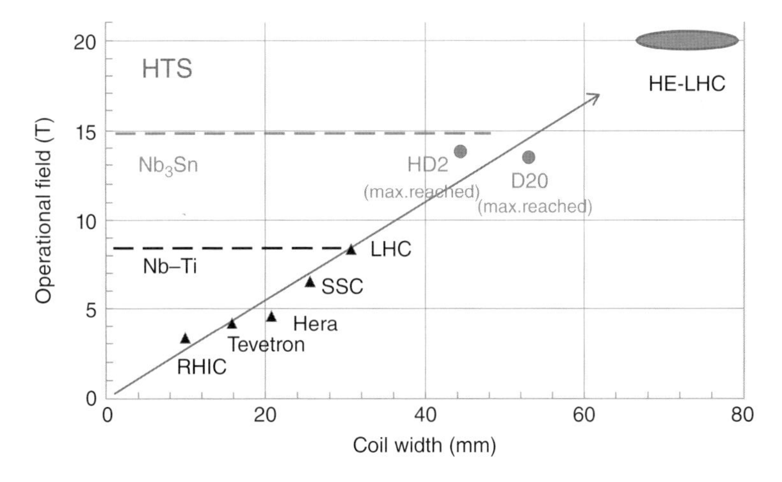

Figure 4.3.7 Plot of the coil radial thickness, or width, versus field for operational (Nb–Ti) dipoles and for record short length (1 m) dipoles (Courtesy of E. Todesco, ref. 8).

Table 4.3.2 Current density $J_{overall}$ and other characteristics of different types of large magnetic systems.

Magnetic system (only dc)	Current density $J_{overall}$ (A mm^{-2})	Operating current (kA)	Typical field range (T)	System stored energy (MJ)
Resistive-air cooled	1–5	1–2	<1	0.01
Resistive-water cooled	10–15	1–10	2	0.05
SC magnets for particle detectors	20–40	2–20	2–6	5–2500
SC Tokamaks for fusion[a)]	25–50	5–70	8–13	5–40 000
SC magnets for MRI	50–200	1	1–10	1–50
SC laboratory solenoids	100–250	0.1–2	5–20	1–20
SC accelerators	200–500	1–12	4–10	1–10 000

a) Top figures refer to ITER (International Thermonuclear Experimental Reactor), under construction.

impractical and too expensive. In Table 4.3.2, the typical range of overall current density of various systems is reported: it is no surprise that accelerator magnets work by far at the highest values.

4.3.3.3 Field Shape

2D magnetic fields are best represented using complex multipoles expansion [9]. Defining the complex variable $z = x + iy$, where the plane (x,y) is that of the magnet cross-section, the function $B_y + iB_x$ of the two components of the magnetic field is expanded in series

$$B_y + iB_x = \sum_{n=1}^{\infty} (B_n + iA_n)\, z^{n-1} = B_{\text{ref}} \sum_{n=1}^{\infty} (b_n + a_n) \left(\frac{z}{r_{\text{ref}}}\right)^{n-1} \tag{4.3.5}$$

The coefficients B_n and A_n of the series expansion, called *normal and skew components*, respectively, are the multipoles of the field, and determine the shape of the field lines. A pure multipolar field has only one nonzero B_n (or A_n) for a given value of n, which is called the *main order of the field*. As an example $n = 1$ is a pure dipole, and $n = 2$ is a quadrupole. Nonzero B_n and A_n for n other than the main order are usually referred to as *field harmonics* or *field errors*. The multipoles are usually quoted as normalized to the main field. In this case, they are indicated with small letters, b_n and a_n, and are always reported in units of 10^{-4}: a b_3 of three *units* in a dipole means that the sextupole error is $b_3 = B_3/B_1 = 3 \times 10^{-4}$. However, as done in Eq. (4.3.5), the normalization requires the definition of a reference radius because the field, except for the dipole component which is uniform in r, increases as a polynomial function of the radius, whose order is $n - 1$. The good choice for r_{ref} is the radius of the largest portion of the cross-section that is actually used by the beam: typically, this is two-third of the inner radius of the coil $R_{\text{in coil}}$. The $r_{\text{ref}} = 2/3R_{\text{in coil}}$ is practically a standard and should be used to compared the harmonic content of magnets with different radius. Normalization to

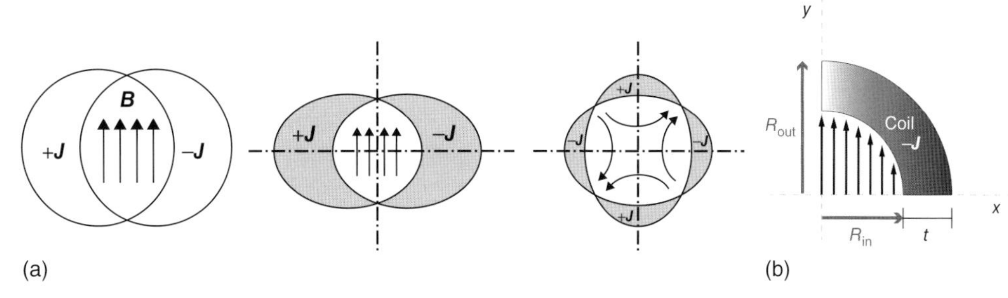

Figure 4.3.8 (a) Overlapping circle and ellipses (dipoles and quadrupoles); (b) dipole generated by a constant thickness cos θ current distribution (one quadrant shown).

the r_{ref} is straightforward, when harmonic (or multipoles) components are measured or reported at a different radius r_0:

$$b_n(r_{\mathrm{ref}}) = b_n(r_0)\left(\frac{r_{\mathrm{ref}}}{r_0}\right)^{n-1} \tag{4.3.6}$$

4.3.3.4 Cos θ Coil

Several ways can be found to generate perfect multipole fields required by HEP accelerator magnets. As an example, we show in Figure 4.3.8a a number of current distributions that generate a perfect transverse dipole field [10], like overlapping circle or ellipses with opposite current. Note how for all arrangements the total surface current $J_s = J \cdot t$, J being the current density and t the coil thickness, is maximum at the mid-plane and zero at the vertical axis, following a "cos θ" behavior. In analogy to the dipole case, the current distribution obtained with two overlapping ellipses, rotated by 90°, produces a quadrupole field. In the above cases, the region of the intersection is not a circle and the minimum diameter is taken as magnet aperture.

As an alternative, a perfect dipole field can be generated by a shell of current, of constant thickness t, in which the volume current density is maximum at the mid-plane and vanishes toward the pole region with a $J = J_0 \cos\theta$ dependence, see Figure 4.3.8b. In this case, the inner region is circular. A $J_0 \cos\theta$ distribution not only generates a perfect dipole field but also is the most efficient current distribution, that is, any other distribution requires more total current (ampere-turns) to generate a given central field, produces more magnetic flux, and has higher stored energy. This consideration is very important for SC magnets where the cost of the conductor is one of the dominant cost factors. It is instructive to compare the central field of the dipole generated by the ideal cos θ distribution, $B_0 = {}^1\!/_2\mu_0 J_0 t$, to that of an infinitely long solenoid of the same current density and thickness, that is, $B_0 = \mu_0 J_0 t$. In practice, the same current density and thickness generates only half the field of the solenoid configuration. Indeed, field levels that are modest in a solenoid can be considered a great achievement for transverse field magnets! In Figure 4.3.9, the progress of high-field solenoid is compared with the one of high-field dipoles, where an approximate factor 2 is clearly visible, as expected from the above qualitative arguments.

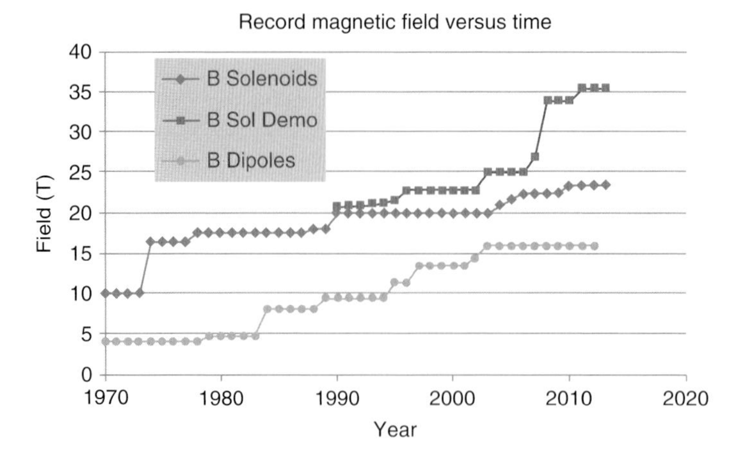

Figure 4.3.9 Progress of central field obtained in solenoids and dipoles (see factor ~2). (Data for solenoids are courtesy of M. Bird, NHMFL, Florida State University.)

Similar to the dipole, a perfect quadrupole field can be generated by a shell of constant thickness with a current density varying as $J = J_0 \cos 2\theta$. In fact, a shell configuration can generate any multipole field of order n with a current distribution $J = J_0 \cos(n\theta)$.

In practice, SC magnets are constituted by shells of constant current density, with spacers, in a way to mimic the cos θ distribution, that is, they are a mix of the two concepts mentioned above. A practical coil cross-section can be approximated as sectors of uniform current density shown in the pictures of Figure 4.3.10a, used also for schematic cross-sections of Figure 4.3.2. The configuration shown for dipole generates an approximate dipole B_1, with higher order field errors. Because of symmetry, the only field errors produced (the so-called *allowed* multipoles) are normal multipoles of order $(2n + 1)$, that is, B_3, B_5, B_7, and so on. Similarly, the configuration of Figure 4.3.10b produces an approximate quadrupole B_2 with normal higher order multipoles of order $2(2n + 1)$, that is, B_6, B_{10}, B_{14}, and so on. In Tables 4.3.3 and 4.3.4, a set of practical formulae for the main field and errors are reported for dipoles and quadrupoles.

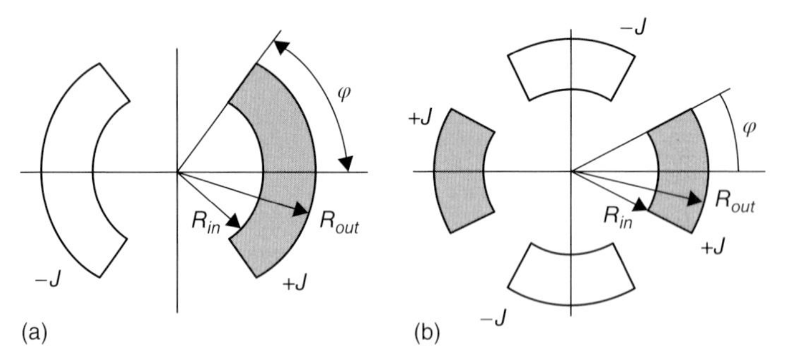

Figure 4.3.10 Principle of sector coils that generate an approximate dipole field (a), and quadrupole field (b).

Table 4.3.3 Practical analytical formulae for the dipole field and field errors for the dipole sector coil configuration in Figure 4.3.14a. The force in a coil refers to a quadrant. The azimuthal stress is intended as average on the coil mid-plane. R_{in} and R_{out} are the coil inner radius and outer radius.

Main field	$B_1 = \frac{2\mu_0}{\pi} J(R_{out} - R_{in})\sin(\varphi)$
Field errors $n = 3, 5, \ldots, 2i-1$	$B_n = \frac{2\mu_0}{\pi} J \frac{R_{out}^{2-n} - R_{in}^{2-n}}{n(2-n)} \sin(n\varphi)$ $A_n = 0$
Force per coil quadrant	$F_x = \frac{\sqrt{3}\mu_0 J^2}{\pi}\left[\frac{2\pi-\sqrt{3}}{36}R_{out}^3 + \left(\frac{\sqrt{3}}{12}\ln\left(\frac{R_{out}}{R_{in}}\right) + \frac{4\pi+\sqrt{3}}{36}\right)R_{in}^3 - \frac{\pi}{6}R_{out}R_{in}^2\right]$ $F_y = \frac{\sqrt{3}\mu_0 J^2}{\pi}\left[\frac{1}{12}R_{out}^3 + \left(\frac{1}{4}\ln\left(\frac{R_{in}}{R_{out}}\right) - \frac{1}{12}\right)R_{in}^3\right]$ $F_z = \frac{3\mu_0 J^2}{\pi}\left[\frac{1}{6}R_{out}^4 - \frac{2}{3}R_{out}R_{in}^3 + \frac{1}{2}R_{in}^4\right]$
Stress in the mid-plane	$\sigma_\theta = \frac{6\mu_0 J^2}{4\pi}\left[\frac{5}{36}R_{out}^3 + \frac{1}{6}\left(\ln\left(\frac{R_{in}}{R_{out}}\right) + \frac{2}{3}\right]R_{in}^3 - \frac{1}{4}R_{out}R_{in}^2\right)\frac{1}{R_{out}-R_{in}}$
Energy per unit length	$\frac{E}{l} = \frac{\pi B^2 R_{in}^2}{\mu_0}\left[1 + \frac{2}{3}\frac{R_{out}-R_{in}}{R_{in}} + \frac{1}{6}\left(\frac{R_{out}-R_{in}}{R_{in}}\right)^2\right]$

Table 4.3.4 Practical analytical formulae for the field gradient and field errors for the quadrupole sector coil configuration in Figure 4.3.14b. The force in a coil refers to an octant. The azimuthal stress is intended as average on the coil mid-plane.

Main field	$G = B_2 = \frac{2\mu_0}{\pi} J \ln\left(\frac{R_{out}}{R_{in}}\right)\sin(2\varphi)$
Field errors $n = 6, 10, \ldots, 2(2i-1)$	$B_n = \frac{4\mu_0}{\pi} J \frac{R_{out}^{2-n} - R_{in}^{2-n}}{n(2-n)} \sin(n\varphi)$ $A_n = 0$
Force per coil octant	$F_x = \frac{\sqrt{3}\mu_0 J^2}{6\pi}\left[\frac{1}{72}\frac{12R_{out}^4 - 36R_{in}^4}{R_{out}} + \left(\ln\left(\frac{R_{in}}{R_{out}}\right) + \frac{1}{3}\right)R_{in}^3\right]$ $F_y = \frac{\sqrt{3}\mu_0 J^2}{\pi}\left[\frac{5-2\sqrt{3}}{36}R_{out}^3 + \frac{1}{12}\frac{R_{in}^4}{R_{out}} + \left(\frac{2-\sqrt{3}}{6}\ln\left(\frac{R_{in}}{R_{out}}\right) + \frac{\sqrt{3}-4}{18}\right)R_{in}^3\right]$ $F_z = \frac{3\mu_0 J^2}{4\pi}\left[\frac{1}{4}R_{out}^4 - \left(\ln\left(\frac{R_{out}}{R_{in}}\right) + \frac{1}{4}\right)R_{in}^4\right]$
Stress in the mid-plane	$\sigma_\theta = \frac{\sqrt{3}\mu_0 J^2}{\pi}\left[\frac{1}{36}\frac{7R_{out}^4 + 9R_{in}^4}{R_{out}} + \frac{1}{3}\left(\ln\left(\frac{R_{in}}{R_{out}}\right) + \frac{4}{3}\right)R_{in}^3\right)\frac{1}{R_{out}-R_{in}}$
Energy per unit length	$\frac{E}{l} = \frac{\pi G^2 R_{in}^4}{2\mu_0\left(\ln\left(\frac{R_{out}}{R_{in}}\right)\right)^2}\left[\frac{1}{8}\frac{R_{out}^4 - R_{in}^4}{R_{in}^4} - \frac{1}{2}\ln\left(\frac{R_{out}}{R_{in}}\right)\right]$

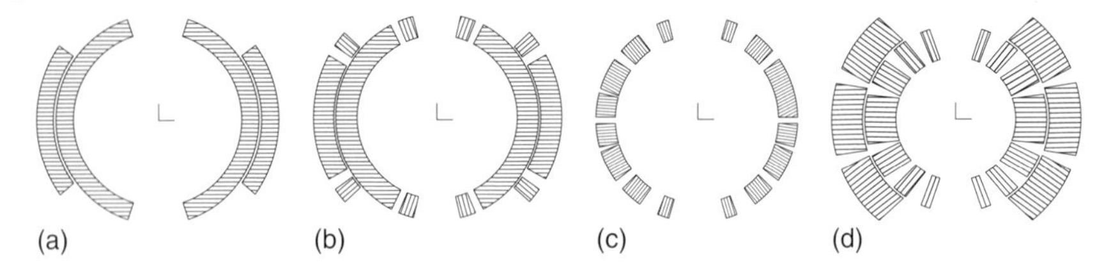

Figure 4.3.11 Cross-section of actual operating dipoles: (a) Tevatron, (b) HERA, (c) RHIC, and (d) LHC.

Table 4.3.5 Scaling of useful quantities versus design goals, kept constant, for dipole magnets.

Design goals	Coil width, w	Force on quadrant, F	Coil stress on mid-plane, σ	Stored energy, E
B, R_{in}	$\approx 1/J$	$\approx$Const.	$\approx J$	$\approx\left(\frac{1}{J}\right)^{\frac{1}{2}}$
J, R_{in}	$\approx B$	$\approx B^2$	$\approx B$	$\approx B^{\frac{5}{2}}$
B, J	$\approx$Const.	$\approx R_{in}$	$\approx R_{in}$	$\approx R_{in}^{\frac{3}{2}}$

Good field quality can be obtained by choosing the sector angles, segmenting the sectors using insulating wedges, and using two (or more) nested layers. This adds degrees of freedom that can be used to improve the field homogeneity, at the cost of an increased complexity of the winding. In Figure 4.3.11, we show the coil cross-sections for a variety of accelerator magnets. It is evident how the coils have evolved in complexity to follow the increased demand of field quality.

Tables 4.3.3 and 4.3.4 report other key quantities for the design of an accelerator magnet, namely the resultant forces in a coil quadrant (or octant), the mid-plane stress, and the energy per unit length. The coil radial width $w = R_{out} - R_{in}$ can be used to estimate the overall coil volume, mass, and material cost. In Table 4.3.5, a few approximate relations, to scale useful quantity based on the design parameters are reported in case of a dipole.

An interesting result of the analysis is that the forces in a dipole scale proportionally to the magnet aperture, and with the square of the field. The accuracy of the scaling is demonstrated in Figure 4.3.12, where we report the horizontal and vertical forces in the coil quadrant of the dipoles of the four large-scale SC colliders, and the SSC. The scaling is reasonably accurate, better than 20%, for magnets that cover a large span of field and aperture. The force scaling explains why the mechanical design of high-field magnets is challenging and why, in general, high-field magnets tend to have the minimum practical aperture. We see, however, that if the coil is designed for a given current density, which is usually the case, the stress in the coil increases only linearly with the field. This is because the width of a coil with given current density also increases linearly with the bore field, thus providing more material to resist the force. In fact, the stress can be lowered by

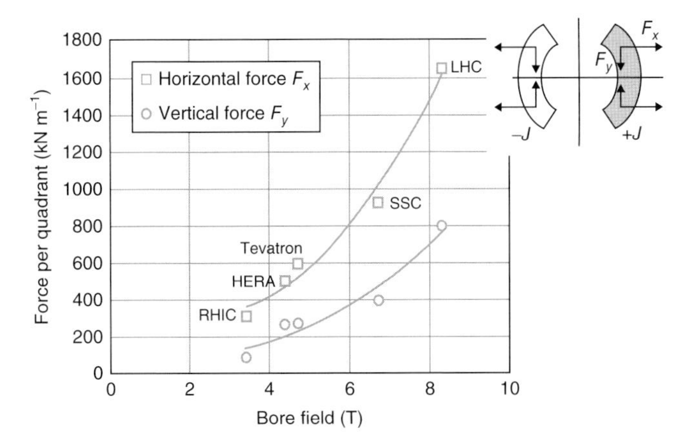

Figure 4.3.12 Force scaling in dipoles, according to formulae of Table 4.3.3.

reducing the current density in the coil and/or R_{in}, as indicated by the scaling relations above.

An interesting result of the scaling analysis is that the magnet energy per unit length is also a strong function of the bore field and aperture, whose consequence will discussed in the section dedicated to quench and protection.

All major projects have used so far the cos θ coil configuration, see Figure 4.3.11. The main characteristics of the various accelerators and of their main dipoles are reported in Table 4.3.6, see Refs [11–14].

4.3.3.5 Other Coil Shapes: Block, Canted, Super-Ferric, Transmission line

The matter of optimal coil shape, especially for high-field magnets, is a subject of debate. There is nonetheless a broad consensus that for dipole magnets with field up to 12–13 T and modest bore (50 mm range), the cos θ coil configuration described earlier is superior, and has the considerable advantage of being well proven. Other coil layouts have been investigated in the past, see as an example Refs [15–17]. In Table 4.3.7, we report the cross-sections of various magnets with coil design other than cos θ, with some comments on their characteristics.

A first alternative for large bore (range of 100 mm) and high field (in excess of 13 T), are coils built as rectangular blocks. These have the advantage of avoiding stress build-up on the mid-plane, a limiting factor important for brittle conductors like Nb_3Sn and high-temperature superconductor (HTS) materials. Indeed, the present record for dipole field in a sizable bore is held by the HD2 dipole model magnet of Lawrence Berkeley National Laboratory (LBNL) [18] as shown in Figure 4.3.7. This is, however, only marginally better than the 15-year older LBNL D20 featuring cos θ coils and manufactured with vintage Nb_3Sn wire (J_c lower by a factor 2 with respect to HD2).

An extension of block-coils to twin aperture magnets (with opposite field) is the common coil design, originated in Brookhaven National Laboratory (BNL) [16]. This option, suited for moderately high (10–12 T) field, has the advantage that the

Table 4.3.6 Main characteristics of large machines based on SC magnets.

Accelerator		Tevatron	HERA	RHIC	LHC

Maximum beam energy	TeV	0.98 + 0.98	0.82 (0.92)	0.25 + 0.25 0.1/n + 0.1/n	7 + 7 (4 + 4)
Injection energy	GeV	150	40	12	450
Ring length	km	6.3	6.3	3.8	26.7
Dipole cross-section	—		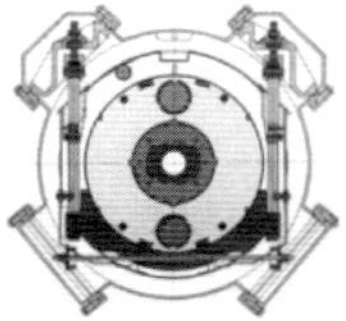	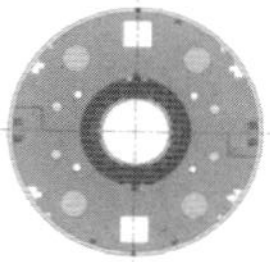	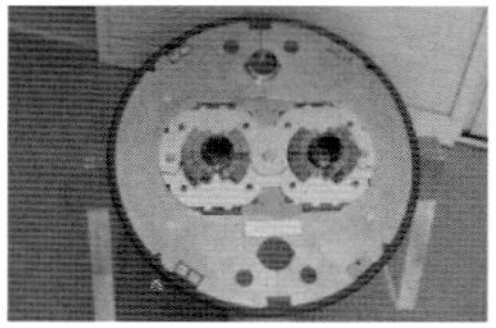

Dipole field	T	4.3	4.65 (5.22)	3.5	8.3 (4.8)
Aperture	mm	76	75	80	56
Magnetic length	m	6.1	8.8	9.5	14.3
Stored energy	MJ	320	250	60	8000
Weight (cold mass)	kg	—	—	3607	27 000
Quantity	#	774	422	264s	1232
Iron yoke	—	Warm	Cold	Cold	Cold
Operating temperature	K	4.2 (He supercritical)	4.5 (He sat.)	4.6 (He forced flow)	1.9 (HEII at 1 bar)
Structure	—	Single bore Warm iron Straight	Single bore Cold iron Straight	Single bore Cold iron Curved	Twin bore Cold iron Curved
Force retaining	—	St. Steel collars	Al alloy collars	Yoke “collars”	St. Steel collars + shrink cylinder via yoke
Operating current	A	—	5027 (5640)	5050	11 850
Physics start	—	1983 (0.512 TeV) 1986 (0.9 × 2 TeV) 2011 (0.98 × 2 TeV)	1990	2000	2010 (3.5 × 2 TeV) 2012 (4 × 2 TeV) 2015 (6.5 × 2 TeV)

Table 4.3.7 Coil shapes alternative to the cos θ layout.

Rectangular coil block	Features	Comments
TAMU	Design for 24 T in 25 mm dual bore. Then designed for 13 and 7 T as demo steps.	Tested on Nb–Ti. Main feature is stress management. By means of ribs and struts and springs (to be demonstrated).
Ellipsis	Designed for 13 T. No demonstrator.	Lower stress on mid-plane, but larger horizontal force than cos θ. Goof field quality, larger amount of SC, large inductance.
Toroidal motor type	Designed for 13 T. No demonstrator.	Very low mid-plane stress, simple winding (flat coils). Very large amount of SC.
HD02	Design for 13–15 T, bore 25 mm. A number of 1 m-long models built and tested. Record field (∼14 T).	Low mid-plane stress. Relatively simple coils (except flare ends). More conductor than cos θ. Very suitable for keys and bladders. Flare ends limitation to be understood.
Fresca2	Designed for 13 T at 70% (for 2015). Large bore of 130 mm.	Similar to HD0 with more margin (large bore). Large amount of SC. Relatively low peak stress and hot spot. Flare ends viability to be demonstrated.
Common coil	A number of coils and assembly for medium field (8–11 T).	Easy winding in the simple approach. For field quality nonplanar coil or open mid-plane is needed which limits the field.
Canted Cosine Theta (CCT)	A few low field (<10 T) models built. In demonstration phase.	Natural strain management; easy to wind with small cable. It required 15–20% more conductor.
Super-ferric		
SIS100	Project under construction.	Suitable to remove large losses at 4 K. Optimized for a large rectangular aperture.
Pipetron	2 m long demo built and operated with Nb–Ti.	Low cost, ideal for $B > 2$ T with 2-bores, for small gap (<30 mm). Use of large compact SC cable (∼100 kA). Suitable for HTS.

coils are simple flat racetrack with large curvature radius. The drawbacks are that the two apertures are strongly coupled and field quality is distorted. Mitigation measures are possible, but make the design more complex or reduce the field level.

An “old” concept [19] experiencing renewed interest is the Canted Cosine Theta (CCT), see [20, 21]. A CCT winding is the superposition of two inclined solenoids, where the current in the azimuthal direction cancels out, and only the longitudinal component remains, thus generating a pure transverse field. The conductors are arranged in a nearly perfect cos θ, in a winding that is easy to manufacture and allows for internal supports, thus avoiding force and stress build-up inside the coil. A CCT requires more conductor than a cos θ coil because of the angle of the current in the winding, and extrapolation to large magnets, requiring wide flat cables, is not evident. Its simplicity is nonetheless a strong motivation for the R&D, especially in view of use of HTS technology, as well as for other applications such as in the field of medical accelerators.

A special case of SC magnet configuration is when the field is dominated by iron magnetization, rather than the conductor current [22, 23]. The focus in this case is to exploit the economic advantage of operating a SC coil, rather than a resistive (copper) coil. In this case, the coil arrangement is usually a rectangular block placed in the most convenient location to magnetize the iron circuit. In these coils, the current density is not a major issue; the main concern is heat removal and cryogenic efficiency.

An effective combination of iron and SC coils is a transmission-line dual magnet, also known under the nickname of “Pipetron” [24, 25]. A single, straight, and large-size SC cable (100 kA range) magnetizes a dual bore iron yoke. The option reported in Table 4.3.7 generates 1.7 T in two 30 mm bore with good field quality controlled by iron shape. This type of magnet can fit very well HTS cables that are widely developed for power transmission lines.

4.3.4 Mechanical Design

4.3.4.1 Collars and Cos θ

As it can be seen from Figure 4.3.5, the racetrack coil shape does not provide any self-support for the lateral forces, which have to be contained by a retaining structure. In addition, always looking at Figure 4.3.5, a compressive force squeezes the coil toward the mid-plane. To avoid cable movement in the azimuthal direction (vertical at the mid-plane), one can insert oversized wedges in the pole region, according to idea of a “roman arch” [26]. An external rim pushes the pole wedge in the final position, generating azimuthal prestress on the coil package, and provides radial prestress; the coupling of these two concepts gives rise to the collar concept, first devised at Fermilab for the Tevatron. In Figure 4.3.13, the illustration is given for the LHC collar (twin aperture). The mechanism is that each pair of half collars is coupled to an adjacent pair of half collars which is mounted upside down (i.e., the collars with long pads is below, while the short collar is on top). The short pad collar is actually a spacer with no mechanical function, while the long

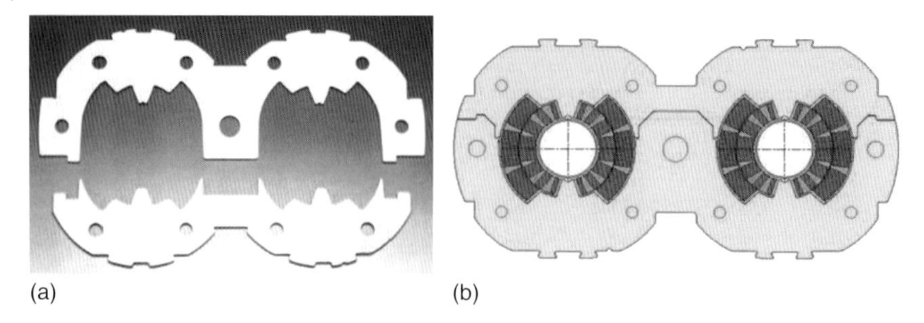

Figure 4.3.13 (a,b) Collar concept for LHC dipoles. The three mid-plane holes are for the locking rods (Source: CERN archive).

pad collar locks with the adjacent, upside-down, collars by inserting very precise rods (tolerance range 10–20 μm) from one end to the other end of the magnet, through the three mid-plane holes. The tolerances required for insertion of the rods cause a loss of the prestress due to a slight opening of the coil cavity, when the collaring pressure is released. For magnets that are collared as single aperture, the collars can be better locked by means of square keys, inserted from the sides in appropriately shaped keyways. The advantage of lateral keys with respect to rods is the fact that they can be force-fit to the mated collars, thus resulting in lower loss of prestress during the collaring.

Preferred option for collars is rigid materials, and foremost austenitic steel. Steel, however, has lower thermal contraction than the coil package, and to compensate for the loss of prestress during cooldown one has to begin with larger values of prestress after collaring than otherwise needed to compensate for key or rod tolerance and collars elasticity only. If the loss of azimuthal prestress is too high, or if the collar rigidity is not sufficient to retain the electromagnetic forces, the collars can be assisted by an external cylinder that pushes the yoke against the collars, contributing to increase the total rigidity of the structure and to reduce the loss of prestress during cooldown. The external cylinder is also prestressed during assembly, for example, by welding shrinkage, or by differential thermal contraction by using aluminum alloy. When the collars are not assisted by the yoke, one speaks of "free-standing collars." This solution is advantageous, given the ease in assembly, and it is routinely used for moderate field dipoles (5–6 T) and for quadrupoles for which forces are lower, see, for example, the LHC main quadrupoles [27].

An alternative way to overcome the problem of prestress loss during cooldown is to use collars made out of aluminum alloy, like in the case of the Hadron-Electron Ring Accelerator (HERA) main dipoles. The concept was taken as the baseline for the LHC dipoles [28]. However, in this case, the low modulus of aluminum may not provide enough rigidity to restrain all movements of a thick coil package and requires a "major" contribution by the yoke. This can only be achieved if the yoke allows movements, for example, being split vertically, and requires tight control of its kinematics during the various phases, a very precise line fitting of the mating surface between collars and yoke components over the magnet length, which is difficult to reach in practical series construction. A statistical analysis of the possible tolerances spread for LHC construction was the base for a change of the LHC dipole design in favor of austenitic steel collars. However, the fee to pay was that despite the 150 MPa peak prestress during collaring, at nominal field the coils are azimuthally unloaded at the pole [6]. In theory, if radial support is maintained, the detachment between pole wedge of the collars and the pole turn of the coil is not a problem, as shown by the fact that LHC dipoles have been routinely tested at 9 T (and even higher fields during prototyping phase). However, because of the presence of friction in a regime of stick-and-slip [29] and in case of inadequate support, considerable training is required to reach this value and the results are sometimes erratic. It should finally be noted that the collaring operation requires heavy tooling. As an example, the LHC dipoles were collared in a press capable of 20 MN per meter length with accuracy of 20 μm. Such tool constitutes a considerable investment for the initial R&D phase of a project.

An alternative concept is to drop the idea of a proximity collar, and use the iron yoke itself as the main support, like it was done for the relativistic heavy ion collider (RHIC) main dipoles. In this case, it is quite natural to place the yoke split in the horizontal mid-plane and to lock the yoke laminations like the collars, with keys that are inserted form the side. Collars are still present, but they serve only as coil assembly features and to transmit the force from the coils to the yoke. This *yoke collaring* concept has been applied in the LHC low-β quadrupoles designed and built by High Energy Accelerator Research Organization (KEK) [30] with very good results. It is instructive to note that similar type of quadrupoles, also installed in the LHC low-β triplet, have been designed and built by Fermilab according to the free-standing collar concept [31]. They met the goal with a coil peak field near 8 T and extremely good field quality. This example shows that different layouts and design concepts can be adopted to reach a certain goal. All can be successful if properly designed and engineered, the choice depending more on the preference of the designer, the available tooling, and other practical considerations.

4.3.4.2 Bladders and Keys

In recent years, a novel concept has proven its soundness and is considered a serious alternative to the collar clamping. The coils are precompressed by thin stainless steel bladders pressurized at 200–600 bar against an external cylinder. Thin keys are inserted between the structure pushing the coils and the cylinders with minimum tolerance, and finally the bladder pressure is released so that the cylinder gives precompression to the coils via the keys, as illustrated in Figure 4.3.14. If the cylinder is made out of aluminum alloy, which is common practice for this concept, the prestress is further increased during cooldown due to the differential thermal contraction of the Al cylinder and the coil pack. The main advantages of this assembly method over the collaring are as follows:

1) The horizontal and vertical prestresses are almost decoupled, allowing a better stress control that is very important for high-field magnets.

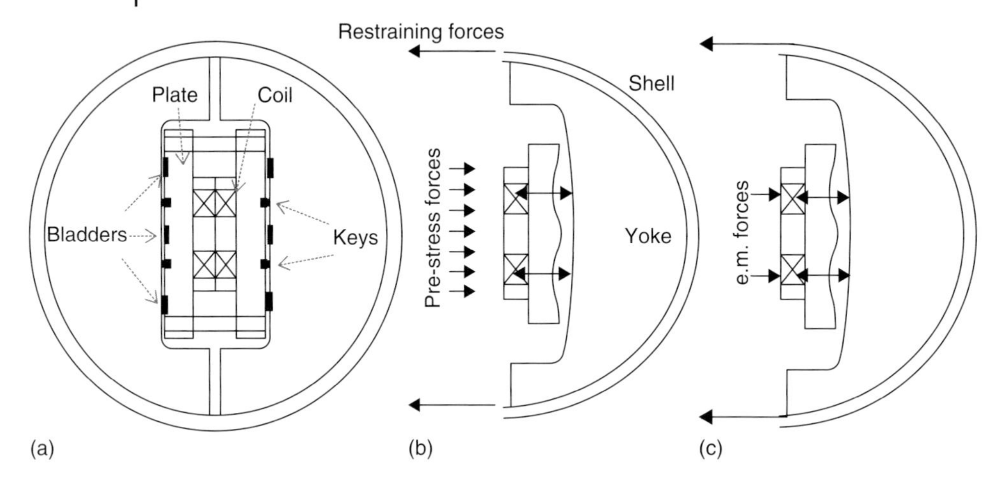

Figure 4.3.14 Functioning principle of bladders and keys: (a) pressurizzation of bladders and insertion of keys; (b) removal of bladders leaving the cylinder opposing the prestress forces via the keys; (c) the prestress forces are replaced by e.m. forces during magnet ramp. (Drawings by courtesy of S. Caspi, LBNL).

2) The stress is given by force applied via the bladder pressure, so it is well controlled, within percents. With collars, the prestress is given by the relative accuracy of the shape of the collars and of the coils, so the geometry is fixed: a coil larger than nominal value may be degraded if it has to fit inside a nominal value collars. The fact of controlling the forces is a real plus for brittle Nb_3Sn and HTS conductors.
3) The bladders and keys structure is also quite easy to assemble and disassemble, requiring a modest tooling in comparison to the collar technique.

A few issues still need to be addressed and demonstrated with this structure. Besides the recurrent desire of pushing magnet performance, engineers are turning their attention to obtaining the field quality by controlling force rather than geometry, and to providing the required alignment over lengths of 5–15 m as typical of accelerator magnets. A few magnets have been built using the bladders and keys technology, such as the 1 m long HD02 and HD03 block dipoles of LBNL (see Figure 4.3.7 and Table 4.3.7; [32]), the series of LHC Accelerator Research Program (LARP) cos θ large-aperture quadrupoles for the High-Luminosity LHC project [33]: Technological Quadrupole with Shell (TQS), Long Quadrupole (LQ), and High gradient/aperture Quadrupole (HQ). Five of these magnets are 1 –m long magnets, while two have a length of 3.6 m. Since the results have been very positive so far, we can say that the bladders and keys technique is now entering in the field as practical alternative to collars: experimented and proven for high-field magnets so far, it may also provide a cost-effective alternative to collars for low- and intermediate-field levels.

4.3.5
Margins, Stability, Training, and Protection

4.3.5.1 Margins and Stability

The operating point of a SC magnet must be chosen below the critical current of the material, as discussed in another chapter [34], to ensure proper engineering margin for stable operation. There are various means to measure the margin, useful in the design and analysis of performance of a SC magnet. We list below the most common definitions:

1) Critical current margin: by calling i the operating fraction of the critical current density $i = J_{op}/J_c(B_{op}, T_{op})$, the critical current margin is defined as 1-i.
2) Margin along the load line: by calling l the ratio of operating to critical current $l = J_{op}/J_{max}$, we define the load margin as 1-l. J_{max} is the critical current evaluated at the intersection of the magnet load line, defined as $t = B_{op}/J_{op}$, and the critical surface. This can be written symbolically as $J_{max} = J_c(tJ_{max}, T_{op})$ where we note the implicit nature of this last relation.
3) Temperature margin ΔT: the difference in temperature from operating conditions T_{op} to current sharing conditions T_{CS}, evaluated at the operating field and current density $\Delta T = T_{CS}(J_{op}, B_{op}) - T_{op}$. We define here the current sharing temperature as the temperature at which the operating current density equals the critical current, or $J_{op} = J_c(B_{op}, T_{CS})$.
4) Energy margin ΔE: the minimum energy that is necessary to induce a thermal runaway of the cable.

The critical current margin, the margin along the load line, and the temperature margin are straightforward and relatively easy to compute. Typical orders of magnitude for the large-scale accelerator dipoles listed earlier are in the range of $i \approx 0.5, \ldots, 0.6$, $l \approx 0.7, \ldots, 0.9$, and $\Delta T = 0.5, \ldots, 1.5$ K. The most useful margin to judge on stable operation of a magnet remains nonetheless the energy margin ΔE. Indeed, a SC magnet is always subject to a series of energy inputs of very different nature, timescale, and magnitude, the so-called *disturbance spectrum* [2]. The energy input causes an increase of the temperature and can bring the SC material above current sharing conditions, where joule heating starts. The enthalpy reserves of the magnet, heat transfer to the surroundings, or a coolant, provide a heat sink, and the final state of the cable (recovery and quench) depends on the balance of heating and cooling. Intuitively, we expect the energy margin to depend on the operating condition, the timescale, and space distribution of the perturbation.

Among the mechanisms that can cause the generation of heat in a SC cable, we can count disturbances that can be extremely localized in space and time, for example, mechanical energy release due to a small motion of a SC wire, a crack in the insulation, or electromagnetic energy release through flux jumps. Other disturbances can affect large portions of a SC magnet and last a significant time, for example, alternating current (AC) loss for pulsed operation or heat deposition from nuclear processes. We do not discuss here magnetization and magnetothermal instabilities that with recent impressive increase of critical current density in Nb_3Sn have become again a major concern [7, 35, 36].

4.3.5.2 **Training**

Magnet training is the consequence of mechanical perturbations that exceed the energy margin while ramping the magnet. Local motion, friction, or cracks release some of the mechanical energy stored in the deformation field associated with the electromagnetic force. At each event, the coil has a possibility to settle, which usually, but not necessarily, reduces the amplitude of the mechanical perturbations at the next ramp. The following quench is hence at a higher current and field, and the magnet *trains* toward increasing performance.

Training is almost inevitable in accelerator magnets, the high $J_{overall}$ and low stabilizer content (50–60%) result in the small (~10–100 μJ) energy margin of Figure 4.3.15 [37], while given the high forces and stresses can produce relatively high energy releases. In Figure 4.3.16, a training curve of an LHC accelerator dipole is shown, which shows a situation shared by essentially all large SC accelerator projects. The quench current must be compared to the nominal value for operation, and the ultimate value, that is, the maximum value for which the magnet has been designed in terms of mechanics, electrics, protection, and so on, which is usually 5–10% higher than the nominal. We also report in Figure 4.3.16 the critical current limit along the load line, often referred to as *short sample limit* of the conductor, corresponding to 9.65 T central field for LHC dipoles. We see that the magnet training starts at a current well under the short sample limit,

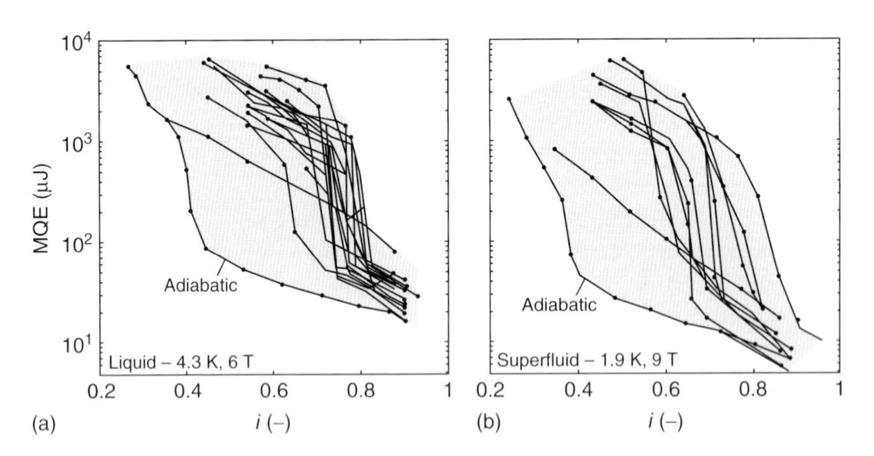

Figure 4.3.15 (a,b) Measured energy margin for LHC cables, and operated at 4.3 K, 6 T or 1.9 K, 9 T, versus *reduced current* ($i \equiv I/I_c$). The shaded area regroups results from cables of different interstrand resistance, average residual resistivity ratio (RRR), and different locations in the cables tested (center, sides). The adiabatic limit refers to a cable impregnated in epoxy. It should be noted that the vertical scale is approximately the same when margin values are expressed in millijoule per cubic centimeter.

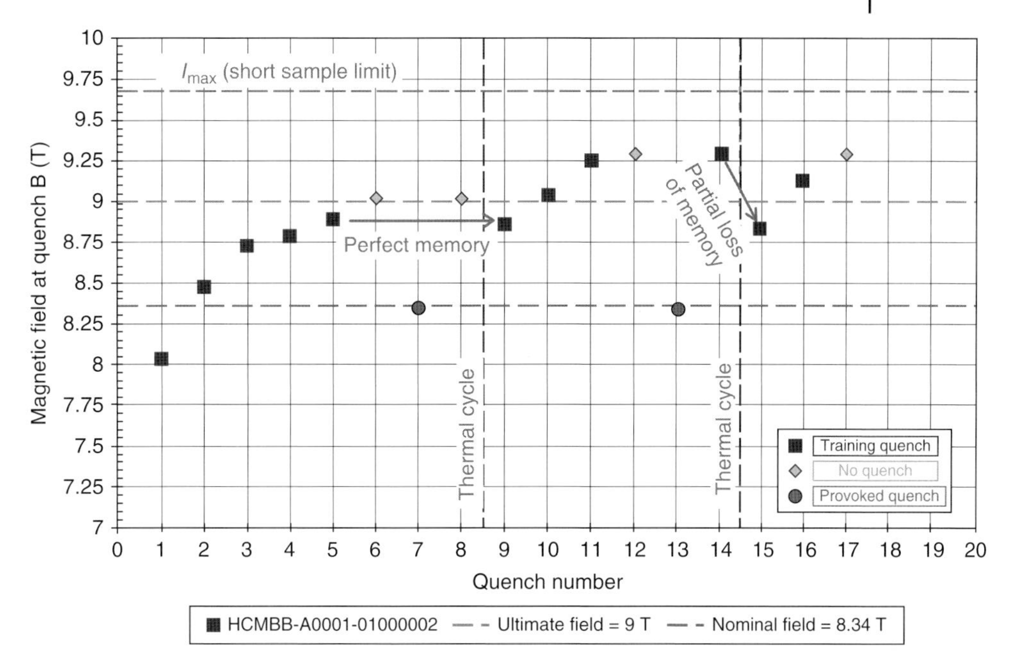

Figure 4.3.16 Quenches of the LHC dipole 1002 at virgin training curve and after first and second thermal cycles (indicated by vertical lines). (Courtesy of A. Siemko, CERN).

at $B = 8$ T in Figure 4.3.16. Training proceeds rapidly toward 9 T, or 93% of I_{max}. After a first thermal cycle, the magnet restarts training at a field value equal to the last one before thermal cycle, which can be seen as a perfect "memory" of the previous powering. Further training brings the magnet to 9.25 T, 95% of I_{max}. After a second thermal cycle, however, there is partial loss of the training memory, and the magnet needs additional quenches to attain once more the previous maximum level. The case of this magnet, which is one of the best in the LHC series production, shows the necessity of margin to cope with a large series production.

This is why the operating point of the LHC main dipoles was designed at 86% of the I_{max}, which is nonetheless in the range of training in the curve reported in Figure 4.3.16. To provide the global picture, we show in Figure 4.3.17 the distribution functions of the first and second quench values obtained at the first magnet cooldown, compared to the distribution of the first quench obtained at the second magnet cooldown, that is, after a full thermal cycle. Training memory is very important, especially when training starts below the nominal operating current of the magnet in the accelerator. Lack of memory is of concern, as it can affect adversely practical operation of a large accelerator. Depending on the circuit, at each quench, a whole string of magnets has to be ramped down, which is very time-consuming and bears considerable risks associated with energy release, quench propagation to neighboring magnets, and helium pressure increase. In practice, one can accept a level of retraining of accelerator magnets not larger than a 10–20% of the total magnet population.

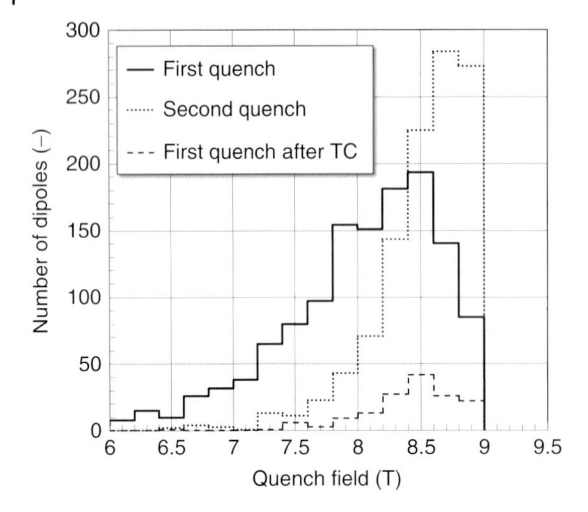

Figure 4.3.17 Histogram of training quenches of the LHC dipole at reception. Out of the 1232 dipoles tested with a first cycle, only about 12% have been submitted to thermal cycle. During test, 9 T was the maximum allowed field.

Table 4.3.8 Summary of training quench fields for the LHC dipoles. Averages and standard deviations in the central column, median values in the column at right.

First quench	8.05 ± 0.61 T	8.15 T
Second quench	8.53 ± 0.44 T	8.61 T
First quench after thermal cycle	8.43 ± 0.35 T	8.45 T

In Table 4.3.8, we give the average and standard deviation for the LHC quench distribution of Figure 4.3.17. Only some 12% of the dipoles have been submitted to one thermal cycle (note that in most cases the dipole undergoing the thermal cycle were the "bad" ones). We see from Table 4.3.8 that the first quench after thermal cycle shows an average loss of memory of 0.1 T: however, this is only 0.1 T above the 8.33 T nominal operating field, while the standard deviation is 0.35 T, so we would expect approximately half of the 1232 to have to quench below nominal operating field at the commissioning of the LHC. The number is actually slightly higher because we have also a small but finite probability to have a second quench below nominal field. So, despite that on average the magnets that have been submitted to thermal cycle during acceptance test have reached 9 T (93% of I_{max}) in less than two quenches, we can state from the acceptance test that reaching the nominal value will require a considerable training campaign. In fact, the retraining of the dipoles during the LHC commissioning was found to be longer than expected from above considerations [38], which suggests that in order to enable a quick commissioning of an accelerator, a safe design margin for large-scale productions such as the LHC is rather in the range of 80% of I_{max}.

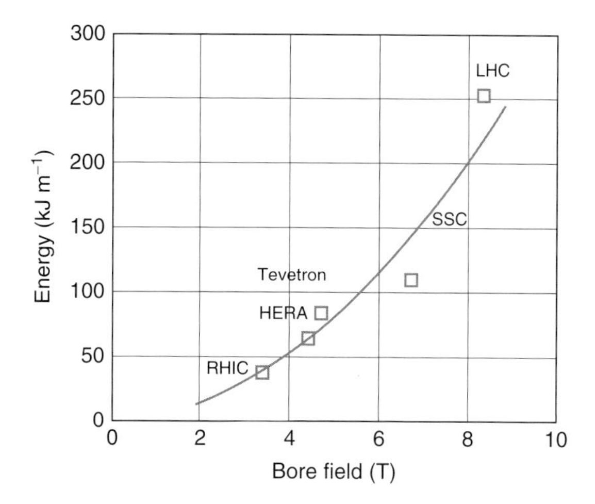

Figure 4.3.18 Stored energy per unit length for accelerator dipoles of previous and existing projects. Each magnet is considered single aperture for this comparison so LHC dipole equivalent length is 29 m.

4.3.5.3 Protection

The main issues for HEP magnets are the large current density in the conductor and the large magnetic energy stored per unit of coil mass, see the energy per unit length for various magnet designs plotted in Figure 4.3.18 together with a fit based on the scaling law of Table 4.3.3. During a resistive transition at nominal operating current in an LHC dipole, the current density in the copper stabilizer is 760 A mm^{-2} in the inner layer cable, and 930 A mm^{-2} in the outer layer cable. This corresponds to a joule power density of 60–80 MW m^{-3} at 20 K, increasing by more than two orders of magnitude at room temperature. Such heating rate is capable to bring copper to melting temperature in ~0.3 s. It is obviously mandatory to switch-off the power supply, typically on a timescale 10 times faster than above, as soon as a quench is detected by electrical means (balanced bridge measurements). This is however not enough, because the magnet, an inductance L, stores a large quantity of energy E that drives the current during the dump process. Accelerator magnets, with well-controlled and highly loaded mechanical structures, can tolerate temperature increases up to room temperature conditions, 300 K, and even higher temperatures in the range of 350–400 K are now being considered for high-field magnets. The hot-spot temperature $T_{\text{hot spot}}$ can be estimated using an adiabatic heat balance, equating the joule heat produced during the discharge to the change of enthalpy of the conductor

$$\int_{T_{\text{op}}}^{T_{\text{hot spot}}} \frac{C}{\rho} dT = \frac{1}{A_{\text{Cu}} A_{\text{tot}}} \int_{t_{\text{quench}}}^{\infty} I^2 dt$$

where C is the total volumetric heat capacity of the superconductor composite, and ρ is the resistivity. We see from the above concept, borrowed from electrical blow-fuses design (whence the improper analysis in terms of *MIIts*), that to limit $T_{\text{hot spot}}$ it is always advantageous to increase the percentage of stabilizer (which has high conductivity) in the composite, which requires higher J_c, if the coil thickness has to stay constant. The speed at which the energy dump takes place, on the other hand, depends on the magnet inductance and the resistance of the circuit formed by the quenching magnet and the external circuit formed by the power supply, switch, and dump resistor, or $R = R_{quench} + R_{ext}$. The characteristic time of the dump is then $\tau \approx L/R$, which can be made short by decreasing the magnet inductance by use of large current cables $L \propto N_{turns}^2 \sim 1/I^2$. Increasing the resistance is advantageous, too. However, from practical considerations of maximum terminal voltage, $\Delta V = RI < 1\ \text{kV}/N_{magnets}$, it follows that in series operation (where $N_{magnets} \cong 40-150$) the external resistance is negligible for the protection of a single magnet. Since the growth of the natural quench resistance is usually too slow, protection ultimately relies on active quench initiation, triggered by heaters embedded in the winding pack, and fired at the moment a quench is detected to spread the normal zone over the whole magnet mass. Active quench initiation is also the only way of making the temperature and voltage distribution more balanced, reducing thermal and electrical stress. In certain cases, the current decay is so fast (dI/dt can range from -5 to $-20\ \text{kA s}^{-1}$) that eddy currents induced in the metallic part of the coil package may generate as much heat as to quench part of the coils in the first part of the decay. This phenomenon, called *quench back*, is usually beneficial to reduce $T_{hot\ spot}$.

In synchrotrons and colliders, the main magnets are powered in series, and the stored energy in a single circuit is much larger than the few megajoule energy stored in a single magnet. In the case of each of the eight circuits formed by the series of 154 LHC main dipoles, the stored magnetic energy attains the gigajoule level. Protection in this case relies on the expedient of subdivision of the circuit, so that each magnet can be discharged independently. Each magnet is bypassed by a SC line (bus bar) with a diode that opens only in case there is voltage across the magnet, see scheme of Figure 4.3.19. The quench induced by the heaters insures that the quench voltage is sufficient to open the diodes (e.g., $\Delta V_{opening} \cong 7\ \text{V}$ at 1.9 K for the LHC diodes). The current through the quenched magnet decreases while its magnetic energy is dissipated in the quench resistance. At the same time, the diode carries the current of the whole line that is dumped with a much longer time constant. In the LHC, the dumping time of a single dipole is about 0.5 s, while the bypass line is dumped with a constant of 100 s. In practice, the diode and bus bar bypass line avoids discharge of the energy of the whole magnet line into the quenched magnet. The above concepts and solutions are commonly applied to large-scale HEP machines, as well as other SC magnet systems. The specific challenges for modern collider magnets is the need to detect and fire the quench heaters rapidly, with times in the range of few tens to a hundred of milliseconds, and the sheer dimension of relevant equipment, e.g., diodes with a 13 kA capacity at 1.9 K.

The situation will become yet more extreme for the new generation magnets, see Figure 4.3.7 and Table 4.3.7, where the stored energy is a considerable fraction of the total enthalpy reserve from operating conditions to the maximum allowed temperature (the hard limit for magnets that dissipate internally most of the

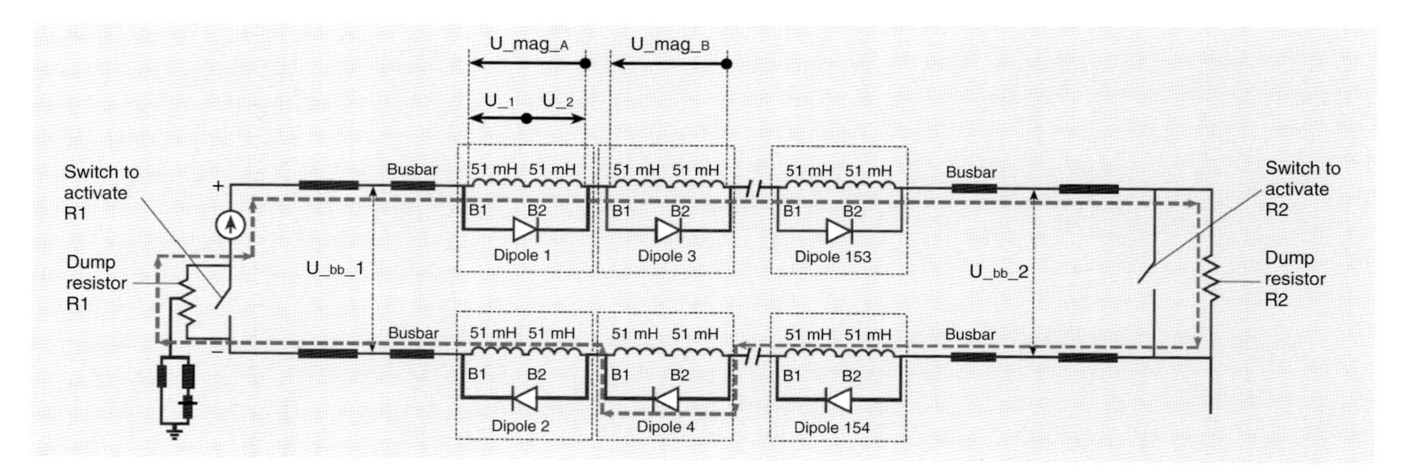

Figure 4.3.19 One of the eight dipole circuits of the LHC, showing the protection elements: the coils of the two apertures (2 × 51 mH), the diode and bus bar bypass, and the current path following a quench in Dipole 4 (dashed line). About 0.3 s after the opening of its diode Dipole 4 is bypassed through the bus bar and diode line, meanwhile the whole circuit is dumped with a 100 s time constant (Source: CERN archive).

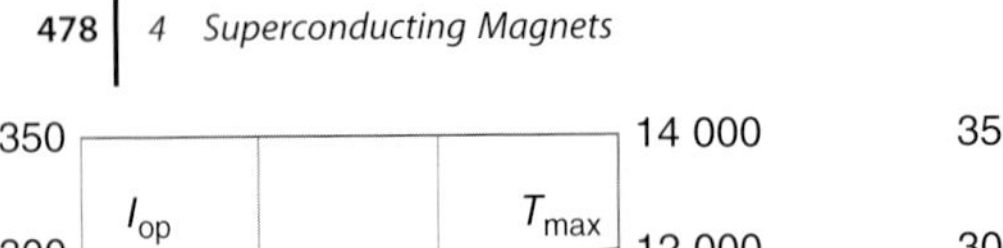

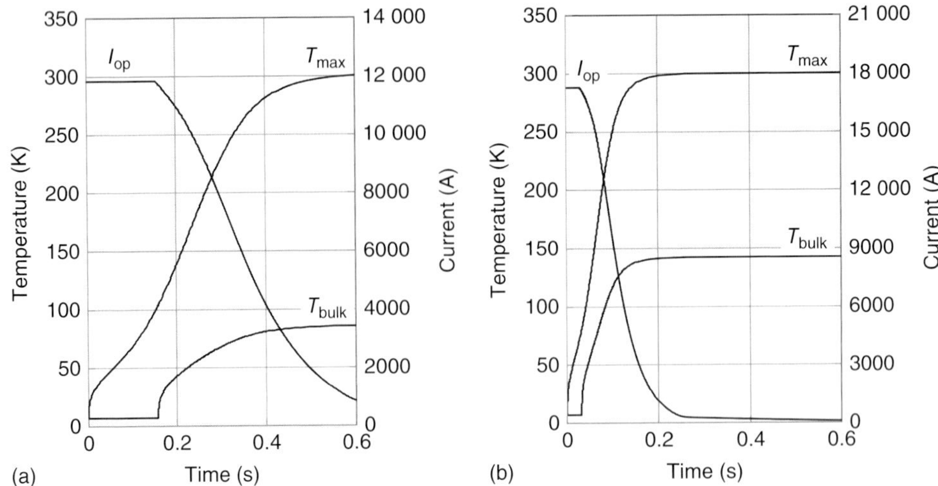

Figure 4.3.20 Quench evolution simulated by means of a simple code for the Nb–Ti LHC dipole (a) and for a Nb_3Sn quadrupole, QXF, under construction for the inner triplet of the HiLumi LHC Project (b) (Figure 4.3.1).

energy). As anticipated, current density and coil thickness for these magnets will be limited by protection considerations, very much like the stress. The high superconductor J_c has to be diluted and coil thickness increased above a theoretical minimum value because copper is needed to "buy time" and heat capacity is needed to absorb the energy. In Figure 4.3.20, the evolution of the $T_{hot\ spot}$ or T_{max} is reported for the LHC dipole in Nb–Ti (a) and for a new generation Nb_3Sn inner triplet quadrupole (QXF) for the HiLumi LHC project (b). One can note that the 450 ms of the LHC dipole to reach 300 K reduces to 150 ms for the 11 T dipole. This calls for improved mechanism of quench propagation, for example, using distributed quench heaters or by enforcing quench-back through induced coupling current [39]. Also, the detection time has to be very short: by looking at the almost vertical slope of the QXF T_{max} in the plot, it is easy to realize that the time left to quench the whole magnets is not more than 20–50 ms.

4.3.6 Field Quality

An almost unique characteristic of accelerator magnets is the necessity of a good field quality. In fact, the typical demand for accuracy at the level of 10–100 ppm is not extraordinary by itself. Other types of SC systems, and especially magnetic resonance imaging (MRI) and nuclear magnetic resonance (NMR) magnets, need homogeneities at the level of parts per million or better. What is extraordinary in accelerator magnets, and in particular for colliders, is that such accuracy is required at a distance of 10 mm or less from the conductor, while in MRI and NMR magnet the parts per million accuracy is required at 100 mm distance and more from main coils. At such a small distance, any coil deformation affects the harmonic content, and the superconductor magnetization has a strong effect.

As discussed earlier, the field in SC accelerator magnets is mainly generated by the current in the coil. Although coils are designed to approximate the theoretical current distributions that produce pure multipole fields, they are invariably subject to construction tolerances and errors. For this reason, the first concern in the production of a SC accelerator magnet is to verify the positioning accuracy of the SC cables. The errors originated by any deviation from the ideal current distribution, including the contribution of a nonsaturated iron, are referred to as the *geometric field errors of the magnet*. Geometric field errors are proportional to the current, and are hence constant when expressed as relative *units* of the main field. As will be shown later, deviations from the nominal coil geometry of the order of 0.1 mm generate geometric errors in the range of few units.

We can classify further the geometric errors in systematic, that is, inherent to the coil design or due to the magnet construction and assembly procedure, and random, that is, originated by variations of the coil geometry due to tolerances in the production components and tooling. Systematic geometric errors may be intentionally introduced in the coil design to compensate partially, or totally, field errors of different origin (in this case, the naming of *errors* is not appropriate). One such example is the geometric sextupole introduced in the LHC dipoles to compensate partially the effect of persistent currents at injection conditions, see Figure 4.3.21 discussed later. In other instances, geometric errors may be due to unexpected coil deformation during its assembly, loading, and cooldown. Systematic geometric errors can be modified most conveniently by making use of shims that introduce a displacement of the pole or the mid-plane of the coils during assembly. This procedure requires some empiricism, to establish the relation between shim position and thickness, and change in multipoles, but is very efficient to correct for drifts in a large industrial production [40], also thanks to the correlation between multipoles measured at low current at ambient temperature and multipoles measured at operational current at cold. To fix the order of magnitude, we have reported in Table 4.3.9 the effect on the low-order allowed multipoles of the LHC main dipole induced by 0.1 mm shims, located either on the coil poles or on mid-plane, as indicated schematically. A shim is conceptually identical to a coil deformation of the same amplitude, so numbers reported there provide a good reference for manufacturing and assembly tolerances. We considered in Table 4.3.9 only symmetric shims, hence only allowed multipoles are affected. When the coil deformations are not symmetric, also non-allowed multipoles appear.

At high field, the iron yoke surrounding the coils saturates, and its contribution to the field is no longer proportional to the current. The effect of the saturation appears as a field error component, which deviates from the linear geometric field. The saturation field error is most important on the main field component (e.g., B_1 for a dipole), with values ranging from tens to hundreds of units. The iron saturation depends on the local field in the iron, and the saturation of the yoke is hence not uniform in the cross-section. For this reason, errors also appear on higher

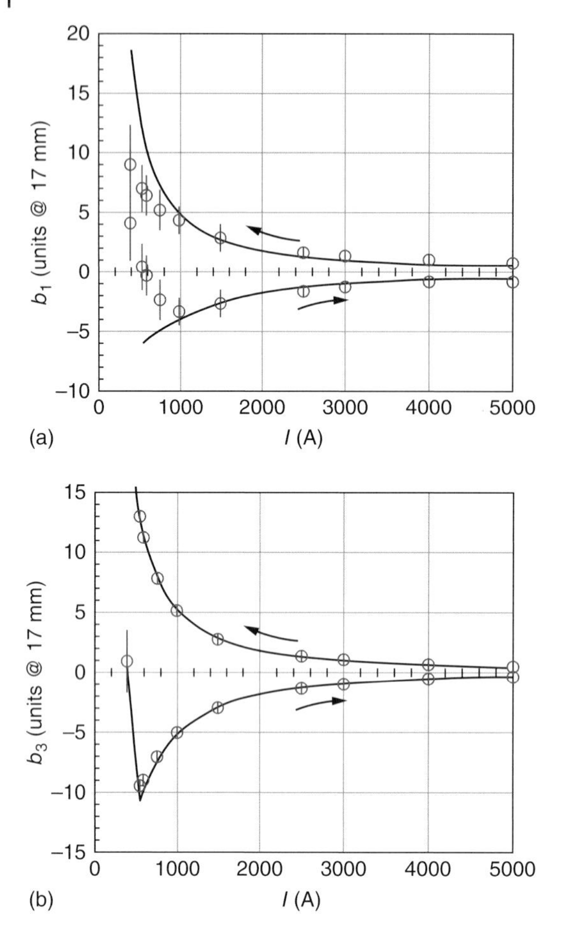

Figure 4.3.21 Variation of (a) main field (b_1) and (b) sextupole (b_3) components, generated by superconductor magnetization (persistent current), during a standard operating cycle of an LHC dipole.

order multipoles. To some extent, it is possible to introduce features in the iron yoke design (holes, shaped interfaces) so that the flux changes driven by saturation can be controlled, and higher order saturation field errors can be reduced.

Finally, the third origin of deviation from ideal field quality is the diamagnetic response of a SC cable to a change in the external field: shielding currents are induced in the SC filaments, also referred to as *persistent currents*, and among filaments in a strand or among strands in a cable, also referred to as *coupling currents*. Persistent currents are hysteretic, they depend linearly on the value of the critical current density, and on the size of the SC filaments. Examples of the contribution of the persistent currents magnetization in the LHC main dipole magnets are shown in Figure 4.3.21. We report there the measured average and spread of the main field b_1 and the sextupole b_3 for all the LHC magnets tested in cryogenic

Table 4.3.9 Calculation of the geometric multipoles (in units) generated in an LHC dipole by a coil deformation (insertion of shims) of 0.1 mm, with the symmetric patterns indicated schematically.

Shim deformation	b_1	b_3	b_5
	9.7	3.0	−0.4
	−5.3	−4.4	−1.1

conditions (about 20% of the total production). In the LHC dipoles, the persistent current errors are of the order of −3 units (b_1) to −7 units (b_3) at injection (760 A, 0.54 T), decreasing as the material J_c drops at higher fields. Recalling our earlier discussion on the geometric field error, a geometric b_3 of about +4 units was introduced in the dipoles to compensate partially for the persistent current error at injection.

Passive magnetic shims can be used to correct or mitigate most of the field errors. Vastly exploited by the RHIC project to correct geometric and manufacturing errors and to steer production, they have also been proposed to compensate for strong persistent current effect in Nb_3Sn magnets, whose high J_c and large effective filaments generates field errors 5–10 times larger than Nb–Ti.

Coupling currents in strands and cables flow in loops formed among the SC filaments, with a return path across the wire matrix, or the strand contacts. The amplitude of the coupling currents, and the associated field errors, are proportional to the field change rate and can be controlled using resistive barriers in the strand matrix, resistive coating to the strands, or inserting resistive barriers (sheets, wraps) in the cable. Indeed, coupling currents were successfully controlled in the large-scale synchrotrons quoted earlier, by acting on the strand surface state. For next generation Nb_3Sn magnet, the issue is however more serious, because of interstrand sintering during heat treatment: in this case, the use of a metallic sheet inside the cable seems the most effective way to drastically reduce cable coupling current. Coupling current control is even more critical in fast-cycled synchrotrons, as discussed in the next section.

A last effect of importance in SC accelerator magnets is a change in the field and multipoles at constant excitation current, conventionally referred to as *decay*, followed by a rapid return to the field and multipole conditions before the drift, which is called *snap-back*. These effects were first observed at the Tevatron during commissioning and fixed target operation, causing significant beam losses, and were subject to intense study at HERA and the SSC. The present understanding is that the change of the persistent current magnetization visible as decay is driven by a variation of the cable self-field associated with a redistribution of the current among the strands. The snap-back, on the other hand, corresponds to the reestablishment of the magnetization state before the decay, driven by the sweep of the background field during a ramp. Intuitively, a relation exists between the amount of decay, the ensuing snap-back, and the background field change necessary to complete the snap-back. Predictive, semi-empirical relations for decay and snap-back, for example, as described in Ref. [41], can be included in the accelerator control system, leading to virtually no beam loss during ramp.

4.3.7 Fast-Cycled Synchrotrons

Injector to high-energy colliders, or accelerators for nuclear physics, require cycled operation at high repetition rate, typically in the range of 1 Hz. AC operation of SC magnets poses a number of challenges, and mainly to achieve the required repetition rate economically and reliably. The following specific issues can be mentioned:

- *AC loss*: The control and reduction of AC loss in the cold mass has foremost importance to reduce cryo-plant investment and operation cost, and to limit the temperature excursions in the conductor.
- *Cooling*: The heat loads on the magnet, and especially those originating from the AC loss and beam heating, must be removed efficiently to warrant a margin sufficient for stable operation.
- *Quench detection and protection*: Protection of SC magnets is especially demanding in case of fast ramping machines due to the relatively high inductive voltages in comparison to the voltage developed by a resistive transition. Voltage compensation and magnet protection must be proven in the presence of an inductive voltage during ramps that can be as large as 1000 times the detection threshold.
- *Field quality*: The contribution of coupling currents in the superconductor and eddy currents in the iron yoke is difficult to predict, control, and measure at the desired resolution during fast ramps.
- *Material fatigue*, over several hundred million cycles, influencing material selection and, possibly, requiring dedicated testing in cryogenic conditions.

None of the above issue is by itself too difficult to address, but their combination has discouraged an extensive penetration of SC technology in fast-cycled synchrotron installations. This may change, as the wall-plug efficiency is becoming an important economic factor in the choice of technology for new installations. This tendency was witnessed as the technical choice fell at the Facility for Antiprotons and Ion Research (FAIR) at the GSI Laboratory in Darmstadt [42], to build the two main rings SIS100 and SIS300 out of SC magnets. The dipoles of SIS100

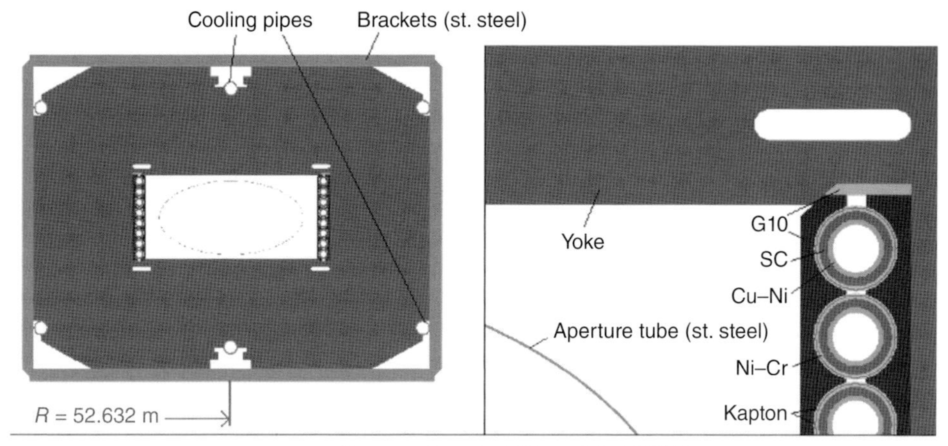

Figure 4.3.22 A cross-section of the super-ferric, fast cycle magnet SIS100 for FAIR project at GSI. (a) Full cross-section; (b): expansion of the aperture–coil interface of top right corner (Courtesy of E. Fisher, FAIR-GSI, Darmstadt-Germany).

are iron-dominated, have a bore field of 2 T, and window frame geometry, see Figure 4.3.22, providing a rectangular bore of 130 mm × 65 mm [23]. The nominal powering cycle lasts 1 s, yielding a nominal field ramp-rate of 4 T s^{-1}. The design is largely inspired by the Nuclotron, which has been in operation at JINR since 1994 [22]. The Nuclotron dipole magnets are operated in the accelerator at a peak field of 1.5 T, ramping at 0.6 T s^{-1}, and have achieved a peak field of 2 T, ramping at 4 T s^{-1}.

The dipoles of SIS300, on the other hand, are of classical cos θ layout, and they shall provide a peak field of 4.5 T in a round bore with a diameter of 100 mm, pulsed up to 1 T s^{-1}. This type of magnet does not have yet a practical benchmark in a working accelerator. An intense R&D has been pursued, supported in a collaboration with BNL in the United States (test of the GSI001 prototype magnet up to 4 T bore field in pulsed conditions up to 4 T s^{-1} [43]), National Institute of Nuclear Physics (INFN) in Italy (test of the DiSCoRaP prototype magnet up to 4.5 T bore field in pulsed conditions up to 0.5 T s^{-1} [44]), and Institute for High Energy Physics (IHEP) in Russia (test of the UNK-inspired SIS300 prototype magnet up to 6 T bore field in pulsed conditions up to 1 T s^{-1} [45]).

These developments bear an important role, beyond the FAIR project. The range of design parameters of the SIS-100 and SIS-300 is typical of injector facilities at major laboratories, such as would be necessary for an upgrade of the PS and SPS injectors at CERN. If properly engineered, SC magnets may offer increased energy, operational flexibility (no limitations on flattop duration), and reduced electric power requirement and operational costs. In addition, this technology may open the door to compact industrial and medical application requiring fast and precise high-field ramping, one such example being hadron therapy. In Figure 4.3.23, a compilation of the various magnets designed and built for cycling operation in

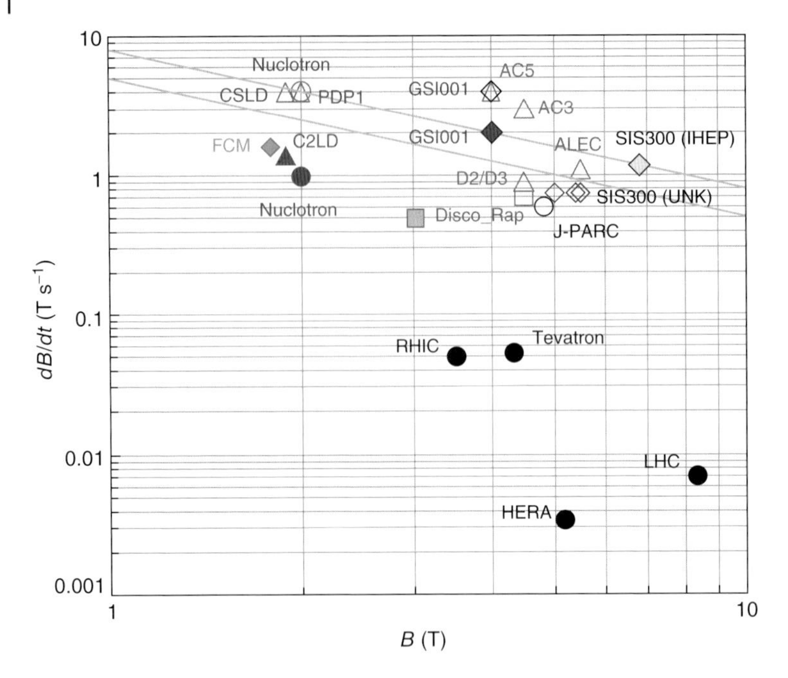

Figure 4.3.23 Compilation of the fast-ramped accelerator magnets in the B – dB/dt space. Also, collider magnets have been also reported for comparison.

the $B - dB/dt$ space is reported. Despite a wide span of technologies and range of fields covered, the magnet behavior tends to group around the line $B \times dB/dt \cong$ const $\approx 7\ \mathrm{T}^2 s^{-1}$ indicate today limit of cooling, because the $B \times dB/dt$ is related to the power dissipated per unit coil volume.

Acknowledgments

The authors thank Ezio Todesco of CERN for the fruitful discussions about magnet design and Paolo Ferracin of CERN for the contribution to Tables 4.3.3 and 4.3.4.

References

1. Wilson, M.N. (1999) Superconductivity and accelerators: the good companions. *IEEE Trans. Appl. Supercond.*, **9**, 111–121.
2. Wilson, M.N. (1983) *Superconducting Magnets*, Oxford University Press.
3. Mess, K.-H., Schmüser, P., and Wolff, S. (1996) *Superconducting Accelerator Magnets*, World Scientific, ISBN: 981-02-2790-6.
4. Ašner, F.M. (1999) *High Field Superconducting Magnets*, Oxford University Press.
5. Tollestrup, A. and Todesco, E. (2008) in *Review of Accelerator Science and Technology*, vol. 11 (eds A. Chao and W. Chou), World Scientific, pp. 185–210.
6. Rossi, L. and Bottura, L. (2012) *Review of Accelerator Science and Technology*, **5**, World Scientific, 51–89.

7. Bottura, L. and Godeke, A. (2012) in *Review of Accelerator Science and Technology*, vol. 5 (eds A. Chao and W. Chou), World Scientific, pp. 25–50.
8. Rossi, L. and Todesco, E. (2011) Conceptual design of the 20 T dipoles for high-energy LHC. Proceedings of the High-Energy Large Hadron Collider Workshop, Malta, October 2010 (eds E. Todesco and F. Zimmermann), CERN-2011-003 (8 April 2011), pp. 13–19.
9. Beth, R.A. (1966) Complex representation and computation of two-dimensional magnetic fields. *J. Appl. Phys.*, **37** (7), 2568.
10. Rabi, I.I. (1934) A method of producing uniform magnetic fields. *Rev. Sci. Instrum.*, **5**, 78–79.
11. Edwards, H.T. (1985) The Tevatron energy doubler: a superconducting accelerator. *Annu. Rev. Nucl. Part. Sci.*, **35**, 605–660.
12. Wolff, S. (1988) Superconducting hera magnets. *IEEE Trans. Magn.*, **24** (2), 719–722.
13. Anerella, M. *et al.* (2003) The RHIC magnet system. *Nucl. Instrum. Methods*, **A499**, 280–315.
14. Rossi, L. (2010) Superconductivity: its role, its success and its setbacks in the Large Hadron Collider of CERN. *Supercond. Sci. Technol.*, **23**, 034001 (17pp).
15. Latypov, D., Jaisle, A., Elliott, T., Diaczenko, N., Gaedke, R., McIntyre, P., Soika, R., Wendt, D., Shen, W., Richards, L., and McJunkins, P. (1997) *Stress Management In High-Field Dipoles*, IEEE, ISBN: 0-7803-4376-X.
16. Gupta, R. (1997) A common coil design for high field 2-in-l accelerator magnets. Particle Accelerator Conference, Vancouver, Canada (Jacow web site).
17. Toral, F., Devred, A., Felice, H., Fessia, P., Loveridge, P., Regis, F., Rochford, J., Sanz, S., Schwerg, N., and Vedrine, P. (2007) Comparison of 2-D magnetic designs of selected coil configurations for the next European dipole (NED). *IEEE Trans. Appl. Supercond.*, **17** (2), 1117–1121.
18. Ferracin, P., Bingham, B., Caspi, S., Cheng, D.W., Dietderich, D.R., Felice, H., Hafalia, A.R., Hannaford, C.R., Joseph, J., Lietzke, A.F., Lizarazo, J., and Sabbi, G. (2010) Recent test results of the high field Nb3Sn dipole magnet HD2. *IEEE Trans. Appl. Supercond.*, **20**, 292–295.
19. Meyer, D.I. and Flasck, R. (1970) A new configuration for a dipole magnet for use in high energy physics applications. *Nucl. Instrum. Methods*, **80**, 339–341.
20. Akhmeteli, A.M., Gavrilin, A.V., and Marshall, W.S. (2005) Superconducting and resistive tilted coil magnets for generation of high and uniform transverse magnetic field. *IEEE Trans. Appl. Supercond.*, **15** (2), 1439–1443.
21. Caspi, S., Arbelaez, D., Brouwer, L., Dietderich, D., Hafalia, R., Robin, D., Sessler, A., Sun, C., and Wan, W. (2012) Progress in the design of a curved superconducting dipole for a therapy gantry. Particle Accelrator Conference 2012, Jacow web site.
22. Kovalenko, A.D. (2000) Nuclotron: status and future. Proceedings of EPAC 2000, Vienna, Austria, pp. 554–556.
23. Fischer, E., Schnizer, P., Mierau, A., Wilfert, S., Bleile, A., Shcherbakov, P., and Schröder, C. (2011) Design and test status of the fast ramped superconducting SIS100 dipole magnet for FAIR. *IEEE Trans. Appl. Supercond.*, **21** (3), 1844–1848.
24. Foster, G.W. *et al.* (2000) Design of a 2 tesla transmission line magnet for the VLHC. *IEEE Trans. Appl. Supercond.*, **10** (1), 202–205.
25. Piekarz, H., Hays, S., Huang, Y., Kashikhin, V., de Rijk, G., and Rossi, L. (2006) Design considerations for fast-cycling superconducting accelerator magnets of 2 tesla generated by transmission line conductor of 100 kA current. *IEEE Trans. Appl. Supercond.*, **18**, 256–259.
26. Perin, R. and Leroy, D. (1998) in *Handbook of Applied Superconductivity*, vol. 2 (ed B. Seeber), Institute of Physics, pp. 1289–1317.
27. Peyrot, M., Rifflet, J.M., Simon, F., Vedrine, P., and Tortschanoff, T. (2000) Construction of the new prototype of main quadrupole cold masses for the arc short straight sections of LHC. *IEEE Trans. Appl. Supercond.*, **10**, 170–173.

28. Billan, J., Bona, M., Bottura, L., Leroy, D., Pagano, O., Perin, R., Perini, D., Savary, F., Siemko, A., Sievers, P., Spigo, G., Vlogaert, J., Walckiers, L., Wyss, C., and Rossi, L. (1999) Test results on long models and full scale prototype of the second generation LHC arc dipoles. *IEEE Trans. Appl. Supercond.*, **9** (2), 1039–1044.
29. Lorin, C., Granieri, P.P., and Todesco, E. (2011) Slip-stick mechanism in training the superconducting magnets in the Large Hadron Collider. *IEEE Trans. Appl. Supercond.*, **21**, 3555–60.
30. Ajima, Y., Higashi, N., Iida, M., Kimura, N., Nakamoto, T., Ogitsu, T., Ohhata, H., Ohuchi, N., Shintomi, T., Sugawara, S., Sugita, K., Tanaka, K., Taylor, T., Terashima, A., Tsuchiya, K., and Yamamoto, A. (2005) The MQXA quadrupoles for the LHC low-beta insertions. *Nucl. Instrum. Methods Phys. Res., Sect. A*, **550**, 499–513.
31. Andreev, N. (2000) Mechanical design and analysis of LHC inner triplet quadrupole magnets at fermilab. *IEEE Trans. Appl. Supercond.*, **10**, 115–118.
32. Hafalia, R.R., Bish, P.A., and Caspi, S. *et al.* A New Support Structure for High Field Magnets. *IEEE Trans. Appl. Supercond.*, **12** (1), 47–50.
33. GL Sabbi, *Proceedings of the High-Energy Large Hadron Collider Workshop*, Villa Bighi, Malta, October 2010, CERN-2011-003 (8 April 2011) (eds E. Todesco and F. Zimmermann), CERN, Genva, pp. 30-36.
34. Wilson, M.N. (2015) Fundamentals of superconducting electromagnets, in *Applied Superconductivity. Handbook on Devices and Applications* (ed P. Seidel), Wiley-VCH Verlag GmbH, Weinheim.
35. Kashikin, V. and Zlobin, A. (2005) Magnetic instabilities in Nb_3Sn strands and cables. *IEEE Trans. Appl. Supercond.*, **15**, 1621.
36. Bordini, B. and Rossi, L. (2009) Self field instability in high J_c Nb_3Sn strands with high copper residual resistivity ratio. *IEEE Trans. Appl. Supercond.*, **19**, 2470.
37. Willering, G. (2009) Stability of superconducting rutherford cables for accelerator magnets. PhD thesis. University of Twente, The Netherlands, ISBN: 978-90-365-2817-7.
38. Lorin, C., Siemko, A., Todesco, E., and Verweij, A. (2010) Predicting the quench behavior of the LHC dipoles during commissioning. *IEEE Trans. Appl. Supercond.*, **20** (3), 135–139.
39. Ravaioli, E., *et al.*, (2013) New, coupling loss induced, quench protection system for superconducting accelerator magnets. IEEE Transactions on Applied Superconductivity, **24** (3), 4, 2014.
40. Todesco, E. *et al.* (2004) Steering field quality in the main dipole magnets of the large Hadron collider. *IEEE Trans. Appl. Supercond.*, **14**, 177–180.
41. Ambrosio, G., Bauer, P., Bottura, L., Haverkamp, M., Pieloni, T., Sanfilippo, S., and Velev, G. (2005) A scaling law for the snapback in superconducting accelerator magnets. *IEEE Trans. Appl. Supercond.*, **15**, 1217–1220.
42. Spiller, P., *et al.*, Status of the fair SIS100/300 synchrotron design. Proceedings of PAC07, Albuquerque, New Mexico, pp. 1419–1421E.
43. Moritz, G. *et al.* Recent test results of the fast-pulsed 4 T Cosθ Dipole GSI001. Proceedings of 2005 Particle Accelerator Conference at Knoxville, Tennessee, 3 p.
44. Sorbi, M., Alessandria, F., Bellomo, G., Fabbricatore, P., Farinon, S., Gambardella, U., Musenich, R., and Volpini, G. (2013) Measurements and analysis of the SIS-300 dipole prototype during the functional test at LASA. IEEE Transactions on Applied Superconductivity, **24** (3), 4, 2014.
45. Kozub, S., Bogdanov, I., Pokrovsky, V., Seletsky, A., Shcherbakov, P., Shirshov, L., Smirnov, V., Sytnik, V., Tkachenko, L., Zubko, V., Floch, E., Moritz, G., and Mueller, H. (2010) SIS 300 dipole model. *IEEE Trans. Appl. Supercond.*, **20**, 200–203.

4.4
Superconducting Detector Magnets for Particle Physics

Michael A. Green

4.4.1
The Development of Detector Solenoids

The discovery of type-2 superconductivity in 1961 [1] was celebrated by the particle physics community. Suddenly, it appeared to be possible to create a large volume of magnetic field at an induction not heretofore considered economical using conventional magnets. In 1960, one of the largest operating particle detectors using a magnetic field was probably the 72 inch bubble chamber at Berkeley. Within days of the announcement of the discovery of niobium tin as a high-field superconductor, particle physicists in Berkeley and other locations set up research groups to make superconducting magnets for particle detectors. There was excitement and constant communication between groups in both the United States and Europe. The cryogenic expertise for the development of the new superconducting magnets came from the people who had developed the cryogenic systems for hydrogen bubble chambers.

The hype and hopes of type-2 superconductivity soon faded with realization that the niobium tin then available was not a usable superconductor for large magnets. Early on, the experimental work shifted to Nb–Ti and Nb–Zr alloys. Before 1964, Nb–Zr was the alloy of choice because it was more stable than Nb–Ti. The alloy superconductors that performed well in short samples would reach only 30% or 50% of their critical current in a magnet [2]. Wires plated with copper appeared to perform somewhat better than bare wire [3]. Coil performance degradation due to flux jumps was the topic of the day. Improvements in magnet performance were not spectacular because there was no general understanding of what was happening within the superconductor. By 1964, the construction of large superconducting particle detector magnets appeared to be nearly hopeless.

4.4.1.1 Early Superconducting Detector Magnets

The paper on cryogenic stability of superconductors by Stekly and Zar [4] caused excitement in the particle physics community. The paper stated that if the superconductor was put in a low resistivity matrix, it did not matter whether the superconductor flux jumped as long as the matrix remained at a temperature below the superconductor critical temperature. The current that had left the superconductor would return to the superconductor once the instability had passed. The discovery of cryogenic stability led to the first large detector magnets being built for particle physics. The first of these magnets was the 12 foot bubble chamber magnet at Argonne in 1969 [5, 6]. The 12 foot bubble chamber was followed by a 7 foot bubble chamber magnet at Brookhaven in 1970 [7], the 15 foot bubble chamber at the Fermilab (1973) [8], a 3.8 m bubble chamber at CERN (Center European Research Nuclear in French) (1973) [8, 9], the LASS magnet at SLAC (Stanford Linear Accelerator Center) (1974) [10], and a number of smaller devices. In 1972, Morpurgo

[11, 12] tested a hollow conductor and forced-cooled cryogenically stable solenoid for the OMEGA experiment at CERN. Cryogenic stability solved the scale problem for large superconducting magnets, but magnets built in this way operated at low current densities (<50 A mm^{-2}). These magnets were far from being thin from the standpoint of particle transmission through the magnet.

4.4.1.2 **Low Mass Thin Detector Magnets**

Truly thin detector solenoids required a superconductor that could operate at higher current densities without flux jumping. Work by Bean [13], Hancox [14], Chester [15], and Smith [16, 17] paved the way to understanding the intrinsic stability of superconductors, which led to the development of modern, twisted multifilamentary conductors with a low matrix metal to superconductor ratio. Increasing the current density in the magnet winding was one way of making the magnet more transparent to particles.

In addition, thin superconducting solenoids had to be cooled in a different way. Helium bath cryostats contain too much material for them to be transparent to particles. In order to reduce the mass of the cryostat, it was found that thin solenoids had to be cooled indirectly by conduction in tubes that contain helium. Experimental work in the 1970s suggested that two-phase helium cooling would result in a lower operating temperature than supercritical helium cooling [18].

The first experiment calling for a thin solenoid was at the ISR at CERN [19]. In 1975, a thin solenoid was proposed for the MINIMAG experiment proposed by the Lawrence Berkeley Laboratory (LBL) in 1975 [20]. This experiment required a 1 m-diameter solenoid that was 0.35 radiation lengths thick, including the cryostat. Two 1 m-diameter test coils were built and tested in 1975 and 1976 [21, 22]. The conductor in the test coils was operated at matrix plus superconductor current densities as high a 1250 A mm^{-2}. The MINIMAG experiment was not built, but a larger detector for an experiment at PEP colliding beam ring at the SLAC was embarked upon. This detector required a clear bore diameter of 2 m with a gap of 3.3 m between the iron poles. A uniform 1.5 T-induction (better than 1 part in 1000 within a 2 m-diameter, 2 m-long volume) was required for the newly developed time projection chamber detector that was developed at LBL. The coil and cryostat had to be <0.7 radiation lengths thick so that calorimeters and muon detectors could be located outside of the magnet. The 0.7 radiation length thick magnet included a 2 m-diameter 1.1 MPa aluminum pressure vessel that made up the inner wall of the cryostat. Work began on a 2 m-diameter test coil in late 1976. This coil was tested in 1977 and 1978 [23]. The thin coil experimental work at Berkeley led directly to the CLEO-1 detector at Cornell University [19] and the PEP-4 detector [24] at the PEP colliding beam facility at the SLAC.

A group at CEN Saclay outside Paris decided to build their detector magnet using a conductor that had a low copper to superconductor ratio soldered to very pure aluminum high residual resistance ratio (RRR) matrix. The advantages of the aluminum matrix were: the minimum propagation zone was lengthened and the coil current density was lowered, so that the energy needed to induce a quench in the magnet was increased by over three orders of magnitude, and the quench propagation velocity along the wire was faster than for a comparable copper matrix conductor. The 2 m-diameter CELLO detector magnet was first tested in 1979 [25]. A conductor made with the copper matrix superconductor co-extruded in pure aluminum (RRR >1000) was developed in a number of locations at about the same time [26]. This type of conductor was used on the CDF detector magnet at Fermilab [27]; the VENUS detector [28], the TOPAZ detector [29], and the AMY detector [30] at KEK in Japan; the ALEPH [31] and DELPHI [32] detectors at CERN; the H-1 (P.T.M. Clee, Rutherford Appleton Laboratory, Didcot, UK, private communication) and ZEUS [33] detectors at DESY (a German laboratory that does particle physics) in Germany; the GSI solenoid (D. Andrews, Oxford Technology Ltd., Oxford, UK, private communication) at Darmstadt; the CLEO-2 detector [34] at Cornell University in Ithica New York; and the CLOE detector at Frascati. A thin solenoid SDC experiment test coil was tested before the SSC was canceled [35]. The Japanese flew a 1 m-diameter balloon solenoid [36, 37] that used a low matrix to superconductor ratio RRR >1000 aluminum matrix conductor to achieve a very low radiation thickness (about 0.25 radiation lengths) for a cosmic ray experiment. Later versions of the magnet achieved radiation thicknesses of the order of 0.12 radiation length because the cryostat vacuum vessel was a thicker honeycomb structure with aluminum surface plates [38, 39]. Detector solenoids for the BaBar [40] experiment at the B factory at SLAC also used an aluminum matrix coil wound within a hard aluminum shell. The BEPCII experiment at IHEP in Beijing, China also used a relatively thin solenoid [41]. The ATLAS experiment [42, 43] and the CMS [44] experiment at the Large Hadron Collider (LHC) use similar fabrication techniques as the previous magnets. All of these magnets will use a pure aluminum matrix superconductor that was wound onto or into a hard aluminum support structure. The large ATLAS and CMS detector magnets will be discussed further in the next section of this chapter.

The use of thin solenoid magnet construction techniques has proven to be less costly even when thinness was not required. As a result, the thin detector solenoid construction techniques were used to build other DC magnets used in physics. An example of this is the two 13.4 m-diameter and one 15.1 m-diameter superconducting solenoids used for the g-2 experiment at the Brookhaven National Laboratory [45]. These solenoids were successfully tested to full field in the summer of 1996 [46]. Table 4.4.1 summarizes the design parameters for a number of superconducting solenoidal-detector magnets and the g-2 magnet [47].

4.4.2 LHC Detector Magnets for the ATLAS, CMS, and ALICE Experiments

The three experiments at the LHC that use detector magnets are ATLAS, CMS, and ALICE. The three experiments are different in concept and in scale. ATLAS is the largest of the three experiments. CMS is a somewhat smaller experiment, but it involves using a single high-field superconducting solenoid that has the largest stored magnetic energy (~2.7 GJ) of any magnet operating in the world until ITER comes on line around 2020. ALICE is a small experiment that uses a large bore

Table 4.4.1 Parameters for various superconducting solenoid detector magnets (1978–2009).

Magnet	Central B (T)	Bore diameter (m)	Magnet length (m)	Matrix material	Coil location	Radiation thickness (radiation length)	Stored energy (MJ)	Conductor J (A mm^{-1})	Cooling type
CLEO-1	1.5	2.0	3.7	Cu	Outside	0.7	10.0	~350	Forced
PEP-4	1.5	2.04	3.84	Cu	Outside	0.85	10.9	645	Forced
CELLO	1.3	1.5	4.02	Al	Outside	0.6	5.0	?	Forced
CDF	1.5	2.85	5.4	Al	Inside	0.84	30	64	Forced
TOPAZ	1.2	2.72	5.4	Al	Inside	0.7	20	56	Forced
VENUS	0.75	3.4	5.6	Al	Inside	0.52	12.0	?	Forced
ALEPH	1.5	4.96	7.0	Al	Inside	1.6	136	30.8	Natural
AMY	3.0	2.39	2.11	Al	Outside	>2	40	50	Pool
GSI	0.6	2.4	3.3	Al	Inside	~1.0	3.4	?	Natural
ZEUS	1.8	1.72	2.9	Al	Inside	>2	16	?	Forced
DELPI	1.2	5.2	7.4	Al	Inside	1.7	108	46.3	Forced
H-1	1.2	5.2	6.0	Al	Inside	1.8	130	46	Forced
CLEO-2	1.5	2.9	3.8	Al	Inside	2.2	25	41.3	Natural
g-2	1.45	14.1[a]	0.18[b]	Al	Inside	>3	5.5	81.8	Forced
KEK-1	1.0	0.852	2.0	Al	Inside	0.21	0.82	241	Pool
KEK-2	1.0	~0.85	2.0	Al	Inside	0.12	0.82	~240	Pool
SDC Test	1.5	1.7	2.4	Al	Inside	1.2	?	63.4	Forced
BaBar	1.5	2.76	3.85	Al	Inside	<1.4	23	37 and 67	Natural
BEPCII	1.0	3.89	3.4	Al	Inside	~0.8	9.5	44.8?	Forced

The conductor location is with respect to the mandrel. A forced cooling system is forced two-phase cooling. A natural cooling system is by natural convection. A pool cooling system is pool boiling.

a) The beam orbit diameter; the outer solenoid diameter is 15.1 m; and the inter solenoid diameter is 13. 4 m.

b) This is the total gap between unsaturated iron poles. The return iron path is C shaped.

(~3.3 m gap) that produces a bending strength of 3 T m. The coils in the ALICE experiment are a rarity is the early part of the twenty-first century in that they are conventional water-cooled coils [48]. The ALICE experiment will not be discussed here.

The ATLAS experiment is huge (over 44 m long and over 20 m in diameter) with a total mass ~7000 tons. The experiment has three types of magnets. There is a relatively thin central 2 T solenoid with a stored energy of 39 MJ. There are two pairs of endcap toroidal coils, which have a total stored energy of 440 MJ. Finally, there are the barrel toroidal coils with an outside diameter of 20.1 m and a length of 25 m and the stored energy of the barrel coils is 1200 MJ. The endcap and the barrel toroidal field magnets have the characteristic of returning their own flux. Close to the coils, the stray field is large, but when moved away from the experiment, the stray field is relatively small.

The CMS experiment is much like a standard thin solenoid experiment. The solenoid magnet is a single superconducting coil with 6.0 m-diameter warm bore and a length of 12.5 m. Earlier magnets had maximum inductions on axis of 1.5 – 2.0 T. The CMS magnet central induction is 4.0 T. The flux is returned through an iron return yoke that is also part of the hadron calorimeter and muon detectors for the experiment. The mass of the iron in the return yoke is about 10 000 tons. The total diameter of the experiment is 15 m. The length of the experiment is 22 m. The total mass for the experiment is ∼12 500 tons.

4.4.2.1 Magnets for the ATLAS Detector

4.4.2.1.1 The ATLAS Central Solenoid

The ATLAS central solenoid was smaller than the CDF solenoid at Fermi National Laboratory or the LEP detector magnets DELPHI or ALEPH or the H-1 detector magnet at DESY in Hamburg, Germany [49]. The central solenoid had a design field of 2.0 T, which is higher than other magnets built until that time. The central solenoid coil inside diameter is 2.4 m and the length is 5.3 m. The radiation thickness of the central solenoid and its cryostat is 0.66 radiations lengths. The total mass of the central solenoid is 5.7 tons.

The central solenoid coil is a 30 mm-thick single layer coil that is, wound the hard way into a 12 mm-thick 5083-H111 cylinder. The total number of turns is 1173. The conductor dimensions are 4.25 mm × 30 mm with 12 strands of Nb – Ti conductor with a copper to superconductor ratio of 0.9. The bulk of the conductor matrix is pure aluminum with an RRR in the range from 520 to 570. The conductor carries a current of 7600 A. The conductor current density is 59.6 A mm^{-2}. The magnet' stored energy is 39 MJ at the full design current. The magnet EJ^2 limit for the central solenoid at its full design current is 1.385×10^{23} J A^2 m^{-4}, which means that the magnet can be protected by a standard quench protection resistor across the leads even without high RRR pure aluminum longitudinal quench propagation strips. (See Eq. (4.4.4) in Section 4.4.5 of this chapter.) In addition to a dump resistor, the quench protection system includes heaters on the coils [50].

The flux for the central solenoid is returned by iron that is part of the hadron calorimeter system for the experiment. The hadron calorimeters and return iron are located between the central solenoid and endcap toroidal coils and the barrel toroidal coils around the girth of the experiment. Half of the ATLAS experiment mass is the iron in the return path. Between the iron return iron and the orientation of the central solenoid with respect to the toroidal coils, there is virtually no inductive between the toroidal magnets and the central solenoid magnet.

The central solenoid must be cooled using a refrigerator that produces at least 130 W at 4.5 K. In addition, there must be a separate circuit of helium from 60 K that produces at least 500 W of cooling (between 60 and 80 K). The central solenoid specification calls for liquid helium mass flows up to 20 g s^{-1}. The gas-cooled leads that connect to room temperature require ∼1.1 g s^{-1} of gas at 4.5 K per lead pair. This gas is returned to the compressor at room temperature. There are no high-temperature superconductor (HTS) leads used in the central solenoid.

4.4.2.1.2 The ATLAS Endcap Toroids

There is an endcap toroid at each end of the ATLAS experiment. Each toroid consists of eight rectangular (almost) coils with an inner diameter of 1.65 m and an outer diameter of 10.7 m [51]. The coil dimension along the axis of the experiment is 5.0 m. The endcap solenoid is designed to capture the particles leaving the collision point at angles between 3.8° and 23.0°. The total assembly mass of each end cap toroid is 239 tons. Each endcap toroid magnet was assembled in 45° segments at CERN [52, 53].

Each endcap toroid coil consists of two double pancake coils of about 200 mm thickness including the 5083-aluminum plate between the coils. The total number of turns per coil is 116. The conductor dimensions are 12 × 41 mm with a 38–40-strand Rutherford cable of Nb–Ti conductor with a copper to superconductor ratio of 1.3. The high RRR aluminum to superconductor ratio is 19. The pure aluminum in the matrix has an RRR >1400. The conductor carries a current of 20 700 A. The conductor current density is 45.1 A mm^{-2}. The magnet's stored energy of each toroid is 220 MJ at the full design current. The magnet EJ^2 limit for each endcap toroid at its full design current is 4.475×10^{23} J A^2 m^{-4}, which means that the magnet can be protected by a standard quench protection resistor system with voltages of ~600 V per toroid across each endcap toroid. (See Eq. (4.4.4) in Section 4.4.5 of this chapter.) In addition to a dump resistor, the quench protection system includes heaters.

The endcap toroids return their flux. There is virtually no coupling between the endcap toroids and the central solenoid. There is some coupling between the endcap toroids and the barrel toroids, but the coupling is relatively small.

The endcap toroids must be cooled from a source of refrigeration that produces at least 330 W at 4.5 K. In addition, there must be a separate refrigeration circuit from 60 to 80 K that produces at least 1700 W of cooling. The liquid helium circuits are designed for a mass flow of 280 g s^{-1}. The coiling for the gas-cooled leads that connect the magnet to room temperature requires ~3 g s^{-1} of gas at 4 K per lead pair. This gas is returned to the refrigerator compressor at room temperature. The endcap toroids do not have HTS leads.

4.4.2.1.3 The ATLAS Barrel Toroid

There is a single toroid around the girth of the ATLAS experiment. This consists of eight rectangular coils 25.3 m long and 5.35 m wide. The toroid inner diameter is 9.4 m. The outer diameter is 20.1 m. The barrel toroid is designed to capture the particles leaving the collision point at angles from about 21° to 90°. The total assembly mass for the barrel toroid system is about 830 tons.

Each barrel toroid coil consists of two double pancake coils that are each 114 mm thick. The total coil package thickness is about 340 mm including the 5083-aluminum plate between the coils and the aluminum covers. The total number of turns per coil is 120 [51, 54]. The conductor dimensions are 12 × 57 mm with a Rutherford cable with 38–40 strands of Nb–Ti conductor with a copper to superconductor ratio of 1.3. The high RRR aluminum to superconductor ratio is 28. The bulk of the conductor matrix is pure aluminum with an RRR >1400.

The conductor carries a current of 20 500 A. At full current, the conductor current density is 30.0 A mm^{-2}. The magnet's stored energy of the barrel toroid is 1200 MJ at design current. The magnet EJ^2 limit for the barrel toroid at full design current is 1.08×10^{24} J A^2 m^{-4}, which means that the magnet can potentially be protected by a standard quench protection resistor system with voltages across the magnet of about 1300 V, without subdivision of the magnet. In addition to a dump resistor, the quench protection system includes heaters. A quench analysis suggests that the peak voltage in the barrel solenoid coils will be about 700 V, even if the quench protection system fails.

The two 25 m-long coils' straight sections are tied together with seven cold ribs. The ribs carry up to 2 MN pulling the straight sections apart, when the coils are charged [55]. The coils are connected to the inside of the vacuum vessel in eight places through flexible tie rods along the coil [56]. The tie rods can flex up to 45 mm to allow for thermal contraction of the coil. There are a number of sliding stops that position the coils within the vacuum vessel. The tie rods and sliding stops are connected to the 60–80 K shield as well. The structure between the coils and the cryostat vacuum vessel was designed to carry the gravity forces from muon detectors, mounted within the barrel toroid [57].

The barrel toroid returns its own flux. There is virtually no coupling between the barrel toroid and the central solenoid. There is some inductive coupling with the endcap toroids.

The magnet design is based on a helium flow of 410 g s^{-1} in the tubes around the coils. The design refrigeration at 4.5 K is 990 W. In addition, refrigeration must be provided from a separate circuit from 60 to 80 K that produces at least 7400 W of cooling for the shields and the cold mass support intercepts. The coiling for the gas-cooled leads that connect the magnet to room temperature requires ~3 g s^{-1} of gas at 4.5 K per pair of leads [58]. This gas is returned to the refrigerator compressor at room temperature. The barrel toroids do not have HTS leads between 4.5 and 60 K. The barrel toroid is shown in Figure 4.4.1.

The ATLAS magnets are cooled from a single refrigerator that produces 6 kW at 4.5 K [59]. The liquid helium flow for all the ATLAS magnet toroidal coils is provided by a helium pump that can deliver up to 1.5 kg s^{-1} from a storage Dewar connected to the refrigerator. In addition, the ATLAS refrigerator produces 30 kW of refrigeration from 60 to 80 K to cool the shields and the cold mass support intercepts for all of the ATLAS magnets plus 12 g s^{-1} to cool all of the gas-cooled leads for the magnets. The basic parameters of the ATLAS detector magnets are shown in Table 4.4.2.

4.4.2.2 **The CMS Detector Magnet**

The CMS magnet is a single large solenoid with an iron return interspersed with a system of hadron calorimeters and muon detector. Within the CMS magnet volume is a vertex detector, a detector that tracks the particles within the 4 T magnetic field and at least some of the hadron calorimetry. The 4 T superconducting solenoid is built like a thin solenoid. The rest of the calorimeters are interspaced within the return iron. The muon detectors are on the outside of the experiment.

Figure 4.4.1 An end view of the barrel toroid being installed in the ATLAS experiment.

Table 4.4.2 The design parameters of the ATLAS experiment magnets.

Parameter	Central solenoid	End cap toroids	Barrel toroid
Magnet inner diameter (m)	2.46	1.65	9.4
Magnet outside diameter (m)	2.63	10.7	20.1
Magnet axial length	5.30	5.0	25.3
Number of magnet units	1	2	1
Number of coils per magnet unit	1	8	8
Central induction of the magnet unit (T)	2.0	Variable	Variable
Number of layers per coil	1	2	2
Number of turns per layer	1 173	60	58
Conductor dimension (mm)	30 × 4.25	41 × 12	57 × 12
Number of copper matrix Nb–Ti strands	12	40	38
Copper based strand diameter (mm)	1.22	1.3	1.3
Aluminum matrix RRR	>500	>1 000	>1 000
Conductor length per magnet unit (km)	10	13	56
Nominal coil design current, I_D (A)	7 600	20 700	20 400
Conductor current density at I_D (A mm^{-2})	59.6	42.1	29.8
Magnet unit stored energy at I_D (MJ)	39	220	1 200
Peak induction in the conductor (T)	2.6	4.1	3.9
Conductor mass per magnet unit (tons)	3.8	20.5	118
Cold mass per magnet unit (tons)	5.4	160	370
Total mass per magnet unit (tons)	5.7	239	830

Most of the muon detectors are within the return iron. The diameter of the experiment is 15 m and the length of the experiment is 22 m.

The standard, thin solenoid technology represented by the central solenoid for ATLAS has a superconductor with a high RRR aluminum matrix with a hard aluminum shell that carries all of the magnetic forces. The electrical resistivity of the aluminum matrix material goes up as it is stained above the yield point. In the ATLAS central solenoid and other magnets such as the Japanese balloon magnets [40], a small amount of other materials such as nickel were added to the pure aluminum alloy to increase its yield strength and yield strain without reducing the RRR of the aluminum very much. This approach breaks down in the CMS magnet because the magnetic forces applied to the conductor package are over an order of magnitude higher for the CMS magnet as compared to the ATLAS central solenoid. Part of this can be taken up by making the coil thicker by a factor of 4 or more, but there is still an increase in the average stress across the conductor by a factor of over 2. As a result, a new type of conductor had to be developed for the CMS magnet.

The CMS magnet conductor is a composite conductor made from a 32 strand Rutherford cable with a copper to superconductor ratio of 1, pure aluminum around the Rutherford cable, and two sections 6082 aluminum for strain control [60]. The Rutherford cable in co-extruded in a 99.998% pure aluminum matrix with an RRR >1500. The finished co-extruded section of Rutherford cable and pure aluminum has the dimensions of 30 mm × 21.6 mm. The co-extruded pure aluminum section with the Rutherford cable is electron beam welded to two sections of EN AW-6082 aluminum that are 21.6 mm × 17 mm in dimension. The maximum tensile force in the composite conductor is 130 kN. The average stress across the conductor at the maximum tensile force is 94 MPa over the entire conductor cross-section. Samples of the EN AW-6082 aluminum had an average yield stress of 187 MPa and an average ultimate tensile stress of 316 MPa when measured at 300 K. At 4.2 K, the yield strength increased to 258 MPa and the ultimate strength increased to 406 MPa. The electron beam weld seam was 2.2 mm wide. The nearest strand of the Rutherford cable was 3.4 mm from the weld seam. The conclusion was that the Nb–Ti in the Rutherford cable was not heated enough to affect the critical current of the superconductor. In the samples measured, the RRR of the pure aluminum was about 3000 at zero field and about 1000 at 5 T. 1950 km of superconducting strand was produced from 148 extrusion billets and cabled into lengths of Rutherford cable that were 2.65 km long. The Rutherford cable was co-extruded with the pure aluminum, which in turn was electron beam welded to the sections of 6082-aluminum. Tests were performed on the conductor during the processing of the conductor wound into the CMS solenoid [61, 62].

The CMS magnet was fabricated in five modules that have an inside diameter of 6 m and a length of 2.5 m [63]. After winding, each of the section was vacuum impregnated with epoxy resin, which was cured at 135 °C. The finished impregnated coil sections were inserted into 50 mm-thick 5083-H321 aluminum shells with 130 mm-thick flanges welded the ends [64]. The helium cooling tubes were welded to the shell before the coils were installed in the shell. After the coil sections were fabricated in Genoa, they were shipped to CERN. At CERN, the five coil sections were connected together using titanium bolts and pins. The peak stresses in the bolts and pins occurred during handling of the magnet. Once the coils were assembled, the electrical connections between the modules were made, cooling tube circuits were connected, the tests were made on the electrical and cooling systems, and coil instrumentation was installed [65]. The magnet shields were installed and the magnet was inserted into the cryostat vacuum vessel. The magnet was then tilted into the horizontal position so that it could be installed in the experiment.

The CMS magnet is kept cold using two redundant thermal-siphon cooling circuits. Cold liquid helium enters the cooling system at the bottom of the coil. Two-phase helium flows up the coil sides and phase separation occur in a tank above the coils. The CMS refrigeration plant has a specified capacity of 800 W at 4.5 K, 4500 W capacity between 60 and 80 K (to cool the shields and cold mass support intercepts), and $4\,g\,s^{-1}$ of helium liquefaction simultaneously [66]. The leads for the CMS magnet are helium gas-cooled copper leads without an HTS lead section at the bottom of the leads [67]. At full current, the leads require about $2.4\,g\,s^{-1}$ to keep them cold.

The CMS magnet and the detector within are located within the iron return yoke that has four layers of iron interspersed with muon detectors. There are layers of iron yoke and muon detectors are in the axial direction as well as in the radial direction. The CMS solenoid cannot be considered thin from a radiation standpoint. In the radial direction, the magnet coil and cryostat are over 4 radiation lengths thick. The hadron calorimeters are within the bore of the solenoid. Particle bending occurs within the hadron calorimeters as well as in the central detector that takes up much of the solenoid volume. The basic parameters for the CMS magnet are shown in Table 4.4.3. An end view of the CMS magnet is shown in Figure 4.4.2.

4.4.3 The Future of Detector Magnets for Particle Physics

This work has described the largest detector magnets of their times. Starting in the late 1970s, larger detector magnets have moved from magnets with conventional water-cooled coils to magnets with superconducting coils. In the larger detector magnets, there are two notable exceptions, the STAR detector solenoid at the Brookhaven National Laboratory. The STAR magnet, which was of a relatively low field (0.75 T), was built with conventional coils because of their lower capital cost. There are questions as to whether the capital cost was really lower. The operating cost since the 1990s would have paid for the extra cost of a superconducting solenoid many times over. The other magnet is the ALICE experiment magnet at CERN. Laboratories such as the Jefferson Laboratory (JLAB) and the NSCL at Michigan State University routinely build dipole and quadrupole magnets with warm iron and superconducting coils that are smaller than any of the

Table 4.4.3 The design parameters of the CMS experiment magnet.

Parameter	Value
Magnet inner diameter (m)	6.0
Magnet outside diameter (m)	~7
Magnet axial length	12.5
Central induction of the magnet unit (T)	4.0
Number of coil modules	5
Number of layers per coil	4
Number of turns per layer	104
Conductor dimension (mm)	64 × 21.6
Number of copper matrix Nb–Ti strands	32
Copper based strand diameter (mm)	1.28
Aluminum matrix RRR	>1 000
Conductor length per magnet (km)	52
Nominal coil design current I_D (A)	20 000
Conductor current density at I_D (A mm^{-2})	14.1
Magnet unit stored energy at I_D (MJ)	2 700
Peak induction in the conductor (T)	4.7
Magnet radial radiation thickness (radiation length)	~4.3
Conductor mass per magnet unit (tons)	~205
Cold mass per magnet unit (tons)	~225
Total mass per magnet unit (tons)	~250

Figure 4.4.2 An end view of the CMS magnet installed in the experiment.

detector magnets at CERN [68]. The newest detector magnets at JLAB will all have superconducting coils [69]. The trend in detector magnets is definitely toward superconducting magnets. There are many good reasons to expect this trend to continue.

The giant detector magnet, such as those at LHC, will be a rarity unless accelerators change to reduce their cost. Magnets for astrophysics detectors may be a future trend. In this case, the universe provides the accelerator and the results of the acceleration are collected by magnets in space or even on earth. Such detector magnets would certainly be superconducting. One such magnet has been proposed for looking for previously undetected axions that could come from the sun [70]. There are two competing trends for detector magnets in the future.

The first trend is toward improved conductors of the CMS type that will permit larger solenoids to be built. Such magnets could have inside diameters that are >6 m and their lengths could be longer than 12 m. These magnets could certainly operate at fields up to 6 T using conductors made with aluminum matrices [71]. One of the great limitations on solenoid detector magnets is returning the magnetic flux. When solenoids are built with a larger diameter and a larger magnetic field, the mass of the iron needed to return the flux becomes enormous. Without iron, one must use coils to return the flux. There is a tipping point where shield coils become the preferred option. An example of such a magnet is given in Ref. [72].

The second trend is toward detectors that use toroidal field configuration. There is some room to grow using toroidal magnets that use aluminum-based conductors and strong aluminum structure, but there is a level of magnetic field where one has to change technologies from the aluminum-based conductor technology to magnets with cable in conduit coils (CICC coils) such as those found in ITER.

4.4.4 The Defining Parameters for Thin Solenoids

In the literature, thinness is defined in terms of interaction lengths, absorption lengths, and radiation lengths. In high-energy physics detectors, there is no single universal definition of thinness. Thus, discussion of interaction lengths must identify which particle, and absorption lengths must identify the particle and its energy. The most common definition of thinness uses radiation lengths as a defining parameter. One radiation occurs where 63.2% $(1 - 1/e)$ of the neutral particles have formed charged particle pairs. This definition is appropriate in many experiments because the calorimeters and muon detectors are the only detectors that are outside the magnet.

The physical thickness of a material that is one radiation length thick is a function of the material atomic number Z and the material specific density γ. In order for a superconducting magnet to be thin, it must be made from low density,

Table 4.4.4 The radiation thickness of various materials.

Material	Z	Mass density (kg m^{-3})	One radiation length	
			(kg m^{-2})	(mm)
Pure elements				
Hydrogen	1	70.8[a]	630.5	8 900
Deuterium	1	163[a]	1 261.0	7 640
Helium	2	125[a]	943.2	7 550
Lithium	3	534	827.6	1 550
Beryllium	4	1 848	651.9	353
Boron	5	2 370	553.9	234
Carbon	6	1 550[b]	427.0	~275
Nitrogen	7	808[a]	379.9	470
Oxygen	8	1 142[a]	344.6	302
Neon	10	1 207[a]	289.4	240
Magnesium	12	1 740	254.6	146
Aluminum	13	2 700	240.1	88.9
Argon	18	1 400[a]	195.5	140
Titanium	22	4 540	168.7	37.2
Chromium	24	7 200	146.7	20.4
Iron	26	7 870	138.4	17.6
Nickel	28	8 902	131.9	14.8
Copper	29	8 960	128.6	14.3
Niobium	41	8 570	~101	~11.8
Tin	50	7 310	88.6	12.1
Tungsten	74	19 300	67.6	3.5
Lead	82	11 350	63.7	5.6
Uranium	92	18 950	61.0	3.2
Compounds, alloys, and other materials				
Water	—	1 000[a]	360.8	360.8
Polyethylene	—	~950	447.8	~470
Epoxy resin	—	~1 450	~406	~280
Fiber glass epoxy	—	~1 750	~330	~189
Carbon fiber epoxy	—	~1 600	~418	~261
Boron–aluminum (45% B)	—	2 550	~381	~149
Mylar	—	~1 370	399.5	287
Sodium iodide	—	3 670	94.9	25.9
Lithium fluoride	—	2 640	392.5	149
304 Stainless steel	—	7 900	137.9	17.4
Nb–47%Ti	—	6 520	132.8	20.4

a) The density applies for the liquid state.
b) The density of carbon is for carbon in the amorphous state.

low Z materials. The radiation thickness of a detector magnet is the sum of the radiation thicknesses of the various elements. Table 4.4.4 shows the radiation thickness of various elements and materials that might be found in particle physics experiments.

The radiation thickness X_0 of a magnet component can be estimated using the following expression:

$$X_0 = \frac{t}{L_r \cos(\alpha)} \tag{4.4.1}$$

where X_0 is the radiation thickness of the magnet component (given in radiation lengths), t is the physical thickness of the material in the magnet component, L_r is the thickness for one radiation length of the material in the magnet component, and α is the particle angle with respect to a line perpendicular to the component. In most cases, radiation thickness is defined when $\alpha = 0$.

The value of L_r used in Eq. (4.4.1) can be used in the following expression:

$$L_r = 158\frac{Z^{-0.7}}{\gamma} \tag{4.4.2}$$

where Z is the atomic number for the heaviest element in the compound that makes up the component and γ is the mass specific density for the material in the component. For pure elements, Eq. (4.4.2) yields a good estimate of L_r, except for ordinary hydrogen, which has no neutrons in its nucleus. For components made from compounds, the use of the Z for the heaviest element in the compound will tend to overestimate radiation thickness, whereas using an average value of Z will often underestimate the radiation thickness. For components made from alloys or composites, the method of mixtures can be applied to achieve a good estimate of L_r.

Since most magnets in a detector and the things that are inside and outside of the magnet are made in layers, one may estimate the total radiation thickness X_{0T} for a particle going at an angle α with respect to being perpendicular to the layer by using the following expression:

$$X_{0T}(\alpha) = \frac{X_{01}(0) + X_{02}(0) + X_{03}(0) + \cdots + X_{0N}(0)}{\cos(\alpha)} \tag{4.4.3}$$

where $X_{01}(0)$ is the radiation thickness of component 1 at $\alpha = 0$, $X_{02}(0)$ is the radiation thickness of component 2 at $\alpha = 0$, $X_{03}(0)$ is the radiation thickness of component 3 at $\alpha = 0$, and so on until one has accounted for all N components.

The radiation thickness at $\alpha = 0$ for various materials can be found in Table 4.4.4, which comes from Refs [73, 74].

4.4.5
Thin Detector Solenoid Design Criteria

The strategy for minimizing the radiation thickness of a superconducting detector magnet requires the following steps: (i) Massive parts such as current bus bars, gas-cooled electrical leads, cold mass support structures, vacuum services, and cryogenic services should be located at the ends of the magnet away from the region that is supposed to have a minimum radiation thickness. (ii) The superconductor should have a minimum amount of copper and niobium titanium. The stabilizer matrix material for the superconductor should be made of a low resistivity, low Z material such as ultrapure aluminum. (iii) The support structure on the outside of the coil, which will carry the hoop forces in the solenoid, should be made from a strong, ductile, low-Z, low-density material with a high thermal conductivity. (iv) The magnet should be cooled indirectly with helium in tubes that are attached to the coil support structure. (v) Intermediate temperature shields for the cryostat should be made of a low Z, low density, and high thermal conductivity material such as 1100-O aluminum. (vi) The inner cryostat vacuum vessel should be made from a strong low-Z, low-density material. (vii) The outer cylinder of the cryostat vacuum vessel should be made from a material with a low-Z, a low-density, and an elastic modulus that is reasonably high. Figure 4.4.3

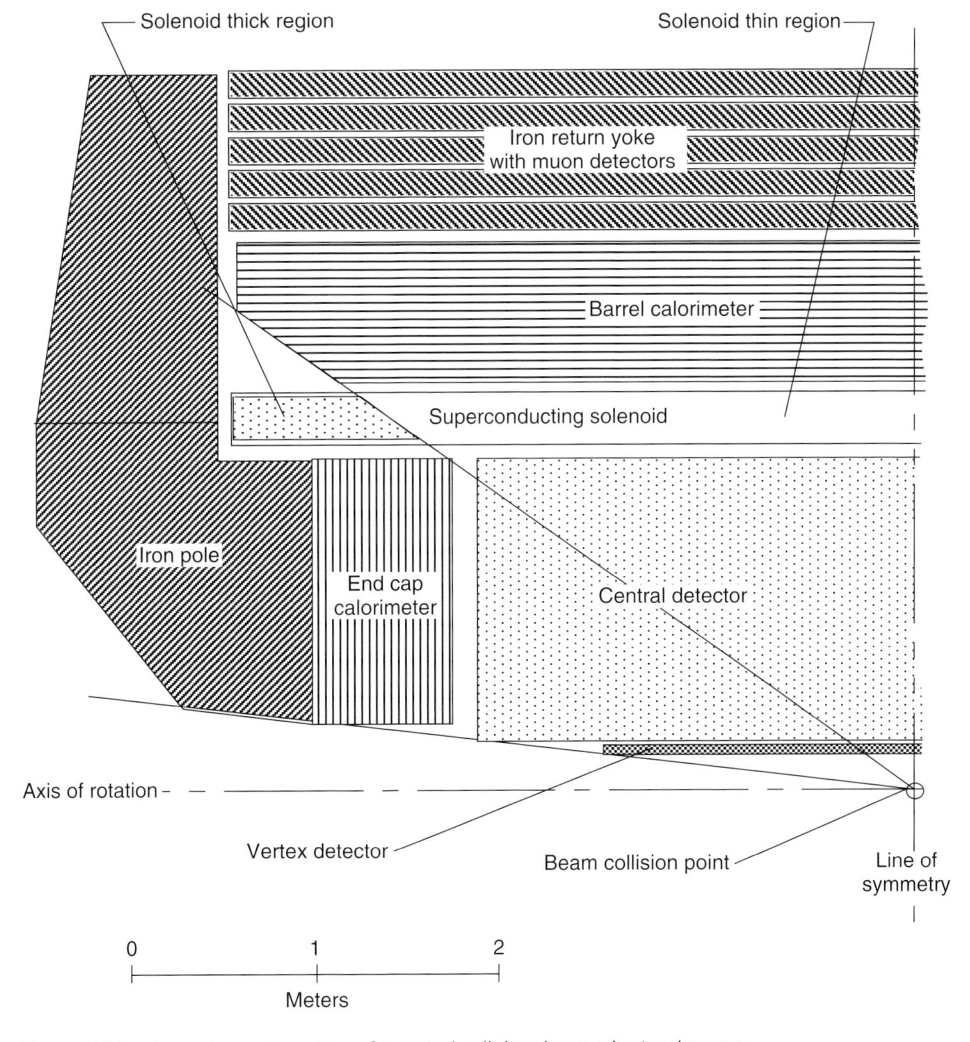

Figure 4.4.3 A quarter section view of a typical colliding beam physics detector.

shows a quarter section of an experiment with a superconducting solenoid that is thin from a radiation standpoint.

The typical physics detector solenoid is usually between two unsaturated iron poles that have an average relative permeability that is >20. The magnetic flux generated by the solenoid winding is returned by an iron yoke that carries the magnetic flux from one pole to the other. The relative permeability of the iron in the return yoke is usually above 50. Figure 4.4.1 shows a typical thin detector solenoid within an experiment located around the collision point of colliding beam storage ring.

The number of ampere-turns needed to generate a uniform magnetic induction within the detector solenoid can be estimated by using the following expression [75]:

$$\mathrm{NI} = \frac{B_o L_g}{\mu_o} \tag{4.4.4}$$

where NI is the total number ampere-turns in the detector solenoid coil needed to generate a magnetic induction B_o in a solenoid that has unsaturated iron poles that are a distance L_g apart. μ_o is the permeability of air ($\mu_o = 4\pi \times 10^{-7}\,\mathrm{H\,m^{-1}}$).

Equation (4.4.3) underestimates the ampere-turns needed to generate the magnetic induction in the solenoid bore anywhere from 3% to 30% depending on the design of the magnetic circuit and the central induction within the solenoid. The equation underestimates the required ampere-turns because the relative permeability of the iron in the poles and the return yoke is not infinite and the iron in the pole pieces is often segmented, with detectors between the segments. Often the extra ampere-turns are put at the ends of the solenoid so that the desired field uniformity within the solenoid can be achieved. Computer codes such as POISSON [76] and OPERA2D [77] or equivalent can be used to determine the number of ampere-turns needed to generate the desired central induction and the desired field uniformity within the detector volume. In more complex cases, a code such as OPERA3D or equivalent may be required.

The amount of superconductor needed to generate the magnetic field is quite small. When Nb–Ti with a critical current density of 2500 A mm^{-2} at 4.2 K and 5.0 T is used, about 0.3 mm of Nb–Ti is needed for every tesla of central magnetic induction produced [78]. The copper to superconductor ratio for the conductor can be as low as 0.8. The amount of stabilizer (usually annealed 0.99999 pure RRR > 1000 aluminum) in the conductor is dictated by the type of quench protection chosen. Quench back may also be a factor in the coil design.

The physical thickness of the superconducting coil is determined by the thickness of stabilizing matrix material in the conductor. The average conductor current density J_m is determined by the safe quench condition for the coil. For safe magnet quenching through a dump resistor, the magnet $E_o J_m^2$ limit can be estimated using the following equation [79, 80]:

$$E_o J_m^2 = V I_o F^*(T_m) \frac{r}{r+1} \tag{4.4.5}$$

where E_o is the magnet's stored energy when it is operated at its design current I_o, V is the discharge voltage for the magnet during the quench (for large magnets, V may be limited to about 500 V), I_o is the magnet design current (I_o is typically >3000 A), r is the matrix to superconductor ratio, and $F^*(T_m)$ is the integral of $J_m^2 dt$ needed to raise the stabilizer adiabatic hot spot temperature from 4 K to a maximum hot spot temperature T_m. (For RRR = 1000 aluminum, $F^*(T_m) = 6 \times 10^{16}$ A^2 m^{-4} when $T_m = 300$ K.)

The magnet's stored energy E_o can be estimated if one knows the solenoid average coil diameter D_c, the central induction B_o, and the gap between the iron poles L_g. An approximate expression for the magnet's stored energy is as follows:

$$E_o \approx \frac{\pi D_c^2 B_o^2 L_g}{8\mu_o} \tag{4.4.6}$$

The stored magnetic energy can be calculated by a number of codes such as POISSON, OPERA2D, and OPERA3D. Other codes such as ANSYS can be used as well.

If the $E_o J_m^2$ limit for the magnet is increased then the magnet design current I_o or the magnet discharge voltage V must be increased as well. Quench back from the coil support structure can be helpful in improving the quench protection for the magnet. Magnets that employ quench back [22, 81] as the primary means for quench protection can be operated at a much higher $E_o J_m^2$ limit, but the typical solenoid that is protected with a dump resistor across the leads has the $E_o J_m^2$ limit given by Eq. (4.4.4).

From Eqs (4.4.4) and (4.4.5), one can determine the thickness of the superconducting coil t_c using the following expression:

$$t_c = \left[\frac{\pi D_c^2 B_o^4 L_g}{8\mu_o^3 V I_o F^*\left(T_m\right)\frac{r}{r+1}}\right]^{0.5} \tag{4.4.7}$$

In order for the coil thickness to be thinner than the value given by Eq. (4.4.6), quench back must turn the whole coil normal in a time that is significantly faster than the L over R time constant of the coil and dump resistor circuit.

In virtually all of the large detector solenoids, the superconducting coil is wound inside the support cylinder [82]. When the coil is inside the support cylinder, the joint between the coil and the support structure is in compression as the magnet is charged. An additional advantage is that the coil package cools down from the outside. Thus, the support cylinder shrinks over the coil. A few of the smaller detector magnets were wound with the coil on the outside of a bobbin or support cylinder. In all these cases, the conductor was designed to carry all of the magnetic hoop forces and the helium cooling tubes were attached to the outside of the coil. The superconducting solenoid coil can be wound in one or two layers. A coil with an even number of layers has the advantage of having both leads from the coil come out at the same end of the coil package. There are a number of accepted ways of winding coils so that they have more current per unit length at the solenoid ends than in the center. One approach is to make the matrix current density higher at the ends by making the conductor thinner along the coil axis.

The thickness of the support shell outside the superconducting coil is governed by the magnetic pressure on the coil windings [75]. Total strain of the coil should be limited to prevent plastic deformation of the conductor matrix. If the conductor has a pure aluminum matrix, the strain limit for the coil should be set to about 0.1% [83, 84]. There are high RRR aluminum alloys that can be strained to almost 0.14%. A conservative view assumes that the support shell carries virtually all of the magnetic forces, and the calculated shell thickness is given by the following expression:

$$t_{\mathrm{s}} = 250\frac{B_{\mathrm{o}}^2 D_{\mathrm{s}}}{\mu_{\mathrm{o}} E_{\mathrm{s}}} \tag{4.4.8}$$

where t_{s} is the design thickness for the support shell, D_{s} is the inside diameter of the support shell, and E_{s} is the modulus of elasticity of the material in the support shell. If the superconductor is included in the overall strain calculation, the thickness of the support shell can be reduced.

The coil cryostat is primarily the vacuum vessel that provides the insulating vacuum for the magnet. The two primary cryostat elements are the outer cryostat vacuum vessels and the warm bore tube. The multilayer insulation and shields make up only a minor part of the cryostat's radiation thickness. A design thickness of a solid outer cryostat wall can be calculated using the following expression, which has been derived from the equation for elastic buckling of a cylinder under external pressure [85, 86]:

$$t_{\mathrm{o}} \approx 1.08\left[\frac{P_{\mathrm{o}} L_{\mathrm{o}} D_{\mathrm{o}}^{1.5}}{E_{\mathrm{o}}}\right]^{0.4} \tag{4.4.9}$$

where t_{o} is the thickness of the outer cryostat wall, P_{o} is pressure on the outer wall of the cryostat (usually $P_{\mathrm{o}} = 0.1013$ MPa is the diameter of the outer cryostat wall), and E_{o} is the elastic modulus of the material in the outer wall of the cryostat.

The minimum thickness of the inner wall of the cryostat can be derived if one knows the design ultimate stress for the material in the inner wall [84, 87]. The margin of safety normally applied to a pressure vessel wall, such as the cryostat inner wall, is usually four (based on the ultimate stress σ_{u}) [88]. However, one should design the cryostat vacuum vessel to the applicable pressure vessel standards. An approximate expression for the minimum inner cryostat wall thickness is given as follows:

$$t_{\mathrm{i}} = 2\frac{P_{\mathrm{i}} D_{\mathrm{i}}}{\sigma_{\mathrm{u}}} \quad \text{or} \quad t_{\mathrm{i}} = \frac{P_{\mathrm{i}} D_{\mathrm{i}}}{\sigma_{\mathrm{y}}}, \quad \text{whichever is larger} \tag{4.4.10}$$

where t_{i} is the minimum wall thickness for the inner cryostat wall, P_{i} is the design internal pressure on the inner cryostat wall, D_{i} is the diameter of the inner cryostat wall, σ_{u} is the ultimate stress for the material used in the inner cryostat wall, and σ_{y} is the yield stress for the material in the wall. Sometimes, the cryostat inner wall thickness is greater than the thickness given by Eq. (4.4.9) so that one can mount particle detectors and other equipment on this wall.

The material thicknesses calculated using Eqs (4.4.6)–(4.4.9) can be used to estimate the radiation thickness of the detector solenoid. Table 4.4.5 compares four

Table 4.4.5 A comparison of four thin solenoids.

Component	Case 1	Case 2	Case 3	Case 4
Central induction (T)	1.5	1.5	0.75	1.5
Solenoid diameter (m)	2.0	4.0	4.0	4.0
Gap between iron poles (m)	3.3	6.6	6.6	3.3
Length of the thin section (m)	3.3	6.6	6.6	3.3
Cryostat inside diameter (m)	1.84	3.8	3.8	3.8
Cryostat outside diameter (m)	2.24	4.28	4.22	4.26
Cryostat overall length (m)	3.85	7.30	7.30	3.85
Magnet ampere turns (MA)	3.94	7.88	3.94	3.94
Magnet stored energy (MJ)	9.28	74.25	18.56	37.13
Magnet design current (kA)	5.0	5.0	5.0	5.0
Magnet self inductance (H)	0.74	5.94	1.48	2.97
Number of coil turns	788	1 576	788	788
Quench discharge voltage (V)	500	500	500	500
Nb–Ti plus Cu thickness (mm)	0.90	0.90	0.45	0.90
Total coil thickness (mm)	9.39	26.58	13.28	18.75
Coil support structure thickness (mm)	12.97	25.95	6.49	25.95
Inner cryostat thickness for 1 bar (mm)	1.24	2.57	2.57	2.57
Outer cryostat thickness for 1 bar (mm)	13.11	25.23	25.23	19.12
Magnet radiation thickness (radiation length)	0.495	0.974	0.589	0.829
Magnet cold mass (kg)	2 120	14 400	6 160	7 000
Magnet overall mass (kg)	~4 000	~23 500	~15 300	~11 800

cases where the coil diameter, the gap between the iron poles, and the central induction are varied.

In all four cases, the cryostat walls and coil support structure are made from aluminum. The superconductor is Nb–Ti with a thick aluminum stabilizer. The assumed insulation system inside and outside the cold mass consists of 60 layers of aluminized Mylar and netting with a single 1 mm-thick aluminum shield on either side of the coil. Larger magnets may have shields at two different temperatures to reduce the refrigerator input power. Figure 4.4.4 shows a cross-section of a coil and cryostat for Case 2 given in Table 4.4.5. In order to make a significant reduction in the radiation thicknesses shown in Table 4.4.5, quench back must be the primary mode of quench protection and the outer cryostat vacuum vessel must be made from a cellular (honeycomb) composite structure that is physically much thicker than a solid aluminum outer cryostat vessel [89, 90].

4.4.6
Magnet Power Supply and Coil Quench Protection

The power supply parameters are set by the coil charge time t_{ch} and the design operating current I_0 for the solenoid. The charge time for a detector solenoid is rarely an issue. Charge times as long as 1 h are acceptable. The charge voltage $V = L_1 di/dt$, where L_1 is the self inductance of the magnet circuit and di_1/dt is the

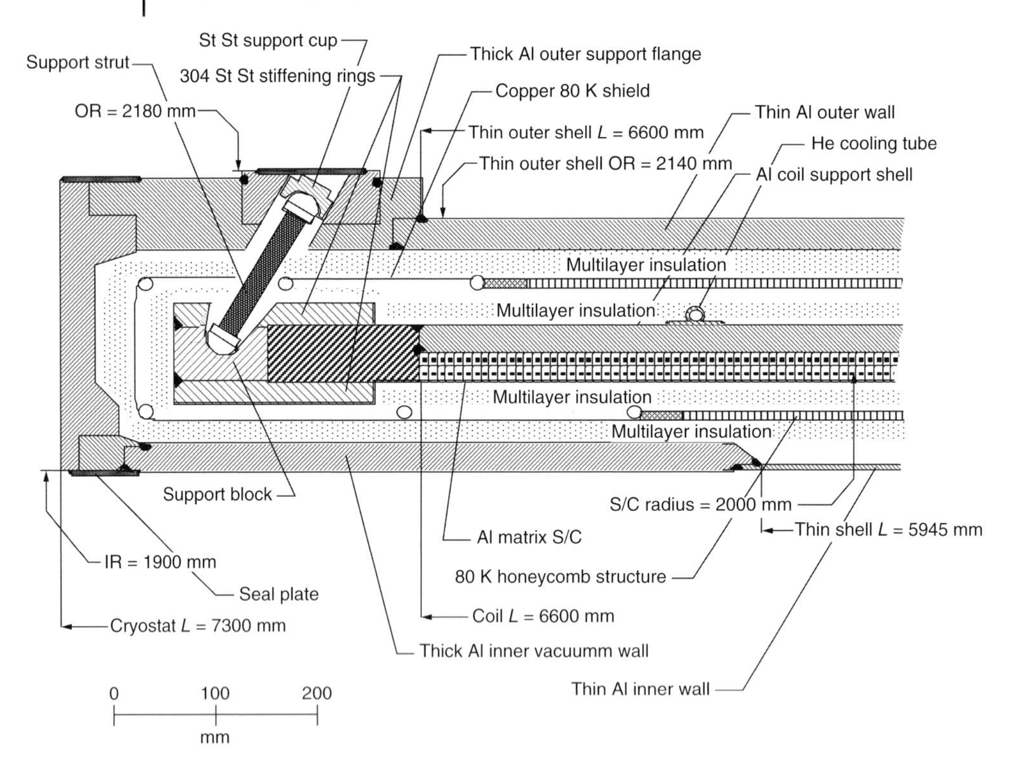

Figure 4.4.4 A cross-section through the end of a 1.5 T thin solenoid with a 4.0 m coil diameter. A self-centering support strut is shown along with the stiff end ring for the superconducting coil package. (See Case 2 in Table 4.4.3.)

magnet current charge rate. (For a typical magnet, $di_1/dt = I_o/t_{ch}$.) To determine the power supply voltage, one must add the IR voltage drop across the gas-cooled electrical leads and the cables connecting the power supply to the magnet. In addition, a voltage drop of 0.9 V should be allocated to the power supply back wheeling diodes and a current shunt.

4.4.6.1 Quench Protection Dump Resistor

Most large detector magnets are protected by a dump resistor across the gas-cooled electrical leads. When a quench is detected, the power supply is disconnected and the dump resistor is put across the leads. The design of a magnet dump resistor circuit is determined by the following relationship [80]:

$$F^*(T_m) = \int_0^\infty j(t)^2 dt = \frac{r}{r+1}\int_{T_0}^{T_m} \frac{C(T)}{\rho(T)} \tag{4.4.11}$$

where $j(t)$ is the current density in the magnet superconductor plus matrix cross-section as a function of time t, $C(T)$ is the conductor volume specific heat as a function of temperature T, $\rho(T)$ is the superconductor matrix material electrical resistivity as a function of temperature, and r is the ratio of matrix material to superconductor in the magnet conductor. T_0 is the starting temperature of the magnet (about 4 K); T_m is the maximum allowable hot spot temperature for the magnet conductor (usually 300–350 K). For a conductor with a very pure aluminum matrix with a RRR = 1000, the value of $F^*(T_m)$ is around 6.0×10^{16} A^2 m^{-4} s when T_m is 300 K.

When the magnet is discharged through a dump resistor, the current decay is exponential with a decay time constant τ_1 ($\tau_1 = L_1/R_{ex}$, where R_{ex} is the resistance of the external dump resistor). The value of $F^*(T_m)$ at the magnet coil hot spot is given as follows:

$$F^*(T_m) = j_o^2 \frac{r+1}{r} \left[\frac{\tau_1}{2} + t_{so}\right] \tag{4.4.12}$$

where t_{so} is the time needed to detect the quench and switch the resistor across the magnet coil (in most cases, t_{so} is <1 s) and j_o is the starting current density in the coil superconductor plus matrix material (I_o divided by the conductor cross-sectional area). If a constant resistance dump resistor is used, the value of the resistance R_{ex} that results in a hot spot temperature less than or equal to T_m can be expressed as follows:

$$R_{ex} = \frac{j_o^2}{2F^*(T_m)} \frac{(r+1)}{r} L_1 \tag{4.4.13}$$

The design value of R_{ex} should be larger than the value calculated by Eq. (4.4.12). For a constant resistance dump resistor, the maximum discharge voltage across the leads $V = R_{ex} I_o$ will occur when the dump resistor is just put across the magnet.

Figure 4.4.5 shows a circuit diagram of the coil, its power supply, and the magnet dump circuit. A quench detection system is also shown. The values for inductances and R given in Figure 4.4.5 would apply to Case 2 in Table 4.4.3. The quench detection system shown in Figure 4.4.3 compares the voltage across the superconducting coil with the *dB*/*dt* voltage due to changes in flux in the coil. If a voltage is measured across the coil and there is no corresponding *dB*/*dt* voltage, there is a normal region in the coil. The normal region detected by the quench detector will open the switch putting the dump resistor across the coil. Other methods can also be used to detect short normal sections within a magnet [91].

4.4.6.2 **The Role of Quench Back**

It has been observed in most of the thin detector solenoid magnets that when the dump resistor is put across the electrical leads, the entire magnet becomes normal through the process of "quench back" [22]. Quench back insures that the coil current will decay faster than is predicted by the L over R_{ex} time constant. As a result, the magnet's hot spot temperature is reduced.

There is a maximum time t_{QB} before which quench back must occur in order for quench back to be a fail-safe method of quench protection [92, 93]. If the resistance of the external resistor induces quench back in a time less than the time t_{QB}, then the hot spot temperature is less than T_m, the maximum allowable hot spot temperature. The maximum allowable quench-back time t_{QR} is the quench-back time required for fail-safe quenching t_{QBR} minus the time required for a heat pulse to cross the insulation between the quench-back circuit (usually the coil

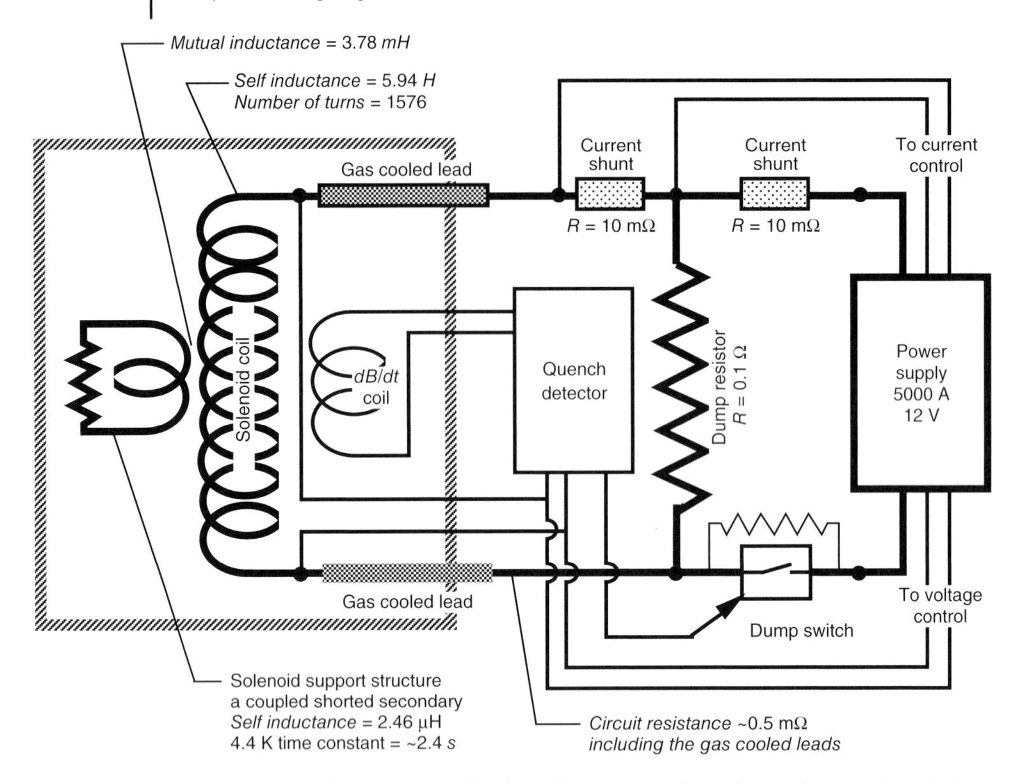

Figure 4.4.5 A schematic circuit for the coil, power supply, and quench protection circuit for a large detector solenoid with a support cylinder. (See Case 2 in Table 4.4.3.)

support structure) and magnet coil t_{H}. (For a layer of ground plane insulation that is 2 mm thick, t_{H} varies from 0.3 to 0.5 s depending on a number of factors.) For large detector solenoids, t_{H} is usually small compared to τ_1. The value of t_{QBR} can be determined using the following expression:

$$t_{\mathrm{QBR}} = \frac{r}{r+1}\left[\frac{F^*\left(T_{\mathrm{m}}\right) - F^*(T_{\mathrm{s}})}{j_{\mathrm{o}}^2}\right] \tag{4.4.14}$$

where r and j_{o} have been previously defined. $F^*(T_{\mathrm{m}})$ and $F^*(T_{\mathrm{s}})$ are defined by the right-hand term in Eq. (4.4.10) for the maximum hot spot temperature T_{m} and the maximum temperature the coil would go to if the entire coil is quenched instantaneously at T_{s}. The value of T_{s} depends on how the stored energy of the magnet is split between the hard aluminum support shell and the coil. For detector solenoids, it is usually safe to split the magnet's stored energy between the support shell and the coil according to their masses.

Once t_{QR} has been determined, it is possible to calculate the resistance of an external resistor needed to cause the coil to quench back from the support tube in a time that is less than t_{QR}. The minimum resistance needed for quench-back R_{min} can be calculated using the following relationship for a solenoid coil that is well coupled inductively to its quench-back circuit [92]:

$$R_{\text{min}} = \left[\frac{L_1}{\tau_2}\frac{N_2}{N_1}\frac{A_{c2}}{I_o}\right]\left[\frac{\Delta H_2}{\rho t_{\text{QR}}}\right]^{0.5} \tag{4.4.15}$$

where L_1 is the magnet coil self inductance, N_2 is the number of turns in the quench-back circuit ($N_2 = 1$ when the support shell is the quench back circuit), N_1 is the number of turns in the magnet coil, A_{c2} is the cross-sectional area of the quench-back circuit, I_o is the coil current, ΔH_2 is the enthalpy change per unit volume needed to raise the quench-back circuit temperature from 4 to 10 K (for aluminum, $\Delta H_2 = 13\,200\,\text{J m}^{-3}$), ρ_2 is the resistivity of the quench-back circuit material, and τ_2 is the L over R time constant for the quench-back circuit (the support tube).

If the minimum quench-back resistance R_{min} is less than the resistance of the quench protection resistor R_{ex}, quench back will always occur during a magnet dump. Therefore, the hot spot temperature of the coil is lower than T_{m}. When quench back is present, one can use a varistor (a diode like resistor where the voltage across the resistor is nearly independent of current) as a dump resistor to speed up the quench process without increasing the coil voltages during the quench [94].

4.4.7 Design Criteria for the Ends of a Detector Solenoid

The previous sections have dealt primarily with the center section of a detector solenoid. Much of the engineering for a detector solenoid is in the ends of the magnet, where thinness is not an issue. For example: (i) the support system for the solenoid cold mass is attached to the ends of the magnet. (ii) The outside ends of the cryostat vessel are often where physical connections are made between the magnet and the rest of the detector. (iii) The current leads and voltage taps into the coil will come out of the coil package at its ends. Gas cooled electrical leads that connect the coil to the room temperature outside world may also be located inside the magnet-insulating vacuum vessel in the end region. (If the magnet has HTS, leads they would be located in this region as long as the field is low enough for HTS leads.) (iv) Cryogenic cooling is usually fed into the solenoid coil from the ends. Cooling should also include the intermediate temperature fluid (either liquid nitrogen at 80 K or helium gas at 40–80 K) used to cool the shields and HTS leads. (v) Cryostat vacuum pumping ports will be located at the ends of the solenoid. (vi) Room temperature ports for voltage taps, quench detection coils, temperature sensors, and pressure transducers will enter the magnet at the ends.

4.4.7.1 Cold Mass Support System

The cold mass supports to room temperature must carry gravity forces, seismic forces, magnetic forces, and shipping forces. Most detector solenoids are designed to be at a neutral magnetic force point when the coil is at its operating temperature, so the cold mass support system must have a spring constant that is higher than the magnetic force constant in the direction when the support system is in unstable equilibrium.

Solenoids that are surrounded by iron are usually, but not always, in stable equilibrium in the radial direction, and unstable equilibrium in the axial direction. In the toroidal direction (about the solenoid axis), there are almost no magnetic forces in a well-built solenoid, although asymmetric holes in the iron can introduce some of these forces. Stable equilibrium indicates that the magnetic forces will act in a direction that reduces a placement error; unstable equilibrium indicates that the magnetic forces will act in a direction that increases the placement error. In the direction of stable equilibrium, the spring constant of the support system is not a critical issue except when determining how the magnet responds to vibration. In the direction of unstable equilibrium (usually the axial direction), the spring constant of the support system must be larger than the force constant for the magnet at its maximum design field. In general, the magnet force constant is linear with the location error and it increases with the magnet current squared. The magnetic force constant is a function of the design of the coil, the iron return yoke, and the pole pieces. Magnetic force constants will change as the iron in the magnetic circuit saturates.

Two types of cold mass support systems are commonly used in detector solenoids. The first is the self-centering support system where the position of the center of the solenoid coil does not change during the magnet cooldown or as the magnet is powered. The second support system carries axial forces with push–pull rods at one end of the magnet while the radial forces are carried by gravity support rods at both ends of the magnet. Both types of support systems must be designed to handle magnet shrinkage during the cool down. A coil that is 6.6 m long and 4 m in diameter will shrink almost 28 mm in the axial direction and the radius will decrease about 8.4 mm. The external cryostat support system should be in line with the cold mass support system in order to avoid bending within the cryostat. The spring constant for the combined internal and external support systems must be greater than the magnetic force constant.

The self-centering support system has several advantages: (i) the position of the magnet center is the same – both warm and cold. The PEP-4 solenoid magnetic center changed <0.3 mm during the magnet cool down. (ii) The radial and axial supports can be combined using either tension or compression rods. Two of the rods can also carry the torsion forces about the solenoid axis. The angle of the support rods can be set so that rod stress is not changed during the coil cooldown. (iii) Since the axial spring constants must be high, the spring constant will be high in all directions. The self-centering support system will have a relatively high first mode vibration frequency. (iv) The self-centering support system is robust in all directions, so earthquake and transportation forces should not be a problem.

The two disadvantages of the self-centering support system are as follows: (i) as the magnet coil cools down, it will move with respect to the ends of the cryostat vacuum vessel at both ends of the magnet. This movement must be considered when designing electrical leads, cryogen feeds, and other attachments to the coil. (ii) Flexure of the coil package (at the ends of the support cylinder) will affect the spring constant of the support system. The stiffness of the ends of the coil package and the number of radial–axial supports are the determining factors for the spring constant of this type of support system. Finite element stress and strain calculations can be used to determine the spring constant of the cold mass support system. A description of the design of a self-centering support system can be found in Ref. [95]. A location of a typical self-centering support compression strut for a detector solenoid is shown in Figure 4.4.2. The strut rotates in its sockets as the solenoid cold mass contracts. The distance between the ball sockets does not change as the solenoid cools down from room temperature to 4 K. The angle of the strut with respect to the solenoid axis changes as the coil end of the strut moves toward the center of the solenoid

4.4.7.2 The Solenoid Support Structure, the Cryogenic Heat Sink

The support cylinder outside the superconducting winding serves the following functions: (i) the outer cylinder carries the magnetic pressure forces that are generated by the coil. (ii) The outer cylinder transfers magnetic, gravitational, and seismic forces from the coil structure to the cold mass support system. (iii) The outer cylinder carries the helium cooling tubes and acts as the heat sink for the coil and all attachments to it. This means that the outer support cylinder must be made from material that conducts heat well in both the radial and the axial directions.

The end ring of the support cylinder should be as stiff as possible in bending. End-ring stiffness can be increased by making the ring thicker, thus increasing its moment of inertia, or one can fabricate a laminated end ring with a high elastic modulus material such as 304 stainless steel (elastic modulus of 200 GPa as compared to 69 GPa for aluminum) on the outside and the inside of the ring with aluminum in the center. The need for stiff end rings on the support cylinder is reduced as the number of cold mass supports per end is increased for a given coil diameter [95].

4.4.7.3 Coil Electrical Connections and Leads to the Outside World

Connections to the superconducting coil that come through the end ring should be mounted on copper bus bars that are electrically insulated from the end rings. These bus bars should be liquid helium cooled in order to avoid heat from outside the coil being deposited directly into the superconducting windings. Heat leaks down pulsed current leads, which are usually not gas cooled, can be particularly troublesome. The cooling circuit used to cool bus bars at the ends of the coil should be part of the magnet helium cooling system. Since much of the cooling circuit is electrically grounded, in-line electrical insulators will be required in the cooling lines that cool the electrical bus bars connected to the superconducting coil.

Most detector solenoids have gas-cooled electrical leads that are fed from a liquid helium pot located somewhere near the solenoid. The current buses between the lead pot and the coil are often cooled by conduction, a practice that has led to a number of failures. All current buses should be helium cooled. The lead pot commonly used in detector magnets can be eliminated by using gas-cooled electrical leads that are attached to the ends of the coil structure. The helium used to cool these leads comes directly from the liquid helium cooling circuit. Gas-cooled leads attached to the end of the magnet are located within the cryostat vacuum, so these leads must be completely vacuum-tight and they must withstand any increase in pressure that might occur in the cooling circuit during a quench [96]. The bundled nested tube leads that were used on the PEP-4 experiment [97] and the g-2 solenoids [98] can be operated at any orientation within the cryostat vacuum vessel. Properly designed gas-cooled leads are stable and they are capable of operating for more than 30 min without gas flow.

4.4.8
Cryogenic Cooling of a Detector Magnet

Most of the detector magnets shown in Table 4.4.1 are solenoids cooled by helium in tubes attached to the superconducting coil or the support cylinder outside the coil. This technique has the following advantages over the bath cooling methods used for early cryogenically stable detector magnets and bubble chamber magnets [99]: (i) tubular cooling eliminates the cryostat helium vessels. As a result, the solenoids are thinner and less massive. (ii) The volume of helium in a tubular cooling system is small. Once this helium is evaporated during a quench, it is expelled from the tube. Large quantities of helium gas are not produced during a magnet quench. The helium expelled during a magnet quench can be returned to the refrigerator where it is recovered. (iii) Tubes can withstand high pressures during a quench. Relief valves for the system can be moved from the magnet cryostat to the helium supply system, which can be outside the detector. (iv) Magnet cooldown can be done directly using the helium refrigerator. (v) Recovery from a quench can be simplified using a well-designed tubular cooling system.

4.4.8.1 Forced Two-Phase Flow Circuits

Detector magnets are cooled with two-phase helium rather than supercritical helium for the following reasons: (i) the operating temperature for the superconducting solenoid is lower [18]. As two-phase helium flows down the cooling circuit, it gets colder, as its pressure goes down. The temperature of a single-phase supercritical helium cooling circuit increases as one goes along the cooling circuit. (ii) The mass flow through the cooling circuit is minimized. As a result, the pressure drop along the flow circuit is lower. (iii) There is no need for auxiliary helium pumping in a two-phase flow circuit. Helium flow can be provided directly by the Joule–Thomson (J–T) circuit of the refrigerator. (iv) A properly designed two-phase helium flow system can be operated at heat loads greater than the capacity of the refrigerator for a period of time. Thus, fluctuations in the heat load can be tolerated by two-phase flow circuits that are designed for the average heat load. The most often stated disadvantage of two-phase cooling is the potential existence of flow and pressure fluctuations in the cooling tube. This has not been a problem in detector magnets when the two-phase helium cooling circuit is properly designed. Experiments with extensively looped cooling tubes that are hundreds of meters long have shown that proper design of the flow circuit can nearly eliminate the flow oscillation "garden hose" slug–plug flow problem [99, 100].

The types of two-phase helium flow circuits that are commonly used in detector solenoids are the forced two-phase flow system and the natural convection two-phase flow system. Forced two-phase flow is appropriate when the flow circuits are long and when the control cryostat is below the top of the magnet. Natural convection two-phase flow is appropriate when there is a large vertical head between the helium cryostat and the load and when there are many parallel flow circuits so that the mass flow in any one circuit can be kept low. Either type of two-phase helium flow circuit can be made to work in most detector solenoids.

The key to stable operation of forced two-phase helium cooling circuits is the control helium vessel and heat exchanger [99, 101]. Flow for the magnet cooling circuit comes from the J–T circuit of the helium refrigerator. Cooling flow can also come from a helium pump, but allowances must be made for the pump work heating generated by such a pump. A positive displacement pump can be used over a broad range of mass flows at high pressure drops [102].

Two-phase flow from the refrigerator J–T circuit flows through a heat exchanger that is cooled in a bath of liquid helium at the suction pressure of the cold end of the refrigerator. The temperature of the helium bath is the lowest temperature in the two-phase flow circuit. Within the heat exchanger, helium in the gas phase is condensed to liquid so that the helium leaving the heat exchanger is either on the saturated liquid line or slightly sub-cooled. As a result, the average density of the helium in the flow circuit is maximized, which will cause the flow pressure drop through the flow circuit to be reduced to a factor of 2 or 3 as compared to the same flow circuit without a heat exchanger in the helium bath. The use of the heat exchanger in the control helium vessel allows the operation of the cooling circuit with a heat flow into the magnet that exceeds the capacity of the refrigerator by as much as 50%. Under this condition, the magnet can be kept cold as long as the heat exchanger in the control cryostat is kept covered with liquid helium. The control cryostat and heat exchanger enhance flow circuit stability.

An alternative to using the refrigerator J–T circuit (or the flow from a two-phase turbine or piston expander) is to use a helium pump. If the pressure drop in the circuit is relatively low (say 5 kPa), a centrifugal pump may be used. If the flow pressure drop is high or if the pump is to be used for a magnet cooldown, one should use a positive displacement pump (a piston pump or a bellows pump). The advantage of using a pump is that the flow does not stop when the refrigerator stops or the power fails. (The pump power must be on some sort of backup power.) The biggest disadvantage of using a pump to circulate the helium is the pump work that must be removed by the refrigerator.

Figure 4.4.6 shows a schematic representation of a forced two-phase helium cooling circuit in its simplest form. The valves in Figure 4.4.6 are shown as they would be when the magnet is operating at 4.5 K. (Un-shaded valve centers are open; shaded valve centers are closed. Dark flow lines carry helium; shaded flow lines are not carrying helium flow.) The refrigeration system shown in Figure 4.4.6 does not include the separate flow circuits within the cold box on the cryostat for cooling the magnet shields, cold mass supports, and HTS leads that may be within the magnet cryostat.

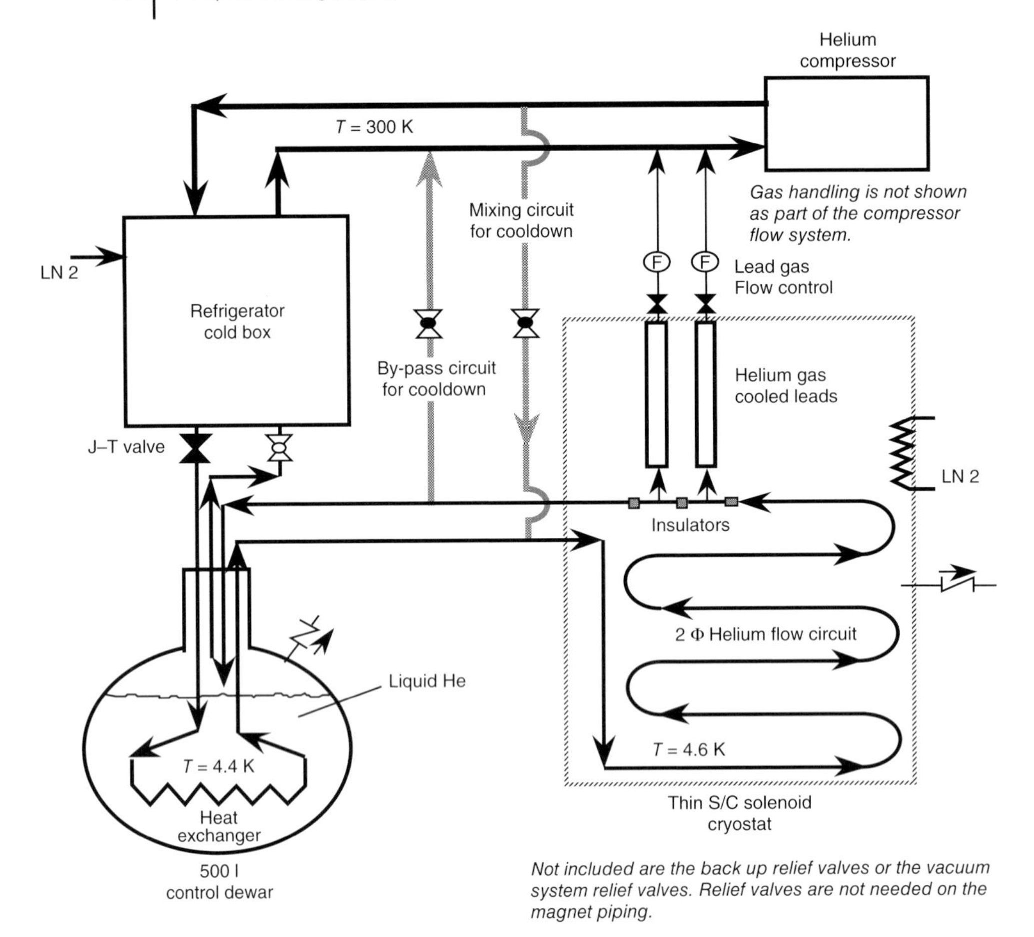

Figure 4.4.6 A schematic representation of a forced two-phase helium cooling system for a large detector solenoid.

The advantages of forced two-phase cooling are as follows: (i) the entire detector solenoid can be cooled using a single helium flow circuit. As a result, the cooldown of the magnet is straightforward because all of the sensible heat of the helium can be employed during the cooldown process. (ii) The operating temperature for a two-phase cooled magnet is lower than it would be for any supercritical helium-cooled magnet. (iii) The control cryostat with its heat exchange can be located flexibly with respect to the magnet coil. Transfer lines to and from the coil can be long, if desired. (iv) Gas-cooled electrical leads and shields can be cooled directly from the two-phase cooling circuit. The connections for leads and shields can be made inside the magnet cryostat vacuum vessel. (v) Since the liquid helium inventory in contact with the magnet is limited to the helium in the cooling tube, the amount of helium gas produced during a quench is small. (vi) The liquid helium in the control cryostat can be used to speed up the recovery of the superconducting magnet after a quench. The primary disadvantage of forced two-phase cooling is that when the refrigerator stops, the magnet cooling stops. The magnet will quench within minutes after the refrigerator stops running. For some users, this is a serious consideration.

4.4.8.2 Two-Phase Cooling Using Natural Convection

The natural convection two-phase cooling system overcomes the primary disadvantage of a forced two-phase cooling system in that the magnet will remain cold and operating even when the refrigerator is not in operation. The cooling for a natural convection cooling system comes from the helium that is stored in a tank that is above the top of the coil. The greater the head between the storage tank and the top of the magnet, the better the natural convection two-phase flow system operates. In order for the natural convention flow system to operate effectively, the following conditions must be present: (i) the pipe from the bottom of the helium storage tank to the manifold at the bottom of the magnet should be short and well insulated. There should be no boiling in helium transferred to the lower manifold on the coil package. (ii) In order to reduce the flow circuit pressure drop, there should be many short upflow circuits in parallel going up and around the coil to the manifold at the top of the magnet. Boiling should occur in these tubes. This increases the helium flow through the cooling system. (iii) The pipe from the manifold at the top of the coil package should dump two-phase helium into the top of the storage tank, where phase separation occurs. This pipe should be insulated from the helium that is in the storage tank. The difference in helium density in the pipe connecting the tank and the lower manifold and the two-phase helium in the cooling tubes circling the magnet provides the driving force for the flow circuit. Figure 4.4.5 shows a schematic representation of a natural convection two-phase helium cooling circuit in its simplest form. The upper part of Figure 4.4.5 shows a schematic of how the refrigerator, its compressors, and the magnet would be hooked up. The lower part of Figure 4.4.5 shows the physical arrangement of the helium storage tank (phase separator) and the detector solenoid. The storage tank shown in Figure 4.4.5 is 500 l, but that tank could be much larger or smaller, depending on the time the magnet must stay cold while discharging during a power failure. The price of helium is high enough that one must not have a tank that is larger than necessary. The loss of hundreds of liters of helium can result in a large expense.

The valves in Figure 4.4.7 are shown as they would be when the magnet is operating at 4.5 K. (Un-shaded valve centers are open; shaded valve centers are closed. Dark flow lines carry cold helium; shaded flow lines are not carrying helium flow.) The refrigeration system shown in Figure 4.4.7 does not include the separate flow circuits within the cold box on the cryostat for cooling the magnet shields, cold mass supports, and HTS leads that may be within the magnet cryostat.

Natural convection flow has some disadvantages, which are as follows: (i) cooldown of the magnet is not as straightforward as with the forced two-phase cooling system. This difficulty can be overcome by having a separate forced flow circuit for the magnet cooldown. (ii) In a natural, convection-cooled magnet

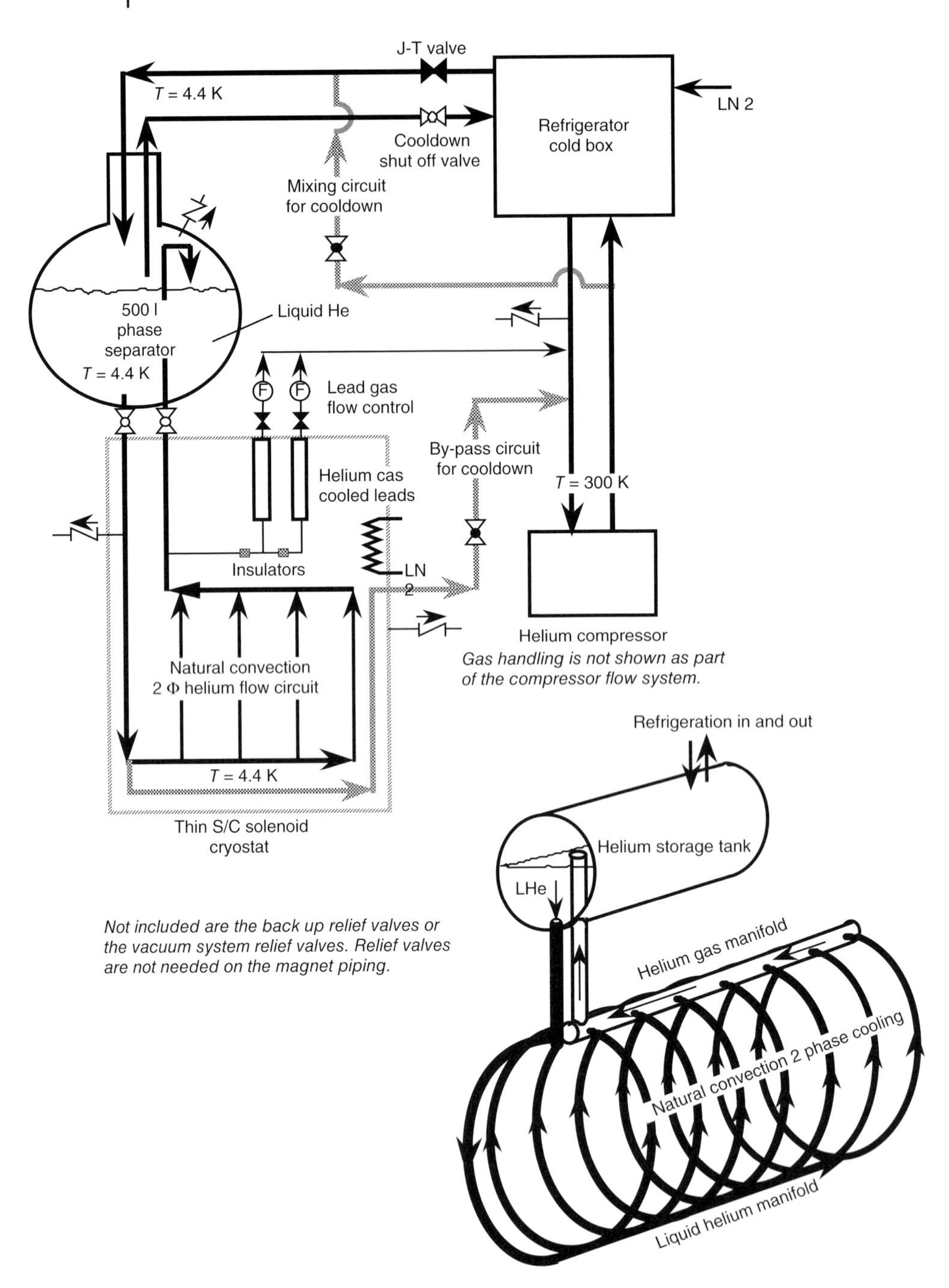

Figure 4.4.7 A schematic representation of a natural convection two-phase helium cooling system for a large detector solenoid.

system, all of the helium that is in the storage tank may be boiled during a magnet quench. This can be overcome by installing an automatic shut off valve, which is triggered by the quench detector, in the pipe between the storage tank and the liquid helium manifold on the bottom of the magnet. (iii) The thin section of the solenoid has a larger radiation thickness above and below the regions where the liquid helium and two-phase helium manifolds are located. (iv) The helium storage tank for natural convection cooling must be located directly above the magnet coil and the transfer lines between the storage tank and the coil should be as short as possible. The physics experiment must accommodate these transfer lines.

4.4.8.3 High-Temperature Superconducting (HTS) Leads

The superconducting detector magnets shown in Tables 4.4.1–4.4.3 use helium gas-cooled leads between 4.5 K and room temperature. The helium mass flow needed to keep the gas-cooled leads cold is a function of the lead current, the IL/A (the lead design coefficient of lead dependent on lead material properties), and the efficiency of the heat transfer between the helium and the lead conductor [103]. Adding a superconductor to the lead does little to reduce the lead mass flow needed to cool the lead [104]. There is an advantage gained, if the temperature of the gas entering the lead is higher (say 40–60 K), because the compressor power needed to produce lead cooling can be an order of magnitude smaller. For example, the equivalent 4.5 K refrigeration is needed to cool and single pair of 20 kA leads can be between 250 and 350 W. The input power for a modern efficient refrigerator (with a Carnot efficiency of 0.35) to provide the cooling would be between 55 and 77 kW. If HTS leads are used so that the gas entering the gas-cooled leads is from 40 to 60 K, the input power needed to cool the leads can be reduced by an order of magnitude. As the total lead current goes up, the cooling of those leads becomes more important. Two papers that describe the melding of HTS leads with gas-cooled leads are found in Refs [105, 106]. The use of HTS leads becomes critical when small coolers are used to cool smaller detector magnets.

4.4.8.4 Detector Magnets Cooled and Cooled Down with Small Cooler

There is a subset of superconducting detector magnets that may, in the future, be cooled using small two-stage 4.2 K coolers. A key to cooling magnets with small coolers is the use of HTS leads that must be cooled using the cooling from the cooler first-stages to intercept the heat coming down conduction-cooled leads from room temperature to 40 or 60 K [103, 107]. A larger detector magnet that is cooled using small coolers will use a thermal-siphon loop to connect the cold head heat exchanger (condenser) to the cold mass [108, 109].

In general, detector magnets cooled with small coolers at liquid helium temperatures will be too large for cryogen-free cooling. Most of the magnets that might be built in 2013 would be made from Nb–Ti, but that will not always be the case. One can make a case for using Nb_3Sn at 4.5 K for a higher field detector magnet. These magnets are in a size range, where it may be attractive to cool them with two-stage 4.2 K coolers using a helium thermal-siphon cooling loop to connect the coolers to the cold mass.

The stray field from a magnet will be an issue when a magnet is cooled using small coolers. Stray magnetic fields can hamper the operation of small coolers [110]. Stray fields will also affect the HTS leads when one operates a magnet that is cooled using coolers [111].

There may be, at some point, magnets made from materials such as MgB_2 and even HTS conductors. Magnets that operate at higher temperature (say in the 15–30 K range) can be cooled using small coolers, by the thermal-siphon cooling loop with hydrogen in loop to transfer the heat from the magnet to the cooler cold heads [112]. When a magnet operates in the temperature range from 15 to 30 K, there can be innovative ways of cooling the magnets down using coolers. This may, in the future, allow more massive magnets to be cooled and cooled down using coolers.

References

1. Kunzler, J.E., Buehler, E., Hsu, F.S.L., and Wernick, J.H. (1961) Superconductivity in Nb3Sn at high current density in a magnetic field of 88 kgauss. *Phys. Rev. Lett.*, **6**, 89.
2. Wolgast, R.C. *et al.* (1962) *Advances in Cryogenic Engineering*, vol. 8, Plenum Press, New York, p. 38.
3. Laverick, C. (1964) *Advances in Cryogenic Engineering*, vol. 10, Plenum Press, New York, p. 105.
4. Stekly, Z.J.J. and Zar, Z.L. (1965) Stable superconducting coils. *IEEE Trans. Nucl. Sci.*, **NS-12**, 367.
5. Purcell, J.R. (1968) The 1.8 tesla, 4.8 m bubble chamber magnet. Proceedings of the 1968 Summer Study on Superconducting Devices and Accelerators, BNL 50155, p. 765.
6. Jones, R.E., Martin, K.B., Purcell, J.R., McIntosh, G.E., and McLagan, J.N. (1969) *Advances in Cryogenic Engineering*, vol. **15**, Plenum Press, New York, p. 141.
7. Brown, D.P., Burgess, R.W., and Mulholland, G.T. (1968) The superconducting magnet for the brookhaven national laboratory 7 foot bubble chamber. Proceedings of the 1968 Summer Study on Superconducting Devices and Accelerators, BNL 50155 (C-55), p. 794.
8. Pewitt, E.G. (1970) *Advances in Cryogenic Engineering*, vol. **16**, Plenum Press, New York, p. 19.
9. Haebel, E.U. and Wittgenstein, F. (1970) Proceedings of the Third International Conference on Magnet Technology (MT 3), Hamburg, Germany, 1970, pp. 874–895.
10. Fields, T.H. (1975) Superconductivity applications in high energy physics. *IEEE Trans. Magn.*, **MAG-11** (2), 113.
11. Morpurgo, M. (1968) Construction of a superconducting test coil cooled by helium forced circulation. Proceedings of the 1968 Summer Study on Superconducting Devices and Accelerators, BNL 50155 (C-55), p. 953.
12. Morpurgo, M. (1970) Proceedings of the Third International Conference on Magnet Technology (MT 3), Hamburg, Germany, 1970, pp. 908–924.
13. Bean, C.P. *et al.* (1965) Research Investigation of the Factors That Affect the Superconducting Properties of Material. AFML-TR-65-431. Air Force Materials Laboratory, Wright Patterson Air force Base, OH.
14. Hancox, R. (1965) Stability against flux jumping in sintered Nb_3Sn. *Phys. Lett.*, **16**, 208.
15. Chester, P.F. (1967) Superconducting magnets. *Rep. Prog. Phys.*, **30** (Pt. 2), 561–614.
16. Smith, P.F. and Lewin, J.D. (1967) Pulsed superconducting synchrotrons. *Nucl. Instrum. Methods*, **52**, 248.
17. Wilson, M.N., Walters, C.R., Lewin, J.D., and Smith, P.F. (1970) Experimental and theoretical studies of

filamentary superconducting composites, part 1, basic ideas and theory. *J. Phys. Appl. Phys.*, **3**, 1517.
18. Green, M.A. (1971) Cooling intrinsically stable superconducting magnets with super-critical helium. *IEEE Trans. Nucl. Sci.*, **NS-18** (3), 669–670.
19. Desportes, H. (1981) Superconducting magnets for accelerators, beam lines and detectors. *IEEE Trans. Magn.*, **MAG-17** (5), 1560.
20. Green, M.A. (1975) *Advances in Cryogenic Engineering*, vol. **21**, Plenum Press, New York, p. 24.
21. Eberhard, P.H. *et al.* (1977) Tests on large diameter superconducting solenoids designed for colliding beam accelerators. *IEEE Trans. Magn.*, **MAG-13** (1), 78.
22. Green, M.A. (1977) The Development of Large High Current Density Superconducting Solenoid Magnets for Use in High Energy Physics Experiments. LBL-5350, Doctoral dissertation at UC Berkeley, May 1977.
23. Green, M.A. (1977) Large superconducting detector magnets with ultra thin coils for use in high energy accelerators and storage rings. Proceedings of the 6th International Conference on Magnet Technology, Bratislava, Czechoslovakia, p. 429.
24. Eberhard, P.H. *et al.* (1979) A magnet system for the time projection chamber at PEP. *IEEE Trans. Magn.*, **MAG-15** (1), 128.
25. Benichou, J. *et al.* (1981) Long term experience on the superconducting magnet system for the CELLO detector. *IEEE Trans. Magn.*, **MAG-17** (5), 1567.
26. Royet, J.M., Scudiere, J.D., and Schwall, R.E. (1983) Aluminum stabilized multifilamentary Nb-Ti conductor. *IEEE Trans. Magn.*, **MAG-19** (3), 761.
27. Leung, E. *et al.* (1983) Design of an indirectly cooled 3-m diameter superconducting solenoid with external support cylinder for the fermilab collider detector facility. *IEEE Trans. Magn.*, **MAG-19** (3), 1368.
28. Wake, A. *et al.* (1985) A large superconducting thin solenoid magnet TRISTAN experiment (VENUS) at KEK. *IEEE Trans. Magn.*, **MAG-21** (2), 494.
29. Hirabayashi, H. (1988) Detector magnets in high energy physics. *IEEE Trans. Magn.*, **MAG-24** (2), 1256.
30. Tsuchiya, K. *et al.* (1987) *Advances in Cryogenic Engineering*, vol. **33**, Plenum Press, New York, p. 33.
31. Baze, J.M. *et al.* (1988) Design, construction and test of the large superconducting solenoid ALEPH. *IEEE Trans. Magn.*, **MAG-24** (2), 1260.
32. Apsey, R.Q. *et al.* (1985) Design of a 5.5 meter diameter superconducting solenoid for the DELPHI particle physics experiment at LEP. *IEEE Trans. Magn.*, **MAG-21** (2), 490.
33. Bonito Oliva, A., Dormicchi, O., Losasso, M., and Lin, Q. (1991) Zeus thin solenoid: test results analysis. *IEEE Trans. Magn.*, **MAG-27** (2), 1954.
34. Monroe, C.M. *et al.* (1988) *Proceeding of the 12th International Cryogenic Engineering Conference, Southampton*, Butterworth & Co., Guildford, p. 773.
35. Yamamoto, A. *et al.* (1995) Development of a prototype thin superconducting solenoid magnet for the SDC detector. *IEEE Trans. Appl. Supercond.*, **5** (2), 849.
36. Mito, T. *et al.* (1989) Prototype thin superconducting solenoid for particle astrophysics in space. *IEEE Trans. Magn.*, **MAG-25** (2), 1663.
37. Makida, Y. *et al.* (1995) Ballooning of a thin superconducting solenoid for particle astrophysics. *IEEE Trans. Appl. Supercond.*, **5** (2), 658.
38. Makida, Y. *et al.* (2007) Cryogenic performance of ultra-thin superconducting solenoid for cosmic-ray observation with ballooning. *IEEE Trans. Appl. Supercond.*, **17** (2), 1205.
39. Makida, Y. *et al.* (2009) The BESS-polar Ultra-thin superconducting solenoid magnet and its operational characteristics during long-duration scientific ballooning over antarctica. *IEEE Trans. Appl. Supercond.*, **19** (3), 1315.
40. Fabbricatore, P. *et al.* (1996) The superconducting magnet for the BaBar detector of the PEP-II B factory at SLAC. *IEEE Trans. Magn.*, **MAG-32** (4), 2210.

41. Wang, L. *et al.* (2008) Superconducting magnets and cooling system in BEPCII. *IEEE Trans. Appl. Supercond.*, **18** (2), 146.
42. Baze, J.M. *et al.* (1996) Progress in the design of a superconducting toroidal magnet for the ATLAS detector on LHC. *IEEE Trans. Magn.*, **MAG-32** (4), 2047.
43. Baynham, D.E. *et al.* (1996) Design of the superconducting end cap toroids for the ATLAS experiment at LHC. *IEEE Trans. Magn.*, **MAG-32** (4), 2055.
44. Kircher, F., Deportes, H., Gallet, B. *et al.* (1996) Conductor developments for the ATLAS and CMS magnets. *IEEE Trans. Magn.*, **MAG-32** (4), 2870.
45. Bunce, G. *et al.* (1995) The large superconducting solenoids for the g-2 muon storage ring. *IEEE Trans. Appl. Supercond.*, **5** (2), 853.
46. Bunce, G. *et al.* (1997) Test results of the g-2 superconducting solenoid magnet system. *IEEE Trans. Appl. Supercond.*, 7 (2), pp. 626–629.
47. Elwyn Baynham, D. (2006) Evolution of detector magnets from CELLO to ATLAS and CMS and toward future developments. *IEEE Trans. Appl. Supercond.*, **16** (2), 493.
48. Swoboda, D., Cacaut, D., and Evrard, S. (2004) Toward starting-up of the ALICE dipole magnet. *IEEE Trans. Appl. Supercond.*, **14** (2), 560.
49. Ruber, R. *et al.* (2007) Ultimate performance of the ATLAS superconducting solenoid. *IEEE Trans. Appl. Supercond.*, **17** (2), 1201.
50. Ruber, R.J.M.Y. *et al.* (2006) Quench characteristics of the ATLAS central solenoid. *IEEE Trans. Appl. Supercond.*, **16** (2), 533.
51. ten Kate, H.H.J. (2005) ATLAS superconducting toroids and solenoid. *IEEE Trans. Appl. Supercond.*, **15** (2), 1267.
52. Baynham, D.E. *et al.* (2006) ATLAS end cap toroid cold mass and cryostat integration. *IEEE Trans. Appl. Supercond.*, **16** (2), 537.
53. Baynham, D.E. *et al.* (2007) ATLAS end cap toroid integration and test. *IEEE Trans. Appl. Supercond.*, **17** (2), 1197.
54. Rabbers, J.J., Dudarev, A., Pengo, R., Berriaud, C., and ten Kate, H.H.J. (2006) Theoretical and experimental investigation of the ramp losses in conductor and coil casing of the ATLAS barrel toroid coils. *IEEE Trans. Appl. Supercond.*, **16** (2), 549.
55. Dudarev, A. *et al.* (2005) First full-size ATLAS barrel toroid coil successfully tested up to 22 kA at 4 T. *IEEE Trans. Appl. Supercond.*, **15** (2), 1271.
56. Mayri, C. *et al.* (2006) Suspension system of the barrel toroid cold mass. *IEEE Trans. Appl. Supercond.*, **16** (2), 525.
57. Sun, Z. *et al.* (2006) ATLAS barrel toroid warm structure design and manufacturing. *IEEE Trans. Appl. Supercond.*, **16** (2), 529.
58. Dudarev, K.V. *et al.* (2002) 20.5 kA current leads for the ATLAS barrel toroid superconducting magnets. *IEEE Trans. Appl. Supercond.*, **12** (1), 1289.
59. ten Kate, H.H.J. (2007) ATLAS superconducting magnet system status. *IEEE Trans. Appl. Supercond.*, **17** (2), 1191.
60. Cure, B. *et al.* (2004) Mechanical properties of the CMS conductor. *IEEE Trans. Appl. Supercond.*, **14** (2), 530.
61. Blua, D. *et al.* (2004) Superconducting strand properties at each production stage of the CMS solenoid conductor manufacturing. *IEEE Trans. Appl. Supercond.*, **14** (2), 548.
62. Fabbricatore, P., Greco, M., Musenich, R., Farinon, S., Kircher, F., and Cure, B. (2005) Electrical characterization of the S/C conductor for the CMS solenoid. *IEEE Trans. Appl. Supercond.*, **15** (2), 1275.
63. Fabbricatore, P. *et al.* (2004) The construction of the modules composing the CMS superconducting coil. *IEEE Trans. Appl. Supercond.*, **14** (2), 552.
64. Sgobba, S., D'Urzo, C., Fabbricatore, P., and Sequeira Tavares, S. (2004) Mechanical performance at cryogenic temperature of the modules of the external cylinder of CMS and quality controls applied during their fabrication. *IEEE Trans. Appl. Supercond.*, **14** (2), 556.
65. Levesy, B. *et al* (2006) CMS Solenoid Assembly. *IEEE Trans. Appl. Supercond.*, **16** (2), 517.

66. Campi, D. *et al.* (2007) Commissioning of the CMS magnet. *IEEE Trans. Appl. Supercond.*, **17** (2), 1185.
67. Fazilleau, F. *et al.* (2004) Design, construction and tests of 20 kA current leads for the CMS solenoid. *IEEE Trans. Appl. Supercond.*, **14** (2), 1766.
68. Bird, M. *et al.* (2004) Cryostat design and fabrication for the NHMFL/NCSL sweeper magnet. *IEEE Trans. Appl. Supercond.*, **14** (2), 564.
69. Quettier, L. *et al.* (2011) Hall B. Superconducting magnets for the CLAS12 detector at JLAB. *IEEE Trans. Appl. Supercond.*, **21** (3), 1872.
70. Shilon, I., Dudarev, A., Silva, H., and ten Kate, H.H. (2014) The superconducting toroid for the new axion observatory (IAXO). *IEEE Trans. Appl. Supercond.*, **24** (3), art. No 4500104.
71. Langeslag, S.A.E., Cure, B., Sgobba, S., Dudarev, A., ten Kate, H.H.J., Neuenschwander, J., and Jerjen, I. (2014) Effect of thermo-mechanical processing on the material properties at low temperature of a large size Al-Ni stabilized Nb-Ti/Cu superconducting cable. *AIP Conf. Proc.* **1574**, 211–218. http://dx.doi.org/10.1063/1.4860626
72. Green, M.A. (1989) in *Supercollider 1* (ed M. McAshan), Plenum Press, New York, p. 627.
73. Particle Data Group (1978) Review of particle properties. *Rev. Mod. Phys.*, **48** (2, Pt. II).
74. Tsai, Y.S. (1974) Pair Production and Bremsstrahlung of Charged Leptons, SLAC-PUB-1365, Table III.6.
75. Smythe, W.R. (1950) *Static and Dynamic Electricity*, 2nd edn, McGraw-Hill Book Company Inc., New York.
76. Los Alamos Accelerator Code Group (1987) *POISSON/SUPERFISH Reference Manual*, LANL Publication LA-UR-87-126, Los Alamos National Laboratory.
77. Vector Fields Ltd. (1991) A static electromagnetics simulation module and the OPERA 3-D system of Cobham plc, Vector Fields Ltd., Oxford.
78. Green, M.A. (1989) Calculating the J_c, B, T surface for niobium titanium using the reduced state model. *IEEE Trans. Magn.*, **MAG-25** (2), 2119.
79. Maddock, B.J. and James, G.B. (1968) Protection and stabilization of large superconducting coils. *Proc. IEE*, **115** (4), 543.
80. Eberhard, P.H. *et al.* (1977) Quenches in large superconducting magnets. Proceedings of the 6th International Conference on Magnet Technology, Bratislava Czechoslovakia, p. 654.
81. Green, M.A. (1984) Quench back in thin superconducting solenoid magnets. *Cryogenics*, **24**, 3.
82. Yamamoto, A. *et al.* (1983) A thin superconducting solenoid with the internal winding method for collider beam experiments. Proceedings of the 8th International Conference on Magnet Technology, September 1983.
83. Yamamoto, A. *et al* (1993) Design study of a thin superconducting solenoid for the SDC detector. *IEEE Trans. Appl. Supercond.*, **3** (1), 95.
84. Roark, R.J. and Young, W.C. (1975) *Formulas for Stress and Strain*, 5th edn, McGraw Hill Book Co, New York.
85. Timoshenko, S. (1940) *Theory of Plates and Shells*, McGraw Hill Book Co, New York.
86. Sauders, H.E. and Wittenberg, D.F. (1931) Strength of thin cylindrical shells under external pressure. *ASME Trans.*, **53** (15), 207.
87. Popov, E.P. (1959) *Mechanics of Materials*, Prentice Hall Inc., Englewood Cliffs, NJ.
88. ASME (2013) Standard that provides Rules for the Design, Fabrication, and Inspection of Boilers and Pressure Vessels. Section 8, Division 1, ANSI/ASME BPV-VIII-1.
89. Fast, R. *et al* (1993) *Advances in Cryogenic Engineering*, vol. **39**, Plenum Press, New York, p. 1991.
90. Yamaoka, H. *et al* (1993) *Advances in Cryogenic Engineering*, vol. **39**, Plenum Press, New York, p. 1983.
91. Wilson, M.N. (1983) *Superconducting Magnets*, Oxford Claredon Press, Oxford, p. 219.
92. Green, M.A. (1984) The role of quench back in the quench protection of a superconducting solenoid. *Cryogenics*, **24**, 659.

93. Green, M.A. (1983) PEP-4, TPC Superconducting Magnet, A Comparison of Measured Quench Back Time with Theoretical Calculations of Quench Back Time for Four Thin Superconducting Magnets. Lawrence Berkeley Laboratory Report LBID-771, Lawrence Berkeley Laboratory, August 1983 (unpublished).
94. Taylor, J.D. *et al* (1979) Quench protection for a 2 MJ magnet. *IEEE Trans. Magn.*, **MAG-15** (1), 855.
95. Green, M.A. (1982) PEP-4, Large Thin Superconducting Solenoid Magnet, Cryogenic Support System Revisited. Lawrence Berkeley Laboratory Engineering Note M5855, March 1982 (unpublished).
96. Green, M.A. (1994) Calculation of the pressure rise in the cooling tube of a two phase cooling system during a quench of an indirectly cooled superconducting magnet. *IEEE Trans. Magn.*, **MAG-30** (4), 2427.
97. Smits, R.G. *et al* (1981) *Advances in Cryogenic Engineering*, vol. **27**, Plenum Press, New York, p. 169.
98. Green, M.A. *et al* (1996) *Advances in Cryogenic Engineering*, vol. **41**, Plenum Press, New York, p. 573.
99. Green, M.A. *et al.* (1980) The TPC Magnet Cryogenic System. Lawrence Berkeley Laboratory Report LBL-10552, Lawrence Berkeley Laboratory, May 1980. *https://publications.lbl.gov* (accessed 28 October 2014.)
100. Taylor, J.D. and Green, M.A. (1978) Garden Hose Test. Lawrence Berkeley Laboratory, Group A Physics Note 857, November 1978 (unpublished).
101. Green, M.A., Burns, W.A., and Taylor, J.D. (1979) *Advances in Cryogenic Engineering*, vol. **25**, Plenum Press, New York, p. 420.
102. Burns, W.A. *et al.* (1980) *Proceedings of the 8th International Cryogenics Engineering Conference, Genoa, Italy*, IPC Science and Technology Press, Guildford, p. 383.
103. Wilson, M.N. (1983) *Superconducting Magnets*, Chapter 11, Oxford Claredon Press, Oxford.
104. Green, M.A. (1990) The role of superconductor in reducing the refrigeration needed to cool the leads of a superconducting magnet. *Cryogenics*, **30** (9 Suppl.), 679.
105. Ballarino, A. (2002) Current leads for the LHC magnet system. *IEEE Trans. Appl. Supercond.*, **12** (1), 1275.
106. Heller, R. *et al.* (2004) Design and fabrication of a 70 kA current lead using Ag/Au stabilized Bi-2223 tapes as a demonstrator for the ITER TF-coil system. *IEEE Trans. Appl. Supercond.*, **14** (2), 1774–1777.
107. Li, W. *et al.* (2008) Design of current leads for the MICE coupling magnet. Proceedings International Conference of Cryogenics and Refrigeration, Shanghai, China, Vol. 22, p. 347.
108. Green, M.A. and Chouhan, S.S. (2013) Cyclotron Gas-Stopper Magnet Cooling Circuit Pressure Drops and the Cool-Down of the Magnet Using a Large Refrigerator or Six Two-Stage Pulse Tube Coolers. Michigan State FRIB Note FRIB-M40201-CA-000107-R001, 10 May 2013.
109. Green, M.A., Chouhan, S.S., and Zeller, A.F. (2014) Design limitations on a thermal siphon 4 K helium loop for cooling-down the cyclotron gas stopper magnet coils. *AIP Conf. Proc.* **1573**, 1543–1550. doi: 10.1063/1.4860890
110. Green, M.A. and Witte, H. (2008) *Advances in Cryogenic Engineering*, vol. **53**, AIP Press, Melville, NY, p. 1299.
111. Green, M.A. and Witte, H. (2008) *Advances in Cryogenic Engineering*, vol. **53**, AIP Press, Melville, NY, p. 1251.
112. Green, M.A. (2014) Cooling and cooling-down a MgB_2 and HTS magnets using a hydrogen thermal siphon loop and coolers running from 15 K to 28 K. *IEEE Trans. Appl. Supercond.*, **24** (3), art. No 501304.

4.5
Magnets for NMR and MRI

Yukikazu Iwasa and Seungyong Hahn

4.5.1
Introduction to NMR and MRI Magnets

To date, nuclear magnetic resonance (NMR) and magnetic resonance imaging (MRI) magnets have been the most successful commercial applications of superconductivity. Proliferation of superconducting NMR and MRI magnets began in earnest starting in the 1980s. By this time, solution to each of the design and operation issues specific and common to superconducting NMR and MRI magnets has emerged to enable them to be profit-making commercial products. These issues include: (i) superconductor itself, particularly technologies of NbTi and Nb_3Sn; (ii) superconducting joint; (iii) adiabatic stability; and (iv) self-protecting techniques.

In this section on NMR and medical diagnostic MRI magnets, after introduction to basics of these magnets (Section 4.5.1), we cover three major topics: Design Issues (Section 4.5.2), Status as of 2013 (Section 4.5.3), and High-Temperature Superconductor (HTS) applications (Section 4.5.4).

4.5.1.1 NMR and MRI

NMR, first measured by Isidor Rabi in 1938 [1], is a quantum effect characterized by the absorption and reradiation of radio waves by nuclei having nonzero spins in a *very uniform* magnetic field [2]. Generally, NMR works better at higher magnetic fields. For most NMR applications, spectral resolution is based on spreading or dispersion by chemical shift differences, which (in frequency units) are proportional to the magnetic field strength.

NMR spectroscopy is used to study, for example, the molecular structures of organic compounds, especially proteins. The gyromagnetic ratio of proton (^1H), 42.576 MHz T^{-1}, is widely used to describe the field strength of an NMR magnet. Thus, the resonant (Lamor) frequencies, proportional to field, at 1 T, for example, are: 42.58 MHz (and 1 GHz at 23.49 T) for protons; 17.25 MHz for ^{13}P; 10.71 MHz for ^{13}C; 3.08 MHz for ^{14}N. The two most compelling arguments for extending NMR to higher field are spectral resolution and sensitivity. And, for generating fields above 2 T, superconducting magnets are the only practical option.

For medical diagnostic MRI, magnetic fields are used to create, through NMR, visual images of a brain and other body parts. To create geometric source of NMR, a spatial distribution in magnetic field, specifically, a "linear" field gradient typically in the range 10 – 100 mT m^{-1} in each of the three orthogonal directions (e.g., x, y, and z) for whole-body MRI, is purposely introduced to restrict resonance to selected subvolume, enabling MRI to map (image) the concentration of nuclei of a selected species [3].

4.5.1.2 **Spatial Field Homogeneity**

Because the absorption and reradiation frequency of radio waves of specific nuclei is quantized to a specific magnetic field, NMR requires, in theory, a *perfectly uniform* spatial field. In reality, however, the absolute spatial uniformity is achievable only over a limited space in which an NMR specimen is placed. Typically, a spatial homogeneity of <0.1 ppm (parts per million) within a 3 cm diameter spherical volume (DSV) is required for high-resolution NMR magnets, while that of <10 ppm within a 45 cm DSV is required for diagnostic whole-body MRI magnets. For a 1 GHz (23.49 T) NMR magnet, 0.1 ppm translates at the extremity of the sphere to a field of 0.024 G, which is ~1/30th the earth magnetic field of 0.7 G. For a 3 T whole-body MRI magnet, 10 ppm translates to a field of 0.3 G. In Section 4.5.2, analytical and manufacturing techniques to satisfy spatial field homogeneity requirements are discussed.

4.5.1.3 **Temporal Stability**

Temporal variation of a magnetic field perturbs NMR signals in a way similar to spatial field inhomogeneity. A typical temporal stability requirement is $<0.01\ \mathrm{ppm\,h^{-1}}$ for solution NMR and $<0.1\ \mathrm{ppm\,h^{-1}}$ for solid-state NMR or MRI. Besides the ability to generate a magnetic field with the electric power necessary only to keep it at cryogenic temperature, the most prominent enabling feature of the superconducting magnet, and perfect for NMR and MRI, is the ease with which the magnet, when operated in *persistent mode* [4], can maintain a temporally constant magnetic field. In a few instances, some are operated in driven mode. Here, we describe the two modes of operation.

4.5.1.3.1 **Persistent Mode**

The absolute temporal stability is theoretically achievable with a *persistent-mode superconducting* magnet. Figure 4.5.1 shows an electrical circuit of a superconducting magnet, in which the magnet is an inductor of inductance L. To energize the persistent-mode superconducting magnet, a persistent-current-switch (PCS), the shaded circle in Figure 4.5.1a, is placed in the magnet circuit, where a heater

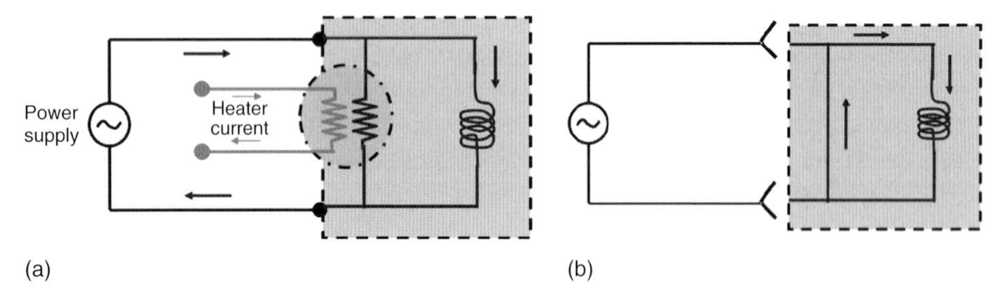

Figure 4.5.1 Circuits for energizing a persistent-mode superconducting magnet [4]: (a) persistent-current-switch (PCS), inside the dashed circle, is "open" (resistive) with its built-in heater (gray resistor) carrying current, to energize the magnet (represented by inductor); (b) magnet in persistent mode, with the current leads decoupled from the magnet terminals. The cold environment is shaded gray.

(gray resistor) maintains the PCS in the resistive state (dark resistor), enabling the power supply to energize the magnet. After the magnet is energized, the PCS heater is turned off, making the PCS superconducting, and the magnet operates in persistent mode as illustrated in Figure 4.5.1b.

4.5.1.3.2 Driven Mode

A driven-mode magnet is operated with its power supply connected across the magnet terminals. Thus, the temporal field instability of a driven-mode magnet depends on the drift rate of a current supply. Currently (2014) all HTS magnets are operated in driven mode. This is because, though a technique to make superconducting splice for REBCO tape has recently been achieved Lee, HTS splices in these magnets are resistive, and likely to remain so for the next several years. To ameliorate the temporal instability of a driven-mode magnet, the so-called external field-frequency lock technique [5–8] is commonly combined with a *stable* power supply having a typical drift rate of $< 0.1\ \text{ppm}\ \text{h}^{-1}$. Active feedback control is one of viable techniques to improve temporal stability. Typically, an NMR probe monitors field fluctuations impinging on the specimen. The field signals are in turn fed to a copper compensation coil, placed in the magnet bore, which generates a required stabilizing field.

4.5.1.4 General Coil Configurations of NMR and MRI Magnets

Figure 4.5.2 shows typical cross-sectional profiles of magnets: (a) NMR and (b) MRI. As illustrated, superconducting coils in NMR and MRI magnets may be categorized: (i) main, (ii) correction, (iii) active cryoshim (persistent), and

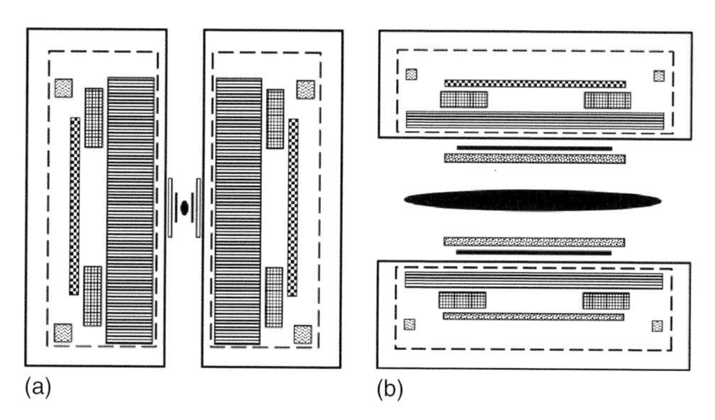

Figure 4.5.2 General configuration and coil components of (a) NMR and (b) MRI magnets: room-temperature (RT) cryostat (outer rectangles); cryogenic chamber (inner dashed rectangles); main (innermost stripe rectangles, represented here by one coil but generally several nested coils); correction (grid rectangle pairs, here one pair, but generally several pairs); active shim coils (checker rectangles); active shielding (chevron pair). Within RT bore, for NMR: a set of RT shim coils, typically copper (white rectangles), and a set of ferromagnetic shims (solid rectangles); for MRI magnet: a set of ferromagnetic shims (solid rectangles) and 3D actively shielded gradient coils (dotted rectangles). Specimen (oval at the magnet center) is small for NMR and can be as large as a person for diagnostic MRI.

(iv) active shielding (persistent). Because of its large size, a medical diagnostic MRI magnet often relies only on ferromagnetic shims to satisfy the field homogeneity requirement. Also, passive shielding, placed at the external of a magnet cryostat, is also common in MRI and low-field NMR magnets. As indicated in Figure 4.5.2, in general the principal field direction of an NMR magnet is vertical, while that of an MRI magnet is horizontal. In an NMR magnet, the specimen and measurement rigs are often introduced into the room-temperature (RT) bore from the bottom.

4.5.2 Specific Design Issues for NMR and MRI Magnets

Five key design and operation issues for a superconducting magnet [4] are: (i) superconductor; (ii) stability; (iii) protection; (iv) mechanical integrity; and (v) cryogenics. Additional important issues, more specific to NMR and MRI magnets, include: field harmonics, field shimming, field shielding, and safety.

4.5.2.1 Superconductor

Figure 4.5.3 shows *approximate* $\mu_0 H_{c2}$ versus T_c plots of 4 HTS (YBCO, Bi2212, Bi2223, MgB_2) and two low-temperature superconductors (LTS) (NbTi and Nb_3Sn). Currently (2013), except Bi2212, the rest is commercially available. Although approximate, these plots indicate which superconductor is most suitable for a magnet to be operated in a given field-temperature regime. Note that, for magnet applications, it is generally agreed that a superconductor must have an engineering critical current density, J_e, of >100 A mm^{-2}. Here, J_c is given by critical current divided by the entire superconductor cross-section, which, except for NbTi, includes those of inherent non-superconducting elements.

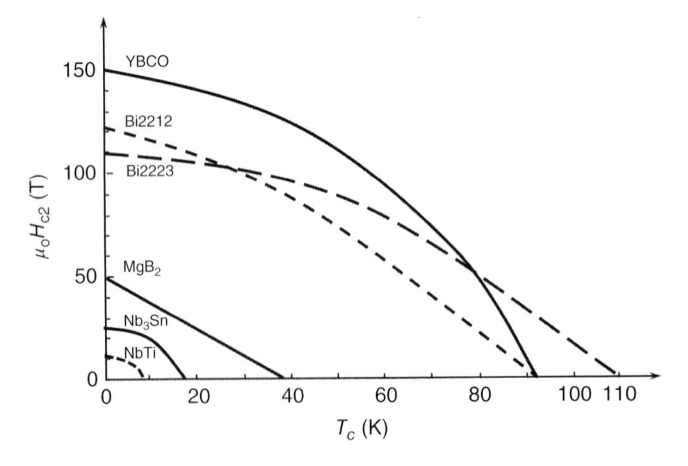

Figure 4.5.3 *Approximate* $\mu_0 H_{c2}$ versus T_c plots of YBCO, Bi2212, Bi2223, MgB_2, Nb_3Sn, and NbTi.

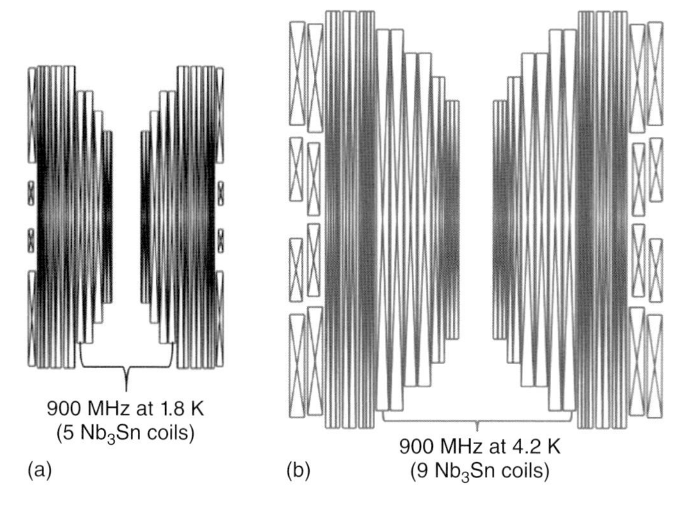

Figure 4.5.4 900 MHz/54 mm bore NMR magnets operating at: (a) 1.8 K, requiring five Nb_3Sn coils; (b) 4.2 K, requiring nine Nb_3Sn coils. (Courtesy: Masatoshi Yoshikawa, JASTEC, 2005.)

Figure 4.5.4 presents to-scale coil cross-sections of two 900 MHz/54 mm bore NMR magnets with operating temperatures at: (a) 1.8 K and (b) 4.2 K. In each magnet, Nb_3Sn coils occupy the inner high-field region, while NbTi coils the outer low-field region. At 21.1 T (900 MHz of ^{1}H NMR frequency), current-carrying capacities of Nb_3Sn wires are below 100 A mm^{-2} at 4.2 K but above at 1.8 K. Therefore, the 900 MHz magnet operated at 1.8 K (a) requires five Nb_3Sn coils, while at 4.2 K, (b) nine coils, as indicated in Figure 4.5.4. Because of its greater size, the magnetic energy of the 4.2 K 900 MHz magnet is more than double that of its 1.8 K counterpart; the superconducting magnet cost is virtually proportional to its stored energy. To reduce the magnet capital cost, superconductors must be chosen to be consistent with their field-temperature-dependent current-carrying capacities. As we shall discuss further in Section 4.5.3.1, currently (2014) the field limit for all-LTS NMR magnet is 1000 MHz (1 GHz).

4.5.2.2 **Stability of Adiabatic Magnets**

Equation (4.5.1), an energy balance equation of a unit superconductor volume, has been widely used in stability analysis of a superconducting magnet [4]. The left-hand side represents the time rate of thermal energy density change of the conductor, where $C_{cd}(T)$ is the heat capacity per unit volume of the "composite" superconductors, comprising superconductor and normal-metal matrix (often copper). In the right-hand side, the first term describes thermal conduction, where $k_{cd}(T)$ is the composite's thermal conductivity. The second term is joule heating, where $\rho_{cd}(T)$ is the composite's electrical resistivity (zero in the superconducting state), and $J_{cd_0}(t)$ is the current density at an operating current $I_{op}(t)$. The $g_d(t)$ describes non-joule-heat disturbance, primarily magnetic and mechanical. The last term represents cooling, where f_p, A_{cd}, and $g_q(T)$ are,

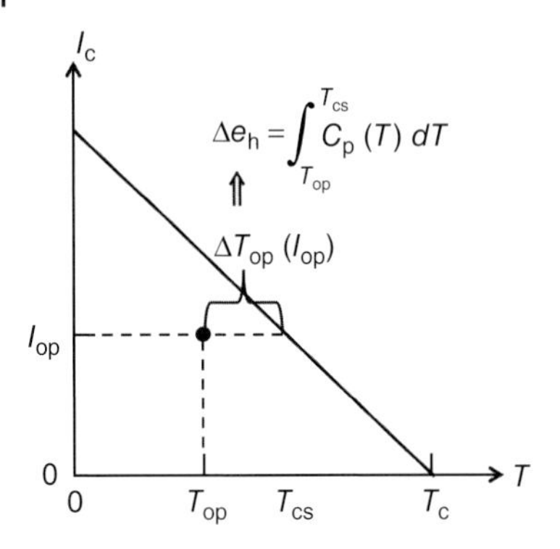

Figure 4.5.5 I versus T plot at constant magnetic field.

respectively, the fraction of the composite perimeter, $\mathscr{P}_D$, exposed to cryogen, the composite cross-sectional area, and the convective heat transfer flux for the cryogen.

$$C_{cd}(T)\frac{\partial T}{\partial t} = \nabla \cdot [k_{cd}(T)\nabla T] + \rho_{cd}(T)J^2_{cd_0}(t) + g_d(t) - \left(\frac{f_p \mathscr{P}_D}{A_{cd}}\right) g_p(T) \quad (4.5.1)$$

Due to the requirements of high-field performance and compact size, most NMR and MRI magnets are designed to be "adiabatic" in terms of stability. The adiabatic magnet, by definition, has no "local" cooling within its winding, that is, $g_q(T) = 0$, although it keeps operated at a target operating temperature with a cooling source. In the absence of this local cooling, the left side of Eq. (4.5.1) in the adiabatic magnet must remain *zero at all times:* in practice, a modest temperature excursion, ΔT_{op}, from the operating temperature, T_{op}, to the current-sharing temperature, T_{cs}, is permitted. In Figure 4.5.5, T_{op}, T_{cs}, and ΔT_{op} are defined for a superconductor's I_c versus T plot under a constant magnetic field, most likely at the highest field point within the winding. Note that, above T_{cs}, a portion of I_{op} begins flowing in the matrix, which generates joule heating; the second term on the left-hand side of Eq. (4.5.1) becomes >0. In the absence of local cooling, the stability condition, $\mathrm{d}T/\mathrm{d}t = 0$, cannot be sustained. Thus, for an adiabatic magnet with its operating point at a dot in Figure 4.5.5, given by the intersection of I_{op} and T_{op}, a disturbance energy density of only up to $\Delta e_h = \int_{C_{cd}}^{T_{cs}} C_{cd}(t)\mathrm{d}T$ is permitted for stability. Often, Δe_h is referred as the *stability margin* of an adiabatic magnet. Substantial disturbances – mechanical (conductor motion, epoxy cracking), flux jumping, or AC losses – that lead to "large" Δe_h are not permitted, particularly in LTS magnets. However, for the HTS magnet that has a much greater ΔT_{op} than that of the LTS magnet, stability is no longer a pressing design/operation issue [4]. Table 4.5.1 lists T_{op}, ΔT_{op}, and corresponding Δe_h for selected LTS (NbTi) and HTS (REBCO, MgB_2) magnets.

Since the late 1980s, mechanical disturbances have been minimized in adiabatic LTS magnets [4]. Still, they afflict *most* of these magnets, in the form of premature quenches, *most* of the time. Although AC losses disrupt most adiabatic magnets,

Table 4.5.1 Selected values of T_{op}, ΔT_{op}, and Δe_h for LTS (NbTi) and HTS (REBCO, MgB_2) magnets.

T_{op} (K)	ΔT_{op} (K)	Δe_h (kJ m^{-3})	Remarks
2.3	0.25	0.1	NbTi 12-T
4.2	0.5	0.5	NbTi 8-T
4.2	1	1	NbTi 5-T
4.2	50	20	REBCO
10	5	12	MgB_2
30	10	4	REBCO
65	10	15	REBCO

they can be managed in NMR and MRI magnets because a "low" charging rate of such a magnet is permissible to thermally conduct away the AC losses to the cooling source outside the winding, while the magnet temperature is kept below T_{cs}.

4.5.2.3 Stress Analysis – Electromagnetic, Thermal, Winding

Governing equation for calculation of stress within a solenoid coil under a magnetic field, self and external, may be given by a force equilibrium equation in Eq. (4.5.2) [4], where $\sigma_r(r, z)$ and $\sigma_\theta(r, z)$ are, respectively, radial and hoop stresses, while λJ and B_z (r, z) are, respectively, overall current density and axial field. The force balance equation requires that, in an arbitrary infinitesimal volume within the winding, the "net force" must be "balanced", that is, zero. In Eq. (4.5.2), the winding shear stress is assumed to be negligible, a valid assumption at the magnet midplane, where in general the stress is at its peak [4, 9]. The inaccuracy, due to the ignorance of shear stress components, appears mainly in the end sections of a solenoid; Bobrov and Williams [9] reported that, for a coil with a length-to-inner-diameter ratio ($\beta = b/a_1$) of 0.53, Eq. (4.5.2) introduces an inaccuracy in stress of <6% in the entire solenoidal winding region. Note that, for a "thin" (theoretically zero radial build) solenoid Eq. (4.5.2) can be simplified to $\sigma_\theta = \lambda B_z r$, so-called BJR.

$$\frac{\partial \sigma_r(r, z)}{\partial r} + \frac{\sigma_r(r, z) - \sigma_\theta(r, z)}{r} + \lambda J B_z(r, z) = 0 \tag{4.5.2}$$

Provided that a winding is isotropic with a Young's modulus of E and a Poisson ratio of ν, Hook's law is applicable as Eq. (4.5.3), where ε_{Tr}, $\varepsilon_{T\theta}$, and ε_{Tz} are total thermal expansions parallel to the respective r, θ, and z directions from a target operating temperature (T_{op}) to a base temperature, typically 300 K [4].

$$\begin{pmatrix} \varepsilon_r \\ \varepsilon_\theta \\ \varepsilon_z \end{pmatrix} = \frac{1}{E} \begin{pmatrix} 1 & -\nu & -\nu \\ -\nu & 1 & -\nu \\ -\nu & -\nu & 1 \end{pmatrix} \begin{pmatrix} \sigma_r \\ \sigma_\theta \\ \sigma_z \end{pmatrix} + \begin{pmatrix} \varepsilon_{Tr} \\ \varepsilon_{T\theta} \\ \varepsilon_{Tz} \end{pmatrix} \tag{4.5.3}$$

The key focuses in stress analysis of high-field NMR and MRI superconducting magnets include: (i) maximum tensile strain on the superconductor, above which the superconductor is degraded [10–13]; (ii) peak stress in epoxy, if applicable, which may cause cracking that may lead to a premature quench or a substantial noise in NMR signal; (iii) potential failure in the winding structure, reinforcement, and/or support structure; and (iv) impact of winding deformation on field homogeneity.

4.5.2.3.1 **Electromagnetic**

Here, we assume an isotropic solenoid, having a winding i.d. of a_1 and o.d. of a_2, with a winding current density of λJ. The axial fields at a_1 and a_2 are defined, respectively, as B_1 and B_2. With $\rho \equiv r/a_1$ and $\kappa \equiv B_2/B_1$, σ_r and σ_θ can be calculated from Eq. (4.5.2) with boundary conditions: $\sigma_r(r = a_1, z) = 0$ and $\sigma_r(r = a_2, z) = 0$. The results are given in Eq. (4.5.4a,b) [4]:

$$\sigma_r = \frac{\lambda J B_1 a_1}{\alpha - 1}\left[\frac{2+\nu}{3}(\alpha - k)\left(\frac{\alpha^2 + \alpha + 1 - \alpha^2/\rho^2}{\alpha + 1} - \rho\right) - \frac{3+\nu}{8}(1-k)\left(\alpha^2 + 1 - \frac{\alpha}{\rho^2} - \rho^2\right)\right] \tag{4.5.4a}$$

$$\sigma_\theta = \frac{\lambda J B_1 \alpha_1}{\alpha - 1}\left\{(\alpha - k)\left[\frac{2+\nu}{3}\left(\frac{\alpha^2 + \alpha + 1 + a^2/\rho^2}{\alpha + 1}\right) - \frac{1+2\nu}{3}\rho\right] - (1-k)\left[\frac{3+\nu}{8}\left(\alpha^2 + 1 + \frac{\alpha^2}{\rho^2}\right) - \frac{1+3\nu}{8}\rho^2\right]\right\} \tag{4.5.4b}$$

In case that σ_r is too large, that is, $\sigma_r > \nu\sigma_\theta + \nu\sigma_z$, ε_r becomes tensile (positive) in Eq. (4.5.3), which implies radial separation of the winding within the coil, the condition not tolerated in the coil. Often used remedy to limit the radial stress is a nested-coil formation that transforms a radially "thick" winding into "thin" coils. Overbanding is another remedy, often effectively deployed to limit the stress [4].

4.5.2.3.2 **Thermal**

Because each constituent of the winding has a different thermal contraction from 300 K to the magnet operating temperature, there will be stresses created by differential thermal contractions. Also, the thermal strain on the superconductor, given by Eq. (4.5.3), must be included in computing the total strain on the superconductor.

4.5.2.3.3 **Winding**

Winding a conductor into a circular turn strains the superconductor by *bending* and *winding tension.* For most superconductors, 95% I_c retention strains are below 1%, for example, Nb_3Sn (0.5%), Bi2223 (0.5%), and ReBCO (0.6%); NbTi may be strained to ~1% though its I_c decreases with strain.

4.5.2.4 **Solenoidal Field**

The center field of a solenoid, having an i.d., o.d., length, and current density of $2a_1$, $2a_2$, $2b$, and λJ, respectively, can be calculated by:

$$B_0 = \lambda T \alpha_1 \beta \ln\left[\frac{\alpha + \sqrt{\alpha^2 + \beta^2}}{1 + \sqrt{1 + \beta^2}}\right] = \lambda J \alpha_1 F(\alpha, \beta) \tag{4.5.5}$$

where, α and β are defined as a_2/a_1 and b/a_1, respectively. $F(\alpha, \beta)$ is known as the *field factor* for the uniform-current-density winding [4].

4.5.2.4.1 **Harmonic Analysis**

Harmonic analysis is a requisite in design and analysis of NMR and MRI magnets that demand a highly homogeneous spatial field [14]. It enables decomposition of an NMR or an MRI field over a target volume, of which the origin is often different from the magnet's physical center, into many gradients. In harmonic analysis, it is common to express the spatial magnetic field, typically in the axial direction (B_z), in spherical coordinates (r,θ,φ):

$$B_z(r, \theta, \varphi) = \sum_{n=0}^{\infty} \sum_{m=0}^{n} r^n (n + m + 1) P_{n|}^m(u)(A_n^m \cos m\varphi + B_n^m \sin m\varphi) \qquad (4.5.6)$$

where $P_n^m(u)$ is the set of Legendre polynomials (for $m = 0$) and associated Legendre functions (for $m > 0$) with $u = \cos\theta$, while A_n^m and B_n^m are constants that are determined by the source, that is, the spatial current distribution. Therefore, the design target of an NMR or MRI magnet, in terms of field homogeneity, is to minimize A_n^m and B_n^m except A_0^0, a target field at the magnetic center.

4.5.2.5 **Field Mapping and Shimming**

Despite the best efforts of designing and constructing a highly homogeneous field magnet, all gradient terms in an "as-wound" NMR or MRI magnet are nonzero in practice. *Field shimming,* therefore, is an essential procedure to minimize these gradients to enable an NMR or MRI magnet to meet its homogeneity requirements [15]. As a first step of field shimming, a technique known as *field mapping* measures magnetic fields at a number of predefined locations; the raw field data are used to derive the target field gradients. For fast and accurate field mapping, which is crucial for effective field shimming, a proper choice of mapping points is important in a way that each mapping point is "orthogonal" to each other in the target spherical coordinates (r,θ,φ); often used is a "helical mapping," where fields are measured along a helical path on the surface of a target cylindrical volume [16, 17]. Table 4.5.2 lists, up to the third harmonic order, the field gradients and the corresponding "symbols" that have been commonly used by NMR/MRI magnet designers.

4.5.2.5.1 **Active Shimming**

In active shimming, a gradient field is generated by current passing through a shim coil, wound with either superconductor and thus operated at a cryogenic temperature or with copper wire and operated at RT. Because a superconducting shim is currently wound with NbTi, which has to be operated in a field below ~10 T, it is placed radially outside of the main–correction coil assembly (see Figure 4.5.2). As discussed briefly in Section 4.5.4.4, HTS shims are now under development. Copper shims, chiefly because of a need to minimize their dissipation, are "small" and placed within the RT bore, as indicated in Figure 4.5.2.

Table 4.5.2 Harmonic orders, field gradients, and shim coils.

Order	Gradient	Shim
First	z	$Z1$
	x	X
	y	Y
Second	$z^2 - \frac{1}{2}(x^2 + y^2)$	$Z2$
	zx	ZX
	xy	XY
	$x^2 - y^2$	$C2$
	$2xy$	$S2$
Third	$z^3 - \frac{3}{2}(x^2 + y^2)$	$Z3$
	$z^2x - \frac{1}{4}x(x^2 + y^2)$	Z2X
	$z^2y - \frac{1}{4}x(x^2 + y^2)$	Z2Y
	$z(x^2 - y^2)$	$ZC2$
	$z(2xy)$	$ZS2$
	$x^3 - 3xy^2$	$C3$
	$3x^2 y - y^3$	$S3$

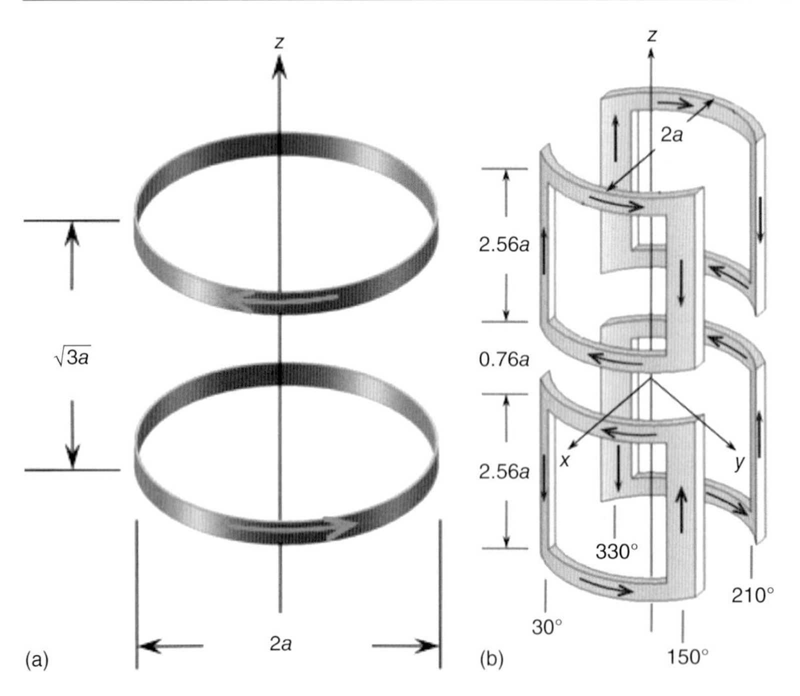

Figure 4.5.6 (a, b) Typical configuration of active shim coils, axial ($Z1$), and radial ($X1$).

Figure 4.5.6a shows a winding configuration of the simplest (and "ideal") $Z1$ shim, comprising two opposing current rings, each of diameter $2a$, separated by a distance of $\sqrt{3}a$; with this separation distance, its third-order harmonics is zero

($d^3Hz/dz^3 = 0$), that is, so-called Anti-Helmholtz pair [4]. Figure 4.5.6b shows the simplest (and "ideal") X shim, which generates a linear B_z gradient field along the x-axis. With a given winding radius of a, the design of these "ideal" shim coils is straightforward and its principle is to minimize the next low and high order terms in the same kind, even or odd [18]. However, in real shim coils, each winding section has both axial and radial builds that introduce unavoidable minor errors. Note that the Y family shim is essentially identical to that of X, except the gradient field is pointed in the y-direction.

4.5.2.5.2 Passive Shimming

Ferromagnetic shimming, an assembly of ferromagnetic tiles, is one form of passive shimming [19]. Each ferromagnetic (usually steel) tile of thickness <1 mm, ranging in size typically from 2 mm × 2 mm to 20 mm × 50 mm, are configured to generate field gradients at the magnet center over a specified volume – it is shown by solid rectangles in Figure 4.5.2. Once assembled and placed in the magnet bore, each tile may be considered as a magnetic dipole and its vector field over the specified volume at the magnet center can be computed. In general, an optimization technique such as linear programming for a given group of ferromagnetic tiles is used to determine their location in order to minimize the target field errors.

4.5.2.6 Field Shielding

Fringing fields – unwanted fields outside a magnet – are dangerous because they can be a safety hazard to those near the system (see Section 4.5.2.7); it may also disrupt or distort field-sensitive equipment. For computing the fringing field $\boldsymbol{H}_\mathrm{f}$ at locations *far* from the magnet, the magnet can be modeled as a dipole with an effective radius R_e [4]:

$$\boldsymbol{B}_\mathrm{f} = B_0 \left(\frac{R_\mathrm{e}}{r}\right)^3 \left(\cos\theta \boldsymbol{i}_r + \frac{1}{2}\sin\theta \boldsymbol{i}_\theta\right) \tag{4.5.7}$$

Equation (4.5.7) indicates that a fringing field is dominated by R_e, implying that fringing field problem is generally more sever in high-field (>10 T) NMR magnets and whole-body MRI magnets, both are now shielded, actively, passively, or by a combination of both. Active shielding uses an electromagnet (shielding coil); passive shielding, a ferromagnetic structure (steel shell).

4.5.2.6.1 Active Shielding

The basic coil configuration of an active shielding magnet is two "ring" coils separated axially, as in a Helmholtz magnet, but in the opposite direction from that of the main/correction coils. However, unlike a Helmholtz magnet in which two "rings" are separated rather close (ideally by a distance of the ring radius), in an active shielding pair, they are separated far apart for two reasons: (i) to maximize the field at each end and thereby cancel out the fringing field of the main/correction coils; and (ii) to minimize its center field because the shielding field subtracts from the main axial field at the center. The active shielding coil pair is schematically shown in Figure 4.5.2 as a chevron rectangle pair.

4.5.2.6.2 Passive Shielding

The advantage of passive shielding over active shielding is generally the cost, but its cost advantage can diminish for "large" and "high-field" magnets. Another advantage, often overlooked, is its "reverse-shielding" capability, that is, shielding the main magnet against a fringe field of a source outside the magnet. Often a thick steel shell is used. Its wall thickness must be sufficient to keep the steel "unsaturated," enabling it to "soak" up a fringing field to prevent it from reaching beyond the shell. If the shell is placed close to the magnet, its wall has to be thick; if placed far from the magnet, the wall can be thin, but generally, the total weight of the shell is nearly independent. The biggest disadvantage of passive shielding is its massiveness.

4.5.2.7 Safety

The biggest safety issue with an NMR or MRI magnet is forces on ferromagnetic objects induced by the magnet's fringing field. Although a shielded magnet has a fringing field small enough in the space where people may move round and field-sensitive equipment may be placed, there have been occasional accidents of large steel objects, for example, tools, chairs, wheelchairs, and gas cylinders, being pulled to a magnet, often one axial end of the magnet cryostat. Other safety issues originate from the time-varying magnetic fields of the gradient coils needed for imaging (nerve, muscle simulation, repeated noise) and pulsed RF field (body heating), but their consequences are not as violent as those of ferromagnetic objects.

4.5.3 Status (2013) of NMR and MRI Magnets

This section describes the status as of 2013 of NMR and MRI magnets, emphasizing those magnets either available from commercial vendors or under development.

4.5.3.1 Solid-State and Solution NMR

A chief difference between solid-state and solution NMR is anisotropic effects, present and absent, respectively, in solid-state and solution NMR. In solid-state NMR, signals are broader, while in solution NMR, they consist of sharp lines. To minimize the detrimental anisotropic effects, the specimen in solid-state NMR, oriented at a "magic-angle" of 54.73° from the main axial field, is spun at a "high frequency" (up to ~100 kHz).

Currently, both solid-state and solution NMR chiefly target protein molecular structures relevant for development of drugs against cancer, viral infection, bacterial diseases, and immune disorders. These include membrane proteins and large protein–protein and protein nucleic acid complexes which, due to their size, demand high NMR resolution and sensitivity. Even at 700–900 MHz frequencies, resolution and sensitivity fall short of that necessary to determine the complete protein structure. It is expected that the development of >1 GHz NMR magnets will enable substantially more complete structural studies of proteins.

Figure 4.5.7 930 MHz/54 mm bore magnet. Cryostat o.d.: 1.72 m. (Courtesy: Tadashi Shimizu, National Institute of Materials Science, 2013.)

4.5.3.1.1 LTS Magnets (400–1000 MHz)

Figure 4.5.7 shows a photo of a 930 MHz/54 mm bore NMR magnet [20] installed in the Tsukuba Magnet Laboratory at the National Institute of Materials Science (NIMS). It is one of the very first >900 MHz NMR magnets developed in the early 2000s by Kobelco. Figure 4.5.8 shows photos of Bruker BioSpin high-field/54 mm bore NMR magnets: (i) an actively shielded 950 MHz and (ii) a 1000 MHz. As indicated in the caption, because of the presence of active shield coils, the 950 MHz magnet's overall outer diameter is as large as that of the 1000 MHz magnet. This all-LTS (NbTi/Nb_3Sn) 1000 MHz magnet has demonstrated one important point about NbTi and Nb_3Sn: they are still the most viable superconductors for NMR magnets up 1000 MHz.

4.5.3.1.2 LTS/HTS Magnets (>1 GHz)

For a >1 GHz NMR magnet, HTS is mandatory. The standard approach is to place an HTS insert in the bore of an LTS NMR magnet. The first systematic approach toward a >1 GHz LTS/HTS NMR magnet was proposed in 1998 by the MIT Francis Bitter Magnet Laboratory (FBML). As of 2014, there are four groups developing >1 GHz NMR magnets: (i) NIMS, with a goal to achieve a 1.05 GHz magnet [21]; (ii) Bruker BioSpin, a 1.2 GHz magnet for completion in 2016; (iii) the National High Magnetic Field Laboratory, an all-HTS 32 T magnet, a precursor to a >1 GHz magnet [22]; and (iv) MIT/FBML. Figure 4.5.9 shows a schematic drawing of a 1.3

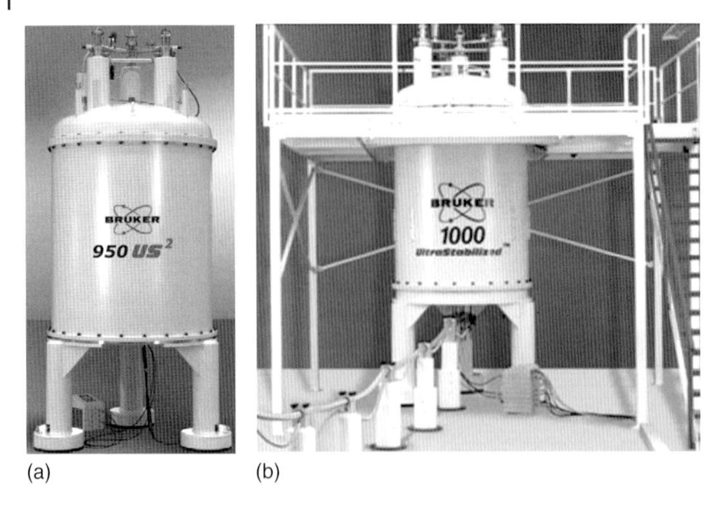

Figure 4.5.8 (a) *Actively shielded* 950 MHz/54 mm bore magnet. Cryostat o.d.: 1.60 m. (b) 1000 MHz/54 mm bore magnet. Cryostat o.d.: 1.60 m. (Courtesy: Gerhard Roth, Bruker BioSpin, 2013.)

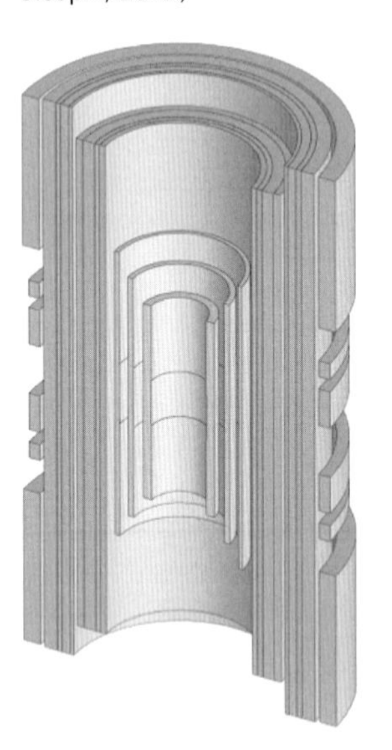

Figure 4.5.9 1.3 GHz/54 mm NMR magnet under development at MIT Francis Bitter Magnet Laboratory.

GHz LTS/HTS NMR magnet comprising a 500 MHz LTS NMR magnet (outer dark gray nested coils) and an 800 MHz HTS insert (inner light gray 3-nested coils) under development at FBML for completion in 2019 [23].

4.5.3.2 **Medical Diagnostic MRI Magnet**

4.5.3.2.1 **Whole Body**

Medical imaging is critical for quality health care for early detection and efficient treatment of disease and injury. The superconducting MRI magnet, operating at

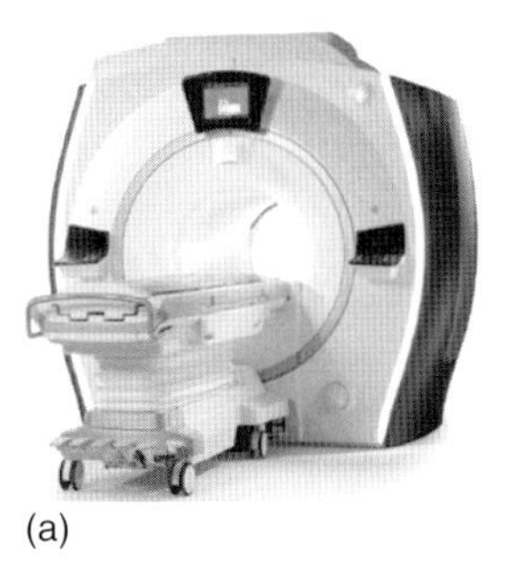

(a)

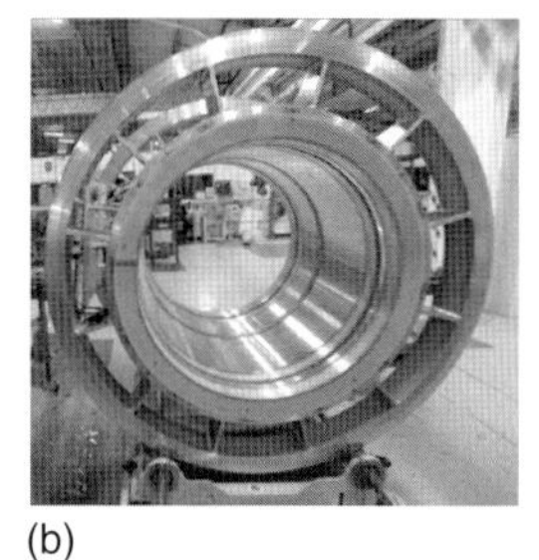

(b)

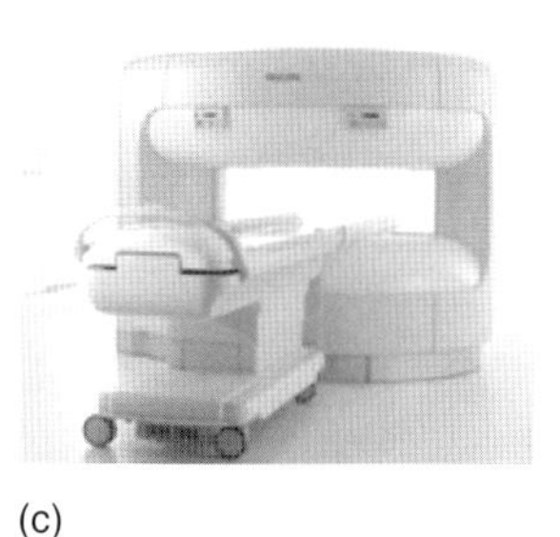

(c)

Figure 4.5.10 (a) 3 T whole-body MRI magnet. Patient bore: 70 cm. (Courtesy: Minfeng Xu, General Electric, 2013.); (b) coil configuration of a 3 T whole-body MRI magnet. (Courtesy: Adrian Thomas, Siemens, 2013.); (c) "open" type whole-body 1.0 T MRI magnet. (Courtesy: Cesar Luongo, Philips, 2013.)

0.5 T, became a clinic tool in the early 1970s. Since then, the field strength has been upgraded to 1 – 1.5 T. A current trend is operation at 3 T, and most recently, a number of 7 T whole-body MRI systems have been put into operation, chiefly for enhanced imaging quality. Virtually all whole-body MRI magnets' operations are now actively shielded.

Figure 4.5.10 shows a photo of (i) a 3 T/70 cm patient bore whole-body MRI magnet by General Electric (GE) and (ii) the coil assembly of a 3 T whole-body magnet by Siemens, of which a pair of outermost coils is for active shielding. Some patients regard these "tunnel" whole-body MRI magnets claustrophobic. To make the environment open, and make the patient readily accessible to the doctor, an "open" type MRI magnet is used. Figure 4.5.10c shows a photo of such an open whole-body MRI magnet by Philips.

4.5.3.2.2 **Extremity**

Another medical MRI magnet being widely used is for imaging body extremities, for example, feet, elbows, hands. Figure 4.5.11 shows a 1.5 T/28 cm bore extremity MRI magnet by GE.

4.5.3.2.3 **Functional**

Functional MRI is for study and examination of brain activity. Currently, fMRI is primarily for research. With head injuries, particularly among athletes and soldiers, more under ages of 40, becoming a serious public health issue, the days when fMRI magnets dedicated to on-site diagnose of head injuries hit the marketplace are not far in the future (Figure 4.5.12).

4.5.3.2.4 **Research**

Research MRI magnets are chiefly fMRI, with field strengths ranging up to 21.1 T. The most prominent is *Iseult* [24], an actively shielded 500 MHz (11.7 T)/90 cm

Figure 4.5.11 Extremity open MRI magnet. (Courtesy: Minfeng Xu, GE, 2013.)

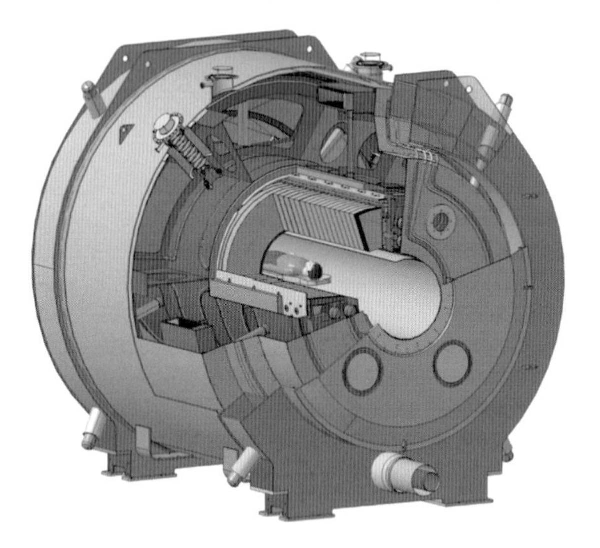

Figure 4.5.12 *Iseult* 500 MHz fMRI magnet. (Courtesy: Pierre Vedrine, Commissariat á l'Energie Atomique, Saclay, 2013.)

bore fMRI magnet under construction by ALSTOM. In 2014, Iseult is expected to become a vital player at the NeuroSpin, Commissariat a l'Energie Atomique, Saclay.

4.5.4
HTS Applications to NMR and MRI Magnets

As already discussed in Section 4.5.3.1, HTS is indispensable for >1 GHz NMR magnets. Other indispensable applications of HTS to NMR and MRI magnets described here, operation at and above 10 K, are impossible with LTS.

4.5.4.1 Annulus NMR

One magnet design option possible with HTS, but impossible with LTS, consists of a stack of many HTS bulk or plate annuli, each capable of supporting a current loop with no joint, and can operate in persistent mode [25–27]. In compact form, they can be desktop NMR magnets.

4.5.4.2 Liquid Helium (LHe)-Free

The price of helium has almost tripled over the last 5 years. NMR and MRI magnets are the largest single user of helium (26%) [28]. One of recent cryogenics trends to reduce LHe consumption in MRI magnets is zero-boiloff (ZBO) refrigeration [29]. Ideally, every superconducting magnet, LTS or HTS, should operate without reliance on LHe. Indeed, GE initiated, in the early 1990s, a trend toward such LHe-free MRI magnets with their all-Nb_3Sn 0.5 T open (doublet) MRI magnet, cryocooled at 10 K, in the early 1990s. Since then, development of cryogen-free magnets for NMR and MRI continues [30, 31].

4.5.4.2.1 MgB_2 MRI

For cryogen-free, low-field ($\leq$1.5 T) MRI magnets, MgB_2 is promising: [32–37] (i) with its critical temperature at 39 K, its energy density margin is approximately two orders of magnitude greater than those of LTS (Table 4.5.1): MgB_2 MRI magnets are immune from those premature quench-causing disturbances that still afflict some LTS MRI and NMR magnets; and (ii) with superconducting splices achievable [38–40], MgB2 magnets may now be operated in persistent mode. Figure 4.5.13 shows a photo of a 0.5 T/56 cm patient gap MgB_2 MRI magnet by ASG Superconductors for Standup Open MRI of Pensacola, FL, USA.

4.5.4.3 No-Insulation Winding Technique

Although insulation is considered indispensable to electromagnets, superconducting or resistive, recent studies, experimental and analytical, have demonstrated, particularly for HTS-tape wound double-pancake (DP) coils operated essentially under DC conditions as for NMR and MRI magnets, that elimination of turn-to-turn insulation from the HTS DP coils dramatically improves performance of the no-insulation (NI) DP coils. The NI winding technique makes the HTS DP coil mechanically robust and improves its field generation efficiency and self-protection [41–46]. An 800 MHz HTS insert [23] currently under development at the MIT FBML applies this NI winding technique.

Figure 4.5.13 0.5 T MgB2 MRI magnet. (Courtesy: Roberto Marabotto, ASG Superconductors, 2013.)

4.5.4.4 HTS Shim Coils

A wide (currently in 46 mm width), thin (<0.1 mm thick) YBCO tape manufactured by AMSC may be cut, with proper dimensioning, into a rectangle loop, folded, with the two vertical ends overlapped, into a $Z1$ shim coil [47]. Unlike a conventional "thick" (of >15 mm radial build) superconducting shim of NbTi, a thin (<5 mm radial build), persistent-mode – because of no joints in the loop-HTS shim – operates in a >12 T field, enabling it to be placed at the innermost region of the main/correction coil assembly. Also, because it can operate at > 10 K, the YBCO shim is ideal for cryogen-free NMR magnets. The HTS shim is now an important design option (Figure 4.5.14).

4.5.4.5 All-HTS 4.26 GHz (100 T) NMR Magnets

Based on a recent design study [48], a 100 T DC magnet, comprising many DP coils, each wound with the second-generation coated HTS tape conductor, is technically feasible. That is, a 4.26 GHz (100-T) all-HTS NMR magnet may become a reality in the future.

4.5.5 Conclusions

As stated at the outset, NMR and MRI magnets have been the most successful commercial applications of superconductivity, chiefly, because they benefit fully from the enabling features of superconductivity. With HTS incorporated, the sky is the limit for these magnets.

Figure 4.5.14 HTS Z1 shim [47].

References

1. Rabi, I.I., Zacharias, J.R., Millman, S., and Kusch, P. (1938) A new method of measuring nuclear magnetic moment. *Phys. Rev.*, **53** (4), 318–327.
2. Hornak, J.P. (1997) The Basics of NMR, *http://www.cis.rit.edu/htbooks/nmr/* (accessed 18 May 2014).
3. Hidalgo-Tobon, S.S. (2010) Theory of gradient coil design methods for magnetic resonance imaging. *Concepts Magn. Reson. A.*, **36A**, 223–242.
4. Iwasa, Y. (2009) *Case Studies in Superconducting Magnet*, 2nd edn, Springer, New York.
5. Paulson, E.K. and Zilm, K.W. (2005) External field-frequency lock probe for high resolution solid state NMR. *Rev. Sci. Instrum.*, **76**, 026104.
6. Yanagisawa, Y., Nakagome, H., Hosono, M., Hamada, H., Kiyoshi, T., Hobo, F., Takahashi, M., Yamazaki, T., and Maeda, H. (2008) Towards beyond-1 GHz solution NMR: internal 2H lock operation in an external current mode. *J. Magn. Reson.*, **192**, 329–337.
7. Yanagisawa, Y., Nakagome, H., Tennmei, K., Hamada, H., Yoshikawa, M., Otsuka, A., Hosono, M., Kiyoshi, T., Takahashi, M., Yamazaki, T., and Maeda, H. (2010) Operation of a 500 MHz high temperature superconducting NMR: towards an NMR spectrometer operating beyond 1 GHz. *J. Magn. Reson.*, **203**, 274–282.
8. Takahashi, M. *et al.* (2012) Towards a beyond 1 GHz solid-state nuclear magnetic resonance: external lock operation in an external current mode for a 500 MHz nuclear magnetic resonance. *Rev. Sci. Instrum.*, **83**, 105110.
9. Bobrov, E.S. and Williams, J.E.C. (1980) in *Mechanics of Superconducting Structures*, vol. 41, Proceedings of the Winter Annual Meeting Chicago, IL, November 16-21, 1980 (ed F.C. Moon), ASME, New York, pp. 13–41.
10. Clickner, C.C., Ekin, J.W., Cheggour, N., Thieme, C.L.H., Qiao, Y., Xie, Y.Y., and Goyal, A. (2006) Mechanical properties of pure Ni and Ni-alloy substrate materials for YBaCuO coated superconductors. *Cryogenics*, **46**, 432–438.
11. Seop Shin, H. and Dedicatoria, M.J. (2013) Intrinsic strain effect on critical current in Cu-stabilized GdBCO coated conductor tapes with different substrates. *Supercond. Sci. Technol.*, **26**, 055 005, (6pp).
12. Osamura, K., Machiya, S., Ochiai, S., Osabe, G., Yamazaki, K., and Fujikami, J. (2013) Direct evidence of the high strain tolerance of the critical current of DI-BSCCO tapes fabricated by means of the pretensioned lamination technique. *Supercond. Sci. Technol.*, **26**, 045012.
13. Kitaguchi, H. and Kumakura, H. (2005) Superconducting and mechanical performance and the strain effects of a

multifilamentary MgB2 Ni tape. *Supercond. Sci. Technol.*, **18**, S284–S289.
14. Garrett, M.W. (1967) Thick cylindrical coil systems for strong magnetic fields with field or gradient homogeneities of the 6th to 20th order. *J. Appl. Phys.*, **38** (6), 2563–2586.
15. Romeo, F. and Hoult, D.I. (1984) Magnet field profiling analysis and correcting coil design. *Magn. Reson. Med.*, **1**, 44–65.
16. Hahn, S., Bascuñán, J., Kim, W., Bobrov, E.S., Lee, H., and Iwasa, Y. (2008) Field mapping, NMR lineshape, and screening currents induced field analyses for homogeneity improvement in LTS/HTS NMR magnets. *IEEE Trans. Appl. Supercond.*, **18**, 856–859.
17. Hahn, S., Bascuñán, J., Lee, H., Bobrov, E.S., Kim, W., Ahn, M.C., and Iwasa, Y. (2009) Operation and performance analyses of 350 and 700 MHz low-/high-temperature superconductor nuclear magnetic resonance magnets: a march toward operating frequencies above 1 GHz. *J. Appl. Phys.*, **105**, 024501.
18. Bobrov, E.S. and Punchard, W.F.B. (1988) A general method of design of axial and radial shim coils for nmr and mri magnets. *IEEE Trans. Magn.*, **24** (1), 533–536.
19. Hoult, D.I. and Lee, D. (1988) Shimming a superconducting nuclear-magnetic-resonance imaging magnet with steel. *Rev. Sci. Instrum.*, **56** (1), 131–135.
20. Kiyoshi, T. *et al.* (2005) Operation of a 930-MHz high-resolution NMR magnet at TML. *IEEE Trans. Appl. Supercond.*, **15**, 1330–1333.
21. Kiyoshi, T. *et al.* (2011) Bi-2223 innermost coil for 1.03GHz NMR magnet. *IEEE Trans. Appl. Supercond.*, **21**, 2110–2113.
22. Markiewicz, W.D. *et al.* (2012) Design of a superconducting 32T magnet with REBCO high field coils. *IEEE Trans. Appl. Supercond.*, **22**, 4300704.
23. Bascuñán, J., Hahn, S., Kim, Y., and Iwasa, Y. (2014) An 18.8-T/90mm Bore All-HTS Insert (H800) for 1.3 GHz LTS/HTS NMR magnet: insert design and double-pancake coil fabrication. *IEEE Trans. Appl. Supercond.*, **24**, 6400205.
24. Vedrine, P. *et al.* (2008) The whole body 11.7T MRI magnet for Iseult/INUMAC project. *IEEE Trans. Appl. Supercond.*, **18**, 868–873.
25. Iwasa, Y., Hahn, S., Tomita, M., Lee, H., and Bascuñán, J. (2005) A persistent-mode magnet comprised of YBCO annuli. *IEEE Trans. Appl. Supercond.*, **15**, 2352–2355.
26. Nakamura, T., Itoh, Y., Yoshikawa, M., Oka, T., and Uzawa, J. (2007) Development of a superconducting magnet for nuclear magnetic resonance using bulk high-temperature superconducting materials. *Concepts Magn. Reson. Part B*, **31B**, 65–69.
27. Hahn, S. *et al.* (2013) Bulk and plate annulus stacks for compact NMR magnets: trapped field characteristics and active shimming performance. *IEEE Trans. Appl. Supercond.*, **22**, 4300504.
28. Banks, M. (January 2010) *Helium Sell-Off Risks Future Supply*, Physics World.
29. Cosmus, T.C. and Parizh, M. (2011) Advances in whole-body MRI magnets. *IEEE Trans. Appl. Supercond.*, **21**, 2104–2109.
30. Terao, Y. *et al.* (2013) Newly designed 3T MRI magnet wound with Bi-2223 tape conductors. *IEEE Trans. Appl. Supercond.*, **23**, 4400904.
31. Good, J. and Mitchell, R. (2006) A desktop cryogen free magnet for NMR and ESR. *IEEE Trans. Appl. Supercond.*, **16** (2), 1328–1329.
32. Iwasa, Y. *et al.* (2006) A round table discussion on MgB2 toward a wide market or a niche production?—a summary. *IEEE Trans. Appl. Supercond.*, **16**, 1457–1464.
33. Yao, W. *et al.* (2006) A solid nitrogen cooled MgB2 demonstration coil for MRI applications. *IEEE Trans. Appl. Supercond.*, **16**, 912–915.
34. Yao, W. *et al.* (2007) Behavior of a 14 cm bore solenoid with multifilament MgB2 tape. *IEEE Trans. Appl. Supercond.*, **17**, 2252–2257.
35. Penco, R. and Grasso, G. (2007) Recent development of MgB2-based large scale applications. *IEEE Trans. Appl. Supercond.*, **17**, 2291–2294.
36. Yao, W., Bascuñán, J., Hahn, S., and Iwasa, Y. (2010) MgB2 coil for MRI

applications. *IEEE Trans. Appl. Supercond.*, **20**, 756–759.
37. Kawagoe, A. *et al.* (2011) Development of an MgB2 coil wound with a parallel conductor composed of two tapes with insulation. *IEEE Trans. Appl. Supercond.*, **21** (3), 1612–1615.
38. Yao, W., Bascũán, J., Hahn, S., and Iwasa, Y. (2009) A superconducting joint technique for MgB2 round wires. *IEEE Trans. Appl. Supercond.*, **19**, 2261.
39. Park, D. *et al.* (2012) MgB2 for MRI magnets: test coils and superconducting joints results. *IEEE Trans. Appl. Supercond.*, **22**, 4400305.
40. Ling, J. *et al.* (2013) Monofilament MgB2 wire for a whole-body MRI magnet: superconducting joints and test coils. *IEEE Trans. Appl. Supercond.*, **23**, 6200304.
41. Hahn, S., Park, D.K., Bascuñán, J., and Iwasa, Y. (2011) HTS pancake coils without turn-to-turn insulation. *IEEE Trans. Appl. Supercond.*, **21**, 1592–1595.
42. Kim, S., Saitou, A., Joo, J., and Kadota, T. (2011) The normal-zone propagation properties of the non-insulated HTS coil in cryocooled operation. *Physica C*, **471**, 1428–1431.
43. Hahn, S., Park, D.K., Voccio, J., Bascuñán, J., and Iwasa, Y. (2012) No-insulation (NI) HTS inserts for >1 GHz LTS/HTS NMR magnets. *IEEE Trans. Appl. Supercond.*, **22**, 4302405.
44. Choi, S., Jo, H.C., Hwang, Y.J., Hahn, S., and Ko, T.K. (2012) A study on the no insulation winding method of the HTS coil. *IEEE Trans. Appl. Supercond.*, **22**, 4904004.
45. Wang, X. *et al.* (2013) Turn-to-turn contact characteristics for equivalent circuit model of no-insulation ReBCO pancake coil. *Supercond. Sci. Technol.*, **26**, 035012.
46. Hahn, S. *et al.* (2013) No-insulation coil under time-varying condition: magnetic coupling with external coil. *IEEE Trans. Appl. Supercond.*, **23**, 4601705.
47. Iwasa, Y., Hahn, S., Voccio, J., Park, D.K., Kim, Y., and Bascuñán, J. (2013) Persistent-mode high-temperature superconductor shim coils: a design concept and experimental results of a prototype Z1 high-temperature superconductor shim. *Appl. Phys. Lett.*, **103**, 052607.
48. Iwasa, Y. and Hahn, S. (2013) First-cut design of an all-superconducting 100-T direct current magnet. *Appl. Phys. Lett.*, **103**, 253507.

4.6
Superconducting Magnets for Fusion

Jean-Luc Duchateau

4.6.1
Introduction to Fusion and Superconductivity

Fusion by magnetic confinement requires large magnet systems to confine the plasma inside the vacuum chamber [1]. The production of the magnetic field with superconducting magnets in the large vacuum chamber of International Thermonuclear Experimental Reactor (ITER) (835 m^3) is one of the main technological challenges, which must be tackled. In the early 1960s, when applied superconductivity was merging with the first small Nb_3Sn magnets, it was quickly identified that this technology was compulsory for fusion [2].

Till the beginning of the 1980s, all the fusion magnet systems were resistive with silver-alloyed copper conductors to improve their mechanical properties and resist the large electromagnetic stresses. This kind of solution was still possible due to the small size of the machines and also due to their pulsed mode operation. The largest machine of this type was JET (major radius of the plasma torus $R = 2.98$ m). In this case (see Table 4.6.1), the required power to energize the toroidal field (TF) system is in the range of 0.5 GW and the tokamak can only be operated thanks to flying wheel generators, a solution which is possible due to the short duration of the JET plasma discharges (10–30 s).

ITER will still be a pulsed machine, but the electrical power necessary to energize the whole system in the case of a solution with resistive magnets ($P = 2$ GW for 500 s) cannot be reasonably obtained from the electrical grid. The high level of required electrical power in case of resistive magnets power, as well as the perspective for future steady state machines, pushed the plasma physics community to develop superconducting magnet systems in their experimental fusion machines.

Superconductivity was introduced at the level of small tokamaks, such as T-7 ($R = 1.22$ m) in 1978 in Soviet Union and TRIAM with Nb_3Sn conductor

Table 4.6.1 Electrical power needed for the tokamak TF system in case of copper magnets for a few fusion machines.

Fusion machine (plasma discharge duration)	Major radius (m)	Plasma volume (m^3)	Plasma magnetic field (T)	Fusion power (MW)	Electrical power TF system(copper magnets) (MW)
TS (1000 s)	2.4	24	4.5	0	~150 (superconducting system)
JET upgrade (10 s)	2.96	100	4	~20	~500 (copper magnets)
ITER (500 s)	6.2	837	5.3	~400	~800 (superconducting system) ~20 in cryoplant

Table 4.6.2 Fusion and superconductivity: the superconducting machines.

Name	Nature	Major radius, R (m)	Maximum field cond. (T)	Stored energy TF (MJ)	Superconducting material/top (K)	Status
Tore Supra	Tokamak	2.4	9	600	NbTi/1.8	In operation since 1988
LHD	Heliotron	3.9	6.9	920	NbTi/5	In operation since 1998
EAST	Tokamak	1.7	5.8	400	NbTi/5	In operation since 2006
KSTAR	Tokamak	1.8	6.7	470	Nb_3Sn and NbTi/5	In operation since 2008
SST-1	Tokamak	1.1	4.2	56	NbTi/5	In operation Plasmas 2013
W7-X	Stellarator	5.5	5	620	NbTi/4	In construction Plasmas 2014
JT-60SA	Tokamak	3.	5.65	1 060	NbTi/5	In construction Plasmas 2018
ITER	Tokamak	6.2	11.8	40 000	Nb_3Sn and NbTi/5	In construction Plasmas 2020

($R = 0.8$ m) in 1990 in Japan. Today, all major fusion machines by magnetic confinement are superconducting machines, as shown in Table 4.6.2. Five systems are presently in operation (Tore Supra (TS) [4] since 1988, LHD [5] in Japan since 1998, EAST [6] in China since 2006, KSTAR [7] in Korea since 2008 SST-1 [9] in India since 2013), and three others in construction (W7-X [8] in Germany, in India, JT-60SA [10] in Japan with an important European contribution, ITER [11]).

The first important introduction of superconducting magnets in magnetic fusion was performed within the construction of the TS tokamak in France ($R = 2.4$ m) (see Figure 4.6.1) and of T-15 [12] in the Soviet Union. In both machines, the magnetic field on the conductor was around 9 T, making the classical use of NbTi conductors at 4.2 K impossible. The debate about two possible choices ran at the end of the 1970s:

1) either the use of forced flow Nb_3Sn conductors at a temperature in the range of 4 K;
2) or the use of bath-cooled NbTi conductors with pressurized helium at 1.8 K through a new technique developed in France at CEA.

The insufficient industrial maturity of Nb_3Sn technology was clearly seen during the acceptance tests of T-15 where resistive parts in the magnets prevented steady-state operation of the tokamak, which definitively stopped operation in 1991. This was further confirmed in a large international fusion experiment at Oak Ridge national laboratory where six large coils were tested introducing forced flow refrigeration, one of them being with Nb_3Sn [13] and also showing resistive

Figure 4.6.1 TS Tokamak: first large superconducting tokamak in operation since 1988.

parts. The main design orientation for the TF system of TS at the beginning of the 1980s was to extend the application of the NbTi superconductor (a cheap material, insensitive to mechanical strain) to a high magnetic field (9 T) by using superfluid helium, at 1.8 K and 1 bar, as a coolant. The industrialization of refrigeration at 1.8 K was really a breakthrough, enabling the production of higher field with NbTi magnets and with Nb_3Sn magnets as well.

TS and LHD have now been in operation long enough to demonstrate that superconducting magnets at low temperature can reliably provide confinement for plasmas at 100 million degree Celsius. Lessons can be drawn from their accumulated experience. This can be of interest at a moment when several superconducting tokamaks are in construction, especially the construction of ITER, a superconducting tokamak reactor with 700 tons of superconductors.

A first international project, the Large Coil Task (LCT), involved the development of a torus with six superconducting D-shaped coils with a bore of 2.5 m × 6.5 m, which were tested in 1986–1987 [13].

By the way, the most emblematic fusion project is ITER by its size and the international teams, which are involved in its construction. ITER is a tokamak but even the teams involved in other fusion concepts (heliotrons, stellarators) consider ITER as a demonstration machine for fusion. The main aspects of superconducting magnets for fusion will therefore be illustrated with this machine.

4.6.2 ITER

4.6.2.1 Introduction

The ITER adventure was initiated in November 1985 when Ronald Reagan and Mikhail Gorbachev met at Geneva and encouraged an international collaboration to take place with the aim of mastering fusion energy. It was however only in 1991 that four parties (Europe, Russia, Japan, and the United States) veritably started a 6 years funded program with a dedicated project team: the ITER program was born. The ITER program has now been extended to three other parties (India, Korea, and China) and the construction of the machine was officially started in 2006 on the site of Cadarache in France.

To prove the feasibility of thermonuclear fusion as a potential source of energy for humanity, ITER is making a major step forward from the most performing fusion machines at present, JT60 in Japan and JET in Europe. JET has a major radius of 3 m, while for the selected configuration in ITER, a major radius of 6.2 m is necessary.

This will be achieved by sustaining stable plasma discharges as long as 500 s for a produced fusion power P_{fus} of 400 MW with an amplification of energy Q of 10 (ratio of the output fusion power to the input heating power). If the extrapolation from JET, based on plasma physics scaling laws, is considered to be adequate, ITER must also handle numerous technological challenges in addition. These challenges regard for instance the plasma-facing components or the high-power plasma heating sources and the very large size components.

Among them, the superconducting magnet system is one of the most remarkable. The superconducting magnet system is the real backbone for the machine representing about 30% of the ITER cost investment. In practice, deeply buried in the very heart of the tokamak, as it is, repairs are hardly envisageable except for a few protruding components like joints for instance. Therefore the quality insurance process during fabrication must be led to avoid any possible fault.

The 838 m^3 ITER plasma (a torus with 12.4 m diameter and 21 m^2 cross section) at 100 million degree Celsius will be confined very closely by the superconducting magnet system, which will operate at a temperature of 5 K.

4.6.2.2 The ITER Magnet System

It is impossible to completely cover all aspects of the ITER magnet system. The presented summary is based on ITER documentation and in particular on [11, 14].

The ITER magnetic field is composed of three major systems (see Figure 4.6.2):

- The TF system (18 TF coils), which produces the main confinement magnetic field for the plasma charged particles.
- The central solenoid (CS), which inductively drives the plasma current (15 MA) for the ramp-up and then the plateau of 500 s, for the reference plasma discharge. The plasma is the secondary of a transformer whose primary is the CS.
- The Poloidal Field system (PF) which plays a role for the plasma shaping, plasma positioning, and its stability.

The main characteristics of the systems are presented in Table 4.6.3. A very important point is the industrialization of the production of Nb_3Sn strands (bronze route and internal tin processes), with a total production of about 500 tons, while the yearly production of Nb_3Sn was around 20–30 tons before ITER. The conductor procurement is presently at a quite advanced stage [14]. Full-size ITER conductor qualification and quality control tests are carried

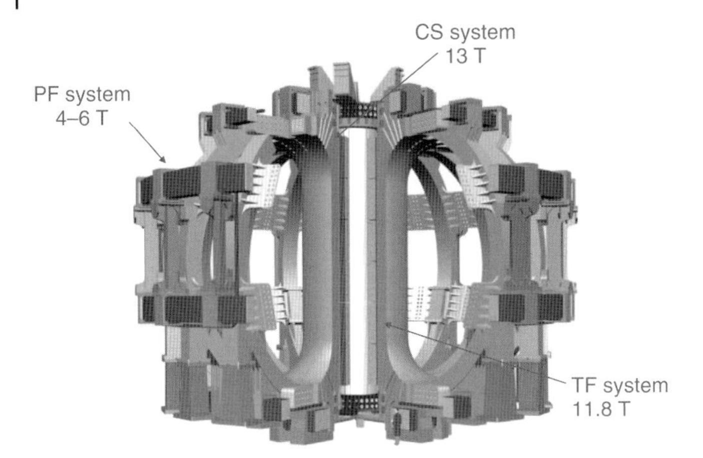

Figure 4.6.2 The three major systems of the ITER magnet system. (Courtesy of ITER.)

Table 4.6.3 Main characteristics of the ITER superconducting systems.

System	Energy (GJ)	Peak field (T)	Total MAt (MA)	Conductor length (km)	Total weight (strand) (tons)
Toroidal field (TF)	41	11.8	164	82.2 Nb_3Sn	6540 (396)
Central solenoid (CS)	6.4	13.	147	35.6 Nb_3Sn	974 (118)
Poloidal field (PF)	4.	6.	58.2	61.4 NbTi	2163 (224)
Correction coils (CC)	—	4.2	3.6	8.2 NbTi	85

out at the SULTAN facility [15]. These tests play an important role to better understand the mechanical role of the strand void fraction and of the twist pitches in mitigating the Nb_3Sn degradation under cycling. The Nb_3Sn double pancakes (DPs) are heat treated for 2 weeks in large ovens. The handling of the reacted conductor after heat treatment in the so-called Wind-React and Transfer fabrication process is a true challenge.

All the ITER magnets systems are wound from Cable in Conduit Conductors (CICCs), which are described in Section 4.6.3. The protection of the systems against quenches is presented in Section 4.6.4. The TF system is a DC system, while the PF and the CS systems are pulsed coils.

4.6.2.3 Main Dimensioning Aspects of ITER

The electrical power associated with the refrigerator to compensate for all losses at cryogenic temperatures can be estimated around 20 MW, a value to be compared to the 2 GW that would be needed for an equivalent resistive system. In a preliminary design approach, the radial extension of the machine and of any

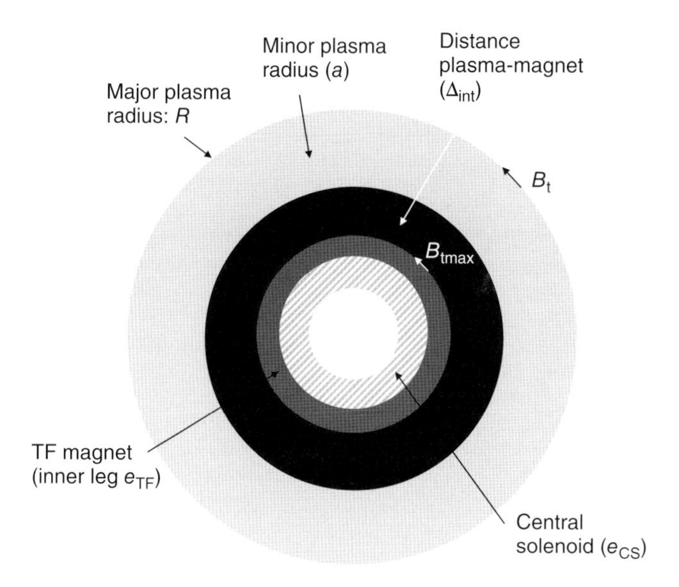

Figure 4.6.3 Main concentric zones of a thermonuclear reactor in the equatorial plane.

thermonuclear reactor had to be built in the equatorial plane such as schematically presented in Figure 4.6.3, starting from the plasma major radius R and progressively building up layers ending with the CS radial extension. In Figure 4.6.3, the main concentric zones are presented starting from the major plasma radius R:

- The plasma of minor radius a
- The blanket-shielding zone of thickness Δ_{int}
- The TF system of radial extension e_{TF}
- The CS of radial extension e_{CS}.

The radial extensions of the TF e_{TF} and of the CS e_{CS} have to be estimated using design criteria related to mechanics and related to the definition of the superconducting cables. The radial extensions of the TF and CS systems are mainly driven in principle by the structures. The blanket shielding zone extension Δ_{int} plays an important role in the magnetic field increase from plasma center to TF conductor. It must be adjusted so as to not only limit the fluence but also reduce the nuclear heating in the TF magnet, aiming at simplifying the cryogenic concept of the TF system. In a future reactor such as DEMOnstration Power Plant (DEMO), the blanket zone extension must be also compatible with a self-tritium sufficiency (tritium breeding ratio (TBR) >1) and to allow heat recovery for electricity production.

The fusion power P_{fus} of a reactor like ITER can be expressed using the engineering parameters R, B_{t} the plasma magnetic field, and A:

$$P_{\mathrm{fus}} \propto \frac{R^3 B_{\mathrm{t}}^4}{A^4} \quad \text{with } A = \frac{R}{a} \tag{4.6.1}$$

The expression of Q cannot be easily analytically expressed and must be numerically calculated.

The selection of the triplet (R, B_t, A) has to be made in tight connection with the questions of cost, magnetic flux for inductive mode, available technology, and of accessibility to the plasma through the ports.

To satisfy the objectives of ITER, the triplet was set at $R = 6.2$ m, $B_t = 5.3$ T, and $A = 3.1$. Owing to the toroidal shape, from the center of the plasma to the conductor on the magnet system, the maximum magnetic field B_{tmax} on the conductor is increased by a factor >2 in comparison with B_t, which imposes to select Nb_3Sn for the superconducting material of the ITER TF system:

$$B_{tmax} = \frac{B_t}{1 - \frac{a}{R} - \frac{\Delta_{int}}{R}} \tag{4.6.2}$$

4.6.2.4 The ITER TF System

The ITER TF magnet system is the most important magnet system of the tokamak. The 18 TF coils are "D" shaped as shown in Figure 4.6.3. Each of them consists of a winding pack (WP) enclosed in a thick steel case. The WP is a bonded structure of seven DP, each inserted within a steel radial plate, which houses the 134 turns of the reacted and insulated CICC. The DPs are reacted within a dedicated mold at 650 °C before being insulated and transferred to the plates. The coil terminals protrude from the TF coils at its lower curved part with the six DP joints (i.e., the joints linking adjacent DPs) and the helium feeder manifolds. The conductor terminations are formed after winding but before heat treatment of the conductor. They consist of the terminals (two per WP) and the DP to DP joints (six per WP). The length of the regular DP is 760 m, and the sides DPs of the WP are shorter.

When energized alone, with no current in the PF and the CS systems, the TF system is submitted to large magnetic forces, mainly a hoop force and a centering force. These magnetic forces induced primary stresses which are contained by a large amount of steel structures (conductor jacket, plates, casings), which result in a low overall current density in the range of 11 A mm^{-2} (see Table 4.6.8). In existing superconducting tokamaks in construction or in operation, the centering force is contained by wedging of the inner legs of the coils, which form a vault such as that presented in Figure 4.6.4. During plasma operation, all TF coils are loaded in addition by out-of-plane forces. These out-of-plane forces are due to the interaction between the TF current and the PF. Secondary stresses associated with out-of-plane forces are induced during plasma discharges, which are contained by wedging and in the outboard region by specific mechanical structures such as those presented in Figure 4.6.4: the outer intercoil structures (OISs). In the curved regions above and below the inboard leg, the coils are structurally linked by means of three upper and three lower precompression rings formed from unidirectional bonded glass fibers that provide compression on inner poloidal shear keys.

The hydraulic length of the regular pancakes is about 380 m long, half of the conductor length, thanks to seven helium inlets per coil. The cooling inlet sections of each DP are located at the inner surface (plasma side) of the coil, at the DP.

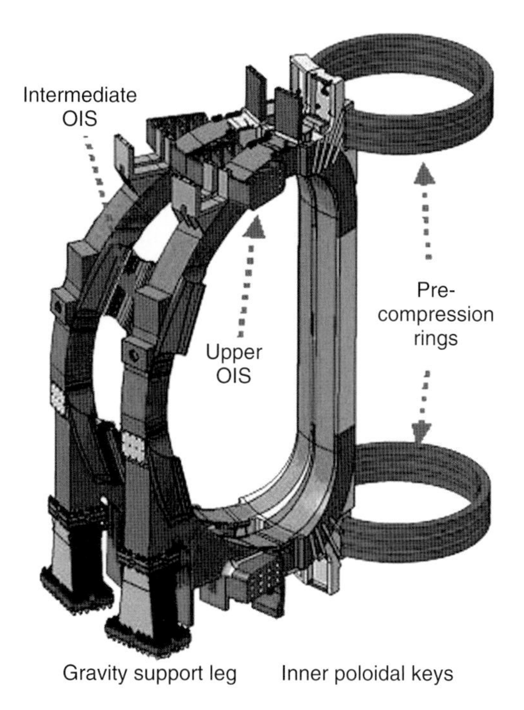

Figure 4.6.4 The ITER TF system highlighting the mechanical structure to resist the out-of-plane loads. (Courtesy of ITER.)

This ensures that inlet (cold) helium is supplied in the high-field region where the nuclear heating is concentrated; it cools the rest of the pancake and exits through the joints located on the outer surface (cryostat side) of the coil at the bottom curved part. This forced flow cooling associated with the corresponding tubing at high voltage is certainly technologically complex. The He massflow circulating in the conductor is 8 g s^{-1}.

4.6.2.5 **The ITER Model Coils**

A very important work was led internationally for the model coils during the preparation phase of ITER (1997–2002) [16, 17]. Two model coils were designed, manufactured, and tested.

- A model coil of the CS, which was manufactured by United States and Japan and tested at Jaeri facility in Japan. Japan, United States, and Europe shared the fabrication of the conductor.
- A model coil of the TF system, which was manufactured and tested at FZK facility in Europe. Europe was in charge of the conductor fabrication.

The model coil experiments (2000–2002) were crucial in testing in real size and relevant lengths the behavior of large Nb_3Sn CICCs in long lengths. Some non-expected degradation in critical performances was found due to the great sensitivity of Nb_3Sn strands to strain. The strain is mainly due to the differential thermal contraction arising between steel and Nb_3Sn from the reaction temperature of 650 °C at which Nb_3Sn is formed to the cryogenic temperature. Some extra strain including bending strain was pointed out in addition during the model coil experiments. This extra strain is related to the Lorentz force loading the strands at nominal current. The drivers of this sensitivity are not yet completely identified (void, twist pitches, cabling patterns) and are still under investigation. The design of the ITER CICCs was later corrected to take into account these effects.

4.6.3
Cable in Conduit Conductors (CICC)

4.6.3.1 Introduction

The DC TF systems of tokamaks are subject to heat deposition due to the field variations caused by the other magnets of the machine, by the plasma itself, and by nuclear heat radiation. In addition, the CS and PF systems are pulsed magnets subject to AC losses. This statement implies to have helium in contact with the conductor to maintain the temperature. Historically, the first two large fusion machines were bath cooled: TS, which introduced superfluid helium as a coolant and LHD, which is not a tokamak and in which heat deposition is therefore limited. In the following machines in operation or in construction, ITER, EAST, KSTAR, W7-X, SST-1, and JT-60SA, the superconducting magnets, are forced flow-cooled magnets. This characteristic is imposed due to the very high voltage, the very high current, and the need to remove important cryogenic losses. The CICC is the selected conductor for the magnet systems of all tokamaks with superconducting magnets except TS. For the ITER coils for instance, the requirement of high current (68 kA for the TF system) and of very high voltages in operation (10 kV to ground for the PF and CS systems in normal operation associated with the plasma discharge) due to the size of the magnetic systems, led to select the CICC as the best choice for the conductors in the present state of the superconducting technology. But any kind of forced flow conductor is in principle able to sustain high voltage to the ground. However, in addition, this type of conductor is well adapted to accept fast heat deposition (such as caused by plasma disruption) with limited temperature increase. The principle of CICC is not recent. M. Hoenig at MIT (USA) introduced it in 1975. Numerous prototype coils have been made with this type of conductor but the experience is not large regarding magnet systems in operation with such type of conductor.

The first coil using this concept with Nb_3Sn was the Westinghouse coil in the "LCT" (see Figure 4.6.5) [13], an R&D program linked to fusion program (1988). The maximum performance of the magnet was however limited due to some spreading out of a resistive phase in the magnet.

The CICC represents a very complex component, integrating a large part of the ITER magnet cost investment (~40%). This type of conductor is rather new and

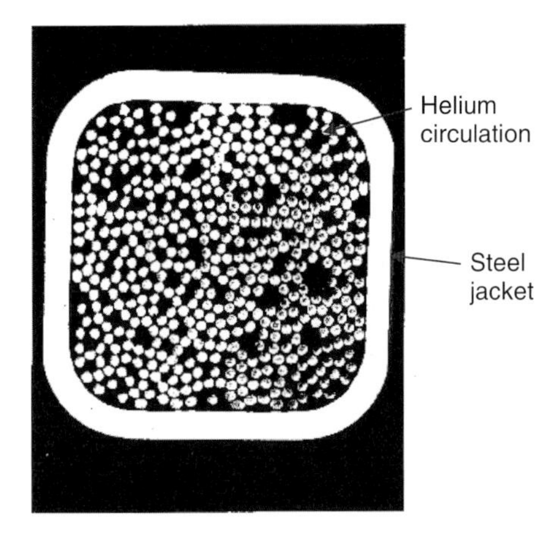

Figure 4.6.5 Airco-Westinghouse Nb_3Sn CICC 20.7 kA, 9 T (20.8 mm^2).

not so many magnets are presently operating using CICC (EAST, KSTAR, SST1, and PF coils of LHD). Practical experience regarding CICC is not large. The main issues, still in discussion, are, for instance:

- Current distribution in the cable
- Stability under fast magnetic field variations
- Degradation of Nb_3Sn strands under cycling due to strain
- Behavior in case of a quench (detection, propagation, maximum He pressure in the conductor).

A CICC is basically made by cabling in several stages superconducting and copper strands and by compacting the cable inside a conduit (most generally of stainless steel). A CICC such as the ITER TF CICC (396 tons of Nb_3Sn strands) or the JT-60SA TF CICC (33.4 tons of NbTi strands) is composed of several components, which are visible in Figures 4.6.6 and 4.6.7: superconducting strands, copper strands, steel bandages, helium, and steel conduit. In a project like ITER, the optimum composition of the conductor components is calculated through design criteria:

- The non-copper section A_{noncu} driving the temperature margin of the conductor and the critical energy.
- The copper section A_{cu} to protect the cable in case of a quench (hot-spot criterion).
- The helium section in strand region A_{He} (in case of Nb_3Sn about 30% to ensure mechanical stability and avoid bending strain).

The heat load from all sources is removed from the coils to keep the temperature constant by circulating a sufficiently high He mass flow through the conductor. This circulation requires a pump work to compensate for the viscous pressure

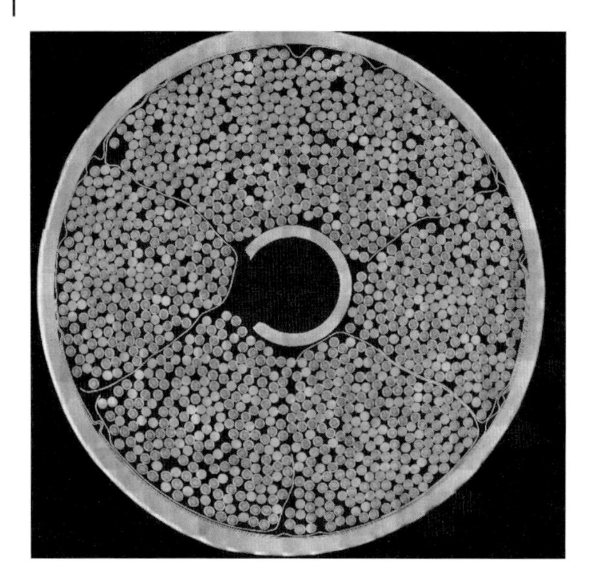

Figure 4.6.6 The ITER TF dual channel CICC 68 kA 11.8 T ($\Phi = 39.7$ mm). (Courtesy of ITER.)

Figure 4.6.7 The JT-60SA TF CICC 25.7 kA 5.65 T (18 mm × 22 mm). (Courtesy of JT-60SA.)

losses, and a heat exchanger where the power is extracted from the system (see Figure 4.6.8).

4.6.3.2 Stability of Cable in Conduit Conductors

The basic principle of the CICC is to take benefit of the very high volumetric heat capacity of helium (about 500 times the volumetric heat capacity of metallic

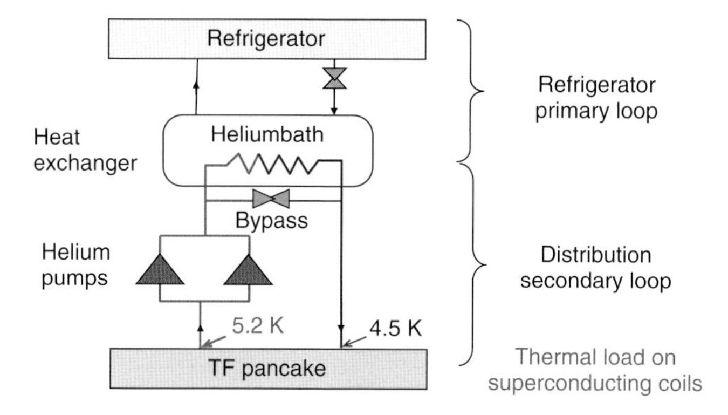

Figure 4.6.8 Cryogenic loop associated with forced flow.

Table 4.6.4 Comparison between enthalpies of metallic materials and helium at 4.5 K.

Material	Enthalpy for a temperature increase of 2 K starting from 4.5 K
Copper	2700 J m^{-3}
Nb_3Sn	7400 J m^{-3}
A316 (steel)	40 kJ m^{-3}
Helium (constant volume)	640 kJ m^{-3}
Helium (local enthalpy)	1660 kJ m^{-3}
Helium (enthalpy at constant pressure)	2270 kJ m^{-3}

Table 4.6.5 Illustration of energy disturbance for the ITER PF CICC.

CICC length affected by the disturbance and origin of disturbance	Time deposition (ms)	Deposited energy (mJ cm^{-3})
Disruption: 20–75 m	100	<30
According to the PF coils		
Mechanical energy: 10 mm	1	100
According to the PF coils		

materials) to limit the temperature increase in case of fast energy deposition (see Table 4.6.4) [18]. This occurs in tokamaks after the very fast decrease to zero of the plasma current in case of a plasma disruption.

During a plasma disruption, a fast magnetic field variation (50–100 ms) affects the whole coil over lengths well above several meters, creating losses in the superconducting strands (Table 4.6.5). This is very specific to fusion magnets.

In the case of the PF and CS systems of a tokamak, the disruption is associated with a current variation and a magnetic field variation. In the case of the TF system, the magnetic field variation is dominant. The time constants of the current

Table 4.6.6 Critical energy of the JT-60SA TF CICC.

h ($\mathrm{W\,m^{-2}\,K^{-1}}$)	τ_{He} (ms)	α_0	E_c ($\mathrm{mJ\,cm^{-3}}$)
600	72	0.59	102
1000	43	0.46	135

and field variations in case of a plasma disruption are driven by the time constant of the vacuum vessel, which filters the variations of the plasma current by developing eddy currents. The critical energy in the CICC is limited during the transients by the simultaneous evacuation of the power deposition into helium, driven by the helium time constant τ_{He}, which can be estimated in case of a square-shaped heat deposition of duration Δt to $\tau_{He} = C_{He}A_{He}/hP$.

C_{He} – volumetric heat capacity of helium
A_{He} – CICC helium section
P – wetted strand perimeter
h – heat transfer coefficient to helium.

An analytical expression of the critical energy can be given, in case of a square-shaped heat deposition, which logically appears proportional to the temperature margin $\Delta T_{margin} = T_{cs} - T_0$, the copper section playing no role. Void is the void fraction (~30%) in the CICC strand region. The critical energy is classically referred to the strands volume ($\mathrm{mJ\,cm^{-3}}$).

$$E_c = C_{He}(T_{cs} - T_0)(1 - \alpha_0)\frac{\text{void}}{1 - \text{void}} \quad \alpha_0 = \frac{1}{1 + \frac{\Delta t}{\tau_{He}}} \tag{4.6.3}$$

A numerical estimation is presented in Table 4.6.6 for JT-60SA TF CICC to illustrate this approach ($\Delta t = 50$ ms, $S = 88\ \mathrm{mm^2}$, $P = 0.68$ m). Two possible values of h have been considered.

It can be seen that the cable in conduit brings an adequate solution to ensure stability and high E_c by providing:

- a local helium reservoir;
- a very high wetted perimeter of the superconducting strands, thanks to the subdivision allowing substantial access to the helium enthalpy reservoir;
- low AC losses for the conductor by controlling its time constants through the contact resistance between strands.

Thanks to the high wetted perimeter, a large heat flux from the conductor to helium takes place during the transients, characterized by the helium time constant of the helium CICC which is in the range of 50 ms, the same order of magnitude as the time constant of the magnetic field variation caused by the disruption. The temperature excursion in the conductor is therefore limited and can be kept under T_{cs} and the conductor stability to face the transient is enhanced. The maximum acceptable energy, named also critical energy, is proportional to ΔT_{margin}.

Table 4.6.7 Stekly parameter in NbTi superconducting fusion magnets.

Fusion project	I (kA)/B (T)	Cu to non-Cu ratio in strand	Stekly parameter, α
W7-X non planar coil	17.6/6	2.7	3.24
EAST	14.3/5.8	1.4	6
JT-60SA	25.7/5.65	1.9	3.3
ITER PF2, PF3, PF4, PF5	55/4.5	2.3	3

Note that the Stekly [19] model at constant bath temperature, adapted to open unlimited bath, is not adequate to model the closed volume of the CICC whose temperature simultaneously increases during heat deposition. By the way, this model is generally not used to dimension the CICC copper section. This is particularly true for NbTi fusion existing CICCs as it is presented in Table 4.6.7, where it can be seen that the Stekly parameter is substantially higher than 1. Respecting the Stekly criterion will have led to copper to non-copper ratios in the range of 6–7.

4.6.3.3 **Current Densities in Cable in Conduit Conductor**

According to the design criteria given in Section 4.3.1, leading to A_{noncu}, A_{Cu}, and A_{He}, it is possible to estimate the achievable current density in a typical Nb_3Sn or NbTi CICC [20]. This current density is logically a function of B_{tmax} and of ΔT_{margin} as presented in Figure 4.6.8 for Nb_3Sn. The change in the slope of $J(B)$ is due to the fact that, for manufacturing reasons, the copper strand section cannot be lower than the non-copper section. At fields higher than this change the cable current density is driven by the temperature margin and no longer by the hot spot criterion. As it has been mentioned, the overall current density in the inner leg of ITER TF is in the range of 10 A mm^{-2} dominated by the structural components (steel); the overall current density in the CICC, as a rule, should be kept above, say 50 A mm^{-2}, without influencing the size of the tokamak. This rule, according to ΔT_{margin}, imposes in the project a maximum field not exceeding B_{tmax}, as shown in Figure 4.6.9.

Based on ITER TF CICC, the relative occupations of the different components of the CICC are shown in Table 4.6.8, highlighting the dominating role of the structures in the overall current density and the relative overall small proportion of the superconducting material (Nb_3Sn): 3.8%.

4.6.4
Quench Protection and Quench Detection in Fusion Magnets

4.6.4.1 **Specific Solution of Quench Protection for Fusion Magnets**

For general considerations about quench and quench protection, see Chapter 4 of the handbook [21]. Although all precautions (margins) have been taken in ITER to avoid any quench during plasma operation, this event cannot be totally excluded. The specific solution used for quench protection in fusion involves dumping the

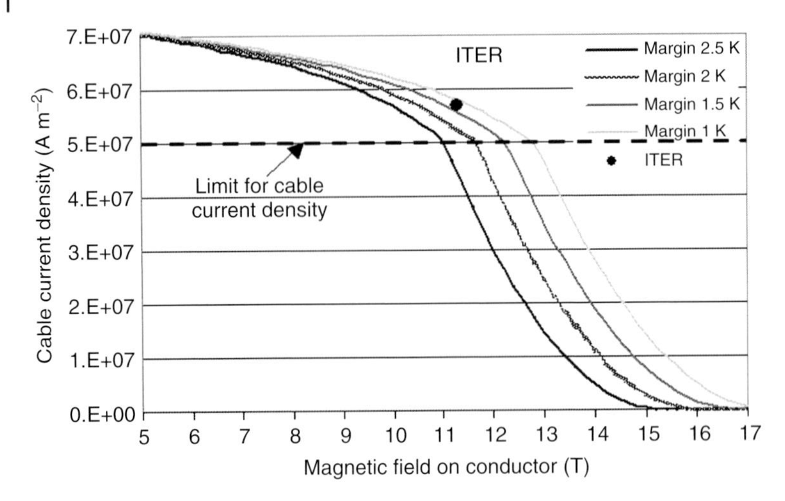

Figure 4.6.9 Influence of B_{tmax} on Nb_3Sn TF cable current density (discharge time constant: $\tau = 11$ s, detection delay $\tau_{da} = 2$ s).

Table 4.6.8 Overall current density in ITER TF CICC inner leg.

Type of material in the cable ($\Phi = 39.7$ mm)	Section in cable (mm^2)	Relative occupation in CICC/TF inner leg
Helium	422.9	30%/6.8%
Total copper	508.3	41%/8%
Non copper	235.3	19%/3.8%
Wrappings, spiral	71.3	6%/1.1%
Total CICC	1238	100%/19.7%
Total structures and insulation	4950	–/20.3%
J_{cable} (A mm^{-2}) 68 kA	55 A mm^{-2}	
J_{noncu} (A mm^{-2}) 68 kA	290 A mm^{-2}	
J_{strand} (A mm^{-2}) 68 kA	91 A mm^{-2}	
$J_{overall}$ (A mm^{-2}) 68 kA	11 A mm^{-2}	

stored energy into external resistors. The resistance can be variable with temperature (ITER, W7-X) [22] to limit the voltage at current-breaker opening. The external resistors are interleaved with coil inductances, the natural subdivision for the TF system being one TF coil or a multiple of TF coils (two for instance in ITER TF). Couples of current leads, generally of high-temperature superconductor (HTS) type, have to be implemented according to the subdivision.

In case of a quench, after detection, a fast safety discharge (FSD) of the current is triggered (time constant τ) so as to extract the magnetic energy of the coil into external dump resistors and protect the coil. The detection and action time τ_{da}, which is the time between the quench initiation and the FSD triggering, must be sufficiently small to limit the temperature increase (hot-spot criterion) in

Table 4.6.9 Typical voltages to ground in ITER magnet system.

Coils	Energy (GJ)	Voltage to ground (normal plasma in operation)	Voltage to ground (safety discharge)
TF	41	0	4
CS	6	~10 kV	5
PF	4	~10 kV	3–5 kV

the coil and avoid any damage. Quench detection using voltage measurements is generally the fastest technical solution available, but, from the resistive voltage, a specific processing is required to discriminate the resistive voltage ΔV which has to be detected from the inductive voltage due to the magnetic field variations. The ITER PF and CS systems are pulsed superconducting systems. There is not much experience about protection of such pulsed superconducting systems.

In ITER and in fusion magnets, in general, the main defaults, which can be envisaged, are related to leaks (associated with the forced flow technology) and to high voltages. In addition, possible degradation under operation and cycling of Nb_3Sn CICC cannot be totally excluded.

4.6.4.2 High Voltages in Fusion Magnets During FSD and in Operation

The values of voltages to ground that are met in ITER are illustrated in Table 4.6.9. For DC magnets, like most of the superconducting systems and for the TF tokamak systems, significant voltages to ground only appear during an FSD. In ITER, and this is quite new, large voltages, even greater than those during an FSD, will exist in normal operation in CS and PF systems, especially during the initiation phase of the plasma.

The most serious electrical failure in superconducting magnets is arcing that can permanently damage part of the magnets. An arc can be induced for instance due to weak insulation, causing a short circuit followed by disintegration of the short circuit due to overcurrent and arc initiation.

The level of damage is related to the stored energy inside the coil: a large part of the energy can be dissipated in the arc. The configuration of fusion coils related to the problem of insulation failure is special and has imposed the use of CICC.

The insulating material around the conductor and around the coil WPs plays a crucial role. All high voltage insulation around the conductor must incorporate a true electrical barrier that could be tested before application of a filler material such as epoxy resin. Glass-epoxy on its own is not adequate as it may contain voids (which cannot be detected during manufacture), and insulation is then provided before impregnation only by the polyimide foils incorporated into the conductor insulation.

The WP is surrounded by insulation and the whole can be taken in addition into a stainless steel casing cooled by He. It is the case for all fusion TF systems but not

Figure 4.6.10 W7-X nonplanar TF coil showing the coil casing and the connection region on the right hand side. (Courtesy of W7-X.)

for the ITER PF and CS systems. For the PF and CS systems, the external insulation is painted with semi-conductive painting ensuring an equipotential voltage.

For ITER, a subassembly, the in-cryostat feeder (CF), connects a coil (electrical power, helium, instrumentation) to the main containment building. The CF contains, in particular, the two NbTi bus bar conductors of the coil. The bus bars are particular sensitive subcomponents operating at low field but generally submitted to high voltage to ground, and therefore special care has to be taken for these components. The bus bars connect the coil to the coil room connections through HTS current leads.

For fusion machines cooled with forced flow helium, the magnets are contained in a cryostat. All surfaces exposed to the cryostat vacuum have a hard ground connection. Currents flowing in potential ground shorts are limited by the grounding system of the power supply. One of the 50 nonplanar superconducting coils of W7-X is presented in Figure 4.6.10.

The casing is visible with its copper bands to improve the casing cooling. On the right hand side of the image, a gray zone is visible, merging from the rest of the magnet, which is associated with the connexions and the helium inlets and will exist also in ITER. The insulation is painted in a semi-conductive gray painting. During the fabrication and tests of the W7-X coils, different kinds of insulation failure were detected and eliminated in this zone after shipping back the magnet to the factory.

The insulating properties of helium are very dependent on the pressure. At very low pressure, typical of cryogenic vacuum, the insulating properties are very good. When the pressure increases, the insulation degrades, reaching the so-called Paschen minimum for a typical value of *pd* (pressure multiplied by the electrode distance).

4.6.4.2.1 **Normal Operation**

In normal operation, the coil vacuum in the cryostat is $10^{-5}-10^{-6}$ mbar. Below 10^{-3} mbar, the voltage breaks down in helium in the range of 10 kV, whatever the distance of the conductor to the ground shield, and an exact number is however difficult to give being dependent on surface properties and shape. This shows that in normal operation with vacuum, there is no risk of arc even in case of a breach in the insulation. In summary, the risk exists only in case of double fault: vacuum degradation + breach in insulation.

4.6.4.2.2 Quality Control During Coil Production

After fabrication, at room temperature, the magnets are generally tested in so-called Paschen condition (degraded vacuum) to detect possible crack or breaches in the insulation and avoid primary insulation defect. This method was pioneered in particular during the toroidal field model coil (TFMC) fabrication [16, 23] and then developed during the W7-X magnets fabrication, illustrating how this bad helium property can be positively used during fabrication.

4.6.4.3 The Quench Protection Circuit (QPC)

The core of a quench protection circuit (QPC) is composed of DC current breakers, see [24] for a review of the existing solutions. The DC current breakers remain closed in normal operation and, when activated, they open, commutating the current into a discharge resistor. The classical solutions are based on the use of existing industrial vacuum current breakers. This type of solution is more and more difficult to implement due to the simultaneous increases of the current and of the voltage.

In TS, the first large tokamak with superconducting TF magnets, the protection system (1.5 kA/3.5 kV during the commissioning) is based on AC three-pole circuit breakers (CB) (six in total) used to divert the DC current into the series resistors. The QPC in this case is a single commutation circuit.

In W7-X, a solution based on a DC industrial mechanical CB from the railway industry was evaluated as the best compromise in terms of availability, reliability, and cost for the design of quench protection units (QPU) rated 20 kA/8 kV. In this case, however, the high current rating requires bypass switches (BPSs) to sustain the continuous current. The QPC in this case is a double commutation circuit such as that presented very schematically in Figure 4.6.11. Before quench detection, the current is circulating in S_1, which is able to sustain 20 kA continuously but not able to open the current. The current in a first phase is commutated in S_2, which is not able to sustain the current continuously, but can open the current at

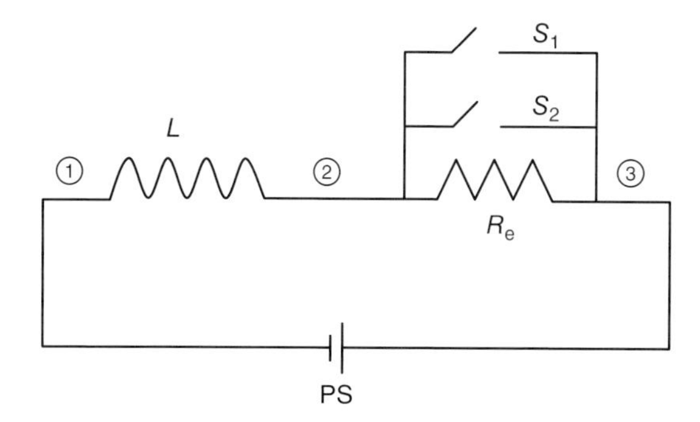

Figure 4.6.11 Double commutation circuit for W7-X.

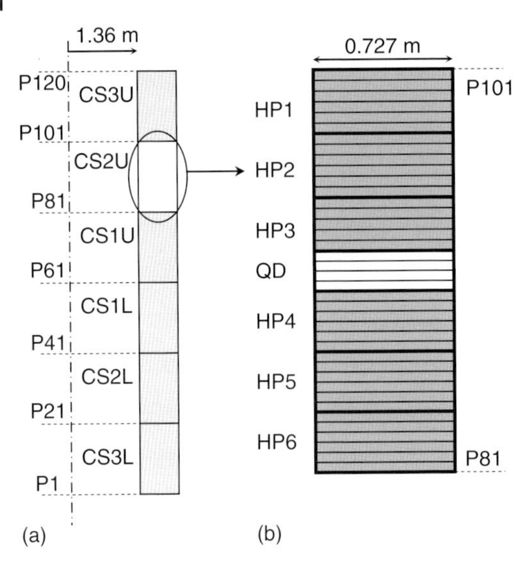

Figure 4.6.12 Overview of the (a) CS cross section and (b) the cross section of CS2U.

the rated voltage. Both switches are doubled in series for redundancy. In case of non-opening, an explosion fuse is activated.

In the case of ITER, the requirement (70 kA/24 kV) has still higher specifications than for W7-X. A current Commutating Unit has been developed based on the same type as the one of W7-X presented in Figure 4.6.12. The current opening is achieved in this case by means of the discharge of a counter-pulse capacitor that creates an artificial current zero in the vacuum circuit breaker (VCB) arc chamber.

For JT-60SA (25.7 kA/3.5 kV), the reference solution is a quite new solution based on CB made with static devices and not based on current breakers. It will use semiconductors controllable both at turn-on and turn-off such as gate turn-off (GTO) thyristors, insulated gate bipolar transistors (IGBTs), or integrated gate-commutated thyristors (IGCTs). The use of these semiconductors is considered as attractive, since current interruption is very fast, arc-less, does not require counter-pulse network, and static CBs are almost maintenance-free.

4.6.4.4 Quench Detection

4.6.4.4.1 Mitigation of the Inductive Part of the Voltage

The voltage across the coil or a part of the coil writes:

$$U(t) = R_{\text{quench}}(t) \cdot i(t) + L\frac{di(t)}{dt} + \sum_{j} M_j.\frac{di_j(t)}{dt} \tag{4.6.5}$$

where R_{quench} is the resistance of the normal zone to be detected, $i(t)$ is the current carried by the quenching sub-element, L is the self-inductance of the coil, and M_j and $i_j(t)$ refer to the mutual inductances between the quenching sub-element and the other magnetic field generating elements with their associated current.

Note that, even if it is more important in this case, the inductive voltage is not restricted to pulsed coils in magnet systems of fusion. Voltages are induced in the TF systems during plasma discharge and other plasma events. To discriminate the resistive voltage from the inductive voltage and obtain ΔV, the usual way is to balance the coil's voltage with another "symmetric coil."

In the TF system of tokamaks, there is an intrinsic symmetry between the coils constituting the TF system. It is possible to use this symmetry to balance the inductive voltage, using simply two TF coils, with a system similar to the classical bridge circuit. In this case, the resistances of the bridge R_1 and R_2 can be very near from each other. Such a system is used for the TF system of TS and of KSTAR and will be used for the TF systems of JT-60SA and SST-1. R_1 and R_2 are adjusted during loading of the TF system in the commissioning phase.

For the CS system of ITER, it is not possible to use this method as the six modules constituting the CS are not magnetically identical, being powered separately and being magnetically coupled differently to the other pulsed systems such as the PF system or the plasma. Several systems are being studied; a possible one consists in opposing the voltage across each of the 60 DPs to the average of the two neighboring DPs (see Figure 4.6.12). Thus, the central difference average voltage (CDA) resistive detection voltage ΔV is associated with the monitoring of each of the DPi. Balance coefficients α and β can be used to compensate for magnetic dissymmetries of sub-elements, which are not negligible among the ITER CS modules.

$$\Delta V_{\mathrm{DPi}} = V_{\mathrm{DPi}} - \frac{(\alpha V_{\mathrm{DPi+1}} + \beta V_{\mathrm{DPi-1}})}{2} \tag{4.6.6}$$

Each module is divided into six hexa-pancakes and one quad-pancake. Using a code providing precise magnetic field calculations, it is possible to predict the residual inductive signal from Eq. (4.6.6) along a reference scenario of ITER. The residual maximum signal during a reference scenario can be estimated, helping to select the threshold level V_{t} (see Section 4.6.4.4.2).

The Co-Wound Tape The most effective system to balance the inductive voltage of the circuit to be monitored is to use a co-wound tape, which follows exactly the same path as the conductor and will see the same flux variations. The co-wound tape has to be carefully insulated from the jacket of the conductor. The principle is shown in Figure 4.6.13. The residual inductive voltage can be very low as the inductive coupling between the conductor and the tape can be considered as nearly perfect. The drawback of such a solution is linked to the difficulties introduced during the conductor fabrication and the risk regarding insulation. It cannot be also avoided that the co-wound tape picks some magnetic signal different from the one of the conductor associated with PF variations during plasma current increase.

4.6.4.4.2 The Main Parameters of the Quench Detection

Quench detection using voltage measurements is likely to be the fastest technical solution available, but, to obtain the resistive voltage which has to be detected,

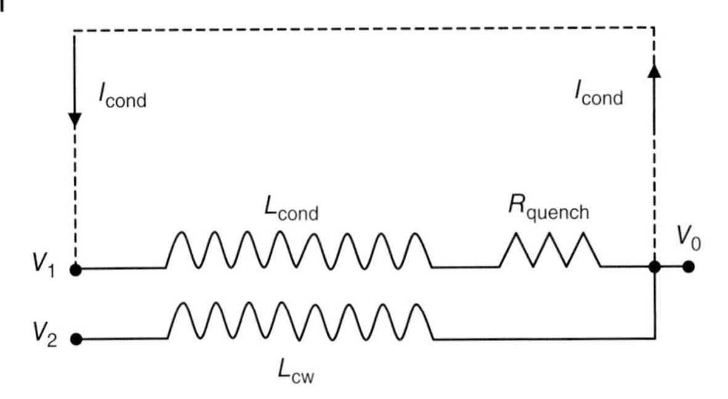

Figure 4.6.13 Monitoring of a quench detection system using a co-wound tape $\Delta V = V_2 - V_1 = R_{quench} I_{cond}$.

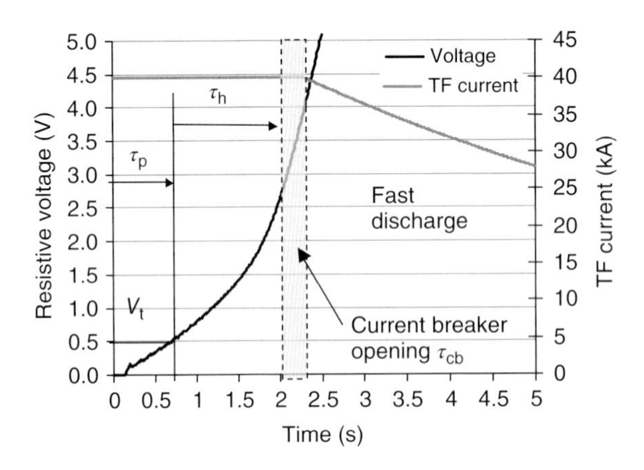

Figure 4.6.14 Main parameters of the quench detection illustrated for a typical resistive signal in ITER TF.

a specific processing is required to discriminate the inductive voltage due to the variations of the magnetic field. The aim is to avoid any false quench detection, which induces a severe perturbation of the system. The problem exists in particular in large pulsed large superconducting systems. This makes the detection particularly difficult.

The main phases of the detection process are illustrated in Figure 4.6.14 for a signal typical of the ITER TF system and starting from a quench initiation, highlighting the values, which have been selected for the detection, which are as follows:

- The voltage threshold V_t above which the quench is detected.
- The holding time τ_h during which V_t has to be continuously exceeded before opening of the current breaker and initiation of the FSD.

The adjustment of V_t is very much linked to the level of the expected residual inductive signal. A very important parameter with this respect is the propagating time τ_p to reach V_t. The total detection and action time, before the initiation of the FSD, is the sum of three terms: τ_p, the holding time τ_h and τ_{cb} the time to open the current breaker. The total detection and action time is: $\tau_{da} = \tau_p + \tau_h + \tau_{cb}$. It corresponds to the time during which the current stays constant in the conductor after quench initiation. Taking into account τ_{da} and τ, it has to be checked for consistency that the classical adiabatic hot-spot criterion is respected. It imposes, for instance, in ITER design, not to exceed 250 K at the end of the FSD, at the point where the quench has been initiated.

4.6.4.4.3 **Quench Propagation in CICC**

Propagation of quench in superconducting magnets has been extensively studied in the "adiabatic" case relevant to magnetic resonance imaging and nuclear magnetic resonance impregnated magnets. However, propagation in an ITER magnet is not "adiabatic." ITER magnets are made of CICC, where the strands are in tight contact of helium, which reduces the propagation velocity and makes the detection more difficult. In the early stage of the propagation, the joule heating generated in the normal zone is transferred into the helium, which consequently expands. This expansion of helium is the vector of the propagation in two directions; a propagation velocity v_p (one front) can be classically associated.

The velocity plays a very important role, as the resistance of the normal zone and the corresponding resistive voltage are the vectors of the primary quench detection. Interesting simplified analytical models have been developed 15 years ago [25], but it is now clear that only dedicated codes such as those given by Gandalf [26] and Vincenta [27] can help to estimate the propagation quench velocity v_p in the early phase of the quench (typically the first second). The propagation velocity is a function of the following parameters, such as the volumetric power in the conductor after quench (ρJ^2), the temperature margin, the initial quenched length L_q, the deposited energy, and power.

The resistive voltage V_t at time t after quench initiation is: $V_t \sim \rho J\ (L_q + 2v_p t)$.

Taking the reference event of a quench initiated in the inner leg of the ITER TF, with 1 m initial quenched length within 0.1 s (duration of a plasma disruption), such a quench is modeled using Gandalf. After 0.8 s, a typical resistive voltage of 0.5 V is reached corresponding to 7 m of quenched length (Figure 4.6.15). This corresponds to a velocity (one front) of 4 m s^{-1} which is slow in comparison with the adiabatic estimation in the range of 11 m s^{-1}.

4.6.5
Prospective about Future Fusion Reactors: DEMO

The next step after ITER will be DEMO, a fusion reactor that should typically provide between 500 and 1000 MW electrical power (P_{elec}) [20]. Its construction could start after ITER has delivered significant results, expectedly 20 years, hence typically in 2030.

There is now an important European activity related to DEMO and coordinated by EFDA. The objectives of DEMO are not yet completely determined and are

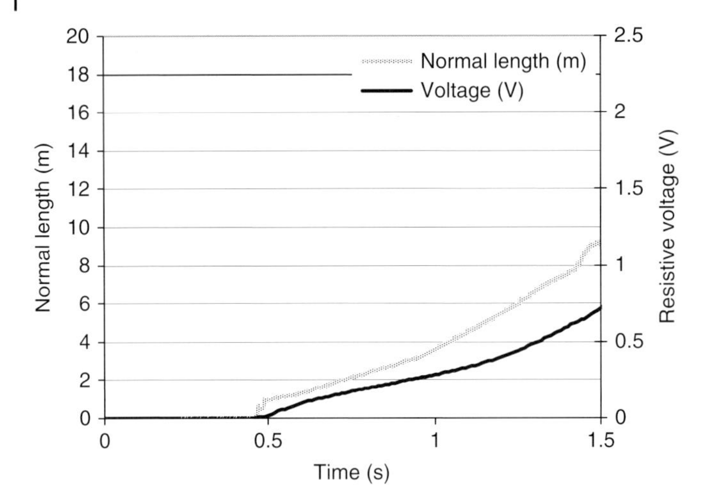

Figure 4.6.15 Quench propagation in the inner leg of the ITER TF system (Gandalf simulation).

in discussion: production of electricity, reactor near ignition, tritium autonomy, limited fusion power (P_{fus} in the range of 1000–2000 MW).

Up to now, the reference design for DEMO studies was a steady-state machine with a noninductive drive of the plasma current. However, in this option, the circulating power, which is the electrical power needed to operate the machine itself, is very high.

In a pragmatic approach, a pulsed version (~2–3 h) of DEMO is now considered with a limited fluence. In comparison with the steady-state machine, the pulsed version gives more importance to the superconducting CS, which has to be larger (see Section 4.6.2.3).

For the pulsed version, the plasma discharge is in inductive mode of the same kind as the ITER reference plasma discharge. Therefore, the observation of ITER operation not only regarding the plasma physics but also regarding the impact of operation on the magnet systems will be crucial for the machine final design. However, the plasma duration is longer than in ITER and the machine should perform near ignition ($Q > 100$). This can be achieved mainly by increasing R up to typically 9 m to increase P_{fus} and Q and accommodate a larger CS for a longer pulse. The final parameters and the objectives are still in discussion, and in particular the aspect ratio A which impacts B_{tmax} (see Eq. (4.6.1)). The present considered B_{tmax} is in the range of 12 T like in ITER corresponding to $A = 3.5$.

4.6.5.1 Which Superconducting Material for DEMO?

HTS materials have been considered for DEMO, but the main option for the TF system is presently Nb_3Sn, like in ITER. The reasons for this choice (in addition to the present nonindustrial maturity of HTS large conductors for fusion) are the following:

- The magnetic field of DEMO TF is in the same range as in ITER.

Table 4.6.10 Expected cryogenic power in DEMO (extrapolation from ITER).

	DEMO (5 K)Nb_3Sn (MW)	DEMO (20 K)HTS (MW)
Cryogenic power magnets	12	2.5
Cryogenic power (thermal shields) 80 K	15	15
Total electrical power for cryogenics	27	17.5
Benefit from HTS	0	9.5

- The estimated electrical power of the DEMO magnet refrigerator is about 27 MW and the expected gain is only about 9.5 MW for a TF system operating at 20 K with HTS (see Table 4.6.10). This gain is small in comparison with the circulating electrical power necessary to cool the blankets with He, which is estimated around 275 MW.
- There is no expected gain in TF current density (and therefore in tokamak size) because the current density is driven by the structures (see Section 4.6.3.3).

In fact, the challenge for DEMO magnets is to achieve a more robust simpler industrial design as for ITER magnets. One of the objectives is certainly to simplify the hydraulics. This can be achieved by reducing all kinds of losses, in particular nuclear losses, by improving the neutrons shielding toward a zero losses magnet playing with the radial extension Δ_{int}. A necessary path is also to increase ΔT_{margin} up to 2 K in comparison with 0.7 K in ITER TF system. On the long term, HTS materials, when mature, could help to progress in this direction. HTS materials could be then envisaged at 20 K in relationship with new types of cryoplants based on conduction cooling where large temperature margins are essential or even with hydrogen cooling.

4.6.6 Conclusion

Today, superconductivity has invaded all large fusion projects, representing typically 30% of the cost investment of the machines. The attention now focuses on the commissioning of W7-X, which will come soon in 2014 and will represent a major step. The conductor rolling has already started in Japan and in Europe for the different magnets of the large fully superconducting JT-60SA tokamak. The cold test of the 18 TF coils will start at CEA Saclay in France in 2015.

But the main challenge is the construction of the ITER magnet system. While the procurement of the conductor is now in a very advanced stage, the construction of the different magnet systems is just starting. This construction is a real challenge by the size, the weight of the components, and the heat treatment, which complicates the fabrication process. The test content for commissioning a large system like the ITER magnet system, minimizing risks, is certainly an issue: insulation, leaks, and quench detection adjustment. The observation of the superconducting machines in operation (EAST, KSTAR) or in commissioning

(W7-X, JT-60SA) is certainly crucial and can help for ITER commissioning preparation.

References

1. Tsuei, C.C. and Kirtley, J.R. (2012) *100 Years of Superconductivity Development*, CRC Press.
2. Rose, D.J. (1960) Energy balance in a thermonuclear reaction. *Bull. Am. Phys. Soc. Ser. II*, **5**, 367.
3. Serio, L. (2010) Challenges for cryogenics at ITER. *Adv. Cryogenic Eng. Trans. CEC*, **55**, 651.
4. Duchateau, J.L., Journeaux, J.Y., and Gravil, B. (2009) Tore Supra toroidal superconducting system. *Fusion Sci. Technol.*, **56**, 1092.
5. Motojima, O. *et al.* (2006) Progress of plasma experiment and superconducting technology in LHD. *Fusion Eng. Des.*, **81**, 2277.
6. Wu, S. and the EAST Team (2007) An overview of the EAST project. *Fusion Eng. Des.*, **82**, 463.
7. Oh, Y.C. *et al.* (2009) Commissioning and initial operation of KSTAR superconducting tokamak. *Fusion Eng. Des.*, **84**, 344.
8. Wegener, L. (2009) Status of Wendelstein 7-X construction. *Fusion Eng. Des.*, **84**, 106.
9. Pradhan, S. and SST-1 Mission Team (2010) Status of SST-1 refurbishment. *J. Plasma Fusion Res. Ser.*, **9**, 650.
10. Matsukawa, M. *et al.* (2008) Status of the JT-60SA tokamak under the EU-JA broader approach agreement. *Fusion Eng. Des.*, **83**, 795.
11. Mitchell, N. *et al.* (2012) ITER magnet design and construction status. *IEEE Trans. Appl. Supercond.*, **22**, 2019808.
12. Chernoplenkov, N.A. (1993) The system and test results for the Tokamak T-15 magnet. *Fusion Eng Des.*, **20**, 55.
13. Beard, D.S. *et al.* (1988) The IAEA large coil task. *Fusion Eng. Des.*, **7**, 240.
14. Devred, A. *et al.* (2012) Status of ITER conductor and production. *IEEE Trans. Appl. Supercond.*, **22**, 4804909.
15. Bruzzone, P., Anghel, A. *et al.* (2002) Upgrade of operating range for SULTAN test facility. *IEEE Trans. Appl. Supercond.*, **12** (1), 520.
16. Ulbricht, A., Duchateau, J.L. *et al.* (2005) The ITER toroidal field model coil project. *Fusion Eng. Des.*, **73**, 189.
17. Ando, T. (2002) Pulsed operation test results of the ITER CS model coil and CS insert. *IEEE Trans. Appl. Supercond.*, **12**, 496.
18. Duchateau, J.L. (2009) New considerations about stability margins of NbTi cable in conduit conductors. *IEEE Trans. Appl. Supercond.*, **19** (Suppl. 19), 55.
19. Stekly, Z.J.J. and Zarr, J.L. (1968) Stable superconducting magnets. *IEEE Trans. Nucl. Sci.*, 367.
20. Duchateau, J.L. (2013) Conceptual design for the superconducting magnet system of a pulsed DEMO reactor accepted for publication in. *Fusion Eng. Des.*, **88**, 160.
21. Duchateau, J.L. *et al.* (2013) Quench detection in ITER superconducting systems accepted for publication in. *Fusion Sci. Technol.*, **64**, 705–710.
22. Monnich, T.H. and Rummel, T.H. (2006) Production and tests of the discharge resistors for wendelstein 7-X. *IEEE Trans. Appl. Supercond.*, **16**, 1741.
23. Fink, S. *et al.* (2002) High voltage tests of the ITER toroidal field model coil insulation system. *IEEE Trans. Appl. Supercond.*, **12**, 554.
24. Gaio, E. *et al.* (2009) Conceptual design of the quench protection circuits for the JT-60SA superconducting magnets. *Fusion Eng. Des.*, **84**, 804.
25. Shajii, A., Freidberg, J.P., and Chaniotakis, A. (1995) Universal scaling laws for quench and thermal hydraulic quenchback in CICC coils. *IEEE Trans. Appl. Supercond.*, **5** (2), 477.
26. Bottura, L. (1996) A numerical model for the simulation of quench in the ITER magnets. *J. Comput. Phys.*, **125**, 26.
27. Takahashi, Y. *et al.* (2007) Stability and quench analysis of toroidal field coils for ITER. *IEEE Trans. Appl. Supercond.*, **17** (2), 2426.

4.7 High-Temperature Superconducting (HTS) Magnets

Swarn Singh Kalsi

4.7.1 Introduction

Large electrical magnets currently are used in a variety of industrial and military settings [1]. The applications range from medical uses to process manufacturing and purification to scientific research. Manufacture of such magnets with high-temperature superconducting (HTS) materials looks attractive. Properties of HTS materials such as BSCCO-2212 ($Bi2Sr_2CaCu_2O$), BSCCO-2223 ($Bi_2Sr_2Ca_2$-Cu_3O), rare-earth-barium-copper-oxide (ReBCO), and MgB_2 (magnesium diboride) are very attractive at low temperatures (<20 K) and ReBCO-coated conductors that are available commercially now are suitable for making useful devices cooled at higher temperature of liquid nitrogen (~77 K). Compared with conventional copper magnets, the HTS magnets have the following benefits:

- *Smaller and lighter* – Reduced size and weight of coils by as much as 40–80% compared to those made with copper wire.
- *More efficient* – Higher energy efficiency and lower operating cost than copper-based systems.
- *Higher magnet fields* – The high current density in HTS wires enables higher magnetic field magnets.
- *Greater thermal stability* – Coils and magnets using HTS wire operate in a stable "cold environment."
- *Longer magnet life* – The "cold cryogenic environment" within which HTS magnets operate eliminates a common source of product failure, that is, heat.
- *Easier cooling* – Because of significantly higher operating temperature of the HTS versus low-temperature superconductor (LTS) materials, the HTS magnets operating at >20 K are much simpler and easier to cool than the LTS magnets operating at about 4 K.

HTS magnets are commercially available today. Experience with the initial applications, which were in the military and scientific domains, is expanding applications to address other markets. Currently available ReBCO-coated conductors provide materials with sufficient critical current density to enable a broad variety of applications.

4.7.2 High-Field Magnets

American Superconductor Corp. (AMSC) [2], Sumitomo [3], and others have built conduction-cooled HTS magnets operating in the 20–30 K temperature range. These systems offer the advantages of high operational stability and the ability to

Figure 4.7.1 AMSC 7 T conduction-cooled magnet.

ramp very quickly. The higher cost and lower performance of HTS material at 20 K compared to LTS material at 4 K is limiting the commercial exploitation.

A 7.25 T laboratory magnet [2] utilizing BSCCO-2223 conductor was built by AMSC for the Naval Research Laboratory in 1998. Operating at 21 K at full field, the magnet provided field homogeneity of ±1% in a 2 inch warm bore. The system was conduction-cooled with a pair of Leybold single-stage cryocoolers that allowed cooldown in <36 h and allowed extended fast ramp operation. The HTS current leads, employed in the magnet, facilitated the operation with a total refrigerator input power of 6 kW. The fully integrated system consisted of the magnet, cryogenic system, control, and protection system, and power supply is shown in Figure 4.7.1.

Table 4.7.1 summarizes the major features of the magnet. It generated a magnetic field >7 T in a 2 inch warm bore. Field homogeneity of 1% was specified within a 2 inch diameter spherical volume (DSV) and 2% within a 2 inch diameter cylindrical and 2 inch tall volume. The magnet HTS winding is conduction-cooled and operates at about 25 K. The magnet employs HTS current leads between 25 K magnet and a 45 K intercept temperature. Conduction-cooled leads are employed between 45 K and room temperature. Two single-stage Leybold cryocoolers driven from a single compressor provided refrigeration at the two temperatures.

The magnet was capable of ramping between zero and full field in 240 s on a continuous basis. It was powered with a four quadrant power supply to permit seamless operation between +7.25 and −7.25 T. A suitable protection system protected it against abrupt quenches and any other unintended operational modes. All components of the magnet system could withstand a shock loading of 6 G in any direction. The magnet cryostat was designed to support experimental equipment weighing up to 250 lb.

Table 4.7.1 AMSC 7 T conduction-cooled magnet features.

Parameter	Unit	Value
Peak field in the bore	T	>7
Field homogeneity	%	1
Useful field volume at room temperature	—	—
Diameter	inch	2
Length	inch	2
Room-temperature bore diameter	inch	2
Operating temperature	K	25
Cooling method	—	cond.
Ramp time, zero to full field	s	240
Experiment weight	lb	250
Shock load withstand capability	G	6
Power supply	—	4 quad

This magnet demonstrated that high-field magnets operating above 20 K could be built using the BSCCO-2223 conductor. It also demonstrated that this conductor provides significant performance advantages for fast ramp magnets or magnets that require high external heat loads.

Sumitomo Electric of Japan also built an 8 T conduction-cooled magnet shown in Figure 4.7.2 with their DI-BSCCO (Dynamically Innovative bismuth-based HTS wire). The magnet had a 200 mm-diameter bore at room temperature and was tested up to 8.1 T. The design study showed that the higher magnetic field (15 T) could be achieved within about the same envelope of the above said magnet. This magnet could be used in various industries such as biomedicine, semiconductor, and environmental industries. Table 4.7.2 lists key features of this magnet.

HTS insert magnets [4] for very high field accelerator magnets are also being considered. Future accelerator magnets will need to reach a magnetic field in the 20 T range. To attain such large magnetic field is a challenge only possible with the use of HTS materials. The high current densities and stress levels needed to satisfy the design criterion of such magnets make ReBCO superconductor the most appropriate candidate.

Brookhaven National Laboratory (BNL) attempted [5] to build HTS coils capable of producing fields >20 T when tested alone and approaching 40 T when tested in a background field magnet. The solenoid was made with ReBCO high engineering current density HTS tape. It had 17 HTS pancake coils and was tested in the temperature range from 20 to 80 K. Quench protection, high stresses, and minimization of degradation of conductor were some of the major challenges associated with this program. The use of ReBCO HTS in high-field magnets is very attractive due to its ability to deliver large engineering current density at very high fields, its ability to handle large Lorentz forces, and its ability to be wound into coils with small radii.

Figure 4.7.2 8 T conduction-cooled magnet built by Sumitomo.

Table 4.7.2 Sumitomo 8 T conduction-cooled magnet features.

Cooling method	**Cryocooler conduction cooling**	**Cryocooler conduction cooling**
Maximum magnetic field	8 T	15 T (example)
Bore diameter at room temperature	200 mm	200 mm
Magnet vessel size (width depth height)	900 mm × 600 mm × 540 mm	1000 mm × 700 mm × 600 mm
Weight	300 kg	500 kg

4.7.3 Low-Field Magnets

Majority of applications of interest are in low-field area. Such magnets have been built for a variety of applications such as minesweeping [6] and degaussing coils [7, 8] for Naval ships, superconducting magnetic energy storage (SMES) for power system stabilization [9–11], magnetic separation [12], beam bending [13–15] in accelerator and synchrotron rings, crystal growth [16, 17], and induction heating [18]. Even HTS materials are being considered for magnetic resonance imaging (MRI) magnets [19], which employed only NbTi so far. Some of these applications are described below.

4.7.3.1 Magnetic Separation

Reciprocating magnetic separators are used in the purification of kaolin clay and titanium dioxide. Kaolin clay is a white pigment with a wide variety of applications and used extensively in paint, paper, and plastics industries. High gradient magnetic separators (HGMSs) have been used in the benefaction of kaolin clay and other minerals for over 30 years. In HGMS, ferromagnetic wire mesh placed into an external magnetic field generates high magnetic field gradients in their surroundings. When magnetically susceptible particles are introduced into such an environment, they move along the gradient toward the wire and are captured at distinct regions on the wire's surface. Detailed discussions and mathematical descriptions of magnetic separation, especially HGMS, could be found elsewhere [20].

A team consisting of DuPont Superconductivity of Wilmington, DE, the Carpco division of Outokumpu in Jacksonville, FL, and the National High Magnetic Field Lab (NHMFL) in Tallahassee, FL designed and fabricated a prototype conduction-cooled HTS coil for use in a reciprocating magnetic separation unit (RMSU) with a warm bore size of 200 mm (Figure 4.7.3). The HTS coil for the RMSU 200 was 0.3 m in length with a 0.25 m cold bore. The central operating magnetic field was a nominal 3.0 T at an operating current of 126 A. The HTS

Figure 4.7.3 The completed HTS pancake coil assembly.

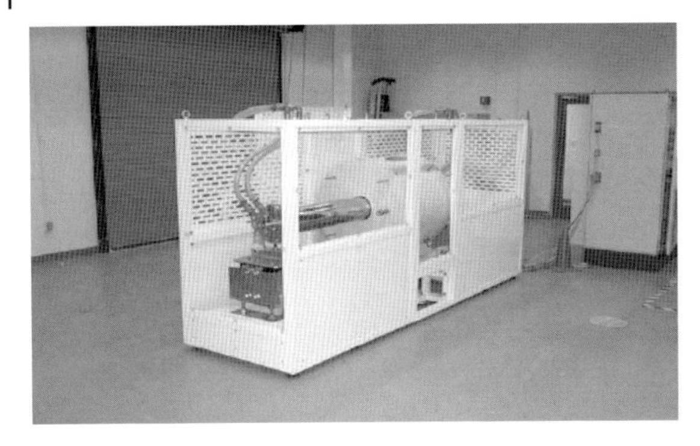

Figure 4.7.4 The reciprocating magnetic separation unit "RMSU 200" at DuPont Superconductivity's magnet test facility in Wilmington, DE, USA.

coil was conduction-cooled with a single stage Gifford–McMahon (G–M) cryocooler, with a nominal operating temperature of 30 K. In terms of combined size and magnetic field strength, this was one of the largest HTS coils produced in 2002. The HTS coil was installed at the DuPont Superconductivity magnet test facility in Wilmington, DE, where it was used as the centerpiece for a pilot-scale test facility (Figure 4.7.4) investigating mineral separations, waste remediation, and other novel chemical processing. This project was successfully finished in August 2002. Fabrication details and test results of this RMSU magnet are presented elsewhere [12, 21].

Later, the team [22] consisting of DuPont, Outokumpu, and the J. M. Huber corporation developed a larger HTS coil for using in a prototype HGMS separator, called *RMSU 500*. The HTS coil was 0.7 m in length and had a 0.5 m inner diameter. The central operating magnetic field was a nominal 3.0 T with a design operating current of 100 A. In terms of combined size and magnetic field strength, this is one of the largest HTS magnets ever fabricated, possessing a stored energy of 0.400 MJ. The HTS magnet was conduction-cooled with a G–M cryocoolers with a nominal operating temperature of 30 K. The HTS conductor used a stainless steel reinforced BSCCO-2223 material. Comparison between the pancake coils used in RMSU 200 and RMSU 500 units is shown in Figure 4.7.5. The RMSU 500 magnet was successfully built and tested [4].

HTS magnets for large-scale industrial processing offer three major advantages over their LTS counterparts:

- Can be operated without the need of a liquid cryogen.
- Can be operated in remote or nonindustrialized regions where access to liquid cryogens requires significant and costly logistics.
- Can be significantly lighter and more robust than their LTS counterparts and are thus road transportable, opening up the business concept of industrial tolling.

A team in Korea [23] also built a magnetic separator based on the conduction-cooled HTS magnet employing BSCCO-2223 pancakes, and had inner and outer coil diameters of 120 and 212 mm, respectively, and coil height of 111.8 mm. The

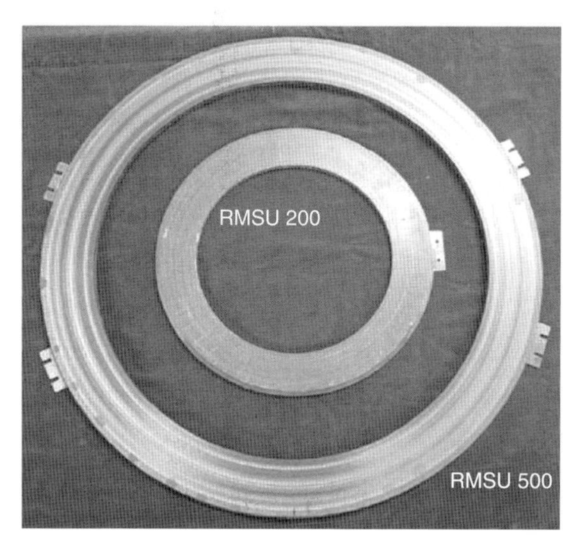

Figure 4.7.5 Comparison of BSCCO-2223 pancake coils for the RMSU 200 and the RMSU 500 units.

magnet generated a magnetic field of 3.22 T in the warm bore with HTS coil operating at a temperature of 10 K. The separating efficiency for wastewater in the iron-steel factory was about 84.1%. This team has recently built another similar magnet [24] with ReBCO conductor.

Another KERI team [25] developed a laboratory-sized HGMS magnet system for cleaning wastewater from steel-manufacturing factories. The coolant of hot roller at steel-making factory includes suspended solid (SS) like steel fines, iron oxide, and emulsified oil. The HTS magnet had eight double-pancake sub-coils and was cooled with a G–M cryocooler. It had a 70 mm room temperature bore and 250 mm of height. This magnet was also used for studying the feasibility of cleaning water from a paper mill [26].

Following the Tsunami in 2004, a Japanese team [27] investigated the possibility of cleaning radioactive soil around the damaged nuclear plants using HGMS magnets. They built a 150 mm diameter and 180 mm long HTS magnet that generated a peak field of 5 T at 20 K. The soil cleaning experiment looked promising and plans were being developed for a large-scale system employing HTS magnets.

4.7.3.2 **Crystal Growth**

Not much literature is available for applications of HTS magnets for crystal growth. In 2001, a cryocooler-cooled HTS magnet model coil was built [15] for Si single crystal growth application for investigating basic characteristics needed for the target magnet design. The following results were obtained:

- In the mechanical characteristic measurements of the model coil, it was confirmed that critical current degradation did not occur at the hoop stress expected in the target magnet.

Figure 4.7.6 An HTS split magnet for silicon single-crystal growth applications.

- The model coil was operated continuously at 290 A, which stored about 50 kJ of electromagnetic energy.
- AC losses of the model coil measured by an electrical method and a calorimetric method were in good agreement with theoretical values. Hysteresis losses dominated during the charging mode.
- This cryocooler-cooled model coil was successfully operated with an AC current of 150 A, 0.1 Hz.

Subsequent to this, a full-scale magnet was built [16] using BSCCO-2223 wire. This was a split coil system (Figure 4.7.6) comprising two coils, each consisting of 18 pancakes. Each coil had an outer diameter of 1.2 m and a thickness of 0.1 m and was operated at 20 K with G–M-type cryocoolers with 3.3 kW compressor for each cooler. The cooldown of the coil was completed within 480 h with temperature difference among pancakes maintained at <18 K. In April 2001, this split coil was successfully energized up to rated current of 210 A at 20 K without a quench. The stored energy of this split coil reached 1.1 MJ at 216 A, 20 K, and the fastest charging time was 1 min.

4.7.3.3 Induction Heating

Conventional AC induction heating has been used in industry since the 1920s. It is widely used [28] to heat up aluminum and copper billets before extrusion. Resistive induction heating systems, which typically have the total efficiency of 50–60%, are generally used with AC current. By utilizing superconductivity, the total efficiency of induction heating system could be increased to ~90%. If the heating power of conventional induction heaters could be reached with a superconducting device, resistive systems could be replaced with superconducting ones, and remarkable savings would develop in the long run. The HTS induction heater concept is based on DC magnets. Billets are rotated in the DC field created with the HTS magnets.

In 1990, this concept got attention [18, 29] for DC induction heating using strong electromagnets. The magnet-wire and motor drive technologies available

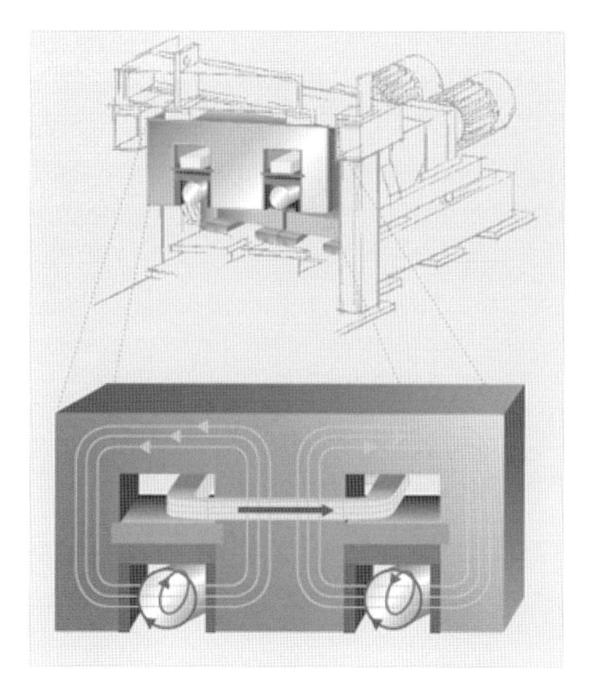

Figure 4.7.7 Induction heater concept with HTS excitation coil. (Courtesy of Zenergy Power.)

at the time, however, did not permit an economical embodiment of the concept. With the emergence of both HTS as a commercially available conductor and advances in solid-state electric motor drive equipment, this almost 20-year-old concept became a commercially viable product [28]. Figure 4.7.7 shows an induction heater concept powered with a DC HTS coil for heating aluminum or copper extrusion billets. The potential for efficiency improvements is substantial because conventional copper coil induction heaters rated for up to around 1 MW operate with an overall efficiency of typically only 50–60%. The efficiency of a HTS induction heater was expected to approach 90%.

The superconductor used for HTS coils was a BSCCO-2223/Ag stainless steel reinforced tape with a cross-section of 0.27 mm × 4.2 mm. The 77 K self-field critical current was 125 A and the critical current at 40 K and with a magnetic field of 1 T applied perpendicular to the surface of the tape was about the same. The tape was wound into a racetrack type coil providing a warm bore of 750 mm × 400 mm. The coil was operated at temperatures from 20 to 40 K, with a normal operating temperature of 22–24 K obtained with two off-the-shelf cryocoolers – one cooler would have delivered the necessary cooling power but the use of two cryocoolers provided an additional safety margin in case of malfunction.

In the first heater, installed at the aluminum extrusion plant, 360 kW motor was employed for rotating the billets. The losses of the frequency converters of the motors were 2–3%, and the power consumption of the cooling system (including the power supply for the magnet) was about 13 kW. In total (including also the energy consumption of peripheral technical devices), the efficiency was >80%. This could be compared with the 50% efficiency for the conventional induction heater.

The convectional copper coil heaters operating from 50/60 Hz power source result in a very shallow skin depth for heating. In the HTS induction heater, a rather low rotational frequency was used, typically down to 4–12.5 Hz (corresponding to 240–750 rpm). Lowering the frequency meant increasing the skin depth, and the heat input became more evenly distributed. For example, a 60 Hz induction heater deposited 20% of the surface power within 15 mm from the surface, but the 4 Hz HTS induction heater deposited 20% of the surface power as far in as 50 mm. The deeper energy penetration and thereby the enhanced temperature homogeneity provided better preconditions for the subsequent extrusion, facilitated a higher processing speed, and eliminated the risk of local surface melting of the billet.

At the Weseralu aluminum extrusion plant in Germany, an HTS billet heater was installed in August 2008. During a period of 2 years, it heated 10 000 tons of aluminum, corresponding to about 350 000 billets. The machine was optimized for billets with diameters of 152–177 mm and a length of 690 mm. The capacity was 2.2 tons h^{-1}. Only 140 s was required for heating 152 mm × 690 mm billets. Since two billets were heated simultaneously, a billet was delivered to the extrusion press every 70 s. The target temperature of the heating process could be lowered due to the better temperature homogeneity. The lower billet temperature in turn enabled making complex profiles. Moreover, the quality of surface finishes was improved. The increase of productivity, directly attributable to the deployment of the magnetic billet heater, was found to amount to an average of 25% across a variety of profiles. The HTS heater was operated for 2 years without severe problems.

In summary, an HTS induction heater manufactured by Zenergy [28] and its technology partner began operation in 2008 and had revolutionized both energy efficiency and process flexibility in industrial aluminum, brass, bronze, and copper processing. In a precision heating process, HTS induction heaters soften raw material billets of non-ferrous metal in order to improve their ductility. The initial motivation for using HTSs in this application was energy savings. The magnetic billet heater met this expectation, with an energy efficiency increase from about 50% for conventional technology to more than 80%. More importantly, however, was a significant productivity increase. The first HTS heater was in operation for 2 years achieving a 25% increase in throughput compared to conventional induction heaters. A payback time of <2 years was estimated using the combined effect of energy savings and productivity improvements.

There has been no news about these heaters since 2010. It is not certain if this technology is still in vogue or has been supplanted by other technologies, like permanent magnet heaters.

4.7.3.4 Accelerator and Synchrotron Magnets

Synchrotron storage rings use a number of very large copper electromagnets, consuming millions of dollars of electricity per year and requiring substantial amounts of cooling water. HTS magnets could be employed to provide significant energy savings: modeling has indicated reduction by a factor of 20 for complete system-wide installations, with equally impressive savings in cooling water demand. The very high current density of HTS wire compared with copper allows more compact coil geometries, leading to greater design flexibility for the magnet, delivering greater optical access to the magnet working area. The HTS magnet advantage is available not just for new facilities: in many cases, existing copper coils can be retrofitted with HTS coils without modification to the iron yoke.

The world's first synchrotron magnet (Figure 4.7.8) fitted with HTS coils was shipped in 2009 from HTS-110, New Zealand to the BNL in New York. This HTS magnet uses less than half of the energy of a copper equivalent, along with substantially less cooling water. Currently, copper coils consume 15 kW of electricity and significant amounts of cooling water during operation. With each synchrotron ring operating 50 or more magnets, the energy usage for an entire copper ring is up to 1 MW.

BNL also attempted building quadrupole HTS magnets [13] for the Facility for Rare Isotope Beams (FRIBs). These magnets, in the fragment separator region of the FRIB, were to be subjected to very large heat loads (over 200 W) and an intense level of radiation (10 MGy per year) into the HTS coils. The prototype magnet concept is shown in Figure 4.7.9. Magnets made with HTS were advantageous over conventional superconducting magnets since heat loads could be removed more efficiently at higher temperatures. The design was based on ReBCO HTS coils operating at 50 K. This HTS material was found to be highly radiation tolerant. The

Figure 4.7.8 Synchrotron magnet employing HTS coils.

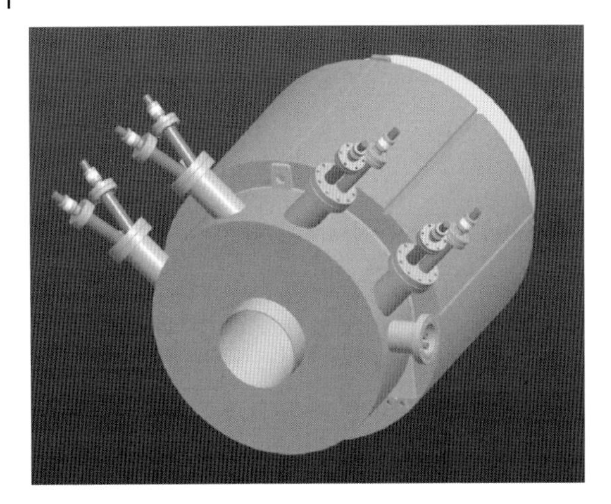

Figure 4.7.9 End isometric view of the assembled magnet showing 300 K yoke, cryostat, and exiting power leads and helium lines.

goal of this R&D program was to evaluate the viability of HTS in a real machine with magnets in a challenging environment where HTS offered a unique solution. This R&D program addressed these issues in a systematic manner.

4.7.4 Outlook

The technology for building HTS magnets is fully developed and ready for employment in many kinds of devices. These HTS magnets have shown to perform the intended function very efficiently in many industrial and research facilities. The capital cost of these magnets, mostly driven by the high cost of HTS materials and the cooling systems, is inhibiting their wider adaptation. More real-life applications will emerge once these magnets meet the economic goals of their users.

References

1. Kalsi, S.S. (2011) *Applications of High Temperature Superconductors to Electric Power Equipment*, IEEE Press\John Wiley & Sons, Inc., ISBN: 978-0-470-16768-7.
2. Snitchler, G., Kalsi, S.S., Manlief, M., Schwall, R.E., Sidi-Yekhief, A., Ige, S., Medeiros, R., Francavilla, T.L., and Gubser, D.U. (1999) High-field warm-bore HTS conduction cooled magnet. *IEEE Trans. Appl. Supercond.*, **2** (Pt. 1), 553–558. doi: 10.1109/77.783356
3. Ohkura, K., Okazaki, T., and Sato, K. (2008) Large HTS magnet made by improved DI-BSCCO tapes. *IEEE Trans. Appl. Supercond.*, **18** (2), 556–559 ISSN: 1051-8223, INSPEC Accession Number:10075303, doi: 10.1109/TASC.2008.920807.
4. Rey, J.-M., Devaux, M., Bertinelli, F., Chaud, X., Debray, F., Durante, M., Favre, G., Fazilleau, P., Lécrevisse, T., Mayri, C., Pes, C., Pottier, F., Sorbi, M., Stenvall, A., Tixador, P., Tudela, J.-M.,

Tardy, T., and Volpini, G. (2013) HTS dipole insert developments. *IEEE Trans. Appl. Supercond.*, **23** (3), 4601004.
5. Gupta, R., Anerella, M., Ganetis, G., Ghosh, A., Kirk, H., Palmer, R., Plate, S., Sampson, W., Shiroyanagi, Y., Wanderer, P., Brandt, B., Cline, D., Garren, A., Kolonko, J., Scanlan, R., and Weggel, R. (2011) High field HTS R&D solenoid for muon collider. *IEEE Trans. Appl. Supercond.*, **21** (3), 1884–1887.
6. Ige, O.O., Aized, D., Curda, A., Medeiros, R., Prum, C., Hwang, P., Naumovich, G., and Golda, E.M. (2003) Mine countermeasures HTS magnet. *IEEE Trans. Appl. Supercond.*, **13** (2, Part 2), 1628–1631. doi: 10.1109/TASC.2003.812811
7. Ige, O.O., Aized, D., Curda, A., Johnson, D., and Golda, M. (2001) Test results of a demonstration HTS magnet for minesweeping. *IEEE Trans. Appl. Supercond.*, **11** (1, Pt. 2), 2527–2530. doi: 10.1109/77.920380
8. Kephart, J.T., Fitzpatrick, B.K., Ferrara, P., Pyryt, M., Pienkos, J., and Golda, E.M. (2011) High temperature superconducting from feasibility study to fleet adoption. *IEEE Trans. Appl. Supercond.*, **21** (3), 2229–2232.
9. Nomura, S., Shintomi, T., Akita, S., Nitta, T., Shimada, R., and Meguro, S. (2010) Technical and cost evaluation on SMES for electric power compensation. *IEEE Trans. Appl. Supercond.*, **20** (3), 1373–1378.
10. Sander, M., Gehring, R., and Neumann, H. (2013) LIQHYSMES – a 48 GJ toroidal MgB2-SMES for buffering minute and second fluctuations. *IEEE Trans. Appl. Supercond.*, **23** (3), 5700505.
11. Zhang, J., Dai, S., Wang, Z., Zhang, D., Song, N., Gao, Z., Zhang, F., Xu, X., Zhu, Z., Zhang, G., Lin, L., and Xiao, L. (2011) The electromagnetic analysis and structural design of a 1 MJ HTS magnet for SMES. *IEEE Trans. Appl. Supercond.*, **21** (3), 1344–1347.
12. Rey, C.M., Hoffman, W.C. Jr., Cantrell, K., Eyssa, Y.M., VanSciver, S.W., Richards, D., and Boehm, J. (2004) Design and fabrication of an HTS reciprocating magnetic separator. *IEEE Trans. Appl. Supercond.*, **12** (1), 971–974.
13. Gupta, R. and Sampson, W. (2009) Medium and low field HTS magnets for particle accelerators and beam lines. *IEEE Trans. Appl. Supercond.*, **19**, 1905–1909 (Accepted for future publication, doi: 10.1109/TASC.2009.2017862, First Published: 2009-06-05, ISSN: 1051-8223).
14. Gupta, R., Anerella, M., Cozzolino, J., Ganetis, G., Ghosh, A., Greene, G., Sampson, W., Shiroyanagi, Y., Wanderer, P., and Zeller, A. (2011) Second generation HTS quadrupole for FRIB. *IEEE Trans. Appl. Supercond.*, **21** (3), 1888–1891.
15. Zangenberg, N., Nielsen, G., Hauge, N., Nielsen, B.R., Baurichter, A., Pedersen, C.G., Bräuner, L., Ulsøe, B., and Møller, S.P. (2012) Conduction cooled high temperature superconducting dipole magnet for accelerator applications. *IEEE Trans. Appl. Supercond.*, **22** (3), 4004004.
16. Tasaki, K., Ono, M., Yazawa, T., Sumiyoshi, Y., Nomura, S., Kuriyama, T., Dozono, Y., Maeda, H., Hikata, T., Hayashi, K., Takei, H., Sato, K., Kimura, M., and Masui, T. (2001) Testing of the world's largest HTS experimental magnet with Ag-sheathed Bi-2223 tapes for Si single crystal growth applications. *IEEE Trans. Appl. Supercond.*, **11** (Pt. 2), 2260–2263.
17. Ono, M., Tasaki, K., Ohotani, Y., Kuriyama, T., Sumiyoshi, Y., Nomura, S., Kyoto, M., Shimonosono, T., Hanai, S., Shoujyu, M., Ayai, N., Kaneko, T., Kobayashi, S., Hayashi, K., Takei, H., Sato, K., Mizuishi, T., Kimura, M., and Masui, T. (2002) Testing of a cryocooler-cooled HTS magnet with silver-sheathed Bi2223 tapes for silicon single-crystal growth applications. *IEEE Trans. Appl. Supercond.*, **12** (1), 984–987.
18. Masur, L., Buehrer, C., Hagemann, H., and Witte, W. (2009) Magnetic Billet Heating. Light Metal Age, pp. 50-55.
19. Parkinson, B.J., Slade, R., Mallett, M.J.D., and Chamritski, V. (2013) Development of a cryogen Free 1.5 T YBCO HTS magnet for MRI. *IEEE Trans. Appl. Supercond.*, **23** (3), 4400405.
20. Svoboda, J. (1987) Magnetic methods for the treatment of minerals, in *Developments in Mineral Processing 8* (ed. D.W.

Furstenau), Elsevier Science Publishers, Amsterdam.

21. Rey, C.M., Hoffman, W.C. Jr., and Steinhauser, D.R. (2003) Test results of a HTS reciprocating magnetic separator. *IEEE Trans. Appl. Supercond.*, **13** (2), 1624–1627.
22. Hoffmann, C.K., Keller, K., Hoffman, W.C. Jr., and Rey, C.M. (2004) Conceptual design of a novel industrial-scale HTS reciprocating high gradient magnetic separator. *IEEE Trans. Appl. Supercond.*, **14** (2), 1225–1228.
23. Wang, Q., Dai, Y., Hu, X., Song, S., Lei, Y., He, C., and Yan, L. (2007) Development of GM cryocooler-cooled Bi2223 high temperature superconducting magnetic separator. *IEEE Trans. Appl. Supercond.*, **17** (2), 2185–2188.
24. Yoon, S., Cheon, K., Lee, H., Moon, S.-H., Ham, I., Kim, Y., Park, S.-H., Joo, H., Choi, K., and Hong, G.-W. (2013) Fabrication and characterization of 3-T/102-mm RT bore magnet using 2nd generation (2G) HTS wire with conducting cooling method. *IEEE Trans. Appl. Supercond.*, **23** (3), 4600604.
25. Kim, T.-H., Ha, D.-W., Kwon, J.-M., Sohn, M.-H., Baik, S.-K., Oh, S.-S., Ko, R.-K., Kim, H.-S., Kim, Y.-H., and Park, S.-K. (2010) Purification of the coolant for hot roller by superconducting magnetic separation. *IEEE Trans. Appl. Supercond.*, **20** (3), 965–968.
26. Ha, D.-W., Kim, T.-H., Sohn, M.-H., Kwon, J.-M., Baik, S.-K., Ko, R.-K., Oh, S.-S., Ha, H.-S., Kim, H., Kim, Y.-H., and Ha, T.-W. (2010) Purification of wastewater from paper factory by superconducting magnetic separation. *IEEE Trans. Appl. Supercond.*, **20** (3), 933–936.
27. Nishijima, S., Akiyama, Y., Mishima, F., Watanabe, T., Yamasaki, T., Nagaya, S., and Fukui, S. (2013) Study on decontamination of radioactive cesium from soil by HTS magnetic separation system. *IEEE Trans. Appl. Supercond.*, **23** (3), 3700405.
28. Hiltunen, I., Korpela, A., and Mikkonen, R. (2005) Solenoidal Bi-2223/Ag induction heater for aluminum and copper billets. *IEEE Trans. Appl. Supercond.*, **15** (2), 2356–2359.
29. Runde, M., Magnusson, N., Fülbier, C., and Bührer, C. (2011) Commercial induction heaters with high-temperature superconductor coils. *IEEE Trans. Appl. Supercond.*, **21** (3), 1379–1382.

4.8
Magnetic Levitation and Transportation

John R. Hull

4.8.1
Introduction

The non-contacting nature of magnetic levitation (maglev) enables the possibility of higher speeds and efficiencies in transportation systems and has attracted a number of engineers and organizations to pursue its realization. In this article, we discuss the principles of maglev and the technology involved in realizing it. We exclude some types of levitation, such as levitation melting, levitation of diamagnetic nonsuperconducting materials and dynamic processes such as those responsible for the performance of the Levitron® toy, and so on, that are unlikely to contribute to transportation applications, at least according to present understanding and development. We follow the discussion of basic concepts with a discussion of various transportation applications and how maglev has been incorporated in their realization, whether in actual projects or in conceptual studies.

A number of past reviews are available on levitation and maglev transportation for the reader who wants more detail than is contained in the present article [1–13].

4.8.2
Magnetic Levitation: Principles and Methods

4.8.2.1 Magnetic Forces

The basic forces involved in maglev are derived from the fundamental laws of electricity and magnetism. The first basic force is the force between magnetic poles. Although isolated magnetic poles are not known to exist in nature, they constitute a convenient model and are discussed in most elementary science texts. For a pair of magnetic poles, opposite poles attract, like poles repel, and the force between point poles is proportional to the inverse square of the distance between them.

The second basic force involved in maglev is the force that occurs when a conductor moves in a magnetic field. This force is derived from Faraday's law and Lenz's law, which teach that an imposed changing magnetic flux produces a voltage in a circuit in such a way that it creates magnetic flux that opposes the imposed change in flux. These laws are responsible for eddy currents.

The response of a material to magnetic fields also produces a maglev force. A diamagnetic material, such as graphite, aluminum, water, or a superconductor, will be repulsed by a magnetic pole, whereas a paramagnetic material, such as oxygen, will be attracted to a magnetic pole. Ferromagnetic materials, such as iron, are strongly attracted to magnetic poles.

4.8.2.2 Static Stability

Earnshaw's theorem is an important theorem that affects the *static* levitation of magnetic systems. The theorem is developed from the property of curl- and divergence-free fields that precludes the existence of local, detached, and scalar-potential maxima or minima. According to this classical theorem, it is impossible to attain stable equilibrium in a system in which only inverse-square-law electrostatic or magnetostatic forces are acting [14]. Braunbek [15] deduced that statically stable electric or magnetic suspension is not possible when all materials have $\varepsilon_r > 1$ or $\mu_r > 1$, but that it is possible when materials with $\varepsilon_r < 1$ or $\mu_r < 1$ are introduced (where ε_r is the relative electrical permittivity and μ_r is the relative magnetic permeability). Earnshaw's theorem is grasped intuitively by most people when they release a permanent magnet (PM) next to the ferromagnetic door of their refrigerator. The magnet either moves to stick to the door or it falls on the floor; it does not hover in space near the point where it was released. We note that Earnshaw's theorem only applies to conditions of static stability and does not apply to dynamic systems.

4.8.2.3 Magnetic Biasing

Although maglev of one PM by another is not stable according to Earnshaw's theorem, this arrangement is still useful if stability is provided by other means. A simple rendition of such a partial levitation, originally proposed by Evershed [16], is shown in Figure 4.8.1. Here, the partially levitated object, that is, the rotor, consists of a PM below which hangs a rigid rod with a point on the bottom. Each PM is magnetized vertically and in the same direction, as shown by the dark arrows in Figure 4.8.1. The PM of the rotor experiences an attractive force toward the

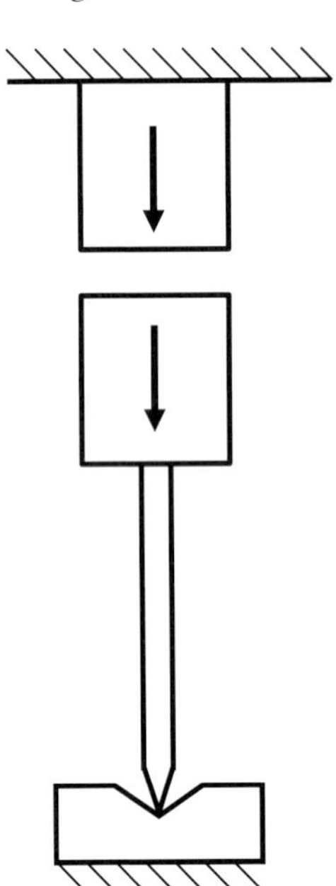

Figure 4.8.1 Evershed bearing design, in which most of the weight is provided by attraction between permanent magnets, and the remaining weight and vertical stability are provided by a small mechanical bearing.

stationary PM immediately above it. This system is statically stable in the radial direction but unstable in the vertical direction. The gap between the two magnets is adjusted in such a manner that the attractive force between the pair of magnets is just $<100\%$ of the rotor weight, so that the rotor would tend to fall. The remainder of the weight is provided by the small mechanical bearing at the bottom, which supplies sufficient stiffness for vertical stability and some additional radial stability.

The Evershed design, either in the simple form shown in Figure 4.8.1 or in some modification, is used in applications ranging from simple toys and watt-hour meters to high-speed centrifuges. This basic design may be combined with other maglev techniques and is often referred to as *magnetic biasing*.

4.8.2.4 **Electromagnetic Suspension**

Electromagnetic suspension (EMS) utilizes the attractive force between a magnetic source and a ferromagnetic material to create the levitation force. Such a system could be easily implemented in an Evershed-type design with either of the systems shown in Figure 4.8.2.

Truly contact-free maglev can be attained by replacing the mechanical bearing in Figure 4.8.2 with a small electromagnet and a feedback system, and one example of many possible embodiments is shown in Figure 4.8.3. In this example, magnetic bias force is provided by a DC superconducting coil wrapped around part of the iron stator. With no current in the stability coils, the system is stable in the radial direction but unstable in the vertical direction. The nonsuperconducting stability coils in Figure 4.8.3 are wired to magnetize the iron in the same sense that the superconducting coil magnetizes it. The feedback system consists of two sensors that detect the top and bottom edge of the levitated iron. For example, the sensors

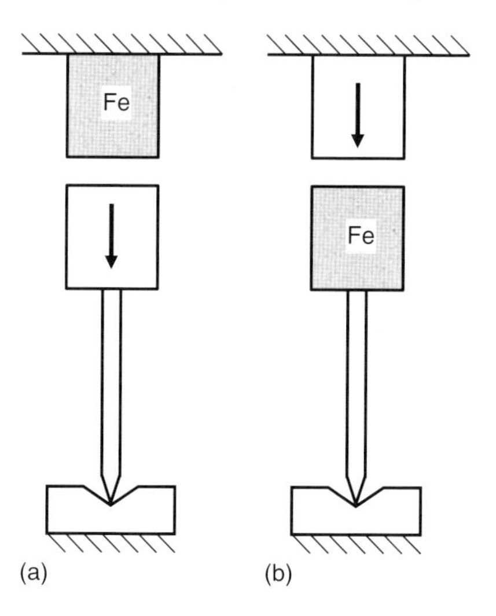

Figure 4.8.2 (a,b) Two examples of an Evershed design using electromagnetic attraction between a PM and ferromagnetic material.

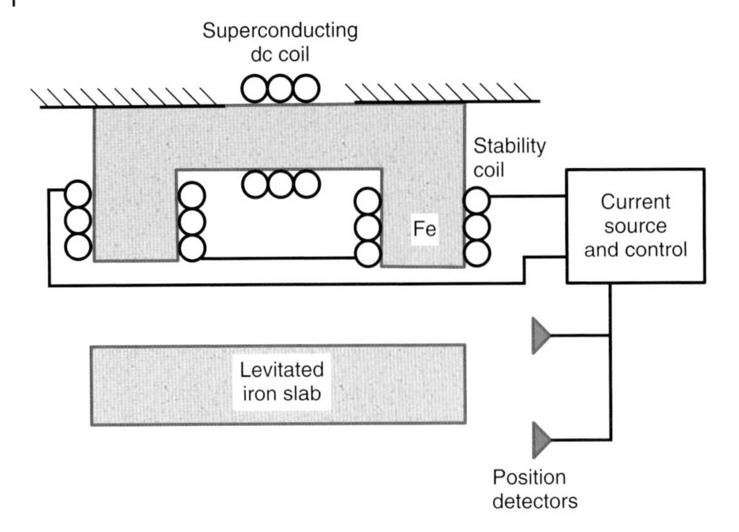

Figure 4.8.3 Attractive levitation of an iron slab with feedback control of an electromagnet.

may be photocells that detect light from a light source to the left of the iron. When light to the bottom sensor is obscured by the falling magnet, the current to the stability coils is turned on, and the attractive force increases. When light to the top sensor is obscured by the rising magnet, current to the stability coils is turned off, and the attractive force decreases. In such a manner, the levitated iron may be stably suspended.

4.8.2.5 AC Levitation

The ability of electromagnetic forces to impart significant levitation forces through Faraday's law is well known from jumping-ring experiments [6] that are performed in many introductory physics classes. Consider the system shown in Figure 4.8.4, in which a coil is stationed above a conducting plate. If the coil is energized with a pulse so that current flows as shown, by Faraday's law, eddy currents will be induced in the plate in such a way that magnetic flux is expelled from the plate.

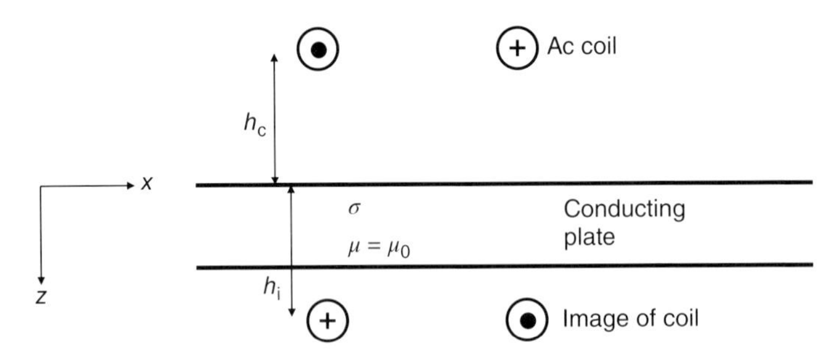

Figure 4.8.4 Repulsive levitation of an AC coil above a conducting sheet using method of images.

The effect of these eddy currents is that the coil will see its mirror image, as shown in Figure 4.8.4, and the interaction between the coil and its image produces a repulsive levitation force. Note that the distance h_i of the image below the top surface of the plate is equal to the height h_c of the coil above the plate. Because of the finite electrical conductivity of the plate, the eddy currents in the plate that arise from a current pulse in the coil will exponentially decay in time, as will the levitation force, after the pulse has ended. In the jumping ring experiment, the conducting plate consists of a ring that sits on top of the coil. When the coil is pulsed, the mutual repulsive force causes the ring to accelerate upwards. To provide a more continuous levitation force, it is necessary to provide another current pulse in the coil before the initial eddy currents have completely died away. This is most conveniently accomplished by supplying the coil with an alternating current.

The general phenomenon can be quantitatively understood in a simpler geometry by considering an alternating magnetic induction, $\boldsymbol{B} = B\boldsymbol{i}$, where $\boldsymbol{i}$ is the unit vector in the x direction, $B = B_0\exp(j\omega t)$, where j is the square root of -1, ω is the radial frequency, and t is time. Let $\boldsymbol{B}$ be incident on the surface of a half space $z > 0$, with electrical conductivity σ and magnetic permeability μ. One then must solve the magnetic diffusion equation

$$\frac{\partial^2 B}{\partial z^2} = \mu\sigma\frac{\partial B}{\partial t} \tag{4.8.1}$$

The solution to Eq. (4.8.1) in the conducting half space is

$$B = B_0 \exp\left(-\frac{z}{\delta}\right) \exp\left[j\left(\omega t - \frac{z}{\delta}\right)\right] \tag{4.8.2}$$

where the skin depth δ is given by

$$\delta = \left[\frac{2}{\mu\sigma\omega}\right]^{\frac{1}{2}}$$

The current density $\boldsymbol{J}$ in the half space, given by Maxwell's equation $\boldsymbol{J} = \nabla \times \boldsymbol{H}$, where $\boldsymbol{H}$ is the magnetic field, is in the y direction; its magnitude is given by

$$J = \left(\frac{B_0}{\mu\delta}\right)(1 + j) \exp\left(-\frac{z}{\delta}\right) \exp\left[j\left(\omega t - \frac{z}{\delta}\right)\right] \tag{4.8.3}$$

Comparing Eq. (4.8.3) with Eq. (4.8.2), we see that current density has the same exponential decay as the magnetic induction but is phase-shifted by 45°. The force per unit volume, given by $\boldsymbol{F} = Re\{\boldsymbol{J}\} \times Re\{\boldsymbol{B}\}$, is in the z direction and its magnitude is

$$F(z) = \left(\frac{B_0^2}{\mu\delta}\right) \exp\left(-\frac{2z}{\delta}\right) \left[\frac{1}{2} - 2^{-\frac{1}{2}} \sin\left(2\omega t - \frac{2z}{\delta} - \frac{\pi}{4}\right)\right] \tag{4.8.4}$$

The force consists of a time-independent part plus a sinusoidal part that is twice the applied frequency. The mean force on the half-space is in the positive z direction, as expected. For the geometry in Figure 4.8.4, there will be a downward force on the plate, with a corresponding force upward on an AC coil above the plate. The pressure P at the surface of the half space is given by

$$P = \int_0^\infty F(z)dz \tag{4.8.5}$$

$$P = \left(\frac{B_0^2}{4\mu}\right)[1 + \cos(2\omega t)] \tag{4.8.6}$$

The average levitation pressure is independent of frequency and proportional to the square of the applied magnetic field. To achieve a relatively constant levitation height, it is desirable that the period of the applied field $\tau = 2\pi/\omega$ be much smaller than the characteristic time of the mechanical motion. However, the frequency cannot be made arbitrarily high because this type of levitation is associated with joule heating; the heating rate Q per unit volume is given by

$$Q = \frac{J^2}{\sigma} \tag{4.8.7}$$

From Eq. (4.8.3), we surmise that the maximum heating rate occurs at the surface and is proportional to the frequency, and the total heating rate in the volume is proportional to the square root of the frequency.

The AC levitation described above is closely related to the magnetic river concept [17], in which the AC coil is replaced by the stator coil of a linear induction motor. Rather than being stationary on the plate, the magnetic field has a horizontal motion which interacts with the eddy currents to provide a propulsion component in addition to the levitation force.

AC levitation can be made passively stable by curving the conducting plate, for example, by making it in the shape of a bowl with coil inside.

4.8.2.6 Electrodynamic Levitation

In electrodynamic levitation (EDL), a moving magnet (PM, electromagnet, or superconducting magnet) interacts with a conducting sheet or a set of coils to produce a levitation force. A drag force, typically much higher than that associated with ferromagnetic suspensions, is associated with the eddy currents. However, above some speed, the drag force decreases as unity divided by the square root of the velocity. The system is passively stable, that is, no feedback is required. The disadvantage is that there is a minimum speed below which the levitation force is not sufficient, so some mechanical support is needed on startup.

The phenomenon of eddy currents caused by a moving magnet can be understood by applying the principle of images, as originally suggested by Maxwell and shown in Figure 4.8.5. In the case of a plane conducting sheet, the imaginary system on the negative side of the sheet is not the simple image, positive or negative, of the real magnet on the positive side, but consists of a moving train of double images [5–7]. The force on the magnet is the sum of forces of the magnet interacting with each of the images. According to this model, when a magnet passes a

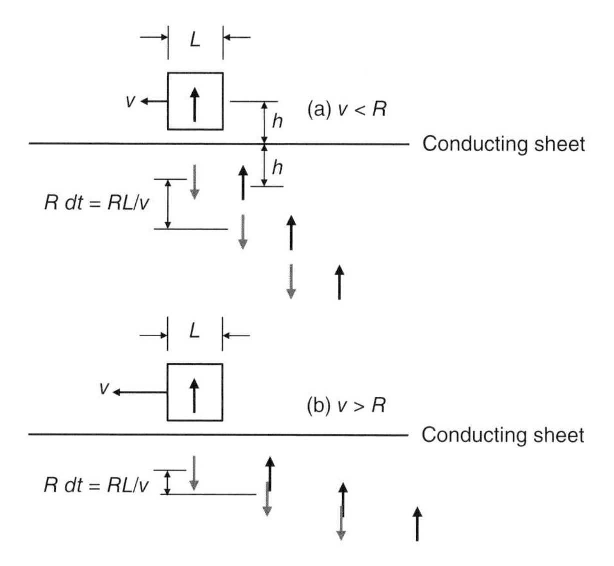

Figure 4.8.5 Maxwell's eddy current model applied to a magnet moving over a conducting plane: (a) low velocity and (b) high velocity.

point on the conducting plane, it induces first a "mirror" image, then a "replicant" image. These images propagate downward at a velocity *R*, which is proportional to the specific resistivity (and to the reciprocal thickness if the sheet is thin when compared with the skin depth). *R* is also the electrical resistance of a square portion of the conducting sheet; its value $R = \rho/2\pi h$ (ρ is the resistivity and h is the sheet thickness) is independent of the size of the square. In electromagnetic units, *R* has the dimensions of velocity.

Two examples at different magnet velocities that apply Maxwell's model are shown in Figure 4.8.5. In the first example (Figure 4.8.5a), the velocity v of the magnet is $<R$. The mirror image has moved down a distance $Rdt = RL/v$ when the replicant image appears at the same horizontal location. Then, as the two images move away tail-to-tail, the induced field falls toward zero. In the second example (Figure 4.8.5b), the velocity is considerably greater than *R*. The mirror image has moved only a small distance RL/v away when the replicant image appears, and the two images nearly cancel each other thereafter. As the magnet velocity increases to a very high value, such that $v \gg R$, there is effectively only one image – a single mirror image under the PM – and the sheet behaves like a perfect diamagnetic material with almost no drag force.

Reitz [18] solved Maxwell's equation for several types of moving magnets with the geometry shown in Figure 4.8.5. In each case, he obtained a "wake of images," similar to those shown in Figure 4.8.5, moving into the plate with a velocity $w = 2\rho/\mu_0 h$, which is Maxwell's *R* expressed in rationalized meter-kilogram-second (mks) units.

The force on a magnet moving over a nonmagnetic conducting plane can be conveniently resolved into two components: a lift force F_L perpendicular to the

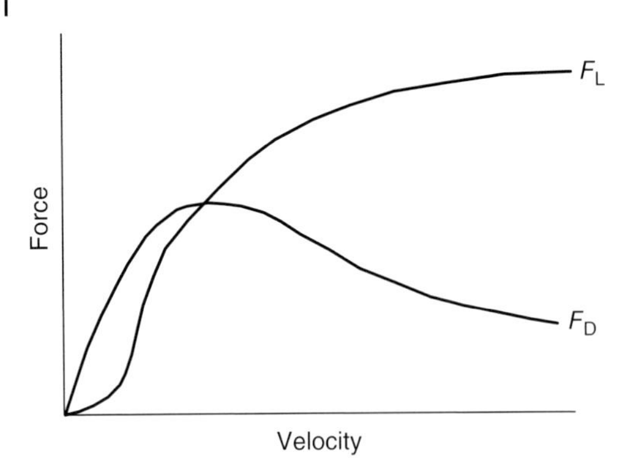

Figure 4.8.6 Velocity dependence of lift force F_L and drag force F_D.

plane and a drag force F_D opposite to the direction of motion. At low velocity, the drag force is proportional to velocity v and considerably greater than the lift force, which is proportional to v^2. As the velocity increases, however, the drag force reaches a maximum (referred to as the *drag peak*) and then decreases as $v^{-1/2}$. On the other hand, the lift force, which increases with v^2 at low velocity, overtakes the drag force as velocity increases and approaches an asymptotic value at high velocity, as shown in Figure 4.8.6. For thin plates, the lift-to-drag ratio, which is of considerable practical importance, is given by $F_L/F_D = v/w$.

Qualitatively, these forces can be understood by considering the diffusion of magnetic flux into the conductor. When a magnet moves over a conductor, the flux tries to diffuse into the conductor. If the magnet is moving rapidly enough, the flux will not penetrate very far into the conductor, and the flux compression between the magnet and the conductor causes a lift force. The flux that does penetrate the conductor is dragged along by the moving magnet, and the force required to drag this flux along is equal to the drag force. At high speeds, less of the magnetic flux has time to penetrate the conductor. At high speed, the lift force that is a result of flux compression approaches an asymptotic limit, and the drag force approaches zero.

The lift force on a vertical dipole of moment m moving at velocity v at a height z_0 above a conducting plane can be shown to be [18]

$$F_L = \frac{3\mu_0 m^2}{32\pi z_0^4}\left[1 - w\left(v^2 + w^2\right)^{-\frac{1}{2}}\right] \tag{4.8.8}$$

At high velocity, the lift force approaches the ideal lift from a single image: $3\mu_0 m^2/32\pi z_0^4$; at low velocity, the factor in the brackets is approximately equal to $v^2/2w^2$, so the lift force increases as v^2.

The drag force, as already pointed out, is w/v times the lift force, so the drag force is proportional to v at low velocity. According to the thin-plate model that we have been discussing thus far, the drag force should fall off with $1/v$ as the lift force reaches its high-speed limit. However, at high velocity, penetration of the eddy currents and magnetic fields is limited to the skin depth, which is proportional to

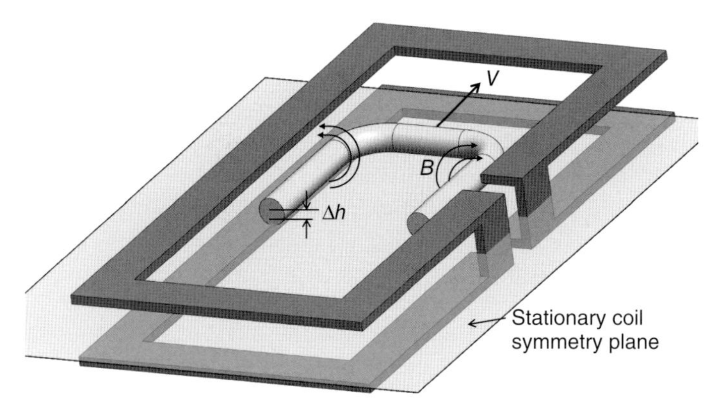

Figure 4.8.7 Null-flux geometry, showing full stationary null-flux coil and one half of levitated coil moving with velocity v and displacement from the symmetry plane by Δh.

$v^{-1/2}$. As a first approximation, one might replace plate thickness by skin depth at high speed. The transition from thin-plate to skin-depth behavior should occur at about 30 m s^{-1} in a 1 cm-thick aluminum plate, for example.

One may improve on the basic system of a magnet moving over a conducting plate by using the null-flux-geometry system [19] shown in Figure 4.8.7. When the moving magnet or coil is in the symmetry plane, no net flux threads the track loop so that lift and drag forces approach zero. Lift forces increase linearly and drag forces as the square of the (small) displacement h from the symmetry plane of the stationary coil. The velocity dependence of null-flux systems is the same as that of the eddy current systems.

4.8.2.7 Levitation by Tuned Resonators

Several levitation concepts involve passive techniques. These concepts have the advantage of simplicity and the lack of a control system. One system achieves a stable stiffness characteristic by using an inductor–capacitor (LC)-circuit excited slightly off resonance [1]. The LC-circuit is formed with the inductance of the electromagnetic bearing coil and a capacitor. The mechanical displacement of the rotor changes the inductance of the electromagnet. The LC circuit is operated near resonance and tuned in such a way that it approaches resonance as the rotor moves away from the electromagnet. This increases the current from the AC voltage source and thus pulls the rotor back to its nominal position. The low forces and stiffnesses of this system, coupled with the necessity for continuous AC energization of the coils, are disadvantages. The system is also subject to a low-frequency, negative-damping instability, and auxiliary damping is usually required for stability.

4.8.2.8 Magnitude of Levitation Pressure

The levitation pressure of a magnetic system is considerably smaller than that of most mechanical systems. The maximum magnetic pressure P between two magnetized objects of magnetization M_1 and M_2 occurs at zero gap between the two objects and is given by

$$P = \frac{M_1 M_2}{(2\mu_0)} \quad (4.8.9)$$

where $\mu_0 = 4\pi \times 10^{-7}$ H m^{-1}. As convenient reference: for $\mu_0 M_1 = \mu_0 M_2 = 1.0$ T, the pressure is 400 kPa. In a sintered NdFeB PM, $\mu_0 M$ is typically between 1 and 1.5 T; ferromagnetic materials may achieve magnetizations up to about 2.5 T.

For a set of DC coils of alternating polarity of spatial period L moving at a height h over a conducting sheet, the maximum levitation pressure is given by the pressure of the image force and is

$$P_{\mathrm{I}} = \left(\frac{1}{2}\pi\mu_0\right) B_0^2 \exp\left(-\frac{4\pi h}{L}\right) \quad (4.8.10)$$

where B_0 is the rms (spatially averaged) value of the magnetic induction in the plane of the magnets. With an NbTi superconducting coil, B_0 can easily be 5 T.

4.8.2.9 HTS/PM Levitation

The development of bulk high-temperature superconductor (HTS) has made possible a robust stable levitation using PM and HTS that can be used in transportation applications. The principles of this technology were discussed in Chapter 4.1.

4.8.2.10 Propulsion

To transport somewhere, one needs to move, and a source of propulsion is desired. When maglev is used for suspension, one usually invokes one of the several types of linear electric motors, although rocket, jet, and propeller propulsion are also possible. A discussion of linear electric machines is beyond the scope of this article, and the interested reader is referred to [17] and similar publications.

4.8.3 Maglev Ground Transport

Maglev ground transport involves the levitation of vehicles by one of the maglev principles discussed earlier. The system invariably restricts the vehicle to movement over a guideway, in the same way that wheeled trains move on tracks. With maglev, there is no mechanical contact between the vehicle to be transported and the guideway that directs its travel. This application has been considered primarily for high-speed transport of people via trains, where damage to the tracks by wear of a high-speed wheel-on-steel-rail system can be significant. Further advantages over wheeled transport are higher efficiency, improved ride quality, and low noise movement above the guideway. Maglev has the potential to be more efficient and affordable than alternative technologies for many transportation applications [12].

4.8.3.1 History

The history of maglev transport has been detailed in [2–4, 6, 9, 12]. As early as 1907, Robert Goddard, better known as the *father of modern rocketry*, but then a student at Worcester Polytechnic Institute, published a story in which many of the key features of a maglev transportation system were described. In 1912, Emile Bachelet, a French engineer proposed a magnetically levitated vehicle for delivering mail. His vehicle was levitated by copper-wound electromagnets moving over a pair of aluminum strips. Because of the large power consumption, however, Bachelet's proposal was not taken very seriously, and the idea lay more or less dormant for half a century.

In 1963, Powell suggested using superconducting magnets to levitate a train over a superconducting guideway. Powell and Danby proposed in 1967 a system that used a less expensive conducting guideway at room temperature. Later, they conceived the novel idea of a "null-flux" suspension system that would minimize the drag force and thus require much less propulsion power [19].

During the late 1960s, groups at the Stanford Research Institute and at Atomic International studied the feasibility of a Mach-10 rocket sled that employed maglev. The maglev principle was later applied to high-speed trains. In 1972, the group at Stanford Research Institute constructed and demonstrated a vehicle that was levitated with superconducting magnets over a continuous 160 m-long aluminum guideway [20].

At about the same time, a team from MIT, Raytheon, and United Engineers designed the magneplane system, in which lightweight cylindrical vehicles, propelled by a synchronously traveling magnetic field, travel in a curved aluminum trough [21]. One advantage of the curved trough is that the vehicle is free to assume the correct bank angle when negotiating curves, but the guideway itself is banked at only approximately the desired angle. The magneplane concept was tested with a 1/25-scale model system that used both PMs and superconducting coils for levitation above a 116 m-long synchronized guideway.

The initial maglev research effort in the United States ended about 1975. Interest revived again around 1989, when four major conceptual designs were funded for several years by the federal government [22], after which government funding again disappeared.

Maglev systems have been studied in several other countries, most notably Japan, Germany, and the United Kingdom. The first maglev train ever in commercial service was a 600 m-long route in Birmingham, England that connected the airport to a conventional rail line. This service operated from 1984 to 1995. Its service was terminated mainly because of lack of parts in a one-of-a-kind technology. The M-Bahn was an elevated maglev train that operated on a 1.6 km guideway in Berlin from 1989 to 1991. Its operation stopped with reunification of Berlin. The Linimo maglev train has operated in Aichi, Japan, near the city of Nagoya, since 2005, on an 8.9 km-long guideway. It has a levitation gap of 8 mm and a top speed of 100 km h^{-1}. In its first year of operation, it handled more than 10^7 passengers.

Research in Germany and Japan has continued to the present, and full-scale vehicles have been tested in both countries. The two main maglev technologies that are being pursued at this time may be broadly classified as EMS, in which active feedback and electromagnets are used, and EDL in which repulsive forces

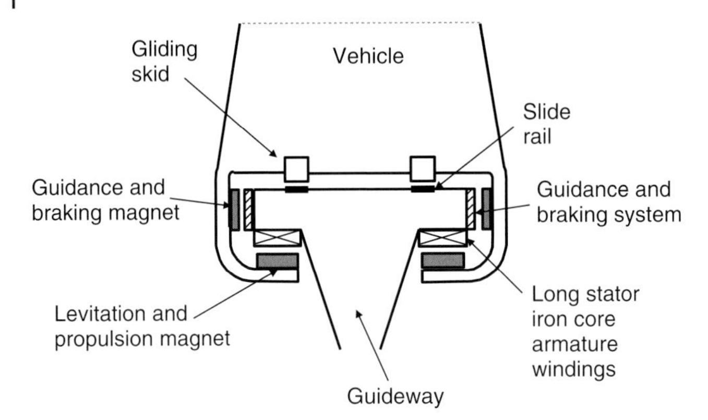

Figure 4.8.8 Schematic diagram of EMS maglev system, similar to that used in the Transrapid.

are generated by eddy currents that are induced in the guideway by the passage of a superconducting magnet.

EMSs depend on the attractive forces between electromagnets and a ferromagnetic (steel) guideway, as notionally shown in Figure 4.8.8. This system has been under intensive investigation in Germany. The magnet-to-guideway spacing must be small (only a few centimeters at most), and this requires that the straightness tolerances of the guideway must be relatively small. On the other hand, it is possible to maintain magnetic suspension even when the vehicles are standing still, which is not true for EDLs. In the system shown in Figure 4.8.8, a separate set of electromagnets provides horizontal guidance force, and the levitation magnets, acted on by a moving magnetic field from the guideway, provide the propulsion force. The German Transrapid TR-07 vehicle is designed to carry 200 passengers at a maximum speed of 500 km h^{-1}. The levitation height is 8 mm, and power consumption is estimated to be 43 MW at 400 km h^{-1}. A full-scale prototype of the Transrapid ran for many years on a 30 km test track in Emsland, Germany, and the TR-07 achieved a speed of 450 km h^{-1} in 1993. The TR-09 operates a 190 MT vehicle at a design speed of 550 km h^{-1} [12]. In 2004, the first commercial implementation of a Transrapid-type system was completed. This maglev train connects the rapid transit network of Shanghai to the Shanghai Pudong International Airport, 30.5 km away with a maximum running speed of 430 km h^{-1} maintained for 50 s.

EDLs depend on repulsive forces between moving magnets and the eddy currents they induce in a conducting aluminum guideway or in conducting loops similar to that shown in Figure 4.8.7. This system has been intensively investigated in Japan. The repulsive levitation force is inherently stable, and comparatively large levitation heights (20–30 cm) are attainable by using superconducting magnets. Various guideway configurations, such as a flat horizontal conductor, a split L-shape conductor, and an array of short-circuit coils on the sidewalls, have been investigated. Each has its advantages and disadvantages. The prototype

Japanese high-speed maglev system involves the use of interconnected Figure-eight ("null-flux") coils on the sidewalls that are cross-connected with wires that go underneath the track. The null-flux arrangement tends to reduce the magnetic drag force and thus the propulsion power needed. A prototype of this system operated for many years on a 7 km-long test track in Miyazaki, Japan. This was followed by ongoing operation at a test track in the Yamanashi prefecture that is 18.4 km long, with extension to 42.8 km expected in 2013/2014. A three-car test train with people onboard achieved a speed of 581 km h^{-1} on this test track in December 2003. The test track encompasses tunnels, two trains passing each other at a relative speed of 1026 km h^{-1}, and trial rides for $>10^5$ passengers. The Yamanashi Test Line operates a 60 MT vehicle with a 100 mm gap [12]. Propulsion power is provided from the track in the form of a linear synchronous motor (LSM). The speed of the vehicle is limited by lack of sufficient power to overcome the aerodynamic drag.

The incorporation of bulk HTS for maglev applications has been studied almost since the discovery of HTS, and HTS could be used in maglev in several ways. First, HTS wire could replace the NbTi wire currently used in the superconducting racetrack coils aboard the vehicle [23]. This would allow cryogenic refrigeration at higher temperatures than the present values near 4 K. Second, trapped-field HTSs could be used to replace the coils [24, 25]. In this case, the trapped-field HTSs would act as very powerful PM analogs that allow much higher levitation heights than can be achieved with conventional PMs on maglev vehicles. Third, the diamagnetic HTSs could be placed over a PM guideway [26]. Fourth, HTS wire or an HTS trapped-field magnet could provide DC magnetization in an EMS system [27]. Arrays of trapped-flux bulk HTS have been magnetized *in situ* by field-cooled and pulsed zero-field cooled techniques [28, 29].

More recently, the option of using bulk HTS over a PM guideway has received attention. One of the advantages of this option is that, like EMS, levitation is possible at all speeds. In 2002, the first person-loading system of this type was reported [30]. Further, real-scale-sized maglev systems followed [31, 32], mostly operating along short distances of a few meters on a double PM track. A maglev facility, called *SupraTrans II*, was started in February 2011 in Dresden, Germany. The double-track guideway is an 80 m long oval and contains a total of 4.8 Mg of PM. A contactless linear motor between the tracks drives the vehicle with a maximum speed of 20 km h^{-1}. The vehicle can transport two people with a total suspended mass of 600 kg at a levitation height above the track of 10 mm.

4.8.3.2 System Technical Considerations

To realize the high-speed potential of maglev in commercial transportation, the system must be inherently safe. There must be sufficient gap and stiffness to overcome the influence of guideway construction imperfections and tolerances, as well as transitory loads such as wind gusts or pressure pulses from a passing train or entrance to a tunnel. Higher speed and small gap require smaller tolerances.

At high speed, the radius of curvature for elevation or direction changes must be sufficiently large such that stressful centrifugal accelerations are not placed on the passengers. This may restrict the number of available routes or limit speed at some route locations to values far below what the maglev technology is capable of.

The largest gaps are realized with high-field superconducting magnets, which at present use NbTi coils. These coils operate in persistent current. They are bathed in liquid helium at 4.2 K. Compared to HTS, the low-temperature superconductors are more susceptible to instability, and minimization of conductor movement with the cryostat and the effect of vehicle vibrations are important design elements for these magnets.

The placement of levitation support on the vehicle is governed by several considerations. Pitch, roll, and yaw stability imply at least four levitation supports, typically at the corner of each individual vehicle. Magnetic field in the passenger compartment must also be kept below certain prescribed limits, as demanded by safety and health requirements.

While much study has been conducted on reducing magnetic drag in maglev systems, it should be appreciated that, at the high speeds, realized by existing prototypes, aerodynamic drag is considerably larger. An EDL has significantly higher magnetic loss than EMS. However, an EDL will have less aerodynamic drag on the levitation components and is usually more efficient at speeds $>450\,\text{km}\,\text{h}^{-1}$ than EMS [12]. Thus, aerodynamic design of the vehicles is as important to maglev as it is to modern-day airplanes. This relationship has prompted several investigators to suggest that, to properly take advantage of maglev, the transportation should be conducted in evacuated tubes or tunnels [33, 34].

In most maglev systems, the largest cost component is the guideway, and in addition to the maglev component, it includes land acquisition, civil construction, and electrical power provisions for propulsion. The cost of the vehicle is usually a small fraction of the total cost. Maglev offers an intrinsic advantage over wheeled systems in this regard. Propulsion is not dependent on friction, and control is not dependent on communication with a moving vehicle. Thus, the reasons that favor long trains in a wheeled system do not apply to the maglev system, and smaller vehicles with closer allowed spacing, requiring lighter guideways, are a result of a system optimization. Maglev has the ability to achieve much higher acceleration and deceleration rates than steel wheels on steel rails. Efficiency improvements derive from the ability to regenerate the braking energy back into the electrical power system.

4.8.3.3 Guideway Design

As we have seen above, it is important to implement an efficient, low-cost guideway. The power system is a key component of the efficiency, and how the design optimizes in terms of energized blocks and sub-blocks will depend on the speed range and the vehicle size.

For EDL levitation, a number of guideway configurations were investigated to reduce cost and improve the magnetic drag characteristics, vehicle stability, and ride quality of the system [8]. Configurations investigated include guideways in geometries of sheet, ladder, discrete coils, and different null-flux concepts. Similarly, for PM tracks, a number of optimization studies have been made, for example, [35–39]. In optimizing PM configurations, one invariably arrives at a Halbach arrangement, which has the advantage of concentrating the magnetic flux on only one side of the PM array.

4.8.3.4 Cryostats and Vehicle Design

The use of superconductivity on a vehicle requires that the superconductors be kept cold. This can be accomplished in several ways. One method is to house the superconductors in a low-loss cryostat and include enough cold thermal mass, for example, liquid helium for NbTi, or liquid or solid nitrogen for HTS, to last until the vehicle can be serviced. A longer lasting approach is to thermally connect the superconductors in the cryostats to a cryocooler. The cryocooler can either re-condense the boiled-off cryogen or directly conductively cool the superconductors. The cryocooler requires electrical power on the vehicle; however, in a commercial system, the power to run the cryocooler should be a small fraction of the normal hotel loads of the vehicle required for lighting and environmental control.

An example for cryostat is the one developed by ATZ for HTS maglev [13]. Each cryostat consists of a stainless steel body with a G-10 plate on the top. A mechanical interface on top is provided to fasten the passenger module. Inside each cryostat, 24 pieces of three-seeded YBCO bulks of dimensions 64 mm × 32 mm × 12 mm are glued and mechanically fastened in a copper holder. The total HTS area is about 490 cm^2 per cryostat. The HTSs are cooled using liquid nitrogen stored in a chamber of the cryostat on the other side of the copper frame by conduction cooling. The 2 mm distance between the YBCO surface and the outer cryostat allows large levitation forces with respect to a high load capacity. The 2.5 l of liquid nitrogen storage enables a 1-day operation without refilling liquid nitrogen. Thermal loss measurements were 2.5–3.0 W per cryostat.

4.8.4 Clean-Room Application

Maglev transport as an application is also being applied in clean-room environments where electronic dimensions are becoming increasingly smaller and contamination caused by rubbing or rolling contact cannot be tolerated [40]. Typically, the component carrier in the pristine environment has a PM, and levitation force is transmitted through the clean-room wall to an HTS. Since the PM and HTS are magnetically coupled, movement of the HTS moves the component carrier.

4.8.5
Air and Space Launch

Use of maglev concepts has also been proposed for assisting the early stages of airplane takeoff and rocket launch. In a maglev-assist launch system, a sled and track act as an electromagnetic catapult to provide initial velocity to vehicles riding on the sled, using electrical power from stationary sources. The launch-assist sled serves as a "virtual first stage" for the vehicle.

A major benefit of maglev assist is that a significant fraction of the energy to achieve vehicle velocity is derived from off-board power, thus increasing the theoretically achievable vehicle dry-mass fraction. Freeing the vehicle of carrying this energy reduces the vehicle size and complexity, reduces the cost to transport the vehicle to its desired destination, and reduces the amount of pollution expended into the atmosphere. The track and sled are infrastructure, and because of the non-contact nature of maglev, they can be reused many times without maintenance to reduce the launch cost.

Airbus has recently publically discussed a maglev-assist concept for airplanes that they call "Eco-climb." Because the power required for sustaining flight is less than that for takeoff, it is claimed that engine size could be reduced, making the planes lighter and more efficient. Further, if the energy required for initial acceleration could be derived from an electrical source, CO_2 emissions could be reduced. Further claimed advantages of this system are possibility of steeper climb resulting in shorter runways and reduced noise.

NASA has funded a number of studies for maglev assist [41–45]. Levitation concepts have included the use of superconducting coils on the sled interacting with a passive null-flux track [42, 44], PMs carried on the sled interacting with passive coils on the track [43], and alternating current in track coils interacting with aluminum conductor on the sled [43].

Maglev-assist launch has many similarities with the maglev and propulsion technologies that have been under development for decades for train transportation. Of all catapult-type launch-assist concepts (e.g., [46–48]), the maglev assist is closest to the required performance requirements in terms of projectile mass and muzzle velocity [41]. The transition from existing maglev train technology to maglev-assist launch is a relatively small leap compared to that required for the other launch-assist concepts.

A 495 kg maglev rocket sled has reached a peak speed of $673\,\mathrm{km\,h^{-1}}$ and an acceleration of 25 g [49]. Electromagnetic propulsion has been developed to help launch airplanes from sea-based carriers. In several Chinese research laboratories, the use of bulk HTS carried in vehicles levitated over a permanent-magnet track has been investigated for train transport and maglev-assist launch [50–54]. The use of maglev-assist launch has several synergies with the reusable launch vehicle, with weight savings that cascade through the design. The benefits that Olds and Bellini [41] ascribed to their Argus concept should apply to most reusable vehicles used in a maglev-assist launch system.

References

1. Frazier, R.H., Gilinson, P.J. Jr., and Oberbeck, G.A. (1974) *Magnetic and Electric Suspensions*, MIT Press, Cambridge, MA.
2. Laithwaite, E.R. (1977) *Transport without Wheels*, Elek Science, London.
3. Rhodes, R.G. and Mulhall, B.E. (1981) *Magnetic Levitation for Rail Transport*, Clarendon Press, Oxford.
4. Jayawant, B.V. (1981) Electromagnetic suspension and levitation. *Rep. Prog. Phys.*, **144**, 411–477; Also (1988) Electromagnetic suspension and levitation techniques, *Proc. R. Soc. London A*, **416**, 245–320.
5. Saslow, W.M. (1991) How a superconductor supports a magnet, how magnetically 'soft' iron attracts a magnet, and eddy currents for the uninitiated. *Am. J. Phys.*, **59**, 16–25.
6. Rossing, T.D. and Hull, J.R. (1991) Magnetic levitation. *Phys. Teach.*, **29**, 552–562.
7. Saslow, W.M. (1992) On Maxwell's theory of eddy currents in thin conducting sheets, and applications to electromagnetic shielding and MAGLEV. *Am. J. Phys.*, **60**, 693–711.
8. Moon, F.C. (1994) *Superconducting Levitation*, John Wiley & Sons, Inc., New York.
9. Hull, J.R. (1999) in *Encyclopedia of Electrical and Electronics Engineering*, vol. **11** (ed. J.G. Webster), John Wiley & Sons, Inc., New York, pp. 740–747.
10. Hull, J.R. (2004) in *High Temperature Superconductivity 2: Engineering Applications* (ed A.V. Narlikar), Springer, Berlin, pp. 91–142.
11. Campbell, A.M. (2004) in *High Temperature Superconductivity 2: Engineering Applications* (ed A.V. Narlikar), Springer, Berlin, pp. 143–174.
12. Thornton, R. (2009) Efficient and affordable maglev opportunities in the United States. *Proc. IEEE*, **97**, 1901–1921.
13. Werfel, F.N., Floegel-Delor, U., Rothfeld, R., Riedel, T., Goebel, B., Wippich, D., and Schirrmeister, P. (2012) Superconductor bearings, flywheels and transportation. *Supercond. Sci. Technol.*, **25**, 014007.
14. Earnshaw, S. (1842) On the nature of the molecular forces which regulate the constitution of the luminiferous ether. *Trans. Cambridge Philos. Soc.*, **7**, 97–112.
15. Braunbek, W. (1939) Freischwebende korper im elektrischen und magnetischen Feld. *Z. Phys.*, **112**, 753–763.
16. Evershed, S. (1900) A frictionless motor meter. *J. Inst. Electr. Eng.*, **29**, 743–796.
17. Laithewaite, E.R. (1975) Linear electric machines – a personal view. *Proc. IEEE*, **63**, 250–290.
18. Reitz, J.R. (1970) Forces on moving magnets due to eddy currents. *J. Appl. Phys.*, **41**, 2067–2071.
19. Powell, J.R. and Danby, G.T. (1971) Magnetic suspension for levitated tracked vehicles. *Cryogenics*, **11**, 192–204.
20. Coffey, H.T., Solinsky, J.C., Colton, J.D., and Woodbury, J.R. (1974) Dynamic performance of the SRI maglev vehicle. *IEEE Trans. Magn.*, **10**, 451–457.
21. Iwasa, Y., Hoenig, M.O., and Kolm, H.H. (1974) Design of a full-scale magneplane vehicle. *IEEE Trans. Magn.*, **10**, 402–405.
22. Coffey, H.T. (1993) U.S. Maglev: status and opportunity. *IEEE Trans. Appl. Supercond.*, **3**, 863–868.
23. Yokoyama, S., Shimohata, K., Inaguchi, T., Kim, T., Nakamura, S., Miyashita, S., and Uchikawa, F. (1995) A conceptual design for a superconducting magnet for maglev using a Bi-based high-Tc tape. *IEEE Trans. Appl. Supercond.*, **5**, 610–613.
24. Sanagawa, Y., Ueda, H., Tsuda, M., Ishiyama, A., Kohayashi, S., and Haseyama, S. (2001) Characteristics of lift and restoring force in HTS bulk – application to two-dimensional maglev transportation. *IEEE Trans. Appl. Supercond.*, **11**, 1797–1800.
25. Fujimoto, H., Kamijo, H., Higuchi, T., Nakamura, Y., Nakashima, K., Murakami, M., and Yoo, S. (1999) Preliminary study of a superconducting bulk magnet for the maglev train. *IEEE Trans. Appl. Supercond.*, **9**, 301–304.
26. Wang, J.S., Wang, S.Y., Ren, Z.Y., Zhu, M., Jiang, H., and Tang, Q.X. (2001)

Levitation force of a YBaCuO bulk high temperature superconductor over a NdFeB guideway. *IEEE Trans. Appl. Supercond.*, **11**, 1801–1804.

27. Wang, J., Wang, S., Ren, Z., Dong, X., Lin, G., Lian, J., Zhang, C., Huang, H., Deng, C., and Zhu, D. (1999) Preliminary study of a superconducting bulk magnet for the maglev train. *IEEE Trans. Appl. Supercond.*, **9**, 904–907.
28. Kamijo, H., Higuchi, T., Fujimoto, H., Ichikawa, H., and Ishigohka, T. (1999) Flux-trapping characteristics of oxide superconducting bulks in array. *IEEE Trans. Appl. Supercond.*, **9**, 976–979.
29. Ishigohka, T., Ichikawa, H., Ninomiya, A., Kamijo, H., and Fujimoto, H. (2001) Flux trapping characteristics of YBCO bulks using pulse magnetization. *IEEE Trans. Appl. Supercond.*, **11**, 1980–1983.
30. Wang, J., Wang, S., Zeng, Y., Huang, H., Luo, F., Xu, Z., Tang, Q., Lin, G., Zhang, C., Ren, Z., Zhao, G., Zhu, D., Wang, S., Jiang, H., Zhu, M., Deng, C., Hu, P., Li, C., Liu, F., Lian, J., Wang, H., Wang, L., Shen, Z., and Dong, X. (2002) The first man-loading high temperature superconducting maglev test vehicle in the world. *Physica C*, **378–381**, 809–814.
31. Schultz, L., deHaas, O., Verges, P., Beyer, C., Roehlig, S., Olsen, H., Kuehn, L., Berger, D., and Noteboom-Funk, U. (2005) Superconductively levitated transport system—the Supratrans project. *IEEE Trans. Appl. Supercond.*, **15**, 2301–2305.
32. Sotelo, G.G., Dias, D.H.N., de Andrade, R. Jr., and Stephan, R.M. (2011) Tests on a superconductor linear magnetic bearing of a full-scale maglev vehicle. *IEEE Trans. Appl. Supercond.*, **21**, 1464–1468.
33. Goddard, R.H. (1949) Apparatus for vacuum tube transportation. US Patent 2,488,287.
34. Forgacs, R.L. (1973) Evacuated tube vehicles versus jet aircraft for high-speed transportation. *Proc. IEEE*, **61**, 604–617.
35. Wang, S., Wang, J., Ren, Z., Jiang, H., Zhu, M., Wang, X., and Tang, Q. (2001) Levitation force of multi-block YBaCuO bulk high temperature superconductors. *IEEE Trans. Appl. Supercond.*, **11**, 1808–1811.
36. Beyer, C., deHaas, O., Verges, V., and Schultz, L. (2006) Guideway and turnout switch for Supratrans project. *J. Phys. Conf. Ser.*, **43**, 991–994.
37. Jing, H., Wang, J., Wang, S., Wang, L., Liu, L., Zheng, J., Deng, Z., Ma, G., Zhang, Y., and Li, J. (2007) A two-pole Halbach permanent magnet guideway for high temperature superconducting maglev vehicle. *Physica C*, **463**, 426–430.
38. Deng, Z., Wang, J., Zheng, J., Lin, Q., Zhang, Y., and Wang, S. (2009) Maglev performance of a double-layer bulk high temperature superconductor above a permanent magnet guideway. *Supercond. Sci. Technol.*, **22**, 055003.
39. Goncalves, G.G., Dias, D.H.N., de Andrade, R. Jr., Stephan, R.M., Del-Valle, N., Sanchez, A., Navau, C., and Chen, D.-X. (2011) Experimental and theoretical levitation forces in a superconducting bearing for a real-scale maglev system. *IEEE Trans. Appl. Supercond.*, **21**, 3532–3540.
40. Minami, H. and Yuyama, J. (1995) Construction and performance test of a magnetically levitated transport system in vacuum using high-Tc superconductors. *Jpn. J. Appl. Phys.*, **34**, 346–349.
41. Olds, J. and Bellini, P. (1998) Argus, a highly reusable SSTO rocket-based combined cycle launch vehicle with maglifter launch assist. AIAA 9801557, AIAA 8th International Space Planes and Hypersonic Systems and Technologies Conference, Norfolk, VA.
42. Dill J. and Meeker, D. (2000) Maglifter Tradeoff Study and Subscale System Demonstrations, NAS-98069-1362.
43. Jacobs, W.A. (2001) Magnetic launch assist – NASA's vision for the future. *IEEE Trans. Magn.*, **37**, 55–57.
44. Schultz, J., Radovinsky, A., Thome, R., Smith, B., and Minervini, J. (2001) Superconducting magnets for maglifter launch assist sleds. *IEEE Trans. Appl. Supercond.*, **11**, 1749–1752.
45. Mankins, J.C. (2002) Highly reusable space transportation: advanced concepts and the opening of the space frontier. *Acta Astronaut.*, **51**, 727–742.
46. (a)Powell, J., Maise, G., Paniagua, J., and Rather, J. (2008) Maglev Launch and

the Next Race to Space. IEEEAC Paper #1536, ver. 7. (b)Also Powell, J. and Maise, G. (2001) Space tram. US Patent 6,311,926.

47. McNab, I.R. (2003) Launch to space with an electromagnetic railgun. *IEEE Trans. Magn.*, **39**, 295–304.

48. Hull, J.R., Fiske, J., Ricci, K., and Ricci, M. (2007) Analysis of levitational systems for a superconducting launch ring. *IEEE Trans. Appl. Supercond.*, **17**, 2117–2120.

49. Hsu, Y., Langhorn, A., Ketchen, D., Holland, L., Minto, D., and Doll, D. (2009) Magnetic levitation upgrade to the Holloman high speed test track. *IEEE Trans. Appl. Supercond.*, **19**, 2074–2077.

50. Yang, W., Wen, Z., Duan, Y., Chen, X., Qiu, M., Liu, Y., and Lin, L. (2006) Construction and performance of HTS maglev launch assist test vehicle. *IEEE Trans. Appl. Supercond.*, **16**, 1108–1111.

51. Qiu, M., Wang, W., Wen, Z., Lin, L., Yang, G., and Liu, Y. (2006) Experimental study and optimization of HTS bulk levitation unit for launch assist. *IEEE Trans. Appl. Supercond.*, **16**, 1120–1123.

52. Wang, J., Wang, S., Deng, C., Zheng, J., Song, H., He, Q., Zeng, Y., Deng, Z., Li, J., Ma, G., Huang, Y., Zhang, J., Lu, Y., Liu, L., Wang, L., Zhang, J., Zhang, L., Liu, M., Qin, Y., and Zhang, Y. (2007) Laboratory-scale high temperature superconducting maglev launch system. *IEEE Trans. Appl. Supercond.*, **17**, 2091–2094.

53. Wang, J., Wang, S., and Zheng, J. (2009) Recent development of high temperature superconducting maglev system in China. *IEEE Trans. Appl. Supercond.*, **19**, 2142–2147.

54. Yang, W., Li, G., Ma, J., Chao, X., and Li, J. (2010) A small high-temperature superconducting maglev propeller system model. *IEEE Trans. Appl. Supercond.*, **20**, 2317–2321.